Technology in Western Civilization

TECHNOLOGY IN

The Emergence of Modern Industrial Society Earliest Times to 1900

WESTERN CIVILIZATION

Volume I

EDITED BY

MELVIN KRANZBERG

CARROLL W. PURSELL, Jr.

EXECUTIVE EDITORS

PAUL J. GROGAN

DONALD F. KAISER

New York
OXFORD UNIVERSITY PRESS
London Toronto 1967

Library of Congress Catalogue Card Number: 67-15129

First Printing, February 1967
Second Printing, August 1967

Printed in the United States of America

Preface

Technology has been one of the major determinants in the development of Western civilization; yet only recently has there been recognition of this fact. As yet this recognition has not been sufficiently incorporated either into general histories of Western civilization, or into those books which deal specifically with the history of technology itself. General history books almost entirely neglect the technological factor, or when they do treat it, it is usually in the form of a chapter on the Industrial Revolution which focuses on the economic and social impact without any serious attempt to explain or understand the interplay of technological factors with the economic and social ones. Histories of technology *per se* suffer from the opposite concern. That is, they address themselves solely to the "internal" history of technology and omit the cultural, economic, social, and other interrelationships of technology. The present book, while focusing on the technological element, still attempts to integrate technological developments with those other aspects of society and culture which have been affected by technology. In a sense, the history of technology is a branch of social history; it also forms an important part of intellectual, economic, political, art, military, and even religious history. At the same time, technology has had its own "internal" history, largely dealing with the "hardware"—the actual technical devices and processes themselves. This book endeavors to present the social and "hardware" history of technology in the broad context of more traditional historical interests.

The survey extends from the very beginnings of man to the present day, and to the problems of the future that our advancing technology is bringing into being. However, unlike other surveys that also attempt to cover the entire sweep of human development, the breaking point between the two volumes is not in 1648, 1715, 1815, or any of the other traditional chronological landmarks. These traditional breaking points have been based largely on political history, but such a division would be foreign to the subject matter of technology itself. Instead, the first volume covers a period from the beginnings of man to the beginning of the 20th century, with emphasis on the period from 1750 to 1900. The second volume deals exclusively with 20th-century developments. The nature of technological history itself has dictated this division. It is only with the Industrial Revolution, beginning in the mid-18th century, that technology became one of the most distinguishing characteristics of our civilization. In our own century, technology has been advancing so rapidly and making itself felt so much in our daily lives—as well as in our social and political institutions—that these recent developments must perforce have a volume to themselves.

In addition, there is a noticeable change in geographical emphasis between the two volumes, and this too is dictated by the subject matter. The first volume is worldwide in scope, although emphasis is upon England in the period after 1750, largely because England captured the industrial leadership of the world in the mid-18th century. The second volume focuses primarily on American developments, not because of any parochial outlook by the authors or editors, but because the industrial leadership of the world passed to America during the 20th century, and because the major technological developments which have had worldwide influence became dominant in America, even when they did not originate there.

In a collaborative work of this sort, there are bound to be differences in individual styles, emphases, and interpretations. The editors have made no attempt to suppress these stylistic or interpretive differences, for they are part of the essence of scholarship. However, the editors have endeavored to provide coherence and unity throughout the book to ensure the integration of the separate chapters into a smoothly flowing, unified narrative.

No comprehensive survey of this nature could ever hope to be a definitive work. The aim has not been to produce an encyclopedia of names and dates of inventions and inventors. Far from that. The effort has been to tell the history of technology with all its various ramifications, and in a unified, integrated manner which will inform and stimulate the reader.

Future historians will undoubtedly come forward with new information that will affect some of the details of technological development presented here; perhaps they will form new interpretations which might revise some of the major outlines. Yet these volumes will stand as a reflection of what historical scholars in the mid-20th century viewed as significant in the development of technology and the first faltering attempts to understand that history in terms of the total social and cultural picture.

The editors wish to thank Professor Paul J. Grogan and Dr. Donald F. Kaiser, who have served as executive editors of the project and who have taken care of the innumerable details inevitable in a large-scale collaborative effort of this type. They have been ably assisted in this work by Mr. Gerald W. Sielaff and Professor Thomas J. Higgins, who were helpful in reading and reviewing the manuscript. This book grew out of a study course jointly prepared by the United States Armed Forces Institute and the University Extension, University of Wisconsin.

June 1967

MELVIN KRANZBERG
CARROLL W. PURSELL, JR.

Contents

Part I
The Emergence of Technology

1 / The Importance of Technology in Human Affairs

MELVIN KRANZBERG AND CARROLL W. PURSELL, JR.

OUR PRESENT TECHNOLOGICAL AGE

In the late afternoon of November 9, 1965, a small electrical relay in a power station in Ontario, Canada, failed. Within a few minutes the flow of electric energy throughout much of the northeastern section of the United States and part of Canada had ceased. Some thirty million people, including those in the great metropolitan areas of Boston and New York, were plunged into darkness. Coming as it did, during the evening rush hour when people were on their way home from work, the shutting off of electric power left hundreds of thousands of New Yorkers stranded in subway trains, confined in elevators stalled between floors of towering skyscrapers, or caught in monstrous traffic jams created by the absence of traffic lights. Even when they finally reached home, many of the now-disconcerted city-dwellers found it to be without warmth, without hot food, and without light. Here was a dramatic demonstration of modern man's dependence on the machine.

Disaster was narrowly averted. Emergency generating equipment allowed essential equipment to function in hospitals and institutions, and with a sense of shared adventure, Americans sought to help their neighbors in a surprising display of good humor and humanity. The great urban centers were able to limp along through the night without many of the technological devices and comforts which characterize life in 20th century America. Yet, had the shutdown of power lasted over a much longer period, it is clear that a considerable disaster could have occurred and that much of civilization as we know it would have been seriously disrupted.

For the fact is that we live in a "Technological Age." It is called that, not because all men are engineers, and certainly not because all men understand technology, but because we are becoming increasingly *aware* that technology has become a major disruptive as well as creative force in the 20th century. The "biggest blackout" of November 1965 gave ample proof of the role of technology in determining the conditions of our life and heightened our awareness of our dependence upon machines, tools, vehicles, and processes.

Equally important, the "biggest blackout" also demonstrated the close relationship between man and his machines from another angle. For while the immediate cause of the power failure was apparently the breakdown of a mechanical component—an electrical relay—this failure might not have occurred had prior decisions been taken to provide "backup" systems, nor would it have extended over such a wide area had the man in Ontario monitoring the power switches acted immediately on the information given him by the dials on his control panel (when he saw the power drop in the Canadian system, he could have switched off the American connection and prevented the power loss in the New York system). Once the blackout had occurred another human failing was revealed: the power company serving New York City was unable to restart its plants immediately because no auxiliary equipment had been provided for that purpose, it being incomprehensible to the engineering mind that such an event could occur.

What distinguishes our age from the past is, first, our belated recognition of the significance of technology in human affairs; second, the accelerated pace of technological development that makes it part-and-parcel of our daily living in ever-increasing measure; and, third, the realization that technology is not simply a limited or local factor but encompasses all men everywhere and is interrelated with nearly all human endeavor.

Man has always lived in a "Technological Age," even though we sense that this is particularly true of our own time. The modern tractor-driven plow represents a higher level of technology than the heavy, crooked stick with which primitive man—or, rather, woman—scratched the soil; and the hydrogen bomb is an infinitely more complex and lethal weapon of destruction than the bow and arrow. Nevertheless, the stick-plow and the bow-and-arrow weapon represented the advanced technology of an earlier era. The heavy stick with which our primitive ancestors prepared the soil for planting enormously increased their ability to wrest a living from an inhospitable and unpredictable nature. Similarly the bow and arrow greatly added to their larder when used to kill game for food. And when used upon their own kind, bow-and-arrow weaponry also gave the first possessors a decided advantage over an enemy who still relied upon rocks and clubs and who could be brought down from afar before their close-range weapons could be brought to bear.

WHAT IS TECHNOLOGY?

While the influence of technology is both widespread and fundamental, the term cannot be defined with precision. In its simplest terms, technology is man's efforts to cope with his physical environment—both that provided by nature and that created by man's own technological deeds, such as cities—and his attempts to subdue or control that environment by means of his imagination and ingenuity in the use of available resources.

In the popular mind, technology is synonymous with machines of various sorts—the steam engine, the locomotive, and the automobile—as well as such developments as printing, photography, radio, and television. The history of technology is then regarded as simply a chronological narrative of inventors and their devices. Of course, such items form a part of the history of technology just as chronologies of battles, treaties, and elections form a part of military and political history. However, technology and its history encompass much more than the mere technical devices and processes at work.

An encyclopedic five-volume work on the history of technology, edited by the late Dr. Charles Singer, defines its subject as "how things are commonly done or made . . . [and] what things are done or made." Such a definition is so broad and loose that it encompasses many items that scarcely can be considered as technology. For example, the passage of laws is something which is "done," but the history of law certainly is not the history of technology.

An element of purpose is stressed in another definition of technology as "man's rational and ordered attempt to control nature." Here the definition is too tight, for while it would include much of technology, many elements would not fit within its limits. The development of certain kinds of toys, for example, does not constitute an attempt to control nature. Furthermore, not all technology exists for the purpose of control, nor, as we shall see in these two volumes, has all past technological endeavor been rational and systematic.

In addition, much of man's technology is devoted to elements which are part of his physical environment but which are not necessarily part of "nature." The various means that man has devised for purposes of controlling the flow of traffic in congested cities are in response to a highly civilized and urban environment which is not a part of the natural environment. Any definition therefore must be extended to include the man-made as well as the natural environment.

To limit the definition of technology to those things which characterize the technology of our own time, such as machinery and prime movers, would be to do violence to all that went before. Indeed, a good case can even be made for considering magic as a technology, for with it primitive man attempted to control or at least influence his environment—a perfectly straightforward goal of all technology. If we now feel that our ancestors used their magic without much success, let us not fall into the error of equating technology only with *successful* technology. The past abounds with failures—schemes that went awry, machines that wouldn't work, processes that proved inapplicable—yet these failures form part of the story of man's attempts to control his environment. Albeit unsuccessful, many of these failures were necessary preliminaries toward the successes in technology.

Sometimes technology is defined as applied science. Science itself is viewed as an attempt by man to *understand* the physical world; technology is the attempt by man to *control* the physical world. This distinction may be briefly put as the difference between the "know-why" and the "know-how." But technology for much of its history had little relation with science, for men could and did make machines and devices without understanding why they worked or why they turned out as they did. Thus for centuries men produced usable objects of iron without knowing the chemical composition of iron and why the various changes occurred in smelting and working it; indeed, they could successfully make things of iron even when they had false theories and incorrect understanding of metallurgical processes. Even today much technology does not represent an application of science, although in such sophisticated technologies as those involving nuclear science, scientific understanding is closely linked with technical accomplishment.

Technology, then, is much more than tools and artifacts, machines and processes. It deals with *human work*, with man's attempts to satisfy his wants by human action on physical objects.

We must use the term "wants" instead of "needs," for human wants go far beyond human needs, especially those basic needs of food, clothing, and shelter. Technology administers to these, of course, but it also helps man to get what he wants, including play, leisure, and better and more commodious dwellings. He cultivates a taste for more exotic foods than those necessary to still the pangs of hunger. He yearns to achieve faster and more lasting communication with others. He wants to travel abroad and be entertained, and to fill his house and his life with beauty as he sees it.

Emphasis upon the "work" aspect of technology shows that it also involves the organization as well as the purpose of labor. For example, the pyramids of Egypt are monuments to the technology of that early civilization. The pyramids demonstrate even today how much can be done with very little in the way of tools but with much ingenuity and skill in the organization of labor. In our own day the efficiency of new tools and processes can only be maximized by utilizing efficient organization. We are increasingly forced to think in terms of "systems," and even decision-making now can sometimes best be done by machines.

THE COMPREHENSIVENESS OF THE HISTORY OF TECHNOLOGY

The nature of invention itself requires that the history of technology be more than a mere tabulation of inventors and their creations. Invention does not come about simply because a creative person decides that he is going to "build a better mousetrap." Invention is a social activity, much affected by social needs, by economic requirements, by the level of technology at a given time, and by sociocultural and psychological circumstances. The fact that some inventions "come before their time" indicates the importance of the sociocultural milieu, and it raises the whole question of the nature and origin of creativity.

Even if we were to try to limit the history of technology to inventions, we would be forced to deal with many social, political, economic, and cultural aspects of civilization. For example, ever since World War II, which stimulated nationalistic feeling in Russia, Soviet scholars have been publishing reports of "firsts" by Russian inventors and scientists. Although most Americans have shrugged off these Russian attempts to claim priorities in inventions, the facts are that some of the Russian claims are well founded and that individual Russian scientists and inventors during the 19th century were the peers of their counterparts in Western Europe and the United States. Yet even if we were to accept all of the Russian claims, we would still face another question: why did Russia lag behind other European nations in industrialization? The answer to that is not to be found in the mental prowess and inventive capabilities of the Russian people, but rather in the complex of social and political circumstances under which invention and innovation thrive.

It is not enough simply to discover who first had the idea for an invention, nor even who first patented the device; we must also see when, why, and how this invention actually came into use. The answers to these questions involve much more than the purely technical factors, which is why the history of technology is such a comprehensive subject. It covers every aspect of human life and must go back to the very beginnings of the human species.

TECHNOLOGY AND THE EMERGENCE OF MAN

Anthropologists seeking the origins of mankind have attempted to differentiate between what constitutes "almost man" and the genus *Homo*, man himself. The chief distinction they have found is that man employed tools, thereby distinguishing him from his almost-human predecessors.

Man, as we know him, surely would not have evolved or survived without tools. He is too weak and puny a creature to compete in the struggle with beasts and the caprices of nature if armed with only his hands and teeth. The lion is stronger, the horse is faster, and the giraffe can reach farther. Man has been able to survive because of his ability to adapt to his environment by improving his equipment for living. As Gordon Childe has pointed out, the specialized equipment man uses differs significantly from that of the animal kingdom. An animal is capable of using only that equipment which he carries around with him as parts of his body. Man has very little specialized equipment of this kind. Moreover, he has discarded some of the organic "tools" with which he started and has relied more on the invention of tools, or extracorporeal organs, that he makes, uses, and discards at will. This invention and use of extracorporeal equipment has enabled man to adapt to nature and to reign supreme among the animals on earth.

Archaeological anthropologists continue to discover older and older fossils of human-like skeletons, almost always surrounded in their graves by primitive tools or implements. It has even been postulated from these findings that technology is perhaps responsible for our standing on two feet and for our being *Homo sapiens*, Man the Thinker. Thus, man began to stand erect so that he might have his forearms free to throw stones; he did not throw stones simply because he was already standing erect. Modern physiology, psychology, evolutionary biology, and anthropology all combine to demonstrate to us that *Homo sapiens* cannot be distinguished from *Homo faber*, Man the Maker. We now realize that man could not have become a thinker had he not at the same time been a maker. Man made tools; but tools made man as well.

TECHNOLOGY AND THE ADVANCE OF CIVILIZATION

The very terms by which we measure the progress of civilization—Stone Age, Bronze Age, Iron Age, and even Atomic Age—refer to a developing technological mastery by man of his environment.

One indication of the start of civilization—the development of settled communities—rests upon a technological innovation: agriculture. In the prehistory before that time, men had been nothing more than hunters; in a sense, they had been parasites upon nature. We do not know exactly how or when agriculture began. Recently there has been found evidence of agricultural communities in the Middle East dating as far back as 8000 years ago. Once men discovered that they could co-operate with nature by sowing seeds and waiting for nature to perform the miracle of growing crops, there arose the possibility of settled and civilized life.

Unlike the hunter, the agriculturist could not afford to live in constant conflict on all sides. Rather, he had to learn to co-operate not only with nature but with other human beings. If he spent too much of his time in fighting, he could have neither the time nor the energy for carrying on his agricultural pursuits. Yet if he ran away from his enemies, his crops would go unattended and he would lose his means of livelihood. With the introduction of agriculture, therefore, civilized society began to emerge. This both spawned and depended upon man's dawning awareness that he must live and work together with others if he was to survive. It is a reasonable, though optimistic, extension of this concern to hope that man has, in the many thousands of years since, begun to realize that he is part of a larger community and that there is a need to co-operate with other human beings in order to advance his control over nature. No longer are his actions, thoughts, or aspirations confined to his immediate locale. Rather he must learn to consider all mankind since he has acquired the skill literally to reshape or destroy the world with the technology at his command.

In terms of energy, there has been transition from human muscle power to that of animals, to wind and water, to steam and oil, to rockets and nuclear power. With machines, we have witnessed change from hand tools to powered tools, from craft shops to mass production lines, from the beginnings of job definitions and quality control to computer control of factories.

The advance of material civilization has not been without interruption, and cannot be portrayed on a graph as a straight line climbing constantly upward through time. Instead, periods of great technological progress have sometimes been followed by eras of relative stagnation, during which time very little advance was made in man's control over nature. Moreover, materialistic techniques may progress while cultural activities such as music, art, literature, and philosophy seem to retrogress. There have been times when religious, philosophical and artistic activities achieved great heights while technology seemed to rest on a plateau.

TECHNOLOGY AND WESTERN CIVILIZATION

Technology and its modern twin, science, are the distinguishing hallmarks of recent Western civilization. The Scientific Revolution of the 17th century was reinforced by a Technological Revolution of the 18th and 19th centuries. These revolutions brought something to our culture that had been unknown to the earlier Western civilization of Greece and Rome or the Eastern civilizations of India and China. Science and technology differentiate our society from all that has gone before in human history and all that has taken place in other parts of the world. While the roots of our Western religious and moral heritage can be found in the Judeo-Christian-Greek tradition, contemporary Western culture is perhaps based more upon science and technology than upon religious and moral considerations.

If we wish to test the hypothesis of the uniqueness and significance of Western civilization, we need merely ask ourselves what "Westernization" means to non-Western societies. To them, it means the acquisition of the products of Western technology, not the political institutions, religious faiths, nor moral attitudes which the West has developed over the centuries. When we speak of the "Westernization" of Japan during the late 19th and early 20th centuries, we refer to the acceptance and the borrowing of Western technology by the Japanese. Similarly, many of the underdeveloped nations of the world want to borrow from the West today. While they often specifically reject Western moral and social attitudes, they want desperately the material advantages which technology can bestow upon them, even though they criticize the West's "materialism." To much of the world, the "American Way of Life" does not mean democracy, much less free enterprise. It means material abundance within the reach of all men; and social and political "isms" become relevant only when they retard or encourage the gaining of that goal.

The attitudes and values of Western man himself have been deeply affected by technological advance. For centuries men thought that it was their lot to earn their living by the sweat of their brows, and there was little hope for material abundance here on earth. In the past, technology was primarily concerned with furnishing the human needs of food, clothing, and shelter. It still serves to fulfill those needs, but now so successfully that modern technology for the first time in history has produced in the United States a society which has not only a surplus of goods but a surplus of leisure as well.

THE HUMAN AND SOCIAL ELEMENTS IN TECHNOLOGY

There have always been those, especially since the Industrial Revolution, who have seen new technologies as a threat to "human values." In the late 18th century the excesses of the growing industrialism in Great Britain—symbolized in William Blake's description of the "dark Satanic mills"—tragically alienated a large and influential segment of our common humanistic tradition. Many within that tradition—including artists, writers, and philosophers—have to this day continued to deplore the Industrial Revolution and our modern urbanized and industrialized society that has issued from it. This estrangement has led in some cases to a failure of the humanities to perform their functions as prophets of mankind. The tragedy is that this alienation, which leaves us all the poorer, seems so unnecessary.

We have come to think of technology as something mechanical, yet the fact remains that all technical processes and products are the result of the creative imagination and manipulative skills of human effort. The story of how man has utilized technology in mastering his environment is part of the great drama of man fighting against the unknown.

Furthermore, the significance of technology lies not only in the uses of technology by human beings, but in terms of what it does to human beings as well. If we regard the telephone, for example, only as a system of wires through which a tiny current passes from mouthpiece to earphone, it would seem to have little interest, except to technicians and repairmen, and virtually none to historians except for the antiquarian desire to discover who conceived the idea and reduced it to practice. But the greater significance of the telephone lies in the newly found ability to transmit voice communication between persons over long distances. It is the communications function of the telephone that gives it importance. The principal significance of this particular bit of technology—as in the function of every technological item—is its use by human beings.

The essential humanity of technology is nowhere better demonstrated than in the fact that it too, like the noble heroes of Greek tragedy, carries within it a fatal flaw which threatens always to lay it low. It is no longer possible, if indeed it ever was, to believe that progress is either inevitable or uniformly beneficent. Granted that technology has contributed to man's material progress, its social repercussions have not always been a boon to all segments of the population. There have been victims of the rapid social readjustment to industrial growth, notably the factory workers in Britain during the early days of the Industrial Revolution. This has led some critics to claim that technology presents two faces to man: one benign and the other malignant. The latter face is most frequently represented today by the destructive potentialities of intercontinental ballistic missiles armed with nuclear warheads.

Yet it advances our understanding very little to say that technology wears two faces, as though one were comedy and the other were tragedy. Technology, in a sense, is nothing more than the area of interaction between ourselves, as individuals, and our environment, whether material or spiritual, natural or man-made. Being the most fundamental aspect of man's condition, his technology has always had critical implications for the status quo of whatever epoch or era. Changes have always rearranged the relationships of men—or at least of some men—with respect to the world about them. Not a few of the historic outcries against technology (or, more properly, against some changes in technology) have been essentially protests against a rearrangement of the world's goods disadvantageous to those who complain.

Some "defenders" of technology claim that it is neutral, that it can have socially desirable or evil effects, depending on the uses which man makes of it. To deny this and to say that technology is not strictly neutral, that it has inherent tendencies or imposes its own values, is merely to recognize the fact that, as a part of our culture, it has an influence on the way in which we behave and grow. Just as men have always had some form of technology, so has that technology influenced the nature and direction of their development. The process cannot be stopped nor the relationship ended; it can only be understood and, hopefully, directed toward goals worthy of mankind.

2 / The Beginnings of Technology and Man

R. J. FORBES

Technology is as old as man himself. Man was evidently a "tool-making primate" from the day when the first human-like creatures roamed on earth, some 25 million years ago. Such very early human remains as that of the "Peking Man" (so called because the fossil bones were found near Peking), dating back about half a million years, are accompanied by stones selected and often shaped to be used as tools. Even when we do find remains of fossil men not accompanied by tools, it is probably because they were trapped by death beyond their usual dwelling site, and therefore without their usual tools.

THE BIOLOGICAL BASIS OF TOOL-MAKING

Although lower animals now and then use a stone or other natural object to break a shell or to help them in some other part of their feeding or breeding habits, only the primates use "tools" in the real sense of the word to cope with unusual situations. But even among these there are differences. While chimpanzees and other apes will sometimes use sticks or stone to perform certain tasks and will sometimes put two sticks together to bring food within reach, this has always remained "a mental isolation of a single feature." The regular shaping of such tools or aids never became a habit with apes; but it did become a habit with man.

Man's earliest natural tools were his hands and his teeth. Biologists tell us that the origin of tool-making is probably related to the advanced functional anatomy of the human hand and its development. Further, though the muscles of the hands of monkeys differ little from those of early man, the nervous mechanism, by which men are able to direct the movements of such muscles is of a

Fig. 2-1. Prehistoric man making first flint instruments. Drawing by Charles R. Knight. (*The Bettmann Archive*)

Fig. 2-2. Stone Age implements, quartz with sharpened edges. (*The Bettmann Archive*)

finer structure, thus making human hands more flexible than those of other primates. The earliest tool-makers were largely carniverous, however, and they needed sharper instruments for hunting, killing, and cutting meat—sharp-edged stones such as could be chipped from pebbles by hammering them with other stones. The larger and more efficiently organized cerebrum of early man allowed him slowly to grope toward supplementing his natural tools.

Not only the development of manual activity but also the development of speech guided man's earliest tool-making activities. The evolution of speech, which "gave everything a name," perhaps helped to differentiate man's world. Different tools of special types came to be used for hunting, fishing, and the making of clothes and shelters.

EARLY STONE TOOLS

Archaeologists identify and classify early human settlements by the shape and type of the tools found there. The first tools were of stone, and the earliest stone tools (called "eoliths," because they date from the Eolithic or earliest Stone Age period over a million years ago) were pebbles already preshaped by nature and simply picked up from a river bed. Very slowly man learned to shape such stones by striking them with other stones, and a limited number of more or less standardized shapes for chopping and cutting were developed. About a hundred thousand years ago primitive man was no longer content with his chopper tool and pointed flakes; he began to make more specialized tools: pear-shaped "hand axes," scrapers, knives, pointed stones, and so forth.

The growing range and complexity of man's tools ran parallel to the development of his activities and achievements in other respects. Archaeological data show that he had now fully become *Homo sapiens,* Man the Thinker. By then speech and language as important adjuncts of thought as well as communication must have been fully developed. His interest, not only in terrestial objects and phenomena but also in those of the heavens, became more and more evident in the drawings he made on his tools, belongings, and later on the walls of caves. These show not only the rudiments of science but also demonstrate his technology, particularly his hunting techniques. The shrewd characterization in these drawings and the clear ordering of his observations which they demonstrate show the development of man's mind, though we are not allowed any intimate contact with his thoughts until the earliest writing appears in the Near East about 3500 B.C.

INVENTION AND DISCOVERY

Two elements governed man's technical progress. Discovery, the recognition and careful observation of new natural objects and phenomena, is a very subjective event until it leads to some practical application shared by others either directly or indirectly. Invention, however, is a mental process in which various discoveries and observation are combined and guided by experience into some new tool or operation. Much experience was needed to lead to truly important inventions, and hence, the material progress of ancient man was very slow.

THE USE OF FIRE

Man's earliest conquest was fire. Ancient myths agree that man was originally threatened and alarmed by forest fires, but he eventually turned this phenomenon into a boon for mankind. The earliest users of fire had to keep going those fires which they might have found in nature, for they had no means of producing it at will. Not until Paleolithic times (Old Stone Age, which lasted about a million years, to 8000 B.C.) in Europe and Asia did man discover the percussion method: lumps of flint and pyrites struck together would give off sparks with which tinder, straw, or other inflammable fuel could be set alight. Other methods, based on the heat of friction produced by rubbing pieces of wood together, as in the fire-saw and fire-drill, seem to have originated in Southern Asia at a somewhat later date.

Fire was the most important discovery of Paleolithic man, who not only warmed his body but also applied fire to the preparation of food. The birth of the art of cooking meant that he could now prepare and "predigest" foodstuffs and augment the diet of fruit, roots, and raw meat on which he had lived for millennia. He not only increased his range of foodstuffs enormously, and was able to vary his diet, but more importantly, by drying meat and other foods with the help of fire, he could now lay in a supply to help him subsist during the lean seasons.

Gradually, various ways of preparing food were developed from the original method of simply holding the item to be cooked over the fire or placing it in the fire. Food could be cooked on top of heated stones and even in glowing embers. As soon as suitable containers could be found or fashioned, boiling, stewing, and frying became regular ways of preparing food; and from primitive means of cooking food in preheated vessels, the art of baking evolved.

Cooking actually led to the invention of suitable containers and other kitchen utensils, of braziers (portable fires) for domestic heating, of grates, and finally of bellows, which were an improvement on the original fan or blowpipe with which the fire was encouraged. It is clear from later documents and from archaeological finds that many later industrial processes involving heat, such as metallurgy, pottery, and brewing, used the accumulated experience of prehistoric cooking. In all early languages industrial operations such as heating, drying, steaming, baking, and washing are indicated by terms derived from the kitchen, and even in such a late growth as alchemical practice (which is hardly older than the eighth century B.C.), it is clear that not only the terms but also apparatus such as filters and water-baths were originally kitchen equipment which was later adopted to chemical use.

Tending the fire required a suitable place in the middle of the hut or cave, usually a mud-plastered, walled spot, the hearth. This was the birthplace of pottery, for prehistoric man soon discovered how hard the mud plaster became after being thoroughly heated. Also from the domestic hearth were developed our kilns, ovens, and industrial furnaces of a later period.

Present evidence points to a fairly late emergence of the art of pottery. The oldest farming villages of the Near East did not contain pots, but by the time of such early urban settlements as Jericho (6000 B.C.), pottery was known. The earliest pottery took the shape of the gourds it was designed to displace, an early example of the tendency of inventors to adopt a natural form in attempting to replace a natural function.

The fuel used by prehistoric man was of the most primitive kind. Dead branches, dry wood, and shrubs were commonly used, but so was such low-grade material as dried bones. In the East, thorny shrubs, withered sticks, twigs, and dried dung were cheap fuels, supplemented by straw and other farming refuse as soon as agriculture was introduced. Only the mountain regions had sufficient firewood. In later times this wood was made into charcoal for industrial purposes. For all ancient industrial processes depended on charcoal, and this charcoal-burning (together with the herds of goats which grazed in the barren mountain region) was responsible for the deforestation of the Mediterranean area in antiquity. Much later, the Greek philosopher Plato (427-347 B.C.) deplored the disappearance of the many forests of Athens due to charcoal-burning and ship-building.

Strangely enough, prehistoric man seldom used convenient outcroppings of coal and lignite. In certain regions of Czechoslovakia, mammoth-hunters seem to have utilized local coal-outcrops for their camp-fires, but this was an exception. By the time (500 B.C.-400 A.D.) of Classical antiquity, however, peat and lignite were well known, although the burning of peat by the early inhabitants of the Low Countries was considered a proof of barbarism and poverty by the Roman author Tacitus (55–117 A.D.) Coal came into more regular use in Roman Britain, but its use in earlier times is hypothetical. This limited supply of second-grade fuel was a major deterrent to the development of industrial processes during the whole of antiquity as well as for many centuries thereafter.

The use of fire also led to the introduction of illuminating devices. In certain cases threading a wick through such oil-rich creatures as the stormy petrel or the candlefish provided a reasonable flame. In most cases, however, the oil was first extracted from the animal or fish, and then the fish-oil, tallow, or other type of fat was used in hollowed-out stone lamps, with a wick hanging over the rim or floating in the middle of the oil. Such stone lamps were in use in Paleolithic times; a specimen was found in the Lascaux caves of France, where it had served the prehistoric painters who decorated the walls there some 12,000 years ago. Sometimes resinous splinters or torches were used. Later, pottery or metal bowls were manufactured, multi-wick lamps being made by forming several spouts along the rim of the bowl to hold more wicks. These gave more stable lighting than the splinter-lights or rush-lights (peeled rushes often dipped in fat or grease to form a primitive candle), which were apt to give off sparks and thus constituted a grave danger-spot in the house. Torches and tapers often consisted of bundles of splinters or rush-lights, sometimes dipped in bitumen or resin. Along the coasts, shells were often used as lamps, for the rim was naturally shaped to hold wicks. For centuries such open lamps were the only form of lighting that mankind knew. Even after the advent of the candle, the vast majority of people usually rose and went to bed with the sun, for oil for illuminating purposes was too expensive for people who were merely eking out a living.

ADVANCES IN STONE TOOLS

As foraging, hunting, and fishing began to provide a wider range of food supply, tool-making outgrew its elementary stage. By 15,000 B.C. more differentiated

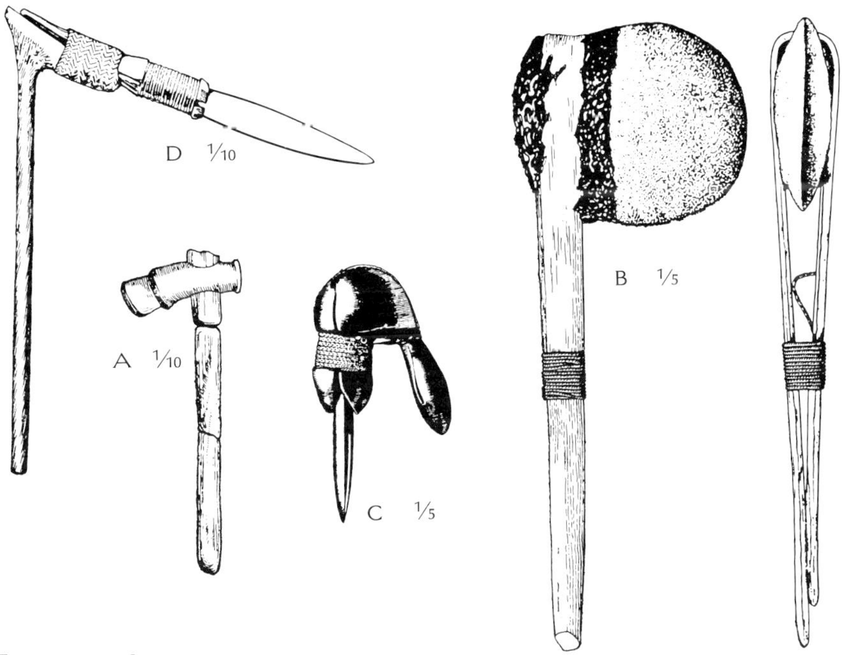

Fig. 2-3. Methods of hafting axes and adzes with stone blades. (A) Axe with blade inserted in a socket of deer antler, which is perforated for the haft. (B) Axe with blade gripped in the bend of a withy, the attachment strengthened with hardened gum and a lashing. (C) Adze with blade bound in a split of the shouldered haft. (D) Adze with blade lashed in a wooden sleeve, which is attached to the shouldered haft by a band of plaited cane. (From Singer *et al.*, *A History of Technology*, Oxford, 1954-58)

and better tools were being made in Europe and the Near East. The earlier technique had been to shape a stone tool by beating it against a hammerstone or between a hammer and a stone anvil, but as more suitable stones for toolmaking were found, flaking techniques were more widely used. Flint, obsidian, or fine-grained lava could be used to produce flakes by applying a stroke with a stone hammer or mallet at certain points of the surface at exactly the right angle. Experience soon taught man to strike his nodule of flint in such a way that the form of the flake detached had an accurately predetermined shape.

It was also possible to produce long narrow flakes or blades for other purposes. This could be achieved by detaching the flakes by means of a wooden or bone punch struck by the mallet or hammerstone. Thus a blade could be shaped into a tool with the two edges sliced down obliquely at one end, forming a narrow chisel edge. This tool is called a burin, of which over twenty types were used by the early nomadic food-gatherers. All these tools, whether hand axes, chopping tools, blades, or burins, could be retouched by secondary flaking and by pressure-flaking (splitting off small flakes along the cutting edge to make it more even) to produce a wide range of special tools for each technical operation such as chopping, cutting, adzing, sawing, scraping, etc.

We must remember, however, that man did not use stone tools only. Unfortunately, remains of the less durable materials such as wood, bone, and ivory have disappeared. Only under exceptional circumstances, such as arise in the salt mines of Carinthia (Austria) or in swamps, do we find such perishable objects as wooden shovels and composite tools consisting of flint pieces with handles, such as sickles. Nevertheless, the few objects found in prehistoric camps and settlements prove that bone tools were in fact used. Splinters could be separated from bones to serve as needles and awls. Wood could be worked by using flint scrapers for chisels, so as to produce dug-out canoes (6500 B.C.) and even, by the end of the Stone Age, to produce primitive tenon (projection cut at the end of piece of material) and mortise (hole cut in second piece to receive projection on first) joints for wooden objects. Antlers were used as picks in the mining of flint by that time. Bone or wooden handles held small flint blades; spears and arrows had hafted heads made of various materials. Hence at the beginning of the New Stone Age (also known as the Neolithic Age, and lasting from about 8000 B.C. to 2500 B.C.) men already utilized a considerable arsenal of specialized tools and weapons. Though our knowledge of these is based on the relatively few examples which have survived, we often have remains of the finished products made with their aid.

THE BEGINNINGS OF MINING

Good flaking stones that could be used for tools and weapons were not too common. Sometimes they were found along the shore where chalk outcroppings had been worn away by the waves and previously embedded nodules of flint had been washed onto the beaches. Flint nodules could also be found inland at several surface locations in Europe and the Near East. The best materials, however, could only be obtained by digging vertical shafts into the limestone to reach the stratum some 30 to 400 feet below the ground which contained flint nodules with better flaking qualities. Prehistoric men probably knew where to dig for flint, presumably by following surface outcroppings.

Many of these Stone Age flint mines have been discovered, and some have even yielded skeletons of miners—with their tools—surprised by the crumbling of their "pipes" or shafts. From these remains we get the impression that originally the tribesmen must have made seasonal excursions to such sites to obtain the flint they needed. Only gradually, by the dawn of the Bronze Age (about 2500 B.C.), did flint mining become a separate profession, with miners living on the spot year round and probably fashioning flint tools to be traded by itinerant hawkers. The same must have happened in the case of the harder, non-flaking stones like diorite and lava which, from Neolithic times onwards, were shaped by grinding and polishing into completely new forms.

Another early product to be mined, apart from the ochres and other colored earths used as pigments for decoration purposes, was salt. This ingredient became very important as man's changing diet—extended by the art of cooking—provided him with more carbohydrates and as the percentage of raw food and meat went down. The diminishing percentage of foodstuffs with a natural salt content had to be supplemented by the addition of other salt. This was sometimes found as deposits of evaporated salt-spray along the coasts, but it was usually obtained by the evaporation of saline waters from such springs as those of the Halle region (Central Europe), or the mining of rock-salt strata, to be found fairly frequently in the deserts of the Near East and the eastern Alps and Carpathians.

TECHNOLOGY AND ASPECTS OF EARLY SOCIETY

All these products—flint, hard stone for tools, and salt (usually traded in the form of "bricks"), together with the semi-precious stones which the prehistoric tribesmen admired, as well as sea-shells from the East—constituted the objects of trade over relatively long distances. We can still trace these trade routes through the river valleys and over mountain passes from the mines to other settlements and even to the coast by the discovery of the treasure and merchandise of peddlers who, when threatened by danger, buried their possessions and often never recovered them.

This developing trade is just one of the signs that society was changing, and with it, technology. For technology is a social product in this sense, that it is one of the interacting factors in a society, which in those early days was still very much limited by its food supply pattern. Its tools and technology were first of all aimed at the support and extension of this food supply, though other factors were also important. Environment and available materials can limit the forms and extension of this technology, but they cannot rob man of his ingenuity. Early food-gatherers and hunters devised a full range of tools directed toward their foraging, hunting, and fishing; these tools have been classified as "crushers," "piercers," and "entanglers." Included among these were the spear and spear-thrower, the simple and composite bow (typical of the hunter of herds), the arrowhead and harpoon, the blow-gun, lasso and bolas, and fish hooks and traps, such as are still used by primitive tribes in Africa and South America.

Nor did the insecurity of daily subsistence prevent early man from creating and cultivating graphic and plastic arts. This was partly because primitive superstition (itself a form of technology) believed that one would be more successful in hunting and fishing if one could draw pictures of the objects for which one hunted and fished. Perhaps also the aesthetic element—the instinct to adorn or beautify—lies deep within man. Using natural pigments crushed in a mortar with pestle (and mixed with animal fats or water in most cases), primitive man painted with his finger tips, brushes, or dry-point crayons. He engraved tools and weapons on the walls of his caves with incised lines or pecking. He also sculpted. Rock or mud were modeled in the round or in relief, and he used ivory, antler, and stone as well. Still, in these early societies there was no specialization or division of labor, and full-time craftsmen were unknown. This holds true even for the larger part of the Neolithic Age, when the gradual development of stock-breeding and agriculture made life safer and more dependable.

THE DOMESTICATION OF ANIMALS

Man gradually tamed such animals as roamed about his settlements or camps and which sometimes may have been scavenging. By Magdalenian times (the late period, 17,000-8000 B.C., of the Old Stone Age) the dog was already domesticated, and probably man had already tamed the first reindeer, goats, and sheep. During the Neolithic Age (Late Stone Age) pigs and cattle joined the range of domestic animals, the horse and the onager (wild ass) being tamed toward the end of that period. From the remains of such tamed animals we may conclude that most domestic animals of Neolithic date had already lost some of the physical characteristics of their wild ancestors and showed distinct traces of domestication. Neolithic and Bronze Age man tamed such animals intentionally. The oldest Egyptian wall-paintings show that even antelopes and gazelles were domesticated, possibly for economic purposes, as their skins and the leather prepared from them were much prized.

The last phase of domestication was characterized by the dominance of such economic purposes. Animals were bred because of their meat, hides, or milk, and man gradually acquired sufficient biological experience to be able to produce more "specialized" animals, the products of which were refined by various techniques. During the Neolithic Age (*c.* 7000 B.C.) we find more and more types of sheep, goats, and cattle bred to suit man's needs. At the same time, the deforestation and cultivation of larger areas of virgin soil began to cause extermination of wild species and to make the domesticated ones more dependent on man's care.

THE BEGINNINGS OF AGRICULTURE

Similar needs led to man's attempts to free himself from the limitations of vegetable food supplied by nature in the wild state. He did not want to be dependent on roots and fruit only, but wanted regular supplies of green vegetables, nuts and oil seeds, cereals, and condiments. In most cases it is impossible to

Fig. 2-4. Plowing and hoeing in Egypt, *c.* 1900 B.C.
From a tomb at Beni Hasan.

determine where or when such cultivated foodstuffs were first discovered and used. In the case of cereals, however, it is fairly certain that they originated from the cultivation and cross-breeding of wild grasses growing in Syria and the highlands to the North, that is, in the Fertile Crescent running from the Egyptian border along the Arabian desert to the delta of the Euphrates and the Tigris.

From 8000 B.C. onwards, the Fertile Crescent played its role as the center of agriculture, whence the knowledge of growing such improved foodstuffs spread to the Mediterranean region and Western Europe. Although most of the cereal grains developed first in the Near East, Europe gave the world two new cereals, oats and rye, which became important crops in Roman times. Rice did not come west from the Orient until about 1000 B.C.

Men also began to cultivate intentionally such plants as flax and hemp and to use their fibers for textiles and rope-making; they also grew plants from which they could extract oils and dyestuffs. The advent of agriculture brought more or less permanent settlements and tended to displace the earlier nomadic way of life. Yet, in these early agricultural settlements there were still few traces of technological specialization; every man had to be a jack-of-all-trades, and every homestead was virtually self-sufficient. Only in a few mining centers do we find signs of craftsmen at work trading their products through peddlers.

Agriculture stimulated technology in various ways. New tools were needed: the hoe, plow, harrow, to prepare and till the soil. Yet, during the thousands of years of prehistory, from the first stone tools to the first cities, technical progress, whether by discovery or by deliberate invention, was very slow. The pace of this development is clearly demonstrated by the late invention of such a simple agricultural tool as the flail, a device for threshing grain.

After the birth of agriculture, prehistoric man was faced with the problem of detaching the grains from the dried heads of the wild grasses he cultivated and improved. He solved this problem by beating sheaves of grain against the compacted mud floor or by spreading it out on the floor and beating it with sticks. Sometimes he had the animals which he had domesticated tread on the grain; sometimes he used the crushing effect of a threshing sledge, studded with flint or iron nails, on the grain spread out over the threshing floor. Much later, even the Romans, with their large farms aiming at mass production of wheat and other cereals, did not possess the flail. It was not until the fourth century in Gaul that the ingenious combination of two hinged sticks produced the flail; yet the technical elements and the need for an effective threshing device had already existed for millennia.

BUILDING

The coming of permanent settlements led to early forms of building. Earlier men had been satisfied with simple windbreaks, seeking more permanent shelters such as caves only for longer stays. The earliest European houses were tent-like constructions, often little more than roofed-over pits and hollows. Gradually pole or frame constructions were developed, which led to more solid constructions made of planks, turf, mud, and adobe. The advent of metal tools in the Bronze Age made possible the building of log houses in the forest regions; while in the south the types of houses with a roofed front porch and a room with a central fireplace prevailed. Longhouses and religious buildings of various forms were also erected, involving a more developed wood-work technology.

In the Near East the earliest farming villages consisted of sapling-supported mud huts, later with wattle (woven branches or twigs) walls smeared with daub (mud). Building construction depended on the local materials available. Reed huts were constructed in the lower river valleys, wooden houses where timber was available. In Jarmo by 6500 B.C. compacted mud-walls were used, and at Jericho we find mud-bricks.

THE URBAN REVOLUTION

Technological developments during the Neolithic Age gradually led to a regular production of surplus foodstuffs, which supported what has been called the "Urban Revolution." In the Near East after 6000 B.C. some farming villages slowly developed into urban centers dominating an agricultural area. Trade was no longer in luxuries alone; the farmers brought their surplus grain and food to the city, where skilled, full-time craftsmen traded the articles which they had produced for the food they needed. This was particularly true after the urban centers began to move toward the river valleys, which were being drained and cultivated but which could not provide the timber, metals, and minerals needed by the craftsmen.

TRANSPORT

Despite an increase in trade, transport was still very primitive. On the water, apart from rafts and inflated skins carrying a timber deck (the keleks of Meso-

Fig. 2.5. Models of Babylonian skin vessels. (British Crown Copyright. *Science Museum*, London)

potamia), there were baskets with a high rim and the coracle (Arabic, *quffa*), a sort of raft made water-tight by a hide covering (in Assyria, by a coat of bituminous mastic). Ships made of bundles of reeds were typical for the Nile Valley civilization. The inhabitants of Mesopotamia, like those of prehistoric Europe, also used the dug-out canoe, the prototype of later plank-built ships. Somehow they learned to avail themselves of the propulsive power of the wind, and their little wooden ships circled Arabia and penetrated the Red Sea, where rock pictures show them to have had square sails. Such ships later inspired the Phoenician ship-builders and became the prototype of the Aegean galley. Despite these early adventurers, most early shipping was merely river transport and seldom risked the dangers even of sailing along the coasts of the open seas.

The technological improvement of land transport was even slower. Early overland trade never involved large volumes of goods, but for a very long time was limited to what men or pack animals could carry on their backs. From 7000 B.C. onward, sledges were in use for heavy loads such as the stones used for later prehistoric (Neolithic, the Late Stone Age, period) monuments. The great megalithic monument at Stonehenge, in England, shows the great feats of land and water transport possible even with the most primitive means; its four-ton stones were carried over land and sea from quarries 150 miles away, its thirty-ton stones were carried some 25 miles by sledge.

For a long time archaeologists have reasoned that such heavy stones must have been moved on sledges placed on rollers and that the idea of the wheel probably derived from these rollers. It is doubtful, however, if wheels evolved from rollers, and we still do not know where, when, and how the wheel was invented.

On early tablets found at Uruk in Mesopotamia (3500 B.C.) we see pictures of sledges on four wheels; a set of solid wheels with their axle were carved from a tree trunk, and the sledge was attached to two such units. True wheeled vehicles, with the wheel rotating about the axle instead of being solidly affixed to it, are not found until the days of the Sumerian royal tombs (after 3000 B.C.).

MAN AT THE DAWN OF HISTORY

Starting as one of the members of the primate group, man possessed certain biological prerequisites for technical progress: the capacity to manipulate with his hands and fingers, the ability to develop speech for communication, and, somehow, a capacity for abstract thought. These had enabled him to develop tools, which in turn enabled him to evolve further.

Although early man undoubtedly first utilized wooden and bone implements, only his stone tools have survived the ravages of time, so human prehistory is divided among the Stone Ages (Eolithic, Paleolithic, and Neolithic), with these tools improving in quality and usability over a long period of time. Suddenly—as such things went in prehistoric times—in the fourth millennium B.C. (4000-3000 B.C.) revolutionary changes took place in man's technology and hence his society.

Man had already discovered fire and had crude weapons. But in the thousand years preceding 3000 B.C. there was a spate of inventions and discoveries. The beginnings of agriculture allowed for settled communities, and man changed from a nomadic hunter and parasite upon nature to become an active partner with nature. He domesticated animals and developed agricultural tools. He made textiles and produced pottery. He invented the wheel and the sail to improve his transportation. And, at the turn of the millennium, he learned how to use and produce copper. Then came the most astonishing invention of prehistory, an invention that divided the very epochs of ancient man.

The development of writing took place about 3500 B.C. The earliest documents were cuneiform clay tablets, forming the archives of the administration of early Mesopotamian (Sumerian) temples. These cuneiform tablets are typical of the way in which language and its earliest written expression, pictographs (that is, ideas expressed in the form of visual pictures), developed. The "naming" of all natural things and phenomena now finds its expression in the word lists (onomastica) in which words for all kinds of animals, plants, minerals, and so forth, are arranged in groups which are supposed to be related in some way. The Sumerian language of the ancient Mesopotamians being agglutinative (that is, forming words by adding prefixes or suffixes to the root forms), it lent itself admirably to such groupings, achieving something like the nature of modern organic chemistry (in which, for example, the addition of descriptive prefixes to a root form such as "benzene" leads to words like "paradichlorbenzene"). Such lists, often designating the objects by their external characteristics which had been observed, led to the "discovery" of the "natural order of things"; they

Fig. 2-6. Cuneiform tablet with account of creation of the world. Babylonian, *c.* 650 B.C. (*The Bettmann Archive*)

also aided budding scientists and craftsmen, for the lists were used in teaching pupils how to read and write. The craftsmen used them to set down their experience and to formulate the recipes and instructions they were gradually producing for technology, new terms being formed and read with ease as necessity dictated.

Cuneiform writing consisted of wedge-shaped marks incised on wet clay, a material in which the Tigris-Euphrates Valley abounded. When dried in the sun or baked in a fire, the clay tablets hardened, and a permanent record was left. Almost at the same time the Nile Valley also witnessed the beginnings of writing by brush dipped in ink or dye on papyrus formed by pressing and drying pithy reeds growing near the Nile delta.

With the beginning of the written record—itself a triumph of technology—we pass from prehistorical to historical times. But in prehistory technology had already provided mankind with the bases of civilization: settled communities made possible through agricultural advance, a wide variety of tools, domesticated animals, means of transportation such as the wheel and sail, and the beginnings of some division of labor. Contemporaneous with the invention of writing was to be the beginning of metallurgy and the accompanying change from the Stone Ages of man to the Copper Age and the Bronze Age in the third millennium B.C.

With the beginnings of metallurgy, the Stone Age of man comes to an end; with the beginnings of writing, prehistory comes to an end; with the beginnings of agriculture, man's parasitism on nature gives way to co-operation with nature. Technology thus made possible the beginnings of civilization in the great river valleys of the Near East, in Mesopotamia and Egypt.

3 / Mesopotamian and Egyptian Technology

R. J. FORBES

When the higher lands of the Near East began drying up with the great climatic changes caused by the end of the last Ice Age (from about 60,000-10,000 B.C.), the population of this area was driven from the once fertile land which is now the Sahara Desert and from the barren highlands of Iran and Arabia into the river valleys. By 3000 B.C. urban civilizations, dependent on agriculture and coincident with the development of metallurgy, had begun to arise in the river valleys. The earliest of these were in Egypt and Mesopotamia (modern Iraq) in the valleys of the Nile and the Twin Rivers, Euphrates and Tigris, respectively, although somewhat later similar cities sprang up farther east in the Indus River Valley (India) and in that of the Yellow River (China).

Once established in the valleys the peoples of both Egypt and Mesopotamia found it necessary to wage an annual war against the floods of the great rivers. In meeting this overwhelming need, each civilization developed social and political systems tailored to the technological necessity of survival. Because of differences in the geographic environment and because the flooding habits of the Nile differed from those of the Tigris-Euphrates, these two societies developed differing technologies, and it is not surprising that their political and social systems also differed considerably. So too did the underlying social attitudes of these regions differ. The people inhabiting the Mesopotamia area were pessimistic in philosophy, and in great fear of demons and evil spirits; the Egyp-

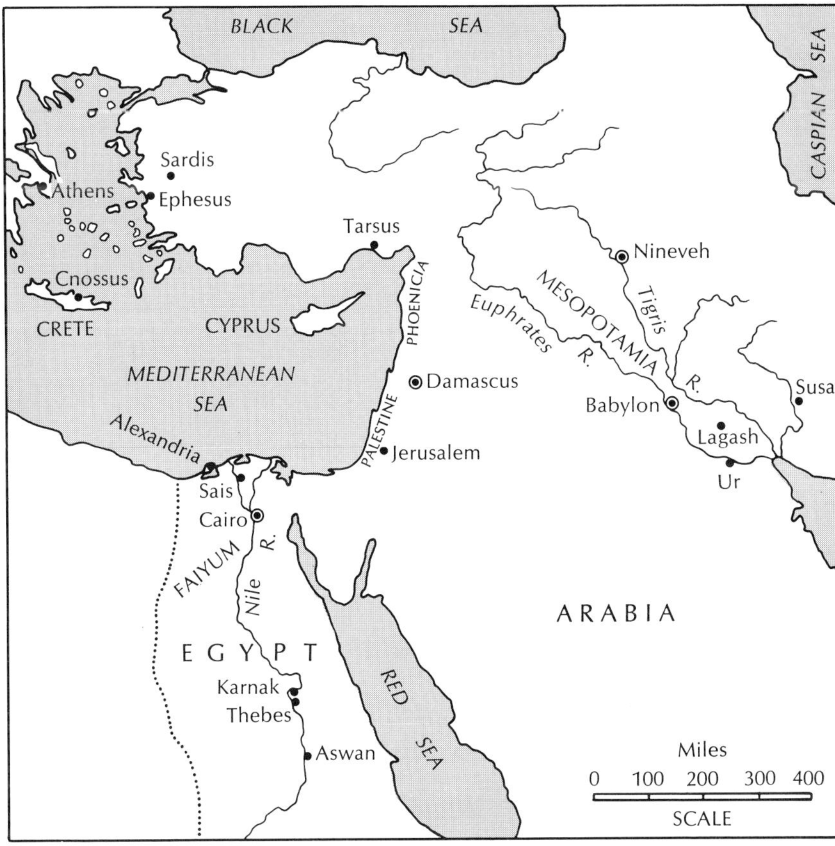

Fig. 3-1. The Near East in ancient times. (From Kirby *et al.*, *Engineering in History*, McGraw-Hill, 1956)

tians, instead of being morbidly concerned with death and the after-life (as some Greek authors of Classical antiquity would mislead us into believing), actually loved the good things in life, which they wanted to enjoy beyond the grave, but without such unpleasant things as taxes and manual labor. These cultural differences were undoubtedly related to the material differences between these two peoples.

THE RISE OF MESOPOTAMIA

The earliest settlements in the land of the Twin Rivers were founded by the Sumerians (about 3000 B.C.), who dominated the southern part of Iraq up to the Persian Gulf. During this Sumerian period, the country knew no central authority and consisted only of city-states (of some 8000 to 12,000 inhabitants), each striving to dominate the others and thereby to capture more of the fertile land. Each city and the land around it were believed to belong to the god or goddess of that city, who had appointed his chief priest to be his representative and shepherd of his people. Thus arose a temple economy, in which the harvest of the land and all the products of the city were delivered to the temple storehouses, whence the priests distributed them to feed and clothe the citizens. Imports of foreign raw materials were handed out to craftsmen, who processed them and were paid for their labor by piece rates.

By 2000 B.C., however, the Sumerian cities to the south had been conquered by Semitic tribes from central Mesopotamia, who united the country, founding the Assyrian and Babylonian empires, known to us through the Bible and archaeological records. Despite the military conquest of the Sumerian cities by these Semitic peoples, the latter assimilated most of the elements of the Sumerian civilization—an early example of the cultural conquest of the conquerors by a conquered people.

Nevertheless, certain elements of Sumerian society were changed. The temple workshops amid the temple precincts, which produced textiles and other goods, slowly gave way to guilds of free craftsmen. These craftsmen produced the articles which necessitated the commercial contact of Mesopotamia with Syria, with Asia Minor and the northern mountains, with Iran, and (by sea) even with Bahrein and the Arabic coast beyond these islands. At the same time, where trade moved, so could armies, and the cities were in constant danger of being invaded from these same regions. Only the rise of powerful military autocracies could secure Mesopotamia from invasion by rapacious neighboring tribes, lacking, as she did, any natural barriers.

THE RISE OF EGYPT

Egypt too was originally inhabited by independent tribes, each of which had established its own "water-province," a union of irrigation and drainage units. Shortly before the end of prehistoric times (*c.* 2000 B.C.) these groups in both upper and lower Egypt had been united under powerful kings (Pharaohs), who ruled the provinces of the land through governors.

Protected by deserts on its western and eastern frontiers, Egypt was safe from invasions for many centuries and suffered only rarely from serious internal strife. The major exception was the Hyksos invasion between 1650 and 1575 B.C.; Egypt reacted by establishing military outposts in Palestine and Syria to avoid repetition of such national disaster, but it never really colonized these foreign parts. Nevertheless, trading contacts existed at an early date with Byblos on the Syrian coast, Cyprus, and Crete; and Egyptian shipping went up the Nile or sailed south along the Red Sea coasts to obtain African and South Arabian products. Egyptian towns never competed economically as did the Mesopotamian towns, nor did the temples dominate the economy until later centuries, when they had grown very rich by royal donations. Both free craftsmen and temple workshops existed and flourished side by side.

LEARNING

In both Egypt and Mesopotamia the only form of education was the temple school, where those who were to become clerks, officials, and priests learned to read and write. Later they were initiated into the various fields of knowledge and became the class of "scribes," looked upon with awe by the common people. In both countries pupils were taught the "order of things as established by the gods in the beginning," in the form of the word lists previously described (Chapter 2). These schools taught their pupils the religious mysteries and other such knowledge restricted to this elite group. Mathematics consisted mainly of simple computations involving the conversion of standard measures or weights and the calculation of areas and volumes of various geometrical figures and bodies, such as the volume of earth needed to build a dike or the area of a piece of land. These were subjects of value to technicians, and we know that the officials who surveyed engineering works often had been pupils in such schools. On occasion technical projects were submitted to the scrutiny and advice of learned bodies of priests who formed advisory boards in the towns of Egypt and Mesopotamia. It was only on these infrequent occasions that craftsmen had contact with the learning accumulated and transmitted in the temple schools. Thus at the very beginning of history, science and technology developed along separate lines with but little contact between the scholarly, clerical caste and the workaday world of the craftsmen.

ROLE OF THE CRAFTSMEN

Sometimes craftsmen were engaged to work the raw materials and products received as tribute or payments in kind by the government, temples, or tradesmen, but most often they plied their several trades on a small scale and in their own shops. Crafts like pottery-making and the manufacture of textiles were often household duties, even in the homes of the rich, whereas the specialized craftsmen worked such things as glass, metals, stone, wood, leather, oils and fats, or luxury foods.

On the large engineering projects of the government and temple authorities, craftsmen were employed as specialists. The common laborers were peasants who paid their taxes by working on such projects during the months their farmlands were under flood water anyway. In Egypt there were very few slaves, only foreign captives or criminals. In Mesopotamia there was a legal class of "bondsmen," who contracted to work for a definite number of years. Slavery played only a very small part in early antiquity; it became a force only in Classical times, that is, during the period of Greece and Rome.

WATER WORKS FOR IRRIGATION

Water supply, irrigation, and drainage dominated this world of early civilization in which the surplus of the farm was used to feed the town. Without the fertile silt deposited on the farmlands by the annual river floods, the soil would soon have been depleted and agriculture made impossible. The prehistoric farming population, which had slowly migrated into the river valleys, had through co-operative efforts drained the originally swampy banks along the rivers in both Egypt and Mesopotamia. Only by co-operation could the dikes be built and maintained, the muddy waters directed over the fields during the inundation, and finally drained off downstream. Flooding problems, and hence their technological solutions, differed appreciably in Egypt and Mesopotamia.

In Egypt the very regular annual rise of the Nile lent itself admirably to basin irrigation. The silt deposited on the land was salt-free, and there was comparatively little silting-up of the canals and ditches which had been dug to spread the waters over the comparatively narrow Nile Valley. The early Pharaohs took over the original local organizations and established "water-houses" in each province (irrigation unit) in order to plan and organize the "cutting of the dikes" (to direct the annual inundation of the fields by the rising Nile) and the constant repair work on both the canals and dikes. They also built Nilometers (graduated wells connected with the river) to measure the rise of the Nile, registered the fields, and set up new boundary-stones when the inundation destroyed earlier landmarks. Many of the inspectors who administered this work were skilled technicians, but in certain cases competent farmers were appointed to survey the work done for the "water-house." Originally, the farmlands reached only as far as the Nile's waters naturally flooded; later, when water-lifting machinery of large capacity had been invented, the area of cultivation was extended to the very rim of the desert.

Mesopotamia was a flat land, and conditions there were very different. The rivers rose early in spring, and water had to be stored then to irrigate the land after the midsummer harvest. Moreover, the Tigris-Euphrates carried about five times more silt than the Nile, and this silt contained salt and gypsum dissolved from the soil in the highlands and carried downstream in fairly large amounts. The large quantity of silt meant that the irrigation canals and ditches had to be cleaned frequently to prevent them from clogging; the high proportion of salt meant that in the course of the centuries several tracts became infertile through the cumulative deposit of salt, a condition which still presents problems for the reclamation of land in modern Iraq.

When Mesopotamia was divided among a number of city-states, these competed for sufficient water to flood their lands, and each city had its own irrigation board. Here too, properly trained officials and experienced farmer-supervisors surveyed the work on the irrigation system, which allowed the farmlands to yield sometimes two crops a year.

It is clear that the care of the irrigation canals and ditches was a major preoccupation in both Egypt and Mesopotamia, and it is small wonder that Pharaohs and emperors were proud of the new works built during their reigns and mentioned them frequently on their monuments. By the same token, when the Mesopotamian warriors invaded enemy country, they regularly singled out their irrigation works for particular destruction.

Outside the great river valleys, in hilly countries like Palestine and Syria, terrace irrigation was practised on a modest scale, the water usually being derived from dammed brooks or rivulets. In more mountainous regions, the inhabitants tapped the water in the foothills of the mountain ranges by driving horizontal tunnels into the hillsides, using vertical shafts for ventilation and inspection purposes. Such tunnel channels, the ancestors of aqueducts, are still constructed and used in Iran and Armenia.

WATER SUPPLY FOR CITIES

The water supplies and sewage systems of the cities were just as carefully regulated as the irrigation system of the farmlands. Most of the cities drew their water from the rivers on which they were located, boiling it when used for domestic purposes. On the other hand, towns in hilly country were supplied by springs and wells, each of which had its own set of problems. Because such towns were mostly built of mud-brick, they tended to rise on the debris of older houses and form a "tell," which meant that the well became deeper and deeper in terms of the ground surface. Springs usually tended to be outside the town's walls, and to ensure a water supply from these springs in case of war, a shaft was dug in the middle of the town to a horizontal sloping tunnel (sinner) which led the spring waters under the walls to the bottom of the shaft. Such water tunnels were typical in Palestinian towns.

The Assyrian king Sennacherib was the first to build a long-distance water supply (691 B.C.) in order to water the fields and palace gardens north of Nineveh. He constructed a dam in the Khosr River near Bavian and conducted its waters through a gently sloping masonry channel to Nineveh, some thirty-five miles to the southeast. This channel, which sometimes bridged valleys on arches, was the earliest aqueduct, and it served as an example to Eupalinos, who built the first Greek aqueduct on the island of Samos (530 B.C.).

Water-lifting devices remained very primitive for centuries. In early days the wooden bailer and the shadoof (a long, pivoted pole with a bucket on one end and a counterpoising weight at the other end) were the only means by which

Fig 3-2. Watering a garden with a shadoof in Egypt, 13th century B.C. (*Science Museum*, London)

the farmer could raise the water by hand to higher ground. Such devices as the Archimedean (or endless) screw, the wheel of pots, or the compartment wheel moved by oxen or by the flowing river did not come into use until about 400 B.C. or later.

PROCESSING OF RAW MATERIALS

The processing of foodstuffs produced in the irrigated gardens and farms involved crushing, pressing, and grinding operations for which several devices were invented. Olives and grapes were originally pressed by treading on them with the feet, but around 3000 B.C. we find pictures of the bag press on the walls of Egyptian tombs. For the bag press a cloth was filled with grapes or other juicy substances and folded in such a way that the two ends on each side enfolded a stick; the two sticks were then turned in opposite directions by four men, and by torsion (or "wringing the cloth" as the texts have it) the grape juice was extracted. A saving of labor was later introduced when one of the sticks was replaced by a noose attached to the upright of the frame in which the bag was hung. The beam press, later to become the major device for extracting juice from grapes and oil from olives, was invented in the Aegean world about 1500 B.C., but it never travelled east.

Grinding grain to make flour was part of the housewife's task; professional millers are not mentioned in Egypt before 1500 B.C., although women sometimes acted as professional millers at an earlier date. The most common milling device was the saddle quern (a saddle-shaped lower stone on which the grain was placed while the woman knelt and rubbed a smaller stone, or pestle, to and fro over the grain). Sometimes the quern was a saucer-shaped basin with a rim to guide the squat and bun-shaped pestle. Saddle querns were developed later which had an upper stone with a hopper and a slit to feed the grinding surface continuously. Toward the end of the second millennium B.C. came the rotary quern; in this the upper stone was rotated by hand by means of a stick fitted into the rim of the upper stone, one of the earliest applications of rotary motion in machinery. The application of small rotary querns went far beyond grinding cereals, and some have been found which were used in grinding colored stones to obtain colors for decorative purposes. During this early period the only implement used for pounding was the mortar and pestle. It was sometimes used for dehusking grain (Egypt), but mainly for such operations as crushing ores and minerals.

Fig. 3-3. Egyptians "wringing the cloth."

Alcoholic beverages were made by fermentation in both Egypt and Mesopotamia. Beer prepared from barley was already a popular drink, although it probably did not taste very much like modern beer, for aromatic herbs were added in some cases. Wine was drunk only by the rich. In Egypt, noblemen cultivated vineyards on the irrigated desert fringe and produced vintage wines named after the vineyard and dated on the seal of the pottery jars in which the wine was bottled and stored in cellars. Mesopotamia had no vineyards of its own and imported its wine from the hillside vineyards of Syria and Palestine. The wines had to be consumed within a year or so, for the containers were not airtight; cork stoppers were unknown, and the jars were closed with a straw plug sealed with clay. Stronger drinks, such as date wine and palm wine, were concocted by adding sugar in the form of honey (the only form of sugar then known) to the fermenting juice.

TRANSPORTATION

Transportation of farm products and other materials was mainly by water, for moving bulky goods by land still constituted a major problem. Although wheeled vehicles were known to have been used by 3000 B.C., their wheels were solid and rimmed with nails or strips of leather. Not until a thousand years later did spoked wheels and lighter wagons begin to appear. War chariots drawn by

Fig. 3-4. Egyptian war chariot, *c.* 1500 B.C. (*The Bettmann Archive*)

Fig. 3-5. Egyptian ship, *c.* 2600 2.3. (*Science Museum,* London)

oxen had for some time conveyed warriors to the battlefield where they then fought on foot, but the light war chariot which now came into use was differently used. Drawn by the horse, it became a new arm of the military, and the charge of these earliest "tanks" created a new strategy. The Egyptian armies adopted the horse-drawn war chariot about 1500 B.C.

War chariots did not affect land transport, largely because there were no properly constructed roads of any length. Few city streets were paved, the major exceptions being those connecting the city-temple with the summer-temple of the god outside the city walls. These processional roads sometimes had ruts hewn into the pavement stones to guide the carts bearing the statues of the gods. Such special tracks or roads were known in prehistoric Europe, too. Most traffic, whether of goods in wagons or on the backs of pack animals, tended to lead to the nearest water, which carried the long-distance transport, a situation which existed in most parts of the West until the 18th or 19th century. The taming of the camel (*c.* 1000 B.C.) made the crossing of deserts possible, thus opening a shorter route from Mesopotamia to Palestine and Egypt. Not until the later Persian Empire, with its centralizing tendencies, was there a proper concern for roads. At that later date also, elaborate bridges were built along with road stations (with garrisons of soldiers and reserve horses for the postal express) and a chain of fire-signal towers. These transportation improvements were short lived, however, disappearing with the decline of the ancient Persian Empire.

River transport on the Nile was relatively simple. Because of the scarcity of timber, which had to be imported from Lebanon or beyond, ship-building started with canoes and rafts made of bundles of reeds, which evolved into larger boats. Ships were constructed with small wooden planks, on the model of the reed boat, and moved up the Nile with square sails stepped forward on a two-legged mast required by the rather unstable construction of the frameless hull. In Mesopotamia, the skin-float (a raft made of hides stretched over a wood frame) and the *quffa* (or coracle, a broad and short boat made of waterproof material stretched over a wicker or light wood frame) dominated the river traffic, as both forms of boats were light and portable and could be carried back upstream by pack animals. Such ships were not fit to sail the seas, however, and sea transport was mainly in the hands of Cretans and Phoenicians who built plank ships, evolved from the dug-out canoe. Such ships could sail the Persian Gulf and Mediterranean Sea with little difficulty.

THE BUILDING ARTS

Timber was scarce in both Mesopotamia and Egypt, so brick and stone were the primary building materials. However, building in brick and stone took very different courses in the two countries. In the early Mesopotamian farming villages walls were first made of rammed or kneaded clay, but this was replaced by sun-dried bricks made in oblong or rectangular open wooden moulds. Sometimes these rude brick walls were decorated with ornamental stone set in various patterns.

Kiln-baked bricks were expensive and therefore they were used for monumental buildings only and were set in bituminous mortar. These bricks were strong enough so that they could be projected over one another, with each layer of brick overhanging the one below; this procedure is known as corbelling, and any piece projecting from a wall and supporting another building member is known as a corbel. When two opposite walls are corbelled out until they meet, a corbelled arch or vault results. Along the great processional roads glazed bricks were used for decorative purposes. In Assyrian times, concrete, consisting of lime, sand, and broken limestone, was used. Natural stone played only a small part in Mesopotamian architecture, for this expensive building material could be obtained from the northern highlands only at great cost and trouble.

Yet the Mesopotamians could build monumental structures. The Tower of Babel, described in the Bible, indicates their ability to erect structures of impressive height. This structure was probably like one of the ancient temple-towers (called "ziggurat") common in Mesopotamia, and consisting of a series of truncated pyramids placed on top of each other, with a shrine on the top platform as well as at its foot. There were several such shrines in Babylon; they were built of mud-bricks, with baked bricks sometimes used as an outer covering, the outside bricks frequently being set in bituminous mastic. Better known as one of the Seven Wonders of the Ancient World was the so-called Hanging Gardens of Babylon. Some scholars believe that these may have been a roof garden planted with trees and shrubs atop a large palace; others, basing their conclusions on the writings of the ancient Greek Herodotus, believe they were simply terraced gardens. Since Herodotus did not provide technical details, but merely spoke of the difficulty of raising water to the gardens, we are still very much in the dark regarding their actual construction.

In Egypt, natural stone abounded in the eastern and western desert regions;

Fig. 3-6. Artist's reconstruction of the city of Babylon. (*The Bettmann Archive*)

hence Egypt developed the use of natural stone in architecture. It was used for major buildings only, however, and ordinary houses were first built of reed bundles or mud-bricks, and later of sun-dried bricks. Kiln-baked bricks were rarely used in Egypt until a much later date. The earliest graves and public buildings were also made of mud-bricks.

Building in natural stone started with the wall and ceremonial buildings encircling the pyramid of King Zoser at Saqqarah (2600 B.C.). His vizier Imhotep is the first engineer and architect known to us by name, and he was later worshipped as the god of wisdom and medicine. Imhotep started using small brick-size blocks of limestone, stacking walls and pillars in small-block masonry and then shaping them on the spot to imitate reed pillars and walls. As the work proceeded, he precut larger blocks of stone to be stacked and thus discovered the elementary rules of building in natural stone, experimenting to explore the possibilities of this new building material.

Fig. 3-7. Pyramids of Gizeh, near Cairo, the largest and among the oldest of architectural monuments. (*The Bettmann Archive*)

In the stone quarries, the exploitation of which was generally a state monopoly, blocks of limestone and granite were detached by drilling holes into which metal wedges were forced. When the blocks split away from the rock, they were dressed with balls and hammers of diorite, a very hard stone, and transported to the building site on rafts and sledges.

The earliest tombs for the Pharaohs were mastabas, rectangular structures with inward-sloping walls set over underground chambers containing the burial pit, the chapels, and rooms for funeral gifts and offerings. The mastabas had originally been made of mud-brick, but they now became monuments of natural stone. The mastaba structure evolved first into the step-pyramid form, and then into the true smooth-sided pyramid.

The greatest of the royal tombs was the Great Pyramid of King Khufu (Cheops) which contains some 2,300,000 four-ton limestone blocks, the joints of which were chiselled on the spot to fit tightly without mortar, and in the core of which corridors and chambers were constructed. This pyramid, which still stands today, measured 756 feet square at its base and rose to a height of 480 feet, making it another of the wonders of the ancient world and indeed of all time. The huge limestone blocks were quarried on the east bank of the Nile and, after crossing the river by barge, were hauled onto the pyramid plateau above the Nile Valley and then up earthen ramps (which finally had as large a volume as the entire pyramid) onto the building level. All this pulling-and-hauling was done without pulleys or cranes; there were no labor-saving devices to ease the strain on human muscle power. The masons and quarrymen were skilled craftsmen, who were paid for their work; the transport was carried out by statute labor, that is, peasants fulfilling their obligations to the state. Marks on the quarried stones show that plans of the work existed, each stone being earmarked for its proper place in the pyramid.

The pyramid was finished off with slabs of polished granite quarried and brought down from Assuan. The building of an entire city of temporary huts to house the workmen, the removal of debris, the supplying of materials and tools, and the feeding of the host of workmen (as many as 100,000 at a time) were organizational problems of enormous proportions. The construction of the pyramids shows what can be done by sheer organization without much in the way of devices; these great constructions represent a triumph of human organization, rather than of machines and tools.

METALLURGY

Many of the techniques used in quarrying and dressing stone were made possible only through the development of metallurgy. At an early stage man discovered and collected such native metals as copper, gold, silver, and meteoric iron. At first he treated them simply as colored stones for ornamentation; then he realized that they could not only be cut, hammered, and ground to shape like other stones, but could also be melted and cast in predetermined shapes as well. By experimenting with the blue and green colored stones (ores) often associated with native copper, man found that these could be smelted with charcoal to yield molten metal. Still other ores when smelted yielded lead, silver, antimony, and tin. Further, several of these could be smelted with unrefined copper or, later, alloyed with molten copper, to produce an "improved" copper: bronze, which was stronger and possessed other special properties which made tools and weapons superior to the older devices fashioned of stone and wood. The Bronze Age, which lasted from about 3000 B.C. until about 1000 B.C. in Egypt and Mesopotamia, used a metallurgy depending mainly on casting and alloying.

The later coming of iron meant the mastering of a completely new set of techniques, for the smelting of iron ores did not yield a molten metal but rather a spongy mass of slag and droplets of iron later called a bloom. This, in turn, had to be reheated and hammered for some length of time to expel impurities and produce a compact mass of wrought iron. This substance had the properties of a metal, though its melting point was too high for the metallurgical furnaces in use in those days. Even though wrought iron was inferior to good bronze for most purposes, by heating it in a charcoal fire and hammering, the

Fig. 3-8. Egyptian metal-workers (above) and vase-makers (below). Wall painting from tomb, *c.* 1475 B.C. (*The Bettmann Archive*)

surface could be made to absorb carbon particles and take on the properties of steel. The hardness of this "steeled" surface could be adjusted by quenching in water and tempering (that is, bringing it to the required hardness by heating and then suddenly cooling it), but these were techniques not yet properly mastered in Egypt or Mesopotamia. Instead, discovery of steel-making was made in the Armenian highlands by the tribe of the Chalybes, and the Greeks were to derive their word for steel from the name of this tribe.

Metallurgy is a complex of techniques involving the choice of proper ores, fuel, method of producing air blast, tools, furnaces, and crucibles. Its earliest home in the Near East lay in the Armenian highlands in Asia Minor. There native gold and copper were certainly in use by 5000 B.C., whence they came to

Mesopotamia where by 2500 B.C. the Royal Tombs of Ur yielded a profusion of objects fashioned from gold, electrum (a native alloy of gold and silver), silver, copper, and various types of bronzes. Mesopotamia had no ores in its plains and hence it depended on trade or conquest for supplies of ores or raw metals from the highlands.

In Egypt the smelting of copper began about 4000 B.C., and copper ores were mined in the Sinai Desert peninsula early in Egyptian history. The government sent expeditions there to obtain copper and copper ores, which were sometimes valued as semiprecious stones. These expeditionary forces were composed of "organizing" officials, skilled miners and smelters, criminals and captives to be used as mining gangs, and bands of soldiers to protect the expedition from marauding desert tribes. The copper obtained from Sinai and other copper mines was used to fashion adzes, axes, arrowheads, chisels, saw blades, drain-pipes (made from strips of copper bent round), and many other items. The hardness of such copper tools was adjusted by cold-working and by annealing (heating and slow cooling).

Although copper ores were abundant in Sinai, Cyprus, Caucasia, and Trans-caucasia, the bottleneck of bronze-making was the short supply of tin, the metal which best alloyed with copper to produce bronze. Originally, tinstone, collected along with native gold in the river beds, had been smelted with black copper to produce bronze (usually containing 8 per cent to 12 per cent of tin). As men learned more about the properties of alloys during the second and first millennia B.C., the tin content of bronze was adjusted to the application, bronze for mirrors and such containing a higher tin percentage than the bronze for tools, where the usual 10 per cent was more proper.

As the meager supplies from Caucasia and Anatolia in Asia Minor began to fail, tin ores and blocks of tin were imported. Phoenician (and later Greek) traders went as far as Brittany in France, Cornwall in England, and even Bohemia in Central Europe to bring tin to the Near East. Egypt did not manufacture bronze on a large scale before 1500 B.C., when tin supplies began to arrive from the West. The shortage of tin, to alloy with copper to produce bronze, was thus one of the major stimuli to long-distance trade and commerce in the ancient world. Here was a forerunner of that cultural intercourse which was significantly to affect the enlargement of man's vision and thought.

Useful as it was, bronze could not always compete with stone. The Egyptian army continued to use flint arrowheads until the fourth century B.C., partly because they had a better penetration power than those of copper or bronze, and partly because bronze was scarcer and more expensive than flint. However, they discarded the stone arrowheads (together with stone mace-heads and sling-stones) when iron became available at low cost. The scarcity and high cost of copper and bronze in the ancient world have caused them to be called the "metals of aristocracy," whereas iron is said to be "the democratic metal."

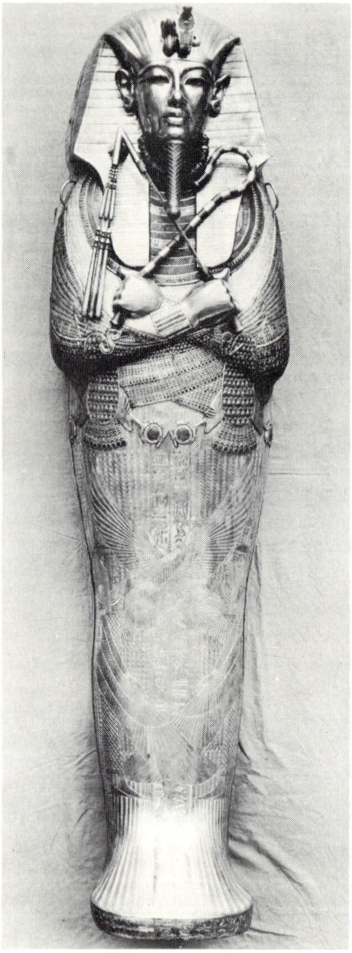

Fig. 3-9. The gold inner coffin of Tutankhamen, *c.* 1350 B.C. (*The Bettmann Archive*)

Practically every country in the Near East and early Europe possessed good and workable iron ores, so iron tools were cheaper and more widely used, once the techniques of manufacturing iron had been mastered.

There is no doubt that the ancient smiths had mastered the various techniques of working gold, silver, copper and its alloy bronze, lead, and antimony. Working sheet-metal, raising, hammering, repoussé-working (hammering on thin metal from the underside to produce a pattern), stamping and engraving, soldering and welding, filigree and granulation, inlay-work and cloisonné held no mystery for these metallurgical craftsmen. Specialization had taken shape during the thirty centuries since the beginnings of metallurgy. Originally the smiths had prospected, mined, and smelted the ores, and then worked and alloyed the metals. Gradually prospecting and mining were left to the miner; the metallurgist smelted the metal from the ore, and the bars and pigs of metal were then refined and alloyed. The metal was now ready for the blacksmith, who specialized in producing such common objects as pots, pans, and ordinary tools, or for the coppersmith (gold-, silver-, tin-, white-smith), who produced smaller objects, art objects, or repaired certain classes of metal products. Metal tools and weapons were still expensive, however, even in the Early Iron Age, so that stone as well as wooden implements continued in use in such areas of production as the textile industry.

TEXTILES

Mesopotamia was the typical wool-producing country of antiquity. The spinning and weaving of wool was originally a female occupation, even in the large temple workshops, but by 2000 B.C. male weavers are mentioned, together with fullers, bleachers, and dyers. At first wool was plucked from the sheep with a knife, but by about 1000 B.C. shears came into use, being made by joining two iron knives with a spring. By completely detaching a fleece, the shears doubled the wool production per animal.

In Egypt, sheep—like some other animals—were considered unclean; hence wool was reserved for outer garments which did not touch the skin. Egypt was the typical linen-producing country, and there the processing of flax was fully known by 3000 B.C. Cotton, though spun and woven in ancient India by 2500 B.C., was unknown in Mesopotamia before 700 B.C., when King Sennacherib had some "trees bearing wool" in his botanical gardens. In Egypt it was not introduced as a material for textiles until the Greco-Roman period, when it was brought from the Sudan.

The fourth textile fiber of antiquity—in addition to wool, linen, and cotton—was silk. A kind of native silk was produced on the island of Cos by the 4th century B.C., but some two centuries later real Chinese silk was imported to Alexandria.

The spinning of fibers was well understood, as was weaving, although no major technological innovations had been made in these processes since prehistoric times. Dyeing and bleaching were practised in both countries, and indeed we often find the production of dyes, such as the famous "sea-purple," concentrated in certain cities which kept herds of sheep in the surrounding hills to produce the wool to be dyed with the special local product.

OTHER CRAFTS

It would be easy to multiply the examples of crafts which contributed to the development of technology in ancient Egypt and Mesopotamia. Pottery and ceramics were produced by craftsmen working in accordance with regional customs, but all contributing to the development and diffusion of better kilns and glazes, and stimulating such other technologies as metallurgy and glass-making.

One of the most important developments in the potter's craft was the evolution of the potter's wheel. The simple clay-disc potter's wheel, such as found at Uruk (3250 B.C.) gradually developed into the foot-wheel (2000 B.C.) and, some two centuries later, the pivoted-disc wheel.

Glass owes its origin to the production of glazes for pottery. Ancient glass-making consisted in fusing the necessary ingredients and casting the molten mass in small moulds, or making small rods of glass, which when still hot were modelled around a sand core and then annealed to form a small pot or flask to contain cosmetics and essential oils or perfumes. However, glass-blowing was unknown to antiquity.

In general, the craftsmen of ancient Egypt and Mesopotamia must have been extremely skilled, for they produced excellent work without the aid of sophisticated tools and machines. Fine woodwork and ivory work show that the ancient craftsman would have little to learn from modern specialists in this field.

Ancient texts mention guilds of craftsmen, but these are not to be confused with modern labor unions. They were first of all religious organizations devoted to the worship of the patron-god. Hence technical operations were still accompanied by religious rites and ceremonies. In many cases, such as the smelting of ores or making glass, the craftsmen believed that they were merely hastening processes which would be accomplished by nature led by the gods anyway. Propitiatory offerings, ritual purity, and prayers accompanied most operations of the craftsman, and sometimes he or his tools, such as the smith's hammer and his anvil, were held to have magic power. Even after many centuries we still find traces of the awe with which the craftsmen looked upon their own achievements.

THE CONDITIONS OF TECHNOLOGY

Among the preconditions of technology are such geographic considerations as the availability of natural resources. The importance of the geographic factor is nowhere more obvious than in the histories of both Egypt and Mesopotamia, with their reliance upon the river valleys which made agriculture possible. It was not just the fact that these were river valley civilizations; the agricultural life of both regions depended upon the fact that these rivers besides providing water for irrigation, periodically overflowed their banks and provided the rich silt which fertilized the land and helped to produce abundant crops. These floodings were so important that both areas had to develop hydraulic-engineering techniques in order to control the waters, since either too much or too little water could be disastrous; irrigation techniques had to be highly developed.

The technological character of irrigation works—the dikes, reservoirs, irrigation canals, and ditches—in turn affected the social and political development of both lands. In order to erect and maintain such large-scale irrigation projects, co-operation among peoples must be assured. Such co-operation could be obtained by democratic co-operation among many different individuals, but it was more likely to be assured, at this stage of political development, through the development of autocratic control by a powerful state which would enforce and supervise the co-operation necessary to build and maintain the dikes, canals, and reservoirs upon which the agricultural subsistence depended. In Egypt the Pharaohs eventually came to control virtually the entire economic life of the nation; in Mesopotamia autocratic rulers allowed for the exercise of strong state control over the economy, although not to the same extent as in Egypt.

This does not mean that technological requirements made necessary the development of autocratic and all-powerful rulers in these areas; it simply means that technology was one of a number of factors which combined to produce certain political conditions. When political conditions deteriorated in both these regions to the point where there was no longer any effective supervisory or enforcing agent to maintain the irrigation works, both the Tigris-Euphrates and Nile valleys declined in productivity and have never since fully recovered economically. The deterioration of public works followed closely the breakdown of public order.

The presence or absence of certain raw materials also had a profound effect upon the development of technologies in both Egypt and Mesopotamia. Although Mesopotamia had little natural stone and few timber resources, it did have ample supplies of clay. Hence the common building material became brick, and clay tablets became their chief writing instruments. In Egypt, natural stone was available in large quantities, so the Egyptians developed the use of stone for building monumental structures. A multitude of associated technologies and techniques were dependent upon the use of these basic raw materials.

As men learned to make and use items of metal, the absence, in certain areas, of tin ores for making alloys of bronze required the development of trading techniques, just as the development of metallurgy itself required the evolution of mining, smelting, and working techniques. Trade and commerce were dependent in great measure upon the evolving technology.

Another important factor affecting the development of technology—and one which we can see at various times throughout history—was religion. It was the religious beliefs of the Egyptians regarding the after-life which caused them to erect the monumental pyramids. The fact that the common people had to sacrifice their energies and labors for this kind of work tells us much about the role and position of the Pharaohs as god-king and also indicates the great social gulf which existed between the rulers and the ruled. Throughout much of history man's religious beliefs have provided an important stimulus to his technology. The building of the pyramids also demonstrates another element within technology itself, namely, the importance of the organization of work as well as the techniques and tools employed. For these great edifices, like the irrigation works upon which the Pharaohs' wealth was based, represent a triumph of human and social organization more than they do the application of developed tools and machines.

The high development of the various crafts in both Egypt and Mesopotamia also manifests another characteristic of technology during most of human history, namely, its separation from science. Both technology and science were related to religion, but in quite different ways. Science and religion were the property of a highly educated caste which had little to do with the work of the craftsman. The crafts followed an empirical tradition, based on experience passed down by oral rather than written means. Few craftsmen could read or write, while the priestly caste alone possessed the esoteric scientific and religious knowledge. Not until much later in human history, as we shall see, did science and technology work together to reinforce one another and increase man's control over his environment.

Nevertheless, both the Egyptians and Mesopotamians could point to significant accomplishments in technological development. They erected monumental structures, some of which were the wonders of the ancient world, and their hydraulic-engineering techniques enabled them to maintain the fertility of large areas for a period extending over a thousand years. Furthermore, their craftsmen produced work of the highest quality. The fruits of their technological advance, however, were largely dedicated to the support of a small group of nobles and priests, with an all-powerful ruler at the head. Technology had not yet developed to the point where man had sufficient control over his environment to provide more than the bare essentials of life for the majority of the population.

Yet, in these early river-civilizations man had already learned to use some elements of nature for his own purposes. Wild animals had been domesticated to serve as carriers for man, to help take some of the burden off men's backs; similarly, the force of the wind had been harnessed by sails, thereby releasing human muscle power. In addition, the great advances in agriculture through the use of irrigation techniques meant that men could work with nature instead of struggling against her in order to provide himself with subsistence. Life was still harsh and work was still overburdening for the vast majority of human beings, yet the beginnings of civilized life in urban communities were to be found, along with the development of writing to extend human powers of memory and speech. Thus, in the very beginning of civilization, technology played an important role and was advanced by the very civilization which it had helped to create.

4 / The Classical Civilizations

A. G. DRACHMANN

THE GRECO-ROMAN BACKGROUND

The culture of the Western world is built upon a double foundation: the Christian religious tradition and the Greco-Roman civilization of Classical antiquity. From about 900 B.C. the Greeks inhabited what is the country of Greece today, plus the surrounding islands in the Aegean and Mediterranean seas and the coast of Asia Minor. They also colonized Sicily and southern Italy, which was known as Graecia Magna (Large Greece). Their contributions to Western culture included a literature that has never been equalled (the epic poems of Homer; the tragedies of Aeschylus, Sophocles, and Euripides; the comedies of Aristophanes); the first rational philosophy (Socrates, Plato, Aristotle); and the foundation of the exact sciences: mathematics, astronomy, and physics.

One thing the Greeks could not do: live at peace with each other. Even though they spoke the same language and worshipped the same gods, the various city-states of Hellas (Greece) were always fighting. For a brief time in the 5th century B.C., Athens, under the leadership of Pericles, almost succeeded in unifying Hellas by bringing the other city-states under the domination of an Athenian "empire." But quarrelling factions within Athens weakened the Athenian control, and the other city-states chafed under their subordination to Athens; Sparta led the city-states in the Peloponnesian war that put an end to the Athenian empire at the close of the 5th century. The city-states resumed their quarrelling, each trying to dominate the others. They proved unable to unite when Philip, king of Macedon (in the northern part of the Greek peninsula), threatened them; he brought unity to Greece by conquest in the middle of the 4th century B.C.

Alexander the Great, Philip's son, went on to conquer the Middle East—Syria, Egypt, Palestine, Persia, and even part of India—near the close of the 4th century B.C. A lover of Hellenic civilization—his tutor had been the great philosopher Aristotle—Alexander spread Hellenic culture among his Oriental subjects; the result was a fusion of Eastern and Western cultural elements, which is known as Hellenistic civilization (to be distinguished from the pure Greek, or Hellenic, culture). But the political unity of East and West did not survive Alexander's death (323 B.C.); the realm split into warring states.

However, the cultural fusion of East and West—Hellenistic civilization—did survive. It was to be assimilated by a people new to Western history, who were again to unite East and West into a great empire which lasted for centuries: the Romans. It is because the Romans adopted and adapted Hellenistic civilization, based on that of Greece, that it is sometimes known as Greco-Roman civilization.

Founded in 743 B.C., Rome was at first but a small city-state among many others in central Italy. The Romans fought their neighbors and eventually conquered them; instead of keeping them perpetually subjected, they gradually extended Roman citizenship to them and changed their former enemies into allies. By degrees all Italy, then all the countries around the Mediterranean, including Greece, were added to the Roman dominions. At its height Roman rule reached from the border between Scotland and England to the Danube, and south to include all North Africa and the Middle East. Inside this enormous territory there was peace from 40 B.C. to the 5th century A.D.; this *Pax Romana* (Roman Peace) was a longer period of peace than has been experienced by the Western world since then. A standing army took care of any fighting that came up along the borders, while law and order reigned in the rest of the empire.

As a city-state Rome had begun as a tribal kingdom, but its first expansion occurred under a republican form of government. However, the quarrels of domestic factions and politicians, and the ambitions of strong men, such as Julius Caesar, put an end to the Roman Republic. From about 40 B.C. under Augustus, the nephew and heir of Julius Caesar, Rome became an empire with a single man, the emperor, at its head. Some emperors were great statesmen, a few were madmen; but to most of the population of the Roman Empire this did not matter. The efficient administration established by the first emperors kept things going, regardless of who was the nominal ruler. Law and order, government and administration, were the main contributions of the Romans to the civilization of the Western world, while at the same time the technical requirements of such an empire led to important mechanical and engineering innovations.

THE BACKGROUND OF GRECO-ROMAN TECHNOLOGY

By the time we first hear of the Greeks, about 900 B.C., most of the early inventions that secured mankind's supremacy over the rest of the animal world had already been made: the use of fire; the domestication of animals; agriculture; the use of metals and iron for weapons, tools, and utensils; spinning and weaving; the potter's wheel and the glazing oven for pottery; the building of houses of wood, brick, and stone; the use of wheels for transport; and the art of writing. In many respects daily life during antiquity already resembled closely that which prevailed in Europe up to the invention of the steam engine: agriculture was the basis of the economy; the home was the center of most production; and almost the only source of power was the muscle power of man or beast.

What we know about the technical achievements during Classical antiquity comes from three sources: there are the writings—a few technical books describing inventions and machinery, and casual mentions in works on other subjects; there are the actual tools, excavated by archaeologists and now preserved in museums all over the world; and, finally, there are pictorial representations in sculpture, in paintings like those preserved to this day on the walls of Pompeii, in mosaics, and in pottery. Greek vases especially have figures of great beauty which illustrate the daily life of the time.

GREEK INVENTORS

Three great inventors, all of them Greeks, are known from antiquity: Ktesibios, Archimedes, and Heron of Alexandria.

It is interesting to note that all these men lived during the Hellenistic period; although none of them lived in Greece proper, they were Greek in culture. Interesting too is the fact that although the preceding Hellenic period (up to about 350 B.C.) had been a period of great cultural activity in art, literature, and philosophy, little was accomplished in the way of technical progress. Some Classical scholars claim that it was the stimulus of the meeting of Eastern and Western cultures, accomplished through Alexander the Great's conquests, which produced the technological advances of the Hellenistic period.

KTESIBIOS

The first, Ktesibios (spelled Ctesibius in its Latin form), lived in Alexandria about 270 B.C. This city, founded by Alexander the Great on the delta of the Nile, although geographically in Egypt was a Greek city in spirit. With its university and its library it became a center of Hellenistic culture. The first three kings (Ptolemy I-III) who reigned in Egypt after Alexander's death promoted the arts and sciences and supported Ktesibios and other inventors.

Ktesibios was the son of a barber, and he put up in his father's shop a mirror hanging on a string with a counterpoise, so that its height could be adjusted to the size of the customer. The counterpoise was placed in a corner, neatly hidden by boards. But as it came down, the air in the enclosed space was compressed and escaped with a loud noise. On the strength of this experience, Ktesibios made the first cylinder with a plunger, and so constructed the first force pump. With his air pump he made the first organ, with many pipes of different lengths and a keyboard for playing them. It was called the water organ, because water was used to keep the pressure of the air constant. Ktesibios also experimented with air and water pressure, studying siphons and founding the science of hydraulics (then known as "pneumatics").

He also invented the water clock. Before his time the sundial was the only means of telling time, and it of course did not work at night or in cloudy weather. Ktesibios constructed his "night clock" or "winter clock" on the following principle: a small stream of water fell into a container called the *klepsydra* (or clepsydra) which had a small hole near its bottom and an overflow hole higher up. As long as more water came in than went out through the hole in the bottom, the water level was constant, and the flow from the small hole was also constant. The water trickled into a cylindrical container with a float that carried a vertical rod and a pointer to indicate the hours on a scale. All reliable timekeepers were built on this clepsydra principle until the invention of the verge-and-foliot escapement in the 12th century A.D.

Ktesibios delighted in mechanical contrivances. Having made a clock that moved by itself, he added a rack and pinion drive to the float and made it perform such tasks as striking the hours and sounding trumpets at noon. (In this respect our cuckoo clock with its performing bird is a direct descendant of

Fig. 4-1. Clock, or klepsydra. (Courtesy of *P. Haase & Son*, Copenhagen)

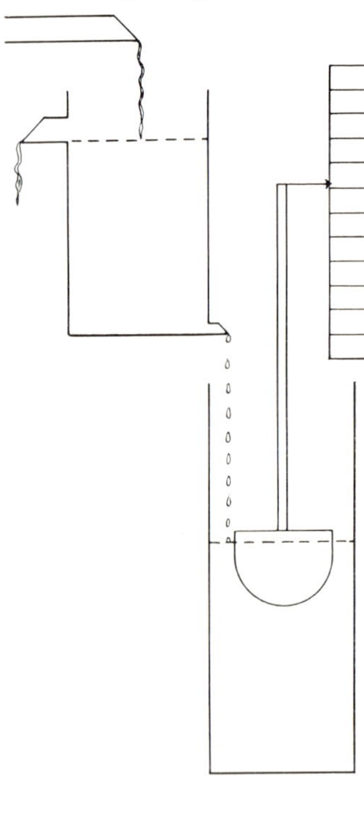

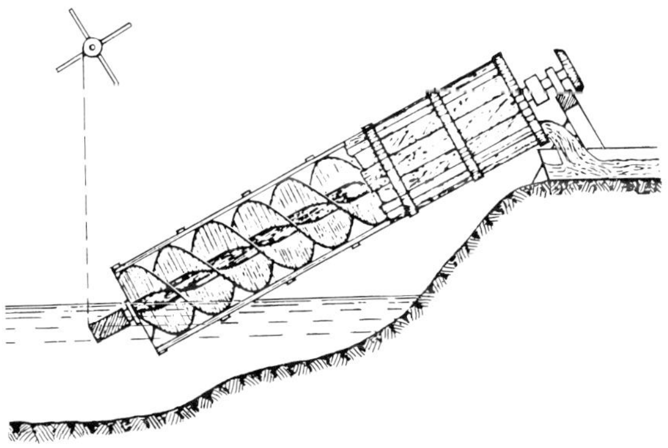

Fig. 4-2. Water-snail. (*P. Haase & Son*)

Ktesibios's water clock.) He was the first to use a toothed wheel, and he also tried to make catapults driven by springs or compressed air. These did not succeed, however, for the technical resources of the time were not sufficient.

ARCHIMEDES

Archimedes (287-212 B.C.) is known as an outstanding methematician, but he was also an inventive genius. Of inventions ascribed to Archimedes, at least four are still in use. He studied the theory of the balance and invented the steel yard (fulcrum scale); he studied the mechanics of the lever and invented a differential gear of ropes and drums; he studied the screw line, and he invented the water snail, which is still in use. He also connected the screw with a gear wheel, making a worm-and-wheel-gear, or endless screw, as it is also called. This is one of the most powerful transmissions ever made, and while it was not much used in antiquity, it is today found in a great number of engines. Furthermore, when his native town, Syracuse, was besieged by the Romans, he kept the enemy at bay by all sorts of mechanical contrivances.

HERON OF ALEXANDRIA

Heron (in Latin, Hero) of Alexandria lived about 60 A.D. (an eclipse of the moon described in his book, *Dioptra*, took place in 62 A.D.). We know nothing of his life, but he was probably connected with the University of Alexandria. Several technical works by him have come down to us: *Pneumatics, Automatic Theater, Dioptra, Belopoiica* (Book of Catapults), all written in Greek, and another, *Mechanics*, has reached us in an Arabic version.

The *Dioptra* describes an invention made by Heron himself, an instrument

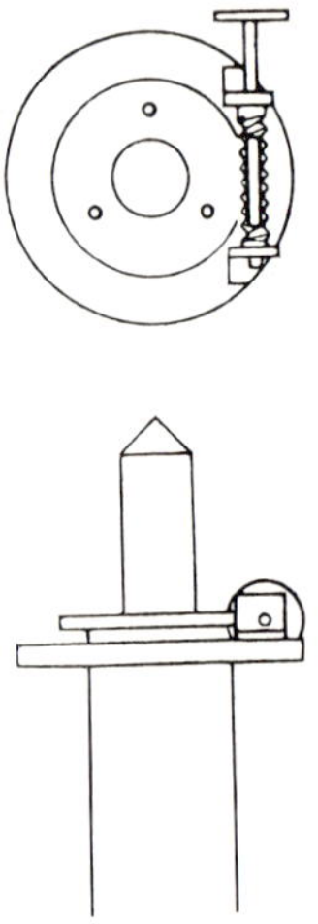

Fig. 4-3. Heron's Dioptra. Around the pivot (attached to the horizontal disc on the upright column) was placed a ring with teeth which were engaged by a screw turning in bearings fastened to the disc. The screw had a longitudinal furrow as wide as the gear-wheel was thick, so the wheel could be turned freely and locked in any position. (*P. Haase & Son*)

for surveyors. It is a combined theodolite (for measuring horizontal and vertical angles) and level, meant to supersede two older instruments, the *groma* and the *chorobates.* (See Figs. 4-3, 4-4, and 4-5.)

The *groma* was used for staking out lines at right angles; its major drawback was that it did not work in windy weather. The *chorobates,* or "land-strider," was a rather cumbersome instrument for taking levellings. It consisted of a 20-foot-long plank with a leg and a plumb line at each end, and for windy weather it had a furrow in the plank to hold water for indicating the horizontal position. No other water-level was known; horizontal beams or walls were tested with a triangle and a plumb line.

Heron's *dioptra* consisted of a stake thrust into the ground on which there was a column with a round disc, and from this horizontal disc a thick pivot stood upright. The rest of the instrument consisted of two different parts, the *dioptra* (theodolite proper) and the level, each of them mounted on a hollow column that fitted the thick pivot. The column carried three pegs to fit into three holes in the ring, and this is the only example known from antiquity of interchangeable parts. The construction and operation of Heron's *dioptra* allowed for much more accurate observations and measurements than earlier instruments.

Heron's instrument shows both the possibilities and the limits of fine mechanical work during antiquity. The brass screws in the level, for example, were inserted into smooth holes and engaged by a small peg driven in from the side. The threaded hole was not introduced until many centuries after the time of Heron, and the familiar pointed wood-screw was not invented until the mid-19th century.

Heron's *Mechanics* was a textbook for architects, contractors, and engineers, and most of the tools described in it were already old before Heron wrote. One thing was new, however: the use of the screw press. Oil and wine were very important products of ancient agriculture; a press was necessary to get the last of the juice out of the grape pulp and to get anything at all out of the olives.

Fig. 4-4. Heron's Dioptra (Theodolite). The dioptra consisted of a vertical half-circle of brass, with teeth on its circumference; it turned on a horizontal axle and was held by an endless screw from below. A round disc was fixed at right angles to the upper, flat edge of the half-circle, and on this disc was mounted the sighting-rod, turning on a pivot in its middle. Two lines at the right angles on the horizontal disc completed the instrument. This was enough for staking out lines at right angles; but if the horizontal disc was divided into 360 degrees, the dioptra could also be used for astronomical observations. (*P. Haase & Son*)

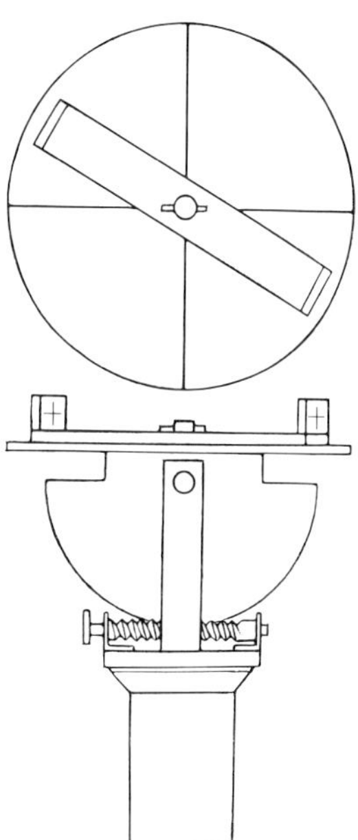

The first presses, using a long lever and a winch, were laborious and time consuming to operate, and they were not very effective in squeezing out the last grape juice or olive oil. The application of the screw to the lever press saved labor, and the direct screw press also saved space. Heron's use of a female screw enabled the screw to press directly on the substance to be squeezed, and hence extracted more juice and oil.

Heron's book *Automatic Theater* describes a plaything, a toy theater, that moves by itself, performing a play without being touched by any human hand. It is driven by a heavy weight resting on mustard seeds that run through a narrow hole, after the manner of the sand in an hour-glass (or the water in a clepsydra). All movements were performed by means of drums and strings, and there were no springs or gear wheels.

Heron's *Pneumatics* also consists mostly of playthings. We find a fire pump, however, and a water organ, but the rest is parlor magic for entertaining: bowls of wine that cannot be emptied, jugs and jars that give out wine or water at

Fig. 4-5. Heron's Dioptra (Level). Heron's level consisted of a very long, horizontal rod of wood, into which was fitted a long tube of brass with upturned ends. These ends carried glass tubes so that when water was poured into this U-tube, the surfaces of the water in the two glass tubes indicated a horizontal line. The sight-holes were made in two small brass plates that were moved up or down by means of brass screws; a slit in each plate was made to coincide with the level of water in the tube. The column with three pegs, which carried the level, was fitted with a plumb-line, for if the "transverse rod," as the level was called, canted to one side, the water levels in the glass tubes could not coincide with the slits. Two targets, half white, half black, were used on the staffs; they were moved up or down by a string over a pully, and carried a pointer to show their height over the ground. (*P. Haase & Son*)

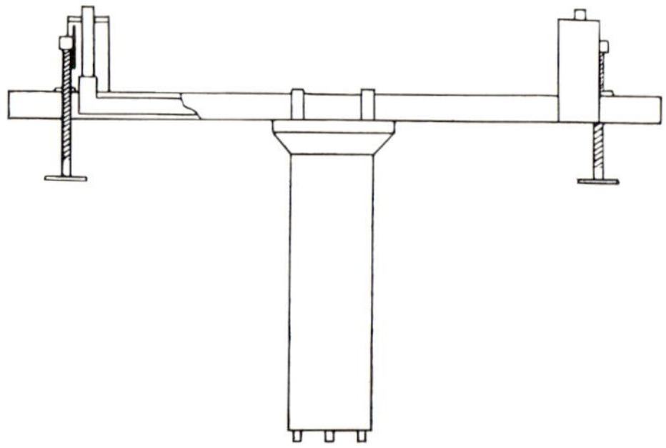

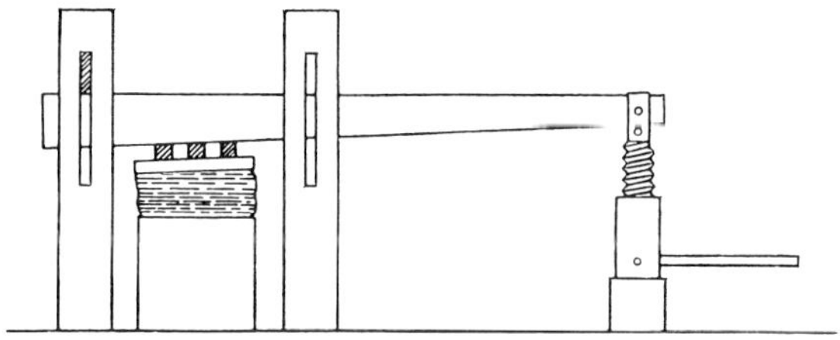

Fig. 4-6. Screw press. (*P. Haase & Son*)

will, a figure of a horse that will drink if offered water, birds that sing or are silent as an owl turns away from them or toward them, and many others. The most interesting feature of these devices is the use of hot air or steam to operate them: a temple door is opened when an altar is lit; in another temple two puppets offer libations when the altar is lit. Both are toy theaters, not life-size. There is a water heater for mixing hot drinks which generates steam for blowing on the charcoal. There is even a small steam turbine: a hollow ball turning on two pivots rising from the lid of a boiling kettle; one pivot is hollow, and the steam coming into the ball escapes through two bent tubes, thus causing the ball to rotate by reaction. This toy was not, however, the forerunner of any real steam engine, then or later.

CONFLICTING INTERPRETATIONS OF HELLENISTIC TECHNOLOGY

Some scholars have argued that Heron's technical devices, despite their ingenuity, constituted no technological advance, that they did nothing to assist man in the performance of work or to further his control over his physical environment. Designed to be used in the temples—if they were actually meant to be put into practice at all, rather than mere doodlings of Heron's imagination—they were meant to amuse, impress, or delude the worshipper, not to lighten man's physical burden. Why, they ask, was so much technical ingenuity "wasted" in this fashion?

One explanation, advanced by the British scholar Benjamin Farrington, is that the ancient Hellenistic economy rested on slave labor. An abundance of cheap human muscle power was available, so there was no need to develop labor-saving devices. And why bother to make things easier for slaves? Even the slaves themselves had little incentive to make their work easier, for they would still remain slaves and be forced to work to the limits of their strength.

In addition, it has been pointed out that there was a gap between the scholarly, or clerical tradition, and the craft tradition. Farrington contends that the ancient world, following the example of Plato, looked down upon manual labor —that was left to slaves—and that men such as Heron, of the intellectual elite,

Fig. 4-7. Heron's owl from his *Pneumatics*. (*P. Haase & Son*)

had little knowledge of technical processes and would scarcely concern themselves with developing devices which would make manual labor easier for the despised lower classes. The workers, who might have been interested in some adaptation of Heron's devices which could make their work less toilsome, had no opportunity to learn of them, for they could not read or write.

The result, according to Farrington, was that few of Heron's devices were translated into devices which could actually be used to perform work. By this interpretation, such devices represent technical ingenuity but not technological progress.

However, a closer examination of the historical evidence would cast doubts upon the validity of Farrington's interpretation of the nature of Greek technology. For example, it would be a mistake to believe that Heron was not interested in practical technical devices designed for the performance of work. Although it is true that in his *Automatic Theater* and in *Pneumatics* Heron was describing toys, not all of them were merely imaginative; indeed, most of the chapters in *Pneumatics* describe existing devices for the entertainment of guests at a feast to which Heron has added improvements. Furthermore, in his other books—*Mechanics, Dioptra, Belopoiica*—Heron deals almost exclusively with implements designed for practical purposes.

Similarly, it is mistaken to state that the abundance of cheap human muscle power hindered the invention of labor-saving machinery. As we have seen in the case of Ktesibios and Archimedes—and Heron's *dioptra* and screw press—many important labor-saving devices were invented during this period. True, slave labor might have been cheap by modern standards; but there was no lack of competition, and such competition would lead to the use of labor-saving machines.

Furthermore, it is questionable if Farrington's interpretation of Plato truly represents Plato's thought. In the *Apology*, Plato's earliest work, Socrates finds that craftsmen know their crafts; in Plato's last work, the *Laws*, he refuses to allow craftsmen to take part in governing the state, for that in itself is a craft, and no man can learn two crafts. Throughout his *Dialogues*, whenever he uses for an illustration a man who knows his job, Plato always refers to a craftsman: a surgeon, a trainer of horses, or the like. Such evidence would indicate that Plato did not lack respect for craftsmen as such.

Nevertheless, the most convincing evidence in opposition to the Farrington thesis of the technological sterility of Classical antiquity is the fact that notable technical advances were made during Greco-Roman times, in order to ease the burden of human labor. A prime example of this was the development of the water wheel.

THE WATER WHEEL

Two of the most laborious tasks in antiquity (and later) were the pumping of water for irrigation and for the city water-works, and grinding grain to make flour for baking. Like most technical problems in antiquity, these were solved not by the inventive feats of such individuals as Heron or Archimedes, but rather through the slow accretion of craftsmen's skills.

Grinding was originally done with a round stone that was rubbed along a hollow in another stone (the quern). Then the grinder (top stone) was made square and provided with a slit for feeding in the corn; the lower stone was flat and was placed upon a table, and the grinder was moved to and fro over it by means of a long wooden arm. This was hard work, and disobedient slaves were sometimes punished by being banished to the mill. By about 150 B.C., two other mills were widely used. One was a rotating hand-mill (or handquern), in which a round millstone was turned backwards and forwards by a handle; the other was a big mill turned by a horse or a mule. In this the grinder was formed like an hour-glass and rested on a fixed lower stone shaped like a cone. The grain was poured in from above, and the top stone was turned by an ass, a mule, or horse harnessed to a long wooden pole attached to the upper stone.

The water mill was not introduced until about a century later, just before the end of the pre-Christian era. It probably had its origin in the water wheel of the period, which was a large wheel with buckets fixed to the circumference, and

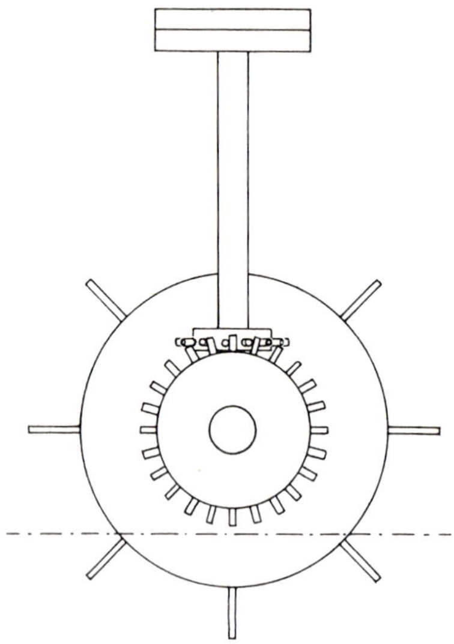

Fig. 4-8. Water mill. (*P. Haase & Son*)

which was set in a stream of water. When turned by manpower, the buckets picked up water from the stream and carried it to the top of the wheel radius, where it was dumped into a reservoir or pipe. At some point it must have occurred to someone that by adding paddles to the buckets, the wheel could be turned by the power of the flowing water itself and thus perform the work of the man.

The next step was to put a gear wheel on the axle of the paddle-wheel and make it turn a vertical axle that carried a millstone. These first gear wheels had round pegs for teeth, which were fitted to the gear wheels set at right angles.

The water mill developed quickly. Rome was provided with ten aqueducts that carried water from mountain springs to reservoirs in the city; the water ran day and night, and the surplus went to rinse the sewers. Where the water came down the Janiculum, one of the seven hills on which Rome was built, it was used for turning mills.

Mills now appeared throughout the Roman Empire, and were applied to several tools, including saws for cutting marble. Although the substitution of water for muscle power represented an enormous stride forward, the new mills had considerable limitations. The most obvious and critical was that they were tied to the geographical accident of sufficient water power. The lack of adequate transportation from the sites of water power (and therefore mill sites) to cities was a drawback which severely limited the wide use of mills. The result was that water mills, while first employed near the beginning of the Christian era, were not to achieve wide use until many centuries later; thus the development of the hammer mill during the early Middle Ages was due to the finding of ore near water power, so that the ore did not have to be transported.

The idea of the water wheel could, of course, be reversed, and the ancients developed a device to that end. It will be recalled that about 150 B.C. the hour-glass shaped mills had appeared whose upper grindstones were attached to a pole; animals were harnessed to the pole and thus turned the mill by going around and around. Since the gearing used in the water mill changes the rotation from vertical to horizontal, the ancients perceived that the horizontal rotation of the horse-walk could be converted to the vertical rotation of a bucket chain, and that animal power could thus be used to perform the laborious task of lifting water. Our earliest evidence of the animal-driven bucket chain for lifting water is to be found in a mural painting dated somewhere between 100 B.C. and 100 A.D., which shows such a device driven by oxen. This device, still in use in Mediterranean countries and the Near East, is now known as a *saqija* (in Spanish, *noria*); it is a striking example of the wish to substitute animal power for human power.

The windmill was not known till many centuries after the Roman Empire. Although a chapter in Heron's *Pneumatics* describes an organ whose air pump is driven by a sort of wind motor, this was just a sketch for a toy, and nothing came of the air pump until the Middle Ages, when a greater variety of power sources was required. Like so much of Heron's work, it remains only a historical curiosity.

MILITARY TECHNOLOGY OF GRECO-ROMAN TIMES

Military needs have frequently stimulated technological developments throughout history. Man's paramount need for security has meant that he lavished care, time, and energy on his weapons. With the Greek city-states constantly engaging in warfare with one another, and with Rome later conquering and maintaining a vast empire, it is not surprising that improvements and some innovations occurred in military technology during this period.

The ordinary hand-weapons—spear, javelin, sword, bow and arrow—had been known, of course, since prehistoric times. Apart from the sword, these were all hunting weapons, which were used also against human enemies. The javelin, or the "hand-flung spear," was greatly improved by the Romans for military purposes, by making it difficult for the enemy to extract it from his shield. Upon attacking, the Roman soldiers first threw their javelins at the enemy, and then closed in upon him with their swords. When the javelin pierced the enemy's shield, its shaft turned at right angles to the point; while this then made the javelin useless as a weapon, it also meant that the shield was useless till the javelin had been removed. While the enemy was thus encumbered, the Roman soldier was upon him. In a later development, the point of the javelin was firmly fixed to the shaft, but it was made partly of soft iron, so that it bent when it struck.

One important invention—the catapult—dates entirely from Classical times. Whereas the ordinary bow sends out its arrow only with the force given it by the strength of the arm of the man who shoots it, the catapult multiplied by many times the power of the bow or the sling. Invented by the Greeks in Sicily shortly after 400 B.C., the first step in the development of the catapult was the *gastraphetes*, or "stomach-bow." This was a crossbow, strung by the archer leaning his weight against the stock while it rested on the ground. In this way a sliding board was driven into the stock, taking the bow string with it; the string was caught by a hook and released by a trigger.

Next came the catapult proper. Instead of a bow made of wood or horn, it was provided with two strong arms whose outer ends carried the bow string, while the inner ends went through two bundles of sinews which possessed great elasticity. The bow string was drawn back by a winch, and the whole device stood on a wooden base. Some catapults shot arrows, others hurled stones. Although the arrow catapults did not work swiftly, they made it possible to shoot at enemies from a safe distance. The stone-throwers were not strong enough to knock down a solidly built wall, but were useful to the besieged for destroying the attacking siege-engineers.

The power of the catapult resided in the two bundles of sinews that activated the arms; their strength depended on their circumference. To build a catapult to a given specification, the diameter of the hole for the sinews was calculated from the length of the arrow or the weight of the stone, and the dimensions of all the other parts of the catapult were given with this diameter as the basic measure-

Fig. 4-9. Gastraphetes, or "stomach-bow." (*P. Haase & Son*)

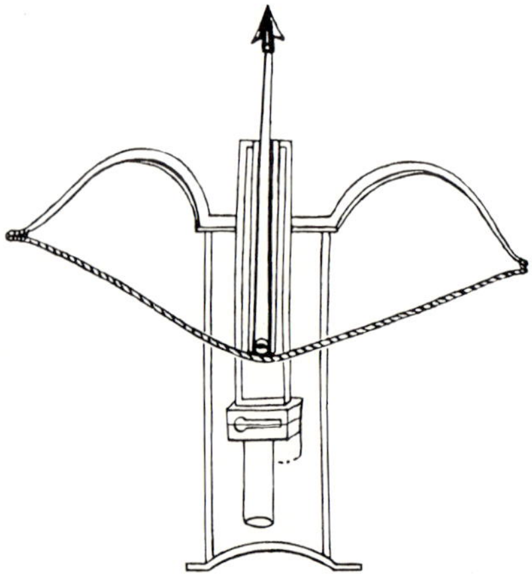

Fig. 4-10. Tortoise. (*P. Haase & Son*)

ment. This is one of the oldest examples of the use of a basic measurement to provide the ratios of the other parts, and it shows that mathematical knowledge had its part in the technical progress of the times.

Pitched battles were the exception rather than the rule in ancient warfare. Most wars involved lengthy sieges against fortified places. The attack against a fortified town was made by "tortoises," battering rams, scaling ladders, and siege towers.

The tortoise was a solid shelter, like the roof of a house, which was moved on wheels toward the wall of the fortress. Under its protection soldiers could demolish or undermine the wall. The battering ram was a huge wooden beam with a bronze head. It was hung from a tall timber structure by chains, so that it could be swung forward against the wall. The scaling ladder was a long ladder wheeled against the wall, and the seige tower was a solid wooden building at least as tall as the wall against which it was wheeled.

All these seige engines had been invented by the Greeks. The Romans merely took them over and improved them.

ROADS

The greatest contribution to military technology in Classical times, however, and perhaps to the advance of civilization itself, was the system of roads developed by the Romans. In Greece the roads had been mere paths or tracks, on which such deep ruts were made by the two-wheeled carts that they could pass each other only with difficulty. Only for the religious processions to the great temples were short paved roads laid down.

The Romans, on the other hand, wanted to move their army to the place it was needed as swiftly as possible, and they built roads everywhere. The roads followed as far as possible a straight line: small hills were cut through, tunnels were made through mountains, ways were cut through forests, and on swampy ground the roads were laid on planks carried by posts rammed down into the mud. The surface of the road was paved with stones, and drainage ditches were dug along it; in some places there were even sidewalks. Rivers were crossed by bridges, some of which still exist and carry modern traffic. All in all, the Romans built some 44,000 miles of road.

These roads made possible swift communication and transportation throughout the vast empire, enabling Rome to maintain its military domination and to administer a large territory. Commerce and culture also benefitted by the roads, many of which are still in use. The roads were the Romans' most valuable contribution to material civilization.

HYGIENE

The Romans also were interested in hygiene, not only personal hygiene but also public health. Life in large cities would have been impossible without adequate water supplies and a sewage system.

A big sewer, the *cloaca maxima,* was built in Rome about 600 B.C., and in 312 B.C. the Censor Appius Claudius built the first aqueduct to carry pure water from a small mountain river to Rome. From that time on, no one in Rome drank from or bathed in the Tiber, which received the water from the sewers. The aqueducts were multiplied, and in 226 A.D. no less than eleven aqueducts carried water to the city. Many of them are still in use.

Most of our knowledge about the Roman water system comes to us from a treatise, *On the Water Supply of the City of Rome,* written by Sextus Julius Frontinus (*c.* 35–103 A.D.). Considered by some scholars to be the greatest engineer in antiquity, Frontinus was commissioner of water works in Rome during the last half-dozen years of his life. His book is a detailed description of the Roman water-works system, including information about the construction of the aqueducts which supplied the city, the breakdown of the distribution of the water, the problems of construction and maintenance, and the laws applicable to the water system. Frontinus proved to be an efficient administrator of the water system, doing away with much fraud and corruption, such as the practice of water-works employees to take on private jobs while on the public payroll and the illegal tapping of water mains by householders. In his treatise, Frontinus pointed with pride to the utility of the aqueducts as compared with the "useless" works of the Egyptian and Greek engineers.

Wherever the Romans occupied or built a town, they took care of the water supply and the sewers, with beneficial effects on the health of the inhabitants. This practice however, fell into disuse during the early Middle Ages, surviving only in a few favored places. As late as the end of the 19th century, epidemics occurred in those European countries where people drank river water contaminated by sewers.

Both the Greeks and Romans were fond of athletics, and the public sports grounds were provided with baths. In the Roman Empire no town was complete without its public bath; in Rome itself they were developed into veritable palaces, with central heating and luxurious appointments, so they became favorite meeting places. During the Middle Ages bathing was frowned upon, and the personal cleanliness that characterized both Greeks and Romans did not come into its own again until our time.

CONSTRUCTION

The great baths and other Roman construction works were built of stone and cement. By a fortunate accident, the Romans discovered and made use of hydraulic cement (that is, cement that hardens under water and is not dissolved by water). Cement until that time had had only a limited use because it could not stand up to the elements, but the Romans fortunately had a supply of *pozzalana* sand nearby, which, when mixed with limestone, formed a cement which was impervious to water. The Romans did not know why this particular sand made such superior cement, but nevertheless they used it in their construction works. After Roman times hydraulic cement disappeared from history, only to be invented in the 19th century. We know it today as Portland cement.

Cement was used both for casting in blocks and as mortar; for common masonry made of sun-dried bricks, ordinary mortar of slaked lime was used. Although hydraulic cement enabled the Romans to bind together the stones of their buildings, bridges, and aqueducts without the need for elaborate mortising of joints, they still bound marble blocks together with iron cramps imbedded in lead, as in the famous Colosseum in Rome.

In addition to the new material of hydraulic cement, the Romans also exploited a new structural device—the arch. Certain forms of the arch and the vault (an extended arch) had been known in Mesopotamia about 4000 B.C.; and a true arch, dating from the 5th century B.C., has been found in the Greek town of Palaera. Although the arch was not invented by the Romans, they developed it and used it, so that it became the characteristic structural member in Roman architecture.

Previous stone construction, under the Greeks, for example, had always employed post-and-lintel construction, that is, columns (posts) supporting a beam (lintel) across them, similar to the way children build with playing cards. Though the Greeks achieved beautifully proportioned buildings (e.g., the Parthenon) with this type of construction, it had its limitations, chiefly the need for many columns to support any long beam, thus making it impossible to roof over any large area without the obstruction of many columns. As we will see later, the arch overcame this limitation; the stress and the thrust of the weight of the roofing was dispersed laterally to the base of the arches. The arch enabled the Romans to cover large areas without columnar obstruction. So successful a solution was this to the problem of spanning large areas that at the beginning of the 20th century the ancient Baths of Caracalla in Rome were copied for the Pennsylvania Station in New York City.

A Roman architect and engineer, Marcus Vitruvius Pollio wrote (25 B.C.) a comprehensive treatise on Roman architecture and building techniques, entitled *De architectura* (On Architecture), and dedicated it to Augustus. The book contains discussions of astronomy, sundials, and others materials, but its language indicates that Viruvius was probably not a highly educated man. Vitruvius' attempts to deal with the scientific principles underlying architectural theory are not convincing, but his detailed description of constructional matters reveals him to be unequalled as a practical builder. In the period following the decline of the Roman Empire, the work of Vitruvius was lost. When the study of Latin authors was revived in the 15th century, a manuscript of *De architectura* was discovered and Vitruvius' instructions on how to build a fortress, a temple, a theater, a seaport, or a house exerted an enormous influence on Renaissance architecture.

"Architect" at Vitruvius' time meant builder, contractor, and engineer, and the two last books of the ten that constitute his work deal with instruments and engines that today would not be considered within the province of the architect. Thus, Vitruvius describes clocks, cranes, and catapults, as well as buildings. So it is not surprising that Vitruvius is our main source of knowledge of the technical achievements of his time.

In the Middle Ages the arched vault of the Romans which developed into the pointed Gothic arch was employed successfully in the great cathedrals of Northern Europe. The dome, another development from the arch (simply an arch turning on its axis throughout its entire circumference), became characteristic of Byzantine architecture; and much modern concrete architecture employs the principles of the Roman arch. Thus the Romans have lasting monuments not only in their own buildings but in the design and techniques of future buildings.

TECHNOLOGY AND SOCIETY

During the thirteen centuries (*c.* 900 B.C.–*c.* 400 A.D.) which we assign to the Greco-Roman period, Western society changed from a conglomeration of small agricultural communities to a worldwide state with great cities and an extensive trade. These cities demanded aqueducts and sewers and the transportation of goods by land and sea. They also required a division of labor, which led to the rise of a class of artisans: smiths, weavers, fullers, bakers, and the like.

There still remained much unskilled, heavy labor to be done, and this was the task of slaves; though slaves were also used for the lighter work of servants, secretaries, or accountants, according to their ability. A slave was not recognized as a legal person, but was the personal property of his owner. Yet, to be a slave was not an inexorable fate, for a slave was sometimes allowed to save up money to buy his freedom, which, in Rome about 50 B.C., might take him some seven years. Once freed, the slave would become a Roman citizen with all the rights belonging to this rank.

Since the owner had to feed and clothe his slaves anyway, he was interested in getting as much work done as possible with the fewest slaves, and undoubtedly some inventions had saving labor as their purpose. Archimedes invented his water snail, differential gear, and endless screw on the strength of his mathematical studies, but they all aimed at making human muscle power more effective. The screw press had the same effect. The water wheel, invented for lifting water, was used for grinding grain and for other purposes.

Thus there was incentive to save human muscle power in antiquity, despite the abundance and cheapness of slave labor. Indeed, the invention of the water wheel—while not fully exploited until the Middle Ages—was a turning point in technical progress, for it introduced the idea of using another source of power in place of muscles. Nevertheless, the chief energy source of antiquity continued to be human and animal muscles; the basic materials continued to be stone, wood, and brick, with metal used chiefly for weapons and sometimes for tools. With the exception of the Roman roads, transportation and communication remained much the same as they had been since the beginnings of historic times. However, the Hellenistic inventors had devised some very ingenious machines, so that the mechanical equipment of Classical antiquity was more than simply a refinement of the tools and implements of pre-Classical times. Thus most of the elements of machinery used to improve technical devices after 500 A.D. had actually been invented before the fall of the Roman Empire.

The Romans are known to history as great engineers. Why is this so? Partly this is owing to the large number and monumentality of their construction works—the great aqueducts, roads, buildings, and bridges which have withstood the ravages of time and still bear testimony to the strength and solidity with which the Romans built. Partly too it is owing to the organizational abilities of the Romans. For engineering consists of more than machines and processes; the task of the engineer is to marshal effectively the resources at his command, to understand the limitations and potentialities of his tools and materials, and to organize his human as well as material resources for the accomplishment of the task at hand. It is perhaps in this last category that the Romans truly excelled.

Using the basic tools which had come to them from previous times and the improvements invented by the Greeks, the Romans nevertheless succeeded in accomplishing engineering feats which had no counterparts until relatively modern times. Roman engineering skill represents a triumph of human organization, and the Romans displayed the same organizing skill in law, government, and military matters. And, while the Roman Empire declined and eventually disintegrated, its engineering constructions remained as a source of wonder and amazement for future generations, a constant reminder of "the grandeur that was Rome."

5 / Technology in the Middle Ages

LYNN WHITE, JR.

The traditional historical picture of the Middle Ages (roughly from the 5th century A.D. to the mid-15th century) has been one of cultural decline, particularly in the early Middle Ages. These centuries, from the 5th to the 9th, have therefore sometimes been called the Dark Ages. Yet such a view of the Middle Ages, and even of its early period, is false when viewed from the standpoint of the history of technology.

Medieval technology continued that of the Roman world. In the eastern half of the Roman Empire, Byzantium, the New Rome established in 330 by Constantine, enjoyed an amazing prosperity and vigor for a thousand years and more. Even when, in the 7th century A.D., the Arabs wrested Syria and Egypt from Byzantium, there was no "decline and fall": on the contrary, the very creative new Islamic civilization incorporated and perpetuated the technical achievements of Greece and Rome.

The idea of the so-called Dark Ages is therefore applicable only to the western portion of the Roman Empire, but again, it is not in terms of technology. In the West the turmoil of the Germanic invasions led to a technological slump only in a few areas. The Romanized Celts of Britain, for example, were pushed into Wales and Cornwall by the fairly primitive Angles and Saxons (who were, however, superb goldsmiths), where they lived in such difficult circumstances that technical rejuvenation could not be spontaneous. Eventually it came from the Continent, where culture never sank so low, despite instability, depopulation, and economic depression. A symbol of the general maintenance of skills in the barbarian kingdoms is the tomb of Theodoric the Ostrogoth (d. 526) at Ravenna: it is capped by a monolithic dome weighing 276 tons which was barged some one hundred miles from Istria and lowered with razor-edge precision onto a masonry drum.

When Roman inventions did pass out of use there was always a good reason. Roman roads were so costly to maintain that even the wealthy Byzantine and

Islamic empires decided that they were not worth the expense. The hypocaust, the Roman system of radiant heating by means of channels through floors and walls, used much fuel in proportion to results and did not respond quickly to the rapid temperature changes typical of Northern Europe; so the Middle Ages invented chimneyed fireplaces and hot-air stoves which were cheaper and more flexible than hypocausts. In Gaul the Romans sometimes used a harvester, pushed by an animal, which chopped off the ears of grain and let them fall into a container; this wasted the straw. When medieval peasants developed a more intensive agriculture which habitually combined stock-raising with cereal production, the straw became valuable and what looked like a sophisticated machine was made obsolete.

Thus, any decline in technology in the early Middle Ages was more apparent than real. As we have seen, a technology is responsive to social needs; the needs of the psychologically urbanized and politically centralized Roman Empire differed from those of the agrarian and politically decentralized states which arose out of the ruins of the Empire in the West. But technical skills seem to have diminished in no significant way. Instead, the changing conditions in the West stimulated technological advance there.

MEDIEVAL EAST AND WEST COMPARED IN TECHNICAL INNOVATIONS

The most curious fact about medieval technology is that while for many centuries both Byzantium and Islam greatly surpassed the West in commerce, political stability, and level of education, nevertheless, it was the West which produced most of the major technological innovations.

There was, to be sure, a technological spurt in Byzantium during the 6th and 7th centuries: the amazing single-shell dome of St. Sophia was designed by the architect Anthemios of Trales in 532, while about 673 a Greek-speaking Syrian refugee from Muslim conquest, Kallinikos, invented Greek fire, a petroleum-based incendiary so efficient in burning enemy ships and siege machinery that its formula was placed under strict security by the imperial arsenal. Considering that Greek fire was in great part responsible for Byzantium's military survival, it is strange that thereafter the medieval Greeks—so vivid and sensitive in many areas of life—showed little interest in technological improvements. Similarly, the Muslims, while they borrowed useful skills from other cultures—paper-making from China in 751, for example—did not make notable contributions, so far as we now know, to mankind's technical repertory.

This is the more puzzling to our 20th-century minds because Byzantium preserved ancient Greek science, and from about 800 to 1200 Islam produced the world's greatest scientists. We today think of technology as applied science, but until the 19th century the connection between science and technology was slight. Science was for intellectuals trying to understand the nature of things;

technology was for workers trying to do things; the notion that knowledge of nature gives us power over nature is old as an aphorism but recent in practice. Medieval Byzantium and Islam produced complex and subtle cultures which focused their energies on art, literature, religion, philosophy, and science, but were little concerned with technical advances.

The contrast with the medieval West is striking and demands explanation. Certainly scientific interests cannot account for it. The early medieval Occident continued the shocking indifference of pagan Rome toward Greek science. Not until the 11th century did Greek and Arabic science become available in Latin, and another 200 years passed before Western Christendom assumed the leadership in science which Islam had held. The West's pre-eminence in technology thus precedes its primacy in science by several centuries.

The victory of Christianity over paganism in the 4th-century Roman Empire had provided an improved psychological basis for technical innovation. The religion of the common man in antiquity was animism: every tree, stream, or mountain had its *genius* or particular spirit which had to be placated before one cut down the tree, dammed the stream, or dug into the mountain for mining. In such circumstances, the Christian smashing of animism liberated artisans and peasants for matter-of-fact exploitation of their natural environment. This change, however, had occurred throughout the late Roman world, and while it helps to account for the eventual speed-up of technological development, it does not explain why the West took the lead.

In Greco-Roman times educated men considered it beneath their dignity to work with their hands. The Jews, however, were an exception: God on Sinai had commanded "Six days shalt thou labor, and on the seventh rest"; the injunction to labor was as binding as that to relax on the Sabbath. Even the most learned rabbis acquired skills at a trade: St. Paul, who studied for the rabbinate, was also a tent-maker. In the 4th century, when Christianity became the official cult of the Roman Empire, it was so corrupted by the influx of opportunists and conformists that monasticism arose as an effort to restore its primitive (largely Jewish) principles. The monks insisted that manual labor is an essential part of the spiritual life, and that "work is worship": *laborare est orare*. Their idealism, intelligence, and energy made monasteries the chief points of cultural radiation during the next seven centuries. The monks were the first intellectuals systematically to dirty their hands with physical work, and we cannot doubt that this combination of brain power and sweat aided technological advance. But since the Greek monks worked as hard as the Latin, this again does not explain the distinctive vigor of technology in the West.

What elements can we identify as peculiar to the Occident? Four things may be suggested.

1. Among the Celts of Roman Gaul there seems to have been somewhat more inventiveness than is detectable in any other part of the Empire. Perhaps this

mood of innovation carried over into the Western Middle Ages and expanded.

2. The Occident was much more deeply shaken by repeated invasions and chaos than were the Byzantine and Islamic regions. There is reason to think that any change aids subsequent change. The greater agony of the West during the early Middle Ages, as the folk-wanderings of the Teutonic tribes gave way to the barbarian kingdoms, may well have corroded traditional ways so deeply that people were generally more open to change, including technological change, than they were in the more "fortunate" Near East, to the eventual great profit of the West.

3. In Greek Christendom an educated laity continued, whereas in the West culture declined to a point where literacy was long a monopoly of the clergy. As a result, the Latin monks came to feel far more responsibility for preserving not only Christian but also pagan or secular culture than was felt by the Greek monks, who could depend on Byzantine laymen to care for the latter. This meant that in Latin Christendom the working monks were closer to worldly concerns than was the case in Eastern Orthodoxy. The Oriental monks have left us nothing comparable to *On Divers Arts* written in 1122-23 by Theophilus, a German Benedictine: this is the earliest European treatise giving specific technological directions for a wide range of complicated processes. Theophilus was not only an expert in metallurgy and glass; he was learned in theology and wrote quite decent Latin. His mentality helps explain the West's advance in technology.

4. Finally, in an effort to understand the technology of an age so permeated with religious attitudes, we must note a basic difference between the theologies and pieties of the Greek Church and the Latin Church. The Greeks have always made right *thought,* or "illumination," central to salvation, whereas the Latins at least since the days of St. Augustine have put greater emphasis on right *will,* or action. The Eastern Church has praised contemplation; the Western, activity. Technology involves doing things, and the mood of the Roman Church fostered it by encouraging activism and practicality in Occidental society.

MILITARY TECHNOLOGY AND SOCIAL CHANGE

If we consider the chronic physical insecurity of life in the early medieval West, it is not surprising that military technology made notable advances there. So long as a horseman had to cling to his steed by pressure of his knees, cavalry was used chiefly for bowmen and the movement of soldiers: the lance could be wielded only at the end of the arm (or with two hands, making use of a shield impossible), because too violent a blow would unseat the rider delivering it. In the early 8th century the stirrup reached the Frankish kingdom, established by the Germanic tribes known as the Franks in what had been the Roman province of Gaul. When combined with a saddle having a high pommel and cantle,

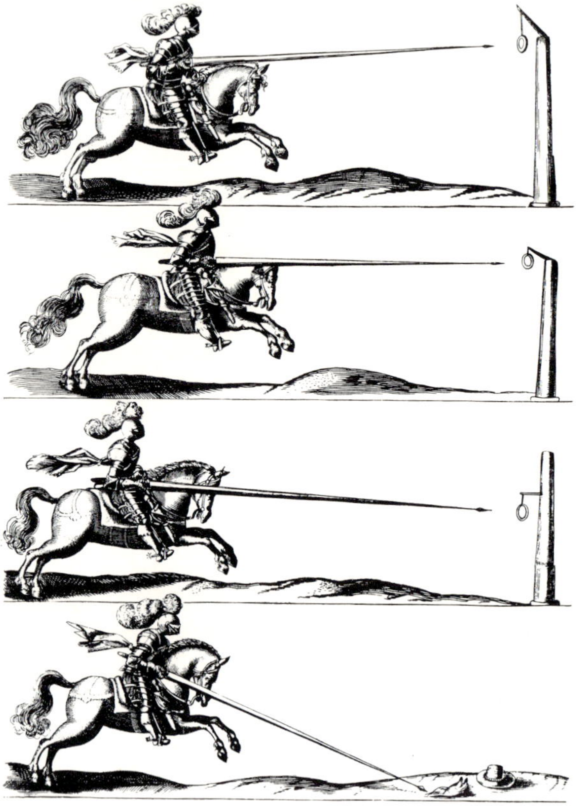

Fig. 5-1. Knight with lance on horseback. When combined with a saddle, stirrups made a single organism of rider and horse. (From John Cruso's *Militarie Instructions*, 1632. Courtesy of *The Huntington Library*, San Marino, California)

stirrups make a single organism of rider and horse. The Franks saw that the lance could now be laid "at rest" under the rider's armpit; the hand merely guiding the blow, which was delivered by the combined impetus of a charging stallion and warrior. The new method of mounted shock-combat involved a great increase in the violence of warfare and indicated a shift from infantry to cavalry as the chief fighting force.

While the Franks of the 8th century were very nearly the last horse-riding people to acquire the stirrup—it had come from India by way of China—they were the first to realize its full implication for battle. There is no absolute determinism in technology: invention is not the mother of necessity. In 732 the

Frankish leader Charles Martel saw a military potential in the stirrup which others had overlooked, and he acted upon his insight. Cavalry is much more costly than infantry, and circulation of coinage in the Frankish realm was insufficient to support an enlarged cavalry out of taxes; so Charles confiscated vast reaches of Church lands and distributed them to retainers on condition that they hold themselves ready at his command to fight on horseback in the new and difficult way. These mounted warriors became the basis on which Charles Martel's grandson, Charlemagne (Charles the Great), enlarged the Frankish domain into the Carolingian Empire at the beginning of the 9th century. When, a century later, the Carolingian Empire disintegrated, this caste of endowed warriors picked up the fragments of political authority and established local rule. Thus the revolution in military technology brought about by the stirrup was the seed of feudalism and of the chivalric culture which the secular aristocracy of the later Middle Ages developed.

The violence of mounted shock-combat led to development of heavier armor, heavier horses, new types of shields, and (in the 11th century) the crossbow designed to penetrate the new armor. (The history of the crossbow is puzzling: the Chinese had it very early; the Romans used it chiefly for hunting birds; but at the time of the First Crusade, 1096, the Byzantines considered it a Frankish novelty.) The new Western military technology was superior to that of the Near East, and elements of it began spreading to Byzantium and Islam even before the Crusades. One of the chief reasons for the eventual failure of the Crusades was that the Muslims learned to fight in the European way.

From the later 11th into the early 13th centuries, military architecture was revolutionized in the West. Often this is credited to Near Eastern influence, but the most careful scholars consider the question still open. One stimulus to better fortification was the development by Europeans, probably on the basis of a Chinese hand-operated rock thrower, of a new and powerful type of counterweight artillery, the trebuchet, which quickly superseded the torsion artillery inherited from the Romans. In the 12th century French engineers produced Gothic architecture, an immensely ingenious system of thrusts and balances using a minimum of masonry to enclose a maximum of space. It was so economical that the most ascetic monastic order of the time, the Cistercians, adopted it and spread it quickly throughout Europe. The rapid and superb expansion of the art of fortification in the West in exactly this period would seem to reflect the same mentality applied to different problems.

THE EXPANSION OF AGRICULTURE

Until very recently agriculture has been the basic form of technology. In antiquity its production of surplus was very low: it is a safe guess that well over nine-tenths of the population had to work the soil to support a tiny fraction of

Fig. 5-2. Wheeled plow. (From Herrad of Landsperg, *Hortus deliciarum, c.* 1170)

humanity engaged in other occupations. Clearly, anything which increased productivity was of major importance.

In late Roman times there were efforts, particularly in the northern provinces, to improve agriculture, but no coherent new system of cultivation emerged. By the middle of the 6th century however, some of the Slavic peasants were using a novel kind of plow very efficient for heavy fertile alluvial soils which were hard to handle with the older, two-ox scratch-plow designed for light soils. The older plow had merely dug the surface of the soil; in order to turn over the soil for planting, it was necessary to cross-plow, that is, to plow the soil twice, the second plowing being at right angles to the first. The new heavier plow had wheels, a vertical blade (colter) to cut the line of the furrow, a horizontal plowshare, and a mouldboard to turn over the sod. Its friction with the dirt was so great that it had to be pulled, at least on newly cleared land or in sticky soil, by eight oxen. It attacked the earth so violently that the cross-plowing required by the scratch-plow was unnecessary, and squarish fields gave way to long strip-fields. Since the mouldboard normally turned the sod to the right and the fields were plowed clockwise, the strips tended to become low ridges favorable to field drainage in the wet climate of Northern Europe. Since few peasants owned eight oxen, co-operative plowing became usual. Likewise, since the fencing of long strip-fields was impractical, villages using the heavy plow divided the arable land into fenced "open fields" embracing many strips which, even though individually owned, had to be cultivated, planted, and harvested on a unified plan. The adoption of the new plow therefore helps to explain the communal pattern of manorial life in Northern Europe.

Starting, it would seem, with the Slavs, the new plow and its related agrarian system spread among the Germans by the early 8th century, and were presumably taken to Britain in the late 9th century by invading Norsemen. Wherever these methods went, their ability to use the heavier and more productive river-bottom soils led to a vast cutting of forests and reclaiming of marshes for agricultural purposes: the face of Northern Europe was changed.

Paralleling and interlocking with the new pattern of cereal-growing was an improved type of cattle-raising. The Romans had not integrated stock-farming closely with agriculture, but had simply pastured their cattle. Proof of this is the scarcity of Roman scythes. Scythes had been used chiefly for cutting grass for hay, which implies an intensive rearing of cattle and sheep, largely in barns, and a concentration of their manure for later systematic fertilization of fields. In the Frankish age, scythes became common, and at the end of the 8th century Charlemagne tried to rename July "Haying Month." In addition to the haying, after the harvest the village herd was turned into the open fields to browse on the stubble, incidentally leaving their droppings to fatten the next crop. Thus the northern medieval peasants worked out a new system of food production more balanced and efficient than anything earlier.

By the later 8th century they had taken another stride, at least in the region between the Loire and the Rhine rivers which was the heart of the Carolingian Empire. Land had normally been left fallow half the time to renew its fertility: the cultivated half of the arable was planted in the autumn with wheat, barley, or rye and harvested in the early summer. But now this "two-field" rotation began to give way to a "three-field" system in which only a third of the land was left fallow. In the autumn another third was planted as before; but in the early spring the remaining third was planted in oats, barley, and legumes to be harvested the later summer. The peas and beans were particularly important, both because their nitrogen-fixing ability strengthened the soil under the burden of this more intensive rotation, and also because they furnished an increased quantity of vegetable proteins for human consumption.

Since the new spring planting required summer rains, it was generally feasible only north of the Alps and the Loire River. Where it could be adopted, however, it was so advantageous that it does much to account for the great vitality of the North in the age of Charlemagne. By providing two sets of crops and two harvests, the three-field rotation much reduced the risk of crop failure and famine. By distributing the work of plowing better over the year, it enabled the plow team to accomplish more. Depending on whether the fallow were plowed once or twice (to turn under the green manure), a community of peasants, with any wasteland to reclaim, by shifting from the two-field to the three-field rotation could increase their production by either one-third or one-half.

The surplus of oats which could be grown in the spring planting of the three-field system is related to another major change in northern agriculture. In antiquity, oxen were adequately harnessed by means of yokes, but the yoke applied to horses is extraordinarily inefficient, both because it strangles the animal as soon as he tries to pull and because the point of traction at the withers is so high that the horse cannot throw his weight into the task of pulling. About 800 A.D. the modern horse harness appeared in the Carolingian realm, consisting of a rigid, padded collar resting on the horse's shoulders and permitting him to breathe, and lateral traces or shafts placed so that the point of traction is effective. With this new harness a team of horses could pull four or five times the load they could draw with a yoke harness.

Hitherto the horse had been valued for its speed; the new harness made horse-power available in conjunction with that speed. The first evidence of habitual plowing with horses, who worked perhaps twice as fast as oxen, comes from Norway in the late 9th century. By 1100, horses were customarily drawing plows, at least in favored regions, all the way from the English Channel to the Ukraine, and throughout the later Middle Ages the horse steadily displaced the ox for farm labor. But this occurred only in Northern Europe, where the three-field rotation made possible, in the spring planting, the surplus of oats needed to feed many horses. The Mediterranean peasants could not shift from oxen to the more efficient horses because, for climatic reasons, they could not produce enough oats.

The early Middle Ages then, witnessed, in Northern Europe, an agricultural revolution unparalleled since the first invention of tillage. Its elements—the heavy plow, open fields, three-field rotation, and horse harness—accumulated and consolidated into a new agrarian system from the 6th through the 9th century. More than anything else the increased surplus of food which it produced accounts for the permanent shift, in Carolingian times, of the focus of European culture away from the Mediterranean to the great plains between the Loire and the Elbe rivers. It accounts for the steady increase of population until the late 13th century, when, because no further agricultural innovations had been introduced, the point of diminishing returns was reached and overcrowding began to worsen the living conditions of the peasantry and undercut the boom in the general economy of Europe which had prevailed from the end of the Viking invasions, *c.* 1000, until *c.* 1300. During these three centuries, the surplus of food permitted an unprecedented growth of cities and an accumulation of capital best symbolized by the enormous Gothic cathedrals which towered over them and which were the pride of the burgher capitalists who created the basic economic and political patterns of the modern West.

TRANSPORTATION DEVELOPMENTS

The Middle Ages likewise revolutionized transport, which made it possible to move the surplus food to the cities where it was needed. Thus technological innovations in agriculture and transportation, along with more settled political conditions, made possible the renewal of town life from about the 12th century on, and hence the foundations on which modern civilization was ultimately to be constructed.

The modern horse-harness, emerging *c.* 800, was essential to the use of horses not only for plowing but also for hauling. However, the wear on a horse's hooves was greater in hauling over roads than it was in tilling fields, and in moist climates horses' hooves grow softer than those of oxen. In the 890's the problem was solved by the daring invention of nailed horseshoes, which appear almost simultaneously in Siberia, Byzantium, and Germany. Iron shoes very quickly became habitual for ridden horses; but there was another problem to be solved before horses could be used for heavy hauling. A horse could plow with lateral traces attached directly to the plow because a furrow is straight. But with traces fastened directly to a wagon, a right turn puts all the strain on the left trace, and vice versa, risking breakage of the harness and overturning the load. The solution, which equalizes the pull on the load, was the whippletree, which appeared in the 11th century. Now, with an efficient horse harness, horseshoes, and the whippletree, heavy hauling by horses was feasible for the first time; and in the early 12th century the horse-drawn *longa caretta* (large cart), holding many people and large quantities of goods, emerged.

About the same time travel was made more comfortable through the development of the springed carriage. Without springs, prolonged speed over rutted and potholed roads is unendurable; the essence of the coach is suspension of the carriage body to cushion the jolts. The germ of this innovation appears among the western Slavs in the 10th century. Four hundred years later this had become a suspended body holding at least six persons. That it moved rapidly is indicated by the fact that a man with bagpipes was perched up on its rear to clear the road ahead: the ancestor of the coach horn and of the modern automobile horn.

Water transportation has always been cheaper than land haulage, and the Middle Ages did not neglect this mode of transport. The essential invention for inland waterways, the canal-lock chamber, seems to have been used at Bruges by 1236. But it was in salt-water navigation that the most significant improvements were made.

As we have seen, man had early harnessed the power of the wind to drive his vessels through sails. But how to go against the wind? Tacking into the wind was a great problem for square-sailed Roman ships. To be sure, fore-and-aft rigs had been applied to small skiffs since the first century B.C., but not to large vessels, perhaps because their keels were not sufficiently deep to prevent lateral drift during tacking. The lateen sail, well adapted to tacking, first appears on merchant ships at Marseilles in the 6th century. Since *lateen* comes from the rare Latin word *latinus* meaning "easy, handy," linguistic evidence would seem to indicate that this new rig was probably developed in the western Mediterranean.

In antiquity, ships were constructed by first building up the shell of the hull

Fig. 5-3. Sailing ship of the end of the 15th century showing rudder. (*The Bettmann Archive*)

out of planks firmly attached to each other and afterwards inserting the skeleton of ribs. This produced a strong vessel, but the process was slow and costly. Skin-diving archaeologists have found the wreck of a Byzantine ship in the Aegean dating from the early 7th century still built in this way. We do not yet know when or where during the Middle Ages our present system of ship-building, by first constructing the skeleton and then nailing on the planks, was developed. Certainly it reduced the costs of maritime commerce notably.

Another great advance in ship-building was the invention of the modern rudder. Early ships were steered by lateral oars which were easily broken in storms and which, when large, were so awkward that they tended to limit the size of ships. In the early 13th century the North Sea area produced a new rudder hinged to the ship's sternpost and operated by a horizontal lever. This was capable of standing the buffeting of great waves, and it could be used on vessels of any size.

Vessels could now be constructed strongly enough to venture into the open seas with safety, and they could be steered against the wind. But how was the navigator to find his way when out of sight of familiar landmarks and when the sky was not clear enough to steer by the stars? Here the East was to provide a technological aid for the West, for the magnetic compass presumably came from China. It reached Europe in the 1190's, and within thirty years was in habitual use even as far as Iceland. Strangely, there is no evidence of it in Islam until 1232. The compass so profoundly affected the art of navigation that, for example, two round-trips annually from the Italian ports to the Near East were now possible, whereas previously only one had been attempted. The returns on capital investment in ships were greatly improved, and the safety of sea voyages enormously increased. By the end of the 13th century, Europe was beginning to contemplate using oceanic sea-routes. In 1291 two members of a great Genoese merchant family, the Vivaldi, led a well-equipped fleet through the Strait of Gibraltar to open the path to India around Africa. The expedition perished, but technological advances by that time had reached such a point that anticipation of Vasco da Gama's historic voyage around Africa in 1498 seemed not impractical.

ARTS AND CRAFTS

Warfare, agriculture, and transport then advanced technologically during the Middle Ages; what of industry?

There is no firm evidence that the water mill was applied in Europe (as distinct from China) to any task save the grinding of grain until about 1000 A.D. The early 9th-century plan of the abbey of St. Gall in Switzerland may indicate water-powered triphammers, but both their identification and their use is uncertain. About the turn of the millennium, however, it is clear that such devices were being employed for fulling cloth, forging metal, and several other industrial processes. Thanks to the same inventions which facilitated hauling, the horse-mill, in which an animal walking in circles turned a vertical shaft to which various types of machinery could be attached, also spread widely and with many applications. Shortly before 1185 the horizontal-axle windmill was invented in the North Sea region, and within seven years it had spread as far as Syria. In the 10th century, vertical-axle windmills, perhaps inspired by Tibetan wind-driven prayer cylinders, had been used in Afghanistan, but these were never diffused in Islam. The windmill was particularly useful in those regions where the flow of streams was so sluggish that water mills were unsatisfactory, or where rainfall was so scanty that streams were scarce. In the 1320's a monk complained that England was being deforested partly because of the search for long timbers for windmill vanes: clearly he lived in a society vividly exploiting power machinery and labor-saving devices.

Machine design was also progressing. While the crank had been known in China since the Han dynasty, it first appeared in Europe in the early 9th century, and by the 12th it had wide application. The compound crank, a combination of the crank and connecting-rod which allows the conversion of continuous rotary motion to reciprocating motion, and vice versa, appeared in

1335. Cams, although known in antiquity, were first generally used in the trip-hammer machines of the 11th century; the groove cam is first found in the 1480's. The flywheel as a regulator of rotary motion in machines is recorded in the 12th century, but, strangely, the pendulum to regulate reciprocating motion is not observable for another 300 years. The earliest machine having two correlated motions is shown *c.* 1235 in a notebook of the French engineer Villard de Honnecourt: a water-driven sawmill which, in addition to the reciprocating action of the saw, provides a rotary feed to keep the log pressed against the saw. The first belt-transmission of power came about 1280 in the earliest spinning wheel, at Speyer, Germany. The 14th century saw an astonishing development of gearing, culminating in 1364 in Giovanni de' Dondi's great planetarium clock. The five centuries following 1000 A.D. greatly elaborated the methods of harnessing and utilizing mechanical power.

THE DIFFUSION OF TECHNOLOGY

The spread of technology in the Middle Ages, as in modern times, knew no geographical barriers. There was nothing self-contained about medieval technology. Norsemen who settled in Greenland taught the Eskimos to make cooperage, and European merchants in the Far East—in the early 14th century the Franciscans maintained a hospice for them in Amoy—showed the Chinese how to distill liquor, an Italian invention of the 12th century. On the western coast of India the crossbow was considered a Frankish weapon, and even in the 15th century a Persian poet knew that eyeglasses—discovered in Tuscany in the 1280's—were European in origin. On their part the Westerners avidly absorbed every item which seemed useful. From sub-Saharan Africa they took sorghum for their fields and the Guinea fowl for their barnyards; buckwheat was brought in from Central Asia by 1396.

Even the distant East Indies made their contribution, thanks to the perennial spice trade. During the 10th century the Javanese fiddle bow reached Europe and eventually became the most important item in Western instrumental music. In the early 15th century the blow-gun arrived, bringing its Malay name with it. Shortly it stimulated interest in air guns, and the air gun, together with the suction pump (an early 15th-century invention), was the chief stimulus to the scientific study of air pressures and vacua.

THE TECHNOLOGICAL PROGRESS OF THE MIDDLE AGES

It has often been said that the greatest of inventions is the idea of invention itself. The European mind first grasped invention as a total project in the later 13th century. For example, at that time the idea of perpetual motion reached Europe from India, and we know that at least two groups were struggling to make perpetual-motion machines. A contemporary Italian surgeon remarks that for the extraction of arrows "every day a new instrument and a new method is invented." The program for producing a weight-driven clock was clearly formulated by about 1271, although the task seems to have required another sixty years or more. Once it was accomplished, a technician in Milan immediately built weight-driven grain mills on the analogy of the new clocks.

As early as 1260 Friar Roger Bacon was looking forward to an age of flying machines, motor boats, submarines, and automobiles. He did not know how all of this was going to be accomplished, but he was confident that it could be done. Not only in its gadgetry but also in its mentality, the later Middle Ages provided the foundation for the subsequent structure of European technology.

Thus the conventional picture of the Middle Ages as a pause in mankind's struggle to conquer environment is inaccurate. Far from stagnating, medieval technology produced a revolution in man's use of energy resources, through the development of water wheels, windmills, and horse-traction; it transformed the art of war by the new power it gave to cavalry and by the development of military fortifications; it increased man's capacity to wrest a living from nature by the use of the heavy-wheeled plow and the three-field system of agriculture; it enabled man to sail afar on the seas through improvements in ships and navigation; and it devised new tools and combinations of tools to make work easier. Above all, it offered a new outlook toward technological innovation, which prepared the way for the mechanical devices of the following period of Western history, known as the Renaissance.

6 / Early Modern Technology, to 1600

A. RUPERT HALL

THE RENAISSANCE

The period from the mid-14th century to the beginning of the 17th was the age of the Renaissance, so called because it represented rebirth (*re-naître*) of interest in the Greece and Rome of Classical antiquity. Rebelling against medieval canons of taste and scholarship, even rejecting in part the claim of religion to dominate all aspects of human activity, the men of the Renaissance sought wider fields of knowledge, keener satisfaction of the senses, a freer range of endeavor. In so doing they took themselves to be re-creating the life of the ancient world, which appeared to them as so much more beautiful, wise, and ingenious than their contemporaneous European world. To the exponents

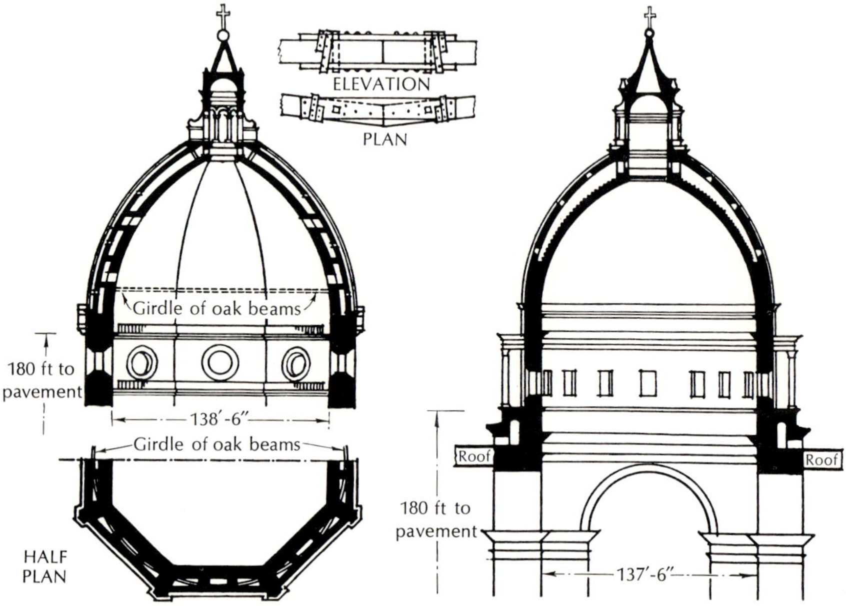

Fig. 6-1. Renaissance dome-construction. *Left*, the dome of the cathedral at Florence by Brunelleschi, 1420, with detail of its oak girdle (center). *Right*, dome of St. Peter's, Rome, by Michelangelo, 1546. (From Singer *et al.*, *A History of Technology*, Oxford, 1954-58)

of this Classical revival, the collapse of the golden civilization of antiquity had introduced an epoch of barbarism from which their own generations were the first to escape. Artists, writers, scientists, and even the more refined craftsmen looked to the past for inspiration and examples on which to model their own work. Latin and Greek were the indispensable keys to style, knowledge, and good taste, assuming a foundational significance in education they were to retain for centuries.

But the Renaissance was much more than the rebirth of interest in Classical antiquity and the revival of the humanistic spirit. Many new and different aspects of Western life showed themselves. For one thing this was the period of the great voyages of discovery which enlarged the horizons of Western civilization, as did the invention of printing, with its incalculable effects upon human communication and the spread of information. The stream of wealth from the New World helped to develop the already-growing economies of Europe. At the same time a new political form, the national monarchy, was emerging, although the Italian city-state remained the center from which wealth and artistic creativity spread to Northern Europe. Feudal warfare gave way to dynastic rivalries, and the nature of warfare itself changed with the introduction of gunpower and cannon.

The incessant warfare—now on an international rather than local scale—did not mean that Europe had lost sight of spiritual values. Indeed, in Northern Europe the Renaissance took the form of a spiritual awakening—an attempt to return to the primitive simplicity and doctrines of the early Church—which led to the Protestant revolt and to a split in Western Christendom. Yet the wealth of the Italian city-states and Papacy itself gave rise to the great works which made the Renaissance one of the most glorious periods in human artistic creativity.

The major influence of the Renaissance on technology was on architecture. The abandonment of Gothic forms by the Italian architect Leon Battista Alberti (1404-72) and his successors, and the gradual spread of the Neo-Classical Palladian style of building from Italy over the whole of Europe involved changes in building technique. The architects and the masons who carried out their designs had to learn how to construct great cathedral domes such as that of St. Peter's in Rome, which Europe had not seen before. This was the most conspicuous instance of the interrelation of style and technique; in less majestic arts, such as those of the joiner and cabinet-maker, the potter, the gold- and silversmith, and the pewterer, the Classical revival, occurring at a moment when the wealth of Europe was increasing rapidly, wrought corresponding changes in craft methods. The Renaissance involved a diversion of money from feudal pomp and power to beauty and luxury; to this extent it both encouraged the artistic craftsmen (goldsmiths like Cellini, painters like Raphael) and modified the objectives of their crafts.

THE ANCIENTS AND THE MODERNS

While most cultural activities of the early Renaissance looked backward to antiquity, technology was in general forward-looking. Niccolo Machiavelli (1469-1527), the Florentine political theorist, was an exception in preferring the weapons and tactics of the Roman legion to those of modern armies. Guido Pancirollus wrote a once well-known book, *The History of Many Memorable Things . . . Now Lost* (1599), in which he contrasted ancient accomplishments with the discoveries of the "modern"; apart from the customs and rituals of antiquity, and its monumental achievements such as the Sphinx and Pyramids of Gizeh, he was able to name only a few lost techniques: the manufacture of Tyrian purple, asbestos garments, and "ductile glass." Marvels attributed to antiquity—Archimedes' burning-mirrors and Archytas' flying models—were not taken very seriously.

On the other side it was argued that the inventions of the modern world demonstrated its technological superiority: this was especially the lesson of

Jan Stradanus' *New Discoveries*, a volume of splendid engravings also produced near the close of the 16th century. The ancients had not commanded the "super-natural" force of gunpowder, nor discovered how to multiply books and pictures by printing. They had found neither the direct route to the East, nor the New World to the West; they had remained ignorant of the use of the magnetic compass and of other navigational aids which had made the 15th-century voyages of discovery possible. The ancients had lacked windmills, iron-shod horses, the art of spectacle-making, mechanical clocks, iron-founding. Only a few years later at the beginning of the 17th century, Francis Bacon was to make these modern technical improvements—the fruit of craftsmen's ingenuity rather than scholar's subtlety—the basis of his argument that progress is both attainable and socially beneficial.

In foreseeing and applauding material progress, Bacon and his 17th-century successors seem to have followed the unconscious instincts of unlettered craftsmen, for whom antiquarianism had no appeal and who tried only to perfect what their predecessors had begun. Craftsmanship is one of the most conservative of all human activities, preserving methods and tools unchanged through many centuries; it is also selective in that when modifications do prevail over tradition they are rarely degenerative. Whatever else may be lost—whether quality is sacrificed to quantity or cheapness—at least there is invariably an increase in mastery over the technique. There is a very clear sense in which technological innovation is progressive, and quite independent of one's personal feelings about the results.

FUEL

Fuel technology in the age of the Renaissance provides an illustrative example of technological progress in the craft tradition. While one way of reducing human effort and speeding operations is to employ machines, another is to consume the energy of fuels. Ultimately, with the steam engine, the two went together; but in the Renaissance the paths of fuel and machine development were still separate.

The most obvious role of fuel in industry was in preparing useful materials: metals, salts and a few other chemicals, bricks, tiles, and pottery, glass, dyes and other materials of the textile and leather trades, pitch, tar, paints, glues, and much else besides. The richer, more spendthrift society of the 16th century demanded an increase of production; when the thriving farmer threw out his wooden tableware in favor of pewter, or the rich bought more silver plate and brilliant earthenwares, the pressure on fuel supplies increased.

But household demands were as nothing compared with the hunger of foundries where cannon—an innovation of the late 15th century—were cast. Several tons of wood were required to smelt one of metal; and the forests suffered. So did the poor man's hearth as the price of timber mounted exceptionally in a time of painful inflation. By the third quarter of the 16th century, dwellers in the iron-working districts of England and northern France were complaining bitterly of the shortage of fuel and demanding checks upon the activities of the ironmasters.

In Northern Europe, and particularly in Britain, coal had been mined for house fuel since the 13th century. In the 16th, coal was adopted for the domestic hearth in regions far from the pits, to which it could be brought by sea—for example, in London, whose air was soon fouled by the smoke. Moreover, as wood grew dearer coal was applied to many simple industrial operations such as brine-evaporating, brewing, brick-making, and sugar-boiling, and was readily introduced into the English glass works. "Charked" into coke, coal was employed for rough smith's work, and in the early 17th century the long search for the secret of smelting metals with coal began.

Now it is true that the products of coal-firing were not necessarily superior to those of wood-burning furnaces, and indeed, it was long before iron made with coal approached the quality of Swedish, charcoal-smelted iron. Moreover, the greed for coal forced millions of men into a dangerous and arduous livelihood in the mines, while it blighted industrial civilization with its own blackness. Yet it is undeniable that the release of the energy contained in coal for lavish use in manufacture for over three hundred years was of enormous importance, for it was the unique basis of industrial technology until late in the 19th century. It was coal mining too that made the steam engine desirable to pump water from the mines. Those parts of the world that lacked deposits of coal were simply left behind in the industrial race, and hence in living standards too, until water power and oil came to their relief. The world before coal was cleaner—but immensely poorer.

THE PRE-EMINENCE OF ITALY

The high Renaissance was a time in which the Mediterranean element in European civilization was strongly dominant; only toward its close at the end of the 16th century did the northern, Protestant, coal-rich fringe of Europe begin to flex its muscles, politically and technologically. The Portuguese had explored Africa and the East; a Genoan, Columbus, had sailed to America in Spanish ships. Aside from navigation and mining, the leadership in techniques unquestionably belonged to Italy. Italian architects framed the Neo-Classical style which was to be imitated elsewhere. Italian engineers, among them Leonardo da Vinci, excelled in the arts of fortification and civil engineering; they constructed the first European canals for inland transport, developing essentially all the techniques ever required.

The most skillful metalwork was done in Italy, and described by an Italian,

Fig. 6-2. A glass works. (From Agricola, *De re metallica*, 1556)

Vanoccio Biringuccio. The Italians were expert in glass-making—Murano, outside Venice, was already old as a center of this art--and perfected (upon Arabic foundations) the gorgeous art of tin-glazed earthenware (majolica), especially associated with the cities of the Romagna and Tuscany. The Italians did the most beautiful work in textiles, in which they employed elaborate silk-throwing machinery and complex (though as yet purely manual) looms for weaving figured fabrics. A few essential materials, including the dyers' mordant, alum, and sulfur were derived almost exclusively from Italian sources. Some of Europe's finest printing was done in Venice; some of its best gunsmiths were in Brescia; the first use of a screw press for striking coins and medals, an art in which the Italians excelled, was in Rome; every traveller returned home from Italy laden not only with pictures and statuary but with as much domestic furnishing as he could carry. The Italian Renaissance was not only a mood, or an intellectual movement; it was the culmination of the development of civilized living which had been taking place throughout the 15th century. These alone made its opulence possible.

THE DIFFUSION OF TECHNOLOGY

Naturally the merits of new techniques and styles were not ignored outside Italy; by the end of the 16th century Italianate products of all kinds were being manufactured in France, the Netherlands, and England. For example, the manufacture of tin-glazed earthenware was established in Holland about 1580 and in England about 1600; whereas both the Italian majolica and its French derivative were chiefly polychrome, these new centers concentrated upon the cobalt-blue decoration upon a white ground typical of "Delft" ware, imitating the blue and white of Chinese porcelain. Similarly, printers like the Plantins of Antwerp and the Elzevirs of Leiden surpassed, in their prime, the by-then degenerate products of Venice and Basle. It seems fairly clear that in most cases in the 16th century—and indeed long afterwards—the diffusion of techniques was chiefly effected by persuading skilled workers to emigrate to regions where their skills were not yet plentiful.

Persecution in this age of religious warfare had the effect of stimulating such migrations of skilled craftsmen, as with the Flemish weavers who settled in the eastern counties of England, or, later, the French Huguenots (Protestants). Occasionally, moreover, large bodies of foreigners were persuaded to come to a country as a deliberate act of state policy: an obvious example is the German community of miners in Cumberland, England, induced by the Elizabethan Company of Mines Royal to settle there for the mining and smelting of copper. Though the diffusion of technical skill by print or systematic teaching was as yet comparatively unimportant, there was in the 15th and 16th centuries—and increasingly later—sufficient mobility of skilled workers to ensure that no technique capable of wide application was long confined to a particular region.

Another example of rapid diffusion was printing. About 1450 Johann Gutenberg in Mainz utilized type cast in an adjustable mould, and within a quarter-century it had spread to all parts of Europe. Caxton began English printing in 1476; the art was already long established in Italy, France, and the Netherlands at many centers. As with printing, so with paper. A large printing trade presupposes an adequate supply of paper; the best and the cheapest had long been made in Southern Europe. The "invention" of paper had passed from China through Islam into Spain, where also the use of a stamp-mill for macerating rags appears to have originated. The water-mark design was introduced in Italy about 1285. In the 16th century the macerating of rags for paper was sometimes performed by water power.

Despite the rapid diffusion of techniques, local methods and regional excellencies could remain significant for relatively long periods. The Mediterranean and the Northern traditions of ship-building long preserved their identities, the latter building more robustly for stormier seas, the former building oared galleys at Venice and Genoa which were rarely seen as visitors to Channel ports. The south German mining region (extending almost from the Adriatic coast into Hungary) showed a unique development of the mining arts and mineralogy, lasting from the 15th century to the latter part of the 17th. England enjoyed a quasi-monopoly of cast-iron artillery, a product of the Weald of Kent that owed much originally to immigrant iron-founders, from about 1540 until 1620, when the rising Swedish ironmasters began to market cannon through Amsterdam. Local textile excellencies—Brussels lace, cambric, Kersey, Venetian point—were as common as local agrarian specialties in food grains, cattle, vines, and olives. Technology, like everything else, was affected by the strong local particularisms which survive today only in such relics as the taste for cider shared by Normans and West-of-England men.

AGRICULTURE

Agriculture still dominated the European economy. The weather, like epidemics, touched the lives of all; population expanded in good years, death by malnutrition and starvation was universal in the bad years. The irregular increase of population from the Black Death of 1348-50 well into the 17th century indicates that agricultural techniques, or the acreage of land under cultivation, or both had made considerable advances.

Throughout this period, however, agricultural practice and the lives of the rural population were at the mercy of outside forces. For example, the depopulation after the Black Death greatly favored sheep-farming, for it took fewer men to herd sheep than to till the fields. This was to give power to the *Mesta* (the sheep-herding syndicate) in Spain, and cause enclosures of the common fields in England. When the recovery of population coincided with the vast inflation of the 16th century, an equally strong advantage was given to cereal farmers in certain regions, as the increased demand from the larger population caused grain prices to rise. The price inflation had been caused first by an expansion of the output of precious-metalworkings in Europe, to which succeeded an avalanche of bullion from Mexico and Peru. Since wages tended to lag behind rising commodity prices, the search for profits was made particularly attractive; moreover, capital for new technical adventures, land improvements, and so forth was relatively easier to come by. Compared with the agricultural changes wrought by economic forces, modifications of technique were insignificant. There were no new crops: some American plants of economic importance were domiciled in Europe from the mid-16th century onwards, including the sweet and common potato, tobacco, and corn. None was yet of great dietetic significance, nor were sugar, the banana, or the tomato.

The economic and social structure which denied ultimate ownership of the land to the vast majority of peasants who worked it, and which was only consistent with the continuation of traditional communal methods, was a great barrier to technical change. Yet there was considerable literary propaganda in favor of improved husbandry, by writers such as Olivier de Serres and Thomas Tusser, and where private ownership permitted and market conditions gave encouragement, changes in method did occur. Local variations in crops and techniques were so extensive that generalization is almost impossible. Probably the best-farmed region in Europe during the 15th and 16th centuries was Flanders, where periodic fallowing of the land was abandoned (the cattle were kept on permanent pasture or fed on fodder crops); peas, beans, and "industrial" plants like rape and flax figured in a fairly elaborate rotation. The sickle was replaced by the more efficient scythe for cutting both cereals and grass; the animals were bedded on the straw. Horses continued to replace oxen for haulage. Elsewhere, in regions of mixed farming, improvement tended to follow this same pattern of intensifying land use and increased yield.

With the general population growth of the 16th century (the population of Castille in Spain, for example, doubled), there was a demand for more food. At the end of the Middle Ages the crop yield per acre was barely half that common in the late 19th century, and cows were equally inefficient producers of milk. One way for increasing production was to enrich the land with animal manure, seaweeds, marl, or lime—a practice used long before but neglected through two or three centuries; or the land area could be increased by breaking new ground or replacing vineyards with grain fields (as happened in some parts of France) and by reclaiming and draining wastes. In the northern Netherlands the mid-16th century was a time of intensive polder-making, that is, the reclamation of low-lying land from the sea by means of dikes. Andries Vierlingh wrote (about 1578) the first systematic account of land drainage, which was to become a Dutch skill above all.

In England the extensive drainage of the eastern Fens was first seriously considered, while in Italy an abortive attack upon the Pontine marshes was launched by Pope Sixtus V. In France an "Association for drying up meres and marshes" worked successfully from its establishment in 1599. In executing all such schemes, it is true, the level of technique required was not high, involving little more than pick-and-shovel work. The skill, which the 16th century began to acquire and the 17th to deploy, lay chiefly in accurately surveying the region to be reclaimed, and then in correctly assessing the engineering remedies to be employed.

WINDMILLS AND OTHER POWER DEVICES

The mechanical element in draining and reclaiming land was small as yet. The great days of the windmill, scoop-wheel, and Archimedian screw were still to come, though a few of these machines were already in use before the end of the sixteenth century. The windmill was likewise used for powering saws and crushers, especially in Holland where the winds were strong and constant. But the windmill was mainly, as it had always been, a grain mill. Although it was formerly held that the windmill came to Europe from the East, we now believe that the European windmill, first recorded in a document of 1180, was very probably *not* derived from the horizontal mills of Asia, but was conceived independently.

The earliest mills were little wooden huts containing the machinery placed high on a strutted post, so that the sails could sweep clear of the ground. In the 14th-century tower-mill the structure was solid, and the shaft carrying the sails was supported by a "cap" on the tower, resting on roller-bearings, which could be swung round by a long pole to bring the sails into the wind. Gearing was of course required in all windmills (as in most water mills) to convert the rotation of the horizontal windshaft into that of the vertical shaft driving the stones, and to increase the latter's speed. This gearing was built of wood. Despite many detailed improvements to the windmill in later times, its essential structure was well established by the end of the 16th century, when it was first illustrated by Ramelli (1588). Inventors proposed modified forms, but none succeeded in practice. The fully developed windmill—capable of a power output up to 50 horsepower—was the most elaborate large-scale machine before the steam engine, and the millwright was the master mechanic of his age.

The windmill, whose history is to be traced in Normandy, Britain, and Holland, became the major prime mover of the flat North European lands exposed to the Atlantic gales. But the windmill was not a dependable power source in hilly districts; there among the hills water power was most abundant and widely used. As early as the 14th century, water power was applied to many industrial uses besides the grinding of grain. What was significant in early modern times was less the devising of new machines to be driven by water, than the increase in scale and number of the machines already known. The largest single topic in the machine books of the 16th century is the application of water power, which could be made available almost everywhere. Agricola, in *De re metallica* (1556), for example, described, in relation to mining and metallurgy, overshot water-driven machines for pumping water (suction and rag-and-

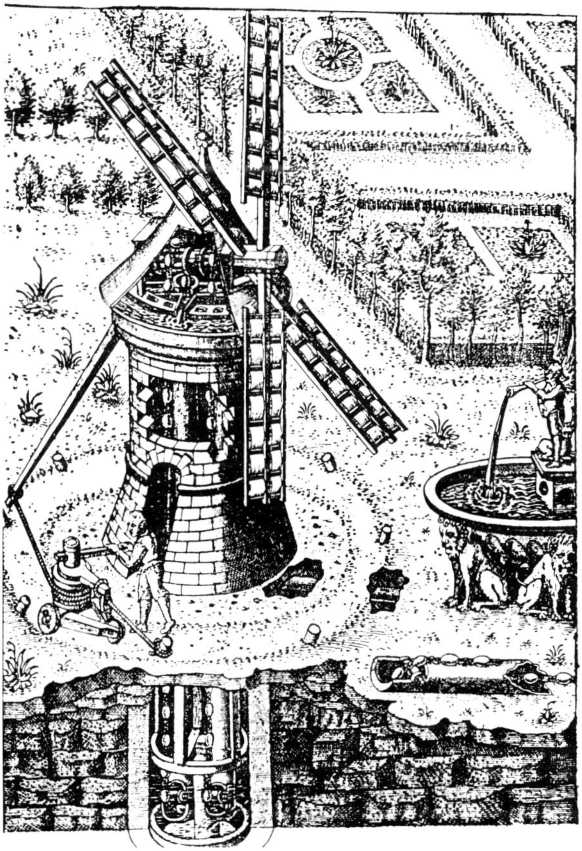

Fig. 6-3. A tower-mill for raising water. (From Ramelli, *Le Diverse et Artificiose Machine*, 1588)

chain), for hoisting spoil, for ventilating the mine, for crushing ore and grinding it, for stirring the mixing tubs, and for blowing the smelting furnaces. For each of these purposes two or three variant forms existed; all were constructed of heavy timber, with massive wheels and shafts, iron being used only for ties and in the bearings. Essentially all such machines, aside from their having such devices as tappets and, more rarely, the crank, were variants on the ancient water mills and windmills for grinding grain. Cranks were still mainly used as handles enabling men to rotate a shaft.

Besides the bellows, the hammers of iron foundries used in working wrought iron, with heads weighing many hundreds of pounds, were driven by water, as were large trueing-lathes, saws for cutting wood, hoists and cranes, fulling mills for the textile industries, and so on. By the late 16th century this use of

Fig. 6-4. Machine for pumping water. (From Agricola, *De re metallica*, 1556)

water power on a relatively large scale was found not only in a few advanced establishments, but was commonplace, indeed almost universal, in certain manufactures. By the early 17th century, blast furnaces, in which water-powered bellows provided the blast, were smelting 200 tons of iron a year each—a quantity as large in relation to earlier centuries as it is small compared with contemporary production. The largest compound water wheels, however, were those devoted to the water supply of large cities. Water wheels in the Thames at London Bridge supplied a part of London's needs; still more famous was the great machine at Marly, on the Seine, built in Louis XIV's reign for the supply of the fountains and other requirements of Versailles. Though less economical, the horse-mill often permitted an equal degree of mechanization where no fall of water was available. The concentration of industry was facilitated accordingly.

There were, of course, limitations on the usefulness of prime movers in early modern times. They had the disadvantages of being perfectly immobile, of being uncertainly available owing to failure of wind or water, and of being expensive both to install and to maintain. It was easier and sometimes cheaper to employ a horse, or even men, when the power was required to be neither very great nor constant. Agricola indicates that water-driven pumps were only required in mining when the depth of the shaft exceeded one hundred feet. In any case, very large increases of power could not be obtained even when they were needed, partly because the machines were badly designed—no attempt to analyze the form of windmill-sails and water wheels with an eye to efficiency was made before the 18th century—and partly because power output was proportionate to size, and the useful limits of size were soon reached. Timber joints would not stand the strain imposed by great power. A new material was required, but not obtained until the 18th century with the large-scale use of iron and steel.

METALS

Iron and steel were indeed becoming much less rare from the 15th century onwards. The key to larger production was the development of the blast furnace. In the early Middle Ages as in primitive times a lump ("bloom") of crude iron was formed by roasting rather than melting ore in an open hearth; the metal was purified and consolidated by further heating and hammering, which removed slag. Temperature was increased by building a furnace and increasing the blast of air, making the bloom purer and larger. When, as in the German *Stuckofen*, the furnace became tall enough and the blast provided was strong enough, the iron actually liquefied inside it. With a furnace having an internal volume up to five cubic yards, it was possible when the iron was liquefied to introduce continuous working, tapping off the molten iron and slag as required. The furnace could then be run for as long as its walls withstood the heat.

The cast-iron product of the blast furnace was handled in two ways; the cast "pigs" were treated much as the blooms had been by reheating and hammering in a second, "chafery" furnace, where sufficient carbon was burned out to produce a malleable wrought iron, or the molten metal was run directly into moulds from which were made cast-iron pots, fire-backs, mortars, cannon and

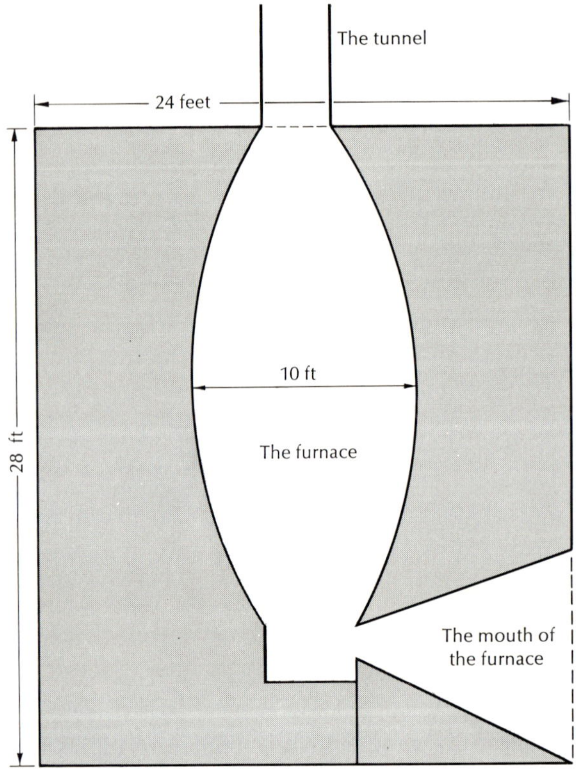

Fig. 6-5. Section through an iron furnace in the Forest of Dean, England, *c.* 1660. (Drawn by the author)

ball, ornamental work, and so on. The use of moulds was relatively minor, since most iron passed through the blacksmith's hands to be made into finished products for the consumer.

English furnaces of the Forest of Dean had in the 1660's the dimensions shown in Figure 6-5; behind the furnace were placed "two huge pair of Bellows, whose Noses meet at a little hole near the bottom. These are compressed together by certain Buttons [tappets] placed on the axis of a very large Wheel, which is turned about by water, and raised again by counterbalance weights." The furnace was filled with alternate layers of ore and charcoal, the tapped metal running into a sand casting-bed before the furnace, being "made so very fluid by the violence of the Fire, that it not only runs to a considerable distance, but stands afterwards boiling for a good while." The furnaces were kept in blast "for many Months together, never suffering the Fire to slacken day nor night, but still supplying the waste of the Fuel and other Materials with Fresh, poured in at the top."

Steel was made by small-scale processes, and was not yet cast in crucibles. One method described by Biringuccio in 1540 was to dip a spongy bloom of wrought iron into a bath of molten iron "often stirring it with a stick as cooks stir food . . . at last, when these pieces are very hot, they are taken out . . . drawn out under the forge hammer, and made into bars." Steel was also made by cementation, that is, heating wrought-iron bars in charcoal. The trick was to raise the carbon content of the iron in a controlled manner, by removing and testing samples until the desired hardness was reached. Iron was also case-hardened by heating in carbonaceous mixtures.

Both iron and steel were largely used for the manufacture of knives, tools, weapons, armor, chains, anchors, screw-bolts, nails, chests, locks, clocks, agricultural implements, and so on, through the various arts of the smith. In almost every case the manufacture of a particular article constituted a distinct craft. Iron was little employed for structural purposes, except as nails (wood-screws were unknown); iron rods and strappings were sparingly employed in masonry work and to reinforce timber. Ships, house timbers, and furniture were pegged together with wooden dowels or "tree nails."

The growing production of iron—about 60,000 tons for the whole of Europe in 1500—was nevertheless sufficient to push copper and its alloys into second place. Bronze, containing from 8 to 20 per cent tin, was melted in reverberatory furnaces and then cast into guns, bells, statues, mortars, and household pieces. Brass was made by melting copper with calamine, an ore of zinc. A good deal of brass was cast into plates to be worked up by artisans, including the makers of scientific instruments. Tin, derived from Cornwall, entered into many alloys, especially solders and the pewter from which domestic goods were made, largely by casting. Lead, also used in alloys, was employed for roofing, lining tanks, and so forth. It is obvious not only from contemporary references but from the wealth of articles surviving today that such artifacts of metal figured a great deal more commonly in the everyday life of the 16th century than in that of the 14th.

The writers of the 16th century who dealt with metallurgy all devoted particular attention to the precious metals, and there were many specialist treatises on them alone, the outstanding one being that of Lazarus Ercker (1574). From the first assay of ores through to the trial of the purity of the finished product, the treatment of the precious metals was an exact art. Quite small variations in the purity of the metal in a coin or a piece of plate could make serious differences in its intrinsic value, and moreover disturb the public confidence in monetary standards. Hence the work of gold- and silversmiths, and of the

Fig. 6-6. Refining silver. (From L. Ercker, *Treatise on Ores and Assaying*, 1574)

coiners, had been very carefully regulated from early times. Gold was obtained by careful washing from its matrix; and by "parting" from silver when the two metals occurred together (it was well known that nitric acid dissolved silver but not gold, and that aqua regia had the opposite effect). Much silver was obtained from lead ores by oxidizing the base metal. The whole series of operations involved was complex—much mercury was mined and distilled for use in amalgamation processes to produce gold and silver—displaying a wide empirical experience of chemistry. Moreover, the assayer's and the refiner's was a *quantitative* craft; profit arose from successful use of the balance, for margins were small. Here, as in navigation, science and craft came close together; but while the navigator was the astronomer's pupil, the chemist descended from the assayer.

THE RISING STANDARD OF LIVING

Agriculture, the omnipresent grain mill, the metal industries represented by a smith in every village—these were part of the technological fabric of life that was changing rapidly in early modern times. Some trades, like those of the tanner and leather worker, or the village wheelwright and carpenter, or the tailor and seamstress, seem to be less obviously in a state of flux; yet when the historical microscope is applied changes are seen to have taken place in all of them.

Tools changed as steel cheapened; the iron spade replaced the wooden spade with an iron tip. Buttons replaced tied "points"; shoes were still made to fit either foot, but in the 16th century knitted silk hose became the rage (formerly hose were cut from cloth and seamed up the calf of the leg). An English clergyman, William Lee, invented a knitting machine (1589), ancestor of the later "stocking-frame." Emigrating to France in search of encouragement and reward, he died at Paris about 1610, neither the first nor the last disappointed inventor. Hops transformed ale into beer, and the first bottled beer was not far off. Wheat and barley began to supplant medieval rye, and by the end of the century men were taking tobacco in long clay pipes, to the disgust of King James I. Not even the most intimate details of life remained impervious to change; the fork began to replace fingers at table, and one's weekly purge came not from the herbalist's, but from the chemist's shop.

TRANSPORTATION EXPANSION

One consequence of the demand for higher living standards and of the increasing technical proficiency that made such rising standards possible was a great expansion of trade and travel in early modern times. It was not mere curiosity that had taken Europeans to the East and West Indies, nor were they inclined to regard Christianizing the heathen as the sole reward their labors deserved. For long-distance trade over many thousands of miles the ship had reached, before the end of the 15th century, almost the form it was to retain until the middle of the 19th: multiple masts and decks, several sails for each mast which could be reefed, sternpost rudder (the wheel for steering was not yet known, however, and there was only one fore-and-aft sail). These vessels were not only gorgeous with carving, gilt, and paint, but serviceable and, already, capacious. Henry VIII's *Great Harry* attained 1500 tons, carrying 195 guns (mostly small) and 90 men; merchant ships were always far smaller, often under the 100 tons displacement (roughly) of Francis Drake's *Golden Hind*, which circumnavigated the world in 1577-81. Nevertheless, techniques of ship-building, rope-making, and so forth seem to have changed only slowly; there was little mechanical aid to craftsmanship except in sawing.

Fig. 6-7. The embarkation of Henry VIII at Dover, 1520. (*Science Museum*, London)

More elaborate port facilities were required for the bigger and more numerous ships; large dockside cranes, sometimes powered by a treadmill, and various types of dredgers both appeared in the 15th century; quays were built with timber piling or masonry; and warehouses were put up for the storage of goods. Inland transport made less progress in early modern times than did transport by sea; private enterprise failed to provide adequate roads and bridges while public authorities hardly attempted to do so. Nevertheless, there was some renewed administrative effort in the 16th century toward the paving of city streets and the provision of sewers. Along country roads wagons began to replace teams of pack horses, and the evolution of lighter carriages permitting faster travel was about to begin, for the well-to-do were starting to demand easier journeying. There were also two important additions to industrial land transport: the introduction of the wheelbarrow, and of wooden tracks in mines along which little four-wheel carts were pushed. This was the birth of the railway. The first true canals were begun near Milan in the 1450's.

TEXTILES

Improved transport had major effects on the one industry that was common to all parts of Europe: the textile industry. Cloth-making for local use was universal, but so was an interchange of high quality fabrics that goes right back through the Middle Ages. England was noted for its high quality woolens;

cotton, silk, and rugs came from the East; fine linen, figured fabrics, tapestries, and so on had their particular centers of manufacture.

All the various stages of cloth manufacture from cultivating the fiber to finishing the fabric were in a state of flux from the late Middle Ages into the 18th century. It would be a mistake to imagine that this great industry was stagnant for centuries until power looms were suddenly invented. Changes contributing both to the quantity and the quality of production were constantly occurring, mostly mechanical but some chemical (relating to dyeing and other parts of the finishing process).

Perhaps the most interesting development, since it for the first time introduced a machine into even humble cottages where the craft was practised, was the spinning wheel. An unknown inventor of the 15th century added the "flyer" to the older spindle wheel so that twisting the yarn and winding it became a single operation rather than two successive ones. But fingers still had to manipulate the yarn. The spinning wheel is a superb example of the application of a relatively complex mechanical device to domestic industry: the sewing machine, later, was another. Neither was employed, as such, in factories. When such devices were supplied with the crank-and-treadle system so that rotary power came from the operator's foot, leaving both hands free, they were exceedingly efficient.

CLOCKS AND INSTRUMENTS

Another fundamentally significant device that early became an accepted part of everyday life was, of course, the clock, which has increasingly ruled human affairs ever since the mid-14th century. Whatever Asiatic and Islamic precursors the mechanical clock may have had, the particular form of balance-escapement that appeared in medieval Europe was an indigenous invention; and it was only in Europe that the clock became a common machine, first on monastic buildings (*c.* 1350), to tell the monks the time of worship, next on town halls as city life became more extensive, then as a bracket-clock in living rooms, and finally as a portable spring-driven watch (*c.* 1500).

Extraordinary craftsmanship was to make the clock into a most beautiful and ingenious device, enriched with automata, astronomical displays, chimes, and alarm mechanisms. The first clocks had been crude products of blacksmith's work, but those of the 16th century show—as do contemporary door and chest locks, and the locks of firearms—the highest skill in working steel and brass. Not only were the working parts accurately made, finished, and polished, but the whole was chased, engraved, gilded, decorated with inlay, and so on.

Along with these crafts of the skilled metalworkers a new one emerged, that of the maker of scientific instruments. This craft really had several branches, for some instruments were made of wood (mariners' forestaffs and quadrants), others of brass (astrolabes, gunners' quadrants, astronomical instruments), and others again of steel (calipers, large drawing compasses). After the introduction of the telescope (1609) the optical lens-maker also entered on the scene. The best of the instrument-makers were splendid craftsmen; some were learned, while others profited by putting into practice the ideas of their learned friends. Their best trade, probably, was in instruments for navigation and surveying, both of which were becoming advanced, mathematical arts.

WARFARE

There was one human activity which, though deplorable, was almost continuous in early modern times and touched on almost every technique: war. The relation of warfare to technological progress will be touched on again; here it is enough to note briefly the unfolding consequences of the discovery of gunpowder. Propellant if not explosive mixtures of combustibles were earlier known to the Chinese and Muslims; whether their knowledge was transmitted to Europe has been much debated. Whether this was so, or whether there was an independent invention in Europe, there can be no doubt that this technique had marked effects on European society.

We might well choose to date the beginning of modern European history from the introduction of gunpowder. By 1325 primitive cannon were in action, and from 1370 mechanical artillery (on the lever principle) was falling into abeyance. By 1450 the hand gun had appeared, beginning the obsolescence of crossbow and longbow. Powder-making became an important industry, along with cannon-founding and gun-making. By 1500 heavy guns, mortars, and explosive mines had made the medieval castle almost untenable. The changes in tactics initiated earlier by the effective use of bow and pike were extended, so that the medieval knight in his panoply of armor withdrew from the field.

Fresh administrative efforts had to be made to equip large armies through long campaigns, or huge fleets like the Spanish Armada (1588), with all the outcry for supplies of every kind—food, clothing, weapons, and munitions—that this entailed. The era of the professional soldier began with the Swiss mercenaries and the Italian *condottieri.* It must never be forgotten that, for the technical machine to be effective, the organizational machine must develop with it. So, in the 16th century, government, society, and human relations began to change from their old pattern in order to exploit the destructive potentiality of new weapons. Even cities changed: the archaic round towers and the high straight walls vanished; in their place came the curtain of geometrically planned defenses, carefully sloped to deflect cannon fire and so arranged that every face could be enfiladed.

Only the wealthiest and most powerful rulers could afford the new gunpowder weapons and the more expensive and larger armies. The rebellious feudal

Fig. 6-8. Early English hand gun (caliver) just after firing. (From Jacob de Gheyn, *The Exercise of Armes*, 1607. Courtesy of *The Huntington Library*, San Marino, California)

nobility in their isolated castle fortresses could no longer withstand the power of the monarch's more powerful gunpowder weapons. Thus the changes in the technology of warfare brought about through the introduction of gunpowder aided in that process of administrative and territorial consolidation which was to give rise to monarchical states and to the nation-state system of Europe as one knows it today.

PRINTING

Just as the invention of gunpowder served to batter down the political positions of feudalism, so did the invention of printing help to remove the barriers of com-

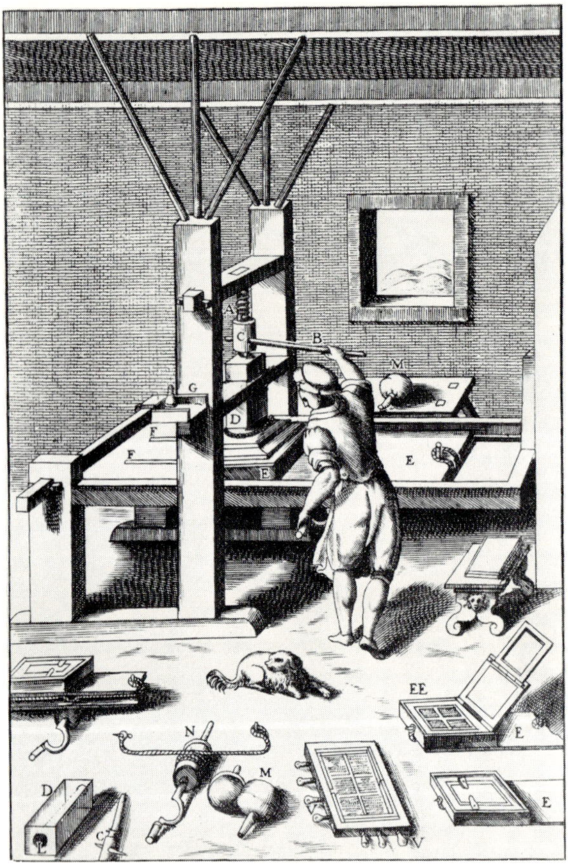

Fig. 6-9. Printing press, *c.* 1607.
(*Science Museum,* London)

munication between men. And, like gunpowder, the invention has had incalculable significance to human history, far beyond its immediate technological effect.

Yet, even treated solely as an episode in the history of technology, the introduction of printing provides an interesting case study of the nature and character of invention. For one thing, there are rival claimants to the title of "inventor": most authorities ascribe the invention of printing to Johann Gutenberg of Mainz, Germany, sometime in the 1440's; however, there are some who claim that Lourens Janszoon Coster, of Haarlem in the Netherlands, printed with movable type as early as 1430. Inasmuch as Coster's process of casting type in sand with wooden hand-engraved originals as patterns was incapable of developing into a practicable method, the honor of actually inventing the process which led to printing as we know it today must go to Gutenberg. Yet the fact that there are two claimants to the invention of printing indicates that the process was "in the air," so to speak, a situation which has frequently led to contemporaneous yet independent inventions of the same device.

There was much more to the invention of printing than merely devising the printing press, and here we find another characteristic of many inventions, namely, the synthesis of several elements, many of which have been known for a long time, to form something new. Other elements were involved in this invention besides the press itself: paper, ink, and movable type. Paper, first invented in China some 1300 years previously, moved very slowly to the west, reaching Baghdad by the end of the 8th century A.D. and Egypt by the beginning of the 10th century. By the middle of the 12th century, paper had found its way to the European continent, where it gradually won acceptance over parchment and vellum. Gutenberg thus had paper at his disposal. The press too was already well known by this time, having been used in Europe for centuries; Gutenberg had only to adapt it for printing on paper.

Another problem, however, presented itself: the ink previously used for printing from wood blocks had been made with a water-soluble base, which ran when used with metal type. However, painters used pigments mixed with linseed oil, and Gutenberg adapted this existing medium to give printing a suitable ink.

Similarly, the use of separate type was already well known to bookbinders during Gutenberg's time, for they used these to stamp titles on bindings of books. What Gutenberg added was to cast this movable type in an adjustable mould. He thus developed a satisfactory method of producing separate type of such accurate dimensions that it could be assembled easily and when assembled could be held firmly together so that successive images could be made from them with reasonable speed and convenience.

The invention of printing, therefore, like that of many other devices, rests upon many prior inventions. Like many other inventors, Gutenberg's contribution was that of a creative synthesis—sorting and selecting the essential elements and combining them into a new form.

THE INDIVIDUAL AND TECHNOLOGY

Individualism and versatility characterized the "Renaissance man." In this context, individualism meant the tendency of men to gratify their senses and to enjoy the gifts of life on earth without undue concern for the life hereafter and without medieval man's burden of spiritual guilt. Such Renaissance individuals as Benvenuto Cellini and Niccolo Machiavelli were concerned with power and glory in this life.

Versatility was also a keynote of the Renaissance. Cellini was famed as a metalsmith who could produce beautiful works of art ranging from small table decorations to large statues, and he thought of himself as a military hero and great lover; his *Memoirs* demonstrate that he at least had considerable literary skill. Michelangelo painted and sculpted with consummate skill, and his architectural designs were equally noteworthy.

These characteristics of versatility and individualism are also found in the leading figures in the technology of the time. Chief among these was Leonardo da Vinci (1452-1519). Today Leonardo is perhaps best known as the masterful painter of "The Last Supper" and the "Mona Lisa"; however, Leonardo earned much of his living through his services as an architect and military engineer.

In his *Notebooks*, Leonardo drew mechanical designs which indicated his recognition of the basic modes of operation of machinery and which were far in advance of their time: a lathe worked by a treadle crank, an automatic file-cutting machine, a rolling mill, a spinning machine, and other mechanical devices. The *Notebooks* also contained designs for some devices which were more utopian than practicable in nature, and which, in any case, could not have been carried out with the technical means and materials then available. Thus he "foresaw" such modern devices as the helicopter, machine gun, airplane, and tank. Despite the virtuosity and mechanical genius displayed by Leonardo in his *Notebooks*, these designs had little influence upon the technology of his time. His notations, recorded in reversed mirror writing (which is not secret, but perfectly easy to read in a glass), and his drawings were little examined by his contemporaries and remained buried and forgotten until resurrected in our own time. While some writers have gone so far as to claim that Leonardo was the "inventor" of the airplane and of other devices, it is obvious that he was not. All such developments came about independently; Leonardo was a lone genius whose immediate impact on the technology of his own time was almost nil.

Unlike Leonardo, who kept secret his speculative notebooks, other Renaissance students of technology made close studies of the devices and processes of their time and incorporated these into handsome books which received wide circulation. Two deserve special mention: Georg Bauer (who used the latinized form of his name, "Agricola," which means "farmer," as does "Bauer" in German) and Vanoccio Biringuccio (1480-1538 or -39). Agricola (1490-1595) was a physician in the mining centers in Saxony, and his work, *De re metallica,* was a summary of the most advanced mining techniques, those practised in the Saxon mines. His clear descriptions and beautiful illustrations of mining machinery have made this work a classic in the history of technology. Biringuccio, an Italian, is famed for his book *De la pirotechnia,* which describes many industrial processes, especially metallurgical operations. His too was a practical work, recording the technical details of applied chemistry in the field of metallurgy.

The pre-eminence of metallurgical technique in the interests of the Renaissance can also be seen in the writings of Lazarus Ercker (1530-93), whose tracts dealt with assaying, mining, and smelting. Other engineers, such as Jacques Besson and Agostino Ramelli, wrote beautifully illustrated treatises showing machine designs and details of building construction.

The beginnings of large-scale publication of books dealing with technology should not mislead one into believing that technology itself was changing in character and becoming the province of highly educated men. True, for the first time the craft knowledge of the past, as incorporated in existing practice, was being written down and published. Nevertheless, the authors of these books themselves made few technological innovations. Technology still remained the province of the craftsman who learned his trade through a long apprenticeship rather than by book-learning. In some cases the powerful guilds resisted changes in long-established processes, devices, and techniques. Thus technology continued to be carried on by traditional means and on a small scale.

THE RENAISSANCE ACCOMPLISHMENT IN TECHNOLOGY

The Renaissance is regarded today as one of the most creative and glorious periods of human endeavor. Yet in terms of the history of technology it perhaps does not rank as high as the Middle Ages with its power revolution and its agricultural innovations. In terms of the basic inventions and improvements made in the Middle Ages, the Renaissance did little more than to increase their size and scope. Machines became larger and more intricate and production increased, but there were few basic innovations.

Yet, as we have seen, there were two major innovations during the Renaissance—gunpowder and printing—which were to have immense consequences for the development of Western civilization.

Just as these technological innovations were to have a profound impact on human history, so did some of the non-technological contributions of the Renaissance make themselves felt upon the subsequent history of technology. This influence is to be found for the most part in the realm of psychology. By turning away from the spiritual emphases of medieval times and by renewing the secular approach which had characterized Classical antiquity, the Renaissance produced a frame of mind which was increasingly receptive to further technological development. Furthermore, the increase in the economic well-being of Europe, the discovery of the New World, and the development of powerful national monarchies provided an economic and political base for further technological expansion, and also widening the horizons of mankind. All these were necessary preliminaries to that great efflorescence of technology in the succeeding century which came to be known as the Industrial Revolution.

Part **II**

Background of the Industrial Revolution, 1600-1750

7/Cultural, Intellectual, and Social Foundations, 1600-1750

A. RUPERT HALL

THE TERM "INDUSTRIAL REVOLUTION"

If we seek to discover the sources of the factory system, the spinning jenny, the steam engine, and the railroad, we must first understand that the phrase "Industrial Revolution"—convenient and conventional as it is—can be misleading. When the term was first used, the character of technological history before 1750 was little understood. Moreover, the original "Industrial Revolution" was taken to be an *economic* event; the name described a rather rapid shift (in England first) of the balance of economic life from agriculture to manufacture, with a corresponding movement of population from the countryside to the towns. And the core of this event was the manufacture taking place in a factory-like establishment, rather than in a home workshop.

No one can doubt that such changes in economic life did occur—in England quite swiftly between about 1750 and 1830, and continuing thereafter—and that they were often (but not always) accompanied by changes in the techniques of making things. When we come to consider the *technical* aspects of the Industrial Revolution, however, we should recall the saying, "nothing succeeds like success." For, as we have seen, technical innovation had been going on for centuries in Europe. Indeed, the pre-Industrial Revolution period was itself an age of rapid technological development, already reflected in the changing economic life of the 16th and 17th centuries.

In short, the "Industrial Revolution" is an age of continued technological *evolution;* it is an age in which earlier tendencies were fulfilled and earlier efforts triumphantly rewarded. At the same time, these technological developments combined with economic, social, cultural, and political changes to produce a revolutionary transformation in Western man's ways of working and living.

THE IDEA OF PROGRESS THROUGH TECHNOLOGY

The first condition of technical accomplishment in the "Age of James Watt," as the early stage of the Industrial Revolution is sometimes called, was obviously an interest in the techniques of agrarian, mechanical, chemical, and other pursuits. A society, like that of India, whose leaders for many centuries had no interest in material techniques is hardly likely to be progressive in its technology. Improvements in technique will occur, but only with a glacial slowness. It is a further stimulus to innovation if technological progress is held to be rationally attainable—that is, if a new invention is regarded as the fruit of careful and patient effort, and not as a kind of miracle—and is considered actually desirable for its own sake.

It seems to be true—though generalizations over such a large area, such diverse cultures, and such a long time-span must be of dubious worth—that there is a long-standing European tendency to regard technological progress as good in itself. To describe the inventor of some useful art as a "benefactor of humanity" is no recent phrase. Nor is the image of a technological Utopia—a society in which men are freed from toil and want by their command over machines and processes—by any means limited to the post-Industrial Revolution era. Men had already entertained this dangerously deceptive dream (deceptive because neither plenty nor idleness is to be equated with contentment) at a time when their actual command over nature was small.

It is certainly true that the ambition to improve society materially was very strong in the 17th century. This ambition derived partly from altruism, and partly from the belief that men allowed to fulfill themselves would be better and more Christian than those condemned to brutish conditions of labor. For the English-speaking world and for many other Europeans this endeavor toward a richer and more just society was expressed by Francis Bacon (1561-1626). Bacon thought that command over Nature enhanced the freedom and dignity of life; that this command was obtained by making new discoveries, like that of printing or the navigational compass; and that such discoveries should arise not from sheer accident (as some had in the past) but from deliberately exploited knowledge. Hence science was the theoretical road to technology. This was not the whole of Bacon's philosophy; he greatly insisted upon the virtue of knowledge for its own sake, and valued science as an aspect of man's human dignity; he was much concerned with the intellectual fabric of science. He was, after all, first and foremost a philosopher. Thus, it is Bacon who contrasts true and false learning in terms of their object:

> But the greatest error of all is the mistaking of the last end of knowledge. For men have entered into a desire of learning and knowledge, sometimes upon a natural curiosity and inquisitive appetite; sometimes

> for ornament and reputation; and most times for lucre and profession; and seldom sincerely to give a true account of their gift of reason, to the benefit and use of men; as if there were sought in knowledge a shop for profit and sale and not a rich storehouse for the glory of the Creator and the relief of man's estate.

Englishmen, and many foreigners, learned from Bacon that the true discoverer is infused with the desire to fulfill God's purpose and to do good by serving humanity. This by no means signified that all sound innovations must be technological ones, nor that those sciences alone should be cultivated that promised immediate utility. But it did exclude merely disputatious or egocentric learning.

THE ROLE OF RELIGION

Bacon's words, and his function as an exemplar throughout the 17th century, immediately suggest two important linkages of technology. The first is with religion.

The question may well be asked: are there elements in the doctrine and morality of Christianity which particularly favored technical progress in Europe? It might be argued, for example, that Christianity was less "other-worldly" than Buddhism, less socially rigid than Hinduism, more practical than Islam. Some historians think that Christianity was neither ignorant nor contemptuous of the hard realities of daily labor, and that consequently the alleviation of its harshness became, as it were, an act of charity for which the alleviator deserved honor. And there is some evidence from the Middle Ages of monastic houses leading technical advance.

In this modern period the question might be raised anew in the form: is there evidence that the religious attitudes prevailing in Protestant countries such as Britain and the Netherlands particularly fostered technological progress there? This is a much more difficult question. For, as has already been pointed out, the Catholic Middle Ages saw the introduction of important inventions, and it was Catholic Italy that excelled in technique at least until about 1500. Even of the 17th century, it cannot be said that technological progress occurred *only* in the Protestant countries, nor that all Protestant groups were associated with such progress. The French, Swiss, and German Protestants—partly because of warlike disturbances—had less success than their co-religionists elsewhere. J. U. Nef has drawn attention to the contrast—almost a Catholic-Protestant contrast—between French technical progress in the manufacture of expensive goods of the highest aesthetic quality, and English technical progress in producing cheaply vast quantities of goods having generally less aesthetic value. The former tended to be brought about through developments in crafts; the latter through developments in the use of machines and in the organization of labor. If this contrast is indeed genuine, the explanation for it must be sought, surely, at the economic and social levels rather than at the religious.

Hence the evidence upon which a contrast between "Catholic lethargy" and "Protestant energy" might be based seems shaky with respect to technical innovation. Moreover, when one reflects upon the differences of a social and economic kind between England and the Netherlands, on the one hand, and France and Italy on the other, it becomes clear that these cannot be ignored, and only the religious ones emphasized. For example, the conditions of the peasantry in England in 1600 were very different from those in France, and so was the state of government. The landed gentry and the upper bourgeoisie triumphed politically in England and Holland; in France they were decisively defeated. Hence it is only fair to recognize that the dissimilarities between these pairs of countries were far from being solely religious.

It is possible to draw a picture of the ideal God-fearing Protestant as the 17th-century preachers often saw him: hard-working, scrupulous, frugal almost to a fault, with a keen eye to material success as a mark of living right. There were such figures in Amsterdam, London, Bristol, Boston, and New Haven, but it is well to remember that the Puritan was as likely to be a die-hard, unimaginative conservative as a man of enterprise; and that one of the keenest proponents of technical invention in 17th-century England was Prince Rupert, most daring (if not most wicked) of the gay Cavaliers, the enemies of the Puritans. The "Puritan ethic" of virtue in hard work with or without profit *sounds* like a good explanation of interest in technology; but the difficulty is to supply instances of its action in practice. On the whole the effects of sermons on business and technology seem to have been negligible at any time, except when they reinforced native self-interest; laws, properly enforced, have always been more effective where self-interest was to be restricted rather than stimulated. Accordingly, it may be more appropriate to see where self-interest led technology in the 17th century.

THE ROLE OF SCIENCE

Technology's other linkage was with science. Did the itch for technical mastery impel men to study Nature? Did scientific insight promote technical mastery? One must be careful not to assume a contradictory attitude. The sensible view seems to be that endorsed by most contemporaries; the advancement of the crafts (into early modern times) owed little to men of learning because the craftsman was much more intimately acquainted with Nature within the terms of his craft and its problems than was the "scientist." In fact it was part of Francis Bacon's argument for the importance of studying the crafts that the "scientist" could learn from them:

> . . . it is esteemed a kind of dishonour unto learning to descend to inquiry or meditation upon matters mechanical, except they be such as may be thought secrets, rarities and special subtilities; which humour of vain and supercilious arrogancy is justly derided in Plato. . . . But if my judgement be of any weight, the use of History Mechanical [accounts of techniques] is of all others the most radical and fundamental towards natural philosophy [science]; such natural philosophy as shall not vanish in the fume of subtile, sublime, or delectable speculation, but such as shall be operative to the endowment and benefit of man's life. . . .

In turn the crafts could benefit from the intervention of learning:

> . . . for it will not only minister and suggest for the present many ingenious practices in all trades, by a connexion and transferring of the observations of one art to the use of another, when the experiences of several [trades] shall fall under the consideration of one man's mind; but further it will give a more true and real illumination concerning causes and axioms than is hitherto attained.

It can scarcely be doubted that the interest, stimulus, and purchasing power of learned or at least semi-educated men had their effects on the crafts through the 17th and 18th centuries. How much the craftsmen actually *learned* from these gentlemen amateurs, however, is more doubtful, and there is very little evidence indeed that craft practice was ever modified as a result of some elaborate theoretical investigation which was scientific in the highest sense. An example of a conspicuous, noble failure is provided by Sir William Petty (1623-87), economist and statistician, who endeavored to introduce a catamaran design into commercial ship-construction; he had only vague theoretical reasons for preferring his novel "double-bottom" design, and the experimental vessels proved no more convincing than his arguments.

The Dutch mathematician and physicist Christian Huygens (1629-95), who introduced both the pendulum and the balance-wheel, had a tremendous influence upon time-keeping and the clock-maker's business. The desire for an adequate marine chronometer was behind Huygens's *practical* interest in horology, though he had in addition purely *scientific* interests in the problems of dynamics that he encountered. It had long been recognized that an exact chronometer furnished one of the most hopeful means of determining the longitude of ships at sea. However, the theoretical considerations which Huygens elaborated upon for the design of such clocks proved quite irrelevant to their mechanism, and the subsequent perfection of the clock and the introduction of an effective marine chronometer are both to be credited to practical clock-makers in France and England. It was also Huygens who first realized that atmospheric pressure could be made to do mechanical work if some cheap and simple way of making a partial vacuum on one side of a piston could be found. It was the

English engineer Thomas Newcomen (1663-1729) who discovered this in his "atmospheric" steam engine.

Examples could be multiplied from military science, surveying (where the roots of modern instruments and techniques were devised by scientists), agriculture, chemistry, optics, and so on to show the constant interaction between science and craft, at least to the extent that educated men devised new things and got them made. Some even passed into general use, thus displacing earlier devices. But it was still very far from being the case that science was governed in its choice of problems by technical needs, or that scientific knowledge was as yet competent to deal with some of the most urgent technical problems. It was Abraham Darby, a practical ironmaster, not a chemist, who first succeeded in making a salable iron with coal; on the other hand, the scientist Réaumur, despite ingenious and indeed illuminating research, failed to clear up altogether the relationship of steel to iron. One has to recall that even in the last years of Sir Isaac Newton (d. 1727), although general scientific attitudes and beliefs were correct, the amount of detailed factual information available was still small, and the ability to make scientific predictions accurately was pretty well limited to astronomy, optics, and some branches of mechanics. Other departments of science were still on the wrong track, hopelessly speculative, or barely susceptible to quantitative treatment. In addition there was virtually no opportunity to apply or do research in engineering and applied science. Thus practical experience allied to native talent could be far more potent than any form of scientific analysis.

Indeed, where the art of discovery in applied science rests wholly with isolated individuals like Newcomen or Marconi, the number of factors that determine whether any given individual shall prove inventive, and furthermore whether his invention succeeds, are so numerous and so complex that the state of his scientific learning is only one minute element in the whole. Even today, when the basic scientific principles underlying a new technique might have long been known, an immense amount of research and development have still to be done before they can be fully exploited. Accordingly, any view that 17th-century scientific developments—magnificent though these were—necessarily and obviously brought out the subsequent Industrial Revolution needs a great deal of careful investigation and qualification. Historical relationships are rarely as simple as they seem.

Yet science contributed to, and was in part a product of, the sense of triumphant progress which was, perhaps, the most significant of all psychological factors behind the Industrial Revolution. Science was a solvent of tradition; it also inspired confidence that technical problems could be solved by patient, systematic experimentation. One may doubt whether the science of his day was adequate to teach Josiah Wedgwood (1730-95), the great English potter, much of direct value about ceramics, even at the beginning of his career; yet Wedgwood undoubtedly regarded himself as having perfected his technique by scientific means (such as pyrometry). Science provided confidence and a method of tackling directly the problems of his manufacture, even though a fundamental understanding of the relevant physics and chemistry was lacking.

James Watt, Wedgwood's friend and contemporary, was another of the same stamp; not a scientist in a refined sense, but a scientifically minded man. It seems doubtful whether Watt gleaned anything of importance from the ideas on heat of the Scottish chemist Joseph Black (1728-99), although Watt never denied that Black's views *could* have been useful; rather, it seems that young Watt apprehended intuitively the simple empirical fact that heat was wasted by an alternate warming and cooling of the steam engine's cylinder. If this is true, it is a very interesting example of the way in which a major technological innovation—the idea of the separate condenser for the atmospheric steam engine—could be approached in two dissimilar ways: by applying the theory of heat, and by direct inference from the engineering situation. By the late 18th century science was far enough advanced to supply a fairly articulate language for the expression of such inferences made by trained, experienced, yet wholly empirical men like Wedgwood and Watt, and to provide a method by which an inference or intuition from the practical state of things could be systematically investigated and perfected.

To take examples of a different kind, the innovations in chemical technique that took place about the end of the 18th century—chief among them the use of chlorine bleaches for textile processing and the synthesis of soda from salt—certainly could not have come about without the prior development of chemical science. The story of the links between "chemists" (those interested in the composition of substances and the processes of chemical change) and practical pharmacists, dyers, acid-manufacturers, distillers, soap-makers, and so on is long and complicated; indeed, science and practice were never wholly divorced from the time of the 16th-century physician and alchemist, Paracelsus, onward. Such an aristocratic and highly theoretical chemist as Robert Boyle (1627-91) was, as a superb experimenter and a follower of Bacon's principles, always conscious of practical interests; in the 18th century a good number of the most eminent chemists occupied posts in "industry." The manufacturing improvements that they and their colleagues introduced were many, especially in England and France; and as the chemists' intellectual grasp on the bewildering detail of chemical phenomena began to tighten, after about 1740, so it seems did their control over manufacturing techniques. Like Watt or Wedgwood, the experimental chemists who mingled in industry (like Black in Scotland and Pierre Joseph Macquer [1718-84] in France) had only slight power to predict what would happen under given industrial conditions; but they had good working means of exploring experimentally certain fruitful situations of industrial significance.

THE ROLE OF SCIENTIFIC SOCIETIES

Obviously, science was an effective, creative element in the technical climate of the Industrial Revolution; its influence, though still slight, had been increasing over several hundred years. This was some (though perhaps not sufficient) justification for the claim that the scientific societies had voiced since the 17th century: that investment in science was investment in prosperity. It was—and still is —one of the best pleas for subsidies and special privileges that can be made. In England the Royal Society (chartered in 1662) adopted a frankly Baconian attitude in its early years, stressing the utility of scientific knowledge. As one of its most distinguished members, Robert Hooke, wrote in 1665: "They do not wholly reject Experiments of meer *light* and *theory;* but they principally aim at such, whose Applications will improve and facilitate the present way of *Manual Arts*."

The Académie Royale des Sciences (Royal Academy of Sciences) in Paris was similarly founded (1666) on the formal assurance that it would serve the public good, and at times its members were directed to technological problems by the king's ministers. Like the Royal Society, the Académie des Sciences aimed at collecting a complete set of descriptions of all trades; this same interest was reflected in the famous *Encyclopédie* of Diderot and D'Alembert (1751-72) which, incidentally, plagiarized the technical plates the Académie had commissioned. In Germany, the philosopher Leibniz, as chief advocate of a (subsidized) German Scientific Academy, urged the practical benefits it would bring, especially in the metallurgical field where Germany had once excelled. Yet it proved easier to promise than to perform; the technical miracles in which statesmen only half-believed never materialized. Fortunately for the progress of science, "mere light and theory" held greater attraction for the best minds than "manual arts," and the growth of scientific organizations did little to alter the fact that it was individual impulse that set men to study this or that.

By 1700 the national scientific societies were firmly wedded to "pure science." As a consequence there was in England (where a sense of responsibility was combined effectively in some members of the upper classes with a belief in material progress) a series of attempts to promote technological advance through other bodies and by other means. The Society of Arts (1754) sought to achieve desirable technical and philanthropic ends—the two were so often combined, as in the efforts to stimulate devices which would ease the lot of the child chimney-sweeps—by offering monetary rewards to inventors. The Lunar Society, so called because it met during the full moon so its members could find their way home, was a small private group in the Midlands. It included James Watt, his partner Matthew Boulton, and Josiah Wedgwood from the world of industry, but also such "pure" scientists as Joseph Priestley, Erasmus Darwin (grandfather of Charles), and William Withering. The Lunar Society—bourgeois, provincial, industrial—was consciously divergent from the grand Royal Society in London. Finally, in the last years of the 18th century, when the Industrial Revolution was well under way, there was founded in London the Royal Institution "for bringing forward into general use new inventions and improvements . . . [and promoting] domestic comfort and economy." Its object was the relief of the poor. Count Rumford (1753-1814), the colonial-born but loyalist refugee from the American Revolution who was the chief instigator of the Royal Institution, asserted that "invention seems to be particularly the province of men of science," and indeed in a short time the Royal Institution, like the Royal Society, had become wholly their province.

It was difficult, in England at any rate, to develop theoretical technology, for each effort seemed to turn into a fresh study of the fundamental scientific issues. The French, in their new institutions founded during the French Revolution, such as the École Polytechnique (the direct ancestor of all modern technical universities), were more successful in sticking to the original plan, and the French correspondingly developed in the early 19th century a more fertile approach to the analysis of prime movers, for example. But even then, science gave at best no more than a hope or a hint of the way to improve techniques; the drive toward improvement had to come from other springs.

THE ROLE OF TRADE

For all except a few idealists, techniques were (and are) means to an economic end. Now the principal economic condition of the 17th and 18th centuries was the expansion of commerce; the little medieval world of Europe had, during the Renaissance, merged with the whole wide world and finally came to dominate it. European rule was already established in Africa, on the Indian mainland, on the islands of Southeast Asia and the Pacific, and on the American continent; the population in Europeanized communities overseas was several millions by 1700, and Europe's control over global trade disproportionately vast. European ships traversed every sea and entered every major harbor. Europe benefitted first from the adventurer's "capital gains"—from the Spanish conquistadors to the Indian nabobs; from commercial profits abroad, made in sugar plantations for example or by participation in colonial trade; from the enrichment of the commerce of Europe by the addition of valuable imported commodities; and finally—though these imported wares were partly paid for by exports of gold and silver bullion to the East—there was a further advantage to Europe in that the empires constituted a vast market for her manufactures. The dependence of the North American colonies upon Britain in this respect is well known; and nearer to London and Paris the economic importance of the subcolonial regions in Eastern Europe, Russia, and the Near East was steadily increasing.

Of course there were failures and periods of economic stagnation. Colonial enterprises swallowed up European capital and population; but by and large it was an age of expansion which brought wealth to the great trading nations, Holland and England. Moreover, the 18th century was one of comparative peace; the destructive wars of religion were over, and the various campaigns of 1740-63 were fought with limited ferocity, for the most part, between professional armies. Britain especially enjoyed tranquility under the Hanoverian kings; security and prosperity effected an unprecedented lowering of the return on regular investment in land, government stocks, or safe lines of trade, thereby both encouraging investment in new ventures of promise (even if the profits were likely to be deferred, as in canal-building) and making the cost of raising capital less. Despite the increasing population it became attractive to put money into capital developments in order to obtain higher, more efficient production with lower labor costs. The great new breweries like Whitbread's that sprang up in London in the later 18th century employed no novel technology (though they later turned to steam pumps), but their size and organization reduced production costs per gallon.

Even without dramatic technical changes—the displacement of the waterwheel by the steam engine, for example, was very gradual—improved industrial organization and the sterner discipline that went with it in the factory could, themselves, yield higher efficiency. Technical modifications permitting a lowering of the worker's standard of skill, or enabling him to control two machines instead of one—typical changes that occurred in the textile industries—obviously increased still further the advantages of the factory system. Moreover in Britain, unlike the United States later, there was a surplus of unskilled population always ready to depreciate the skilled craftsman's labor, for the rural population was expanding as opportunities for employment on the land diminished, partly as a consequence of the enclosure of the common fields and concomitant changes in the use of agricultural labor.

The standard of living of the English peasant, fast becoming a landless laborer, declined steadily during the Industrial Revolution. Bad as were the new industrial towns of northern England from a modern point of view, jerry-built and unsanitary, and harsh as were the conditions of work, country-dwellers moved into them readily as an alternative to cottage starvation. The phenomenon has often been repeated: the rural idyll is too often only the poet's vision and the commuter's dream. There can hardly be any doubt that the Industrial Revolution came just in time to save the British population—and indeed the whole of Europe—from the tragedy of overpopulation and starvation that struck Ireland in 1845. Thus the 18th century was not so far wrong in finding charity in the employment of children in factories; those for whom there was no work could not eat. Realities were harsh in this society of extremes which was still, overall, close to the subsistence level.

In short, economic and social conditions provided most markedly in Britain—but in varying degrees all over Europe and North America—an opportunity for the full transforming application of the technical resources and skills that had been slowly built up during the previous centuries. There was ample cheap capital; an abundance of mobile labor and the new ability in the countryside to feed the growing towns; there were markets ready to absorb the new production. If its quality was not the very highest this was not necessarily important, so long as it was cheap. To these conditions, war in the late 18th century brought the element of boom to hasten some developments even more. Once the occasion thus offered, a tremendous potentiality was released; latent, unrewarded inventiveness found its sudden expression and its profitable rewards. And once this innovation was rewarded, a regenerative process operated. The enterprising manufacturer was able to make enough profit from his first successful ventures—not all were successful at once—to finance an expansion of his operations, and so on, creating fresh industrial capital by the growth of the business.

Similarly, invention generated fresh invention: an obvious example is the mutual interaction of spinning and weaving capacities in the textile industries. The steam engine did not remain fossilized in the state to which Watt brought it with the separate condenser; it became in rapid succession a prime mover (rather than a pump, as it had previously been), a high-pressure engine, and a locomotive—thus also illustrating diversification of a machine by invention. As some aspects of technology became stronger they pressed heavily on others that were weak, and so enforced further changes; moreover, technology itself gave rise to new industrial and economic activities, such as machine-building and exporting.

Ultimately a new stability was to be reached on a higher plateau about the mid-19th century, when the technical possibilities of steam, iron, and coal had been fully exploited and a still newer technology was in the process of being born. But this renewed transformation could not have taken place had it not been for the Industrial Revolution, whose technical, cultural, and social foundations were firmly laid in the century and a half preceding 1750. Yet these technical, social, and cultural developments were not sufficient by themselves to prepare the groundwork for industrialization; to them were added economic and political forces which strengthened the basis for a major industrial transformation.

8 / Economic and Political Developments, 1600-1750

SHEPARD B. CLOUGH

In all the annals of history, no area and no period ever experienced such profound changes as Western Civilization has undergone in the last 350 years. In this period the population of the Western world, including Europe and the Western hemisphere, increased by an estimated one thousand per cent; the average age of man, more technically known as the expectation of life at birth, went up from 25 or 30 years to 70 years; and the goods and services available for human consumption have become more varied and perhaps some eight times more abundant per capita.

Basically these changes which have so altered man's life would have been impossible without what has come to be known as the Industrial Revolution—a movement which became accelerated in the 18th century and which in a sense has continued to the present time. This great revolution made possible extraordinary increases in material well-being by developing means by which raw materials could be processed by machines driven by mechanical power. Furthermore, this revolution included the exploitation of materials stored up in the earth's crust over eons of time and thus gave man an infinitely greater supply of natural resources than could be produced by growing plants and animals. Moreover, improved transportation and communications permitted the tapping of resources in what had been exceedingly remote and inaccessible parts of the globe, whether they were mountain tops, holes in the earth, or lands at the antipodes. Lastly, it made possible a division of labor—a specialization by men and by areas—that destroyed individual and regional economic self-sufficiency and that resulted in greater output and also greater inter-responsibility of groups and peoples than had ever before existed.

THE ROOTS OF REVOLUTION

Preparations for the enormous changes connected with the Industrial Revolution extend back a long way, even to the discoveries of fire, of metal-smelting, of the wheel, and of money as a medium of exchange. Significantly enough, as we have seen, knowledge of industrial techniques, of the mathematics of building, and of the physical world generally was not lost with the decline of one culture and the rise of another, but was preserved and passed on to the others in an accumulative process. By the end of the 15th century, Western Europe seemed to be gaining a position of primacy as the repository of this great body of knowledge about the physical world and how to make things useful to man. Also, people in Western Europe were astoundingly curious about the physical world. They wanted to extend the knowledge which they had and to put it to practical use in order to improve their well-being with less human effort. The *Notebooks* of Leonardo da Vinci (1452-1519), which offer a microcosmic view of the technical European world at the beginning of modern times, bear witness to the extensive range of Europe's interests from wheelbarrows to flying machines, from firearms to spinning machines, and from the anatomy of man to the wing structure of birds.

Neither a store of knowledge about the physical world and about productive techniques nor the desire for more knowledge of a similar kind was, however, sufficient to effect the great breakthrough to the Industrial Revolution. In fact, no single explanation of a complex phenomenon is ever adequate. Many factors —social, economic, political, and intellectual—always come into play in any important change. Old rigidities in the manner of doing things must be broken down and greater flexibility shown in the goals for which men strive, in the organizations which regulate societal affairs, and in the leadership of the society in question. Moreover, surpluses beyond what is needed for immediate consumption must be built up in order that men may divert some of their energies and resources away from traditional productive techniques to experimentation with new methods of turning out goods. In brief, the range of opportunities for alternate decisions by members of a society must be greatly enlarged, for change arises from the freedom to make other than traditional decisions.

THE EXPANSION OF TRADE

In the first two centuries of modern times, that is, from approximately 1500 to 1700, Western Europe developed in such a way as to provide conditions for fundamental economic and technological change. One of the most important contributing factors to this development was the so-called overseas expansion of Europe. The explorations and discoveries by such intrepid men as Vasco da Gama, Columbus, Magellan, and a host of others, opened up to traders new routes to the East and brought into Europe's sphere of economic exploitation two new continents of great wealth in the Western hemisphere. The Europeans' vast superiority over all other peoples in seamanship, navigation, and in weapons, especially in firearms, made it possible for them to use the new finds to their advantage.

The actual amount of overseas trade along the new routes, and for that matter along old ones, is impossible to measure accurately, but all authorities agree that it was large and grew amazingly fast. In the case of England, foreign trade doubled from the decade 1701-1710 to the decade 1731-1740, and it doubled again from the latter decade to that of 1761-1770. Increased water-borne foreign trade meant an expansion of ship-building in port areas, and this expansion resulted in a stimulus to other industries such as lumbering, rope-making (including the growing of hemp), tapping pines to get pitch for caulking seams, sail-making, the making of navigational instruments including clocks, the forging of ship's hardware, and the construction of port and harbor installations. Furthermore, the demands of shippers for goods which they could take abroad were so concentrated in the area near ports that they provided a great incentive for a few producers to improve their techniques in order to furnish more goods at lower costs and thus to improve their competitive position. Builders' hardware such as nails, screws, latches, and hinges were in heavy demand and came to be produced in more abundance. Similarly, textiles were much sought after by shippers for their clients overseas, as were trinkets, firearms, and "fire-water."

THE INTRODUCTION OF NEW PRODUCTS

The reverse side of the foreign-trade coin was that shippers were bringing into Europe new products, or old products in new quantities, the use and processing of which created new industries, new habits, and new social institutions. One of the most important of the new products was the potato. This lowly tuber has the virtue of producing more calories per acre than any other easily grown crop, is adaptable to a great variety of soils and climatic conditions, and can be eaten day after day without fatiguing the taste buds of the human—or at least without fatiguing them very much. Introduced into Spain from South America in the 16th century, the potato soon spread to Ireland, where it gained the name of Irish potato, and subsequently went to the rest of Europe and to North America. It increased enormously the food supply of these areas and thus made possible an extraordinary growth of population.

Sugar from cane was another important product which came into Europe from the New World in the first two centuries after the discoveries. Although cane sugar had been known in Europe for some time, it was relatively rare before the 16th century; it was used largely for medicinal purposes by the wealthy, while honey continued to be the chief sweetener of the poor. The importation of large quantities of cane sugar resulted in this product's coming within the price-range of the masses, who added it to a great variety of foods and drinks for taste and energy. Its by-product molasses was used in fermenting a most extraordinary distilled drink, rum. Thus the refining of sugar and the distilling of rum became large industries in port areas such as London, Amsterdam, and Nantes.

Cocoa, tea, and coffee also opened new taste experiences to the Europeans. All three products were used for drinks, and cocoa was used extensively in candies. The new beverages gave rise to special shops where people came to partake of them, to converse, and to exchange political and commercial information. Furthermore, tobacco was introduced into Europe from America and by making smoking and snuff-taking possible, gave people entirely new sensations. Quinine, or Jesuit's bark, was still another new product which proved to be extremely useful in the relief of fever, especially malaria.

Most important of all the new products for consumption, so far as the mechanization of its manufacture was concerned, was cotton. Cotton prints, imported at first from Persia and the East and later made in Europe as fustians (partly of linen), created a style range which reached its height in the reign of Queen Anne in the early 1700's. Superior to wool for underwear and for summer dress, and eventually much cheaper than linen, cotton products were in such demand that textile-makers were drawn into the cotton industry and then vied with one another to lower costs. They were unhampered by guild regulations, for guilds were not organized in this trade, and being free from prescriptions of methods of manufacture, they could turn their minds and talents to finding new and faster methods of spinning and weaving.

Of all the products which came from the New World to the Old none were more important than gold and silver. These two precious metals came into Spain and Portugal in such volume that it is said Europe had by the end of the 16th century three times more bullion than it had at the beginning. It has been estimated that the Spanish colonies sent the homeland some 16.8 billion grams of silver and 181.3 billion grams of gold between 1503 and 1660.

THE MOBILITY OF MONEY

No matter what the actual amount of gold and silver was that Europe received from overseas in the 16th, 17th, and 18th centuries, the volume was great enough to augment markedly the supply of money. Money, which is at once a measure of value, a store of wealth, and a medium of exchange, is one of the most important inventions of all times, and should be ranked along with fire and the wheel. By its use man has been able to exchange goods readily, even with total strangers; and such exchange in unknown and uncertain markets has permitted a high division (or specialization) of labor, of producers, and of areas.

Moreover, an increase in the supply of money was accompanied by a more extensive use of credit instruments, particularly the bill of exchange, and by a rapid increase in banking. The bill of exchange, which is something like a modern check, was an order by an individual businessman on a debtor to pay a creditor a certain sum on a specified date. It was thus a way of settling accounts between persons and was very useful when debtors and creditors were in distant places, for it removed the necessity of actually transporting precious metals. Moreover, the bill of exchange could be used to extend credit to a borrower, for payment of the bill could be in the future. It could also be for a sum that included interest, for the interest charge would be easily hidden in the rate of exchange of one currency for another mentioned in the bill. In fact, lending for interest became ever more common and the Church, which had forbidden the charging of interest in the Middle Ages, frowned on it less and less, if the money was to be used in a socially or politically desirable cause. Thus capital, which may be defined as anything having exchange market value that can be employed in the production of other things which have exchange value, became more mobile. It could be amassed more easily in the form of money than in the form of land; it could be transferred from place to place; and it could be split up so that investments of one person could be in different enterprises. Indeed, it was the mobility of capital that made possible the development of the capitalist system and the accumulations of capital which could be invested in industrial enterprises.

THE RISE OF BANKING

The growth of banking accompanied the extension of the use of money, of credit instruments, and of mobile capital. Banks provided an exceedingly important means of bringing together the resources of many people for making loans to entrepreneurs who wanted to extend their activities. Their services seem to have been demanded by the business community and to have been only incidentally stimulated by the Protestant ethic (hard work, thrift, sobriety, and relentless pursuit of one's calling), as some historians have maintained. Many important banks and many banking practices were developed in areas which were Catholic and were to remain Catholic—Augsburg in Southern Germany, Antwerp in present-day Belgium, Barcelona in Spain, and Florence and Genoa in Italy.

The Bank of Amsterdam, founded in 1609, was brought into being at the insistence of the city in order to prevent the gouging of legitimate businessmen by free-lance money-changers. By limiting the number of money-changers much of the exchange of currencies was taken over by this institution. Furthermore, the bank was given a monopoly of paying bills of exchange over 600 florins; it transferred funds on its books from one customer to another on order; it issued "deposit slips" which came to be used as a medium of exchange and hence as money by their holders; and it made loans to the city, to the Dutch East India Company, to the municipal pawnshop, and later to other borrowers.

The Bank of England, created in 1694, played a somewhat different role in its economic life than the Bank of Amsterdam, although the functions of the two institutions ultimately came to resemble each other more than they did at first. This bank was chartered to accept deposits, to buy and sell precious metals, to borrow, lend, and transfer funds, to discount bills of exchange, and to issue its "promises to pay on demand," or banknotes. By the control of credit and the issuing of banknotes it came to perform the function of a central bank. It could do much to regulate the money market and to help with loans those banks which got into trouble.

THE PRICE RISE

With the increase in metallic money, in bills of exchange, and in banknotes and other "bank money," the supply and rate of circulation of money appear to have grown faster than the supply of goods in the market. At any rate, European prices rose by about three times from 1500 to 1640. Then as the inflow of gold and silver from the New World temporarily abated, and before bank money had acquired much volume, prices declined somewhat from 1640 to approximately 1720. After the latter date they entered a new period of rise which continued through the French Revolutionary and Napoleonic wars to about 1819.

The argument has been advanced that generally rising prices in the 16th and 18th centuries stimulated entrepreneurial activity, because wages did not keep pace with prices; and with lower labor costs, entrepreneurs were encouraged to establish or expand productive plants. Although businessmen may not have taken full advantage of the "profit inflationary" conditions, rising prices and lagging wages were at least temporarily a boon to them and rising prices relieved their burden of debt.

On the other hand, very sudden increases in prices, which frequently accompanied famines, usually indicated periods of economic crisis. Under famine conditions, workers had to spend a larger proportion of their incomes on food and did not have much to spend on other things; peasants had less to sell in the market and hence less money for their purchases; and lords got smaller deliveries of goods in kind from their dependents and therefore had smaller incomes for their own use.

THE DECLINE OF THE MANOR

The long-term price rise had a definite impact on the feudal-manorial system, which was showing signs of being upset also by the growth of trade. During the Black Death in France and England (in the 14th and 15th centuries), and in areas like East Prussia (14th century), the labor supply was short; hence lords attempted to keep their peasants from running away and also to attract new ones by making the conditions more favorable. One of the things they frequently did was to commute (change) certain manorial obligations, such as working for the lord or making deliveries in kind to him, to money payments. Consequently, as prices rose, the value of the money payments declined. Peasants found it easier to raise the money which they had to put out, but lords discovered that their incomes from this source went down. Under such circumstances the lords were tempted to raise surplus goods to sell in the market rather than just to meet the needs of the manor.

A dramatic example of this kind of shift was the large scale on which lords in England gave up regular farming in the 16th century and turned to the raising of sheep for wool. In this change-over, lords found it necessary to "enclose" their fields for pasture and thus to deprive peasants of the right to till strips of land in the big fields. Once the peasants lost all or most of their lands, they had to seek other employment, which meant that there were fewer hands on the enclosed estates for the cultivation of traditional crops. Thus the self-sufficiency of the manor declined; and large landowners grew some produce, if not wool then wheat, flax, or the minor grains, for sale in the market.

Other important developments were also making severe inroads into the institution of the feudal manor. In England, for example, lords discovered mineral wealth upon their lands and turned from agriculture to mining. A case in point was the discovery of coal near the sea at Newcastle. Coal was in demand for domestic heating, brewing, salt-refining, and alum-processing, since wood was becoming scarce around large cities, smelters, and ship-building centers.

Then, too, younger sons of lords who would not inherit the family estates and who previously had sought refuge in the Church, in governmental posts, in military exploits, or in overseas discoveries and trade, now realized that they had another career outlet for their talents—they could go into industry or commerce. Similarly, the younger sons of peasants, who would not inherit the father's right to work specified parcels of land on the manor, might easily be enticed into working in new trades, for life in the towns had attractions rarely found on the farm. The 16th and 17th centuries heard few cries from employers that they could not obtain the workmen they needed.

THE GROWTH OF CITIES

The development of trade, especially overseas, and the expansion of industrial production, even at the handicraft stage, meant a growth of cities during the period from 1500 to 1750. There were, of course, temporary setbacks. Italian cities were hard hit by plagues in the fourth decade of the 17th century, but then recovered. Some German cities suffered from the Thirty Years' War (1618-48). Antwerp was hit by the war with Spain, when it was sacked (1576) and its trade down the Scheldt River prevented. Cities in France and England did not undergo such setbacks.

Nevertheless, the trend was toward greater urbanization, and with this growth, more people perforce had to rely upon the market for the satisfaction of their wants rather than upon their own production of everything. Although census data for the towns and cities are few and far between for the period under consideration, we know that most cities were getting larger and were constantly extending their walls to accommodate more people. London grew from 130,000 in 1631 to 403,000 in 1662 before the Great Fire, fell to 384, 000 in 1676, and reached over 900,000 in 1801. The population of Paris rose from some 200,000 in 1600 to perhaps 600,000 in 1789; that of Amsterdam went up from 100,000 in 1600 to 221,000 in 1795; and that of Berlin increased from 8000 during the Thirty Years' War to 20,000 in 1700, or 56,000 for Greater Berlin.

Within the same two and a half centuries, Europe's population enjoyed a general increase. Although population estimates are made by calculating the number of families and the number of houses and multiplying both by assumed averages, it is believed that the population of Europe went up from some 100 million in 1650 to about 140 million in 1750 and to 187 million in 1800. Obviously this increase in population, which was particularly marked in Western Europe, meant a demand for new housing, more clothing, more food, and, indeed, for all articles of human consumption. This demand, added to that which came from the growth of towns and cities, resulted in added pressure upon producers to turn out more goods. And this pressure had the effect of encouraging businessmen to improve their productive techniques.

POLITICAL FACTORS IN RISING PRODUCTION

Demand for goods was, furthermore, augmented by at least two other significant changes. One was that, with the growth of centralized, political states, rulers were impelled to build imposing palaces to impress their subjects, and to tame their rivals by a round of luxurious court life. Their example was followed by many lesser personages—clerical, noble, and bourgeois. Perhaps the best known example of the extravagant chateau and ostentatious court life was that of Louis XIV (1638-1715) at Versailles, with its Grand Lac, its fountains, its gardens, its lesser buildings in the park, and its incessant "gaiety." In fact, so much money was spent on this lavish establishment that the public was much irritated and the Treasury of France embarrassed.

The other development which led to a greater and more concentrated demand for goods was war. In fact, incessant wars among the rising states of Western Europe during the 17th and 18th centuries were probably the largest single enterprise before the building of railways. With the centralization of political power in states, armies provided by leaders (*condottieri*) of mercenary, military gangs were no longer satisfactory, for they were not to be trusted to do what their employers wanted. Consequently, states began to create their own military establishments and to keep standing armies. They gave their soldiers uniforms, provided them with arms, furnished them food and housing, and equipped them with means of transportation. Furthermore, such efforts at organizing military units were supplemented by the formation of companies and regiments of men, by the establishment of hierarchies of command which have come down to the present time, and by rigorous and elaborate training. All these changes required the expenditure of large sums of money, especially in times of actual war when much equipment was used up and had to be replaced.

Naval warfare also became more important. As the European states competed for the wealth of overseas colonies, naval power became the key to military conquest and national power. The merchant ships carrying overseas commerce had to be protected, while those of enemy nations were attacked for gain, just as the English freebooters plundered the galleons carrying gold from the New World to Spain.

Although warfare and the destruction accompanying it probably had a retarding effect upon European economies, the fact remains that the need for war materials put pressure upon producers to turn out more goods, led to the greater organization of men in combined efforts, and had a profound impact upon public finance and banking.

The need of governments to meet the costs of war led to the imposition of heavier and more regular taxes. And the collection of large sums from a great many individuals resulted in the amassing of mobile capital which was then spent on relatively few industries. Tax collecting resulted, in turn, in tax collectors, tax farmers (private individuals who paid large sums immediately for the privilege of collecting certain taxes and were then allowed to keep all the tax money in excess of their original payment), and government treasuries having funds which might be spent for goods or used to help existing industries.

Furthermore, states frequently got into financial situations where they could not meet their needs by taxation and had to borrow. Thus public debts came into being; and the managing of public debts was instrumental in the founding of banks, as in the case of the Bank of England.

THE ORGANIZATION OF PRODUCTION

With the increased demand for goods and services, with the development of trade, and with a rise in the use of money, existing organizations for controlling production and commerce (the craft and commercial guilds) were hard pressed to maintain their practices and their power. They had come into being primarily to establish monopolies over the production and sale of specified goods and services in a given area, usually a town or city. Guild masters banded together not only to keep all the business for themselves but also to limit the number of persons who might enter their trades. Ultimately they became so exclusive that they let into their monopolies only members of their families and their friends.

Gradually they did less of the work themselves, and hired journeymen to work for wages. They sought to expand their businesses and to produce for unknown buyers, sometimes in distant markets, and at unknown prices. They thus became a capitalist oligarchy.

In spite of their power, guildsmen were not able to extend their control over production outside the towns or cities or over new trades, like cotton textiles, which were developing. Hence the position held by the guilds began to decline as bold and energetic men circumvented the rules of business which the masters had tried to lay down. These non-guild entrepreneurs frequently bought raw materials and distributed them to home workers for whom they sometimes also provided machinery, as in the case of looms or spinning wheels. In this "domestic" or "putting-out" system, workers processed these raw materials in their own homes during whatever hours they had free from work in the fields. Subsequently the entrepreneur collected the finished products which he sold in the market for profit and paid the workers at piece-rate for their labor. It was only a short step from this arrangement to the factory system in which laborers went to the entrepreneurs' buildings to process raw materials and to operate machines.

THE DEVELOPMENT OF ECONOMIC THEORY

Lastly, economic theory began to change and to reflect many of the economic shifts which were taking place. Earlier notions that economies were mostly static and that an individual or a state, in order to get more, had to take from someone else were gradually altered. Evidence showed that economic growth was possible—that more economic activity could supply a greater abundance of goods and services.

The attitude toward gold and silver and toward money and interest-taking also began to undergo fundamental changes. The theory that the wealth of an area depended upon the amount of gold and silver which it could amass (the bullionist theory) gave way before the lessons of experience. In fact, much evidence existed to indicate that a great increase in bullion in proportion to the amount of goods and services available might not result in greater wealth, but in higher prices. Slowly the view became clear that the volume of goods and services available and the ability to reproduce them was the proper way to gauge wealth.

Moreover, the newly unified political states had a body of economic thought, known as mercantilism, which purported to explain the ways to become economically strong. This theory included such ideas as the limitation of imports in order that the homeland might have all the business of production for itself, and the restriction of exports of raw materials in order to prevent rivals from developing their productive capacities. Mercantilist theory also embraced the doctrine that national shipping should be limited to the bottoms of the home country (the Navigation Acts) in order to develop the home merchant marine. It held, too, that colonies existed for the good of the motherland and that they should restrict exportation of important materials to the homeland, should not manufacture themselves what the home country could furnish, and should buy whenever possible from the mother country.

The theory and practice of mercantilism, so well illustrated by the policies of Colbert in France during the reign of Louis XIV, constituted the economic side of state-building. They indicated that the state was prepared to aid businessmen in their quest for greater economic activity. At the same time, however, critics of man-made restrictions on economic processes became more numerous and vehement. Often they directed their barbs at guilds, for the critics sought greater freedom in matters of production and trade. They sometimes appealed to laws of nature in their efforts to break down existing rigidities and one school, the Physiocrats, led by Dr. François Quesnay (1694-1774), physician to Madame de Pompadour, pleaded for a return to a "natural order" of economics —laissez-faire—in which men would not interfere.

By the early 18th century many changes had come about, both in practice and in theory, to permit and to encourage individuals to increase production, to broaden their market activities, and to make more profits. Thus economic leaders were stimulated to find new methods of production. Necessity may well be the mother of invention, but the father is undoubtedly social and economic opportunity.

9/The Agricultural Revolution, 1600-1850

G. E. FUSSELL

THE SLOWNESS OF AGRICULTURAL CHANGE

Revolution is one of those forceful words that carry a sense of immediacy; an event that happens with an explosion of uncontrolled force, cataclysmically. Developments in farming practice are rarely of this nature, but the term has become accepted as a description of a lengthy period in modern history, and must be used here in that sense, although there was in fact only slow, intermittent change of great variety in widely separated, often rather isolated, areas. Indeed, the tempo of improvement from 1600 to 1850 was at least five times slower than that introduced during the first half of the 20th century. Yet the changes of the earlier period were along the same lines as those of our contemporary era, although done without the benefit of modern technical and scientific knowledge.

The similarities are owing to the fact that organic and biological processes, which are the fundamentals of farming, are difficult if not impossible to change drastically, and those that can be slightly modified must be tackled with extreme care if disastrous results are to be avoided.

Success in farming depends on the sensible use and co-ordination of a number of interlocking factors. This was true in the times before science was systematically applied to agriculture—during the greater part of the time of the so-called Agricultural Revolution.

THE OPEN-FIELD AND THREE-FIELD SYSTEM

The first essential in farming is the right to occupy and cultivate a specific area of land for the production of arable crops, and to use the grazing land available outside the area for maintaining livestock.

At the end of the 16th century the greater part of Western Europe still used the thousand-year-old system of open arable fields, with grazing on the surrounding waste lands or in the neighboring forests. "Open" field meant that no fences or barriers separated the plot of one farmer from that of his neighbor. The working farmers occupied scattered strips of land, varying in area up to an acre or so and usually much longer than broad, in each of the village fields, which totalled two, three, or even more in number. The crops cultivated were selected and controlled by the community, and a sequence of winter grain, spring grain, and fallow (that is, no crop at all) was followed in rotation. For example, one year Field A was planted for winter grain, Field B for spring, and Field C allowed to lie fallow; the following year Field B had the winter grain, C the spring, and A was fallow; and the third year C had the winter grain, B was fallow, and A was planted for spring grain. This was called the three-field system. This rotation was not completely rigid, though a farmer who did not observe the rules ran the risk of providing feed for his neighbor's livestock.

As soon as the harvest was in, the arable fields were thrown open for grazing by the animals, which consumed the stubble and weeds and fertilized the soil with their excreta. When not grazing on land cleared of crops the beasts fed in the fallow field, or upon the poor grazing of the unreclaimed grassland and the woods that separated village from village. Some meadows, usually near a brook or river, were kept closed for a hay crop from April or so until it was mowed, at which time the animals were let in again to find what feed they could. Often these meadows were divided into lots or strips for individual mowing.

Widespread as these arrangements were, there remained broad areas where they had not been adopted. Along the Mediterranean basin in France and in many parts of Italy and Spain, a system of land use derived from Classical times continued, and indeed continues today in some measure. Under this system the fields are small in size, square or roughly rectangular in shape, and worked on a two-cycle crop and fallow rotation. Usually these fields were worked by individual farmers, although some were affected by rights of common grazing after the crops had been removed.

In the wilds of northwestern France, as in some of the more remote and hilly parts of Great Britain, convertible husbandry was practised. This involved the plowing up of a relatively small area, enclosing it, and growing crops on it for several years in a row, often with oats or rye; after that it was allowed to fall down again to rough grazing. Some fields in these areas were cropped continuously; though in most of these places livestock was more important than crops, both for supplying the farmer and his family with food and providing some income.

It was largely on the open fields that the Agricultural Revolution took place, beginning in the 17th century. The revolution consisted of three items: (1) the introduction of new crops to the open fields, often after enclosure; (2) the improvement of implement design and the invention of new ones; and (3) the improvement in livestock. The new crops were the artificial grasses and roots which could provide fodder for livestock. Fed on the new fodder crops the livestock increased both in size and productivity; they also supplied more and better organic manure.

To enable these things to be done the rearrangement of the scattered strips of the open-field farm was fundamental. In spite of some exceptions, it was widely accepted by farming theorists from the 16th century on that only with fenced fields free from communal grazing rights could farmers make the best use of the soil.

THE NEW CROPS

It was not on the open fields north of the Alps but in the Mediterranean lands that the cultivation of alfalfa, clover, and other artificial grasses was first practised. Actually, these "new crops" were not really new, for they had been known to the Romans, and Classical farming textbooks, such as that of Varro, had advised their cultivation. Farmers in Piedmont and Lombardy (in northern Italy), in the south of France, and perhaps in Spain, must have utilized these forage crops throughout the vicissitudes of the medieval period, though there is little or no concrete evidence of it, and some scholars have suggested that alfalfa was brought to Spain by the Arabs. We do not know when and where the field cultivation of clover originated, although it was already grown in Upper Italy and Brabant in the second half of the 16th century.

These fodder crops and the roots, such as turnips, were the pivot on which the Agricultural Revolution turned. The larger and more certain supply of feed they provided enabled more stock to be kept, and these livestock produced more and better manure so that the arable land could be more effectively fertilized in the days before chemical fertilizers had been discovered. Moreover, the cultivation of roots and grasses alternately with cereals, in a four-course rotation or some modification of it, added to the total of cultivated areas; and deeper plowing and cultivation (hoeing and weeding) between the rows of roots combined with the chemical reactions of the legume plants to make the seedbed more productive. The advantages of these new crops had been recognized by the Venetian Torelli and by the Frenchman Olivier de Serres in the 16th century.

The spread of crop-rotation systems employing fodder crops was hindered by the restrictions of the open-field system. Nevertheless, this crop system took root in Flanders, and thence spread to other portions of Western Europe. Flemish husbandry, the intensity of which so impressed travellers in the first half of the 17th century, became renowned, largely owing to their enthusiastic writings on the subject. A little later England took the lead.

For centuries before 1600 there had been close contact between the Low Countries and Great Britain in business relations and in intellectual and military exchanges. When the novel crops and methods were introduced in England they were taken up enthusiastically in the expanding literature devoted to agriculture, and to some extent by the landowners and farmers themselves. An abortive survey of English farming made in Charles II's reign (1660-85) shows that already at this date the new crops were being grown in widely separated districts and by 1700 were fairly well known to farmers in many different parts of England.

ENCLOSURES

The very first English textbook on farming, Fitzherbert's *Husbandry* (1523), emphasized the desirability of "several" or enclosed fields for the progressive farmer, and this was constantly repeated by later writers on farming. By 1700 a large proportion of landowners favored this rearrangement of the farmlands because they expected larger rents to accrue, and because they, like the most foresighted of their tenants, wanted to adopt new crops which would promote increased yields, improve soil conditions, and provide the additional fodder required for the stock. Both technical and economic considerations combined to stimulate this movement toward enclosure of open fields and caused it to be carried out at an increased rate as the 18th century advanced. Although the change had been going on slowly for years, a new legal method for obtaining enclosure of open fields was devised in the late 17th century which speeded up the enclosure movement. Private Acts of Parliament now enabled those in favor of enclosure, provided they were powerful enough or owned enough of the village lands, to get the work done with greater facility.

Nevertheless, English farmers were not encouraged to develop their systems along the new lines in the first half of the 18th century, for there was not sufficient pressure for increased agricultural production. There was little overseas trade in grain, and the government attempted to control its import by means of Corn Laws, which levied duties on imported grain in order to maintain the profits of domestic farmers. When the farmer had produced more than enough to meet his own requirements, he found it difficult to find a market for his surplus grain. For one thing, there was the problem of transport. The cost of carrying bulky goods was high; roads were poor, and the canals had not been built yet. Coastal shipping could carry some goods to London, the largest market, but the grain first had to be brought to the coast. Livestock could of course travel on the hoof, and came south to London from Scotland or east from Wales.

In the half-century between 1750 and 1800 there was increasing pressure for greater agricultural production partly from a growing population, partly perhaps from a rising standard of living. The number of people who ate wheat was 4.1 million in 1700, 5 million in 1750, 8 million in 1800. About 20 per cent of the people lived on lesser grains such as rye, oats, and barley in 1700; only 10 per cent in 1800, so on both counts the pressure on the farmers for increased production was much heavier between 1750 and 1800 than before.

This conclusion is supported by the rate of enclosure authorized by Private Acts of Parliament:

	ACTS	COMMON FIELD AND SOME WASTE ACRES	ACTS	WASTE ACRES ONLY
1700-1760	152	237,845	56	74,518
1761-1800	1,479	2,428,721	521	752,150
1801-1844	1,075	1,610,302	808	939,043

Nearly twelve times as much common field with some wasteland was authorized to be enclosed in the last forty years of the 18th century as in the first sixty, and about ten times as much wasteland. A smaller area, one-third less, of open field was authorized between 1801-44, and nearly another million acres of waste. This was only a small part of the total area of the country, but it was, it may be assumed, a very large part of the arable land even excluding the 1.25 million acres of wasteland that were also enclosed. The newly enclosed fields were cultivated with new machines and improved implements as well as being planted with novel crops in new rotations.

IMPROVEMENTS IN FARM IMPLEMENTS AND CROP SYSTEMS

The first true machine with internal moving mechanism that was introduced on the land was the seed drill. What is thought to have been a seed dropper fitted on a plow had been used in ancient Babylonia, but this had been lost; in

Fig. 9-1. Jethro Tull's seed drill. (*Science Museum*, London)

all of Western Europe seed was normally sown by hand until the 18th century, and later. In the last years of Elizabeth I's reign (1558-1603) two people suggested dibbling seed (making a hole in the ground and putting the seed in it) by hand, with the idea of economizing on seed, and perhaps to encourage tillering (that is, increasing the number of stalks growing from one seed).

Jethro Tull was the first Englishman to make a successful seed drill and also to develop a system of horse-hoeing with crops planted in rows for that purpose. Although Tull's work was done in the late 17th century, his book on the subject was not written until 1731. It was another thirty years before his principles were generally accepted by other theorists, and even longer before they were adopted by the ordinary farmer.

The men who became the "heroes" of the 18th-century changes in farming practice in England were "Turnip" Townshend and Coke of Norfolk. Charles, Viscount Townshend (1674-1738) had had an active career in politics before he retired to the farm where he dedicated himself to increasing the productivity of his land. His staunch advocacy of turnips in the crop-rotation systems won him the nickname of "Turnip" Townshend. Coke of Holkam (1752-1842), in Norfolk, was a rather late advocate of the four-crop rotation. Coke was very successful with this rotation practice, which came to be called the "Norfolk system," because of its association with him and with other landowners in Norfolk. The light lands of West Norfolk were reclaimed by marling heavily, and by introducing alternate crops; the rotation was wheat, roots, barley, with

Fig. 9-2. Plowing with two-wheel plow in Hereford, England, *c.* 1800. (From a painting by an unknown artist. Courtesy of *R. W. Stow*)

clover and rye grass undersown, and left for one or two years: it was then plowed out and the rotation repeated. Known as the Norfolk four-course rotation, this system spread over a large part of the country, coming to full fruition in the period of "high farming" between 1840 and 1880.

The value of the four-course system is immediately apparent. Instead of one-third of the arable land lying fallow, the whole of the cultivated land is cropped, the root break and clover ley providing larger supplies of fodder than ever before. The better-fed cattle, sheep, and pigs supplied more meat, milk, wool and mutton, and pork and bacon, not to speak of organic manure that maintained a higher degree of fertility. The Norfolk system became deservedly famous and was copied throughout Western Europe, and even in America.

Growing turnips and similar crops involves a good deal more work than do cereals, though even these were sometimes hand-weeded in earlier times. Root crops have to be sown in drills or rows. At an early stage they have to be bunched by hoeing out the plant, leaving groups or "bunches" to grow. A little later the "bunches" have to be "singled," that is, one plant only, preferably the strongest, must be left in the ground, the rest being chopped out with a hoe. This work, which must be done by hand-hoeing, no doubt gave seasonal employment to more farm workers, their wives, and families than cereal growing, except perhaps during harvest when labor was always scarce and everyone, even village tradesmen, lent a hand. The root crops, too, were hoed between the rows until too leafy. This is why Tull's horse-hoe and its innumerable successors of varied design became popular—they saved labor in addition to keeping the root land free from weeds.

The seed drill was a new machine, the horse-hoe a new implement. Although very ancient, the plow too was susceptible to improvement. Mainly it was of two types, the essential difference between them being that one, the swing plow, had no wheels, while the other was fitted with wheels at the fore end of the beam. Some of these were one-way plows, that is, they turned the soil only to the right or to the left; double-furrow plows and ridging plows with a double mouldboard were also in use. There were heavier and lighter designs, but on the light lands in Norfolk the wheeled plow with a high gallows drawn by two horses was favored. It was effective in those soils, and did the work rapidly, the estimated output being an acre a day. It was not a new implement in the 18th century, for it had been seen in Norfolk by a traveller in the late 17th century, and was known a century before that in Gelderland and Cleves in the Low Countries and in the neighborhood of Cologne.

It was the swing plow, greatly praised by Arthur Young, the 18th-century English traveller and writer on farming, and frequently recommended by him to farmers all over the country, often unjustifiably, that engaged the attention of inventive men. Much of their inspiration came from the Low Countries, but it was in Britain that the principles of plow design were worked out by James Small of Blackadder (Scotland) and others. In America Thomas Jefferson, too, worked out the proper design of a mouldboard on mathematical principles.

Iron was used for the mouldboard and coulter in these plows. In 1785 Robert Ransome took out a patent for tempering cast-iron plowshares, and in 1803 another for a chilled share hardened so that it remained sharp in use. The plow, the major implement in 18th-century farming, was thus made more effective, enabling the soil to be more pulverized and thereby providing a better seedbed.

Although there were improvements in plowing, hoeing, and harrowing, harvesting remained as laborious as ever, the grain being cut by a sickle or a scythe. At the end of the 18th century, some efforts had been made to produce a reaping machine, but it was not until the 1820's and 1830's that Bell's machine in Scotland proved workable, as did the famous reaper designed by Cyrus

Fig. 9-3. James Small's plow, designed in the 18th century. (From J. Allen Ransome, *The Implements of Agriculture*, 1843)

Fig. 9-4. Reaping with sickles, *c.* 1800.

McCormick in America (1834). A winnowing machine of great simplicity was also designed: how many farmers used it is a question. A threshing drum that did the job was produced in the later years of the 18th century, and was used by a few farmers in northern England and Scotland.

Thus by the end of the 18th century English and Scottish farmers had brought the Norfolk four-course rotation to a degree of perfection that was the admiration of Western Europe, and the inventors had provided new machines and improved implements that enabled them to do their work more effectively. In addition to the roots and clover with the rye grass that gave them larger supplies of fodder, sainfoin (a bean-like plant used for fodder) was grown along the limestone and chalk escarpment that stretched across the country from the Suffolk ridge to Wiltshire, and some alfalfa was also grown here and there. The construction of irrigated water meadows along some streams added to the yield of grass feed and hay there. In other areas substantial quantities of potatoes were grown; these had been introduced into Northern Europe from the New World, and some people still questioned the use of this new tuber as human food. Yet Ireland by this time was using potatoes as a dietary staple.

IMPROVEMENTS IN STOCK-BREEDING

The ample supply of forage produced by the new methods was the foundation upon which the improvement of livestock was built. Here again a few men made outstanding contributions, the most famous of the English breeders being Robert Bakewell of Dishley Grange, Leicestershire (1725-95). Bakewell's greatest success was in sheep-breeding. Using quite unpromising material, perhaps the long-legged Lincolnshire sheep and the Old Leicester, he bred the New Leicester sheep; this new breed carried a lighter fleece but gave the greatest weight of mutton for the least expenditure of feed in the least possible time. Although the mutton was so fatty that few modern townsmen would be able to stomach it, it was avidly demanded by the rapidly increasing manufacturing population. Bakewell hired out his New Leicester ram for very high prices, and the breed was widely used for crossing to improve other breeds; it helped to produce the Border Leicester, for example.

Bakewell is said to have imported horses from the Low Countries to use as breeding stock, and he succeeded in producing a very fine type of black Shire horse. With cattle he was less successful. He chose to develop the local Longhorn breed, but though he made some improvement in meat production, possibly at the expense of milk yield, the Longhorn gave place a little later to the Shorthorn dual-purpose beast, known as the Durham, or Teeswater, breed. Bakewell worked on pigs as well, but without notable success.

His stock-breeding made Bakewell famous, and his farm was very up-to-date in other respects. He not only grew roots and cabbage as well as the grasses, but he made some water meadows as well. Farmers from all over the known world visited Dishley to admire and learn his methods.

Other farmers were also successful in improving livestock. The Southdown sheep that grazed on the short sweet grass of the Sussex Downs were improved beyond recognition by John Ellman of Glynde in that county (1753-1832). Native pigs were successfully crossed with Chinese and Neapolitan animals, and heavier, fatter, and more meaty bacon produced.

All over England the better farmers followed these outstanding examples and were encouraged to do so by the great "Sheep Shearings," which were really general agricultural shows where the new machinery and the champions of the improved breeds of livestock were exhibited. Societies of landowners and farmers, great and small, and a myriad of small, local county and village gatherings offered prizes for crops and stock, and to competitors in plowing matches. The outbreak of war against France in 1793 led to even more strenuous efforts to improve the yield of foodstuffs. To promote the improved farming methods, a Board of Agriculture was set up, partly financed by a government grant.

There can be no doubt that some larger and more productive animals were bred in some parts of England at the end of the 18th century, but it is difficult to estimate the degree of improvement. Beeves weighed at Smithfield in 1710, according to one account, are said to have weighed on the average 370 lb. and calves 50 lb., whereas in 1785 they weighed 800 and 148 lb., respectively. Other equally optimistic figures vary somewhat. Estimates of the size and fleece of the sheep show equally large increases, but it is really impossible to accept such averages in the absence of any accurate knowledge of the distribution of the breeds and the ratio of the number of each to the total population of cattle and sheep. All that can be said is that better animals were bred in some parts of England, or by particular farmers, and that the progress made was continued during the 19th century.

EXTENT AND SPREAD OF THE AGRICULTURAL REVOLUTION

The so-called Agricultural Revolution in Britain was not complete; that is, it was not country-wide. Its methods were continued, perhaps intensified, under the stimulus of wars, arrested to some degree by depressions after the wars; nevertheless, the revolution expanded until the coming of overseas competition in foodstuffs in the second half of the 19th century and the disastrous animal diseases and frightful weather of the 1870's and 1880's.

The main constituents of the Agricultural Revolution were the introduction of novel fodder crops on the arable land in a four-course rotation in fenced fields that had been substituted for the great hedgeless expanses of the open fields. Clover and rye grass, sainfoin and alfalfa, roots, turnips, swedes and manfolds, cabbages and carrots were tried. Water meadows were constructed. New machines—the seed drill; the threshing drum; the winnowing machine; chaff cutters; improved plows, swing and wheel, single-and-double furrow, and with a double mouldboard for ridging—were made, and many different patterns of horse-hoe, harrow, scarifier, and cultivator were designed. Careful breeders developed fine breeds of livestock, and were followed by worthy successors.

But there is no way of gauging the progress. The area under the new crops or the new rotations cannot be assessed; the yield of each crop can only be guessed at in the light of the most incredible contemporary estimates, or a few scattered figures obtained from the rare surviving farm accounts. Much more would need to be known about the animals before any likely measure of their improvement could be calculated. Yet for all this, the system was everywhere admired, and attempts to emulate the successful innovations of the British farmers were extensively made.

Only a trifling change took place in French farming in the 18th century, and the improvement made was confined to narrow limits. Most farmers were prevented by their poverty from exercising much initiative, and they were handicapped by lack of sufficient horse-power. However, the work done on sheep was, if not outstanding, at the lowest estimate, effective. It was only after 1815 that the new implements and machines began to be imported.

Several of the French nobility and gentry set out to emulate "Turnip" Townshend and Coke of Norfolk, from 1760 until the outbreak of the French Revolution (1789), but their efforts were cut off by that event. The writings of the *Encyclopédistes* and the textbooks on farming did have some effect, but what improvement there was in French agriculture was mainly the work of the practical farmer. Most of the French farmers, however, continued to follow traditional practice. The Loire River, which divided France geographically between north and south, also served as a demarcation line between farming practices. In the south the new crops from overseas—corn, tobacco, and potatoes—were adopted. This was all the simpler because alfalfa and sainfoin had been grown in the enclosed fields there, and buckwheat also, perhaps introduced from Spain. Mulberries, too, flourished, and, of course, the vine. As in Italy, the potato proved popular throughout the 18th century, especially because of the crusading efforts in its behalf by a man named Parmentier.

The fodder crops were not known in northern France until 1750, but thereafter some of the open-field communities made arrangements to grow alfalfa and other crops on the fallow. It would be too much to say that alfalfa and clover became common, but these crops were grown in some places. In northeastern France, crops were grown for industrial purposes; wood, madder, and saffron, for dyeing; cole, rape, and linseed for oil. Some roots were cultivated, and when Napoleon's continual wars cut off imports of cane sugar, he encouraged, nay commanded, the farmers to grow sugar beets, and to produce chicory as a substitute for coffee. In France there was little if any change in the agricultural implements in general use. They retained their ancient form, and on some farms the grain was still trodden out by the oxen as it had been in Biblical times.

In stock-breeding, France, like Germany, Sweden, and later the United States, had imported Merino sheep in an attempt to produce the fine wool on which Spain had the monopoly. Spain forbade the export of these sheep, but Louis XVI was able to obtain some Merinos in 1786; he maintained this flock in a pure strain at Rambouillet, and Napoleon secured a supplementary flock in 1801. The character of French sheep was modified by careful crossing. Earlier than this, Merino stock had been imported into Sweden and bred both pure and crossed with the native breed, so that by 1765 a substantial number of both were kept there. Frederick the Great of Prussia did the same thing in 1747, in spite of the Spanish embargo, and gave some of his best sheep to the Elector of Saxony in 1751-52 with such pleasing results that the Saxon Merino became renowned. In 1775 Austria also obtained breeding stock, and toward the end of the century "Farmer" George III of England also experimented with Merinos. The breed was not successful in Britain, but in northern Germany and Saxony proved profitable. After the 1840's the competition of Australia, where the Merino flourished, proved fatal to the breed in Europe, for the Australians could outproduce and sell wool more cheaply than the European sheep-breeders.

In Germany the Thirty Years' War (1618-48) had arrested whatever progress was being made before that time. Cultivation of the fodder crops, artificial grass and roots, legumes, and rape had begun in the Lower Rhine area, but between the Elbe and the Rhine three-course open-field farming had remained the general rule. The three decades of war devastated and laid waste a large part of East Prussia, Silesia, and other parts of Middle Europe. Frederick the Great of Prussia set himself to repopulate and reclaim these parts of his dominions. His object was to restore his realm to prosperity, and he may or may not have been inspired by the advanced farming that was then already being practised in the Netherlands and coming into vogue in England. He drained marshes, built new villages, and encouraged immigration from the Netherlands and southern Germany. The redistribution of the scattered strips of the open fields was made by his orders, and common grazing on the fallow abolished in many villages. He not only encouraged better animal-breeding, but he set the example, particularly for horses, since he wanted chargers and horses for military transport. He supplied pedigreed bulls and rams to the landowners for breeding improved cattle and sheep on their estates.

Other German states also were the scene of agricultural transformation. Where the open-field system continued, the traditional three-course rotation was changed to the "improved" three-field system between approximately 1750 and 1800. By this "improved" system some 50 to 75 per cent of the fallow land was used for growing fodder crops, red clover, potatoes, roots, and pulse. Some tobacco and industrial crops were grown. Credit for the widespread adoption of the "improved" system belongs mainly to the princes of the small German states, who simply ordered their subjects to adopt agricultural improvements. The advantages were obvious, and were the same as those secured in other countries by adopting fodder crops. Where formerly cows were fed on straw in winter and grass in summer so that at the end of the winter they were in very poor condition, clover, hay, and roots made it possible to keep them in good fettle during the winter with possibly an average rise in live weight and some increased milk yield. Stall feeding was introduced even in the summertime, partly because more grassland was being plowed up for crops, so there was less land available for grazing. At the same time stall feeding allowed for the conservation of organic manure where it had formerly been dissipated on the wasteland. Agricultural change came later in Germany than in Great Britain. Nevertheless, the improvements called the Agricultural Revolution had been made by about 1850, though, of course, adaptation continued as it did in other countries.

In Italy the impact of agricultural change was felt in the north, but not in the south. Northern Italy had been a country of rather intensive cropping, irrigated meadows, and good cattle-breeding for centuries, but large parts of southern Italy were very backward, working along traditional lines under great hardship and difficulty. Potatoes, corn, and tobacco were introduced and spread during the 18th century. Spain, famous for its Merino sheep, was not receptive to changes of any sort, and the same held true for Portugal.

It will be recalled that many of the innovations of the Agricultural Revolution had their start in the Low Countries. It was from the Low Countries that England had learned to grow the fodder crops, to use a modified rotation system and better plows on enclosed fields. The far-reaching changes in farming methods thus set in motion spread from England to Scotland, France, Germany, and other parts of Europe. Not surprisingly, they finally returned to the Low Countries in a slightly changed form. Dairy farming in the Netherlands was intensive, and some new things, the churn-mill, for example, were produced there.

VETERINARY MEDICINE

Animal disease was rampant all over Europe. Veterinary medicine was elementary, and when disease reached the scale of an epidemic the village cow doctor or farrier was overwhelmed. Throughout the 18th century there were epidemics of cattle plague, which were transferred rapidly from one country to another by military activity and dealings in livestock. Figures of enormous total losses between 1700 and 1770 are stated in contemporary accounts.

When cattle plague broke out the only effective means of control was to slaughter all the animals, and this was widely adopted. These terrible conditions underlined the pressing need for more adequate knowledge. The earliest veterinary schools had been founded in Spain under Ferdinand and Isabella, but were prescientific. A school for the study of veterinary science was established at Lyons, France, in 1762, with a second French school in 1767. Denmark followed the example of France, establishing a veterinary school at Copenhagen in 1773.

Indeed, societies for the improvement of farming, and educational institutions were being set up in most countries toward the end of the 18th century. Experimental work, too, was being done (throughout the Western world) in the realms of plant and animal physiology, soil conditions, and plant growth. But these experiments were not yet sufficiently simplified for use by the ordinary farmer, and had not yet arrived at a point where they could be.

THE AGRICULTURAL REVOLUTION AND THE SOCIAL STRUCTURE

Society itself had been strongly affected by the agricultural changes, perhaps more especially in England where the Industrial Revolution was leading to large concentrations of a rapidly rising population. There had, of course, always been landless men in the countryside, but the enclosure movement and the intensive farming requiring large capital is believed to have increased their number. By the end of the 18th century the threefold organization of country-dwellers into landlords, tenants, and laborers had become, generally speaking, the normal state of rural society. There were still small holders and rural tradesmen, but the majority fell into these three strata. On the Continent even before the French Revolution there had been sporadic measures to modify servile tenures and serfdom; the Revolution completed this change in France, and Napoleon spread it throughout Europe. The Agricultural Revolution produced more food, but there were many more people to eat it, large numbers of whom were employed in industry. Yet for a large majority of the population the new processes and the new supplies did nothing. The poor were still sunk in the deepest poverty in spite of the comforting belief that the general standard of living was rising. For the landowners and the farmers of large holdings worked on the modern system, however, there were great rewards. The innovations of the Agricultural Revolution—the enclosed fields, the new crops, the new rotation systems, the new and improved machines and implements, and the improved stock-breeding—opened the road that was to lead to the great development of food production in lands such as America and Australia that were to be only fully explored in the 19th century.

10 / Metallurgy in the Seventeenth and Eighteenth Centuries

CYRIL STANLEY SMITH

In the 16th century, for the first time in history, practical metallurgists began writing down the technical details of their craft in order to instruct others. The most important of these metallurgical authors are Biringuccio (1540), Agricola (1556), and Ercker (1574). Their books excel, respectively, in the alloying and applications of metals; in mining, ore-dressing, and smelting; and in assaying. So good were these books, and so advanced were the techniques described therein, that they continued to be reprinted for nearly 150 years.

This 16th-century burgeoning of books is partly a result of the technology of printing, but even more it reflects a change of mind. Previously, literary men were uninterested in technology, and technologists had no urge to write and would have found few readers if they did. But just because there were few written records in this earlier period does not mean that metallurgy had been lagging. Surviving objects from the hands of metal-workers often provide a better record than written treatises; and jewelry, implements, and weapons, to say nothing of cathedrals, testify to a continuing and increasing skill during the period when written records are sparse.

We must remember that metals are of no use until they have been reduced from their ores. Although many of the more useful metals are present in the crust of the earth in extraordinarily small amounts, much of the task of concentrating them has already been done by geological processes, which concentrate minerals in ore veins. The chemical and physical properties that cause geographic segregation of some metals also make their subsequent processing by man easy. Absolute abundance and discovery are quite unrelated: gold, for instance, one of the earliest metals to be collected and used, makes up only 10^{-8}th of the earth's crust, while titanium, one of the latest, is 1,000,000 times more abundant! Even the so-called rare-earths are ten times more abundant than "common" lead.

MINING OPERATIONS

The utilization of all metals begins with mining, for the first step in the metallurgical process is to obtain the metallic ores from the earth.

Ancient mines typically began at the surface where mineral was exposed and worked down following the vein; better practice, commonly followed by the 15th century, was to begin at the bottom and work up. Whenever possible, a tunnel would be run in from the side of a hill with a slight upward slope so that both drainage and ore removal would be facilitated. By that time, too, miners had found that the straight shaft and tunnels, not deviating to follow every little vein of ore, were preferable. Starting from level ground a shaft was sunk vertically branching out with horizontal tunnels at intervals into the ore body. In soft rock these were laboriously lined with heavy timber to prevent collapse.

Though rock was often cracked with the aid of fire and water, ore was usually removed by the use of brute force and, according to Agricola, by "certain iron tools which the miners designate by names of their own, and besides these there are wedges, iron blocks, iron plates, hammers, crowbars, picks, hoes and shovels." Every mine had a smithy nearby to repoint such tools.

Work at the rock face continued with the methods and even the very implements unchanged from Classical antiquity, but methods of draining and ventilating the mines, of carrying out the ore-bearing rocks, and of crushing and washing the ores had all been partly mechanized by the 16th century. Technological needs and capabilities had developed together in symbiotic relationship, each depending to some extent on the other. Mechanization decreased the unit cost of product, but it required large amounts of capital. The organization needed both for operating and for financing large mining and smelting operations benefitted from the rising trade and wealth of the Renaissance.

Much of our information regarding mining techniques of the 16th century comes from the *De re metallica* (1556), published a year after the death of the author Georg Bauer, who wrote under the name of "Agricola." A physician in Joachimsthal, a mining and smelting town in Central Europe, Agricola began a systematic study of minerals and mining. His books on rocks and minerals have won him the title of "father of mineralogy," for he was the first to describe rocks and minerals in terms of their observable, physical properties rather than of their supposed magical and philosophical properties. More important to the history of technology was his *De re metallica,* a systematic and comprehensive treatise on all aspects of mining and smelting; its excellent illustrations shed much light on the devices and techniques of his times.

Mine surveying (important both for controlling operation and for establishing ownership) was done with simple methods using stretched cords for angle or length measurement, and levels and magnetic compass with wax circles for recording directions and angles. According to Agricola, the results commonly were reproduced at natural size in an adjacent field instead of being plotted on a reduced scale on a map.

Agricola borrowed some of the passages in his *De re metallica* from the *De la pirotechnia* of Vanoccio Biringuccio, which had been published some sixteen years earlier (1540) in Venice. Biringuccio's treatise was a practical manual, setting down actual accounts of metallurgical processes—assaying, smelting, alloying, casting—as he himself had viewed or practised them. The birth of a new scientific spirit in the mid-16th century is shown by his emphasis on actual observations and experiments; his *Pirotechnia* represents the first description in full working detail of many metallurgical processes. These accounts reflect Biringuccio's observations in his wide travels through the metallurgical centers of Italy and Germany; in 1538, a year before his death, he was made head of the papal foundry in Rome, and his treatise was published a year after his death. Practical metallurgists must have made much use of Biringuccio's text, for it went through many editions. However, it has been overshadowed by Agricola's later and more famous—albeit less comprehensive—book, and it was not until 1942 that an English version of the *Pirotechnia* appeared.

In discussing the conditions necessary for a successful mine, Biringuccio wrote:

> But of all the inconveniences, shortage of water is most to be avoided, for it is a material of the utmost importance in such work because wheels and other ingenious machines are driven by its power and weight. It can easily operate large and powerful bellows that give fresh force and vigor to the fires; and it causes the heaviest hammers to strike, mills to turn, and other similar things whose forces are an aid to men.

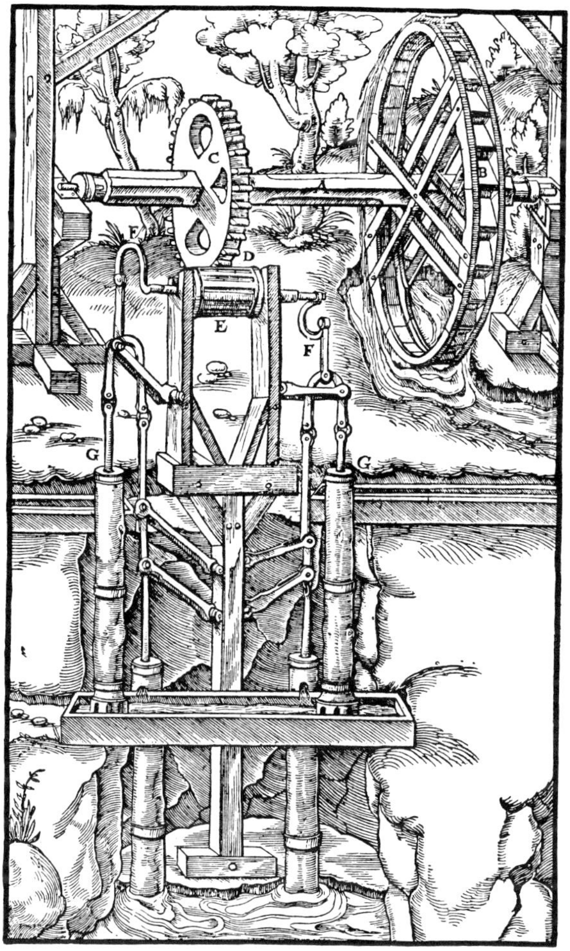

Fig. 10-1. Mechanical pumps for mine draining. (From Agricola, *De re metallica*, 1556)

The use of water power for pumping water out of the mines and for hoisting the ore greatly extended the depth to which an ore body could be profitably worked. Many ingenious pumping devices were developed specifically for mine use, and in later times it was the mine pumping problem that led to the development of the first steam engine, Savery's "Miner's Friend" in 1699, and was later the incentive for the improvements of the steam engine by both Newcomen and Watt.

After mining, the ore was carefully selected by hand, crushed by hand, or in larger plants by water-driven stamp-mills or rotary stones. It was then concentrated by washing, the separation depending only on the differing ease with which particles of varied density would be washed away. This was sometimes done by hand, in pans or troughs, but in larger mines more elaborate separating tables with transverse grooves were used.

Ores containing sulfur, arsenic, or antimony—"harsh" ores, as they were called—were then roasted, an operation that was often done on raw ore also in order to facilitate crushing. The roasted ore, now a concentrated form of metallic oxide, though usually containing some residual sulfide, was then reduced to the metallic state by heating it with charcoal. The metallurgists of the time, of course, recognized the presence of sulfur which could be removed as a visible, stinking smoke on roasting, but the role of oxygen and the fact that charcoal played a double role as fuel and as chemical reducing agent was not seen until the "Chemical Revolution" at the end of the 18th century. Here is an example of how technology developed quite apart from science during this early period, for an understanding of the chemical changes involved was not necessary for the discovery of the efficiency of the roasting and smelting operations.

SMELTING OF METALS

Bismuth and antimony sulfide were extracted from their ores simply by heating them red hot and allowing the metal to drain off, the rocky matter being left behind unmelted. Copper and lead ores, usually after roasting, were mixed with charcoal and smelted in small vertical shaft furnaces equipped with water-driven bellows to force an air blast through the mixture and so produce a hot enough fire. Since many ores, even after washing, contain a considerable amount of infusible siliceous or calcareous rock, it was common to add additional rock or to mix ores of different kinds as fluxes—that is, to make everything fusible and to allow the liquid metal to separate from the slag.

The furnaces were commonly built in groups of three or four against a permanent wall behind which were the water-driven bellows providing the blast. A typical furnace for smelting copper would be built of blocks of refractory (difficult to fuse) sandstone to give a shaft about 18 to 24 inches square internally and about 5 to 6 feet high, with at least the front part at the bottom easily replaceable. Iron-smelting furnaces were larger and free standing. Before use, the furnace was lined with a thick layer of clay or a mixture of clay and charcoal powder which could later be replaced with comparative ease.

Baskets containing weighed charges of ore, flux, and charcoal were dumped into the top of the furnace. When everything was melted, the products were run into a forehearth for separation of the slag and the metal, which was ladled into the moulds. Copper needed repeated passage through the furnace to remove sulfur and iron, and the last stage involved a refining hearth wherein a blast of air from the bellows was directed upon the surface of a pool of molten

Fig. 10-2. Blast furnaces for smelting copper-lead ores. (From Agricola, *De re metallica*, 1556)

metal to remove the last of the impurities in the form of a slag. The final step was to stir the metal with a wooden stick (which had the effect of reducing the excess of oxygen) until the appearance of a sample that was removed and solidified showed that the metal had the desired gas content, whereafter marketable cakes of copper were obtained by pouring water on the surface of the molten bath and removing a sequence of solid crusts.

An extremely complicated and interesting process known as liquation was

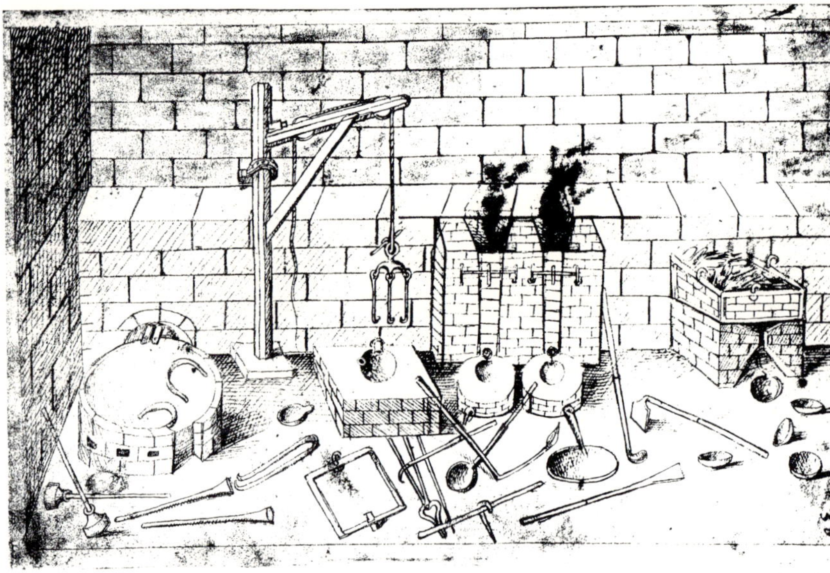

Fig. 10-3. Interior of a German smelting house, *c.* 1480. Mixed copper and lead ores were smelted in the blast furnace (center). The resulting copper-lead cakes were then heated at a red heat in the liquation furnace (right) to allow the molten lead to run off with the precious metals in solution, while solid copper remained behind. The lead was then cupelled in the hearth at the left to give pure auriferous silver (litharge was returned to the blast furnace). (From *Mittelalterliche Hausbuch*, *c.* 1480)

developed in Saxony for the treatment of complex copper and copper-lead ores containing enough gold and silver to warrant their extraction. In this process, use is made of the fact that copper and lead are completely miscible with each other when liquid at high temperatures, but that on cooling the copper crystallizes out long before the lead solidifies and that the precious metals have a much higher affinity for the lead than the copper. When a cake of 75/25 lead-copper alloy was cast from a mixed charge smelted or melted in the blast furnace, then reheated to a dull red heat, the lead drained away carrying the gold and silver with it; it left behind a porous mass of copper which needed only fire refining to be ready for the market. The gold and silver would be obtained from the drained-off lead by cupelling. Cupellation simply involved using a blast of air to oxidize the lead to molten lead oxide (litharge), which flowed away (to be reused as a source of lead with a new copper charge) leaving the pure unoxidizable precious metal behind.

It is uncertain when the liquation process first becomes important. Although cupellation was used in ancient Greece, liquation probably was not. There is a hint of its use in the 12th century, but the first technical description dates from near the end of the 15th century, by which time the various furnaces involved in the operation had already reached an advanced state of development. The whole process was under such good control that it did not change until it was supplanted by electrolytic methods in the 19th century. The silver obtained from German copper ores by the use of this method enriched the mine-owners of southern Germany, such as the famous Fugger family, which provided the finances for the Holy Roman Emperor. Indeed, the wealth derived from silver mining provided much of the resources for trade and banking in both southern Germany and northern Italy during the Renaissance period.

ASSAYING

An important aspect of some metallurgical processes during this period was the technical, practically scientific, control that was applied, especially in assaying. Assaying refers to the examination or testing of ore—especially gold or silver ore, or an alloy—to analyze the nature and proportion of its ingredients and to determine its purity. Although the operator of any mine or smelter would use crude tests to find the yield of metal that could be expected of his ore, assaying with high precision was carried out only on the precious metals, gold and silver. Though the chemical theory was, of course, not understood at the time, the processes depended on the facts that (1) molten metal will settle out when a roasted ore is melted with a flux and a reducing agent; (2) if lead oxide is added to a charge, during its reduction it will collect and carry down with it even traces of gold and silver; (3) all common metals, except gold and silver, will oxidize when heated in air and the oxides will dissolve in molten lead oxide; and (4) nitric acid will dissolve silver away from a silver-gold alloy leaving pure gold behind—the operation known as parting.

Methods that had been originally developed for the assaying of gold and silver were used to find quantitatively the traces of these metals in ores of copper and other metals. Frequent assaying gave the manager of a smelting or refining plant a check on the effectiveness of each stage of the operations, and told him whether a given mass of ore, copper, or lead would repay treatment.

The assaying operations were intended to reproduce quantitatively, but on a miniature scale, the operations of smelting. They produced tiny but weighable beads of pure metal, using mainly pyrotechnical methods, quite different from most of today's chemical analyses, which involve the precipitation of chemical compounds from aqueous solution.

The assayers were perforce educated men, and it was natural that they would write descriptions of their operations: the first printed book on metallurgy (1517) was *Probierbüchlein*, a manual or handbook for the assayer. Agricola had a whole long chapter on assaying, and Lazarus Ercker published in Prague

Fig. 10-4. A 16th-century assay laboratory. In the back (left to right) are a cementation furnace, wind furnaces for crucible melts, and a muffle furnace. The little blast furnace in the right foreground is for copper assaying and is blown by a steam injector (aeolipile). The athanor (middle foreground) is making acid for parting. The towers of the athanor and the cementing furnace are full of charcoal, which burns only at the bottom and falls down as it is consumed, thus maintaining a steady fire without attention for as much as a day. (From Ercker, *Treatise on Ores and Assaying*, 1574)

in 1574 a lavishly illustrated book of 288 folio pages devoted entirely to the subject. There is no trace of theory in these books, but they reflect an empirical knowledge of chemical substances and their reactions that is both sophisticated and accurate.

Ercker was another product of the rich mining area of Central Europe. By the age of twenty-five he had become an assay master (1555), and the following year he published a tract on assaying. He eventually became chief superintendent of mines and comptroller of the Holy Roman Empire and the kingdom of Bohemia, where there were rich mining areas in the Erzgebirge, on the Saxon-Bohemian (today's German-Czechoslovakian) border. Ercker's important book, entitled *Beschreibung allerfürnemsten Mineralischen Ertzt und Berckwercksarten* (Treatise on Mineral Ores, Mining, and Smelting Operations), had even more beautiful woodcut illustrations than did the texts by Biringuccio and Agricola, and like them, it was much used as a practical guide to assaying by master craftsmen in their shops.

The assayer's techniques, the reactions that they discussed, and particularly their viewpoint regarding the quantitative immutability of substances through many solutions and changes of state, provided the background for the later development of quantitative chemistry on a theoretical basis. Modern science owes quite as much to the assayer's discoveries of the chemical properties of matter as it does to the better known early measurements of the movements of the stars, which eventually led to mathematical astronomy and mechanics.

ALLOYS AND APPLICATIONS

Relatively pure copper was used to some extent for decorative work and utensils, but its main application was in the form of alloys. In the case of bronze (a mixture of copper and tin), the alloy gave a material of greater strength and improved castability; in the case of brass (a mixture of copper and zinc), both these effects and also lower cost were obtained. At the time of its discovery in the fifth millennium B.C., bronze was probably made by the use of mixed ores, but it was thereafter made by intentionally adding separately smelted tin in known amounts to the copper. Brass was made by heating small bits of copper with roasted zinc ore mixed with charcoal. The zinc was never seen as a separate metal in the process, but was directly absorbed by the copper. The brassmakers developed special furnaces tall enough to obtain a good natural draft, long before the draft-inducing effect on a separate chimney, such as later used, had been discovered.

One effect of alloying, the tendency to lower the melting points of metals, was used particularly in making solders. Soft solders (tin-lead alloys) and hard solders (gold-copper, silver-copper, silver-copper-zinc, and copper-zinc) were widely used in making and repairing jewelry, cooking utensils, and tools.

Various combinations of gold, silver, and copper were used for coinage, combining beauty, tarnish resistance, and value in varying degrees. Some ingenious chemical methods were developed to produce a gold-rich or silver-rich surface on alloys of lower value—antecedents of the bimetal (clad) coinage of the United States in 1966.

Pewter was extensively employed in making a kind of plate for the growing lower-middle class. It consisted of an alloy of tin with from 10 to 25 per cent lead, hardened by natural impurities and small additions of mercury, brass, or (from the 16th century) bismuth. The pewterers' guilds established elaborate methods of control over their members, for tin was an expensive metal and there was strong temptation to use more lead than was proper. Since an excessive amount of lead increased the density of the alloy, the guild could detect adulteration by weighing a ball of the pewter cast in a standard mould.

Pewter which melts very easily, lent itself particularly well to the casting of numerous identical objects in permanent moulds made of iron or, more commonly, of carved soft stone as well as in temporary moulds of clay or wood.

Although many Roman brass pots were obviously turned, the first written description of a metal-turning lathe (1125 A.D.) was in connection with the finishing of pewter vessels. The pewterer's easily castable alloy and his permanent moulds with their removable parts carrying decorative detail may have provided the background for the development of type-casting in the 15th century.

Printing provides an early example (for some centuries almost an isolated one) of the modern way in which materials of new composition are developed in specific relation to a desired application. The Chinese, beginning in the 11th century, were the first to use movable type: this was made of a fired ceramic, carved wood, or cast bronze, the last made from individual patterns in clay moulds by conventional methods. Once Europeans started to adopt the use of movable type in the middle of the 15th century, they devised methods far outstripping those of the Orient and marking the beginning of mass production of identically dimensioned objects. Though the essence of the typecaster's technique probably came from the pewterer, pewter would wear rapidly (as the Chinese had already found), and printers soon hardened their metal with greater amounts of antimony or perhaps bismuth, for type could be of a more brittle material than was suitable for domestic utensils. By the 17th century, typecasters had adopted the cheaper, harder alloys of lead and antimony containing only minor amounts of tin; essentially that used today. For the type there had to be developed an ink quite different from that used for writing, and, of course, the presses and the production of paper also needed development. Not until the advent of the automobile was any engineering development so dependent upon specialized new materials for its realization as was printing.

IRON-WORKING

There are three main forms of iron: (1) *wrought iron*, melting only at a temperature above any achieved before the 19th century, forgeable and weldable when hot, ductile when cold; (2) *steel*, with a lower melting point, capable of being forged at a red heat, relatively soft when slowly cooled but spectacularly hard after quenching (cooling quickly by plunging the hot metal into water or some other bath); and (3) *cast iron*, which is easily melted and cast to shape, relatively soft and easily machined, although quite brittle. All these materials, with their great differences in quality and nature, come from the same iron ore and fuel. They all contain over 95 per cent iron; their differences arise mainly from different furnace designs and processing procedures which affect the amount of carbon that is absorbed from the fuel and remains in the final product. Wrought iron contains no carbon; there is between ¼ to 1¼ per cent in steels; and 2½ to 3½ per cent in cast irons.

The presence of these differing amounts of carbon, however, was unsuspected until the end of the 18th century. Although the distinction between bronze and copper was clearly due to a difference in composition (for the making of bronze needed a physical mixing of ores or metals), carbon as an alloying element in steel was overlooked. This was partly because such very small quantities of it had such a profound effect, but mainly because its role as an alloy was obscured by its role as the fuel used in the smelting process. Thus for centuries, Aristotle's belief that steel was a more highly purified form of iron persisted; this seemed reasonable, for the purifying effect of fire was well known, and steel was produced by prolonged heating in a charcoal fire. Although the element carbon was not definitely identified until the 1780's, this did not prevent metallurgists from making steel with some degree of reproducibility.

Because of some imagined crudeness of the material, iron fared particularly poorly in the early written record. Though iron has been both quantitatively and qualitatively by far the most important metal for about 3000 years, knowledge of its early use is still fragmentary, and the technical records before the 18th century mostly deal with the non-ferrous metals (that is, metals *not* containing iron).

Iron differs from the principal non-ferrous metals in having a melting point so high that it cannot, when pure, easily be extracted from its ores in the molten state. If it contains enough carbon, it will melt at a temperature only slightly above the melting point of copper, but it is then brittle and cannot be forged. Although in China such cast iron was used extensively for agricultural tools and other utensils from the 6th century B.C. on, in Europe it was virtually unknown until the 15th century A.D.

Until that time all iron was made by direct reduction of the ore to a soft metallic sponge; this was never melted, but rather was compacted by forging at a white heat to give a strong ductile metal in shapes for use. This product—wrought iron—could be of high purity but it was usually heterogeneous, containing varying amounts of slag and rocky particles that drastically modified its properties. The reduction operation was carried out using bellows and charcoal fuel in relatively small hearths, and the product was highly dependent upon the quality of ore and the skill of the operator.

This direct production of wrought iron from the ore continued to the end of the 19th century in some parts of Europe (even today in remote parts of Africa), but in the beginning of the 15th century, it began to be displaced as a major production process by the indirect method of producing iron. In this, the ore was smelted in a tall blast furnace wherein the reduced iron remains in intimate contact with charcoal, so becoming further carburized and eventually melting so as to be easily separated from the slag formed by fluxing the rocky parts of the ore. The resulting cast iron was then converted to wrought iron by a separate refining. This last operation used a hearth somewhat larger but not otherwise greatly different from the earlier reduction hearth, though the chemical aim, practised though not yet understood, was decarburization rather than reduction.

In both cases the product was a spongy mass of solid iron which had to be

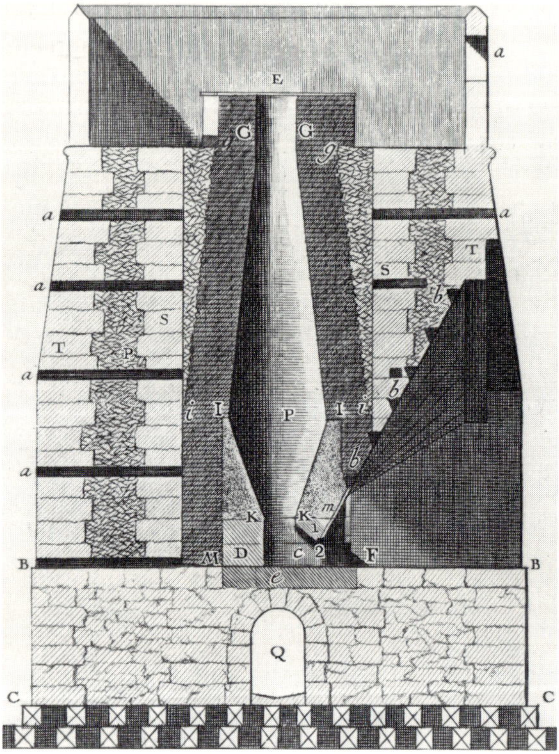

Fig. 10-5. A blast furnace for smelting iron. The furnace was about 25 feet high, charged by hand. The blast was provided by two large water-driven bellows. Some of the iron was run directly into finished moulds for casting cannon, firebacks, and the like, but most of it was run into grooves cut in a sand bed, providing the large "sows" shown in the picture. These were used for subsequent remelting in the foundry, or to a larger extent, for refining to wrought iron. (From Diderot, *Encyclopédie*, Paris, 1765)

freed from slag as well as possible by hammering. The product could be shaped by a blacksmith by forging into a variety of tools and hardware. Wrought iron is unmatched for forgeability when hot and has considerable strength when cold. Moreover, iron parts are easily welded together simply by heating them to a white heat, placing the parts to be joined in contact and rapidly hammering to consolidate them. Welding together of this wrought iron was used to build up the great iron bombards (cannon) of the 15th century and architectural hardware, notably the magnificent wrought-iron gates of Spain.

As the scale of iron-working operations increased, large water-powered hammers were used in the refinery to produce merchantable rods, but final shaping was still done largely by hand. As time went on, cast iron, as the cheapest of metals though relatively brittle, found a considerable use of its own for casting pots and cannon, pipes, rails, machine parts, and, in the 19th century, as a major constructional material in bridges and buildings.

Charcoal (made by a partial burning of wood, usually in piles covered by turf or clay, to drive off the volatile matter) was for a long time the only metallurgical fuel. Although the purity of charcoal gave a quality product, at least with selected ores, its low resistance to crushing limited the height of smelting furnaces. The pressure to find some other fuel came mainly, however, from the increasing cost of charcoal as forest areas decreased in the 16th and 17th centuries. The deforestation came from the expansion of agriculture, from the increasing use of wood as a structural material and as a fuel, and from the growing demands of the smelting furnaces themselves.

Mineral coal was an obvious substitute and was abundant enough, but its impurities, especially sulfur, rendered iron hopelessly brittle. Furthermore, the presence of volatile matter in the coal, as well as the softness of coal on initial heating, interfered with furnace operations. Thus the early experiments on smelting iron with coal, notably by Dud Dudley in England around 1620, met with failure.

By 1700 coal fuel was successfully used in operations which needed a long flame and avoided direct contact between the product and the coal, such as in glass kilns and in reverberatory furnaces (where the heat from a flame is deflected downward from the roof of the kiln) smelting copper and lead. However, it was not until 1709 that smelting of iron with mineral fuel was successfully achieved—by Abraham Darby at Coalbrookdale in Shropshire. He had previously used coke in malt kilns, and he found that coking (precooking of the coal) removed most of the unpleasant qualities that had defeated earlier smelting experiments with coal. He was lucky too in the particular bed of coal that he had available for his experiments.

With strong coke as a fuel, the blast furnace was capable of almost unlimited expansion in size. Coke-smelted iron was, however, more impure than the charcoal product and harder to work to good wrought iron in the refinery hearth. Output continued to be limited until economic pressure forced the adoption of lower quality standards. The situation was saved by the replacement of the refinery by the puddling furnace, which simultaneously increased production and gave a better product. The puddling furnace, introduced by Henry Cort in 1784, was to prove a major factor in the Industrial Revolution. As we shall see later, this new and cheaper process for producing iron provided much of the impetus for changing the basic technological material used by man for his tools and instruments from wood to iron. When carbon and silicon were removed from molten cast iron, crystals of more or less pure iron separated out from the

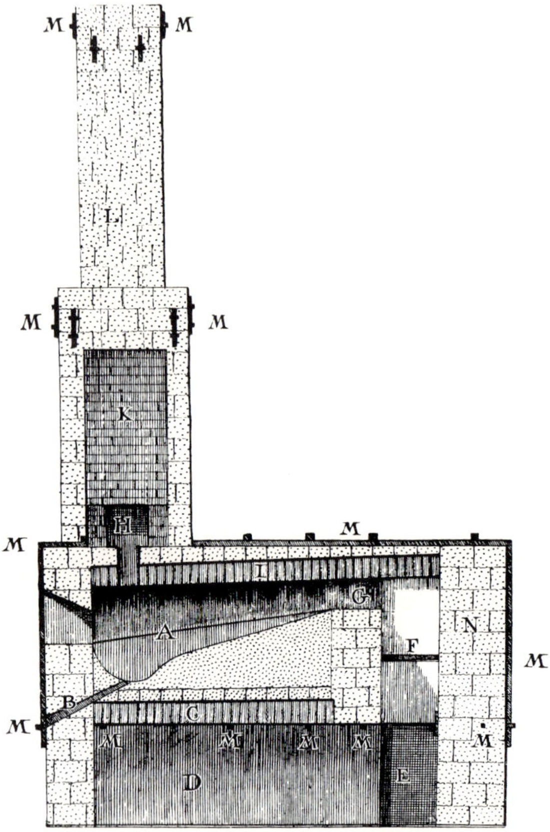

Fig. 10-6. Perspective view of reverberatory furnace for melting cast iron in the foundry. This furnace was sketched by Gabriel Jars on a visit to England in 1765 and is reproduced from his famous book, *Voyages métallurgiques*, Paris, 1774.

liquid. Impurities became concentrated in the residual liquid and eventually were removed by oxidation; but inevitably there was some slag left in the final product, the quality of which was highly dependent upon the workman's skill. The earliest iron-workers used large hammers to expel this slag. Soon, however, water-driven helves permitted the scale of operation to increase and enabled iron-makers to improve the metal by more extensive forging.

After the first hammering, the resulting "loupes" were reheated and further forged into bars or sheets for use. Beginning early in the 17th century, small rods for nail-making and the like were produced in the slitting mill (an ingenious contrivance with rotary cutters slitting a forged flat strip into as many as 12 square-sectioned strips), but larger sections had to be forged. The rolling mill with cylindrical rolls was used in flattening iron for slitting and had previously been used for making uniform narrow strips of coinage metals and wide plates of soft lead for roofing purposes.

The precise steel regimen used to produce early steels is unknown, but it probably involved simply keeping the reduced iron sponge in contact with hot charcoal long enough to allow absorption of carbon to occur, sometimes, per-

Fig. 10-7. Puddling furnace for making wrought iron. Note the similarity to the reverberatory melting furnace (Fig. 10-6), but the much larger hearth in relation to the work area. (From DuFrenoy and de Beaumont, *Voyages métallurgiques*, Paris, 1827)

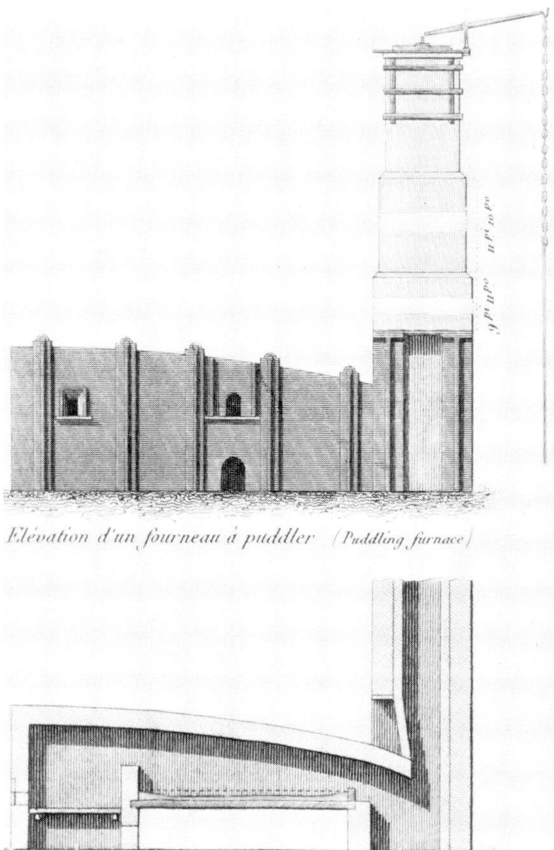

Fig. 10-8. Large water-driven hammer for working wrought iron blooms into merchantable shapes. (From Diderot, *Encyclopédie*, Paris, 1765)

haps, via the intermediate formation of transitory amounts of molten cast iron. Reasonably reproducible steel, as distinct from iron, had appeared as early as about 1200 B.C., but for many centuries the products varied greatly in carbon content and uniformity. We can best trace the history of steel by following the development of swords through the centuries.

Quite early it had been discovered that steel made the best sword blades. Some early swords of about the 8th century B.C. have handles of forged iron with blades of hardened steel. Later weapons had conspicuous patterns forged into their blades by hammering layers of steel and iron together. The most famous of such textured metals is the Damascus sword blade, which appeared in the Middle East early in our era and spread over the whole Islamic world in the days of its supremacy.

There were several different kinds of texture in Damascus blades, but the most interesting ones owed their beautiful appearance as well as their effectiveness as weapons to a duplex laminated structure of iron carbide and steel. This

Fig. 10-9. Short sword found in Luristan (a mountainous region in Western Iran) dating from *c.* 800 B.C. The blade is of carbon steel, the handle assembled from several pieces of forged iron accurately fitted together. (Courtesy *Metropolitan Museum of Art*)

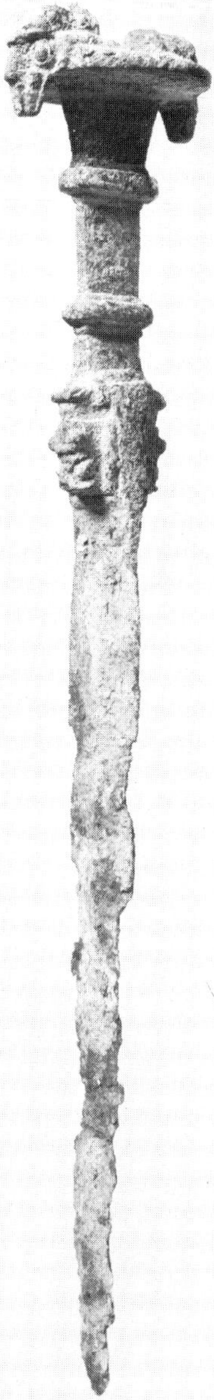

was produced by melting a very high-carbon steel, solidifying it slowly so that it developed large crystals, and then forging at a relatively low temperature to avoid destroying the segregation pattern. India had a virtual monopoly on the raw material for these swords for a long time, but the Persians were most renowned for the forging and use of them. The swords of medieval Japan constitute the supreme metallurgical art, though they were without influence on Western metallurgical or military history. The Japanese blade was a complex forging, composed of a soft wrought iron interior with the outside surface and cutting edge made of steel forged in such a way as to control the carbon content exactly and to obtain a very fine dispersion of slag particles. Although the cutting edge was extremely hard, an ingenious differential quenching treatment left the body of the blade soft and the sword as a whole was not brittle. Careful polishing technique gave the blades a velvety matte surface with subtle differences in light reflection from the hard and soft areas, and displaying rich beautiful detail originating in the welding and forging technique.

While the Oriental world was so advanced in its metallurgical processes, it should be noted that steel was not melted on any significant scale in Europe until 1740 A.D., and that the crystallization of Damask steel was not duplicated in Europe until 1821.

The spectacular properties of steel develop only when it is properly heat treated. The extreme hardness of rapidly quenched (cooled) steel containing over about 0.4 per cent carbon is accompanied by extreme brittleness. Today this brittleness is relieved by the operation called "tempering," which simply involves reheating the quenched steel at an intermediate temperature to relax its hardness and restore some measure of toughness as required for each type of application. In earlier times, however, the steel was quenched in such a way as to come out of the cooling bath at the right hardness, an operation requiring considerable dexterity on the part of the smith. It was this quenching process that originally was called tempering, and "tempered steel" used to be synonymous with "hardened steel." Just when the change in technology and terminology took place is uncertain. Sixteenth-century writers refer to both methods, but it was probably not until after the 17th century that the quench-and-reheat method became almost universal.

Even before the 17th century, the smith knew that success depended upon a succession of colors: the radiant red or white heat of the steel as it was removed from the fire for quenching, and the temper colors (interference colors due to the film of oxide on the surface) which would form on a de-scaled surface in the quenching bath or on later reheating. The relation between the different temper colors and the hardness appropriate to various types of service is mentioned by most writers of the 16th century and later. The experienced smith could tell when the metal had reached the right color for removal from the quenching bath for the purpose for which the metal was intended.

No metallic alloying elements were involved in any of the earlier steels: the properties depended mainly on the control of carbon by furnace regimen, though minor amounts of phosphorus and nitrogen had perceptible hardening effect, and these elements with sulphur and slag could cause embrittlement. Variations in ore and skill made the steels from different regions very different in quality. Those of Spain and Styria (near the present-day Austro-Yugoslav border) were most renowned until after 1740; then English cast steel, based on Swedish iron, outranked all others for cutlery and fine tools because of its uniformity and hardness. This was the famous Sheffield steel.

At least in medieval times and perhaps much earlier, case-carburization was used to harden the surface of a tool or weapon. This involved prolonged heating of the finely shaped iron at a good red heat when packed in charcoal, often in a sealed container or a clay jacket. The hardening effect was sometimes enhanced by the use of organic substances containing nitrogen. Beginning very early in the 17th century, this method was used to convert whole bars of iron into steel which were sold to cutlers and others for subsequent forging to knives and tools. The product, "cementation steel" or "blister steel," was more reliable than natural steel coming from the hearth. The name blister steel comes from the fact that the bars of iron being slowly heated in the charcoal developed blisters in the course of their conversion from iron to steel; the reaction of the oxide and slag with carbon gave carbon monoxide gas which formed the blisters. Cementation steel was the basis for the pre-eminence in steel for many decades of Sheffield, England, and it remained generally the raw material for cast steel. The best grade of cementation steel was made by welding two or three bars of blister steel together and forging them out to a product called shear steel, supposedly named from its use in making shears.

The scale of production of steel during this period was inevitably small. Most steel for agricultural implements—picks, hoes, and the like—was of a relatively crude quality made by direct reduction of the ores. The smith using it learned to select the quality he needed for a given job by the appearance of a fractured surface. Very often a tip or a thin layer of good steel was welded to the service end of a tool body of cheaper local iron.

The quality and uniformity of steel for special tools and springs was greatly improved after the invention of the crucible melting and casting process by Benjamin Huntsman around 1740. A watch-maker living in Sheffield, he was led to formulate this new technique because of the lack of uniformity of the material available for his watch springs. His success was owing to a combination of several factors. There was a good local sandstone for his furnace, and good local clay for his crucibles, of a type already extensively used in glass furnaces. In his furnace design Huntsman exploited the relatively new use of tall chimneys to obtain a strong draft in place of the older cumbersome bellows with their uneven blast. In addition, he used hard coke, which gave, with an adequate

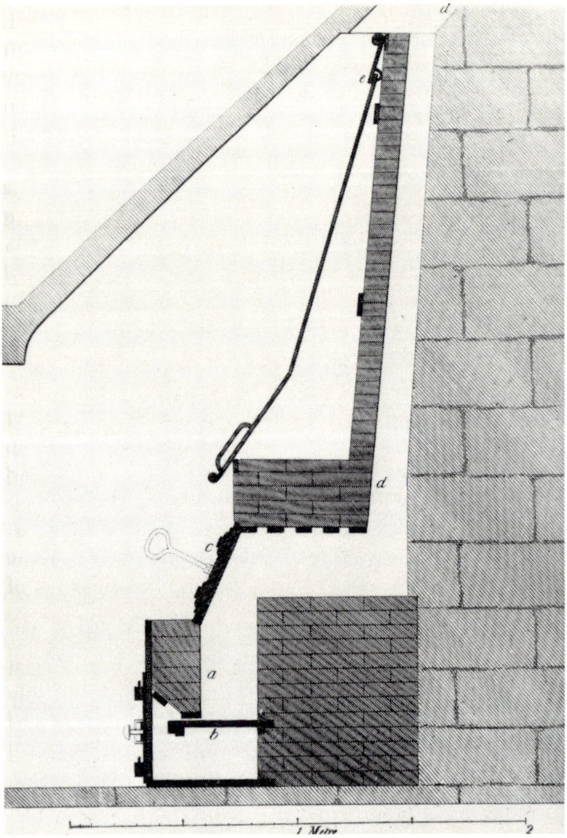

Fig. 10-10. Furnace for melting crucible steel. Coke was used as the fuel and the draft produced by the tall chimney enabled the necessary high temperature to be reached. (From *Annales de chimie*, 1798)

draft, a higher temperature than charcoal, and was then common as an industrial fuel. Even with all these factors, it was still not possible to melt anything but the higher carbon tool steels, though there would have been little market for the softer steels. Huntsman tried to keep his process secret, but competition slowly arose and by 1787, shortly after Huntsman died, there were eleven steel "refiners," that is, crucible steel melters, in Sheffield alone.

THE CASTING OF METALS

Making a casting involves primarily the shaping of a refractory (heat-resistant) material into a receptacle (mould) in which molten metal can be poured and allowed to solidify without changing its shape. The first moulds were simply depressions in the ground, but metallurgists in the ancient Near East learned very early how to shape negative cavities in stone and clay, then to assemble moulds from several pieces to form haft holes and lighten the casting, and to carry decorative detail. Innumerable objects in the world's art museums testify to the skill of early founders. The most magnificent products ever made by piece-mould technique are the ceremonial bronze vessels of Shang dynasty (China, 1527-1027 B.C.). Later came "lost wax" (*cire perdue*) casting in which a model of wax is buried in a sandy clay, the wax melted out, and molten metal poured into the residual space, which can be of very complicated shape.

Bronze (an alloy of copper and tin) has been a favored material for castings from early times. It was melted at first perhaps in a kiln such as those used by potters, or by heaping charcoal over it in a clay-lined depression in the ground and blowing with a pipe; eventually the crucible itself was set in an outside fire urged by draft or blast. In medieval times, perhaps much earlier, the bronze for large objects, such as bells, was melted in large pots on top of which a shaft was built to hold the metal in contact with the fuel, urged by a blast from the bellows. The arrangement was much like the modern cupola, but it was displaced in making the largest castings by a reverberatory furnace which separated the fuel from the metal and which, by the end of the 18th century, was capable of melting large amounts of metal.

A vividly detailed description of the whole process of casting a large bell is found in the *Treatise of Divers Arts* of Theophilus (*c.* 1125 A.D.); 600 years later nearly identical descriptions were given for the casting of bronze cannon, for the gun-founder took over most of the bell-founder's techniques. In making a large bell, a clay core was turned on a lathe, then coated with a layer of tallow of the thickness that the metal was to be, and turned to the correct contours, then surrounded again with clay and reinforced with iron bars to complete the mould. (Meltable patterns of tallow or wax were later replaced by clay which could be stripped out, though decorative details were always of wax.) In guns, the central hole was shaped by a long clay core built up on an iron rod; iron chaplets were embedded in the mould near the breach to keep the core straight and axial. The assembled mould was baked out with a fire, placed in a pit, surrounded by tamped earth, and the molten metal run into the mould.

The core-supporting rods remained embedded in the bronze and can be seen today in most surviving guns of the 16th century and earlier. After the middle of the 18th century, cores were omitted, for boring machines, which had initially been used simply to make the cast hole truly round and of even dimensions had become powerful and rigid enough to bore the hole in solid metal.

The making of cast iron in the blast furnace, which began mainly as a stage in the making of wrought iron, involved the running of the molten metal into depressions in the ground—the pig bed, hence the term "pig iron"—which soon

Fig. 10-11. View in an iron foundry. Cast iron was here cast directly from the blast furnace. Castings are made in sand moulds in boxes (on the left) or in loam moulds (turned on the lathe, in the upper right corner), buried in the ground. At the right is a cast section of a cast-iron water pipe, a very important product at that time. (From Diderot, *Encyclopédie*, Paris, 1765)

suggested the direct casting of simple objects in open sand moulds. Cheap sand soon displaced expensive loam as the principal material for moulds, and with experience larger and more complex items could be cast. Soon even cannon, the largest iron products of their time, were being made of cast iron.

The huge bombards, in which gunpowder was first effectively used in warfare, had been made of longitudinal wrought-iron bars welded together into a cylinder and surrounded by circumferential iron hoops. At first these complicated forgings, with their weak welds which frequently burst when the cannon discharged, were reinforced with external cast iron, but soon they were displaced entirely by cast cannon. The making of cast-iron cannon began in France, but by the mid-16th century was a very important business in Sussex, England, as well.

Although cast iron became an important material for making cannon from the 15th century onward, the best cannon were still made of the more reliable bronze, which was not entirely displaced for this purpose until the middle of the 19th century. If iron was essentially the smith's metal, bronze was, through most of European history, the material for the founder.

An important development of the founder's art came with the application of cast iron to engineering purposes: the large cast-iron water pipes for the royal fountains of Versailles in 1664 were forerunners of more democratic water systems. The use of cast-iron members in the famous bridge at Coalbrookdale was followed by increasing use of cast iron in architecture; it also prepared the way for the use of cast iron for machine and engine frames, having rigidity and dimensional permanence far exceeding that of wood, the basic material for machinery since its beginnings.

THE JOINING OF METALS

The ease with which metal parts can be securely attached to one another in construction and repair has been responsible for many of their applications. The great freedom of design which modern welding methods have given to the engineer has not always existed, however.

Lead was used in Mesopotamia as a kind of solder to join copper before 3000 B.C., though most joints at the time were done by riveting or crimping. Alloys approximating the low melting point in the gold-copper series were effectively used as solders in the manufacture of jewelry by 2600 B.C. (The alloy of about 20 per cent copper melts at 889° C, 174° below the melting point of pure gold.)

No joining technique today surpasses in elegance that used by the Etruscan metal-workers in their granulation work of the 7th century B.C. This involved successive oxidation and reduction to produce an extremely thin layer of the copper-gold solder to join decorative granules to the surface of gold. In China, bronzes with abnormally high tin were sometimes used in a pasty condition (like plumber's wiping solder) to join together parts of bronze vessels. The commonest joining method used in antiquity was simply the "casting-on" process, in which a precast part was set into a mould and fused into the main casting, or else molten metal was run through a locally moulded channel to form a ring around parts to be joined. This technique was often used to join dissimilar metals, as in the casting of a decorative bronze handle onto an iron blade. From its earliest days, iron was joined by hammer welding at a white heat, a process applicable to no other metal. The blacksmith found that copper made a strong solder for clean iron, and he later used brass—that is, brazing solder.

The source of heat used in these operations was at first just a melting furnace or forge fire. Local sources of intense heat, such as the oxy-gas torch and the electric arc, appeared only in the 19th century. The former had a remote antecedent in the mouth blowpipe used from Egyptian times to urge a charcoal fire locally into incandescence; a small blowpipe was used in the 17th century, also, with an oil lamp flame for soldering jewelry and for localized heating in shaping glass, enamelling, and making artificial gems. Such blowpipes were worked with bellows in the 18th century. Immediately after the gases had been discovered, the intense oxy-hydrogen flame was used for scientific research. Platinum was commercially melted with an oxy-coal gas flame in 1857, but the oxygen torch did not become an important workshop tool for joining metals until the 20th century.

METALLURGICAL SCIENCE IN THE 18TH CENTURY

Although the principal improvements in production practice—the use of coke, the crucible and puddling processes, and cast steel—occurred in England, the main additions to understanding occurred in Sweden and France. In those two countries occurred the great events of the 18th-century "Chemical Revolution," in which oxygen was recognized and a new meaning given to the word "element," but it is interesting to note that the new-found understanding of chemical reactions had a background in practical metallurgy. For example, most of the reactions that appeared in the first tables (1718) of chemical affinity were, in fact, the reactions that assayers and smelters had been using to separate metals for some centuries. Similarly, the phlogiston theory (which held that burning objects gave off an immaterial substance called "phlogiston"), which inspired such fruitful chemical argument and experiment in the 18th century, was a direct outgrowth of the observation of spectacular physical changes that occur when ores are heated with charcoal. This theory was completely abandoned only when it was recognized that atmospheric oxygen and charcoal were needed not only to provide the source of heat in metallurgical operations, but that they were even more significant as chemical reagents.

Swedish chemists with a metallurgical background developed new methods of chemical analysis based on selective precipitation from aqueous solutions which far outstripped the limited range of capability of the old assayers. Analytical chemists soon discovered new metallic elements in old minerals and disclosed the presence of definite chemical impurities associated with the variable behavior of the products of different mines and smelters. Analysis provided the basis for a new scientific approach toward metallurgy, though the effect on practice was not to be felt until the next century.

A great triumph of the chemists was the recognition in 1773 to 1786 that carbon was responsible for the age-old distinction between wrought iron, cast iron, and steel. Back of it was the important work of the great French metallurgical scientist, R.A.F. de Réaumur, who in 1722 had developed a new type of malleable cast iron. He did this by first making castings of brittle white cast iron and then annealing them for a long time packed in iron ore or in bone ash. (Today we know that this resulted in removing the carbon entirely or converting it to graphite in the form of microscopic spheres that are mechanically harmless, quite different from the flakes in gray cast iron.)

Réaumur's work was inspired by his original theory that the differences between the different forms of iron were due to the presence of some "sulfurs and salts" in steel, which modified the structure. Though he was unaware that steel was microcrystalline in structure, he did study the change in its texture as revealed by the fracture and was quite modern in his belief that behind texture lay the key to the properties of metals. He also worked on the crystallization of metals from the liquid state and pioneered in studies of high-strength fibers and of crystallized glasses.

Also in the 18th century there were several systematic studies of some physical properties of alloys (including thermal expansion and especially density) and the beginning of the study of mechanical properties, though engineers were not yet ready to use strength figures effectively in design.

SUMMARY

The period from roughly 1600 to 1750 was one which saw considerable growth and improvements in all areas of metallurgy—mining, smelting, alloying, analyzing, and shaping metals.

At least as important as these metallurgical advances was the increasing pressure on the whole of metal technology from the spheres of economic, political, and social life during these same years. The rapid depleting of forests forced the use of mineral coal as a major fuel; it was used in the form of coke, which was also better adapted to use in larger furnaces. The increasing scale of warfare put a continuing strain on the founder to produce more, and more reliable, cannon. And, finally, the quickening of the Scientific Revolution, which itself had important roots in the arts and mysteries of the craftsmen, began to have increasing relevance to the improvements of those same processes.

Between the practical triumphs of the great metallurgical commentators—Agricola, Biringuccio, and Ercker, whose books are still ornaments of the 16th century—and the leaders of the chemical and industrial revolutions of the late 18th century, lie the years that witnessed the beginnings of modern metallurgical thought, but they are mainly noted for the practical adaptation of the older processes to production on an ever larger scale.

11 / Instrumentation

SILVIO A. BEDINI AND DEREK J. DE SOLLA PRICE

By the beginning of the 17th century, many of the tools and production techniques basic to the period 1600-1750 had already been established in major European centers. In fact, the great tradition of fine brass instruments produced for princely patrons and wealthy amateurs by skilled artisans of the south German states and of Flanders had already reached its climax and was on the wane by the end of the Thirty Years' War in 1648. Difficult living conditions during those years caused many of the German artisans to seek employment elsewhere, especially in France and Italy, where the names of German craftsmen in the major cities of both countries became increasingly prevalent.

A gradual shift in the need for scientific instrumentation, however, was noticeable from the beginning of the 17th century as a result of several causes. Primarily responsible for the shift was the changing practice of scientific research, resulting in the movement toward a wide variety of more precise standards of measurement and observation necessitated by the new emphasis on and interst in the study of natural philosophy and medicine. Other factors contributing to the change were the important new inventions that were being made and the re-examination of old inventions in the hope of adapting them to the needs of the times. Consequently, the change was not so much the result of the development of new tools and techniques but rather the reverse, for new tools and techniques were developed to keep pace with the new needs.

NEW INSTRUMENTS

The end of the 16th century witnessed a decline in the use of ancient and once-popular scientific instruments—sundials, astrolabes (computing devices for calculating the positions of stars), quadrants (for finding altitudes or angular elevations in astronomy and navigation), and armillary spheres (astronomical models showing the great circles of the heavens). As noted, new studies in natural philosophy and medicine brought with them the need for greater precision in observation and measurement, and attention was directed to old instruments which already existed and which could be revised to serve new purposes.

For example, the gradual disappearance of the medieval manorial system and the resulting changes in land ownership during the Renaissance exerted considerable influence on the science of surveying by the end of the 16th century, and the improvement of instruments kept pace with the change. Astronomical instruments of the past were modified for more practical purposes; from the dorsum (back part) of the astrolabe came the alidade and the circular angle scale which were mounted on a tripod as a socketed instrument for the use of surveyors in the field. When the ordinary magnetic compass was added for taking directional bearings, a new type of instrument was born. In England an instrument combining circumferentors in vertical and horizontal planes was developed by Leonard Digges in 1571; it became known as the theodolite and was used to measure horizontal and vertical angles in surveying. Although the theodolite was widely used in England and subsequently exercised great influence on the development and refinement of surveying instruments, it did not achieve the same degree of popularity in continental European countries. In France another form of the instrument, called the graphometer, was invented by Philip Danfrie in 1597. This combination was a direct adaptation of the torquetum, an astronomical instrument first introduced in the 13th century; the graphometer spread from France to other countries, and variations of it were soon being produced in Germany and Italy. The third form which followed in due course was the circumferentor, or Holland Circle, devised by Jan Dou in Amsterdam in 1612.

Similar changes took place in navigation instruments. The most commonly used of these was the magnetic compass, which—although undoubtedly of much earlier origin—was first described in the 14th century. The magnetic properties of the lodestone were known and applied in the 12th century, when a needle rubbed with the stone and suspended on cork or straw in a container of water—in conjunction with a compass card—was used by navigators to find their bearing. The compass continues to serve the navigator in the same manner today.

Another common instrument of the navigator was the forestaff or cross-staff, incorporating a linear scale, used by seamen to take the altitude of the sun or a star and thus determine the ship's latitude. First described in the 14th century, the cross-staff remained in use until superseded by the Davis Quadrant, or backstaff, which was introduced in the 1590's by Captain John Davis. This instrument was based on the principle of the cross-staff, but it enabled the navigator to sight on a shadow cast by the sun, instead of having to sight directly into its glare. The backstaff became standard mariners' equipment until replaced by the octant in the mid-18th century.

Another adaptation from the astronomer's tools was the "nocturnal," which was known in the Middle Ages but did not come into common usage until the second half of the 16th century. It was used to tell the time or the latitude at night by sighting on either the Little or the Great Bear (the Little or Big Dippers). To these must be added the mariner's astrolabe, a short-lived but well-known adaptation of the astronomer's astrolabe, made by the Spanish or Portuguese during the first half of the 16th century. It survived for about a century thereafter, the first scientific instrument designed exclusively for navigational use.

Attention was next directed to instruments for mathematical computation. The most important of the new types to be developed was the sector, a pair of hinged and graduated rules, which—based upon the theory of similar triangles—permitted a great variety of computations to be made. Among the early forms of the sector were those designed by the Englishman Humfray Cole and by the Italian engineer Federico Commandino, both in the second half of the 16th century. The latter ones were improved by Guidobaldo del Monte, and it was with this instrument that Galileo Galilei (1564-1642) first entered the field of scientific instrumentation. Realizing the sector's potential, Galileo perfected it for a greater variety of uses and popularized it among his pupils, disciples, and patrons. So great was the demand for his "geometrical and military compass" that he employed an instrument-maker to produce them on a full-time basis. The records indicate that at least 300 of these instruments were made and distributed by Galileo at Padua before he moved to Florence in 1610. Galileo's sector was capable of such diverse uses that it remained popular into the 19th century, long after other and superior instruments had become available.

The significance of Galileo's employment of a full-time instrument-maker for manufacturing on a commercial scale must not be overlooked. His only predecessor in this was Tycho Brahe, who, besides patronizing German instrument-makers, had also employed a craftsman at Hven to produce instruments for his observatory. These two examples did much to popularize instrumentation for serious scientific purposes as opposed to instruments designed for mere casual interest or diversion.

Another instrument to achieve popularity at about the same time was the hodometer (or odometer) which measures the distance travelled by a vehicle. Although the hodometer was attributed to Ctesibius (300-230 B.C.) and Heron of Alexandria (c. 62 A.D.), there is doubt that it was used in practical form before the late 16th century. The hodometer and kindred mechanical devices were popularized by the 16th- and 17th-century editions of Vitruvius' treatise on architecture, one of the most famous technical works of Classical antiquity.

Early use of the hodometer in Europe is attributed to Jean Fernel in 1528 and others. Perhaps the earliest sophisticated form in which a recording mapmaker combined the hodometer and compass appeared at the end of the 16th century and was described by Anselme Boetii di Boodt in 1609. Prior to this date Christoph Schissler, one of the foremost instrument-makers of Augsburg, had incorporated a magnetic compass with an hodometer, and another well-known maker, Thomas Ruckert, had devised a method of automatically recording compass bearings on a paper tape as early as 1575. The hodometer exerted a strong influence on the growing preoccupation with map-making in Europe during the late 16th and early 17th centuries; it reflected the growing importance of a new craftsman in the field of instrumentation, the mechanician and clock-maker.

NEW MATERIALS AND SKILLS

The new scientific instruments brought forth new artisans, working with other materials and different tools. As brass instruments were gradually augmented by new instruments made of glass, the skills of the turner (the lathe operator) and the glass-blower replaced those of the brass-worker and the engraver. With this change, the centers of instrument-making moved away from the traditional centers in south Germany and Flanders, and the accompanying skills were dispersed to other parts of Europe.

As the shift came about from brass to glass as the material for instrument construction, new skills were required. Consequently the instrument-makers became most closely allied with the other craftsmen whose techniques they needed for the new instruments. In mid-17th-century England, for instance, instrument-makers became closely associated with the spectacle-makers and clock-makers, as well as with cabinet-makers and joiners, glass-blowers and makers of measuring rules. In 1667 the major makers of mathematical instruments formally joined the company of clock-makers, and the makers of optical instruments soon affiliated with the company of spectacle-makers. Inevitably there was dissension between the various groups, and especially between the makers of telescopes and microscopes on the one hand and the instrument-makers who had joined the clock-makers' company on the other. Those who worked independently added, of course, another dimension of controversy. In addition, rivalry and friction arose between instrument-makers who worked in wood and those who worked in metal. These differences led to the increase of specialization in these crafts rather than the establishment of single shops producing a variety of instruments. The position of the instrument-maker in society changed accordingly. No longer allied to one or two patrons, they developed into dealers as well as makers, and served many instead of a select few. Their shops became centers where both amateurs and serious scientists could meet and exchange ideas.

The first glass instrument to receive the attention of the scientific researcher was the thermometer. Its earliest form was the thermoscope, first developed by Galileo and then improved by Santorio Santorio in the last decades of the 16th century. Basically it was a revival of a device invented by Philo of Byzantium in the 2nd century B.C. which Galileo modified for possible application to the practice of medicine. He experimented with the thermoscope to investigate the degrees of heat and cold by the application of a scale. Early in the 17th century, Santorio, a professor of medicine at the University of Padua, applied a form of the thermoscope for indicating the changes in the heat of the human body during illness and produced what was in effect the first clinical thermometer.

Others soon became interested in the thermoscope. Francis Bacon (1561-1626) in 1620 described a similar instrument to which a paper scale was applied. Some twenty years later Archduke Ferdinand II of Tuscany, Galileo's patron, developed the instrument further by using colored alcohol instead of water as the liquid in a hermetically sealed tube. The scale was divided on the glass by means of minute enamel beads. At least four types of these Florentine thermometers, as they were called, were developed for use in experimentation by the members of the Accademia del Cimento (one of the first great scientific academies, established in Florence in 1651). Their success was a triumph for the superb skill of the glass-blowers of Florence. It was not until the early 19th century, however, that the thermometer achieved its major importance as a scientific instrument of precision.

The barometer had a similar, but briefer, evolution. It was invented in 1643 by one of Galileo's pupils, Evangelista Torricelli (1608-47), to demonstrate the existence of atmospheric pressure. Several decades passed before the correlation between climate and atmospheric pressure was recognized, a development which added greatly to the usefulness of the instrument.

THE TELESCOPE

By far the greatest revolution in 17th-century scientific thought was brought about by the telescope. The origin of this instrument is obscure, and although there is evidence that its first form may have been developed during the last quarter of the 16th century in England or Italy, for practical purposes the invention is better ascribed to Holland during the first decade of the 17th century. The most popular contender for the honor of inventor is Hans Lippershey (?-1619), a spectacle-maker of Middleburg.

At first the telescope was considered to be little more than a toy but it was soon applied usefully to making observations at sea. It was for this purpose that the first telescope was brought to Venice in 1609 by an anonymous Frenchman; he hoped to interest the Venetian leaders in the instrument because of their city's noted maritime superiority. A report of the telescope came to Galileo's attention in June 1609, and he constructed one of his own. It was Galileo who first opened up an entire new world of science by turning the telescope skyward.

Galileo's first telescope—makeshift at best—was made with a tube of lead. His second, which he presented to the senators of Venice for testing, was an im-

Fig. 11-1. Lens-grinding lathe of Ippolito Francini. From a 1666 illustration. (*The Smithsonian Institution*)

proved model contrived from a rolled sheet of tinned iron plate, soldered at the edges and covered with crimson cotton. The tube was fitted with a concave lens at the eyepiece, and a concave lens was used for the objective. The lenses were undoubtedly ready-made and acquired by Venetian friends from the glass-makers of Murano. The success of Galileo's telescope was immediate, and he produced instruments for wealthy patrons from all parts of Europe as well as for the city government of Venice. He soon encountered serious difficulties, however, in production of the instruments.

The art of optical glass-making was still in its infancy, and Galileo had to rely on the mirror-polishers and glass-makers of Murano and Venice for his supply of lenses. Because existing methods did not produce glass of a quality consistently good enough for lenses, it was a hit-and-miss proposition, and the discards were far more numerous than the acceptable pieces. Lens-grinding was accomplished on the rudimentary lathes of the mirror-polishers, and it was a long time before satisfactory techniques were developed to meet the demand for suitable lenses.

The first major improvement in the production of optical lenses for scientific purposes was made by Ippolito Francini (?-1653), a young lens-grinder employed in the Medici palace workshops at Florence. During the last decade of

Fig. 11-2. Lens-grinding lathe invented by Guiseppi Campani, *c.* 1660-64. (From *Memoirs de l'Académie Royale des Sciences,* 1764. *The Smithsonian Institution*)

his career he produced lenses for Galileo's instruments. Francini modified the ordinary lens-grinding lathe into a more sophisticated machine tool which produced lenses of greatly increased accuracy. Another pioneer in the development of optical glass for scientific purposes was Eustachio Divini (1620-95) of Rome, who improved lens-grinding techniques and made substantial improvements in the lens-grinding lathe.

It remained for Guiseppi Campani (1635-1715) of Rome to develop the art of lens-grinding into a science, with the invention of a lathe which ground and polished lenses directly from the glass, instead of moulded discs. Prior practice had been first to cast the glass into moulds and then grind and polish the casting into final form. The quality of Campani's lenses and telescopes was superior to any others produced in Europe to the beginning of the 18th century. He experimented with and perfected telescopes of considerable focal length (and therefore high magnification) which were sought and purchased by amateur astronomers and scientists alike from all over Europe.

Long before Campani, Christian Huygens (1629-95) of Holland had produced lenses for telescopes, but his work was experimental in nature and not commercial. Although Huygens made many contributions to science, the quality of his lenses and telescopes never equalled those of Campani. Others attempted to emulate Campani's techniques, but without success. Robert Hooke (1635-1703), upon reading of Campani's invention of the lens-grinding lathe, announced a similar invention of his own, and published a drawing of it in his *Micrographia* in 1665. No record has survived, however, of any lenses produced with this equipment.

The telescope required other improvements, in addition to better lenses, during this early period. At first lenses were enclosed in thin, turned wooden cells with coarse threads, which were inserted into cardboard tubes covered on the outer surface with cloth, decorated paper, or leather. Later Divini and Campani both experimented with the production of large telescopes with tubes of wood, and several examples of their instruments with octagonal tubes have survived, but they are unwieldy. Nevertheless, telescopes with octagonal wooden tubes continued to be used at sea throughout most of the 19th century.

It was not until the last two decades of the 18th century that sheet brass was generally available for use in production of telescope tubes. The first one to employ this material successfully was John Dollond (1706-61) of London. Thereafter it was possible to produce telescope tubes of any diameter or length, and there came into being the instruments which are familiar to us today.

THE MICROSCOPE

There were similar problems with the compound microscope. Anton von Leeuwenhoek (1632-1723) in Holland and others had made progress in the grinding of small lenses for simple microscopes which permitted the examination of bacteria and spermatozoa. But the increasing requirement for precision in medicine demanded the development of instruments with greater adaptibility.

The compound microscope (using two lens systems to achieve greater magnification) had its earliest development in Italy, where Divini and Campani were among the pioneers who brought the instrument to a stage of greater usefulness. As with the telescope, the first compound microscopes produced in the mid-17th century were made of cardboard tubes covered with decorated cloth, paper, or leather. The inner tube was covered with marbled paper, presumably because it was smooth and permitted easier sliding. The lenses for the eyepiece and objective were encased in turned cells of hard wood, and the assembly was supported on a wooden or metal tripod. Focusing was accomplished with sliding tubes which permitted only coarse adjustment. The next step in the evolution of the compound microscope was the invention of the screw-barrel instrument by Campani some time after 1665. He replaced the sliding tubes with a turned hard-wood tube having coarse wooden threads, which permitted more precise focusing.

To summarize the individual contributions to the microscope's evolution, Divini first applied the doublet lens—consisting of two plano-convex lenses combined at their convexities—a focusing arrangement by means of sliding tubes, adjustment of focus between the objective lens and the object to be examined, and the first attempt to develop a screw adjustment for focusing. Campani's major contribution included the invention and perfection of the screw-barrel instrument, which permitted relatively fine focus between lens and object, and between the eyepiece and the objective; reduction in size of the instrument

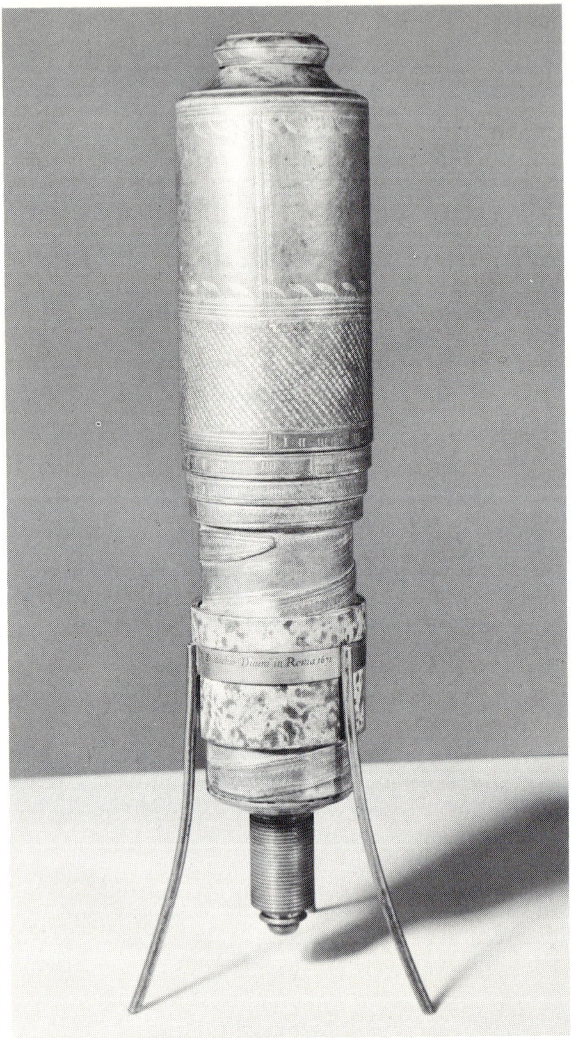

Fig. 11-3. Divini sliding-barrel microscope.
(*Instituto di Fisica*)

which made it easier to handle; and development of a slide-holder as part of the instrument, permitting the incorporation of a controlled light source.

As the compound microscope developed, increasing reliance was placed upon the turner, or lathe operator, for providing accessories as well as suitable mountings for the optical parts. Wood turnings were required for parts of telescopes as well as for microscopes, and for the construction of numerous parts of other scientific instruments which became popular during this period. One of the best known of the early microscope-makers was John Marshall (1663-1725), who had received his early training as a turner and later produced microscopes commercially. The compound microscopes produced in England during the second half of the 17th century, for instance, used cardboard tubes covered with decorated leather and strengthened at each end with large wooden or ivory turnings which housed the lenses. The tube was supported by a brass or wooden rod attached to a turned wooden base. A coarse threaded screw from a lateral brass arm permitted the instrument to be raised or lowered appropriately in relation to the object being studied. This type of support was further refined by the use of a ball-and-socket joint by which the instrument could be inclined to take best advantage of the light. This joint was a feature of the instruments produced by Marshall from the design of Robert Hooke.

Quite a different type of support was employed by Edmund Culpeper (1660-

Fig. 11-4. Campani screw-barrel microscope. (*Medical Museum of the Armed Forces Institute of Pathology*, Washington, D. C.)

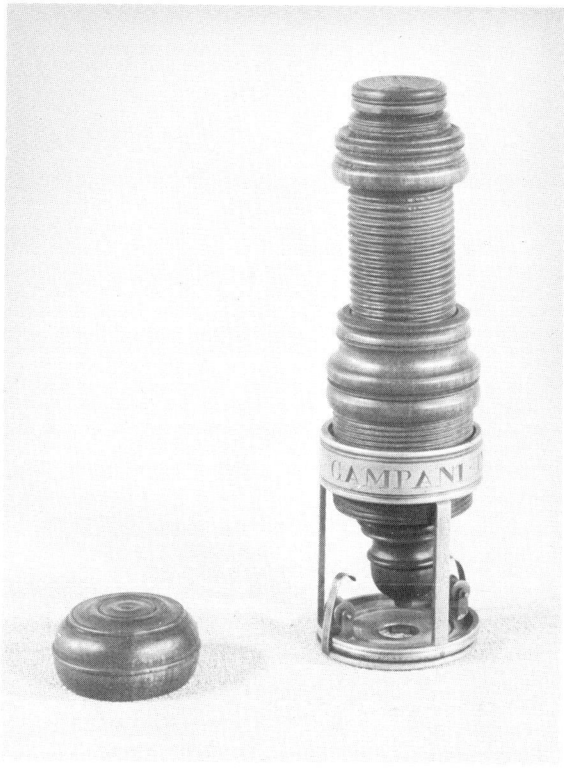

1738) and the later English makers who copied his instruments. Culpeper adapted the tripod support which had been a feature of the earlier Italian sliding tube microscopes, and he added a stage and a base with a tripod support made of turned wood. He later improved his instruments by making these parts of turned brass, but the eyepiece and the objective were still housed in wood turnings.

The basic elements and the appearance of the compound microscope evolved slowly with changes in construction materials. Early in the 18th century the Marshall type of instrument was being copied by Culpeper for John Yarwell, dealer in instruments, on a smaller scale and made of brass. The manner of attaching the instrument to a support was improved but led to the suppression of fine adjustment. These adaptations could only be accomplished with the skills of the turner in metal. The general design of the compound microscope during this period closely resembled a form of candlestick which was then popular, suggesting that both may have been produced by the same artisans.

The major faults of the compound microscopes of this period were mechanical rather than optical; although these faults resulted primarily from design features and the materials employed, the method of preparing and selecting specimens to be examined was also to blame. Precise scientific instruments were produced from such relatively fragile materials as cardboard and wood in this period because means for producing them in more permanent materials were not yet developed.

In both the telescope and the microscope the basic problem was weight, which was a major consideration with any instrument that had to be lifted, carried, or held aloft. Cast-brass tubes were being made in this period, but were unsuitable for the purposes required, since the weight was considerable and borings in brass were not accurate. Tubes could also be produced by soldering sheets of rolled brass or iron, but again the thickness of the metal would make the resulting instrument too heavy and cumbersome.

OTHER INSTRUMENTS

Two other major instruments developed in the 17th century were the air pump and the electrical machine. They were first produced by Otto von Guericke of Magdeburg. The vacuum pump was the first of the complex and large machines to be developed for laboratory use, beginning the long road to "big science" and its cyclotrons. Von Guericke's early experiments with the air pump were probably made before 1654, and the form of the instrument developed rapidly from a well-caulked cask filled with water evacuated by a brass pump to a copper sphere from which air was removed directly without first being replaced by water. In 1654 before the Imperial Diet at Ratisbon he performed an impressive and famous demonstration in which two carefully-fitted hollow bronze hemispheres were emptied of air. Eight horses harnessed to each hemisphere were unable to separate them until a cock was opened to permit air to enter the vacuum.

Von Guericke's air pump was considerably improved by Robert Hooke while he was working at Oxford in 1658-59 for Robert Boyle, one of the most competent scientific investigators of the 17th century. Boyle produced three versions of the air pump, each an improvement over its predecessor, and he and Hooke made numerous experiments with this equipment. Others who became interested in the air pump and made substantial contributions to its development included Christian Huygens of Holland and Denis Papin of France. By the beginning of the 18th century Francis Hauksbee (1687-1763) of London was producing air pumps commercially.

Another major contribution of von Guericke to scientific instrumentation was the frictional electrical machine which he developed around 1660 and first described in a book published in 1672. The original form of the machine was relatively simple, consisting of a globe of sulfur mounted on an iron axle between two supports. As the globe revolved, the operator's dry hand was pressed against the surface to generate static electricity in the sphere which would attract paper and other light objects. Others, including Isaac Newton, experimented with the electrical machine, and by the beginning of the 18th century the sulfur globe was replaced with one of glass and other improvements followed. Francis Hauksbee, formerly assistant to Robert Hooke, was the key figure in the continuing experimentation with electricity. Two other major contributors to the development of this device were Stephen Gray and Charles F. Dufay.

As new scientific instruments were developed in the 17th and 18th centuries, refinements were required for greater precision. Foremost among these was the use of the screw in many instruments for fine focusing and adjustment, as in the microscope, micrometer, and artillery instruments, as well as for sight adjustments on observatory instruments. The mandrel lathe and the ornamental turning lathe served prominently in advancing the development of screws with fine and precise threads.

As the methods improved for making scientific instruments of greater accuracy, it became possible to improve the design and the construction of the instruments themselves to provide greater precision in results. Features which are now common to many instruments—such as cross supports, accurate centering, and precision gearing—were the result of long development and slow adaptation. The application of the telescopic sight as a component of surveying and navigational instruments was another important improvement, deriving from the success in measurement and observation achieved in the use of the astronomical telescope.

THE CLOCK-MAKERS

The need for increased precision in other parts of instruments was partly met in the shops of the clock-makers, whose contributions were fully as great as had been those of the founder and the turner in the past. Whereas earlier clock-makers had relied most heavily on handicraft, by the beginning of the 18th century machine-tool methods were introduced to meet the increasing market for timepieces and instruments.

The application of the pendulum to clockwork in the mid-17th century resulted not only in more accurate time-keeping, but led to the improvement of the mechanics of clockwork. The clock-maker evolved from the guild artisan to a mechanician as his trade imposed more rigid requirements and increased skill. With the advent of the pendulum clock, new escapements were devised, and gear teeth had to be cut with greater precision and to better design. The superiority of epicycloidal teeth was researched and demonstrated for the first time by Ole Roemer and Christian Huygens in 1674-75. Further studies initiated by Camus in 1735 and by Thiout in 1741 led to improvement of clockwork and its application to scientific instrumentation.

This change was not rapid, and it depended on the improvement of the tools and the techniques of the clock-maker. The simple "turn" held in a vise, which had served the clock-maker for centuries, was replaced by a more elaborate and adaptable tool early in the 18th century. Another natural development was the fusee engine, made entirely of metal and having a lead screw and gears to control the cutting tool which produced a spiral thread. By the second half of the 18th century the manually operated in-feed was replaced with an automatic feed in a fusee engine designed and developed by 1763, and illustrated by Ferdinand Berthoud. This machine was, in effect, a specialized all-metal lathe for producing precision metal parts by means of adjustable automatic mechanisms.

THE LATHE

One of the most basic tools which was adapted for new uses during these years was the lathe, which played a significant role in the evolution of other tools and of precise instrument-making techniques. The earliest form of the lathe, powered by a spring pole and treadle drive, had developed into a far more useful tool by the 16th century; additions included a continuous drive mechanism, a mechanical tool holder, and a live spindle (that is, the shaft-like part which rotates while holding the piece to be turned on the lathe). In the next century the lathe acquired a sturdier bed and stocks which extended its utility. The transition from the larger and less accurate type of lathe for cutting screw-threads, delineated in Leonardo da Vinci's drawings, to the more sophisticated

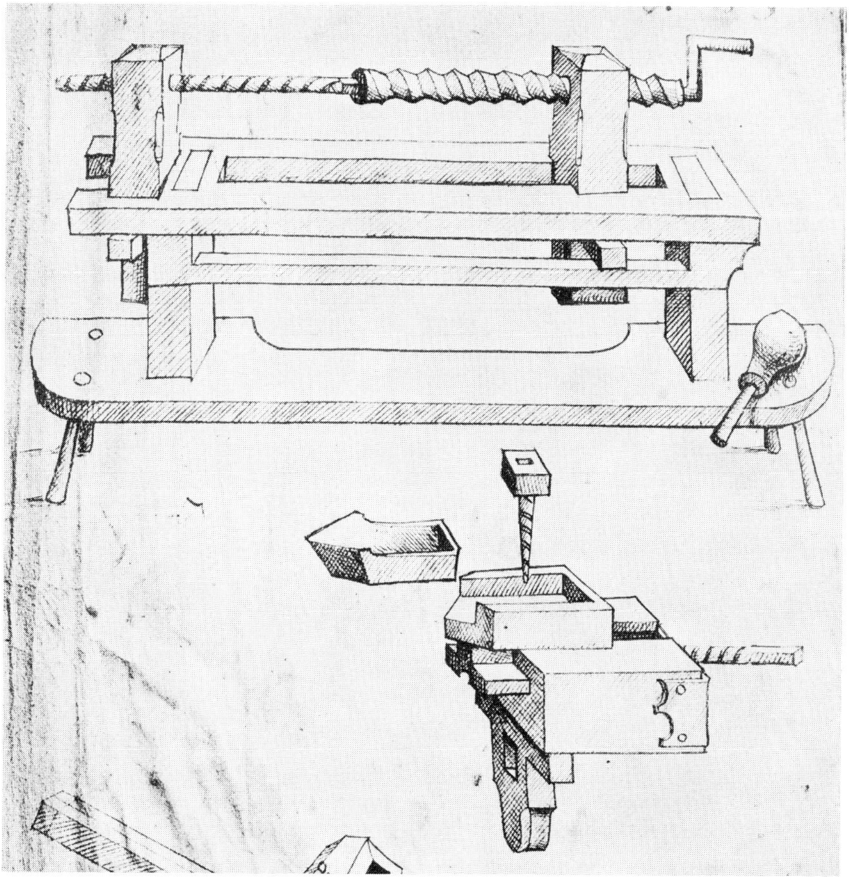

Fig. 11-5. Screw-cutting lathe with cross-slide, 1480. (From *Mittelalterliche Hausbuch*, c. 1480. *The Smithsonian Institution*)

form can best be seen in the works of Jacques Besson, engineer at the court of King Charles IX of France.

Besson is credited with the development of the lathe for ornamental turning, and his book, published at Lyon in 1569, illustrated and described this type as well as an early screw-cutting lathe. Clearly apparent in Besson's drawings is the progress made in the drive of the workpiece as well as in the longitudinal feed of the tool.

The importance of lathe developments in this period cannot be overstressed. The lathe was of great significance, not only in the production of 17th-century scientific instruments, but also for the turning of parts and for its use in developing the screw for scientific instruments and for other tools. Besson was largely

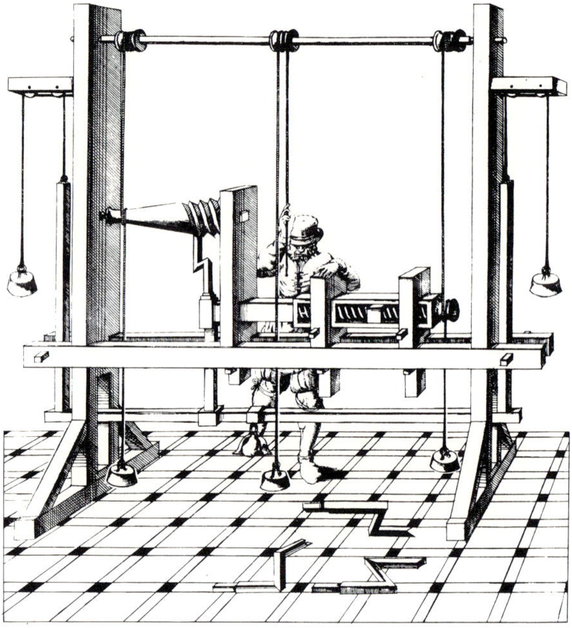

Fig. 11-6. Screw-cutting lathe by Jacques Besson.
(From a French edition of 1578. *The Smithsonian Institution*)

responsible for the development of the lathe along two separate lines of evolution, both of which were important in the new technology of instrument-making. On the one hand, as the screw-cutting lathe was improved and finally evolved into the industrial lathe, it had great impact on the production of other machine tools. On the other hand, the ornamental turning lathe, which had been primarily a plaything of the wealthy, developed into a tool of critical importance to the Scientific Revolution of the 17th century.

Most important was the introduction of cams and templates (patterns, usually in the form of thin plates), which permitted more intricate motions in the production of the workpiece as well as increasing the accuracy. These made feasible the control of the tool as well as of the workpiece, and mechanical control of the tool became an important factor in precision work accomplished by the industrial lathe.

After Besson, the next lathe improvement was recorded by Salomon de Caus, engineer to Henry, Prince of Wales, and later to King Charles I. De Caus illustrated and described an oval turning lathe in a book published in 1615.

Although metal had long been worked on wood-turning lathes, using different tools, metal-cutting as a specialized lathe technique was described for the first time by Charles Plumier in a work which appeared in Lyon, France, in 1701. Both the method and the tool were crude, compared with later, more sophisticated versions, but Plumier's machine is nevertheless significant. The turning of iron on a lathe was a rare art at this time. Plumier mentioned that he knew only of two other men capable of turning satisfactory mandrels of iron on a lathe.

Although the screw was known in early times, it had limited application until

Fig. 11-7. Metal-cutting lathe by Charles Plumier, 1701.
(*The Smithsonian Institution*)

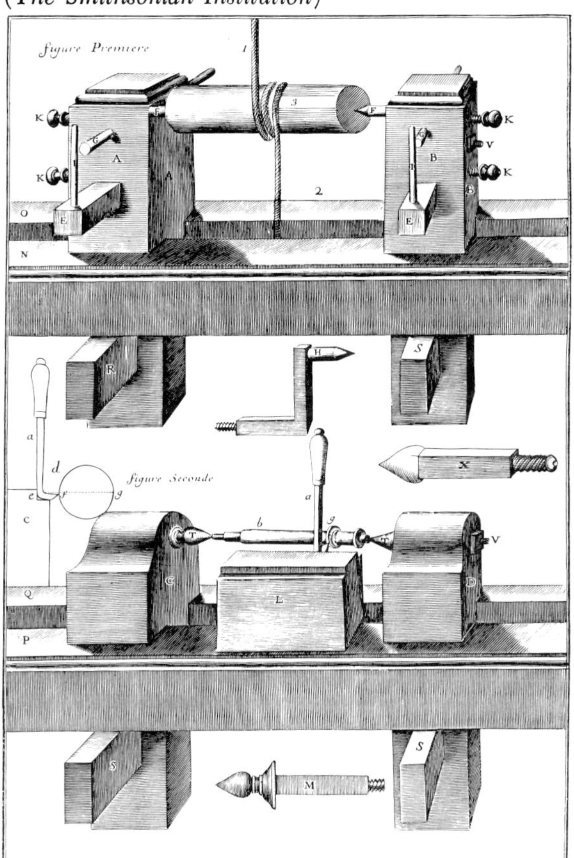

the 16th century because it lacked precision. Screws with both fine and coarse thread, for example, were produced in wood and in metal in the 16th century, but they were not sufficiently refined and accurate for application to instruments of measurement.

Despite the availability of lathes such as Besson and Plumier described, they were not precise enough, and both metal and wood screws continued to be fashioned by hand with chisel and file. Because these hand-made screws were expensive, however, an alternate method was sought and finally found in the mandrel lathe.

Screws produced on the master lead-screw type lathe had wide application in the printing and minting crafts and in tools such as vices, where precision was not essential. The earlier press screws made of wood were replaced by copper press screws, as seen in a Nuremberg printing press produced around 1550 by Danner, and they were later made of iron. Similar screws of much smaller size were occasionally used in scientific instruments. Steel fixing screws for assembing armor in the late Middle Ages were being applied to scientific instruments in the early 17th century. The development from these crude screws found in the fittings of 16th-century armor to sophisticated adjustment screws can be easily traced. Adjustment screws were first introduced late in the 16th century by the Danish astronomer Tycho Brahe (1546-1601) on some of the astronomical instruments in his observatory at Hven. The earliest examples of fine precision adjustment screws were used by Christoph Trechsler (*fl.* 1571-1624) of the Dresden armory. He applied the threaded screw as a micrometer device in a gunner's level dated 1614, which has survived.

PRECISION SCIENTIFIC INSTRUMENTS

Machine tools for the production of screw threads and precision parts for instruments evolved simultaneously with the advancement of clock-makers' techniques. The clock-maker's lathe for screw-cutting which was illustrated in 1701 in Charles Plumier's work on the art of turning, employed a sliding spindle with a set of master threads. The same mechanism was constructed by Thiout in France just before the middle of the 18th century.

The next major advance came with the development of the dividing engine for graduating scientific instruments. In France, the duc de Chaulnes (1714-69) attempted to produce a mechanism for graduating the circle by mechanical means at the same time that Jesse Ramsden (1730-1800) was developing a machine for the same purpose in England. While the work of Chaulnes was experimental, Ramsden produced tangible results with his dividing engine of 1763 for the accurate subdivision and graduation of circles. A side product of Ramsden's accomplishment was his invention and construction in 1777 of a special screw-cutting lathe to produce the extremely accurate tangent screw re-

Fig. 11-8. Ramsden's circular dividing engine.
(*The Smithsonian Institution*)

quired for his dividing engine. Ramsden's engine revolutionized the production of precision instruments and greatly improved the accuracy of observation in many sciences. With this new tool it was possible to make astronomical observations of longer duration, using clock-driven telescopes of improved design and accuracy. The dividing engine could also be used for the graduation of scales on surveying and navigational instruments.

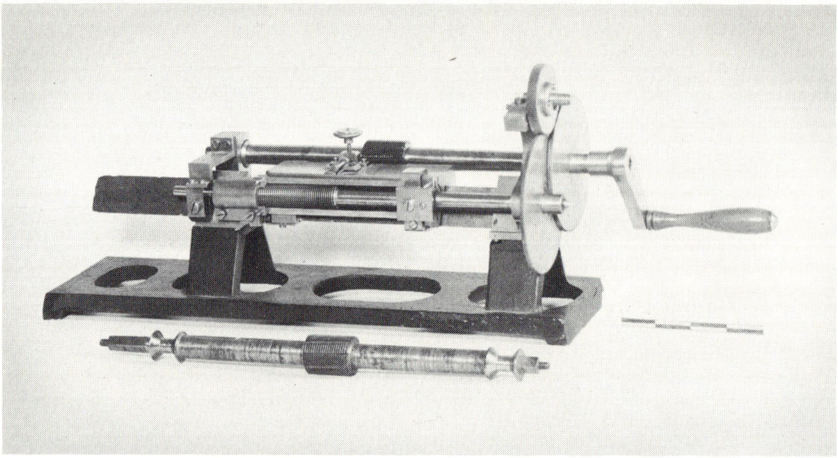

Fig. 11-9. Ramsden's precision hobbing or screw-cutting machine. (*The Smithsonian Institution*)

CENTERS OF PRODUCTION

It was not until late in the 16th century that the craft of the instrument-maker was transmitted from Flanders and the south German states to other parts of Germany, to France, England, the Low Countries, and Italy. In each of these centers the craft grew around a master or school, and sometimes it was the result of encouragement by a wealthy patron, but the spread of the craft was not accidental.

For generations, Nuremberg had remained the traditional center of the instrument-maker's craft, largely because of the high quality of the instruments produced by Regiomontanus (1436-76) and by succeeding master makers of that city. Nuremberg was subsequently rivalled by Augsburg, partly as a result of Tycho Brahe's numerous important orders for special instruments to equip his observatory at Hven, and partly because of the fine work produced by Christoph Schissler the Elder (*c.* 1561-86) and his son.

During the same period instrument-makers and mathematical practitioners responded to the patronage of Emperor Rudolph II (1552-1612) at Prague. Most notable among these craftsmen was Erasmus Habermehl (?-1606) a native of southern Germany, who produced many ingenious instruments of original design and with superb engraving. Thomas Gemini (alias Lambritt, *fl.* 1524-62) who had been trained in the famous center of the Arsenius family at Louvain (Belgium), came to England for refuge. Like many other instrument engravers, he was a maker of plates for printing books and maps. It was Gemini who produced the pirated edition of Vesalius. At that time sheet brass was first being produced, and he initiated the use of this material in a great craft tradition continued by Humfray Cole (1530-91) and other outstanding English artisans who followed, often in master-apprentice succession.

CHANGING DEMANDS

With the changing scientific scene in the 18th century, the demands made on the instrument-maker became greater and greater, not only for the production of many different devices for the practitioner, but for the construction of precision tools for the researcher and apparatus for physical experiments. By this time the vacuum pump and the electrical machine had become standard laboratory equipment, and these were followed by stills, thermostats, and similar laboratory equipment. The active scientist required new instruments which had to be constructed to his own specifications for the accomplishment of special purposes. This necessitated increased skill of the instrument-maker as well as greater flexibility and accuracy of his tools. Another need that found expression during this period was for instruments which could be combined for a variety of purposes. Finally, as the research front advanced, a need arose for the standardization of instruments and their components. This could be accomplished only through the combined skills of numerous specialist craftsmen to produce the ultimate equipment and apparatus. This new experimental laboratory equipment and apparatus involved new principles of design and construction, and called upon the instrument-maker to devise tools, machines, and techniques which proved to have great impact on other aspects of the Scientific and Industrial Revolutions.

12 / Transportation and Construction, 1300-1800
The Rise of Modern Civil Engineering

JAMES KIP FINCH

The awakening of trade and commerce and of town life in the late Middle Ages had revived the need for transportation and construction, and a new generation of workers emerged to carry forward the services of the ancient Roman *architectus*. Little is known of the remarkable group of gifted craftsmen who first met these needs, but during the following centuries their successors were again to attain professional status. At the same time, broadening activity was to lead to increasing specialization and ultimately to a division of interests. Three countries played a part in this evolution of an ancient profession. The change began in the Italy of the Renaissance, passed to French leadership in the 17th and 18th centuries, and following the French Revolution, was carried forward by Britain.

THE ITALIAN RENAISSANCE

The Italian Renaissance is usually thought of as marking man's discovery of his past and a turning back, especially to earlier Roman culture, for inspiration and guidance. Actually it was a strange blending of the old and the new in a land lacking any political unity, torn by constant strife between the military bosses who temporarily controlled its northern city-states, and subject to foreign invasions that made it one of the major battlegrounds of Europe. Nevertheless, increasing urban and commercial needs demanded attention, and even the military struggles of the day stimulated, as war sometimes does, technical advances. Exploration was to reveal a new world, and natural science, throwing off the shackles of earlier Greek philosophy and of religious inhibitions, turned to observation and experiment. It has, in fact, been suggested that the diversified life of the day stimulated rather than retarded man's search for improved means of meeting pressing material needs and wants.

The major technical advances of the period were made in its earlier years. It was not until the so-called High Renaissance of the 16th century that the "Classical revival," which was to have such a profound and long-lasting effect on art, architecture, literary interests, and learning, became the dominating influence of the day.

THE ARTIST—ARCHITECT—ENGINEER

Few of the earlier Renaissance workers are known by name but they apparently were, as in the Middle Ages, gifted craftsmen, capable planners, and expert practical builders. Even as late as 1420 the designer of the great dome of the Cathedral of Florence directed its construction himself. About 1415, however, the papal secretary, Poggio, discovered a manuscript of *De architectura,* the famous treatise by Vitruvius, the Roman architect of about 25 B.C. First printed in 1486, this became the architectural bible of the day, and interest turned increasingly to art and decoration inspired by the achievements of the past. The outstanding designers of later years almost invariably began their careers as assistants to painters or sculptors. This led to an increasing division between the designer and the practical builder. The artist-designer left not only the task of carrying out his plans but many structural and constructional problems to the *campo mastro*, or construction superintendent. Architecture began to be regarded as a "fine" rather than a "practical" art, thus associated with painting and sculpture, but with little interest in manual skills and construction techniques. Nevertheless, the designer and builder still remained one person; he was artist, architect, and engineer all rolled into one. He was an independent worker, regarding his skills as part of his stock-in-trade, and offering his services on a competitive basis to a prospective client, or rather patron. This client was normally one of the political-military bosses of the day, but sometimes it was a city requiring a bridge or other public work.

BRIDGES

For many of the bridges needed by the revitalized cities of Italy, Renaissance builders abandoned the semi-circular arched form of Roman times. The large piers needed to distribute the load of this type of bridge obstructed the flow of the rivers, and as a result, closed almost two-thirds of the natural flood channel. Such bridges became, in effect, dams with holes in them. Floods greatly increased the velocity of flow through their restricted openings; and the scouring action which followed tended to undermine the shallow piers.

Renaissance builders therefore adopted the segmental, circular arch of larger radius which eliminated much of the obstruction to the water flow. Two especially notable bridges of this form, built by gifted mason-craftsmen, survive today: the Ponte Vecchio, or "Old Bridge," over the Arno River at Florence (built in 1341-45) has spans of about 96 feet with only a 15-foot rise, a ratio of rise to span of a little over 1:7; and the Ponte Castelvecchio (1354-56) over the Adige at Verona (destroyed by the retreating Germans in World War II but since carefully restored). This latter bridge, of somewhat greater rise, has a major span of 160 feet.

Since the thrust of an arch more or less follows its curvature, these segmental spans reduce both the dead weight of the arches and the vertical component of their thrust, and thus the load to be carried by the piers. Those of the Ponte Vecchio are notably slender—under 20 feet—still further increasing the waterway for flood flows. Later, artistic considerations led to the abandonment of this form of arch.

The Ponte Santa Trinita of 1566-69, just below the Ponte Vecchio, is a "basket-handle," or elliptical arch. The designer was a well-known architect of the day, Ammanati, whose work reflects in its deeply moulded arches, panelled piers, and scrolled keystones, the decorative flair of the High Renaissance. (Also destroyed by the Germans in World War II, it has since been carefully rebuilt.) This elliptical form was regarded as more pleasing, for a smooth curve joined arch and pier, thereby avoiding the abrupt angle of the segmental form. The segmental arch does appear in a few later works, such as the famous Rialto bridge of 1588-91 in Venice; but the compromise form, the elliptical arch with

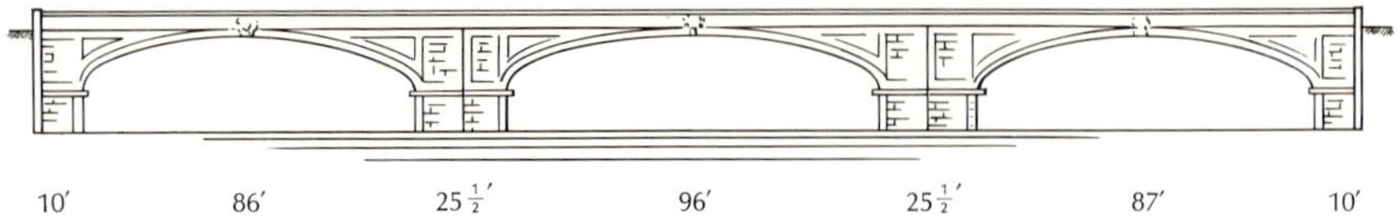

Fig. 12-1. The Ponte Santa Trinita, Florence, 1566-69. Note the "basket-handle," or elliptical arch.

piers about one-fourth of the span in width, became practically standard, and the segmental form was not revived until late in the 18th century in France.

BRIDGE CONSTRUCTION

In the area of bridge construction, as compared to design, no notable advances either in methods or equipment took place during the early years of the Renaissance. As in earlier times, cofferdams and piles were employed for foundations. There appears to have been some improvement in using enclosing walls of sheet-piling, but the builder was still limited to the man-operated pile-driver, pump, and hoist of Roman antiquity. Not until the advent of steam-driven equipment in the 19th century was it possible to found bridge piers more than 8 or 10 feet below water level.

THE BRIDGE TRUSS

A later and very important development in bridge construction—the truss—was carried over from work on buildings. The truss in its simplest form undoubtedly was used in the roofing of buildings as early as ancient Greek times. Early trusses built of wood have perished, but the use of a horizontal tension tie to take the outward thrust of two sloping rafters must have been adopted early. In the roofing of the Gothic cathedral and other later medieval buildings, other members were added, frequently of little structural value and reflecting, it is said, wooden ship-building practices. A remarkable early metal truss of bronze had also been used in covering the portico of the Pantheon in Rome about A.D. 130, but it had been melted down to make cannon in 1625. It was not until the late Renaissance, however, that the truss bridge appears in the famous *Four Books on Architecture* of Andrea Palladio (1518-80), published in Venice in 1570.

Palladio does not claim to have originated this form and notes that similar works had been seen in Germany by a friend. Earlier timber bridges were usually simple beams supported by timber pile piers. Palladio's simplest bridge, that over the Brenta near Bassano, had such piers supporting beams of about 34-

foot span, but with the added support of strut braces at each side sloping upward from the single bent piers. A later work over the Cismone, however, was a true "through" truss of quite modern form, thus providing a clear waterway below the floor level for the passage of flood flows and debris—solid, beautiful, and convenient. Both these forms reappear, notably in later Swiss bridges, but it remained for early American engineers of the 19th century, working in a land of plentiful timber but limited funds, to carry forward the development of the truss bridge.

THE DOME

The Renaissance also witnessed the development of another notable construction form: the dome. The dome had earlier been developed to a high degree by the ancient Romans. The Pantheon in Rome (built *c.* 130 A.D.), with its great dome 140-feet in diameter and with its 28-foot circular opening at the top, still remains one of the world's earliest and most outstanding domed structures. True domes exert a radial outward thrust; in the Pantheon this thrust was resisted by massive low encircling walls.

A notable advance in retaining the outward thrust of a dome was in the great church of St. Sophia (Sancta Sophia), built in Constantinople in 537 A.D. There, an interlocking stone course at the base of the dome held its thrust and made it possible to raise the dome to a great height while using only relatively slender supporting walls. Italian builders were to carry forward dome design,

Fig. 12-2. Palladio's Bridge of Cismone. A true timber-truss bridge that contains all the elements of later designs and, unlike the stone arch, left a clear waterway below for the passage of floods and debris.

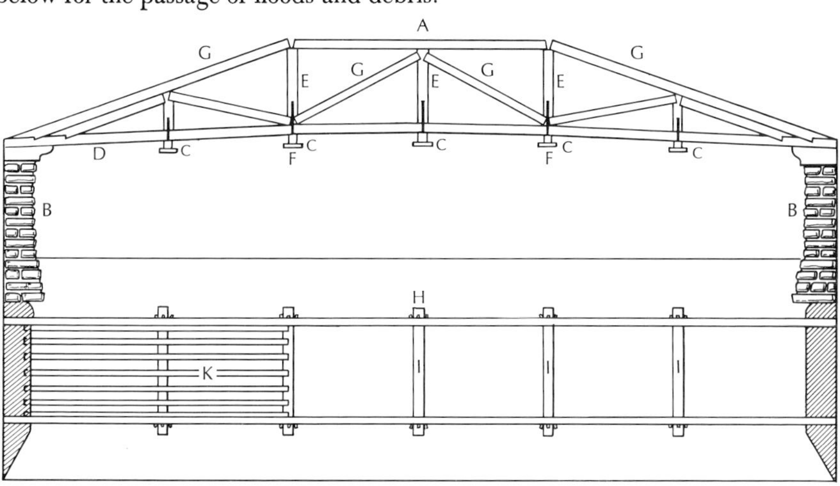

but it was almost 900 years before the next great advance took place in dealing with the problem of the thrust exerted by a dome.

In 1420, Brunelleschi (1377-1446), one of the great artist-sculptor-architects of the Renaissance, began the dome of the cathedral of Santa Maria del Fiore in Florence. Construction of the cathedral had begun in 1296, and it still stands as one of the few examples of Italian Gothic. It has a great octagonal, pointed, ribbed dome 143 feet in diameter and 105 feet high, with its base 200 feet above the cathedral floor. The dome's lower portion is of solid stone but the upper is a double brickwork shell, the inner part supporting a lantern (that is, a tower placed at the apex of the dome). The lower stones were bound together with iron cramps (pieces of iron, bent at the ends, to hold blocks of stone together). Header stones tied to the upper brickwork, and so-called cinctures of wood, about one foot square, bolted and with an encircling iron band, were also used. Its great height, however, made its thrust largely downward rather than outward, so these cinctures were probably not needed.

The dome of St. Peter's Cathedral in Rome came over a century later. Having the same diameter but being lower than that of Santa Maria del Fiore, its dome exerted more outward thrust which, in this case, was met by wrought-iron bars which hooped the dome.

Although domes are interesting and impressive structures, they are of architectural rather than engineering interest, and are not widely used in modern construction. Nevertheless, domes are among the greatest works ever erected by man.

CANALS

Some of the many bridges built during the Renaissance were erected to span the canals which were increasingly coming into use. Canals had been dug in ancient times, serving chiefly for irrigation and domestic water supply; canals primarily for navigation date only from the 12th century. The canal slowly evolved to become of vital importance in the days before the advent of the railroad, when road traffic was limited and waterways were the major transportation arteries of trade and commerce.

First exploited in Northern Italy, navigation canals were based on a sequence of developments rather than a single invention. In 1167, Milan and other northern cities organized the Lombard League for mutual protection, and new city walls were built at Milan. Water, secured for a wide moat from nearby streams, was soon used for irrigation. The need for more water followed, and a canal was built some 16 miles to tap the Ticino River. First known as the Ticinello, it was used to bring building stone to the city; as it was extended and put to more uses, it was in due course called the Naviglio Grange, and became the first of a network of canals in the basin of the Po and its tributaries.

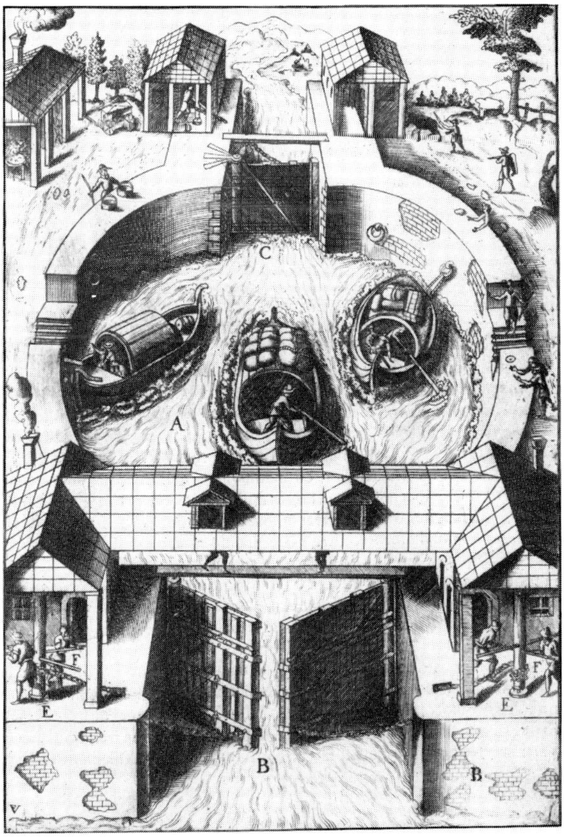

Fig. 12-3. The canal lock in its developed form of the 15th century, showing both single leaf and mitering gates.

This increasing use of canals for transportation made necessary the development of the canal lock, which although sometimes hailed as the greatest hydraulic invention of the ages, appears also to have been the result of evolution rather than pure invention. The early Italian canals were primarily of the slack-water type: that is, barriers were built at intervals along a stream to hold back the stream flow and create slack-water of sufficient depth to float boats. In earlier Italian works, as in China, these barriers were temporarily removed to permit boats to pass from one level to another, upward movement against the current being made by haulage ropes.

The first improvement, dating from the 15th century, was the use of gates to hold back the flow. The term "lock" refers to such gates. Spacing two such locks closely enough to form a boat basin, or lock chamber, permitted a boat to be raised or lowered by simply filling this chamber from the upper or emptying it to the lower level. Vertical lift-gates were used early as locks, and single gates turning horizontally like doors on vertical hinges were also adopted. In order to permit the easy opening of these horizontal locks, a small gate was provided in the main gate; the small gate could be opened first in order to relieve the pressure on the main gate. The earliest description of such a canal lock appears in 1485 in the first printed book on architecture, *De re aedificatoria,* by Leon Alberti (1404-72).

The next step was the double mitering gate, one leaf swinging from each side, meeting at an upstream angle, supporting each other and reducing the span and size of the gate. Illustrated by Leonardo in his famous Notebooks, this is usually credited to him and is probably his major engineering contribution; for, in spite of his fascinating mechanical and other speculations, his notebooks did not become available until much later, and few actual engineering works can be attributed to him.

MILITARY ENGINEERING AND THE EARTHWORK FORT

The term *ingeniator* had come into use for the early workers who built the ingenious fortresses and engines of war of the feudal era. In spite of diverging interests, what we know today as architecture and civil engineering long remained practically one profession. However, the new interests and techniques following the advent of gunpowder ultimately led to the recognition of military engineering as a specialized branch of the profession. It was not until the mid-18th century that John Smeaton in England adopted the title civil engineer (that is, one concerned with civilian projects) to distinguish his interests from those of the expert in military construction.

Gunpowder and cannon, first used in battle in France at Crécy in 1346, slowly brought to an end the age of armor, the medieval castle, and the walled city. When the first printed technical book, *De re militari* by Valturius (1413-83), appeared in 1472, it covered the techniques and equipment of the pre-gunpowder era such as catapults, as well as containing material on cannon. The famous *Trattrato di architecttura* of Francesco di Giorgio Martini (1439-1502), a contemporary of Leonardo da Vinci, came later. It is a transitional work, a mixture of classical architecture with more modern fortress plans. Martini is usually credited with originating the star-shaped, bastioned fortress of earthwork and low walls which replaced the earlier fortress and city walls.

Another work, *Della architecttura militare* (published in 1599) by Francisco de Marchi (1504-77), completed the divorce between architecture and military engineering. Its author, a practical military engineer who lacked both the classi-

Fig. 12-4. The bastioned earthwork fort. Gunpowder put an end to the stone medieval fortress and walled city and led to the development of the star-shaped fort with cannon emplacements. (This illustration is one of over 150 such plans from de Marchi, *Della architecttura militare,* 1599.)

cal training and aesthetic interests of the architects of the day, devoted this book to the techniques of the new era. Over 150 plates with full *expositioné* describe the selection of sites and fortress plans for mountain locations, for island and seashore protection, and for towns on plains. Their design provided inclusive lines of fire for cannon and defenders and the avoidance of dead areas (that is, areas where the defenders could not aim their weapons effectively against attackers).

The impact of the new techniques of military engineering was also reflected in the first printed books on surveying, which were devoted to military problems. Likewise, the making of cannon stimulated metallurgical developments, while the problems of trajectory gave rise to new scientific speculations. Italy, however, became a major battleground of Europe, and for further advances in transportation and construction and what became known as civil engineering, one must turn to France and her great period of leadership in the 17th and 18th centuries.

THE FRENCH PIONEERS OF MODERN ENGINEERING

By the close of the 17th century the French nobles had become mere courtiers, and Louis XIV could claim, "I am the state." In sharp contrast to the political disunity which had characterized Italy during the Renaissance, the new unity of France made it possible for engineering to be pushed forward under an absolute monarch, and for the foundations of modern civil engineering to be established in France.

Characteristically, limitations of resources and the various political and economic interests of the nation determined the scope and character of French works. Her workers still relied, as had those of the Renaissance, on timber, stone, and brick for materials, and on man-powered equipment. Coal and other resources that later furthered Britain's industrial development were not exploited. Foreign trade also was a minor factor in a largely self-contained and self-sufficient agricultural economy. In contrast to the dominant role private enterprise was to play in the development of British engineering, French works were primarily governmental or public undertakings and remained so even in the 19th century. The kings of France took a particular interest in palace-building, and in military activities, which involved the construction of roads and bridges for troop and other movements. French engineering advances therefore centered primarily on construction and transportation.

PONTS ET CHAUSSÉES

There had been little major construction during the Hundred Years' War, but a new generation of builders emerged in the 16th century and, gaining experience through practice, built a few notable works. In the 17th century, during the long reign of Louis XIV (1643-1715), more construction was undertaken, and as in Italy, a number of independent architect-engineers emerged. By 1672, Vauban, the famous military engineer, who had served as a personal consultant to individual commanders, suggested the organization within the army of an engineering corps (*Corps du Génie*).

Louis XIV's famous finance minister, Colbert, besides trying to advance the economy of France, also helped organize its cultural interests. With his support an Academy of Sciences was created in 1666 "to investigate Nature and perfect the arts." An Academy of Architecture followed in 1671. Colbert also was the first to suggest the term *ponts et chaussées* (bridges and roads) for an engineering organization. By the close of the 17th century the growth of governmental construction activities and the need for organizing their planning and direction led to the founding in 1716 of the famous Corps des Ponts et Chaussées. The creation of this "national highway department" was a notable dividing point in French engineering.

The new organizational lines frequently crossed: architecture, for example, as in the past, remained closely linked to engineering. Many engineers of the Corps des Ponts et Chaussées were also members of the academies of architecture and of science. In their earlier years these groups had given some attention to engineering problems, and scientists had been called upon to advise on some engineering projects; but the establishment of the Corps brought together under one head a group of noted engineers with opportunities for specialized conferences and exchanges of experiences, and was a notable step in securing professional recognition. Most of all, it provided the essential organization for the remarkable growth of works in the 18th century which made France the leading engineering nation of the world. Under the Minister of Finance and headed by its First Engineer, with a staff of *inspecteurs-généraux* contacted by engineers-in-chief in each province, the Corps assured effective design, direction, and supervision of governmental engineering works.

ENGINEERING AND NATURAL SCIENCES

Throughout the 18th century there was increasing interest among engineers in developing more exact, more fully quantitative, mathematical understanding of the problems of design to supplement the earlier reliance on judgment, experience, and "rules-of-thumb." The liaison between science and engineering, which was to become so important in the 20th century, was beginning to develop, however imperfectly.

With few exceptions scientific interests centered in astronomical and other "pure" mathematical pursuits. Gautier in his pioneer text on bridges (1716) complained that *les savants* (the "scholars") had no interest in the mechanics of arches, while the architects were absorbed in the aesthetics of design. In certain respects the engineering needs and resources of the day offered few strong incentives for scientific endeavor. After all, there seemed little point in striving for more exact knowledge which would lead primarily to a saving of materials that were already relatively plentiful and inexpensive. On the other hand, a practical, useful technique of design required not only an understanding of the mechanics of such constructions as beams, columns and arches, quays, locks and retaining walls, but also an appraisal of the forces involved and a testing of materials to determine the safe loads to which they could be subjected. Hence these problems received increasing attention from those engineers who sought to perfect their art.

A military engineer and outstanding writer of the day, Belidor (1697-1761) published *La Science des Ingénieurs* (1729) in which he took up the mechanics of the simple beam, earth pressure, and retaining walls. (By the close of the century Coulomb [1736-1806], another military engineer whose name is recalled in the electrical unit named in his honor, had corrected and clarified these earlier ideas.) The great engineer, Perronet (1708-94), devised a testing machine to aid in selecting the strongest building stone to resist the high thrust of his flat arches. In 1789 Girard published the first text on the *Résistance des Solides*. Thus, modern structural mechanics, a blending of theory and practice, was born.

Similarly, French interest in canal and water supplies led to notable advances in practical hydraulics. In the later 17th century it had been shown that the velocity of water flowing from an orifice was substantially the same as that of a body falling freely through a height equal to the head, or depth of water, at the opening. Scientists surmised that the velocity of flow in a canal or open channel similarly increased with the depth of flow. In 1730 Pitot (1695-1771), who was later to build a notable water supply for the city of Montpellier, devised the Pitot tube. With an opening facing the current, flow could be measured by noting the velocity head, that is, the height to which water was forced up in the tube. Earlier ideas proved erroneous since, due to frictional drag, the maximum flow was near the surface and smallest at the bottom of the channel.

In his studies on the water supply of canals, Chézy (1718-98), Perronet's right-hand man, developed the famous *formule d'hydraulique* (hydraulic formula), $v = C\sqrt{RS}$, which bears his name and which has been applied to flow in both small ditches and great rivers. The low slopes necessary for canals also demanded accurate levelling in location and construction. To this end the French contributed the vernier, the modern bubble-tube levelling device, and the surveyor's level with a telescopic sight. Bion's famous treatise of 1723 described practically all the tools of the profession, from drafting to surveying instruments, which are used today.

ROADS

While French engineers were not lacking in mechanical abilities, their major area of interest was in the field to be designated as civil engineering. In this area their contributions to road-building, to bridge construction, and to canals were outstanding. The interests of the French kings, especially Louis XIV, emphasized the construction of main roads for troop movements and royal tours, and their finance ministers encouraged the construction of these arteries to help stimulate internal trade and commerce.

Remains of Roman roads appear to have influenced French practice, although in a modified form, for their heavy foundations and paving, "a cut-stone wall laid on its side," was too costly for general use. In planning their roads, French engineers also followed Roman practice, locating them as straight as

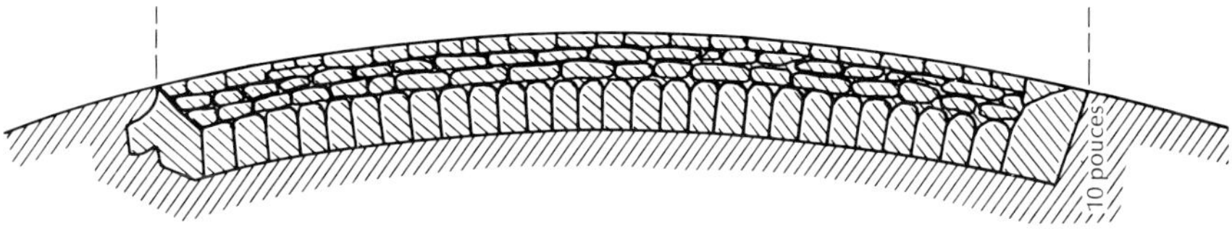

Fig. 12-5. A new form of road construction appeared in 1764 when Trésaguet developed a base of flat stones set vertically.

possible between towns with little attention to grades. A new art of practical road-building had to be created.

Gautier, who published the first book on bridges in 1716, also wrote on roads; he advised surfacing with local materials, such as gravel, sand, and earth, and to leave compacting to rain and traffic rather than employ women and girls for this work. Over a century was to pass before road rollers and other equipment were devised for compacting road materials. Main routes came to be built on a base of flat stones with a surface of broken or whole stones. A new form appeared in 1764 when Trésaguet (1716-96), of the Corps des Ponts et Chaussées, developed a base of flat stones set vertically in a rough voussoir-arch shape, which Telford later adopted in Britain and which is usually known as a Telford base.

One element in the reform of roadways was constant maintenance and repair, which in earlier days was often neglected. The *corvée* plan, using the forced labor of those living in the area served by a road, had earlier appeared in some of the frontier provinces, but in 1738 it was made general. Local workers were taken from their farms for *journées de travail* (workdays), and were also subject to the *taille* or royal tax, two burdens which escaped the nobility and the clergy. These obligations of the common people, mostly the peasantry, compared with the exemptions of the first two estates—clergy and nobility—became another source of the growing resentment which led to the French Revolution of 1789.

Yet the *corvée* system was effective in getting roads built and maintained, and it is estimated that before the Revolution, France had a network of main highways totalling some 30,000 miles, or nearly that of the present day. By 1789, on the eve of the French Revolution, a keen and observant British visitor, Arthur Young, characterized French roads as "stupendous and magnificent, we have no idea of what such a road is in England." Actually, he was comparing French main routes with the washed out and rutted roads of Britain; many of the local roads in France were impassable for three or four months of the year and were little better than their English counterparts. Nevertheless, it was well into the 19th century before any other nation had a highway system comparable to that of France.

BRIDGES

As in Italy, the revival of road-building and the beginning of canals led to the construction of many bridges. In replacing timber bridges, subject to fire and decay, by stone arches, French engineers faced the same problems that had perplexed their predecessors: the obstruction of the normal flood channel of the river by the bridge piers, and the limitations of available equipment in securing safe pier foundations.

The first notable modern bridge, the Pont Notre Dame in Paris, replaced an earlier timber span between the island, la Cité, and the north bank of the Seine in 1500-1507. Its design is attributed to an Italian expert, Fra Giovanni Giocondo, and its six practically circular arches of moderate span (56 feet) recalled earlier Roman designs. Native French talent, however, slowly developed, although it was almost a century before other notable bridges were built.

Two later bridges, the Pont Henri IV at Châtellerault, and the famous Pont Neuf, the oldest bridge still standing at Paris, are architecturally pleasing structures of modest, substantially circular spans. The former, begun in 1576, was not completed until 1611, and the latter, begun in 1578, was opened in 1607. A third work, and a far bolder structure, followed the elliptical form of the Italian Renaissance. This was the bridge spanning the Garonne at Toulouse. It had seven relatively low rise *anse de panier* (basket-handle) arches with a maximum span of 104 feet. Although the bridge was begun in 1542, work ceased during the religious wars in France, and it was not completed until 1611. All three of these bridges show a typically French innovation in design.

Because the flow through an opening is substantially increased by rounding its inner edge, the French bridge-builders secured a similar effect by bevelling the opening of their arches starting at the crown and increasing to the springing where the bevel blended into the cutwaters of the piers. Shorter piers also could be used and this tapering form, known as the *corne de vache,* or cow's horn, was widely adopted.

The pier foundation problem, however, still remained critical. The *battardou,* or cofferdam, consisting of two rows of sheet-piling surrounding the pier site, could not be driven deeply into the river bed with man-operated pile-drivers. Although the cofferdam was filled with earth and clay, inflow of water was inevitable. Archimedean screws plus batteries of *chaplet* (chain and bead) pumps, operated by men working in shifts, were unable to cope with the water seepage, with the result that piers could be carried down but little below stream bed level.

In 1685, a new method for constructing pier foundations was introduced in building the Pont Royal. This famous Paris bridge was built for Louis XIV in order to join his palace of the Tuileries with the south bank of the Seine. Al-

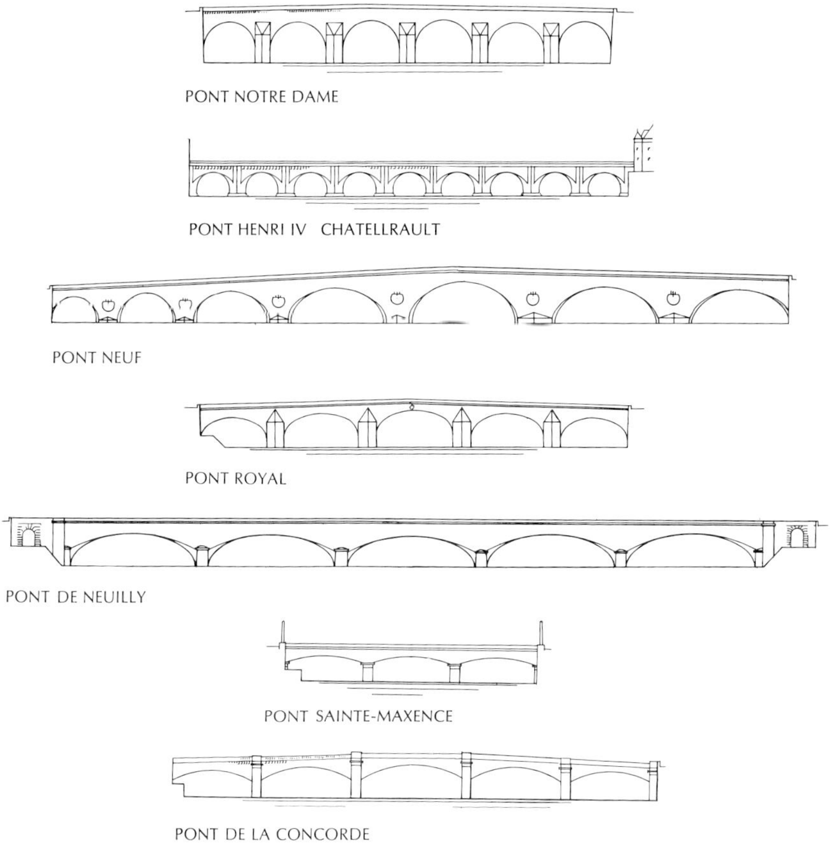

Fig. 12-6. French stone-arch bridges, illustrating their evolution from the Roman form of the Pont Notre Dame of 1507 through the earlier elliptical types to Perronet's daring revival of the segmental arch in the late 18th century. *Top to bottom*: Pont Notre Dame, 1507 (note the circular arches); Pont Henri IV, 1611; Pont Neuf, Toulouse, 1611 (note the elliptical form of the arches); Pont Royal, 1685; Pont de Neuilly, 1772; Pont Sainte-Maxence, 1786; Pont de la Concorde, 1791.

though its five elliptical arches are not especially notable in span (68 to 77 feet), its symmetry and excellent proportions, with the floor rising slightly to give clearance for the larger central arch, give it a feeling of repose and stability. The first men who were called in to design the Pont Royal—two famous architects, Mansard and Gabriel—apparently failed to solve its foundation problem. Then the Belgian priest, François Romain (1646-1735), who had built bridges in the Low Countries, was called in. He adopted a new and daring foundation technique. Using a man-operated treadmill dipper dredge, developed in the Low Countries, he first dredged the pier site. A large Marne river barge, loaded with and surrounded by loose stone, was sunk on this base. This formed a *caisse*, a box or caisson, inside which the first lower outer course of pier stones was placed. Surrounded by a line of piles, the pier was then built up with rubble, using Roman pozzuolana cement mortar (which would set without contact with the air)—for the first time in a French work. The method was a success, and Father Romain was hailed as "First of Engineers of the Century."

When the Swiss-French engineer, Labelye, was called to London (1738) to build the Westminster Bridge over the Thames, he adopted Romain's caisson plan using an open box with sides that could later be removed. In 1753-57, de Voglie used the *caisse* in building the Pont de Saumur over the Loire on a foundation of piles driven in the dredged area and cut off under water. Today, a more modern form, the "open" caisson with dredging wells, is a widely used construction device.

PERRONET

In 1744 a bureau of designers was created in the Corps des Ponts et Chaussées to plan main routes, and in 1747 a school was established to train men for the Corps. After an active career Jean Rodolphe Perronet (1708-94) was appointed director of this pioneer venture in engineering education, the École des Ponts et Chaussées. In 1764 he also became chief of the Corps and after serving for 30 years, was hailed as father of the École and the outstanding French bridge engineer.

In 1772, he built the Pont de Neuilly over the Seine below Paris. With several low rise elliptical arches combined with circular face openings providing the bevelled *corne de vache*, and with relatively narrow piers, its graceful and simple form made it perhaps the most beautiful stone arch ever built. It has unfortunately been replaced with a modern span. Telford in Britain was to copy it in 1827 in his Gloucester arch over the River Severn.

At the same time, Perronet was engaged on a bridge over the Oise about 35 miles north of Paris, in which he revived the Italian segmental arch of the Renaissance. His Pont Sainte-Maxence marked a new level of skill and daring in stone arch design. Its three 72-foot segmental spans with an extremely low rise, just over 6 feet, reduced both the dead weight of the arches and the downward component of their thrust, making it possible to support the structure on four slender 9-foot diameter columns instead of the earlier massive and solid piers. This tour de force of bridge construction thus provided a remarkably free waterway for the passage of floods.

Perronet had planned a similar design for his last and best-known work, the Pont de la Concorde in Paris. Because his design was regarded as too daring, he was forced to use arches of somewhat greater rise than he had originally proposed. With five spans, a center arch of 102 feet and rise of 13 feet, with slender columnar piers and open, ballustrade railings replacing the earlier solid parapet walls, this bridge still remains a most graceful and aesthetically pleasing memorial to its builder. Completed in the early years of the Revolution, it has been widened in recent years, closely following its original form.

Another noted engineer, Gauthey, who built the Canal du Centre, adopted an oval, streamlined form for his piers in order to reduce their resistance to flow. This, however, involved the difficult problem of building arches of varying span, so it was not followed in later works. Unlike Perronet, Gauthey added various decorative features to his bridges, in keeping with the over-ornamented architectural style of the day.

In these works the stone arch had reached a high level of development. But at the same time in Britain an old material, iron, was becoming available in quantities that made feasible its use for bridges. The stone arch was no longer the only long-life type of bridge; although it continued to be used for the earlier railroad bridges and some later spans, it gradually gave way to new forms.

CANALS

Great French rivers—on the east, the Rhone flowing south to the Mediterranean; and the Seine, Loire, and Garonne, penetrating the center of France eastward from the Atlantic—provided a unique opportunity for inland water transportation. At their mouths great port cities had developed, and by joining their headwaters with canals, a complete system of inland navigation could be provided. Although a few notable works had been built earlier, it was not until well into the 19th century—when railroads were replacing canals in Britain and the United States—that France finally met the many obstacles retarding canal-building and completed one of the world's greatest systems of inland waterways.

As early as 1560 the Canal de Craponne, some 40 miles long and named after its builder, a military engineer, had been constructed in the south of France, north of Marseilles. A pioneer venture, it had no locks and served only for irrigation and to power some small water mills. But Sully, the chief minister under Henry IV (1589-1610), had supported a project to join the Loing, a tributary of the Seine, with the Loire in the center of France. The resulting Canal de Briare was a pioneer work, the first canal to cross a mountain divide. Started in 1604 by contractor-promoters, work on the canal ceased with Henry's murder in 1610; but plans were improved, and it was resumed by new promoters in 1638 and opened in 1642. Forty miles long with 41 locks, its water supply proved inadequate. Projected in 1692, the Canal d'Orléans, 75 miles long and leaving the Loing at Montargis, provided a more direct, competing route to Orléans with a much-improved water supply.

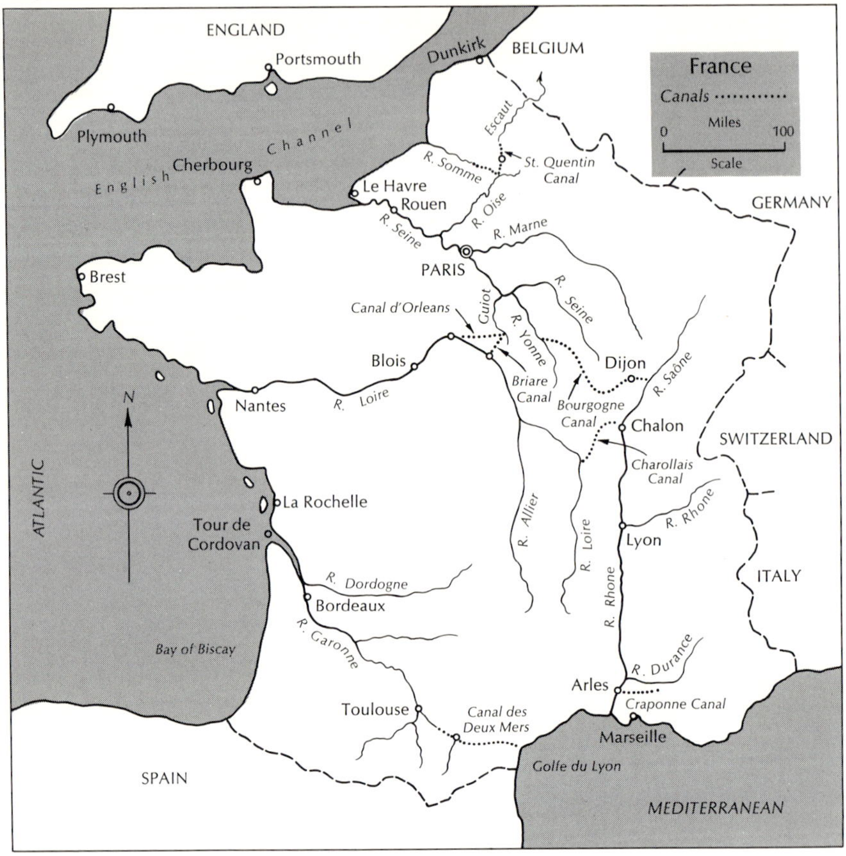

Fig. 12-7. Early French canals.

The outstanding French canal work, however, was the Canal de Languedoc, also known as the Canal du Midi or Canal de Deux Mers, which joined the Atlantic and Mediterranean. By uniting Toulouse on the upper Garonne with Cette (Sète) on the Gulf of Lyon, this canal enabled shipping to avoid the long sea route around Spain. Although considered at an earlier date, the project had been abandoned as too costly and impractical. Finally in 1661, this great work, 148 miles long with 120 locks and a summit level 621 feet above the sea, was undertaken by Pierre-Paul Riquet de Bonrepos (1604-80). Riquet first secured the support of Colbert, Louis XIV's famous finance minister. Some limited royal funds were promised, and Riquet, who was appointed to collect local taxes, finally persuaded the provincial authorities to invest some funds. Eventually he poured his own family fortune into the work.

In addition to numerous locks and crossings, a reservoir to store the flow of high mountain streams for the water supply and a pioneer tunnel at Malpas were required. Riquet had models made, directed the estimated ten thousand workers (of whom 600 were women), and met the criticism of those who doubted that the canal could be built or would work. The canal was completed in 1681 at an estimated cost of over six million dollars. His son finished the canal, for, worn out by his labors, Riquet had died a year earlier. His heirs were forced to give up a large part of his interests, and waited 40 years before receiving a very substantial return on a work which was said at the time to be "the wonder of Europe," the greatest engineering work of its day.

Canal-building, however, faced many problems which long retarded further developments. The number of lockages possible depended on securing and storing water from the variable flow of stream headwaters higher than the summit level. This was later to lead to canal lifts, inclines, and even pumped supplies. Because traffic developed slowly, profits were small, thereby discouraging private investment in canals. Moreover, canals were little related to military and other interests of the king, so there was only limited royal support.

Two other major projects received early attention but only one was completed before 1800. Both were in the center of France: the Charolais Canal or Canal du Centre, to unite the Loire with the Saône and thus the Rhone; and, to the north, the Bourgogne connecting the Yonne, a tributary of the Seine, with the Saône. The far less difficult Centre project was built first. Seventy miles long, with a two-mile summit level reached by 30 locks rising 250 feet from the Loire, it descended by 50 more locks the 430 feet to the Saône. Begun in 1783, it was completed in 1792 by a force of soldiers under the direction of an eminent engineer of the Corps, Gauthey (1732-1806). A later lateral canal of the upper Loire completed the route. The Bourgogne project, favored because it led to the Seine and Paris, was not completed until 1832.

It was thus not until the 19th century that other major works were built by the French government. As traffic developed, the early small sailing boats gave way to larger barges, and over 3000 miles of canals, with some 2000 lateral canals on the upper rivers, plus 1800 miles of river navigation, gave France close to 7000 miles of waterways.

HARBORS

With a rapidly growing trade and an increase in sea-going traffic caused by the acquisition of overseas colonies, such shipping aids as lighthouses and harbors came to be of critical importance. As early as 1611 the Tour de Cordovan, an early lighthouse over 150 feet high, had been built at the mouth of the Gironde River, about 60 miles below Bordeaux. Although regarded as the first modern

Fig. 12-8. Vauban's harbor at Dunkirk before its destruction in 1713. (From Belidor, *Architecture Hydraulique*, 1737-53)

lighthouse, it was a Renaissance-type building, little resembling modern works. First lighted by wood fires, coal followed in 1782, and after several alterations, lamps with reflectors were installed.

Another early work at La Rochelle is referred to as the first breakwater of Western Europe. It was a *dique,* or dike, of loose stones built in shallow water to close the harbor during the siege by the King in 1627 and prevent the entry of a relief fleet from England.

When Louis XIV determined to build up a French fleet, he called upon one of the greatest French military engineers to develop and fortify a harbor at Dunkirk on the English Channel. Sébastian le Prestre de Vauban (1633-1707) had a long and famous career. "Builder of Fortresses" for the king he served for 55 years, he was made a Marshall of France in 1703. It is said that St. Simon coined the title *patriote* for him. Vauban's engineering contributions included not only Dunkirk but other notable works, and also the initiation of the Corps of Engineers.

With few exceptions, notably Brest where deep water prevailed, harbors in northern France presented two major problems: a shallow, sloping shore and a tidal range frequently reaching 20 feet or more. Vauban constructed two parallel jetties to confine the flow of the streams entering the sea at Dunkirk and thereby to produce a scouring effect and keep the harbor from silting up; since normal flow was insufficient, storage reservoirs were built to release water to aid this scouring action. A depth of 15 feet was secured for entrance to the inner basins, with locks to maintain the water level at the several dock, repair, and other facilities. With its famous island fort, the Rixban, Dunkirk was a remarkable fortified harbor. It so threatened British vital interests that in 1713, fortunately after Vauban's death, the Treaty of Utrecht forced France to demolish this masterpiece.

French engineers also undertook one of the world's greatest harbor works, the great *dique* or breakwater at Cherbourg. Two and a half miles long in deep water (40 feet), it was begun in 1781 but not completed until 1858. Harbor-building became, in fact, a continuing problem of improvement and replacement.

REVOLUTION AND ECLIPSE

The French Revolution brought about political and social reforms but did little to change France's engineering interests. A few engineers left the country, but new works were built. Under Napoleon the difficult Canal de St. Quentin was completed, and also the famous road through the Simplon Pass in the Alps. The Revolution emphasized science and arts and trade. It established a new and more comprehensive scientific and engineering college, the École Polytechnique, which threatened for a time to supplant the more specialized École des Ponts et Chaussées and École des Mines (School of Mining Engineering). The École Polytechnique stressed mechanical and mathematical techniques and theory in engineering design, and its professors and graduates made notable contributions along those lines. The specialized schools also began to place more emphasis on theory.

BRITISH LEADERSHIP

While France made remarkable engineering progress during the 18th century, engineering was also developing in Britain. There the embryonic Industrial Age stimulated engineering to new heights of innovation and creativity, in transportation and construction as well as in mechanical and metallurgical fields.

Samuel Smiles, in his *Lives of the Engineers* (1861-75), remarks, "Although the inventions of British engineers now exercise so great an influence all over the world . . . most of the continental nations had a long start on us in art, in science, in mechanics, in navigation, and in engineering." Many British industries, he observed, had been started by foreigners: cloth-making by Flemings, paper-making and mining by Germans. And foreign influence had been important in other engineering areas: Holland provided water and windmills; a Dutchman, Vermuyden, first drained the Great Level of the Fens; and the English fortresses and cathedrals of the Middle Ages probably owed much to French talent.

There had been some engineering works in the Elizabethan Period, the most

Fig. 12-9. The first iron bridge was built at Coalbrookdale, England, in 1779. (*Science Museum,* London)

notable of which was the New River water scheme for London. Using ditches and bored-log pipe, it had been completed in 1613 by Sir Hugh Myddleton (1555-1631), a London goldsmith, financier, and mining promoter. This pioneer venture was not a government undertaking but was projected and carried through by private enterprise, a procedure that was to become especially characteristic of and basic to British economic growth. Thus, while French works were government projects, Britain's remarkable industrial advances and her major public services—from canals, railroads, and water supplies, even to lighthouses—were carried forward under the stimulation and challenging competition of private initiative. Some six or seven private companies were later to serve London's growing water needs, and it was not until 1902 that the Metropolitan Water Board, a governmental organization, took over Myddleton's New River enterprise.

Little engineering advance was made in the 17th century as Britain passed through political revolution and civil war, and power was transferred from King to Parliament under a constitutional monarchy. In the 18th century, however, construction played a major role in opening interior areas for the new industrial growth of Britain and in providing for the essential transportation needs of industry, commerce, and trade.

Unlike France, Britain had plentiful resources of coal and iron which were a major factor in her industrial and engineering development. The Industrial Revolution ushered in a true Age of Iron. The first iron bridge was built at Coalbrookdale in 1779; construction and transportation, after centuries of limited means, were provided with new materials and tools with which to meet the changing needs and wants of an emerging era. The combined needs of manufacture and trade so stimulated building and construction that British civil engineers soon came to lead all others.

The engineers who were to meet the challenge of the new industrial age were generally men of humble station, practical planners and builders. Indeed the pioneer inventors of Britain's textile industry were men who cannot be classed as engineers, while those who carried forward Britain's construction and transportation works were, by modern standards, both civil and mechanical engineers. Unlike French practice, there was little or no liaison between architecture and engineering. True, some attention was given to architectural taste in design; even the railway pioneers, in an effort to placate those who opposed the advent of the "Iron Horse," gave special attention to aesthetic values in designing viaducts, stations, and bridges. Nevertheless, the British engineer of the 18th century was neither architect nor engineer in the sense that his earlier Italian or French counterpart had been. Almost invariably, the British engineer had begun his career as an apprentice to a mill-wright—the practical planner and builder of water- and windmills, the machines of the pre-industrial era.

We thus find the early leaders of British engineering interested in both the older arts of construction and the problems of a new mechanical era. When the British Institute of Civil Engineers was founded in 1818, the majority of its members were connected with mechanical pursuits. With the rise of new interests and techniques, specialization was inevitable, and an Institute of Mechanical Engineering was finally established in 1847. Britain thus took over engineering leadership under quite different conditions and with quite different interests from those which had prevailed in France.

BRINDLEY

A small canal and lock are said to have been built in the reign of Elizabeth I, but the British canal era really began through the efforts of James Brindley (1716-62). Born of poor parents and lacking formal education, Brindley proved to be an able and resourceful builder. Early apprenticed to a mill-wright, he later set up his own shop and achieved a reputation as an ingenious worker. Coming to the attention of a progressive property owner, the Duke of Bridgewater, he was engaged to plan a canal to bring coal from the Duke's mines at Worsley to Manchester.

Work was begun in 1759, and in 1762 the project was extended some 24 miles to reach the Mersey River and thus join Liverpool and Manchester. The final locks were completed in 1773. Brindley had advised carrying the canal across the Irwell River on a canal aqueduct which avoided the use of locks up and down from this intervening stream. Drainage conduits had earlier been used in France, but Brindley's Barton Aqueduct, referred to as his "Castle in the Air," was the first of a type of British canal structure later developed on a much larger scale.

While this work was under way, Brindley was asked to plan and direct a far more formidable project, the Grand Trunk Canal, running south and east from the Mersey to the Trent and, with its extension the Birmingham Canal, to the Severn. Over 140 miles long, with 75 locks, several aqueducts, 5 tunnels, and over 100 highway bridges, it was a notable and daring venture. Josiah Wedgwood, the pottery manufacturer, joined with the Duke of Bridgewater and Earl of Gower in supporting this project. But Brindley died before it was completed in 1777.

This was the canal that opened to economic growth and development the great industrial center of Britain, the Manchester-Sheffield-Birmingham area. By the close of the 18th century Parliament had approved over 80 acts authorizing canal works, and canal-building became a major activity of early British engineers. Roads were yet to come, preceding railroads by only a few years.

SMEATON

In sharp contrast to the self-taught Brindley, John Smeaton (1724-92), the son of a lawyer of Leeds, had an excellent education. Expected to follow law, he early developed mechanical interests and served for a time as apprentice to an instrument-maker in London.

The Dutch had early turned to windmills to meet their power needs, and in 1755 Smeaton travelled in the Low Countries, noting especially the mills used for pumping water, grinding grain, making paper, and for other purposes. He also was interested in the Royal Society and in 1759, received its Gold Medal for a paper on water and windmills. As a result, when the Society was asked to name someone to design a lighthouse, Smeaton was suggested—not because he had any experience in such works but because he had acquired a reputation as an ingenious and thoughtful engineer who would not undertake any work he could not carry through. He succeeded in building the lighthouse at Eddystone (in the English Channel, 14 miles south of Plymouth) after two earlier lighthouses on that site had ended in disaster.

The first attempt had been as early as 1696, when an eccentric mechanical genius, Winstanley, built a curious wooden structure on the dangerous Eddystone ledge. This was carried away in a great storm in 1703. Another designer, Rudyard, then built a new tower of quite modern form in 1709. Unfortunately, it was of wood and, with early lighting methods, caught fire and burned down

Fig. 12-10. Smeaton's Eddystone lighthouse. Built on an exposed ledge in the Channel and completed in 1759, it was a "welcome beacon to mariners who had earlier faced the dangers of a hidden reef."

in 1755. The proprietors, however, still had some years under their 99-year lease, and they turned to Smeaton on the advice of the Royal Society. Smeaton visited the exposed ledge and began his plans in 1756.

He pointed out to the proprietors that a wood tower would be least expensive and would probably last to the termination of their lease. The leaseholders, however, preferred the more costly but permanent stone. Smeaton describes this first of modern lighthouses in his *Narrative of the Construction of the Eddystone Lighthouse* (1791), an engineering classic. The dangerous Eddystone ledge was cut to receive the interlocked bottom stones which were cut and fitted in a yard on shore during the winter months, when it was not possible to work on the exposed ledge. In calm weather between storms, the tower, "shaped like the lower trunk of an old oak tree," was carried up to the lantern level with its top almost 100 feet high. First lighted in 1759, it was replaced in 1877 because of the undermining of the ledge on which it rested and the need for a higher light.

Smeaton later was active in or consulted on many other works, including harbors, dams, canals, drainage projects, and bridges. He was the first to adopt the title civil engineer, to indicate the professional who worked on engineering projects which were not military in nature or purpose. When in London he met with his engineering friends in a group known as the Smeatonian Society, a forerunner of the Institute of Civil Engineers of 1818. He also made notable contributions to mechanical as well as civil engineering; he was a jack—and master —of all trades.

RENNIE

John Rennie (1761-1821), born in Scotland like Watt and Telford, worked for a famous mill-wright and also studied three years at Edinburgh University. Visiting James Watt at Birmingham, he was asked to supervise the construction of the Albion Mills near London in 1784-88, a pioneer application of steam power in the operation of a flour mill with novel sifting, fanning, and other devices. Achieving a reputation as an able engineer, Rennie was called upon (1791) to undertake a canal project. His later career centered in civil engineering works.

One wonders how Rennie and his contemporary, Thomas Telford (1757-1834), found time for the great number of works on which they were consulted. Smiles lists twenty or more of Rennie's canal projects; at the same time he worked on drainage plans south of Ely and on the construction of fen works from Lincoln to the sea which required detailed surveys and studies of rainfall, flood, and channel flow. In addition, he is said to have been consulted on at least 50 other works including bridges.

Rennie's first bridge was a stone arch over the Tweed in 1799, but like Telford, he pioneered the use of the new construction material, cast iron. Rennie's first bridge over the Thames—Waterloo Bridge—had been an elliptical stone arch of 9 notable spans of 120 feet. However, in 1815 he turned to cast iron to build Southwark Bridge, spanning the river in three record-breaking arches of 240 feet each. Unfortunately, he did not live to see the completion of his third bridge over the Thames, the New London Bridge. Replacing the six-centuries old famous bridge of nursery-rhyme fame, this bridge was completed by his sons, John and George, in 1831. For this work Rennie had turned again to stone, using five arches similar to Waterloo but of somewhat greater span. Although tidal currents so undermined the Thames bridges built by others that they had to be rebuilt, Rennie's bridges still remain.

In his harbor works Rennie included harbors of refuge, those for commerce, and also naval bases. His plans (1806) for the notable breakwater at Plymouth were carried out by his son John and completed in 1848. A careful and thoughtful engineer, Rennie has been criticized as an over-conservative, cautious builder. British works in general, however, were solid and long-lasting, built to endure rather than with possible future alteration or extension in mind.

TELFORD

Another outstanding engineer of the earlier period was Thomas Telford (1757-1834), who was born in a remote shepherd's hut in the valley of the Esk in Scotland. Through sheer determination and self-education, he became noted not only for his outstanding engineering works, but for his literary interests, well-written reports, and personal qualities. He began life as a mason but, "turning from hand to brain work," he set out for London when only twenty-five, made friends, and was soon engaged in building and engineering works. One of his first major undertakings was the Ellesmere Canal, projected by some of his friends in the difficult, rough borderland of Wales.

Like Rennie, Telford was a leader in the use of the new construction material, cast iron. The Ellesmere Canal involved two major aqueduct structures, the Chirk and the Pont-Cyssylte. For the latter, he adopted a new design: the piers were stone but the supporting arched ribs and the water channel were built in sections of cast iron bolted together. This was the first use of cast iron on a major work, and Sir Walter Scott described it as the most impressive work of art he had ever seen.

In 1802, attention turned from canals to the long-neglected problem of roads. Called upon in 1802 to report on roads in Scotland, Telford directed the construction, over eighteen years, of an estimated total of 920 miles with 1200 bridges. These roads brought new life to the Scottish Highlands, and Telford became widely known as an expert, "the Colossus of Roads."

Among the many roads upon which he was called to advise was the main route from London and central England to Holyhead on the Isle of Anglesey off the north coast of Wales, the point of embarkation for Ireland. Here, as in his other road work, he adopted the Telford base, or foundation course, which had been first used by Trésaguet in France more than forty years earlier. Unlike French practice, he sought not only to save distance in locating his roads but to reduce their grades so that heavier loads could be moved over them. Thus, in place of grades of 15 per cent then commonly encountered, he sought a maximum of 5 per cent.

Telford's road-making in Wales did not get under way until 1815, just at the time when another famous road-builder, John Loudon McAdam (1756-1836), became the overseer of the Bristol roads. McAdam condemned both French practice and that of Telford; he argued that their cost was excessive and that

Britain's severe frost conditions inevitably led to the lifting of the large base stones to the surface. In their place he recommended a more "elastic," compacted, broken stone: the "macadamized" road. This type of construction became standard for the best roads of Britain. Furthermore, whereas British roads had been built on a toll basis by Turnpike Trusts, with managers who served without pay and engaged local labor for construction and repairs, McAdam insisted that trained and experienced personnel were essential. Slowly, his ideas were adopted.

Telford was a great builder of bridges, as well as of roads. One wonders, as with Rennie, how he found time for his many engagements, for he had also constructed harbors in Scotland while engaged in road-building there. A number of Telford's bridges were stone structures, more or less following French examples. Like Rennie, he built cast-iron bridges. However, when wrought iron, with its reliable tensile strength at last became available in quantity, Telford began using it for suspension bridges.

Suspension bridges using wrought-iron links, or chains, had been built in the United States as early as 1801. In 1826 Telford completed his record-breaking span of 580 feet over the Menai Straits, replacing a ferry on the route to Holyhead. The bars he used were carefully tested for strength, and the design was checked using the structural mechanics of the day. Many of the early suspension bridges provided only a light, suspended timber roadway platform, and a number of these spans failed because of wind action. Although Telford's bridge later suffered from wind action too, and had to be strengthened and repaired, it remains a pioneering masterpiece.

SUMMARY

The engineering activities of these years were clearly and directly tied to the major concerns of the times and places of their undertaking—the renewal of city life in Renaissance Italy, the martial and ceremonial activities of Louis XIV of France, the industrial and commercial ventures of Great Britain in the 18th century.

First tied to military necessity, some engineers began to specialize in harbors, canals, and other "civil" pursuits. Armed with new theoretical and mathematical knowledge, given new materials and machines with which to work, and patronized by public or private interests, a new breed of men arose who could, as a matter of routine, accomplish large works to match the growing concerns of the age. They bridged the rivers, built the canals, and paved the roads which were to provide the transportation arteries for the Industrial Revolution.

Part III
The Industrial Revolution, 1750-1830

13 / Prerequisites for Industrialization

MELVIN KRANZBERG

When we think of the Industrial Revolution we usually think of the steam engine, the railway locomotive, and the factory system. But these are merely the best known of a series of fundamental technological, economic, social, and cultural changes which, taken together, constitute the Industrial Revolution. From the very beginnings of civilization the hearth and home had been the center of production and of life. The Industrial Revolution transformed all of society by taking men away from the traditional agricultural pursuits which had formed their main occupation throughout history and introducing them to novel ways of working and living in factory and city.

DEFINITION OF THE INDUSTRIAL REVOLUTION

The term "Industrial Revolution" was first used by some French writers at the beginning of the 19th century. These men, much impressed by the French Revolution with its transformation of political life, and seeking to characterize the revolutionary changes then taking place in industrial and economic life, coined the phrase. However, the term did not acquire much currency until popularized by Arnold Toynbee (1815-1883), uncle of the famous 20th-century historian of the same name, who used it to describe England's economic development from 1760 to 1840. The years mentioned by Toynbee have since been frequently employed as the "dates" of the Industrial Revolution. Actually, these dates are misleading. Neither the beginning nor end of such a deep-seated and complex social and economic transformation is capable of sharp definition. We must regard an industrial revolution more as a *process* than as a distinct period of time. And this explains why the Industrial Revolution can take place in different places at different times—why some countries such as Russia and China did not begin their industrial revolutions until the 20th century.

Some scholars deny that the Industrial Revolution was very revolutionary.

They correctly point out that many of the elements entering into the Industrial Revolution reach far back in history and had a very slow evolutionary development. For example, the power revolution in medieval times laid the basis for the changes in energy sources which took place in the 18th century and many of the other technological elements in the Industrial Revolution began as much as a hundred years earlier.

Now if we think of the word "revolution" as implying only a quick and sudden overthrow, then the Industrial Revolution was certainly not so much revolutionary as it was evolutionary. However, in the sense that the process of industrialization thoroughly transformed every aspect of society, then it is proper to describe it as a revolution: it certainly revolutionized men's ways of living and working, and in the process gave birth to our contemporary civilization. Furthermore, while certain elements of the Industrial Revolution did appear long before 1760, it was only in the closing decades of the 18th century that these and other factors came together to produce these climactic changes in life and society. Hence, if we didn't have the term "Industrial Revolution" to describe the conjunction of the various technological and socio-cultural phenomena which changed the character of life in the closing decades of the 18th century, we would have to invent some other phrase to describe the process whereby an agrarian society becomes transformed into a modern industrial society. It is simpler to continue to call it the Industrial Revolution.

TECHNOLOGICAL AND SOCIO-CULTURAL CHANGES

The technological changes involved in the Industrial Revolution included the use of new basic materials such as iron and steel. These had been used previously, of course, but now they became the primary materials which men used for machines, for buildings, and for any durable product. There were new energy sources—that is, new fuels and motive power—such as coal and the steam engine, and, at a later date, electricity, petroleum, and the internal-combustion engine. The invention of new machines markedly increased the production of goods with a smaller expenditure of human energy. Production was also increased by a new organization of work in the factory system with its division of labor and specialization of function. Raw materials had to be brought to the factories and the manufactured goods sold in a wide market, making necessary important developments in transportation and communication. These rested upon the new fuels and new machines and included the railroad, the steamship, and later the automobile and airplane as well as the development of the telegraph and wireless. Later, in the 19th century, was to come the increasing application of science to industry. The Industrial Revolution also included a series of technological transformations in agriculture which made possible an adequate food supply for the large section of the population now working and living in cities and hence unable to grow their own food.

All the changes involved in the Industrial Revolution were not simply technological; they were also social, economic, cultural, and political. A series of economic changes which profoundly affected the technological changes and were affected by them involved a wider distribution of wealth, the transfer of the source of wealth from land to industrial production, the growth of large-scale international trade, and the advancement of capitalistic growth. Political changes reflected the economic shifts. New institutions were developed and novel governmental policies were employed, corresponding to the new needs developed by a society which was in the throes of industrialization. There were also sweeping social changes, including urbanization, the rise of working-class movements, and the emergence of new patterns of authority. Cultural transformations of a broad order also occurred. The worker, for example, acquired new and distinctive skills, and his relation to his work shifted: instead of being a handicraft worker with tools, he became a machine operator subjected to factory discipline. Finally, there was a deep-seated psychological change in man's view of his relations with Nature: the Industrial Revolution heightened man's confidence in his ability to use the resources of nature and to master her.

WHY DID THE INDUSTRIAL REVOLUTION BEGIN IN BRITAIN?

A major problem in any discussion of the Industrial Revolution involves its causative factors. After all, we want to know what factors will stimulate technological growth today; perhaps if we knew what caused the industrial growth in the 18th century, we might learn some lessons which would be applicable to our contemporary situation. In terms of the technological transformations of the later 18th century, the question is usually asked in this form: Why did the Industrial Revolution occur first in England?

England and France, which were locked in combat for world supremacy during most of the 18th century, were, as the most advanced and powerful nations in Europe, the only two countries in which the Industrial Revolution could possibly have begun. To answer the question about England's priority in industrialization, therefore, we must compare England and France in the 18th century to see why England took the lead when France might have been expected to become the leading industrial nation.

Throughout most of the 18th century, France was the greatest power in Europe politically, militarily, economically, and culturally. Despite some military defeats, the loss of her overseas colonies in mid-century, and the approach of royal bankruptcy, which was to eventuate in the French Revolution, France was still the wealthiest nation in Europe. French trade and industry were rich.

For example, in 1789, on the eve of the French Revolution, France's foreign trade amounted to 200 million dollars, while England's was 160 million. Furthermore, France had a population of 26,000,000 while England had only 9,000,000. France possessed not only raw materials and significant natural resources, but she had an energetic middle class. In addition, France in the 18th century had the cultural leadership of the world. The Enlightenment, which was the primary cultural movement of the 18th century, had its home in France; throughout Europe, the thoughts and attitudes of the French philosophers were copied. And the rulers of the other European states sought to imitate the splendid court of the French monarchs at Versailles.

Despite these evidences of France's premier role in Europe, we know that the Industrial Revolution occurred first in Britain. We must therefore analyze the factors which went to make up the Industrial Revolution in the mid-18th century to see how certain of these prerequisites of industrialization were better fulfilled in England than in France. At the same time it must be realized that both these countries met these prerequisites for industrial growth better than any other European states.

CAPITAL FOR INDUSTRIAL DEVELOPMENT

Capital for investment in machinery is necessary for industrialization. In addition to the cost of the machinery itself, there is also the expense of bridging the gap between invention and innovation, that is, the development of a basic idea into a usable process or piece of machinery. Many inventions would have been stillborn had not capital been found to make their application effective. For example, the industrial application of Watt's fundamental invention of the steam engine was delayed for almost two decades until Matthew Boulton provided the capital and the drive which made the steam engine commercially successful.

In the 18th century, both France and England possessed large accumulations of capital which might have been invested in industrial advance. In the case of France, however, much of this capital was devoted to unproductive uses, such as forced loans to the state, the upkeep of a wasteful court and bureaucracy, and war expenditures; and it was later dissipated by the royal bankruptcy which precipitated the French Revolution.

England's capital, although perhaps not so great as France's, was rapidly increasing during the 18th century through agricultural improvements and commercial expansion. The lowering of the British interest rate in the first half of the 18th century made cheap capital available for industrial investment. At the same time, a profit inflation during the first half of the 18th century occurred as a result of wages lagging behind prices; this gave the British capitalists additional funds for investment in industrial machinery and expansion. In addition, the English monetary system was far sounder than that of France. England possessed the Bank of England since 1694, which made it possible to mobilize British capital readily. Furthermore, a widespread network of country banks throughout England allowed the diffusion of capital to burgeoning industries outside of London. France, on the other hand, had no solid or adequate banking system until Napoleon established the Bank of France, after the Bank of England had been in operation for a century.

It is not simply the quantity of capital which is important but also its liquidity or mobility, in other words, its *availability* for investment. Although France had more capital funds, these were not as available for investment in industrial enterprise as were the British funds. With the availability of capital there must also be a willingness to use it. Otherwise, capital might flow into the hands of those who will sterilize it by hoarding, by luxury consumption, or by low productivity investment outlays. Socio-cultural forces thus enter into the availability of capital for technological expansion. France had more capital, but British capital was more mobile and hence more available for industrial investment. And, as we will see, social forces gave Britons more opportunity to invest their funds in industry.

LABOR FOR INDUSTRY

As with capital, so with labor. A supply of laborers is necessary for industrialization, but here again it is not simply the quantity of workers but rather their ability to play an effective role in the technological process. In the mid-18th century, France's population was almost thrice that of England. But mobility and adaptability, as well as the size of the labor force, determine a country's potentiality for industrialization.

Serfdom and the guild system, two obstacles to the workers' freedom of movement and occupation, had been almost completely broken down in England by the beginning of the 18th century. They were not overcome in France until the French Revolution at the end of the century. Hence English workingmen, many of whom had been forced off the soil by the enclosure movement, were freer to move from farm to factory, and English merchants were less handicapped in what they might do and whom they might employ. When the French revolutionaries finally removed many of the legal restrictions from French labor, they established small peasant proprietorship as the characteristic form of French landholding, a move that *kept* workers on the farm rather than dispossessing them into factories.

Availability of labor also requires adaptability, that is, the education of the workers to a level of technical knowledge necessary for industry and their willingness to accept the rhythm and discipline of factory life. For example, one of the chief obstacles to Watt's steam engine coming into general use more rapidly seems to have been a chronic shortage of skilled labor. The problem of maintaining or repairing engines which were in use seems to have been beyond the abilities of most mechanics until the new machinery had been around for some twenty years. Because the machinery was developed first in Britain, British workers acquired a head start in the acquisition of technical skills. The factory organization of labor in England had begun even before Watt invented the steam engine; hence the British worker was already becoming accustomed to the pace of factory production when the introduction of machinery made the factory the characteristic working place. Here again, as in the area of mobility, England's labor supply was ahead of France's.

MARKETS FOR MASS-PRODUCED GOODS

Because industrial machinery involves a large outlay of capital, it is not worthwhile to invest in it without access to large markets where the mass-produced, manufactured wares can be sold profitably. How did England and France compare in regard to markets in the mid-18th century?

The dominant system of economic practices and theories in the 18th century was called mercantilism. Utilizing high protective tariffs, mercantilism endeavored to stimulate a nation's manufacturers by discouraging the import of goods manufactured abroad and by seeking colonial markets for goods manufactured in the home country. Both England and France pursued mercantilist policies, which meant that they had their home markets at their command. Since France's population was thrice that of England, she had a greater home market than did England. In the matter of colonial markets, however, England was forging ahead. France and England competed for imperial domains throughout the 18th century, in both North America and Asia. In this duel for empire, France lost out to England, and Britain monopolized trade with her colonies in India and America (including Canada).

With markets, also, it is not simply a case of quantity, or the number of possible customers. We must think of the character of the market, especially its ability to buy goods, and in particular, the kind of goods which could be manufactured by machinery. The Industrial Revolution opened an age of large-scale production for the needs of the masses. This kind of production requires standardization and repetition—just the type of job which the machine can do best. Since the English could not hope to compete with the already well-established French trade in hand-made, deluxe articles, they sought to satisfy the demands of the larger mass market whose needs could be met only by cheaper goods, which in turn could be produced only by machines. Once England began to manufacture and supply low-quality goods in quantity, her industrial leadership inevitably rose.

An increase in population during the latter part of the 18th and throughout the entire 19th century meant a growing market for mass-produced goods. Because England's population was growing at a faster rate than France's—and eventually outstripped the French population—the country was provided with a powerful stimulus to boost manufacturing.

RAW MATERIALS

Raw materials and natural resources are essential for industrialization. Britain and France both possessed coal and iron, the prime raw materials of the Industrial Revolution; however, England's coal and iron were readily available and in close proximity to one another, whereas France's resources, though abundant, were not comparable in quality or in location. The French ore deposits in Lorraine, for example, contained much phosphoric material. These ores could not be exploited until later in the 19th century when metallurgical science developed a process to separate the phosphoric adulterants from the iron ore and produce usable steel.

It should be noted that natural resources are not a fixed quantity, for certain resources are not available until technology reaches the level which enables them to be exploited. The uranium-bearing ores of the Belgian Congo, for example, became valuable natural resources only when the scientific technology of the 20th century found a large-scale use for uranium. Economic need and technological level are closely intertwined in determining the role of natural resources. Coal was not important in English industry until the supply of wood (for charcoal) began to give out in the 17th century. France did not have a similar charcoal shortage because it had forests to supply adequate metallurgical fuel. In Britain, the use of coal, which arose from necessity, became the basis for further technological advance.

THE TRANSPORTATION LINKAGE

Raw materials, producers, and markets must be brought together. Transportation is basic even for the most primitive production; the wheel, the ox-cart, and the pack ass, for example, played an important role in the prehistoric transition to the Bronze Age. In an industrial age secondary benefits derive from the provision of transportation facilities: the development of engineering skills is stimulated, and transportation itself becomes a major industry. The building of the railroads, for example, encouraged the growth of the iron and coal industries.

At the very beginning of the Industrial Revolution, transportation was recognized as the hinge upon which the development of industry turned. Here again, England was ahead of France during the crucial period of preparation for the Industrial Revolution.

The 18th century was a period of great road- and canal-building activity. In

England privately owned turnpike companies built a wide network of roads. In France too there was much road-building activity through the application of a road-work tax and the beginning of training engineers for the construction of roads. The École des Ponts et Chaussées, founded in the mid-18th century, was the first engineering institution ever established, so that France for a time was ahead of England in the design of its roads. However the road network in France was developed to fit the needs and desires of the absolute monarch. The French monarchy wanted to dominate the Continent, so roads were built for purposes of military strategy. At the same time, the monarchy desired straight roads radiating from the seat of government in Paris and Versailles. Unfortunately, the needs of the military and of courtiers do not always correspond to the needs of commerce and industry. England, free from the threat of invasion by virtue of her insular position, built its roads to follow routes of commerce rather than for military or political considerations.

There are also, of course, geographical differences between the two countries. England is a "tight little island"; France is a much larger country with an extraordinary number of geographical and geological variations which add to the difficulties of building an adequate transportation system. The shorter distances in England meant that it was commercially feasible to develop canal and road links between points of production and distribution, whereas the greater distances in France meant that in many cases it was not.

In the matter of internal politics, England possessed fewer hindrances of prejudices, traditions, and institutions to the movement of raw materials and goods. Under the Old Regime (that is, the government of France before the Revolution of 1789) French commerce was hampered by local tolls, internal customs duties, and by carrying monopolies. These obstacles to the free flow of trade were not wiped out until the French Revolution, whereas England had done away with most of these ancient encumbrances at a much earlier date.

In foreign transport England was also ahead. Britain had for a long time depended upon overseas commerce while France was not nearly so dependent upon foreign supplies or markets. Hence the British merchant marine was larger and more efficient than the French throughout the 18th century.

INVENTORS

Because of the spectacular inventions which ushered in the Industrial Revolution and because most of the inventors were British, it would be simple to attribute Britain's priority in the Industrial Revolution to the fact that its inventors were more ingenious than their French counterparts. But such a simple explanation shows how easily factual data can be misinterpreted. While there was a rich supply of inventive geniuses in England and Scotland in the 18th century, Frenchmen were just as smart and inventive, and we have some very great French inventors of this period—Vaucanson and Jacquard—to prove it.

However, as in the case of so many other prerequisites of industrialization, qualitative factors predominate over the quantitative. France and Britain were perhaps equal in the mass of brainpower, but here again, the economic, social, and technical opportunities available in Britain were more favorable to technological innovation. Hence British inventors had more profitable opportunities to turn their ideas into actuality than did their French counterparts.

THE ROLE OF INDUSTRIAL TRADITION

Paradoxically enough, the opportunity for the exercise of the profit motive in England might have been greater because of her lack of industrial tradition in the past. Rigidity in business methods and in production techniques can be a retarding factor in industrial development, and this was to prove one of France's failings in the mid-18th century. Ever since Colbert, Louis XIV's Minister of Finance, had established fine luxury manufacturing at the close of the 17th century, French manufacturers were loath to depart from those business and technological practices which were so highly profitable.

England, on the other hand, had been backward in luxury manufactures during the 17th century and still employed low-profit craft techniques. Therefore when it was possible to make profits by the introduction of new machines and new methods, British entrepreneurs were quick to grab at the opportunity. French manufacturers saw little need to jeopardize their high profits by industrial innovation. Later in the 19th century the British, enjoying industrial leadership and high profits, saw little need to modernize their machinery and techniques of production. The result was that industrial leadership passed to the innovative manufacturers of Germany and the United States.

Without the breakdown of traditional attitudes and without the disintegration of rigid behavioral patterns from the past, industrial innovation cannot take place. Had there been an unwillingness to break the chain of routine, had the workers accustomed to agricultural pursuits remained steadfastly attached to doing the same kind of work in the same old ways, they could never have developed new technical skills nor could they have adapted to factory work.

ENTREPRENEURS

Equally important was the breakdown of traditional business patterns, stimulating 18th-century entrepreneurs to invest in new enterprises and to foster technological growth. Medieval merchants had an "artisan mentality"; they thought "small." With the Commercial Revolution of the 17th century, we find the formation of a class of entrepreneurs who were prepared intellectually and technically to exploit the potentialities for profit through new enterprises and industrial innovations. This new entrepreneurial spirit was of prime significance, for much of the technological advance of the 18th century lay not so much in entirely new techniques as in the harnessing of the traditional skills of clock-makers, mill-wrights, blacksmiths, and the like to novel productive organizations and new needs envisaged by the entrepreneurs.

When we compare 18th-century France and England in respect to entrepreneurial advantages, we certainly find no differences in the profit-seeking tendencies of their businessmen. English merchants were no more greedy than businessmen of France or elsewhere. However, in terms of the opportunities to exercise that desire for profit, English capitalists had greater opportunity for giving vent to their profit-making urge than did their counterparts in France. For one thing, the English social system and scale of social values did not discourage industrial and commercial activities to the same extent as in France. Although the landed aristocracy was still the uppermost social class, business participation in England did not carry quite the same social stigma which it did in France. Furthermore, English commercial magnates found it easier to rise in the social scale and to amalgamate with the landed aristocracy. In 1688, for example, the Glorious Revolution in England had brought together the commercial and landed goups to form a ruling elite.

This was not the case in France. Even though French financiers found it possible to buy titles of nobility, they were still sneered at and despised by the nobility with older claims to titles. While English aristocrats could invest in commercial enterprises without suffering social ostracism, it was beneath the dignity of the French aristocracy of the Old Regime to dabble in business. This is one example of the degree to which the social attitudes of the British were more favorable toward business activities than were those of the French until the French Revolution gave the businessman a larger role in society.

AGRICULTURAL CHANGE

Another prerequisite for an industrial transformation, which is often neglected and yet is of profound importance, is the need for advances in agriculture. An agricultural revolution played as large a part in the great industrial change at the end of the 18th century as did the engines and machines to which attention is more usually directed.

An advancing agriculture was indispensable for technical improvement, even in prehistory. At the end of the Neolithic period, for example, the development of agriculture made it possible to support those technological specialists who were not directly engaged in food production and who ushered in the Bronze Age. The growth of great factory centers in the 18th century would have been impossible if agricultural production had not advanced sufficiently to provide for the needs of a large industrial population.

As we have seen, Britain outstripped France in agricultural advance during the 18th century. The intoduction of the Norfolk (four-field) system, the use of nitrogen-restoring crops, the development of scientific stock-breeding, and the enclosure of the open fields and commons—all helped to augment English agricultural production and to enable Britain to feed its growing population of factory workers. France under the Old Regime lagged behind Britain in introducing the new agricultural techniques and hence did not possess the essential agrarian basis on which to found an industrial revolution.

GOVERNMENT POLICIES AND INDUSTRIALIZATION

Industrial advance is much affected by governmental policies and institutions. During the 18th century the English Parliament was controlled by an oligarchy of the commercial magnates and landed aristocracy. This group was consistently devoted to the expansion of commerce and to the prosperity of agriculture, as demonstrated by the Enclosure Acts (inspired by the capitalistic development of agriculture) and the Navigation Acts (resulting from the pressure of commercial and shipping interests). As a rising industrial bourgeoisie began to rival the power of these special interests, it too began to find the government responsive to its demands. Its primary demand was that the government pursue a policy of laissez-faire; that is, keep its hands off business and allow industry to develop as it pleased. Such policies made a substantial contribution to industry in England at this stage of its development.

While England was encouraging industrialization, the French monarchy was doing little in its behalf. This was in striking contrast to the attitude of the French government at the end of the 17th century. Then, under the leadership of Colbert, France had pursued mercantilist policies which had encouraged industry. After Colbert, however, the French government scarcely concerned itself with the welfare of business, and when it did do so its clumsy interference hampered industrial expansion.

England's industrial progress under a laissez-faire policy led Arnold Toynbee to state that the "essence of the Industrial Revolution is the substitution of competition for the medieval regulations" which had controlled the production and distribution of wealth. Although free competition, as advocated by Adam Smith and other laissez-faire economists, helped to stimulate British industry in the 18th and 19th centuries, in other times and places government regulation, the very opposite of laissez-faire, has helped to stimulate economic and industrial growth. The point is that the government must provide a milieu favorable to industry either by direct action or by being purposefully passive. It must be responsive to the needs of industrial growth, as was England in the crucial middle decades of the 18th century when the Industrial Revolution had its beginnings.

SOCIAL ATTITUDES TOWARD INDUSTRIALIZATION

Not only the government but the entire society must develop values, attitudes, and institutions favorable to industrialization. Specifically these would include a desire for material progress, the approval of social mobility, a willingness to accept new ideas and techniques, and an appreciation of technological advance as leading to material betterment. These attitudes cannot be developed overnight.

The change in values and attitudes requisite for industrialization had arisen slowly during the early modern period as Western Europe moved from the medieval emphasis upon the spiritual aspects of life into a secular and quantitative approach to life. By reviving pagan notions concerning the value of life here on earth, the Renaissance stressed material goals. Despite its emphasis upon Classical antiquity, the Renaissance produced an open-minded attitude toward novelty which helped to break the crust of custom and to encourage the reception of inventive ideas which affected industrial change.

The transformation in values also owed much to the Protestant Revolution. Some scholars, particularly Max Weber and R. H. Tawney, have claimed that the Protestant ethic—with its emphasis on hard work, sobriety, thrift, and the relentless pursuit of one's calling—fostered the growth of capitalism in Protestant countries. This in turn led to a frame of mind which hastened the oncoming of industrialization. Although the Weber-Tawney thesis has been disputed, there is no doubt that the Protestant Revolution marked a major step in the secularization of society. At the same time, in line with the emphasis on material betterment, production became directed toward utility as well as aesthetic and spiritual satisfaction.

In England during the 16th and 17th centuries, many developments helped to destroy the power of old institutions and traditional habits of mind and to foster a frame of mind favorable toward industrialization. First was the break with the Roman Church, leading to the dissolution of the monasteries and the creation of a class made wealthy by the acquisition of former Church properties. Then came the parliamentary revolutions and religious struggles of the 17th century. These developments helped to pave the way for new ideas, new processes, and a new society.

Concomitantly, a Scientific Revolution was taking place. Although it is difficult to detect any direct and immediate influence of scientific discoveries upon technological developments during this time, the Scientific Revolution, exemplified by Sir Isaac Newton, helped to diffuse a general attitude of receptiveness to new ideas. By the mid-18th century, Dr. Johnson was moved to state, "the age is running mad after innovation."

TECHNOLOGICAL LEVEL AND SOCIAL NEED

Many discoveries have undoubtedly perished with their discoverers because they did not fit the needs or techniques of the times. For example, the Greeks of the Hellenistic period knew something of the energy of steam, but they utilized this principle only in the making of a toy because the social and economic conditions of Hellenistic life did not favor its development as a machine to do work. Even if social and economic conditions had been favorable, Hellenistic metallurgical techniques would have been inadequate to produce a steam engine sufficiently powerful to do meaningful work. Similarly, Branca designed a steam turbine in 1629, but nobody was capable of actually producing one until two centuries later.

Had James Watt's invention come earlier, it is possible that the techniques and machines to produce the metal shapes basic to his engine would not yet have existed. What is more important, it might not have attracted sufficient capital for manufacture on a commercial scale, for there would have been little economic need for it. The steam engine was an outgrowth of the need to pump out mines; and 150 years earlier there would not have been the same incentive to develop a machine which could perform that task.

There are many examples of the retardation of technological advance because of the inadequacy of contemporaneous technical means. In 18th century Britain socio-cultural factors, economic needs, and technological capabilities came together in a happy combination to foster industrial advance.

BRITAIN'S LEADERSHIP

In short, there was no single factor which can account for Britain's leadership in the Industrial Revolution. Instead, it was a multiplicity of factors—technological, social, economic, political, and cultural—which came together in the mid-18th century to provide the stimulus for industrial advance.

In all these factors, Britain had a slight advantage over France. But the advantage was qualitative rather than quantitative. France had more people, more capital, more trade, more roads—more of everything—but British labor, financial resources, market opportunities, transportation network, etc. were more available and adaptable to industrial production.

Not the least of the factors which favored Britain's industrial leadership were the socio-cultural elements: social attitudes toward change, governmental response to business requirements, industrial tradition, and so on. The great engineering inventions which characterized the Industrial Revolution would have been impossible without the dynamic interplay of a great many forces which brought together social stimulus, technical level, economic need, and inventive genius.

Most, if not all, of these elements came together first in textile production in Britain, the first manufacturing enterprise to undergo industrialization.

14 / The Textile Industry, 1750-1830

ABBOTT PAYSON USHER

The introduction of power machinery into the textile industries of Great Britain, Europe, and America between 1750 and 1830 was fundamental to the success of the Industrial Revolution and the cause of enormous changes in Western society. These years saw the first phase of a transition from industries organized on the merchant-employer system to those organized on the factory system. Decentralized capitalist control was supplanted by fully centralized control accompanied by new and often onerous discipline and supervision. These new conditions of work (and therefore of living) were so distasteful to the workers that they were accepted only when mechanization had proceeded so far that decentralized production could no longer compete in the market place. The introduction of power machinery into the textile industry on a comprehensive scale was the underlying factor in the change.

THE OLD ORGANIZATION OF PRODUCTION

Under the old system of textile manufacture the merchant employer purchased raw materials which were then given out to workers in farms or in town cottages. Weavers sometimes owned their own looms, and their cottages, but in many instances they rented fully equipped cottages from the employer. In some regions, notably in Norfolk, England, the merchant employer confined his activity to the production of yarn, which was then sold to local weavers or exported. If finished cloth was produced, finishing processes were carried out in a central shop directly controlled and supervised by the employer.

The details of this oganization of production varied considerably from region to region and from period to period. The capitalist financed the system of production and disposed of the product. The actual work was done on farms and in cottages, so that there was no possibility of the employer exercising much supervision over the work. It was also difficult to enforce standards of quality, or to maintain a schedule for the completion of the work given out. Embezzlement of raw materials was common.

The factory, as a centralized establishment, offered opportunities for supervision and discipline that were clearly advantageous to the employer. Attempts were made to establish centralized workshops even before any changes were made in the technology of production. A number of large cloth-making establishments were set up in the west counties of England in the 16th century, and a number of privileged establishments (state-chartered monopolies) were set up in France in the latter part of the 17th century. But these factories showed little vitality and did not compete successfully with the merchant-employer system. One abortive effort was made in the silk industry, where a power-driven throwing-machine was developed in Italy toward the end of the 13th century. On this basis, silk-throwing became a factory industry. The machine was introduced into England by Thomas Lombe in 1721, but only one factory was established before 1750.

EARLY MECHANICAL DEVELOPMENTS

The history of the textile industries after 1750 is marked by the development of power machinery and by the gradual emergence of the factory system as a dominant form of organization. The rate of development closely followed the advances in the mechanization of production. An understanding of the broad outlines of the technology of the textile industries is, therefore, essential to any accurate appreciation of the economic and social changes.

At the beginning of the 18th century the primary processes of preparation, spinning, and weaving had been only slightly mechanized. Carding and combing for preparation of the fibers remained hand processes. Spinning the fiber into yarn had been significantly improved by the addition of the flyer to the spindle in the 15th century, but the looms for weaving the cloth had not been notably improved since the beginning of the Christian era.

Then the cloth had to be finished. The disagreeable process of fulling (shrinking and consolidating the cloth) had benefitted by the introduction of the fulling mill early in the Middle Ages. Formerly the fulling had been done by treading on the cloth in a vat filled with a strong solution of soap and urine; this task was now taken over by a simple mill which pounded the cloth with pestles moving vertically, or hammers set at an angle. The fulling machines were driven by water power or by winches turned by animal power. In the 16th century gig mills were introduced to raise the nap (surface) on woolens by working over the surface with teasels (the barbed fruit of certain plants). But this machine did not come into general use in Europe.

This thin record of technical improvement in the textile industry was not due to lack of imagination, nor to failure to realize the advantages which might be achieved by mechanization of these laborious and time-consuming processes. The development of the flyer covers about a century (1430-1530) between the earliest illustrations of the device and the mature spinning wheel with a treadle drive. In this period, Leonardo da Vinci sketched devices for the application of power to a spindle and flyer, and his notebooks also contain drawings for improved gig mills and for a power loom. There were actual attempts to develop a power loom in France. The draw loom, so important in the silk industry, was progressively improved in France in the course of the 18th century, and the Jacquard action (1804) successfully mechanized the greater part of the process, though power was not yet applied.

The Dutch ribbon form was the most important achievement prior to 1700 in the field of woven textiles, although the knitting frame invented by William Lee in 1589 must also be counted as a major invention. It was a significant mechanical advance, and it gave a new impulse to the mass production of an important series of consumer goods.

The relatively late achievement of mechanization of spinning and weaving was due to the essential difficulties in the production and control of the motions underlying these processes. The carded and combed yarns were fragile and incapable of taking much strain until considerable twist had been imparted to them by the spindle and flyer. For its part the loom presented a series of problems in the control of tension. The mechanical achievements of the 18th century in textile manufacturing were dependent on the great advances in light engineering at a craft level in clock- and watch-making, in lathe work in wood, and in the various crafts working with non-ferrous metals.

The importance of clock- and watch-making is frequently underestimated. All the problems of the construction and control of geared mechanisms were involved. The higher levels of work in England and Europe embraced a full mathematical analysis of the pendulum, and of the design of gear teeth. Problems of compensation for temperature and position were fully studied. Through the work of Huygens, Clement, Hooke, LeRoy, Harrison, and Berthoud, both the pendulum clock and the spring-driven chronometer became instruments of precision. Over a period of three centuries, the clock- and watch-makers provided the best instruction in engineering and made outstanding contributions to the formulation of principles.

SPINNING

The early efforts to mechanize spinning have traditionally been attributed to the imbalance between weaving and spinning created by the invention of Kay's flying shuttle. The resulting increase in the efficiency of the loom is presumed to have created a greater demand for yarn so that there was a new incentive to improve the productivity of the spinners. This rationalization is improbable, because the flying shuttle was not adopted generally until after the initial inventions in spinning were well launched. So the spinning inventions can best be treated as an independent episode. The breakthrough to a solution of the problems of spinning is to be found in the work of John Wyatt and Lewis Paul. Unfortunately, the relations between the two men are not fully known, and the record of their earliest work on the spinning machine is incomplete.

John Wyatt was born near Litchfield, England, in 1700. Trained as a ship carpenter, he soon became interested in the metal trades. He was an inventor with a fertile imagination and was continuously occupied with the development of new projects. His invention of a machine for turning and boring metals proved important for his career, because after he failed to get expected financial aid from an armorer in Birmingham, he turned for help to Lewis Paul. Rights were made over to Paul to satisfy debts which Wyatt had incurred.

Lewis Paul was the son of a French refugee to England. A protégé of the Duke of Shaftesbury, he had contacts with well-known personalities, including Dr. Samuel Johnson, the lexicographer. Paul seems to have been an inventor in his own right, but few of his papers have survived.

Wyatt's son claimed that the spinning machine was invented by his father about 1730, and that a small model was actually operated at Sutton Coldfield in 1733. This reminiscence is not now accepted. The association of Wyatt and Paul began about 1733, but the earliest patent was taken out by Paul in 1738. There are passages in Wyatt's letters which acknowledge that the essential concept of the machine was the work of Paul.

There is no drawing of the device in the patent records, but the surviving description is important because it specifies the use of more than one pair of rolls, thus making it possible to spin more than one thread at a time. The carded cotton or wool was formed into a continuous sliver and "put between a pair of rollers, cylinders, or cones, or some such movements, which being twined around by their motion, draws in the new mass of wool or cotton to be spun, in proportion to the velocity given to such rollers, cylinders, or cones; as the prepared mass presses regularly through or betwixt these rollers, cylinders, or cones, others, moving proportionally faster than the first, draw the rope, thread, or sliver into any degree of fineness that may be required." From these rolls the thread passed to the flyer and spindle. The twist was imparted by the flyer as in hand spinning, and the finished yarn was wound up on the spindle. If the spindle revolved faster than the rolls, the yarn was built up on the bobbin under tension; if the spindle revolved more slowly, the thread was deposited on the bobbin without tension.

Their first machine was set up in London in 1740, and other mills were set up at Birmingham in 1741, at Northampton in 1742, and at Leominster in 1744. Certain passages in the correspondence between Paul and Wyatt show that there was much uncertainty in their minds about the operation of the rolls. They had doubted the wisdom of using two or more sets of rolls, and of subjecting the carded sliver to any drawing out before it had been given any twist by the spindle and flyer. A second patent taken out by Paul in 1758 suggests that the attempt to draw out the sliver between the rolls was definitely abandoned. But it should not be presumed that the idea of mechanical spinning was given up.

Work on a carding machine was begun by Bourne at Leominster in 1743, and by Paul in 1748. Both men took out patents in 1748 on carding machines operating on different principles, and it is clear that both men assumed that the entire process of yarn production could be mechanized. Paul's correspondence shows that he believed that the vital problem was to determine the proper rates of revolution for the spindles and the rolls. These difficulties were ultimately surmounted by passing the carded slivers through a preliminary spinning process in a machine with rolls and flyers, but designed to produce only coarse rovings (extended, twisted fibers). It is dangerous to presume that Paul had achieved full success in the management of the rolls, and the spindle and its flyer, but his accomplishment was certainly greater than is commonly recognized.

About 1750, the cotton industry in Lancashire began to develop rapidly. Calico printing made much progress, and high grades of fabrics were added to the well-established type called fustians (coarse cloths of cotton and linen). There was, thus, a strong demand for yarn, so that there was a great interest in the improvement of spinning. In 1761, the Royal Society of Arts offered "rewards" for the best invention to spin six threads of wool, flax, hemp, or cotton at one time "and that will require but one person to attend it." Several machines were submitted to the Society. One award was made in 1763 for a six-thread machine, but nothing more was heard of it.

The next few years witnessed the invention of most of the famous spinning machines of the era. James Hargreaves's spinning jenny was a machine of this type, but we do not know if the reward offered by the Society of Arts stimulated his work. It is alleged that he invented the jenny in 1764, though not much is known about it until 1767, and it was not patented until 1770. Arkwright's first patent for the water frame was taken out in 1769. His second patent, for the carding engine, and the use of rolls in preparing rovings, followed in 1775. Crompton's work on the mule began about 1774, but the machine was not completed until 1779. These machines (see Figs. 14-1, 14-2, and 14-3 for a description of their mechanical features and operation) were complementary rather than directly competitive, and the mule combined features of both the jenny and the water frame (hence its name, indicating a hybrid).

The carding engine was no less important than the spinning machines. It supplanted a laborious hand process and was essential to the full mechanization of processes from the raw cotton to the finished yarn. An incomplete machine, patented by Lewis Paul in 1748, was coming into use in Lancashire about 1760. The calico printer Robert Reel employed James Hargreaves to make improvements in the carding engine, but we hear no more about the engine until Ark-

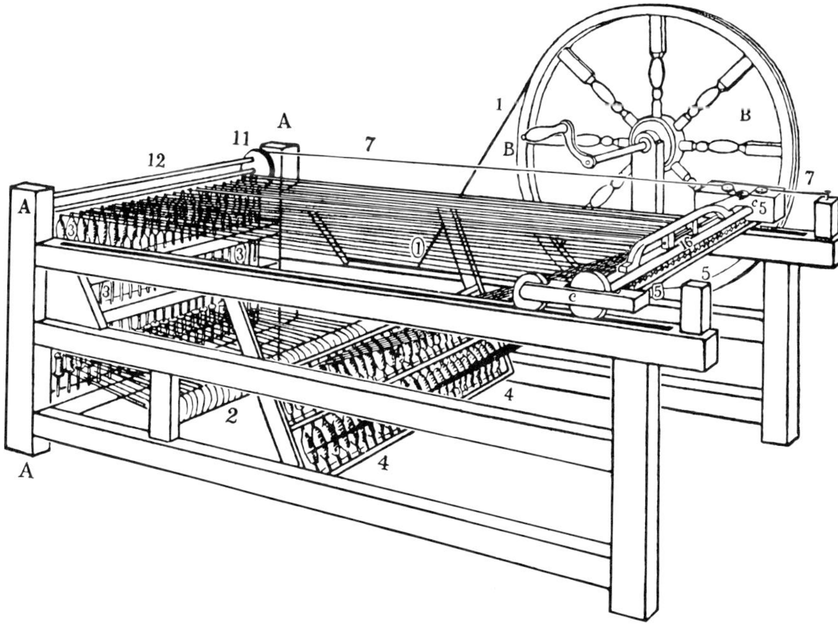

Fig. 14-1. Hargreaves's spinning jenny. The jenny is shown in the illustration in a slightly improved form. There is a box (4-4) underneath, which contains rovings, previously prepared on a spinning wheel. The carriage (5-5) with a moveable clasp bar firmly holds the rovings which pass through to the spindles (3-3). The spindles are rotated by means of the large wheel (B-B). The jenny was worked by one person, who took up his position in front of the frame. The rovings were then drawn between the "clove" of clasp bars of the carriage, having first been placed in position for commencing work at the end of its traverse, nearest the spindles. After the bottom bar was lowered, the carriage was drawn away from its position, until enough rovings to form one "draw," or length of yarn, had been given out, the amount of which was regulated by a mark on the side of the frame. The lower bar was then raised, the rove held, and the spindles set in motion by the spinner turning the wheel (B), at the same time commencing to draw the carriage further out from its position near the spindles. The attenuation and twisting of the rove went on simultaneously until the requisite degree of fineness was attained; when the outward traverse of the carriage was stopped, the spindles were in operation for a short time longer in order to impart sufficient twist to the thread. In the yarns for warps this was much more than for wefts, which did not require the same degree of strength. When this twisting had been completed, the carriage was slightly backed, and the guide or faller wire (12) was gently brought down upon the threads, by means of the cord (7) depressing them to the required level; the wheel (B) was then turned slowly round, causing the spindles to wind up the thread as the carriage returned to the first position.

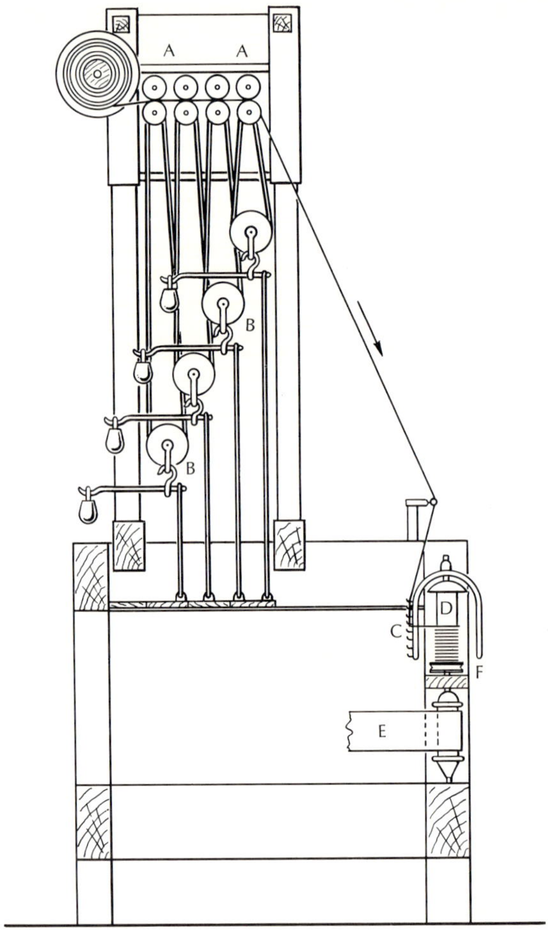

Fig. 14-2. Sectional drawing showing the principle of throstle or water frame. There are four pairs of rollers at the top; the upper rolls are kept in contact with the lower rolls by weights. Supposing the rolls numbered from left to right, the second, third, and fourth pairs revolve more rapidly than each preceding pair. The roving was thus drawn out, in the course of its passage through the rolls, to a degree of fineness that could be regulated by variations in the speed of the rolls. Leaving the rolls, the roving passed downward to the flyer and bobbin, receiving the twist necessary to form it into yarn by the rotation of the flyer (C). The finished yarn was reeled upon the bobbin (D) because of a difference in speed between the flyer and the bobbin: this could be accomplished either by driving the bobbin faster than the flyer, or by driving the flyer faster than the bobbin.

wright patented (1775) a machine with a crank and comb to take the carded sliver from the rolls. The sliver was delivered in a continuous strip into a tall can. In other machines the sliver was taken off in short strips that were later put together. There exists no satisfactory drawing of the carding engine in its early state.

The mule was never patented, for Crompton believed that it would not be possible to enforce patent rights over a machine that could be so easily produced. It seems likely that the use made of the rolls would have been an infringement of Arkwright's patents of 1769 and 1775, so that a license would have been necessary. The throstle was adopted to the production of the coarser yarns, in counts up to 40, but notably in the 20's or lower (the lower the count, the coarser the yarn, a system of designation still preserved on spools of thread). The mule was capable of spinning the higher counts used in the muslins and fine calicoes, so that it was competitive with the fine hand-spun Indian yarns, which commonly ran as high as 80, or even 120. For exhibition purposes, counts as high as 350 were produced on the mule.

Although the jennies were ultimately supplanted by the mule, for more than a generation they afforded support to the smaller enterprises and gave a longer lease of life to the independent weavers. Hargreaves secured a patent on the jenny, but he was never able to enforce his right to compel users to pay license fees. Soon after the patent was granted, he offered a reward of ten guineas for information about illegal use of the jenny. At a meeting with manufacturers at the Bull's Head Inn, in Manchester, October 2, 1770, the manufacturers offered £3000 for the right to use the machine. Hargreaves asked £7000, but later reduced his demand to £4000. Preparations were made to bring suit, but it appeared that jennies had been sold to manufacturers before the patent was granted, so that there was no hope of enforcing exclusive rights under the patent.

Arkwright was not much more successful. No suits were brought to defend either of his patents until 1781, when a suit was brought to prevent the use of the carding patent without a license. The defense claimed that Arkwright's patent specifications were incomplete, and the judgment accepted this claim. Arkwright then petitioned Parliament to have the patents of 1769 and 1775 consolidated and to extend their life for the normal term of the second patent. The incomplete specifications, he explained, were simply a means of preventing foreign manufacturers from using the patent. When no action was taken in Parliament Arkwright collected evidence to show that the specifications of the carding patents were sufficient and brought suit in February 1785 on the issue of the adequacy of the specifications. The defense was not prepared to meet this issue, and made no attempt at this trial to challenge the originality of the inventions. Arkwright won this case.

The verdict in favor of Arkwright filled many manufacturers with dismay, since they had projects for substantial extensions of both carding and spinning. A writ was secured to test the validity of the patent of 1775 on three primary

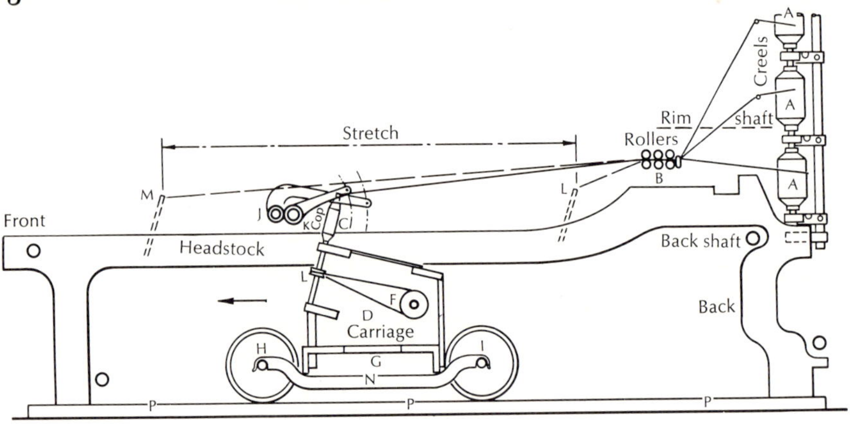

Fig. 14-3. Crompton's mule. The rovings held on the creels (A, A, A) were passed through the rolls below and thence to the spindles when the carriage was nearest the rolls (at L). The cycle of operation began by setting the rolls in motion and moving the carriage away from the rolls toward the front of the frame. The spindle revolved to give a twist to the thread. Shortly before the carriage reached the limit of its run the rolls stopped, so that the yarn was drawn a little more after the twisting process was nearly complete. The yarn could bear tension at this point, and as a consequence, fine yarns could be produced.

issues: that the roving machine was a repetition of the throstle for which the patent had expired; that Arkwright was not the inventor of the spinning machine patented in 1769; and that the improvements in the carding engine were the work of Hargreaves. At another trial in June 1785 the defense produced an imposing array of witnesses, whose testimony left Arkwright without the slightest claim to the invention of any major feature of either the spinning machine or the carding engine. An unfavorable verdict was inevitable.

After the trial, Arkwright secured new evidence for his claim to the invention of the crank and comb on the carding engine, and attempted to refute the claims of others regarding roller spinning. The court, however, refused to grant a new trial. The son of Hargreaves's partner later declared that the damaging testimony of Hargreaves's widow and son about the carding engine was in error and that Hargreaves had learned about the crank and comb from Arkwright.

Wholly apart from the patent contests and the questions of priority of invention, the importance of spinning mechanization for the development of the factory system is best shown by the bare chronology of its development. Factories on the basis of Paul's machine were set up in London (1740), Birmingham (1741), Northampton (1742), and Leominster (1744). After the early start in Derby in 1720, silk-throwing developed as a factory industry in Congelton (1752), Stockport (1752), Macclesfield (1756), and Milford (1775-76).

The early Arkwright mills were: Nottingham (1769), Cromford (1771), Belfer (1775-76), Birkacre (1777), Holywell (1777), Taplow (1780). In 1788, a pamphlet on the cotton industry gives the distribution of 120 spinning mills as follows: the Lancashire group, including Lancashire, Cheshire, Flintshire, and Westmoreland, 57; Derbyshire, 17; Yorkshire, 11; Staffordshire, 7; others in England, 9; Scotland, 19.

WEAVING

Almost all weaving was done on hand looms until after 1810, even though in many establishments the weavers were collected in large buildings with fully centralized organization. The development of the power loom marks the transition to another phase of the Industrial Revolution in which all the processes were competently mechanized.

The primary inventions in weaving cover the long span between Kay's flying shuttle of 1733 and the decisively successful loom of Roberts that was patented in 1823. During this critical period, machine-building came to be the work of skilled workers in iron and steel using precision lathes. The craftsmen of the clock and watch trades were thus supplanted by men who can properly be described as professional engineers. There were no formal schools, but the shops of men like Maudslay, Clement, and Roberts furnished the type of apprentice training which was equivalent in character to that of the scientists of the period who were trained in the private laboratories of chemists and physicists. Cartwright was the last of the great inventors who belong to the craft period.

John Kay had been a reed-maker. (A reed is a device on a loom, by means of which threads are drawn between the separated threads of the warp.) He made a number of inventions of minor importance; he substituted wires for split cane in the reed, and improved the making of a special twine for incidental work with the reed. In 1733 he patented a carding engine and the famous flying shuttle. This device increased the speed of weaving even on the narrow goods (18-30 inches), and dispensed with the second weaver that had been necessary on the broad cloths. This important invention came into use very slowly, obstructed by both indifference and violence.

The flying shuttle provided a solution to the essential problem of applying power to the loom, but though it established the method of producing the necessary motions, it left serious problems in their control. The force required to drive the shuttle into the shuttle box was as great as the force needed to carry the shuttle through the web. These forces must also be checked by the springs and other arresting features of the shuttle box. It was essential too that a power loom be stopped if a thread broke in the warp or in the shuttle. The piece of cloth in the loom must be kept at a constant width. In the hand loom simple laze rods could be kept in the warp, to be moved forward at intervals by the

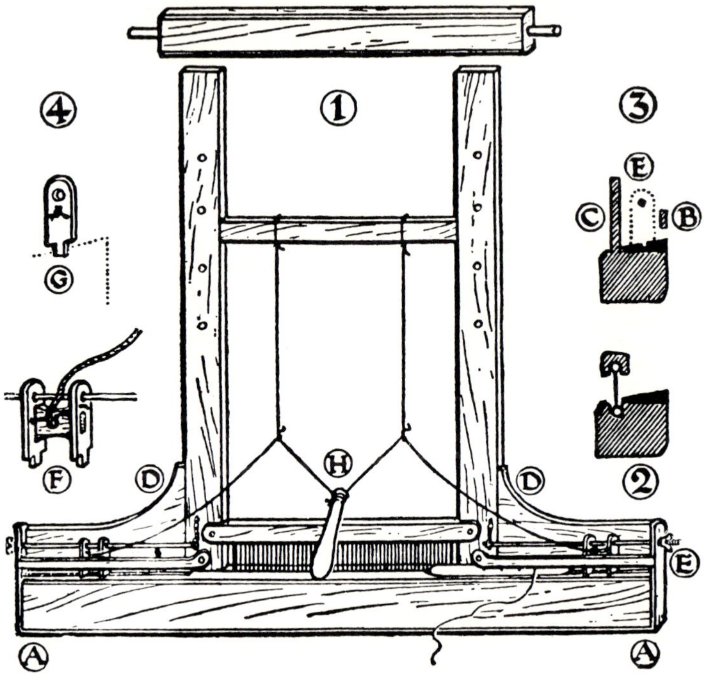

Fig. 14-4. Kay's flying shuttle. In its original form the reed batten performed only one function, the beating of each new weft thread into the cloth. It was suspended from the top of the loom and controlled by the weaver by hand or by treadles. Kay built out the ends of the batten to provide boxes (3) at each end for the shuttle, and a runway (2) was made behind the reed so that the shuttle could pass through the open warp and come to rest securely in the box toward which it was moving. The boxes were tapered so that the shuttle would not rebound. The shuttle was put in motion by the pickers (F), which could be given a sharp flick by the cords connecting them to the handle (4).

weaver. In the power loom, temples were set at the sides of the loom to hold the web at a proper width. Even the breakthrough achieved by Kay's flying shuttle did not quickly inspire serious work on the loom, though some new types were produced and patented.

Major work on the loom was initiated by Edmund Cartwright, an Oxford man who had prepared for the Church. Though he was made a fellow at Magdalen College, he accepted instead a post as a country clergyman: first at

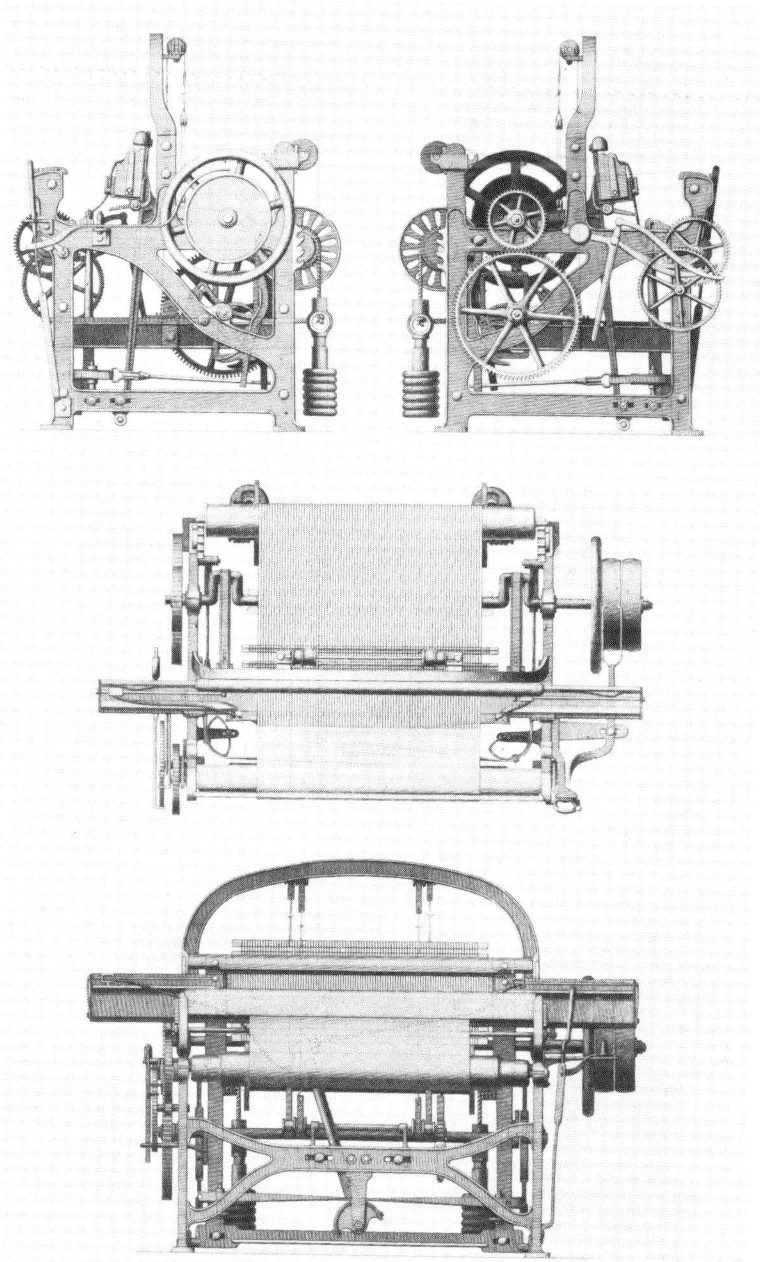

Fig. 14-5. A power loom in 1835. *Top*, end views; *center*, top view; *bottom*, front view. (*Science Museum*, London)

Brompton in Derbyshire, later, at Goadby Marwood in Leicestershire. Once established in his rural parish he took an interest in medicine and agriculture, and was ultimately drawn into an interest in industry. On a visit to Manchester he heard much about the new spinning machines which would produce more yarn than the weavers could weave. Henceforth Cartwright applied himself to the problem of weaving, although he had never seen a loom in operation before. "It struck me that as plain weaving can only be three movements which were to follow each other in succession, there would be little difficulty in producing them and repeating them. Full of these ideas I immediately employed a carpenter and a smith to carry them into effect. As soon as the machine was finished, I got a weaver to put in a warp which was of such material as sail cloths are usually made of. To my great delight, a piece of cloth, such as it was, was the product. The reed fell with the weight of at least a half a hundred weight and the springs which threw the shuttle were strong enough to have thrown a Congreave rocket. In short, it required the strength of two powerful men to work the machine at a slow rate and only then a short time. I then secured what I thought was a most valuable property by a patent on April 4, 1785."

Cartwright's model was developed substantially by improvements covered by patents in 1786, 1787, and 1788. In 1787, a small factory was set up at Doncaster equipped with 20 looms; 8 for calicoes, 10 for muslins, one for cotton checks, and one for coarse linen. Although this factory was not successful, in 1791 the Grimshaws of Manchester planned a larger factory with 400 looms. However, before it was brought into operation, the buildings were attacked and burned to the ground by a mob of weavers fearing the loss of their jobs. There were other attempts to use Cartwright's patents, but the general adoption of the power loom waited upon the development of models by other inventors. New devices for sizing and dressing the warp proved to be essential.

The type of loom long dominant in England was first patented by Thomas Johnson in 1803 and 1805, and improved by Horrocks in two stages, 1813 and 1821. The loom was still unsatisfactory, so Richard Roberts undertook further improvements. Roberts was an outstanding representative of the new generation of machine-builders and inventors. His early training was at the Bradley Iron Works under John Wilkinson, and he worked for two years with Maudslay. In 1814 he set up his own shop at Manchester, where he built a precision lathe, a planing machine, and a punching machine. In the textile field, his primary achievements were the self-actor mule (1825) and the power looms (1822, 1830). The partnership of Roberts and the Sharp Brothers of Manchester—formed in 1828 under the name Sharp and Roberts—introduced these textile machines to the industry.

The early state of the power loom is illustrated by one developed by John Austin of Glasgow. He began work in 1789 and planned to take out a patent, but later gave up the idea. In 1796, his loom with improvements was demonstrated before the Glasgow Chamber of Commerce, and two years later 30 looms were installed in the spinning mill of Monteith at Pollackshaws, whose weaving shed was later expanded to accommodate 200 looms.

The importance of the improvements in the power loom is best shown by the numbers of power looms in use at different dates.

NUMBERS OF POWER LOOMS

	1813	1820	1829	1833
England	2,400	12,150	45,500	85,000
Scotland	?	2,000	10,000	15,000

THE DRAW LOOM

The development of the draw loom, used in the production of figured fabrics, was the work of a succession of French inventors. The French had always specialized in fine and expensive cloths of high artistic quality; hence they took the lead in mechanizing the weaving of patterns in cloth.

The draw loom is a complex mechanism because each thread in the warp must be controlled separately. In its early forms, the loom required several workers. The earliest modification known to us is a device called the "draw boy," which was introduced in the 15th century by Jean le Calabrais. Groups of cords required to produce a pattern in the fabric were connected with a single master cord so that the manipulation of the warp was simplified. The device was adequate for narrow fabrics, but impractical on broad webs, for the working of the cords was beyond the strength of the worker. This problem was solved by an arrangement made by a Lyonese workman Danjou in 1609. Another notable step was taken in 1725, when a punched cylinder of paper was applied to the task of selecting the threads of the warp to be raised or lowered. Work was done on this concept of the loom action by several inventors. In 1747, Vaucanson endeavored to improve these devices, but no complete machine was produced. The model was deposited in the Conservatory of Arts and Crafts, and about 1800 J. M. Jacquard (1752-1834) was urged to work over the model.

Jacquard built his first action in 1804, combining much of Vaucanson's work with the work of Falcon (1737), and it proved to be a complete success. There is still much discussion of the originality and importance of Jacquard's work. Some writers feel that he added nothing of importance to Vaucanson's action; others recognize him as a talented and resourceful inventor. The issue involves the concept of the process of invention. Many believe that the task of the inventor is completed when the underlying concept of a train of mechanism has been conceived. In the most extreme cases the bare idea is counted as an invention, even when no model had been made. When invention is understood to be a continuing process, any given train of mechanism will be regarded as a cumulative achievement that involves many separate and independent acts of insight, each of which must be considered as a distinct invention.

THE TEXTILE INDUSTRY IN 1830

The family of textile machines was still incomplete in 1830. The combing machines were far from satisfactory. The power loom was capable of much improvement, and the application of power to the Jacquard action was not perfected. Mechanization proceeded more slowly in the woolen and worsted industries than in cotton. The combing machines developed more slowly than the carding engine, so that the preparatory work was not mechanized as rapidly. The power loom made its way more slowly, because it was most economical for plain goods. Fancy-weave patterns needed in smaller quantities could be produced with economy on the hand loom. Indeed, the hand loom maintained itself with great tenacity for specialized types of weaving in England and Scotland, and on an even larger scale in the rest of Europe.

The new technology was more efficient in the cotton industry than in woolens and linen. The cotton industry was also favored by the changes in the supply of raw cotton. The introduction of Eli Whitney's cotton gin (1793) made the upland cotton of the United States a major source of supply at much lower price levels. Wool remained in short supply with prices at relatively high levels until Australian wool became a predominant factor in the market in the 19th century. The price of flax for making linen was also high. Thus, among the textile fibers, all the advantages were held by cotton. The extent of the change in the different textiles is shown by the data on the values of exports and home consumption in Britain, and the proportions of new raw material used.

RELATIVE POSITION OF THE BRITISH TEXTILE INDUSTRIES

	VALUES IN THOUSANDS OF POUNDS STERLING		
		EXPORTS	HOME CONSUMPTION
1783	Wool	3,700	13,100
	Cotton	360	600
	Linen	700	3,300
1859-61	Wool	15,041	24,959
	Cotton	49,000	28,000
	Linen	6,119	9,381

For other dates, we have the proportions of new materials consumed by the various industries.

	COTTON	WOOL	FLAX	TOTAL
1798-1800	16.08	42.15	41.77	100
1829-31	41.47	25.48	33.05	100
1859-61	68.40	17.42	14.18	100
1880-82	66.35	20.90	12.75	100

With the establishment of inspection under early factory acts statistical material that is extensive and trustworthy became available. For 1833, 237,000 operators are reported in cotton factories: 200,000 in England; 32,000 in Scotland; 5000 in Ireland. There were 9,333,000 spindles in operation, and 100,000 power looms. In cotton, there were 250,000 hand-loom weavers. The factory workers received £6,044,000 in wages; the hand-loom weavers, £4,375,000.

Although the record of the textile industry of Great Britain, taken as a whole, was not uniform during the years 1750 to 1830, the industry was sufficiently homogeneous to be distinctive. Where inequalities still existed—such as the amount of mechanization in the cotton as opposed to the woolen branch—the differences were neither radical nor permanent. The path of improvement laid out by the cotton industry was soon to be followed by manufacturers of wool and the other textile fibers.

The cotton industry influenced not only the rest of the textile industry but the entire Industrial Revolution as well; for it was in this industry that a small group of craftsmen-inventors, responding to the changing tastes and economic patterns of the nation, created the first and most influential modern industry. It was from this industry that manufacturers in other areas and in other fields drew their inspiration, and social critics, their polemics. The "dark Satanic mills" of the Romantic poets were cotton mills, and they owed their special character to the machine and its pervasive influence.

15 / The Steam Engine Before 1830

EUGENE S. FERGUSON

Before the 18th century, the industrial activity of Great Britain, like that of other countries, depended on the use of four basic sources of power: human, animal, wind, and water. Of these, water was the most desirable for many types of industry, especially those requiring large amounts of power. The number of mill sites suitable for such use was, however, strictly limited by the number of streams available and the fall of water along these streams. When the fall of water was used as efficiently as possible, and when the water-power sites were all taken up, no further growth was possible. By 1700 industrialization had progressed far enough that some pinch was being felt; in mining areas too, as the mines went deeper, there was a need for more power for pumping water from mines. Another source of power was much desired—one which could deliver large amounts of power per unit and which could be made available anywhere. The answer proved to be the steam engine.

The first full-scale steam engine, that is, one built for sale and not for the laboratory, appeared in England in 1699; the low-pressure steam engines of Newcomen (1712) and Watt (1775) dominated the 18th century; and the high-pressure steam engines of Trevithick and Evans were operating before 1802. Thus in a span of just over a century a new prime mover had appeared which would have an important effect upon the relationships between men and upon the way men lived and worked.

MINE DRAINAGE

The first use of the steam engine was to pump water out of mines, taking over the earlier work of men, horses, and in a few cases water power. As mines became deeper, horses had replaced men, but the number of horses that could be used depended upon the extremely limited access-opening to the mine shaft. Furthermore, horses had to be permitted to rest frequently, and the cost of continuous drainage in many mines crept closer and closer to the value of the coal or ore being mined. Water power could be used only when both flow and fall were ample—an exceptional set of circumstances at most mines. Thus there was a strong economic incentive to introduce an alternate device.

In 1698, Thomas Savery (?1650-1715), a native of a mining district of Devon and a prolific inventor, patented a water-pumping device that consisted essentially of a vessel (cylinder) that could be alternately evacuated and put under pressure. (The vessel was of moderate size, perhaps 50 gallons.) A vacuum could be produced by first filling the vessel with steam and then condensing the steam by pouring water over the outside; the pressure could be supplied by admitting steam from a high-pressure boiler. When the vessel was evacuated, water would be forced up from a lower level by the atmosphere to fill the vacuum, and when put under pressure, the water would be forced into a pipe extending upward from the vessel.

It was because of the high-pressure requirement that the Savery engine was unsuccessful in mine-pumping. Although the vacuum process raised water about 20 feet, to force vertically another 180 feet, for example, would require a pressure of about 80 pounds per square inch, which was well beyond the safe capacity of boilers that Savery could build. Mine depths of several hundred feet were at this time not uncommon; the Savery engine in any case would

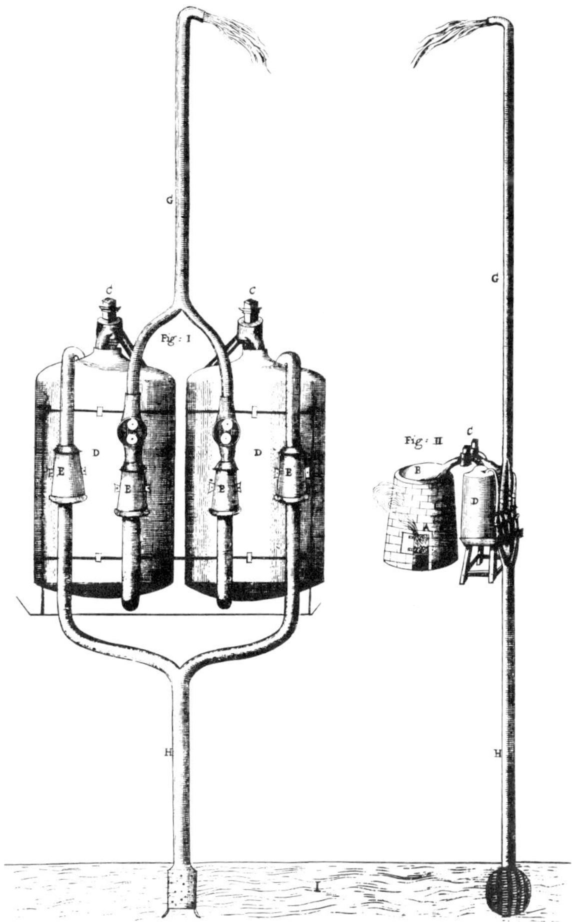

Fig. 15-1. Savery steam engine, 1698. When steam from the boiler (B) is condensed in the cylinders (D), water is drawn from the pool below and partially fills the cylinders; when the valves are operated and steam is admitted under pressure to the cylinders, water is ejected through the tops of the pipes (G). (*Science Museum*, London)

have had to be located within 20 feet of the lowest water level in the mine. While the raising of water in stages, using more than one Savery engine, and the placing of boilers underground were not unthinkable, the engine apparently had not sufficient promise in its original form to be tried in many mines. After the appearance in 1712 of the Newcomen engine, which met the objectives of the Savery engine, there was no further development of the Savery engine for mine-pumping. Nevertheless, it remained useful for pumping to heights under 20 feet and was so used throughout the 18th century.

THE NEWCOMEN ENGINE

In 1712, Thomas Newcomen (1663-1729), an ironmonger of Devonshire, put to work his first successful atmospheric steam engine at a coal mine near Birmingham. The engine provided power for a vertical reciprocating lift pump. The whole assemblage looked like a familiar well pump, greatly enlarged. The pump rod was hung on one end of a heavy, horizontal working-beam, pivoted at the center; the steam cylinder of the engine was located beneath the other end; a movable piston within the cylinder was attached by a chain to the beam end. When at rest, the pump end of the beam was down and the engine end was up. The steam cylinder was then filled with steam. Next, a water spray, injected into the cylinder, condensed the steam, forming a vacuum within the cylinder. The atmosphere, acting on the upper side of the piston, pushed it downward into the vacuum, dragging with it one end of the beam; thus the pump end of the beam was raised and a pumping stroke took place. The cycle could be repeated as often as 14 times per minute.

The Newcomen engine was a phenomenally successful machine. It could be built in sizes large enough to cope with great quantities of water; pump rods could be extended to any reasonable depth while the engine remained above ground. Although it may appear to have been designed in the only possible—or at any rate, feasible—way, a close examination of Newcomen's engine reveals it as one of the great original synthetic inventions of all time. As an ironmonger, Newcomen was of a class of highly skilled mechanics, generally of wide experience. It is assumed that he was also familiar with the latest scientific publications, because elements of his engine exhibit similarities to published drawings and descriptions that can hardly be fortuitous. Taking elements that were in existence, some in the laboratory and some in a shape quite different from that which he finally adopted, Newcomen unerringly selected the right elements and put them together in such a way that it was instantly clear to an observer that this was the way a steam engine should be built. The unique character of the engine can be better appreciated when we recognize that no other engine was brought forward to replace or compete with it until sixty years later, when the Watt engine appeared.

By 1725, the Newcomen engine had come into general use for mine-pumping in the United Kingdom; in 1723 an engine was taken to Austria, where it was erected by an Englishman; in 1732 an English engine was erected in France; before 1734 a Newcomen engine was in use at the Dannemora Mines in Sweden, having been built by a native of Sweden who had obtained his knowledge of

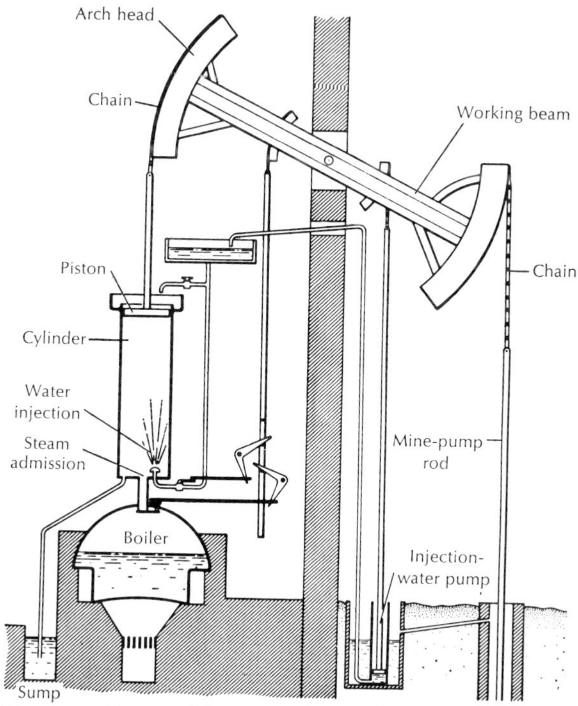

Fig. 15-2. Successful engine (1712) by Newcomen introduced steam into a cylinder that was then cooled with the injection of water, creating a partial vacuum. Atmospheric pressure forced the piston down to achieve pumping; the weight of the mine-pump rod and equipment then raised the piston for a new cycle. (© 1964 by *Scientific American, Inc.* All rights reserved)

engine-building in England; and in 1753 the first steam engine in the Americas was an English engine erected in New Jersey by a young immigrant whose father was an engine-builder in the mining district of Cornwall. There was never any question as to where the steam engine originated; it was solely a British innovation.

At mid-century, a few engines were being used by the wealthy to supply water for private domestic water systems; a handful had been applied to the filling of water-mill reservoirs; but the great majority of all engines were used for their original purpose, which was to pump water out of mines. In design, arrangement, and even in performance the Newcomen engine had not changed substantially since 1712. It was this performance—and specifically, the problem

Fig. 15-3. Newcomen engine at a mine in Dannemora, Sweden, *c.* 1734. (*Science Museum*, London)

of economy or thermal efficiency—that in the late 1760's interested John Smeaton (1724-92).

Following the thread of pioneering that he had done in theoretical and practical testing of water wheels and windmills, Smeaton in 1769 built a relatively small test engine for the purpose of investigating steam-engine performance. He also obtained performance data for a number of engines then operating in the Newcastle-on-Tyne district, learning incidentally that since their invention a hundred Newcomen engines had been erected in the district and that fifty-seven were still in operation. In 1772 Smeaton published tables, based upon his investigations, of optimum dimensions of the various parts of a Newcomen engine.

He found, for example, that the engine reached maximum thermal efficiency when the vacuum was approximately 8 pounds per square inch, a little more than half an atmosphere. Where the vacuum was higher or lower, the efficiency of the engine decreased. Making practical use of his studies, which were the only systematic investigation of the Newcomen engine in the sixty years following its inception, Smeaton built a number of large engines, with cylinder diameters up to 72 inches and stroke lengths to 9 feet. The thermal efficiency of these engines was nearly twice that of other engines of the period.

It should be noted that the work of Smeaton, which brought the Newcomen steam engine to its zenith of perfection, occurred after Watt had, in 1769, taken out his patent for a steam engine that would make the Newcomen engine obsolete.

JAMES WATT

James Watt (1736-1819), a Scotsman, was instrument-maker to the University of Glasgow in 1763 when he was asked to repair a teaching model of a Newcomen engine. The curious phenomena that Watt observed—and which he could not explain—turned his attention to the steam engine and determined the object to which his genius would be applied.

Watt's invention of the separate condenser illustrates clearly the way a new idea may enter a mind. Nearly two years after Watt had begun to study the problem of condensation in the Newcomen engine—specifically, why several cylinder-volumes of steam were required to fill the cylinder once—the idea of the separate condenser occurred to him; the details were worked out within hours and a model was tried within days. The patent for a separate condenser was taken out in 1769, after Watt had obtained a sponsor to pay for the development of a full-size engine. The sponsor was John Roebuck, a businessman who was operating the Carron Iron Works, in Stirlingshire, and trying to develop a coal supply nearby. Roebuck had known of Watt's ideas and experiments for nearly three years when, in 1768, in return for a two-thirds interest in the patent, he agreed to pay the patent fees and repay debts that Watt had incurred in his experimental work—in all, some £1200 (about $6000)—and to provide materials and workmen for future experiments. Actually, Roebuck paid Watt about £1000 before he found himself in financial troubles that prevented further support.

Roebuck was impatient for results, but the workmen that he furnished could not meet the demands of precision that the engine required if it was to be successful. Even Watt, who was contributing his own time to the project, was unsure of its ultimate success; after an unsuccessful trial of the engine in 1770, he accepted an offer of employment as a civil engineer to survey and plan a new canal. When, after a few months, he returned to the engine, he was hustled off on a day's notice to make another canal survey. He had to decide whether to continue the engine experiments without pay, or to work for the canal projectors who would pay him. The developmental work on the steam engine languished. Ironically, had Roebuck not completely failed in his business, it is uncertain when, if ever, Watt might have picked up the steam-engine work.

Fortunately, in 1774 Matthew Boulton (1728-1809), a creditor of Roebuck and a prominent Birmingham manufacturer of metal novelties—buttons, buckles, and the like—accepted Roebuck's interest in the Watt steam engine patent as settlement of his account. "None of his creditors," Watt had reported earlier to Boulton, "value the engine at a farthing." Boulton took Watt as a partner, paid him a salary to cover his living expenses, and what was equally important to a man of Watt's temperament, encouraged him to work despite failures, and rejoiced with him in his successes. The flame of Watt's genius was easily buffeted by adversity; it might have gone out if Boulton had not acted as its protector.

While the inventive genius of James Watt is of central importance to the steam engine, the ability of Matthew Boulton as an entrepreneur must not be overlooked. When Watt came to him as a partner, the steam engine had not yet operated well enough, in full size, to permit building an engine for sale. The time yet required to complete the development was uncertain; the fourteen-year patent had eight or nine years to run, and Boulton doubted whether he could recover his investment during that period. Therefore, he encouraged Watt to petition Parliament to extend the 1769 patent. Watt had believed he might obtain a new patent, but he was advised by London solicitors that an Act of Parliament extending the patent would be less costly than a new patent. Accordingly, in 1775, an extension was granted for a twenty-five-year period. Before discussing the effects upon rival builders of this patent extension, usually spoken of as the period of the Boulton and Watt monopoly, let us review the principal modifications of the steam engine during the time it was thus protected.

THE WATT ENGINE

The contributions of Watt changed the fundamental character of the steam engine by introducing new and original elements that had not been anticipated in any form. After nearly twenty productive years of invention and development (1765-85), Watt left the steam engine substantially in its final form. While the reciprocating steam engine has been superseded in the 20th century by the steam turbine, two basic elements of the modern steam plant—the separate condenser and the speed governor—were introduced by Watt before 1790.

Watt made three fundamental changes in the steam engine. As in the Newcomen engine, he employed a vertical steam cylinder and a horizontal, pivoted working-beam. Instead of injecting water and condensing the steam in the cylinder, however, Watt added a separate vessel, somewhat smaller than the cylin-

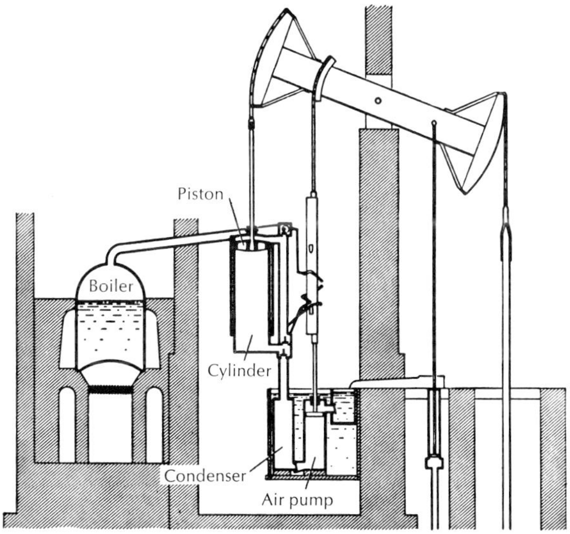

Fig. 15-4. Watt steam engine, 1775. Contributions by Watt to the steam engine included the development of a separate condenser, shown here. (© 1964 by *Scientific American, Inc.* All rights reserved)

der, in which condensation could take place. The result was to maintain the steam cylinder at high temperatures and the condenser at low temperatures. The saving of steam, since the steam cylinder did not require reheating after each condensing stroke, was on the order of one-half. Thus the Watt engine had a thermal efficiency twice that of the best Newcomen engines, as built by Smeaton.

The second innovation was the double-acting feature, which made practical the rotative engine. In the Newcomen engine, and in Watt's earliest engines, the working stroke occurred in one direction only. By introducing steam into the cylinder first at one end and then the other, force could be exerted alternately on each side of the piston. The chain connection between the piston and the end of the working-beam had to be modified, because it was now used to push as well as pull. Watt's solution to the problem was typically brilliant. The upper end of a solid piston rod was made to exert force on one end of the working-beam through a straight-line linkage that he devised. By attaching a connecting rod and crank on the other end of the working-beam, the reciprocating motion of the piston was thus converted to rotary power output.

Watt's third innovation was the speed governor. The familiar fly-ball governor, whose weighted arms swing farther away from the vertical axis of a rotating shaft as the speed increases, had been used for many years to regulate the

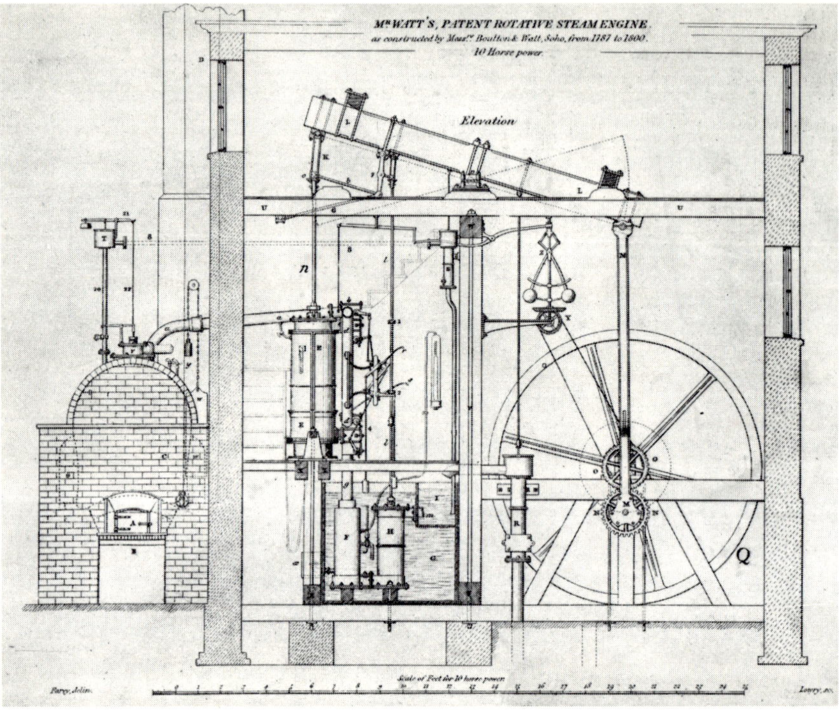

Fig. 15-5. Watt engine depicted in an 1826 illustration. The engine transformed the vertical action of the piston rod (*n*) into rotary motion through the flywheel (Q). (From John Farey, *Treatise on the Steam Engine*, 1827. *Science Museum*, London)

distance between the millstones of a grain mill. As the speed changed, the gap between the stones was changed. In the Albion Mills, a large flour-milling concern near London, Boulton had noticed the governor and had called it to Watt's attention. Within a few months the fly-ball governor had been fitted to the Watt steam engine. However, in adapting the governor to his engine Watt changed its action essentially and thus laid the foundation for modern control systems. Whereas the millstone governor simply changed the gap as the speed changed but had no control over the speed, Watt used his governor to regulate engine speed by regulating the steam supply to the engine. If the speed of the engine began to increase, for example, the steam throttle valve would be closed sufficiently to prevent the speed from increasing further. In effect, the governor received information about engine speed while regulating engine speed. The process of informing the governor of the status of the quantity being governed is now called feedback. This far-reaching principle made possible self-regulation rather than merely automatic machines. The Newcomen engine, for example, was automatic but not self-regulating.

During the twenty-five years of the Boulton and Watt "monopoly," in 1775-1800, the firm was responsible for the construction of nearly 500 engines, with about 200 being pumping engines which performed the same work as Newcomen engines, albeit more efficiently, and 300 rotative engines which supplied power to rotating shafts. During almost the entire period, the engines were assembled on the customer's premises by men sent out by Boulton and Watt; however, the customer purchased most of the materials for the engine from various suppliers and in addition paid Boulton and Watt a royalty. The valve assembly, which was separate from the engine cylinder, was made in Boulton's works in Birmingham; the engine cylinders were cast and bored in John Wilkinson's foundry at Bersham, ninety miles from Birmingham; and air-pump parts could be supplied by any one of a number of founders. The royalty was initially set at one-third of the coal-saving, as compared to a Newcomen engine, but this was soon reduced to a formula based upon the size of the engine and the number of hour's operation.

The demand for steam engines to turn rotating machinery was very great. Matthew Boulton, in urging Watt to get on with the rotative engine design, had

Fig. 15-6. Boulton and Watt's rotative beam "lap" engine, 1788. (British Crown Copyright. *Science Museum*, London)

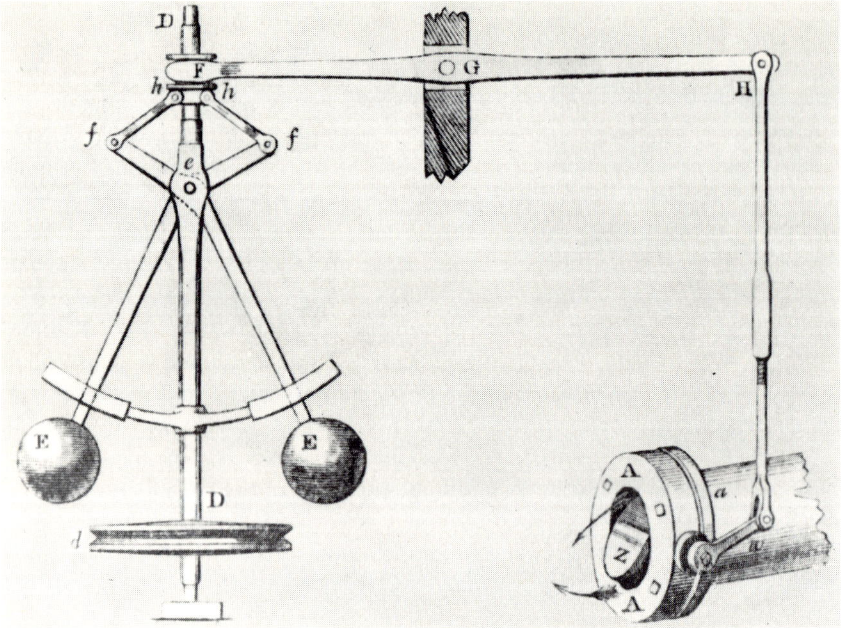

Fig. 15-7. Fly-ball governor, 1787. The ball spindle (DD) is rotated by a cord passing around the pulley (*d*). As speed increases, the balls (E, E) swing outward, the bar (FGH) is pulled down at the left end and raised at the right, and the steam valve (Z) is partly closed in order to keep the engine from a further increase of speed. (From Farey, *Treatise on the Steam Engine*, 1827)

described the industrial community as "steam-mill mad." Recent research has shown that the demand for engines was not in fact satisfied by Boulton and Watt, and that many other steam-engine builders were active during the "monopoly" period. Even aside from the direct piracy of the Watt design (Wilkinson, for example, built more than twenty engines without obtaining licenses from Boulton and Watt, while Bateman and Sherrat, of Manchester, supplied close to forty Watt-type engines in Lancashire), a large number of Savery and Newcomen engines were supplied to customers who could not obtain Watt engines or who preferred the lower cost of other, less efficient engines. Joshua Wrigley, a successful pump-maker of Manchester, his successors, and others built Savery-type engines to pump water for water wheels at least until 1850. Probably fewer than half of the engines built during the 1775-1800 period were built under Watt's patent.

In mine-pumping, the Watt engine made real inroads on the Newcomen engine wherever the fuel costs were high, for the efficiency of the Watt engine, as mentioned above, was about twice that of a Smeaton-Newcomen type. In the 1790's, however, barely half of the pumping engines in Cornwall—where coal was as expensive as in any place in England—were Watt engines. In the Derbyshire coal fields, a visitor noted as late as 1811 that he "met with no Pumping Engine on Boulton and Watt's principle, at a Coal-Pit; the old Atmospheric Engines, well-contrived and executed, being thought to answer better in such situations."

THE NATURE OF STEAM-ENGINE DEVELOPMENT

Although the original steam engine was not created to do more than solve a limited problem, which was that of pumping water out of mines, it would be difficult to overestimate the influence of the steam engine upon the whole Western world. We find in the minds of its original inventors and entrepreneurs no notion of the ultimate effects their work would have on other men's lives and thoughts. Through no overt act of theirs, the steam engine took on an identity quite independent of its creators. The possibility of using a steam engine for other purposes began to dawn upon later innovators; being eminently adaptable, it was adapted to the particular task at hand. "I wish to God," exclaimed Charles Babbage in 1820 while correcting errors in a newly prepared mathematical table, "these calculations had been executed by steam"; to which a colleague replied, "it is quite possible."

Each step in the development and adaptation of the steam engine to new uses appears, by hindsight, so logical and natural that the steps are frequently thought to have been inevitable. Even as early as 1798, when from our viewpoint the capabilities of the steam engine had scarcely been explored, the path of innovation already appeared to be a smooth one. Robert Livingston, a widely travelled American who was later to be associated with Robert Fulton, complained that "the slow steps by which this engine has advanced to its present state of improvement are really astonishing, considering how naturally most of those improvements suggest themselves."

A brief review of the manner in which the steam-engine developments were carried out leads to a quite different conclusion. The 1698 engine of Savery was unsuccessful in mine-pumping, but was widely used to raise water short distances, as in water-supply systems. The 1712 engine of Newcomen, which was successful in mine-pumping, remained substantially unchanged and unmodified for over fifty years. Furthermore, the physical arrangement of Newcomen's engine was followed by James Watt, even though his condenser, speed governor, and some lesser modifications changed the engine in ways that had not occurred to others. Watt was never accused of infringing upon another inventor's patent, nor of having stolen his ideas. The high-pressure engine, which might have competed with the Newcomen engine as early as 1725, if the proper person had recognized its possibilities, waited instead until after 1800 to be successfully employed.

It seems clear that the contributions of Newcomen and Watt determined, in many detailed ways, how the steam engine must further develop, and it is to the genius of these two men that we owe the steam engine as we know it.

AN ALTERNATE SOURCE OF POWER

When, in the latter part of the 18th century, the steam engine was used to provide rotary power, it replaced both horses and water wheels. Horses, up to a practical maximum of about eight, could be used to turn heavy vertical shafts by walking round at the ends of radial, spoke-like bars. The term "horsepower" was first defined by Watt in order to provide a uniform basis for deciding how many horses could be replaced by a given steam engine. It was desirable to replace water wheels not only because of the wheel's inflexibility of location and the frequently limited power available at a single mill site, but also because mill dams and sluices, so necessary for an adequate fall of water for a water wheel, prevented the navigation of a stream or river. Windmills in England were never effectively employed either in mine-pumping or in large-scale manufacturing.

It will be noted that in every case the steam engine replaced an existing power source. As the power of the steam engine increased, however, the process of replacement became irreversible. When a steam engine doubled the capacity of a mine-pumping system, for example, and permitted the mine to go deeper, there was no returning to the former power source, because its maximum capacity under given conditions had been exceeded. The steam engine, as it came to be built in larger and larger sizes, tended to combine operations that had formerly been dispersed. When the direct economic advantages of such concentration had been recognized, the trend to centralization was clearly established.

HIGH-PRESSURE ENGINES

A new competitor of the Watt engine appeared about 1800, just as numerous builders entered the steam-engine field to take advantage of the expiration of the basic Watt patent. Richard Trevithick (1771-1833) in England and Oliver Evans (1755-1819) in the United States built high-pressure steam engines, dispensing with the condenser entirely and exhausting the spent steam directly to the atmosphere.

Trevithick, a native of Cornwall, was the son of a mechanician who repaired —and on occasion improved the performance of—Newcomen mine-pumping engines. Thus the younger Trevithick had grown up in the midst of steam engines; at the age of nineteen he was working as an engine operator, and eventually he followed his father's trade as engine mechanician.

The possibility of using steam at high pressure was not new when Trevithick seized upon the concept. A high-pressure steam engine had been illustrated in a German book of 1724; in 1770, a steam carriage which employed a very similar high-pressure engine had been tried without conspicuous success on the streets of Paris; and a high-pressure water engine using water in place of steam as the working substance had been built in England and illustrated in an English journal as early as 1787. Finally, in 1800, about the same time as he was experimenting with a high-pressure steam engine designed to propel a carriage, Trevithick also built a high-pressure water engine at Wheal Druid Mine. In 1802, he patented his high-pressure steam engine.

The application of high-pressure steam to an actual engine was not as simple as might appear, for reasons very similar to those that delayed Savery a century earlier. Although the techniques of fabrication were advancing rapidly in 1800, safe high-pressure boilers were still beyond the capabilities of nearly all metalsmiths. In spite of a disastrous explosion of one of his boilers, which killed four men in 1803, Trevithick in 1804 went on to demonstrate the first clearly successful locomotive engine in the world. On an iron railway connecting the Welsh Penydaren Iron Works with a seaport nearly ten miles away, his single-cylinder steam locomotive successfully hauled a loaded train. His second locomotive was demonstrated in 1805 on a wooden-track railway at Wylam colliery, near Newcastle. While his second locomotive was too heavy for the existing railway (railways for hauling wagons drawn by horses were widely used in England before 1750) and his first locomotive was used only for a short time, the seed of steam locomotion had been firmly planted. A steam engine could be used for locomotive power only if the engine was powerful, light, and compact. Only the high-pressure steam engine met these requirements. Trevithick was the man who saw the connection.

Oliver Evans was forty-six years old when he built his first steam engine. He had already worked out a revolutionary concept in four-milling machinery and had published a pioneering handbook on mill-wrighting. Vertical bucket elevators and horizontal screw conveyors were employed to make Evans's flour mill completely automatic, requiring no manual lifting or transfer of grain in any part of the mill.

In 1801, Evans built in Philadelphia a small, stationary, high-pressure steam engine which he employed to turn a rotary crusher that produced agricultural-grade limestone. Over the next several years he developed a stationary high-pressure steam engine that competed successfully with Watt-type engines then being built in the United States. In 1804 he had obtained a patent on the high-pressure engine, and in 1815 his patent was extended by Congress. The engine that became associated with Evans's name was a highly original adaptation of the conventional vertical-cylinder beam-engine. Instead of placing the engine cylinder under one end of the beam and the crankshaft under the other, as in the Watt engine, Evans put both cylinder and crankshaft at the same end of the beam, thus drastically reducing the strength required of the beam. His solution

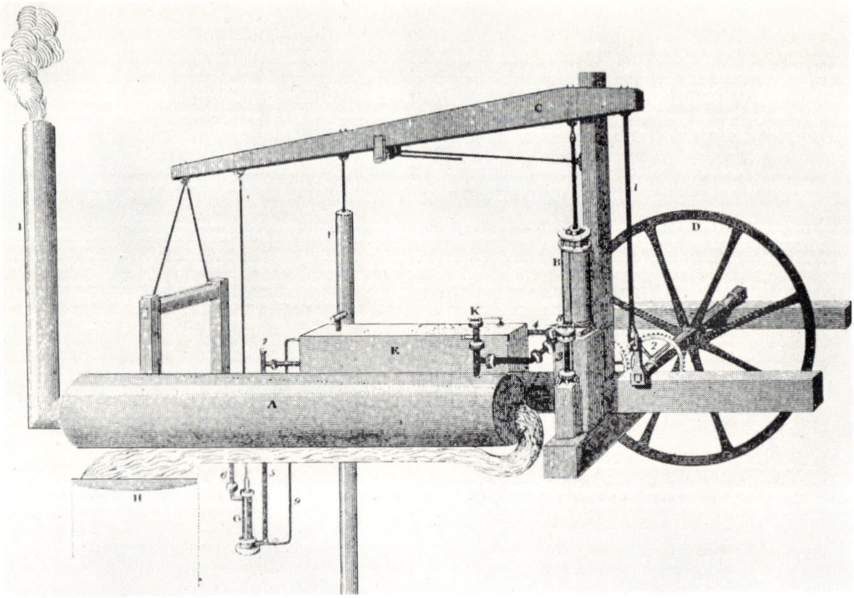

Fig. 15-8. Evans's "Columbian" steam engine, 1813. By placing both cylinder and crankshaft at the same end of the overhead beam (C), the size of the beam could be reduced to less than that of the Watt and Newcomen engines. (From *Emporium of Arts and Sciences,* 1814)

of the problem of moving the upper end of the piston rod in a vertical line was a linkage that became known throughout the Western world, including Russia, as the Evans straight-line linkage. Evans saw immediately the possibility of using his engine to drive a carriage on a smooth paved road and he wrote often about the future of locomotive power, but his first engines were used to propel boats and to drive saw mills, flour mills, and boring machines. Before 1812 Evans had ten engines in operation and at least six more under construction.

Trevithick's knowledge of the principles on which he had built was rudimentary, but the understanding of thermodynamic phenomena was still in a qualitative state. Trevithick saw only the gain of force against a piston as the steam pressure was increased to several atmospheres. He eliminated the condenser without qualms because he reasoned that the loss was merely one atmosphere of force, or less. He was not aware of the great advantages of the condenser in (1) increasing the thermal efficiency and (2) increasing the work output of an engine for each pound of steam generated.

In Cornwall, the pioneering high-pressure work of Trevithick led eventually to the very efficient Cornish pumping engine as it was developed by Arthur Woolf (1776-1837) and others. Taking in steam at about 50 pounds per square inch above atmospheric pressure, the Cornish engine exhausted into a condenser maintained at very low pressure. The result was an efficiency almost three times as high as that of the best Watt engines.

Evans, on the other hand, thought he had discovered a new quantitative principle of steam utilization. Noting the rapid increase of pressure as the temperature of the steam increased, Evans proposed an attractive conclusion; he stated that doubling the temperature of the steam would increase the effect of the engine by a factor of 16 while requiring only double the fuel. Although his reasoning was quite in error, there was nobody at that time who could tell him exactly why, and his erroneous conclusion had the positive value of sustaining his efforts to promote the use of high-pressure engines.

The advantage of the high-pressure steam engine was its light weight and small size for a given power output. Thus the high-pressure engine was necessary to the success of the railway locomotive. On the western rivers of the United States—the Ohio, Mississippi, and Missouri rivers system—the simplicity of operation and low first cost made the high-pressure engine a universal favorite, in spite of its dismal explosion record. Forty steamboat boiler explosions, at least, occurred during the first fifteen years following the first appearance, in 1816, of a steamboat on the western rivers. During the next ten years ninety more explosions occurred. The loss of life, though appalling, was somehow accepted, even as it is today, as the price of progress.

EFFECTS OF THE STEAM ENGINE

The spread of steam engines used in manufacturing processes was rapid and widespread. In Birmingham, for example, the number of engines in use in 1820 was 60, developing an estimated 1000 horsepower. By 1835 this number had risen to 169 developing some 2700 horsepower, and by 1840 it had risen again to 240 engines developing 3436 horsepower. Glasgow saw its first engine applied to manufacturing in 1792, and by 1825 there were 310 at work producing 6406 horsepower. A census of engines in France in 1833 discovered 947 in use, of which 759 were manufactured in France and 569 were using high-pressure steam. In the United States, in 1838, the Secretary of the Treasury reported to the Congress that there were 3010 steam engines at work in the nation, of which 800 were used on board steamboats, 350 in locomotives, and 1860 in various manufacturing establishments and public works. It was an impressive record.

Based on this record, four far-reaching effects of the 18th-century steam engine can now be recognized. First, the factory system of manufacture, in which machine tenders were brought together, usually in a single building, around a central power source. When the prime mover was a steam engine, the factory might be located at the convenience of the owner, being no longer dependent upon water power, and its size could be increased indefinitely because there was no natural limit to available power. An important advantage of a central power source is that all the machines required for a given product can be located in one place, thus eliminating the need for excessive transporting of components and partly finished products. From the beginning, however, the principal advantage of the factory system lay in its discipline that brought workers to a central location for regular periods of work. The steam engine simply reinforced this discipline.

Second, the steam engine gave transportation systems a power source of undisputed merit for more than a hundred years. Third, the disadvantages of steam engines—bulk, danger of explosion, and the troublesome latent heat which appeared to be a loss, until thermodynamic analysis was possible—drew forth a sustained effort to find a prime mover that would replace the steam engine entirely. Some investigators tried alternate working fluids, such as ether and mercury, in place of water; others picked up the thread of explosive internal-combustion engines which in the 17th century had been dropped in favor of steam. Although the steam turbine, which is the successor to the steam engine, is presently unchallenged in very large sizes, the internal-combustion engine has replaced the steam engine in railway locomotives, agricultural tractors, and many boats and ships in all but the largest sizes.

Finally, the Watt steam engine was the first large machine to be built of metal in considerable numbers that required precise and consistent dimensions of its working parts. The novel boring mill of John Wilkinson, which enabled him in 1775 to make a cylinder that "erred [from a true circle] not more than the thickness of a sixpence," was the key to Watt's first successful full-size engine. As the advantages of precision in machinery became manifest, an array of basically new machine tools—lathes, planers, and the like—was developed over the next seventy years. Thus the steam engine set in motion a chain of circumstances that has conditioned how we live and view the world today.

We may reasonably wonder why the steam engine appeared first in England rather than in another country, particularly when we see that the important scientific groundwork for both the Savery and the Newcomen engines was of continental origin—French, Italian, and German. The existence in Germany and France of the *Stangekunst,* a push-pull system of rocking bars that would transmit reciprocating power over distances up to a mile or more, probably delayed serious work there on a prime mover such as the steam engine. The power of water wheels could be thus transmitted overland to mine pumps, apparently in amounts large enough to keep even the deeper mines free of water. From the Seine River, near Paris, where large quantities of water were pumped to a level nearly 400 feet above the river, in three stages, to supply the palace at Versailles, a veritable forest of *Stangekunsten* marched up a hillside from the water wheels to the two intermediate pump-houses. The great Marly machine, as it was called, was one of the wonders of the age, outlandish as it may appear to modern eyes.

It is not difficult to see why continental mine operators, not panting after change merely for the sake of change, would retain their systems of water-powered *Stangekunsten.* The English, on the other hand, had no satisfactory systems of mine-pumping. The economic incentive was no doubt a powerful one, paticularly in the expanding economy of the late 17th century. However, the absence of the *Stangekunst* in the United Kingdom is striking and, thus far, quite unexplained. What set of circumstances in England led a man like Newcomen to spend ten years in developing a completely untried device rather than import a well-proven system from the Continent? An answer to this question will have to wait until we know a great deal more about Newcomen than we do today.

Because the steam engine has long been in the Western world a symbol of progress as well as power, it is well to examine the conditions that it imposes upon those who would employ it. The managers of industry adopted the steam engine with no conscious reservations. The machine multiplied man's ability to bend nature to human ends, even as it augmented the economic power of its owners. Andrew Ure, the Scottish chemist who in 1835 set forth with unbridled enthusiasm the manifold advantages of the factory system in his book, *The Philosophy of Manufactures,* wrote: "The steam engine is, in fact, the controller-general and main-spring of British industry, which urges it on at a steady rate, and never suffers it to lag or loiter, till its appointed task be done."

A few years later, a writer in *Blackwood's Edinburgh Magazine* reflected upon a railway locomotive, which someone had called an "iron slave." Noting the number of people who were employed in tasks such as greasing the machinery, trimming the lamps, watering the boiler, and tending the fire, he wrote: ". . . methinks there is something hypocritical and deceptive in this obedient engine of yours. Goes of itself, you say. Does it? Your iron slave wants many other slaves, unfortunately not of iron, to attend on it; on this condition only will it serve you. No despot travels with so obsequious a train, and so subservient, as this quiet-looking engine." In this way did the steam engine both free and enslave the Western world.

It is fitting that we should honor Thomas Newcomen, James Watt, and the others who were responsible for the origin and development of the steam engine. It has contributed in many subtle as well as obvious ways to the material plenty that is technically possible for all men. It will be noted, however, that each new or continued application of steam power has required of the living generation a value judgment as to the need for or wisdom of that application. Difficult as it may be for most men to comprehend, power carries no imperative independent of human judgment.

16 / Metallurgical and Machine-Tool Developments

EUGENE S. FERGUSON

The technologist's consideration of iron as an industrial material was vastly changed during the Industrial Revolution. In 1750 iron was used in machines and structures only where wood or another cheaper and more easily wrought material simply would not do. By 1830 iron was the first material considered by engineers and mechanicians for a wide range of uses. For the most part, iron in 1750 was worked up by a blacksmith, shaped and welded by his hammer to form brackets, straps, wagon tires, nails, and a variety of other ironmongery. This malleable metal, called wrought iron, contained a very low fraction—on the order of 0.1 per cent or less—of carbon. Cast iron, which was hard, brittle, not at all malleable, contained a high fraction—2.5 to 4.5 per cent—of carbon. Perhaps as little as 5 per cent of all iron was made into castings. In 1830, on the other hand, cast iron was widely used for machine frames, water and gas pipe, and building members; wrought iron was being rolled into rails; and other new uses for iron multiplied rapidly.

This enormous difference in the employment of iron came about through a complex of interacting innovations. The supply of iron was increased when the steam engine multiplied the ironmaster's supply of power; the rapidly increasing use of steam engines in turn increased the demand for cast iron; new techniques of iron-making further increased the quantities that could be made economically; and the increased supply of iron was rendered more useful by a new class of tools, called machine tools, that could cut the hard metal, both in its cast and wrought form.

THE FOUNDRY TRADE

In order to convert iron ore to iron, the ore must be heated in a reducing atmosphere. The ore (an oxide of iron) can be most conveniently reduced (that is, brought to a metallic state by removing the non-metallic elements in it, in this case, oxygen) by placing it in contact with a burning fuel which is supplied with insufficient oxygen for complete combustion. Under these conditions the oxygen in the ore will combine with the fuel. The ore, now robbed of its oxygen, becomes iron. From ancient times, the fuel thus used in the smelting of ores was charcoal, which was produced from wood.

When, in the 17th century, industrial requirements of wood—for glass-making, soap-making, and brewing as well as for smelting—had increased to the point where depletion of English forests became alarming, there was constant pressure to substitute coke (obtained from coal) for charcoal wherever possible. Coke, however, contained certain impurities, particularly sulfur, which contaminated the materials with which it came into contact. Melting chambers in glass furnaces could be enclosed so that the molten glass was not thereby contaminated by the products of combustion of the coke, and malters could use coke for drying their malt; but early attempts to use coke for smelting iron had failed because the coke usually contained sulfurous and phosphoric compounds, and the iron, being extremely sensitive to very small concentrations of these compounds, was spoiled by contact with the coke.

In 1708, Abraham Darby I (1676-1717), became proprietor and ironmaster of Coalbrookdale, on the Severn River, west and north of Birmingham, England. He had earlier been in business in Bristol, first as a malter then as a copper smelter, and already knew coke as a well-tried fuel. Thus, when he went to iron-making at Coalbrookdale, it is not surprising that he experimented with coke in the smelting of iron ore. By a fortunate coincidence, both the ore and the coke that were available to Darby were unusually low in sulfur and phosphorus, and the cast iron of Coalbrookdale soon became well known in the Birmingham district.

Coalbrookdale was responsible during the 1720's for a second major innovation. The steam cylinders of the Newcomen steam engines in their early years were made of brass or bronze, which were the only economical metals that could be cast in the large sizes required by the engines. As early as 1722, however, iron cylinders for Newcomen engines were being cast and finished at Coalbrookdale by the successors of the elder Darby. In the years from 1722 to 1738 a total of 38 steam-engine cylinders, as small as 17 and as large as 38 inches in diameter, were made at Coalbrookdale. During the next ten years, after Abraham Darby II (1711-63) had taken over the management of the firm from the interim proprietors, some 43 cylinders were finished. In the mid-1750's Darby was able to cast cylinders up to 48 inches in diameter, and in 1763 he furnished to a Newcastle colliery a 74-inch cylinder, 10½ feet long, weighing over 7 tons.

Coalbrookdale was easily the most prominent foundry at mid-century. Other iron works were slow to adopt coke for smelting because they were not as fortunate as Coalbrookdale in the quality of their supplies. Nevertheless, the undoubted success of the Darbys was constantly before the trade, and as might be expected, more and more iron works were established where conditions of coke and ore were promising.

Another impetus to the growth of the foundry trade was given by John Smeaton, a prominent English engineer who in 1754 specified that cast iron be employed instead of wood for the main shaft of a windmill. Within a few years Smeaton was using cast iron for both water-wheel shafts and for gears. From Carron, the celebrated Scottish iron works that came into being during the Seven Years' War with France (1756-63), cast-iron cannon and naval mortars —the latter known as carronades—were supplied to the Royal Navy and to the governments of Russia, Denmark, and Spain.

In 1779, a cast-iron arch bridge of 100-foot span was thrown across the Severn River at Coalbrookdale, where it yet stands. Abraham Darby III (1750-91) was the designer and builder. In a kind of virtuoso performance, Darby made the main ribs 70 feet long and proceeded successfully to cast directly from the blast furnace ten ribs of this incredible length. Around 1786, John Rennie, another prominent British engineer, employed cast-iron gearing and machine-framing throughout a grain mill, and before the end of the 18th century cast iron was being used in mill buildings as a structural material, for beams as well as columns.

THE INFLUENCE OF THE STEAM ENGINE

Partly as a result of such foundry techniques as those pioneered by the Darbys, innovations in iron-making and in steam-engine building combined to produce an accelerating advance of techniques, augmentation of power, and mastery over materials. In this clear-cut example can be seen the interaction of one branch of technology with another. An understanding of this process, whose complexity and impetus toward change are increased exponentially as more and more branches of technology interact, is central to a comprehension of the way in which technological change occurs. The Industrial Revolution is only a convenient label for the period in which the process began to alter radically, within a single lifetime, man's material environment.

The employment of the Newcomen and Savery steam engines in pumping water to turn a water wheel has been described in Chapter 15. Using this technique, the steam engine was thus able to keep iron works in operation during the seasons when natural water supplies failed. From the year the Watt engine was introduced, however, the spectacular effects of the steam engine upon iron production first began to be felt. In 1776, just a year after he had supplied to Boulton and Watt the first accurately bored steam-engine cylinders, John Wilkinson installed a Boulton and Watt steam engine in his own iron works to operate directly the blowing cylinders of his blast furnace. A blast of air at 4 pounds per square inch pressure was supplied continuously to the furnace; the higher temperatures made possible by the powerful blast simplified the problems of using coke for smelting. So successful was this application of the steam engine to the blast furnace that within a few years Wilkinson had equipped four more of his furnaces with steam engines, and Boulton and Watt engines had been installed in Coalbrookdale and in several other iron works.

In 1783, the steam engine was employed to drive a forge hammer directly, again replacing an intermediate water wheel. A Boulton and Watt rotative engine was installed in John Wilkinson's works at Bradley; through a cam it lifted the forge hammer-head, weighing 800 pounds, a distance of two feet. Again, the steam engine answered admirably for this task, and within a year at least three other iron works had adopted the new plan. As the steam engine demonstrated its ability to turn a camshaft, it became evident that it might be employed also to turn a rolling mill. Probably it was Wilkinson who first made this innovation, in 1786.

The sharp increase in the available quantities of hammered and rolled plates encouraged the substitution of iron for copper in the making of steam boilers. Although it is conceivable that high-pressure boilers for stationary and locomotive service might have been made of copper, the high costs of copper would certainly have retarded the growth of the steam-engine industry. Even the low-pressure copper boilers of the Boulton and Watt power plants, built before iron plates were available in adequate quantities, were twice as costly as iron. In any case, rolled-iron plates were widely employed shortly after 1800, although a dogged preference for hammered sheets over rolled sheets persisted for some time.

WROUGHT IRON

In 1750, only a very small part, perhaps 5 per cent, of iron went into castings. Although over the next fifty years cast-iron production increased many times and a larger portion of iron was used as cast iron, wrought iron was in 1800 still by far the most important product of the iron works, a primacy that was only to be eclipsed by Bessemer iron (steel) nearly sixty years later. The contributions of Henry Cort (1740-1800) to the techniques of refining and working wrought iron, particularly his substitution of coke for charcoal as the fuel, were essential to the continued pre-eminence of the material.

The puddling process, which Cort introduced in 1784, was different from the earlier refining processes, but the chemical changes to be effected were the same. In the earlier process, in which charcoal was the fuel, cast iron in the form of "pigs" was heated in a very hot hearth-fire while a constant current of air was blown over the pigs. The excess air burned the carbon in the iron to carbon monoxide or carbon dioxide, thus freeing the iron of its carbon. As more and more carbon was oxidized, the iron became malleable and could more easily be worked. In Cort's puddling process, on the other hand, the pig was melted in a chamber of a coke-fired reverberatory furnace which kept the fuel separate from the iron. Air was brought more quickly into contact with the molten metal by stirring the pool with an iron "rabbling" bar. As carbon was removed, the melting temperature of the iron increased. The successful removal of carbon was marked by a general loss of fluidity of the pool; the limits of attainable furnace temperature were eventually reached; and the iron mass, now low in carbon, could be gathered together in a sort of pasty ball, called a puddle ball. The puddle ball was then fished out of the furnace and squeezed into shape in a grooved rolling mill. Cort patented his process in 1784.

Many objections were raised immediately against the new puddled and rolled wrought iron. James Watt, for example, thought at first it would be brittle when cold because its "crystals [were] spun out by the rolling." He wrote, just after the puddling patent was awarded, "it is tender to the file and soft to the hammer, rusts very readily and ought never to be used where it is subjected to any strain, as it is very weak, therefore useless for engine work, ship work, etc., but good for nails because easily wrought." Some of the difficulties in using coke as fuel were the same as those encountered in the coke-smelting of cast iron. Sulfur, in particular, was an annoying contaminant. Nevertheless, the process was attractively economical, and it was susceptible to great improvement. When the pig iron and fuel were properly selected, the resulting wrought iron compared favorably with Swedish charcoal iron, which for generations had had a reputation for excellence.

CRUCIBLE STEEL

One more important innovation in 18th-century iron technology must be mentioned because of its ultimate effect upon the machine-tool industry. During the decade before 1750, Benjamin Huntsman (1704-76), a clock- and instrument-maker of Doncaster, near Sheffield, tried to make a more uniform steel for clock springs than was available. In no place was steel-making better known than in Sheffield, which from late medieval times had been the home of cutlers and edge-tool makers. To make steel, bars of good wrought iron, preferably Swedish charcoal iron, were packed in charcoal, usually in a closed box; the iron, placed thus in close contact with carbon, was heated at a high temperature —over 1000° F—for several days. Under these conditions, some of the carbon passed into solid solution with the iron, making the bar hard near the surface by increasing its carbon content. The shell-hardened material was called "blister steel" because of the appearance of a bar after this carburization process. The process was also called "cementation," and the steel so produced was cemented steel.

Huntsman's improved process consisted in breaking blister bars into short lengths, adding a flux (a substance to help the metal fuse together), and melting the heterogeneous material in a clay crucible. Steel of sufficient hardness, uniform in composition, retaining some of the toughness of the blister bars, was the immediate contribution that Huntsman made to the cutlery trade; the steel that he taught the world to make was, after his death, a necessary ingredient of the new machine tools, for which it provided the actual cutting edge.

THE EXPANDING MARKET FOR IRON

By 1790 the impressive repertoire of the ironmaster in England included coke-smelting, coke-fired puddling, Huntsman's steel, and steam-driven rolling mills and forge hammers. Through the use of these new techniques and devices, the output of the English iron industry was doubled in the eight years between 1788 and 1796 and was doubled again in the following eight years.

Although the expansion of iron production occurred during and perhaps largely because of the extraordinary demands of the Napoleonic Wars, which ended in 1815, it is significant that after a short postwar decline in 1816-17, the iron industry was urged on to greater exertions by new demands, especially the demand for rails by the growing railways. In 1820, for example, John Birkinshaw, of Bedlington Iron Works, near Newcastle-on-Tyne, successfully rolled wrought-iron rails for railways. Designed to replace the earlier cast-iron fish-bellied rails, the Birkinshaw rail was rolled in 15-foot lengths and weighed 28 pounds per yard. When, in 1830, the Liverpool and Manchester Railway was opened, the superiority of wrought- over cast-iron rails had been conclusively shown and cast iron had been driven from the field. The wrought-iron mills in South Wales, many of which had been established on the heels of Cort's first successes, supplied wrought-iron rails to the burgeoning railway systems of England and of the United States. The familiar American rail pattern, whose broad base flange permitted spiking of the rail directly to wooden ties, was first rolled in Wales in 1831 at the request of Robert Stevens, who had journeyed from the United States in order to buy rails for a New Jersey railroad.

Another source of demand for iron came from ship-building. The first iron steamship was built in 1822 by an Englishman, Aaron Manby, for a projected freight line on the Seine River, in France. The rolled wrought-iron plates used in the vessel were ¼ inch thick, similar in every way to boiler plates.

In North America a similar though slower development was taking place. The techniques of iron-making had been brought over with the colonists; following English precedent a fully developed iron works, consisting of smelter, finery, forge, and slitting rolls, had been operated at Saugus, Massachusetts, in the late 17th century. The charcoal-iron industry was vigorous throughout the 18th century; small furnaces were widely scattered throughout New England and the Middle Atlantic states, particularly in eastern Pennsylvania and New Jersey. Like continental European countries, the United States retained charcoal as the fuel for iron-making long after the British iron industry had turned almost entirely to coke.

In 1830 the United States produced some 165,000 tons of iron, about one-fourth as much as the United Kingdom. While a very considerable part of American production was in bars and sheets, heavy imports were received of Swedish and Russian sheet and bar iron, the Russian polished sheets being particularly esteemed for their appearance and malleability.

The uses of such materials multiplied rapidly. Cut nails (rather than hand wrought), a distinctly American innovation of about 1800, were sheared from rolled nail plates. The West Point Foundry, in Cold Spring, New York, was able in 1822 to furnish E. I. du Pont with a pair of roller wheels for a powder mill; these wheels were over six feet in diameter and each weighed more than seven tons. Henry Burden (1791-1871) of Troy, New York, started in 1825 to make railroad spikes by machine and by 1835 he had mechanized the making of horseshoes. Shovels and spades, made of American rolled iron, were produced from 1826 in large quantities in Alan Wood's Delaware Iron Works, a water-powered rolling mill located near Wilmington. Also in 1826 the Collins Company was formed in Connecticut to mass produce axes. Boiler-makers in the United States made their boilers by riveting rolled wrought-iron sheets in much the same way as their English counterparts. Here as there, boiler-makers took the lead in developing iron boats.

However, no serious attempt appears to have been made in the United States to roll wrought-iron rails until after 1840. The problems to be overcome in rolling a long, thin, iron piece of irregular section were not inconsiderable, and iron-makers were still wrestling with them more than twenty years later. It should also be pointed out that the tariff on railroad iron was so low that iron from Wales could be delivered in the United States at a price below that charged by a local rolling mill.

THE MACHINE-TOOL INDUSTRY

Remarkable as were the various processes and techniques that have already been described, the full use of iron as a constructive material depended upon the development of machine tools. As soon as a designer or machine-builder could obtain the same shapes, and put the same holes, slots, and recesses in iron as in wood, he could add the strength and rigidity of iron to his machines and thus increase indefinitely their power, capacity, and durability.

The idea of guiding a cutting tool precisely to cut away specific portions of the workpiece (the piece to be cut) was demonstrated by instrument-makers' tools throughout the century before 1775; in 1775 the first large and powerful machine tools began to transform the process of machine-building. In the clock-maker's gear-cutting machine of 1700, for example, the workpiece, a round disc, was notched by a rotary cutter to form teeth; the cutter could be moved parallel to the axis of the workpiece; between cuts, the workpiece was indexed (rotated and accurately positioned by a perforated index plate).

Fig. 16-1. Gear-cutting machine, *c.* 1672. The gear being cut is at the upper end of the vertical spindle; the circular cutter is on the horizontal shaft just behind the gear. The large circular dividing plate at the bottom of the spindle positions the gear as each successive tooth is cut. (British Crown Copyright. *Science Museum*, London)

THE BORING MILL

The possibility of building large machines whose parts were as nicely fitted to each other as those of a clock or a scientific instrument was glimpsed in 1769 by Matthew Boulton, the future partner of James Watt, who recognized even more clearly than the inventor the need for care in building the new style steam engine. Your engine, he told Watt, will require "very accurate workmanship . . . with as great a difference of accuracy as there is between the blacksmith and the mathematical instrument-maker."

The wisdom of Boulton's statement was shown six years later, in April 1775, when Boulton and Watt were partners. An accurately bored cylinder was of critical importance: Watt had since 1769 tried unsuccessfully to obtain a cylinder that would be steam-tight as the piston moved up and down inside it,

Fig. 16-2. Model of Wilkinson's boring mill, 1775. The cylinder to be bored is shown secured by chains in the center of the picture. The cylinder is cut away to show the boring head inside the cylinder, midway between the chains. The boring bar is supported at both ends of the cylinder, thus providing greater rigidity than in earlier boring mills. (British Crown Copyright. *Science Museum*, London)

and in the absence of such a cylinder, the development of his engine was at a standstill. Just such an iron cylinder, slightly over 18 inches in diameter, was finally cast and bored at John Wilkinson's iron works.

Wilkinson's boring mill, to which the successful cylinder owed its existence, differed fundamentally from earlier cylinder boring mills, though by hindsight the essential change appears to be minor and obvious. In earlier mills, the boring head was a disc with several cutting tools fastened around the periphery. The boring head was supported and turned by the extension of a water-wheel shaft. Wilkinson's contribution was to extend the supporting shaft all the way through the cylinder (which was cast hollow), encasing the shaft in bearings at both ends of the cylinder rather than at one end only. In so doing, the tendency of the boring head to droop and to be deflected by soft and hard spots in the iron was effectively curbed. Wilkinson's mill could bore a cylinder 50 inches in diameter which, in Boulton's words, "doth not err the thickness of an old shilling. . . ." (A worn shilling might be 0.050 inches thick, which would mean that the maximum error of the boring machine was approximately 1/1000 inch per inch of cylinder diameter; an ordinary boring mill of 1900 might err by half as much.) If Boulton's statement is correct, the precision of Wilkinson's machine is wholly remarkable. Just as Coalbrookdale fifty years earlier had furnished cylinders for Newcomen engines, so John Wilkinson's works now furnished the cylinders for Watt engines. Until 1795, when Boulton and Watt finally built a boring mill on their own premises, Wilkinson furnished all but three or four of the several hundred cylinders required for Watt engines.

It would be a mistake to dismiss John Wilkinson's accurate boring of a steam-engine cylinder in 1775 as a timely but relatively trivial accomplishment. Twenty-five years after Wilkinson had shown the way, the boring of a 64-inch cylinder in Boulton and Watt's new foundry still required 27 working days. In the United States, where machine tools were nearly unknown, the problem of boring a 40-inch steam-engine cylinder, about 6½ feet long, for the new Philadelphia Water Works pumping engine was a major one. This cylinder was bored in Nicholas Roosevelt's Soho Works, near Newark, New Jersey. Named after the Birmingham shops of Boulton and Watt, Roosevelt's machine shop was as advanced as any in America. Nevertheless, a contemporary description of the boring operation in 1800 points up the primitive state of the machine arts in the United States.

Roosevelt's boring mill, arranged horizontally, consisted of a boring head, driven by a water wheel, and a movable carriage to which the cylinder was

Fig. 16-3. Model of Smeaton-type boring mill, 1770. The cylinder, secured to the small carriage, is advanced against the boring head (shown near center of cut-away cylinder). The boring bar is not supported at its outboard end as in the Wilkinson boring mill. (British Crown Copyright. *Science Museum*, London)

fastened and which served to advance the cylinder against the boring head. About ¾-inch of material had to be removed from the inside surface of the cylinder. The account of an observer, who represented the customer, was as follows:

> Two men are required. One almost lives in the cylinder, with a hammer in hand to keep things in order, and attend to the steelings [the cutting tools]; the other attends to the frame on which the cylinder rests which is moved by suitable machinery; these hands are relieved, and the work goes on day and night; one man is also employed to grind the steelings; the work is stopped at dinner time, but this is thought no disadvantage as to bore constantly the cylinder would become too much heated; the work also stands whilst the steelings are being changed, which requires about ten minutes time, and in ten minutes more work they were dull again. . . . The workmen state that the boring was commenced the 9th of April and had been going on ever since, three months, and about six weeks more will be required to finish it.

As we shall see shortly, the machine industry of the United States grew very rapidly in the generation after 1800. It was during this period that the idea of special-purpose machine tools, designed to perform limited operations on particular components of a manufactured device such as a musket or a clock, was advanced. By 1850 the idea was being so eagerly pursued that the resulting system of manufacture became known in Europe as the "American system."

THE ENGINE LATHE

A characteristic of machine tools that encourages their proliferation is the ability of one tool to machine the parts of another and later tool. In the Philadelphia Water Works, for example, a rolling mill and a boring mill were installed, to be driven by the steam engine during slack periods when it was not required for pumping water. The boring mill, a considerable improvement over Roosevelt's, enabled Philadelphia mechanics to enter the steam-engine trade.

The basic machine tool of any workshop was and still is the engine lathe, on which a workpiece is fastened and rotated by a spindle axis; the cutting tool, rigidly supported, operates upon the workpiece to produce a cylinder or other figure of revolution. The ordinary wood-turning lathe, in which a chisel-like tool is held by the operator against the workpiece, is of ancient origin. Numerous precedents for the industrial engine lathe can be found throughout the 18th century, particularly among the scientific-instrument-makers' tools, the most notable being the little (12 x 8 x 6 inches overall) screw-cutting lathe of Jesse Ramsden, the celebrated English instrument-maker, which was described in detail in a book of 1777.

The heavy industrial lathe appeared around 1780 in France, but because of the much more vigorous development of the lathe in England, the French antecedents are frequently overlooked, nor can their influence upon English makers be more than guessed at, in the absence of any serious study of the question. Nevertheless, the existence of two medium-sized, sturdily constructed lathes in the museum of the Conservatoire Nationale des Arts et Métiers, in Paris, shows that the mechanical skill if not the economic demand were present on the Continent as well as in the United Kingdom. The earlier of these two lathes, built between 1770 and 1780 by Jacques de Vaucanson (1709-82), has a sliding tool-carriage, advanced by a long screw, called a lead screw, which runs parallel to the axis of the workpiece. The later lathe, a machine of 1795 attributed to Senot, about whom nothing else is known, had in addition to the sliding tool-carriage a system of gearing, called change gears, which connected the lathe spindle with the lead screw, thus controlling the rate of advance of the tool-carriage in such a way that a helical screw-thread could be cut on the workpiece. Vaucanson built also a drilling machine in which the positioning and advance of the drilling bit were controlled much as in the modern drill press. The axis of the bit in Vaucanson's machine was horizontal, while most later drilling machines (from about 1800) were vertical.

HENRY MAUDSLAY

The name of Henry Maudslay (1771-1831), of London, has become inseparably tied to the industrial lathe, for good but complex reasons. He was by no means first to design a screw-cutting lathe: Ramsden and Senot were earlier. Long and heavy workpieces—six feet long, a foot or more in diameter—could be turned in sturdy lathes, constructed of heavy timbers laid in solid masonry foundations. Such lathes, which had rudimentary but effective sliding tool-carriages, could be found before 1815 in England, Germany, and the United States. Even Maudslay's tool-carriage, when one was received in Philadelphia shortly after 1820, was modified by Mason, Tyler, and Baldwin, and a substantially more rigid and workman-like tool-carriage resulted. Furthermore, Maudslay made no particular effort to sell lathes. He preferred, for example, to make lead screws for other machine-tool builders rather than build the entire machine himself. Nevertheless, Maudslay is entirely deserving of his great reputation.

There appear to be two principal reasons for the fame of Maudslay's lathes and for the influence they undoubtedly exercised on later machine tools. The first reason is best expressed by John Farey, Jr., a perceptive contemporary writer on mechanical subjects. In 1810, reflecting a widely held view, Farey described Maudslay's lathe as "the most perfect of its kind." The second reason was the remarkable group of English machine-tool builders who worked in Maudslay's shop at one time or another.

Maudslay, a native of Woolwich, near London, was from the age of twelve

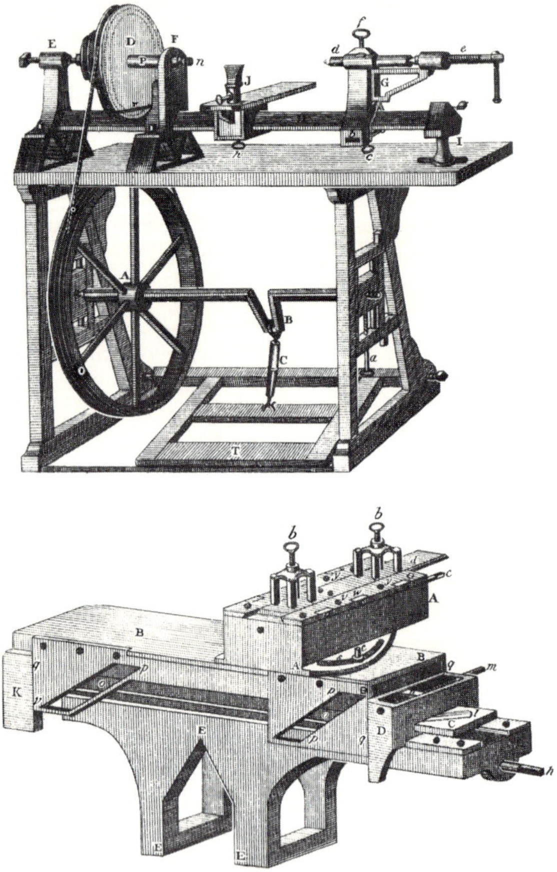

Fig. 16-4. Maudslay's lathe and slide-tool, *c.* 1807. The slide-tool, *lower* (much enlarged), was fitted over the inverted-V lathe bed, replacing the conventional tool rest, shown installed, *above*. The cutting tools were secured under the four-legged yokes by thumbscrews (b,b). (*The Smithsonian Institution*)

employed in workshops. He spent some eight years, between the ages of eighteen and twenty-six, in the shop of Joseph Bramah, who made locks but whose most significant contribution was his hydraulic press. After setting up his own shop, Maudslay was engaged for several years in making the well-known Portsmouth block-making machines. Marc Brunel and Samuel Bentham, who knew how the Taylors of Southhampton had before 1780 partly mechanized the production of ships' blocks (pulleys), had supplied plans to Maudslay for completely mechanized production of blocks, to be carried out in the Portsmouth dockyard. Special-purpose machines, using precisely the system that fifty years later would be known as the "American system" of manufacture, were designed to carry out sequential operations, starting with the raw materials, of wood

Fig. 16-5. Scoring machine, *c.* 1810. The two wheels have cutting tools on their peripheries. Revolving at high speed, these cutters cut a groove in the pulley-block; a block is shown being cut at left. This is one of the 43 machines, known as the Portsmouth block-making machinery, built by Henry Maudslay for the Admiralty. (British Crown Copyright. *Science Museum*, London)

and metal, and ending with finished blocks. In all, Maudslay built 43 machines, some of which still survive in the Science Museum, London. Although in successful operation before 1810, widely celebrated in print, and in operation for more than a century, the Portsmouth block machinery made an unaccountably slight impression upon British gun-makers and others who might have adapted the system to their manufactures. Nevertheless, Maudslay's reputation as the leading machine-builder in England grew.

The opportunity to work in Maudslay's shop was eagerly grasped by four young men who became leading machine-tool builders of the next generation. His influence upon the thinking of the machine-tool industry was thus felt for many years after his death in 1831. Joseph Clement (1779-1844), best known for his work on Charles Babbage's mechanical computers, came to Maudslay in 1815 as chief draftsman; in 1817 he set up his own shop in London, where he remained for the rest of his life. He made substantial though detailed innovations in lathe and planer design; their influence upon the trade was enhanced by the publication of drawings and descriptions.

Richard Roberts (1789-1864), a Welshman, worked for Maudslay from 1814 to 1816, then established a shop in Manchester. An original and prolific inventor, he built in 1817 a metal-planing machine, then a new if not unique machine tool. He improved the engine lathe also, supplying the "back gear" arrangement of shifting the spindle-drive from high speed to low, which remained the standard design for over a hundred years. He also built many drilling machines.

Joseph Whitworth (1803-87), whose name was to become the most famous in machine-tool circles, was born near Manchester and entered Maudslay's London shop in 1825. After working in the shops of Joseph Clement and John Jacob Holtzapffel, another tool-builder, Whitworth returned to Manchester to found his own machine-tool works. While neither Clement nor Roberts regularly made machine tools for the trade, Whitworth operated probably the largest—certainly the most prestigious—works devoted to making machine tools for sale. Once established, Whitworth was responsible for introducing a standard screw-thread, and by manufacturing and selling standard gauges his influence in length-measurement was felt all over the world.

James Nasmyth (1808-90), son of a successful Scottish portrait painter, came to Maudslay's shop only three years before Maudslay's death. An able and original mechanical designer, Nasmyth devised for Maudslay a special-purpose milling machine for milling the flats of hexagonal nuts. He later designed a popular version of the planer, known as a shaper. Nasmyth's shaper was dubbed "steam arm," duplicating as it did the manual operation of grooving with a cold chisel. His best-known innovation was the steam forge hammer, which remained a standard and powerful iron-forming tool for over a century. He established large machine shops near Manchester.

Nasmyth, in his *Autobiography*, described his experiences in Maudslay's

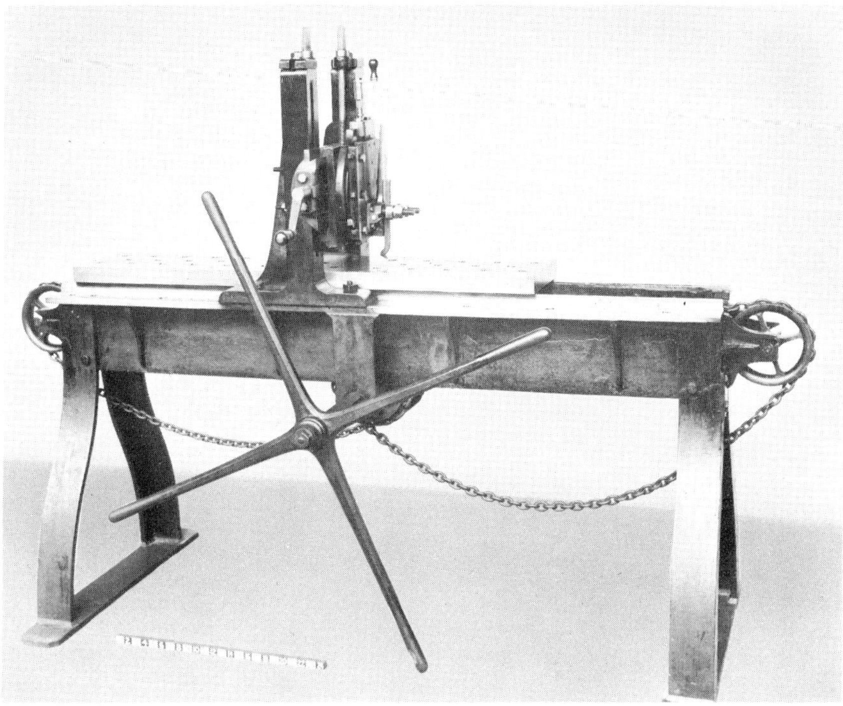

Fig. 16-6. Metal-planing machine, *c.* 1817. Built by Richard Roberts, this planing machine is one of the earliest built. The workpiece (not shown) was fastened to the perforated table and pulled against the cutting tool (set vertically above the table) by the chain. (Lent to the *Science Museum*, London, by the late R. Roberts, Esq., Manchester)

shops. Here we can glimpse Maudslay's abundant talent and traits of character that so powerfully imprinted his way of doing things on the whole British machine industry. His "innate love of truth and accuracy" led Maudslay to have surface plates—precisely planed, flat metal surfaces—placed throughout his shops, so workmen might test their own work for precision as they proceeded. His bench micrometer (for extremely accurate measurements) was probably the first to be used in a machine shop; he called it the "Lord Chancellor," after the one from whom there was no appeal. His own skillful use of tools increased his standing among mechanicians and brought him impressively close to the perfection he strove for.

"To be permitted to stand by," wrote Nasmyth, "and observe the systematic way in which Mr. Maudslay would first mark or line out his work, and the masterly manner in which he would deal with his materials, and cause them to assume the desired forms, was a treat beyond all expression. Every stroke of the hammer, chisel, or file told as an effective step toward the intended result. It was a never-to-be-forgotten practical lesson in workmanship in the most exalted sense of the term." An old hand in Maudslay's shop recalled, many years after the master's death, "it was a pleasure to see him handle a tool of any kind, but he was *quite splendid* with an eighteen-inch file."

OTHER MACHINE-BUILDERS

Maudslay's shops were by no means the only important machine shops in England, but his unusual and well-documented influence upon other builders has called for special mention. The steam-engine shops of Boulton and Watt in Birmingham employed some of the most advanced techniques of metal-cutting; Matthew Murray (1765-1826) of Leeds was an original and, within his lifetime, celebrated builder of steam engines and textile machinery; and there were many others.

Bryan Donkin (1768-1855), a very able mechanician, in whose London shops the Fourdrinier paper-making machines were built, was a close friend of Maudslay, consulting with him frequently on mechanical problems. Donkin, who has been curiously neglected by historians of technology, spent a major part of his career in developing and bringing to practical perfection the great continuous-web paper machines that so radically changed the paper-making industry. Many of these machines, which were 6 feet or more in width and between 25 and 40 feet long, served for fifty years or more. The essential arrangement of the Fourdrinier machine remains yet unchanged. In 1832, a young American technologist, George Escol Sellers, visited Donkin's shops, where he was suitably impressed by machine work of the highest order. While Sellers was with Donkin, a messenger brought word of an accident to one of the Fourdrinier machines that Donkin had built. An operator had dropped a wrench into the machine; before the machine could be stopped it had been seriously damaged. To the visiting American's amazement, Donkin went to his storeroom, took from the shelves replacements for the parts that had been broken, and dispatched a workman who, he said, would have the machine back into operation before midnight. This was the first time Sellers had seen reduced to practice the notion of interchangeable parts for large machines. Not only did this show the precision of the machine tools on which Donkin had made the machinery, but this whole concept of interchangeable parts was the basis for mass production which was to revolutionize industry and society.

Donkin and Maudslay had taken out a minor patent together when both were young men; throughout their lives both were obsessed with the pursuit of perfect accuracy in screw-threads. Both men contributed ideas and skill to the development of workshop precision and accuracy. Both men were of that remarkable generation of mechanicians who, within less than fifty years, transformed machine building from a branch of blacksmithing to an art in which man's mastery over metals was fundamentally changed.

By 1830 all of the principal machine tools—with the exception of the surface grinder which came a few years later—had been brought to a form that would be instantly recognizable to a machinist of today. Most of the basic innovations had been made by men directly influenced by the Maudslay school. The planing machine emerged after 1810 and probably before 1820. Claims of priority have been made for several machine-builders, including Matthew Murray and James Fox (1789-1859) of Darby; the planing machine of Richard Roberts, carrying a date of 1817, whose authority has been insufficiently questioned, exists in the Science Museum. The shaper, a modification of the planer, was made popular by James Nasmyth around 1835.

One form of the milling machine, which was similar in principle to the clockmaker's gear-cutting machine, was built by Nasmyth when he was scarcely twenty-one years of age. Much of the further development of the milling machine, however, occurred in the United States, and until late in the 19th century the milling machine was not a popular tool in British shops.

AMERICAN DEVELOPMENTS

The backward state in 1800 of American machine shops has been noted. Many primitive practices, such as the turning of metal in a lathe with a long-handled, manually guided cutting tool, persisted in small machine shops until after 1840. In the absence of planing machines of sufficient capacity to finish very large castings, American machinists resorted to the cold chisel and hammer. In 1839, for example, a marine steam-engine cylinder, 6 feet in diameter, was seated on an iron bed-plate casting by finishing the casting with hand chisels and files. On the other hand, American practice was by 1830 in many ways coming rapidly up to British standards of excellence, and in some respects had already leaped beyond.

George Escol Sellers, the Philadelphian whose visit to England has already been mentioned, was in the Maudslay shops in 1832, about a year after the founder's death. He knew well the reputation of the master, and had been anxious to visit the shops, particularly because he had heard that the largest marine engine cylinders that Maudslay's had ever made were then being machined.

"Here I must confess to a feeling of great disappointment," he recalled, "for the cylinder struck me as a mere pigmy compared with the cylinders of the North River and Long Island Sound boats of that period. . . . There were at that time boats on the American rivers with condensing engines, whose cylinders would cover two of the one on the boring machine placed one on top of the other." In his opinion, Roberts's lathes, in Manchester, were far ahead of Mauds-

Fig. 16-7. Blanchard's gunstock machine, 1821.
(*The Smithsonian Institution*)

lay's; and he found no planing machines at work in the Maudslay shops, flat surfaces being wrought by chisel and file. He noted before leaving England that he had not seen in London or Birmingham or Manchester a lathe as large as the one that he and his brother were then building in Philadelphia.

Perhaps the most original tool to emerge in the United States before 1830 was the gunstock machine of Thomas Blanchard (1788-1864), a Massachusetts machinist. Blanchard's machine, which from 1821 was used to carve out wooden gunstocks in the Springfield Armory, employed a rotary cutter; a tracing wheel of the same diameter as the cutter was held against a master gunstock pattern; through a rectangular linkage the movement of the tracing wheel was transferred to the cutter, which produced a stock identical with the pattern. A generation later the Blanchard duplicating principle was applied to metal components as well as wood.

In the United States the concept of using special-purpose machines to reduce the requirement of skilled labor was accepted by laboring men as well as by management, and the "American system" of manufacture, in which complex mechanical devices were produced in a series of sequential machine operations, flourished after 1830 in many industries besides gun-making. It was probably the "go-ahead" attitude, held by all classes of Americans and so often noted by European visitors to the United States after 1830, that permitted the "American system" to flourish in America while it withered in Europe. It will be recalled that the Portsmouth block machinery exhibited the necessary elements of the system, but that its influence in England was slight.

Another significant characteristic of American machine-building, in contrast to British practice, helps to explain the rapid strides taken by American builders to overcome the lead of English industry. In American machines, design was based upon function. No unnecessary parts were tolerated; no decoration was acceptable. Only the working surfaces were to be finished; all other parts of the machine could be left as-cast or as-forged. A young Scottish engineer, who in 1837 visited the locomotive works of Matthias Baldwin, in Philadelphia, reported that he had seen "no less than twelve locomotive carriages in different states of progress, and all of substantial and good workmanship. Those parts of the engine, such as the cylinder, piston, valves, journals, and slides, in which good fitting and fine workmanship are indispensable to the efficient action of the machine, were very highly finished, but the external parts, such as the connecting rods, cranks, framing, and wheels, were left in much coarser state than in engines of British manufacture." Bleak as this appeared to many a European craftsman, who lavished care upon the appearance of finish and decoration, it will be recognized by technologists as a labor-saving means to a purely functional end.

SUMMARY

The decades between 1750 and 1830 were of critical importance for the art of iron metallurgy. Both the making of iron and its working were vastly improved by developments during these years. The production of cast iron was improved by the use of mineral coal and coke, and castings of larger and larger size were successfully attempted. The production of wrought iron was revolutionized by the substitution of the puddling process and rolling mill for the earlier hearth-fire and forge hammer. The use of steam engines to power blast engines and other mill machinery greatly benefitted the whole iron trade.

The increased demand for and increased supply of iron, both wrought and cast, was paralleled by growing sophistication in the handling of metals. The blacksmith was reduced from the primary fabricator of iron products to something more like the kindly shoer of horses known to later generations. The development of machine tools—which were in themselves machines and were used to make other machines—such as the engine lathe, drill press, planer, and so forth, completed the round of innovation which spiralled upward during these years.

Increased demand for iron forced the use of steam engines to help increase production; the demand for steam engines greatly stimulated the market for iron and at the same time demanded new and more exact machine tools for the working of iron; these new machine tools, in turn, enormously increased the usefulness of iron for a whole host of purposes previously reserved for the less durable but more easily wrought wood. And so the spiral continued until the entire industry, and other industries and, indeed, all of society, was transformed.

17 / Steam Transportation

EUGENE S. FERGUSON

Perhaps more than any other single factor, improved transportation was the key to the success of the Industrial Revolution. In Great Britain the division of labor and increased production which resulted from the introduction of new machines and the factory system depended upon an ability to move goods and raw materials (as well as people) freely and economically about the island kingdom. In Europe, Napoleon's efforts to unite the entire Continent into one Grand Empire were frustrated almost as much by a lack of transportation facilities as by the rival ambitions of captive peoples. In the United States, adequate transportation was absolutely imperative to the new nation which occupied the narrow coastal plain of the Atlantic seaboard, and dreamed of filling all the west as far as the Mississippi River, and perhaps even beyond. For all of these, the steam engine, which had liberated manufacturers from the tyranny of mill streams and hand power, was now to break the restraints of seacoast and navigable stream with the steamboat and the steam locomotive.

Although the voyage in 1807 of Fulton's steamboat, popularly known as the *Clermont,* marked the beginning of commercially successful steamboat operation, this success was based upon the intensive work of many steamboat inventors over a twenty-year period, which was itself the outgrowth of a much longer period of steam-engine development. As early as 1788, when the steam engine had become an accepted power source for turning machinery as well as for pumping water, the steamboat was shown to be at least technically possible if not yet economically feasible. In the United States, John Fitch demonstrated on the Delaware River his steamboat, which proved a principle but made no money; in the south of Scotland, William Symington demonstrated his steamboat in a lake near Dalswinton.

As in the case of the steamboat, the first steam locomotive had not appeared without predecessors. The steam locomotive that Richard Trevithick in 1804 put on the Penydaren colliery railway, in southern Wales, was the first locomotive engine to operate on rails, but he, among others, had already built more than one steam carriage, intended to move on common roads. The exact date of commercial success of the steam locomotive, therefore, is much less distinct than that of the steamboat; however, there was no longer a reasonable doubt after the opening in 1825 of the Stockton and Darlington Railway in northeastern England. It was a day of unqualified triumph for George Stephenson's locomotives. In coal cars temporarily fitted with seats, well over 300 passengers were carried in a single train for more than eight miles while a throng of perhaps 40,000 or more watched the spectacle.

THE TRANSPORTATION PROBLEM

A nation can never know, in advance, what the final solution of a pressing national problem will be, particularly if the technical means that will in time make the problem soluble have not yet emerged. Thus it was in England and in the United States during the period of the Industrial Revolution. That a need existed for better transportation was evident to thoughtful men. In 1776, the great English economist, Adam Smith, declared that "good roads, canals, and navigable rivers, by diminishing the expense of carriage [of goods, thus opening new markets to producers] are on that account the greatest of all improvements."

The prominent English potter, Josiah Wedgwood, who was a promoter of the Trent and Mersey Canal, had in his youth seen pack horses, in files of forty or fifty, heavily laden with coal or potter's clay or crates of the potter's ware, floundering knee-deep in muddy roads that were all but impassable, urged on by the whips of impatient drivers. Such sights were common in many parts of England before 1750. Up to 1760, although English vessels out of Liverpool called at ports around the world, there was no road for carriages into Liverpool. There were no coaches and no wagon trade even to Manchester, less than forty miles away. In the United Kingdom, as in the United States, the seaports in the most distant parts of the country had cheaper and easier transportation than inland points only a few miles apart. Coal from Newcastle was carried to London in a fast fleet of coasting vessels; coal in the interior often arrived, as noted above, on the backs of pack horses.

Internal improvements in England lagged behind those on the Continent, particularly in the Netherlands. As on the Continent, the first responses to the need for better transportation both in the United Kingdom and the United States, were the improvement of rivers and the construction of canals. In the United Kingdom, the building of canals employed much money and talent from about 1760, and by 1830 the system of inland waterways was essentially complete. In the United States, on the other hand, the first long canal—the Erie—was started only in 1817, and the canal age in the new country did not get under way until after 1820, partly because of this reliance on both inland river and coastal water routes.

EARLY STEAMBOATS

The first clear idea of a steamboat occurred in 1736 in England, when Jonathan Hull published a description and pictorial sketch of a boat he proposed to use to tow sailing craft into and out of harbors. The engine was of the Newcomen type; the propulsion was by paddle-wheels. There is no evidence that such a craft was built, and it was not until the last fifteen years of the 18th century that the half-formed ideas and speculations concerning the steam engine as a prime mover for boats began to be put into actual use in building.

In the United States the earliest credible steamboat builder was John Fitch (1743-98). In many ways he was, as he described himself, the "most unfortunate man of the world"; he had no money, and there is little doubt that his personality repelled those who might have supported his efforts. Fitch was one of that peculiar class of innovators that has been fairly numerous in the United States: unlettered, often quite ignorant of the physical principles on which their innovations rest; persistent, yet seldom successful either in a conceptual or commercial way. They may, like Fitch, have been "ahead of their time," but it is questionable whether they possessed the level heads that are capable of making a practical contribution at any time. Writers have either ridiculed or glorified men like Fitch; yet none has been seriously studied.

In 1787, in the Delaware River at Philadelphia, Fitch placed in a boat perhaps 30 feet long a steam engine that he and Henry Voight, a Philadelphia clock-maker, had made. The steam engine, which employed a separate condenser, drove paddles, probably at the stern of the boat. On August 22, in the presence of several statesmen attending the Constitutional Convention then in session, Fitch piloted his boat while the steam engine propelled it against wind and tide. In the following year, more successful attempts were made by Fitch. In 1790 he established service between Philadelphia and Burlington, some twenty miles away, but it was not sufficiently regular for commercial acceptance.

Fitch demonstrated that his approach to the steamboat was not chimerical, although he was unable to build a boat that was commercially successful. He spent a great deal of time and effort petitioning the legislatures of the various states, asking for exclusive rights to steamboats on their rivers, and in proposing to the national Congress that he be given money to develop a suitable craft to stem the current of the Mississippi. He spent energy also in contesting the competing priority claims of James Rumsey, of Virginia, who late in 1787 had built a steamboat propelled by a jet of water issuing from its stern.

In Scotland, the steamboat that William Symington (1761-1831) was associated with in 1788, was a double-hulled catamaran (a boat with two parallel hulls). The builder of the hull, Patrick Miller, was a wealthy man of leisure who already had worked out a manually operated paddle-wheel propulsion system. He employed Symington to adapt a steam engine to his existing system. The Miller-Symington boat of 1788, employing a Watt-type condensing engine, probably travelled at about four miles an hour. However, Miller apparently favored his manually operated boats, and steam experiments were accordingly dropped. In 1801, Symington built the more famous *Charlotte Dundas,* intended for towing boats through the Forth and Clyde Canal. None of his steamboat projects was profitable to him, however, and Symington is remembered, like Fitch, as a man who was not quite able to exploit ideas that in other heads and under other circumstances were to prove economically profitable.

There were two parts to the problem that the early steamboat builders were unable to solve satisfactorily. They could not, or at any rate did not, build a steam engine that could be depended upon; and it was not apparent that any one system of propulsion was inherently superior to alternative systems. Since the prime mover and the propulsion system had to work together, the undetermined variables were numerous and refractory. Fitch considered at various times such propulsive devices as a chain of paddles, located at either side of the hull; a mechanism that paddled the boat in the fashion of a war-canoe, with six vertical paddles on each side; a screw propeller; and a set of three duck-feet paddles hanging over the stern of the vessel, which is the system he used in his partly successful boat of 1790.

Benjamin Franklin, however, who was supporting Rumsey's trials, favored reaction propulsion using a pump to squirt water rearward through a stern-facing tube, believing, erroneously, that a paddle-wheel was impracticable because so much force was expended in depressing and lifting slugs of water. He had seen in Paris a boat being propelled across the Seine by an Archimedean screw mounted in the air, on the boat's deck. Thomas Jefferson, who succeeded Franklin as American Minister in Paris, had seen it also. He described the screw, about four feet in diameter and three or four feet long, and its action, which appeared to him to be one of advancing in air as a wood-screw advances in wood, though with more slippage. Jefferson thought that the effect of the boat screw would be increased by locating it underwater rather than in air. The builder of the screw, he noted, had no idea of the principle he was employing.

John Stevens, another American inventor of the time, used a screw propeller, in appearance very much like a modern boat propeller, but later abandoned the screw in favor of paddle-wheels. Thus it may be supposed that the correct solution of this part of the propulsion problem was not self-evident. Similarly, the steam engines used ran the gamut from Newcomen-type atmospheric engines to high-pressure non-condensing engines, and were as often as not built on the design and under the direction of the experimenter himself. The art of steam-engine building was not widely diffused in the latter part of the 18th century, particularly in America, and most of these engines were neither practical nor dependable.

ROBERT FULTON

It has been argued that the steamboat was brought into being by an economic need, but there was a great deal more to meeting this need than simply recognizing it. The passengers had been waiting on the wharf for twenty years when Robert Fulton's *North River* made its maiden voyage up the Hudson from New York City to Albany in 1807.

Robert Fulton (1765-1815), a native of Pennsylvania, had at the age of twenty-one gone to London to become an artist. The high excitement of canal-building and mechanical innovations in England was an attraction that eventually took him from his painting; between 1793 and 1806, when he returned to the United States, he had been active in many schemes involving canals, boats, and submarines. He invented a submarine and tried to sell it first in France, then in England. After long and unsuccessful flirtations with the French and British governments (then engaged in the Napoleonic Wars), he gave some serious thought to a steamboat that he agreed in 1802 to build for Robert Livingston, the American Minister to France. Chancellor Livingston, who fancied himself quite an inventor, had a monopoly on steamboat navigation in the Hudson River, granted by the State of New York. His attempts, before going to France, to build a steamboat had resulted in dismal failure, so he enlisted Fulton's aid. Starting with full knowledge of the work of Symington and others in England and France, Fulton made some further experiments on the Seine River, in Paris, then ordered a steam engine from Boulton and Watt, in Birmingham, to be shipped to New York.

In the summer of 1807, even while he prepared his boat in New York for its first passage up the Hudson, Fulton's thoughts were far away on the Mississippi. To him, as to Fitch before him, the vast opportunities on the western rivers were more attractive than the route for which Livingston had a monopoly. In anticipation of these opportunities he had built his boat with a flat bottom, drawing only two feet of water, suitable for the western waters. After Fulton's European years, in which failure had been a constant companion, it is hardly surprising that he could not foresee the remarkable success of his boats in the Hudson River and on Long Island Sound.

Fulton's steamboat, called by him "the North River steamboat" but later

Fig. 17-1. Fulton's "North River steamboat" (later known as the *Clermont*) on the Hudson River, 1807. (*The Bettmann Archive*)

known as the *Clermont*, after Chancellor Livingston's estate, departed from New York on August 11, 1807, at 1:00 p.m., and made the run to Clermont in 24 hours. There it stayed until the following day and then continued on to Albany. The total running time was 32 hours, the distance 150 miles, the average velocity about five miles per hour. The boat returned to New York in 30 hours running time.

The North River steamboat was long and narrow, 150 by 13 feet; two paddle-wheels each 15 feet in diameter, located on the sides of the hull, made the boat appear wider and better proportioned; the paddle-wheels were driven by a Boulton and Watt vertical double-acting steam engine. The drive arrangement, which was a unique bell-crank device, probably was designed by Fulton himself.

Holding to a regular schedule, the dependability of Fulton's steamboat service slowly became obvious in 1807 and 1808. By 1811, Fulton had a steamboat running on the Mississippi River; in 1812 he established a machine shop in Jersey City to build steam engines. Before his death in 1815 he had steamboats running on Long Island Sound in addition to those on the Hudson, and he was pushing to completion the first steam warship, a twin-hulled battery for coast defense.

While Fulton's commercial success is easily seen, his substantive contribution

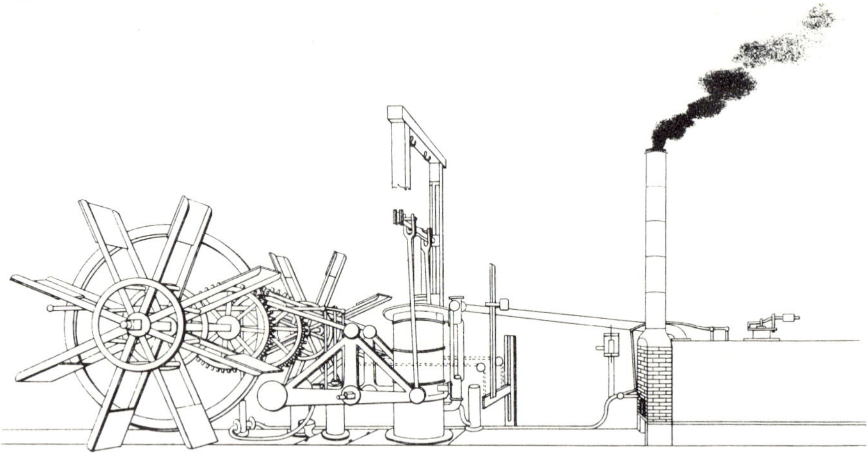

Fig. 17-2. Machinery of Fulton's steamboat, 1807. The vertical cylinder transmits force through the vertical connecting rods, triangular bell-crank, and horizontal connecting rod to the paddle-wheel shaft. The engine and boiler were built by Boulton and Watt of Birmingham, England. (*Science Museum*, London)

to steamboat technology is perhaps not so obvious. His hull design, which he based upon the results of towing models in a test tank, was highly unorthodox, consisting merely of a straight box chopped off on straight lines at each end to form a 60-degree bow and stern. The hull, then, of which he was so proud, was a definite hindrance. Before the North River boat's second season, the hull had been rebuilt along more traditional lines.

Fulton's solution of the twin problems of prime mover and propulsion system was the sole basis for his success, but in view of the earlier gropings of many steamboat enthusiasts, his success was definitive. The paddle-wheel was unquestionably the most efficient propulsive device available when the prime mover was a low-speed steam engine, and the steam engine he obtained was from the most able steam-engine builders in the world.

In the United States, the western river steamboat was a type that developed rapidly during the 1820's. In 1830, there were nearly 200 steamboats operating in the Mississippi watershed, a number two-thirds as large as that of the entire British merchant fleet. Despite many differences of detail, the western boat as a type was distinguished by a high-pressure, non-condensing steam engine which drove a stern paddle-wheel. While boiler explosions were not confined to high-pressure engines—heavily barricaded "safety barges" for passengers were towed by steamboats on the Hudson River—the conditions under which western high-pressure engines operated were particularly conducive to disastrous explosions.

Boilers were cylindrical, about 4 feet in diameter. Pressures of 150 pounds per square inch were common, and boiler settings were unsubstantial, resting as they did upon flat, flexible boat hulls. Boiler water came unfiltered out of the river, and mud settled in the boiler immediately above the fire. Being a good insulator, the mud caused overheating of the boiler shell with consequent weakening. Even if a safety valve had never been tied down during an exciting race (and they often were), the record of explosions would have been nearly as dismal. Thirty-five explosions occurred between 1815 and 1830, killing 250 people; during the 10 years following 1830 another 70 explosions occurred. Nevertheless, the western steamboat helped greatly to populate the east central section of the United States, and the eastern river steamboat increased the tempo and regularity of life in a nation becoming rapidly industrial.

Steamboat development in Great Britain was only a few years behind. In 1812 Henry Bell's *Comet* was successfully operated on the Clyde River, in Scotland; by 1816 steamboats were plying the Thames, Mersey, Trent, Tyne, Ouse, and Humber Rivers in England, and steam ferry service had been started between Scotland and Ireland.

The Atlantic Ocean was crossed in 1819 by the American steamship *Savannah*. Like all ocean-going steam vessels before 1838, the *Savannah* carried a full suit of sails. The captain was expected to use steam power only when the wind failed, and the paddle-wheels were collapsible so that they could be stowed on deck when not in use. In a passage of 29 days, the *Savannah* used its engine for about 90 hours (less than 4 full days). The British steamer *Enterprize*, sailing out to India in 1826, used her engine on 63 days of her 103-day passage.

EARLY LOCOMOTIVES

Like Fulton's *Clermont*, the first successful locomotive followed decades of trial and innovation. The first steam locomotive, which was built in Paris in 1770, was a generation ahead of the successful English development of the locomotive, but it apparently had no direct effect upon later investigators. Nevertheless, the machine itself was a remarkable *tour de force* and deserves mention as an example of the highly sophisticated concepts and workmanship that were then employed.

In 1770, Nicolas Cugnot (1725-1804) demonstrated in the presence of French army representatives a three-wheeled self-propelled steam wagon, which may have been intended to carry a burden on its own chassis or, as tradition has it, to drag artillery pieces behind it. A copper boiler, which held perhaps 400 gallons of water, supplied steam to two high-pressure steam cyclinders, which were mounted vertically above the single front driving wheel. The piston rods, extending downward from the cylinders, turned the driving wheel by

Fig. 17-3. Cugnot's steam carriage, 1770. The boiler is at left; the vertical engine cylinders are shown to the right of the boiler. (*Science Museum*, London)

means of a ratchet device. As a piston rod moved downward it engaged teeth on the driving wheel, thus urging it onward; the two cylinders alternately supplied impetus to the wheel.

Aside from the obvious excellence of workmanship represented by this machine (which still exists and can be inspected in the Conservatoire Nationale des Arts et Métiers, in Paris), there are two other remarkable features. First, the high-pressure two-cylinder reciprocating engine is so nearly like a drawing that had appeared first in a German reference book of 1724 that the similarity can hardly be merely coincidental. Still, the Cugnot engine is the only high-pressure engine that is known to have been built during the 18th century. Second, the Cugnot carriage exhibits clearly a device for converting the reciprocating motion of the steam engine to the rotary motion of a wheel. The problem of converting reciprocating to rotary motion had been attacked, on paper, before 1769, but again Cugnot's device is among the very first to be reduced to practice. While the crank and connecting rod, which to a modern mind is the obvious solution of the problem, were by no means unknown to Cugnot, their superiority over alternate devices was not immediately evident, even to so acute a mind as James Watt's.

In every way, then, the Cugnot steam carriage was remarkable. Even if no direct lineage can be shown to Trevithick's first locomotive, Cugnot's machine demonstrates to us the capabilities of European machinists, metalsmiths, and innovators, and makes more understandable the work of Trevithick and the other builders.

In 1784 James Watt took out a patent for a steam-driven road carriage. The carriage was but one item in a patent that contained several diverse claims, and it was added as an afterthought. William Murdock (1754-1839), a Scottish mechanician who was then Boulton and Watt's erector and representative in the Cornwall mining region, had had that summer an idea for a steam carriage which employed a low-pressure condensing machine. In his enthusiasm, he considered abandoning his position with Boulton and Watt in order to obtain a patent and pursue the idea of the carriage. By offering to include the carriage in Watt's current patent, Murdock saved the expense of a separate patent; and Matthew Boulton was able to convince him to continue his job in Cornwall. In 1786, Murdock built a model of a steam carriage, on which he used a steam cylinder ¾ inches in diameter and a 1½ inch stroke, probably operating without a condenser. Murdock set off to London to exihibit his model, was met on the road by Boulton, and again was induced to return to stationary-engine building in Cornwall. Nothing more of significance was heard about Murdock's carriage.

In 1801, Richard Trevithick built a steam carriage that he operated briefly in Camborne, in western Cornwall. Trevithick had a year or two earlier built a successful high-pressure stationary steam engine, without a condenser, that he used for winding (lifting) ore from the underground workings. Trevithick journeyed up to London to take out a patent on his steam carriage. (It should be noted that while the steam carriages that had appeared thus far were all intended to run on common roads, the railway was by no means unknown when Trevithick first put a locomotive on rails.) His patent, granted in 1802, also covered the use of high-pressure steam engines to propel vehicles on railways. The railway he used, in fact, had been used for years as a wagon way.

The idea of a planked wagon way (the prototype railway) came out of the needs of mines. In Agricola's mining book of 1556, there is shown unmistakably a plank way laid underground for a four-wheel cart. Furthermore, the miner who pushed the cart did not have to steer it, for a stud on the cart, extending downward between the parallel planks of the wagon way, kept the cart from running off the planks. In the Newcastle coal-mining district, wagon ways were used extensively above ground to haul coal from the minehead to piers on the Tyne River, where it could be loaded into coasting vessels. As early as 1734, flanged iron wheels were being used on wooden railways.

Horses were employed, from the first, to pull the railway wagons. Rails were improved from time to time, first by fastening a series of cast-iron plates, each several feet long and 3 or 4 inches wide, to the wooden stringers of the railway. Later, cast-iron angle bars provided standing flanges on either side of the iron tracks, thus eliminating the need for flanged wheels on the railway wagon. By the time a steam locomotive was considered seriously as a device to pull the wagons, a great deal of experimenting with tracks had taken place. Nevertheless, the vicissitudes in determining a satisfactory solution to the problems of putting a locomotive on rails had only just begun.

Trevithick's first successful locomotive was operated on the wagon way of the Penydaren Iron Works in southern Wales. The rails of the wagon way, some ten miles long, were cast-iron angle plates with standing flanges supported on stone blocks. In February 1804, Trevithick successfully hauled a load of ten tons of iron and seventy men over a distance of more than nine miles. One of his passengers was, according to Trevithick, "the Gentleman that bet five hundred Guineas against it." This bet by Anthony Hill was, in fact, the reason for the trial. The locomotive was actually too heavy for the road, and the ironmaster of Penydaren works happened to need a stationary engine; so the wheels were removed from the locomotive and the engine was set to a task in the iron works.

This first locomotive employed a single-cylinder horizontal engine, 8¼ inches in diameter and with a stroke of 4½ feet. The piston rod, working through connecting linkage, turned a large flywheel, of perhaps seven or eight feet diameter, which was in turn connected to the wheels through an intermediate idler-gear. The steam pressure is unknown but the standard working pressure in locomotives up to about 1830 was between 30 and 50 pounds per square inch.

Trevithick's next locomotive was operated in 1805 on a coal railway near Newcastle. Although this second locomotive, like the first, was too heavy for the rails, it brought to the attention of coal operators the possibilities of steam locomotive power, and over the next several years the ideas planted by Trevithick took root as many other enterprising men built locomotives, and eventually the locomotive-powered railway flourished in the coal country.

One of the difficulties that plagued locomotive builders until after 1835 was that of obtaining sufficient friction between the driving wheels and the rails. That much of this difficulty existed merely in the minds of the builders made it no less real. Before Trevithick built his first steam carriage, in 1801, he carefully determined whether a carriage could in fact be propelled by an engine that turned its wheels. Ordinarily, carriage wheels turned because a horse drew the vehicle. It was not self-evident that the horse could be dragged by reversing the procedure. Mounting an unhitched cart, Trevithick and a friend, by dragging on the spokes rotated the wheels by hand, and determined that the cart would thus climb the steepest hill they could find.

In 1812, John Blenkinsop built a rack-and-pinion geared locomotive to obviate the supposed traction difficulty. A pinion (gear wheel), turned by the steam engine, dragged the locomotive along a toothed rack. At about the same time, William Brunton proposed, in all seriousness, a steam locomotive that would be pushed along by mechanical feet, mounted on jointed walking rods which extended like legs from the locomotive carriage. He obtained a patent on such a machine in 1813.

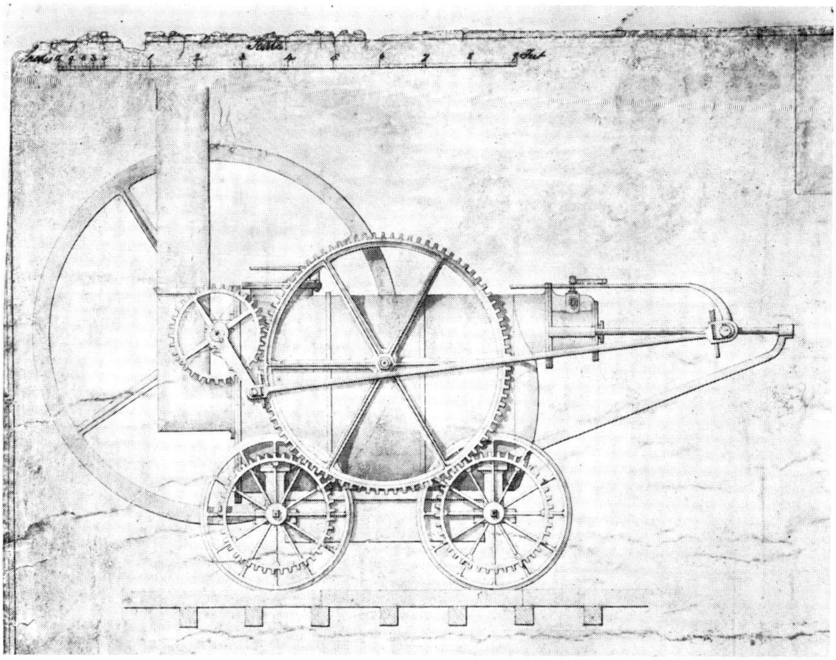

Fig. 17-4. Original drawing of a Trevithick locomotive.
(*Science Museum*, London)

Anyone who has seen a 20th-century steam locomotive spin its wheels in trying to start a heavy train of cars knows that the tractive effort of smooth wheels on smooth rails is limited. The knowledge that emerged from the railway experience up to 1836 was that the limit was much higher than almost anybody thought it could be. The issue was partly obscured by complicating factors: the locomotives were relatively light and the required tractive effort increased sharply as the grade (or slope) of the road increased. A car, weighing one ton and moving on a level rail requires a tractive force of about 10 pounds; when ascending a grade which rises one foot for each 100 feet of horizontal travel, the required tractive force is trebled. Despite a series of tests that were made around 1820, a general doctrine was diffused that simply was not in accordance with the facts. It was generally believed—and this belief was imported into the United States from England—that locomotives could be used only when railroads were nearly on a dead level. In England, on grades as slight as one foot per 100, cars were dragged up the incline by ropes, using stationary steam engines for motive power. It was not until 1836, when an American locomotive builder challenged the doctrine directly and dramatically, that its hold upon the engineering mind was shaken. William Norris, of Philadelphia, demolished the belief by driving a locomotive and a loaded passenger car up a long inclined plane whose vertical rise was over 7 feet in a 100. This incident anticipates the end of our story, but it illustrates how even a limited number of interrelated variables can so complicate a purely technical problem that any attempt even to analyze it into its essential elements is discouraged.

GEORGE STEPHENSON

George Stephenson (1781-1848), whose name is generally associated with the locomotive, built his first engine in 1814, taking full advantage of the experience of earlier builders. Over the next fifteen years he made numerous basic improvements, but his fame rests not so much upon originating radically new devices as upon uniformly selecting the right device to perform the task at hand.

It was the potential of the railroad itself, rather than the locomotive, that gave Stephenson his first opportunity. In the first dozen years of locomotive development an increasing number of people were made aware not only of locomotives but of railways in general, all of which were industrial railways and each of which served only a single mine or iron works. Long before the locomotive appeared to be more than an experimental machine, possibilities were being explored of building horse-powered railways to act as common carriers. As enthusiasm mounted, the scope of the horse-powered railways expanded from mere canal feeders, bringing freight to a canal, to clear alternatives to canals. Certainly the first cost of a railway was smaller than that of a canal, and the surprisingly bulky loads that horses could haul on level rails encouraged optimistic organizers and investors to find ways to minimize the rather limited carrying capacity and high operating costs per ton of freight.

When, in 1821, the Stockton and Darlington Railway was chartered, more than a score of railways had already been authorized. On none of these was it intended to use locomotives. The Stockton and Darlington, located in the coal fields thirty miles south of Newcastle, probably would not have considered locomotive power either had not George Stephenson been called in as a civil-engineering consultant to locate the line and to decide how the railway should be built and operated. In response to Stephenson's request, the charter was modified in 1823 to permit locomotives to be used; in 1824 two locomotives were ordered; and when it commenced operations in 1825, the Stockton and Darlington became the first common carrier in the world to employ locomotive power. This railway was in an important sense the prototype of all railways that were built within the next generation, for the practices adopted by Stephenson were widely copied by visiting observers from elsewhere in the United Kingdom, from the Continent, and from the United States. Within the year following the opening of the Stockton and Darlington, more than a dozen new railway companies were chartered. Among these was the Liverpool and Manchester

Railway, which became the theater for the next significant act in demonstrating in definitive terms how railway systems should be designed and operated.

George Stephenson was appointed chief engineer by the company in 1826. Although by the time the railway opened for business in 1830, Stephenson's already enviable reputation had been permanently established, his first experiences were not encouraging. Despite the success of Stephenson's locomotives on the Stockton and Darlington, the majority of directors of the Liverpool and Manchester Railway tended to favor stationary engines and rope haulage for the entire thirty miles of their line. Stephenson planned to use three inclined planes on the line, each more than a mile long. One was to overcome a grade of 1 in 48, the other two a grade of 1 in 96. Yet even on the levels connecting the planes, Stephenson's recommendation of locomotive power was questioned. Over the protests of a minority of directors, led by the treasurer, Henry Booth, the directors engaged two other civil engineers to study the situation. Not surprisingly, these engineers reported in favor of rope haulage. Each section, to be powered by a stationary steam engine, would be about 1½ miles long.

In view of the counter-arguments of Stephenson and the higher cost of the rope-hauling system, the directors finally decided to arrange what was to become the most famous contest in railroad history. The Rainhill Trials, which were held in October 1829 were for the purpose of discovering and testing "the most improved locomotive engine." On the Rainhill Level, between Rainhill and the Sutton planes, some nine or ten miles east of Liverpool, a 1½-mile course was measured off. At minimum, a six-ton locomotive was required to draw a load of 20 tons at the rate of 10 miles per hour. To demonstrate steady dependability, each locomotive was required to make forty trips over the course, totalling the distance from Liverpool to Manchester and return.

About ten entries were expected, but only three locomotives were actually on hand for the trials. Robert Stephenson (1803-1859), son of George, entered the *Rocket,* which had been built in the shops of Robert Stephenson & Co., the "& Co." meaning George and two financiers. Timothy Hackworth, master mechanic of the Stockton and Darlington Railway, entered the *Sans Pareil;* and John Braithwaite and John Ericsson entered the *Novelty.*

The *Novelty,* though light, elegant, and speedy, had some new mishap at every trial. An observer explained it thus: "first an explosion of inflammable gas which Burst his Bellows then his feed pipe blew up and finally some internal joint of his hidden flue thro his boiler so that it was no go." The *Sans Pareil,* which according to the same observer "rumbles and mumbles and roars and rolls about like an Empty Beer Butt on a rough Pavement," gave out before going half the required distance. Stephenson's *Rocket,* awkward though it appeared, ran the required distance at an average speed of 15 miles per hour, reaching on the last lap nearly 30 miles per hour.

Between 10,000 and 15,000 spectators were present for the first day of

Fig. 17-5. Stephenson's *Rocket*, 1829. A contemporary drawing from *Mechanics Magazine*, October 24, 1829. (*Science Museum*, London)

the trials; and while delays, breakdowns, and modifications of rules stretched the affair over several days, it seems certain that the best locomotive won. The *Rocket*, a joint design of Robert Stephenson, his father George, and Henry Booth, who suggested the small fire-tubes in the boiler, was the embodiment of the many advances that had occurred since George Stephenson had built his first locomotive fifteen years before. Again the master had provided the prototype locomotive, whose essential features were still recognizable in the last of the 20th-century steam locomotives.

In an equally remarkable though less rational manner, Stephenson was responsible also for the 4 feet 8½ inch gauge, or distance between rails. This gauge was selected because it was the gauge of the Killingworth colliery wagon way, for which Stephenson built his first locomotive.

THE SPREAD OF RAILROADS

The success of English railroads whetted the interest of the French. Following the pattern of the Stockton and Darlington, a short, horse-drawn coal line was opened in 1818. In France, unlike England, the government took direct control of railway development, and the growth of the national railway system was inhibited until nearly 1850 by the government's inability to ignore the various pressure groups whose interests could not be equally served by any single system.

In the United States, the absence of an adequate transportation system between the eastern seaboard and the interior was brought home during the War of 1812, when a British blockade of the eastern seaboard cut normal lines of transport and taxed both ingenuity and strength to haul cannon and other war supplies to the scenes of action on the Great Lakes. The first substantial result of the agitation for better transport was the Erie Canal, started in 1817 and completed in 1825, extending 365 miles across the particularly favorable terrain of upper New York State from Albany, on the Hudson River, to Buffalo on Lake Erie.

The state of Pennsylvania looked with an anxious eye upon the Erie Canal, which threatened not only to arrest the growth of Philadelphia's carrying trade but to cut off the dribble of eastern-bound wagon-freight that struggled over the Allegheny Mountains. In 1824, when the success of the Erie Canal appeared certain, an association of Philadelphians, whose business interests and civic pride would benefit by the development of a transport system across Pennsylvania, sent a representative to England to learn everything he could about canals, railroads, and highways. In its instructions to the emissary, William Strickland (1787-1854), the association reflected on the state of railroad building in the United States. "Of the utility of railways, and their importance as means of transporting *large burdens*, we have full knowledge," advised the sponsors, but "of the mode of constructing them, and of their cost nothing is known with certainty." Anything at all regarding locomotives was "entirely unknown within the United States."

The Philadelphia and Columbia Railroad, part of a mongrel system of railroads, canals, and inclined planes that eventually crossed the state of Pennsylvania, was opened in 1831 as a public way, on which any citizen, upon payment of tolls, might place his own freight wagons and haul them with his own horses. The impracticability of this system became immediately evident, and within a short time the railroad itself owned and operated rolling stock. Locomotives were placed on this railroad in 1834.

Railroads spread rapidly in the United States, beginning in the late 1820's. When Andrew Jackson entered the White House in 1829, no railroad employed locomotive power. Eight years later when Jackson finished his second term as President, a total of 1450 miles of railroads, using locomotives, had already been completed, and the growth of the railroads had only just begun. In the United Kingdom, on the other hand, 1600 miles of railroad had been completed by 1841.

Although a number of English locomotives were imported, starting in 1829 with the *Stourbridge Lion* for the Delaware and Hudson Railroad, a locomotive-building industry sprang up in the United States almost at once. The West Point Foundry machine shops, in New York City before its operations were consolidated

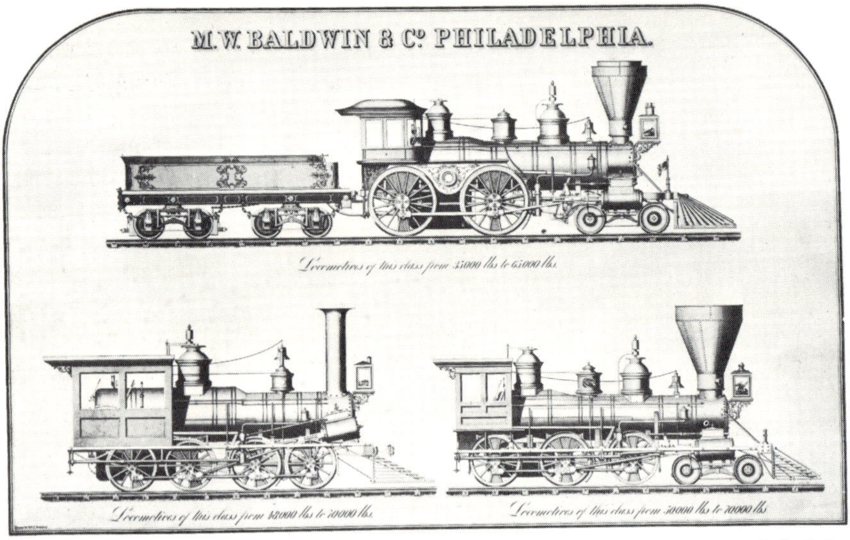

Fig. 17-6. American locomotives built by M. W. Baldwin & Company, Philadelphia. (*Science Museum*, London)

at Cold Spring, built one or two locomotives in 1831; Matthias Baldwin (1795-1866), a Philadelphia engraver and machinist, turned to locomotive building in 1832 and devoted the rest of his life to the trade; William Norris, whose audacious hill-climbing locomotive has been mentioned, was a well-known builder for a time; and there were many others who entered the field. About 350 locomotives were at work on American railroads in 1838; of these not more than 75 were of foreign make. During the 1840's a distinctively American style of locomotive—long, rakish, elegant if a bit spindly—emerged as a result of an accretion of the ideas of many builders, reflecting perhaps in its developed form an American point of view toward mechanical construction in general. When once the railroad had entered the United States, there was little opposition to its extension. The conflicts that developed with canal companies and riverboat interests were local and intermittent. In a country expanding as rapidly as the United States, there was business enough for all. Objections to railroads on other than economic grounds were thought to be frivolous.

In England, on the other hand, the arguments against railroads and locomotives were numerous, even if in the end they were declared irrelevant. Landowners who had leased rights-of-way to horse railways suddenly found their estates crossed by noisy, smoky locomotives. Landowners who were also highway commissioners objected strenuously and sometimes physically to the invasion of their lands by the Liverpool and Manchester surveying crews. George

Stephenson was threatened with ducking in a lake on one estate; on another he conducted furtive surveys while the owner was at dinner; a third he could not enter at all because a continuous watch was set by day and at night guns were fired across the grounds at irregular intervals. Parliament, exercising the right of eminent domain, eventually overcame these difficulties for the railway companies, but only at a price: as a concession to objectors, a change was included in railway charters requiring that locomotives must not emit smoke.

While in railway circles the final triumph of the steam locomotive was being celebrated, particularly after the Rainhill Trials of 1829, a sanguine group of enthusiasts was hard at work on the older problem of establishing the steam carriage as a proper vehicle to carry people on common roads. The best-known name in steam-carriage development was Goldsworthy Burney, who from 1826 built a series of carriages that would carry fifteen or twenty passengers at the rate of six to ten miles per hour. The steam carriages were not, in general, excessively heavy, weighing usually from 1½ to 2½ tons (comparable to modern American automobiles). Gurney used a water-tube boiler, which reduced the danger of boiler explosion. Some of the coaches made rather impressive records in giving dependable service on short runs. The steam carriage was expected by its advocates to compete not only with stage coaches but also with railroads. One glowing argument suggested that steam carriages would eliminate the "inconvenience to the public of abandoning their present habits [of using stage coaches] and adjusting themselves to the monotonous and limited accommodation offered by railways." However, the steam carriage was effectively taxed out of existence shortly after 1830; a resurgence of interest in the 1860's brought forth in 1865 the famous Red Flag Act, which required a flagman on foot to precede each steam vehicle. It was not until 1896 that British roads were at last successfully invaded by steam-driven vehicles.

SUMMARY

The impact of the steamboat and railroad was enormous both in England and the United States, but their results were somewhat different. Steamboats multiplied with wonderful rapidity in the United States after 1807; American railroads before 1837 had outdistanced the English systems they were patterned after. While in the United Kingdom the new systems of transportation were connecting already established centers of population and industry, those in the United States were designed to probe the sparsely settled territory to the west, bringing to large sections of the new continent not only whole populations but also the avenues of trade that permitted rapid development of centers of industry.

The steamboat and railroad not only encouraged commerce and industry but also the industry that built transport systems. Enlarged physical plants and refined techniques were required in the construction of boilers, the rolling of rails, and the casting and finishing of engine parts. Bridges in unimagined numbers were required by the railroads; as wooden bridges were replaced by iron ones, industrial capacity was of necessity expanded.

Finally, the steamboat and railroad were both cause and effect of increased commercial and industrial activity. Called into existence by pre-existing needs, they in turn stimulated further developments in these fields. Much of the increase in the iron trade of the 19th century, both in Europe and in North America, was dependent upon the enormous demand for rails created by the transportation revolution. The call for ever larger steam engines and boilers forced the development of many new and larger machine shops and foundries, especially in the United States. Heavy rolling stock in turn forced improvements in bridge design and construction techniques. In the United States the Federal Government became heavily involved in surveying railroad lines and clearing obstructions from the western rivers. In this country, and to a lesser extent in Great Britain, improved transportation was, in the late 19th century, the gateway to future greatness. The steamboat and railroad proved to be the key to the gate.

18 / The Social Impact of the Industrial Revolution

ERIC E. LAMPARD

THE INDUSTRIAL REVOLUTION: TRIUMPH OR CATASTROPHE?

The Industrial Revolution was for long represented by its historians (and by many of its contemporaries) as a "sudden and violent" change which inaugurated an age of economic exploitation and social unrest. The revolution, they say, began in the late 18th century when steam engines were harnessed to new machines in the textile mills of central and northern England. It led inexorably thereafter to the crowding of scores of thousands of men, women, and children into slum quarters of the bleak towns which mushroomed about the factories. There the workers were driven to labor long hours under abominable conditions that were not materially remedied before the mid-19th century through the mounting pressures of trade unions and the passage of benevolent factory acts. "The effects of the Industrial Revolution," said Arnold Toynbee in 1881, "prove that free competition may produce wealth without producing well-being. We all know the horrors that ensued in England before it [the revolution] was restrained by legislation and combination."

For Toynbee and most other critics of the early industrial society, it was the factory-owner who appropriated the economic benefits, while the worker and "the community" bore the costs. When the Industrial Revolution later spread to continental Europe and North America, it appeared that many of the same processes were repeated in circumstances which, however much they might differ in local color and detail, gave rise to similar forms of exploitation and the same kinds of social abuse. In these later industrial revolutions, likewise, time and suffering were required before public-health and factory regulations, backed by a more enlightened and democratic public opinion, could prevail against the factoryman's greed and the cruel indifference of town governments.

During the present century, this almost wholly pessimistic view of early industrialization has come under critical review. As a result, the optimism expressed by some contemporary observers in the late 18th and early 19th centuries found new support in the writings of 20th-century scholars. Historians were able to show that many of the technological and organizational novelties commonly associated with the Industrial Revolution had, in fact, been unfolding over previous centuries on the Continent as well as in the British Isles. This held for the engines utilizing atmospheric pressures and even for the so-called uniformity system of interchangeable-parts manufacture which underlies modern mass production. Likewise, factory organization did not break in upon a stagnant world of industry, but had its prototypes and forerunners dating back at least to the Italian Renaissance. Many economic and social abuses of "the bleak age," moreover, were found to antedate the steam engine and the textile mill.

Social reformers, such as the German businessman and revolutionary propagandist, Friedrich Engels, had clearly romanticized the realities of a bygone rural life and industry in which "workers enjoyed a comfortable and peaceful existence" and where "children grew up in the open air of the countryside . . . and there was no question of an 8 or 12 hour day." Hours of toil in household and workshop had most likely been long and arduous and, according to a contemporary spokesman for the factory interest, William Cooke Taylor, child labor in Lancashire "was at its worst and greatest height before anybody thought of a factory."

The researches of economists have also revealed that, notwithstanding the possibility of dislocation and hardship in areas of industrial decline, the workers who migrated to new areas of industrial revolution received rising real wages and could expect to enjoy a higher level of living. Finally, it was suggested, the shocking squalor of the manufacturing towns—which had alarmed reformers and reactionaries alike at the time—was to be better understood in terms of the circumstances rather than the rank cupidity of the capitalist class.

Neither the original pessimistic view nor the more recently found optimism of the revisionists, however, can be taken as altogether single-minded efforts at historical description and analysis. Nor should they be dismissed out of hand as mere political or class prejudice. Both positions contain the not-uncommon mixture of history with advocacy which marks the treatment of any really *live* issue—from the Reformation to the causes of world war. The bitter indictment of an Engels from "the Left" and the complacent apologetics of a Cooke Taylor from "the Right" refer, in fact, to much the same bodies of historical evidence and rehearse many of the same arguments that have served pamphleteers and scholars down to the present day. Revived interest in the historiography of industrial revolution is, indeed, related to the current phase of "competitive coexistence" in which the leading Cold War protagonists recommend their own particular styles of industrialization and social change to newly developing countries in Asia, Latin America, and Africa.

The pessimistic left focuses its critical attention upon the immediate experience of industrial revolution and dwells upon the "full" costs of the economic and social upheavals that have characterized the early stages of industrialization carried out under capitalist auspices. The optimistic right addresses itself rather to the ultimate outcome and insists that the benefits from economic growth and rising levels of living accrue to the entire population and not to the owners of capital alone. Moreover, they remain unimpressed by the more recent performances of the collectivist alternative. A critical view of the Industrial Revolution is put forward by the socialist and, oddly enough, the old-style conservative, while a favorable judgment is usually rendered by the classical liberal and his latterday descendent, the new-style conservative, who has made himself the ideological custodian of industrial achievement. Whereas the prosecution's frame of reference is generally social responsibility and the idea of "community," that of the defense is liberty of the individual and "free" enterprise in the pursuit of wealth.

Since the historical specialists have so far tended to adopt opposing views on the social impact of the Industrial Revolution, it is not surprising that the general reader has difficulty making up his mind on the subject. When most of the historical evidence is, so to speak, short-run and parochial, concerned only with a few decades in time and with one or two localities, it is not easy to develop any adequate sense of a prolonged revolutionizing process. Also, where the established frameworks of interpretation are either too partial (narrow and biased) or too abstract (theoretical), it is hard to see the longer-run process unfolding in any authentic context of history. The divergent views on the Industrial Revolution outlined above serve to underscore the hazards involved in generalizing about a many-faceted process of social change, the character and effects of which *in the event* were highly diverse. Nevertheless, within the brief scope of this chapter, the next task will be to construct a socio-economic framework in terms of which the historical experience of industrial revolutions can be more fully understood.

THE ECONOMIC BENEFITS AND COSTS

From an economic standpoint, any industrial revolution may be regarded as the initial reorganization of a society's productive resources that gives rise to a rapid and "self-sustaining" increase in per-capita incomes. The metaphor of "revolution" denotes a social departure in which the total output of goods and services begins, virtually for the first time, to grow at a discernibly faster rate than population. From the purely economic standpoint, therefore, industrial revolution is but another term for accelerated economic growth. But it is also more than the phenomenon of economic growth since it comprehends both the departure from some traditional order of society characterized by, among other things, little or no economic growth, *and* the inauguration of a new "industrial" order characterized by, among other things, continuous and cumulative increases in output and income. The long-run tendency for per-capita output and incomes to rise is to be included then among the major consequences of a social movement or transformation called "industrial revolution."

Industrial revolution is thus a particular form of social change. Its occurrence transcends explanation in purely economic terms. For example, changes in technology and social organization—which may occur independently of changes in economic variables such as incomes and investment—are often among its mainsprings. Hence, our concern here is the *social matrix* itself, whose changing contours, taken in conjunction with dependent economic movements, embody the revolution's social impact.

Requirements for economic growth may be baldly stated as continuous increases in the per-capita supply of productive resources, that is, the labor and capital (including raw materials) from which all goods and services derive. In short, growth depends on the *supply capacity* of the economy which governs potential output at any time. The means to achieve rising output per capita are twofold: (1) an absolute increase in the quantity of resources utilized per head of population by the addition of more labor and/or capital (including materials) than heretofore; (2) an increase in the relative *efficiency* with which available resources are utilized by altering productive methods of organization so that any given output per capita is obtained from a smaller input of resources. The labor and/or capital withheld from the production process by more efficient operations may then be utilized in raising output in some other line of activity. Thus a unit of resources spared through greater efficiency (2) is equivalent to a unit of resources added to the per-capita supply (1). By whichever means or combination of means the supply capacity of the economy is enhanced, output is potentially increased. In short, the "problem" of economic growth resolves itself into the problem of developing in the population a style of behavior that will enlarge the society's capacity for production.

The concept of growth as a supply function must be distinguished from the concept of the determination of total product. Total product at any time is normally dependent on total demand, within some upper limit set by the output potential of the economy. The analysis of the determination of total product is thus the analysis of demand. Growth analysis, on the other hand, is concerned precisely with the upper limit of supply potential and, more especially, with the rate at which the upper limit of supply changes over time. Our analysis of industrial revolutions, therefore, is concerned not so much with actual output or total demand as with the changing supply capacity of the economy.

That economic behavior which is most appropriate to enlarging the supply capacity is called saving and investment. The act of diverting a part of the existing resource supply from current consumption and replacement activity is termed *net* saving; the act of converting such saving into the enlargement of labor and capital stocks in the future is *net* investment. To augment productive capacity and supply potential, therefore, requires saving beyond mere replacement levels and investment beyond mere replacement levels. Simply to replace existing stocks of inputs would represent a condition of zero net saving and investment, a condition that Adam Smith in the 18th century already recognized as a "stationary state."

Just how rapidly, and in what measure, a rising rate of net investment will return higher incomes per capita in the future depends on: (1) the proportion of total income that is saved—the saving ratio; (2) the amount of investment required on average to produce a unit of output—the so-called capital-output ratio; and (3) the rate of growth of population. Given the level of saving (1), the growth of product and income will vary inversely with (2) the capital-output ratio. The higher that ratio, the more capital that is used up on average in producing a unit of output, the slower will a unit of investment pay off in added output (and vice versa). Hence the critical relevance in all growth calculations of (3) the population factor, since it represents the denominator that is divided into the output numerator at any time to give output per capita.

This highly abstract and artificial framework of economic and demographic (population) variables permits a closer understanding of the "problem" of industrial revolution. Unless a country can borrow or otherwise obtain *extra* resources from abroad, it must "squeeze" its saving out of resources that are currently available at home. When a large proportion of the resources that are withdrawn from current consumption and replacement are used not so much for the provision of new capital goods (the celebrated textile machines, for example) as for the production of necessary "social overhead" (roads, bridges, canals, harbors, warehouses, hospitals, sewage facilities, business and residential construction, and so on), the incremental yield of goods for personal consumption by workers and other members of the population is likely to be low. The additional income earned by workers in the installation of the social overhead, in fact, will have no corresponding or "off-setting" equivalence in increments of consumer goods. Prices of the necessities that loom large in the budgets of workers' families are likely to rise at least until such time as higher prices serve to attract resources into enlarging the economy's supply capacity for consumer goods. Thus, at the very moment when a society is first preoccupied with expanding its overall supply capacity (and largely because of the preoccupation), prices may rise and *real* incomes per capita fall.

From the standpoint of the mass of working consumers then, the achievement of economic growth under the constraints of industrial revolution may impose considerable hardship for an indefinite period. This would almost certainly be the case if the job opportunities connected with the installation of the social overhead were to encourage couples to marry younger and have larger numbers of children. The increase in population would, in the short run, heighten pressure on the already limited supplies available for consumption.

But one important mitigating circumstance in many low-income pre-industrial situations is the likelihood of a degree of "disguised unemployment," or more accurately, *underemployment* of existing resources. In some activities, for example, the marginal product of additional units of input may be zero or actually negative in the sense that employment of additional units of input would add less to income than to costs. Hence, the successful deployment of resources from such typical lines of underemployment as traditional agriculture into activities where marginal products are likely to be positive, such as commerce or industrial manufactures, would in itself be tantamount to an absolute increase in the society's total resources—simply because labor productivity is higher in the modernizing sectors. An alternative strategy would be to keep existing labor and equipment at work for twelve instead of ten hours a day and, without any evident loss or gain in efficiency, total resource input per capita would nonetheless be increased by the additional man-hours of labor input.

The central question—whether or not industrial revolutions necessarily impose hardship on the mass of the population—probably turns on: (1) the degree of underemployment, and (2) the rate of population increase. If underemployed resources abound, if there is "slack" in the traditional system, the necessary movements of economic variables need impose no severe deprivation in consumption terms. In such an instance, the population variable would be decisive.

THE INDUSTRIAL REVOLUTION AND "THE SOCIAL QUESTION"

It is not inconceivable, therefore, that per-capita incomes might actually rise during the course of an industrial revolution. Yet even in that best of all possible worlds, the broad masses of the population and their spokesmen might still find the social costs of economic change excessive. A society is always more than a market place, and a livelihood is never the whole of life. Almost any of the economic growth possibilities considered above would have entailed profound alterations and disturbances in the inherited pattern of everyday life—regardless of changes in average levels of living.

In social terms, the transition from a traditional or pre-industrial society to an industrial society may be characterized as the disintegration and "break up" of an old order and its gradual replacement by a new order which is organized and integrated along different lines. What might at first appear to be rather commonplace matters (to 20th-century people) of changing jobs, shifting from one type of occupation to another, moving away from home or other familiar scene, getting poorer or richer, would, under conditions of industrial revolution, turn out to be nothing short of the disintegration and displacement of an older way of life.

During the first phases of the Industrial Revolution, the distress and bewilderment of populations in towns and in the country was, if anything, aggravated by the very gradualness of change. Aware that the old order was in flux, few could yet be sure that they or their children would find a place in the new "dis-order," or of what the outcome would be. Certainly the traditional agrarian society, with its high incidence of poverty, hunger, and disease, and its legal and social discriminations, was not Utopia. The idealized image of America held by Europeans from the 16th century was itself a significant rejection, in some ways, of the world that was known. Yet to many caught up in the throes of industrialization, the old order acquired a sufficiency and equity, a benign corporateness, as it receded further into the past. For a country like England with relatively low density of population, commercial vitality, and "ancient" liberties, even the recent past would seem to be "Merrie England"; but in France or Germany the golden age was still further back in the lost order of the medieval world. By the second quarter of the 19th century, contemporaries already spoke of the need for social reconstruction as "*the* social question."

The "social question," the "social problem," or the "labor problem" represented efforts to comprehend and cope with the gathering impetus of industrialism. Contemporary diagnoses and prescription, of course, varied with the observer's position and outlook in life, and with the place and date of his observation. Nevertheless, the posing of the question at any time revealed the growing concern among different segments of society that the new and potentially more productive ways of livelihood were contingent upon the acceptance of far-reaching alterations in the conduct and conditions of everyday life.

Under industrialism the focus of the social question changed almost as rapidly as the environment itself. The age-old phenomenon of pauperism and the poor was gradually transmuted into the more pressing "social problem" of a self-conscious working class or "industrial proletariat"; while the response to that issue in turn brought the entire social order into question. Under industrial conditions it was increasingly difficult to accept poverty as part of some inscrutable God-given natural condition; it was truly a social question. With the further unfolding of industrial development in the later 19th century, however, it was apparent that larger numbers were adjusting to the routines and discipline of the new order, reconciled that for them at least, the benefits in income and acquired status, on balance, outweighed the heavy costs. Changes in social organization and individual behavior tended to assume a more regular form such that, underlying the politicized "problems" of adjustment to social change, some observers could discern the formative societal processes: the matrix of industrial society.

It is possible for the historian of the Industrial Revolution to define the essential contours of these underlying movements in the changing context of the social question. The problems identified by those who experienced the revolution in their own lives—problems of population, the factory, the city, class conflict, the right of association, illiteracy, and persistent poverty—furnish important clues to the processes that were actually shaping the industrial order.

Conspicuous among the problems of early industrialization in a number of countries, for example, were those relating to: (1) the recruitment and training of workers for specialized installations of manufacture such as mills and factories; (2) the adjustment of workers to the routine of factory operations and to impersonal "industrial" relations with their employers in matters of work discipline, hours, conditions, and wages; also (3) the adjustment of working populations to the housing, public health, congestion, rules and conditions of changing town life; (4) the political integration of the greatly enlarged working classes into the emerging social structure. This meant achievement of full civil rights, political recognition, and the right of association. Other problems included: (5) the formation and integration of new middle classes of talent and education—the independent professionals, public servants, and higher grades of technical and clerical "white collar" workers—into a social hierarchy with the older middle classes of property-owners, merchants, and industrialists; (6) the adjustment of personal behavior to continuing alterations in the social environment, or the accommodation of inherited rules and mores to new patterns of individual behavior; and (7) development of the means, such as enlargement of communications and other service facilities, provision of educational and training systems, inauguration of forms of "social" insurance against periodic economic distress, physical disability, and old age, and so on, to institutionalize social change.

Among these "typical" problems of early industrialism—freely adapted from a more comprehensive list suggested by the German scholar Wolfram Fischer—item number (7), purposeful and controlled adaptation to industrial change, clearly overlaps and even subsumes some of the others. Indeed, insofar as it raises the larger issue of the overall coherence and direction of industrial change, the question of adaptation becomes virtually "*the* social question" itself. Underlying the many issues of personal and institutional adjustment was the pervasive tendency toward greater specialization and differentiation of functions in almost all aspects of life. As a consequence, the Industrial Revolution involved greater functional interdependence in the economic sense among more specialized groupings of workers and machines in different regions and countries, at the same time as the older hierarchical forms of personal and social interdependence were dissolved. Many of the changes, costs, and benefits, of early industrialism were rooted in this intensified socio-economic division of labor. Inherited relationships and structures that proved to be too deeply embedded in small market towns and traditional agrarian communities disintegrated.

POPULATION MOVEMENTS

The long-run tendency of industrialism toward greater specialization and differentiation of functions manifested itself in three characteristic population movements: (1) occupational differentiation, involving the shift of workers into full-time, specialized employments, notably in non-agricultural industries but in agriculture also; (2) spatial differentiation, or territorial redistribution of population, especially in the form of urban concentration; and (3) social-structural differentiation, involving upward social mobility and fuller participation in shaping the social order, through the achievement of civil rights for all citizens and expansion of opportunities for attaining "middle class" status. To be sure, the organization and structure of traditional societies are also characterized by varying degrees of functional-structural differentiation. But the effect of the Industrial Revolution was to break up the prevailing order insofar as it was characterized by smaller and more evenly distributed populations, comparatively undifferentiated rural-agricultural occupations, and by the ranking of classes or "estates" in established hierarchies of status.

The Industrial Revolution put mobility in the place of stability. As a consequence of progressive economic specialization, industrial populations tended to become, on average, more skilled and productive, more urbanized in residence and, with rising levels of education and per-capita income, more "bourgeois" in their social orientation and style. In short, the industrial regime required occupational, residential, and social mobility.

The transition from the traditional to the industrial, however, was a long drawn out, uncertain process. At any time institutional and structural rigidities in the old order could, and in some countries did, arrest the course of social renovation and economic progress. In the late 18th and early 19th centuries—the period with which we are presently concerned—the shape of things to come was everywhere unsure. Under the technological and organizational conditions of the Industrial Revolution, moreover, the very same occupational and residential shifts that were a condition of economic progress were also a primary source of dislocation and stress. In conforming to these structural movements individual members of the population felt the social impact of the revolution in the pattern of their own lives.

THE SOCIAL MATRIX

The interrelation and interplay of economic, technological, and other social facets of the early industrial revolutions may now be stated more boldly. The growth of output and income can be represented, from the economic standpoint, as the conjoint effect of: (1) increased inputs of productive resources, and (2) greater efficiency in combining such inputs into a given output. Hence any long-run decline in ratios of input to output (measured output divided by measurable inputs) is attributable to innovations in productive methods that prove to be more economical in their use of resources than any previously tried methods. Greater efficiency in resource utilization—as indicated by rising productivity per unit of capital and labor employed—stems, therefore, from advances in men's knowledge of ways and means to combine labor and materials in the production process; in short, from improvements in technology and productive organization. The development and application of relevant knowledge to economic ends, however, involves distinctive cultural and social attributes in a population as well as highly sophisticated economic traits.

The variable quantities of labor and materials thus comprise only a part, albeit the most obvious and measurable part, of the total input at any time. As such, the tangible inputs contribute very little to increased productivity. Additional allowance must be made for the other part of input, which includes: the layout of work tasks, equipment, and buildings; capacities of transport and communications systems; the quality and flexibility of financial organization; the availability of relevant systematized knowledge; the liberal or illiberal character of law and order, and the like. Much of this part of total input is social and cultural rather than material or economic in nature, and a great deal of it is not subject to direct measurement; yet this cultural and social endowment, perhaps, contributes the larger part of the productivity increment.

The productivity gains that follow directly from the installation of improved machinery and fuels in well-designed factory buildings are obvious. Indeed, the disintegration of rudimentary, undifferentiated work processes and their reintegration in more specialized factory sequences is commonly thought to be the essence of the Industrial Revolution; notwithstanding the well-known instances of factory organization long antedating the power-driven machinery of the 18th century. But, as indicated, changes in social and political organization, not to mention intellectual concerns, also contribute to rising productivity and may well form the *institutional* matrix for many of the "feedbacks" upon efficiency and growth that are manifested in the behavior of entrepreneurs.

The migration of "surplus" population from the countryside was an example of this interchange, dependent upon social factors as well as economic ones. Such migration was a logical requirement both for raising farm output and efficiency, and for augmenting and maintaining the potential labor supply in more productive non-farm pursuits. The legal emancipation of the land and its peasant workers is a necessary condition for the realization of the economist's notion of a "market" for labor and for many kinds of agricultural, mineral, and forest products. Agrarian reorganization by one or another means has come to be regarded, in Alexander Gershenkron's phrase, as "a major prerequisite of modern industrialization." Without social renovation of the countryside, price changes and resulting movements in the terms of trade between country and town cannot lead to a more rational (economical) allocation of productive resources.

Only less significant for the cause of industrialism than agrarian reform is the universal need for education. By the 18th century, it was already an axiom among most liberal-minded reformers in Europe and America that, without opportunities for education, neither the material nor the moral promises of emancipation could be fulfilled. During the 19th century, education and democracy came to be regarded as twin solvents of "the social question" raised by the disintegration of the traditional agrarian order. Efforts to decrease illiteracy and promote useful knowledge among the poor, although requiring the diversion of scarce materials and talents from immediate goods production, were, nevertheless, expected to yield increasing returns to society as well as economy from the improved quality of human resources entering upon employment and citizenship.

As a consequence of the twin processes of education and urbanization in the 19th century, sons ceased to follow their fathers' trades. The family and the institution of apprenticeship no longer determined one's occupation. Eventually the school rather than the family or the neighborhood came to be the principal agency governing a youth's chances for success in the world.

There is a certain irony in the fact, therefore, that by the criterion of educational opportunities, England in the 18th century would not appear to have been likely to inaugurate either the Industrial Revolution or the political democratization of an industrial society. Yet during the slow build-up of commercial energies from Tudor times, the pre-industrial institution of apprenticeship and other forms of on-the-job training, augmented by the influx of artisan refugees from abroad and supplemented by sectarian educational concerns of dissenting groups at home, evidently sufficed to produce small cadres with the practical skills and entrepreneurial drive to carry England over the threshold of industrialism. As for political democracy, the "rule of law" and the formalities of representation were well-established even before the full impact of the Industrial Revolution was felt. If English society embarked upon its industrial transformation almost inadvertently, it had already been preoccupied for the better part of two centuries with laying the institutional foundations of a liberal state.

Economy and polity alike are thus deeply embedded facets of society. Agrarian reorganization, emancipation, and the spread of literacy represented instances of the ways in which political, legal, and educational processes, interact and thereby contribute to greater efficiency in the economic system. Such feedbacks from outside the market facilitate and expedite the essential transfer of resources from traditional to growth sectors. Needless to say, comparable impulses from the economy are, in turn, passed on to political and intellectual, facets of society. In these continuing interchanges the phenomenon of social change has its momentum.

A striking instance of such interchange is the role of scientific knowledge in shaping the industrial-urban society. Even more than political or legal processes, the intellectual process of science and its projection in technology impinge directly upon the course of economic change. Since the 17th century, there can be little doubt of the increasing involvement of the marketplace in intellectual and institutional tendencies of science. But from the economic standpoint, science, even technology itself, counts only when it can be harnessed to the commercial and profit-making ends of the enterprise. Potent as such economic incentives were to become during and after the Industrial Revolution, at no point in the industrial era did business firms, with one or two notable exceptions, ever produce the basic knowledge that was to be adapted to their use. Historically, the task of "inventing" knowledge has fallen to private individuals, institutions of higher learning, philanthropically endowed foundations, and, increasingly, to governments. The accretion of knowledge is thus a socio-cultural process; its patenting, where possible, a socio-legal process; and its application to commercial ends, a socio-economic process.

In no two countries was the course of interchange among different facets of society ever alike in all respects. The institution of a market, the establishment of constitutional government, the expansion of knowledge, indeed, the setting of the stage for so many elements that were to combine in industrial revolutions, were the achievements of particular populations at different moments in history. Nevertheless, in every case, these varied elements comprised part of the necessary social and cultural overhead which is not produced or paid for by the decisions of entrepreneurs. They comprised, in short, the "other part" of total input which contributes so greatly over time to incremental productivity.

If so great a part of the Industrial Revolution is not the product of the market, but rather of the whole social process, why is it that pessimistic historians have blamed the industrialists and "the curse of Midas" for its social costs, while optimists have praised them for the benefits? An answer to this question properly lies within the field of intellectual history, but the economic historian may, at least, indicate that the distribution of "praise" or "blame" no longer seems pertinent to his primary task of identifying the social matrix of industrialism and of delineating the forms and direction of its trend. In this respect, for all the censure and celebration of the Industrial Revolution, we still know very little, for the study of social change has not until recently been a disinterested focus of historical inquiry.

Social change in a population takes the form of cumulative alterations in the daily routines and life cycles of individuals, the trend of which is persistent and largely irreversible. Since the end of the 18th century, we have come to identify a major strand of social change as industrialization. In all of this, businessmen or industrialists have played a leading role, but they do share the billing with many other innovating actors, and in no historical sense have they monopolized the roles of either heroes or villains.

THE TREND TOWARD RATIONALITY IN INDUSTRIAL SOCIETY

If the most conspicuous form of innovation under industrialism has been specialization and differentiation of functions, the prevailing trend has been toward greater rationality in the structure of the resulting interdependence. The evaluation of persons, places, and things in terms of their relative capacities to perform "utilitarian" tasks rather than according to terms of inherited custom or ascribed status was central to the late medieval "spirit of capitalism." But, the word *Rationalität,* as Max Weber used it, applied not only to the economic order of Western Europe, but increasingly to its law, politics, science, and technology—even to its music and architecture. Finally, in the form of the Calvinist ethic, it permeated religion itself. This critical principle of rationality was already inherent in the centralizing tendency of the absolutist state and the capital-based business system that emerged from the disintegrating medieval order of the 14th and 15th centuries. But whereas rationality had been a disruptive principle under the old order, it was to provide a basis for re-integration under the new.

From the standpoint of nascent industrialism, mercantilist policies of the nation-state—which aimed to build up the power and prestige of the monarch or republic—had the special virtue of encouraging merchants and artisans to link technological innovations with their business operations in more systematic efforts to increase production. The wealth and power which accrued to the business community in certain countries, moreover, gave it an important share in dismantling the older structure of mercantile laws and restrictions and in establishing the more rational structure of the "free" market.

During the 18th and, more especially, the 19th century, the introduction and large-scale application of fuel-burning machines to production and communications, under more competitive conditions, imposed a more radical order of rationalization on people and place. The population and its economic activities were increasingly concentrated in an array of urban centers; the industrial city was re-organized and re-shaped into spatially segregated areas of work and residence, with more sharply differentiated hours for labor and domestic life and with requisite changes for the population from productive to consumptive roles at particular times in its daily and weekly routine. The centripetal regimen of industrialism was, to be sure, already immanent in the commercially oriented social order that had emerged in certain parts of Europe and North America before the perfection of the steam engine and the promise of large-scale organization. Nevertheless, the novel and exacting regimen of technical efficiency, based upon the comparative job performances and operational requirements of machines but translated into the older efficiency calculus of capitalist bookkeeping, greatly amplified and accelerated the movements of population and productive activities (other than a rationalized agriculture) out of the countrysides and into the towns.

THE FIRST GENERATIONS OF INDUSTRIAL WORKERS

Everywhere, the first generations of industrial workers were born in the countryside but migrated to the towns. In the year 1800 only Great Britain and the Netherlands had more than fifteen to twenty per cent of their populations living in cities and in most countries the proportion was under ten. Meanwhile, since more of those born in country villages or on farms were surviving into adult years, the countrysides of Europe and North America were becoming more densely populated. Rates of natural increase among rural populations were usually higher than those of urban populations, owing to the higher mortality rate that long prevailed in the congested and insanitary towns. Nevertheless, it was in the towns that the industrial revolution in manufactures and communications was generating most of the new jobs, while in the countryside, where population multiplied, the agricultural revolution tended to reduce the growth of job opportunities on the land.

Most opportunities for a livelihood before the Industrial Revolution—even many of those not directly involved in husbandry—were to be found in small villages and country hamlets rather than in towns. But improvements in mechanical manufacture and the growing resort to water and, later, to steam power, tended to concentrate the leading sectors of industrial manufacture—notably textiles—closer to the sources of their energy, the swift-flowing streams and the dank coal fields. With the installation of power-driven machinery in factory organizations, the localization of manufacturing activity gathered momentum.

In the early industrial society, there was thus a contradiction of sorts between the different rates of natural increase of rural and urban populations and the locus and rate of growth in opportunities for employment. From the late 18th century, this contradiction could only be resolved by people *moving to* jobs, by migration of the "excess" country labor force to the new industrial towns. By 1850 there were half a dozen countries, perhaps, with more than fifteen to twenty per cent of their growing populations living in towns—Belgium, Saxony, and Rhenish-Prussia were significant newcomers—and even in North America, where vast areas of cheap agricultural lands were still available, in the manufacturing states of Rhode Island and Massachusetts almost half the population resided in urban areas.

For a long time, historians debated whether the first generations of industrial-urban populations were attracted to the town opportunities or driven there by force of circumstance. While both "forces" seem to have contributed, there is no longer much doubt that: (1) laborers in manufacturing were better off financially than those in most agricultural and traditional sectors of the economy, or that (2) this advantage reflected part of the higher productivity of industrial organization and technology. Moreover, most historians now recognize that the cruel pressure of a rising population on the land, rather than any special vindictiveness of capitalistic landlords, accounted for most of the rural exodus.

In parts of England and some German states, the absolute numbers of small holders of various tenures increased somewhat and, for a period at least, "enclosing landlords" even created jobs for farm labor by investing in land improvement. More than three-quarters of the numbers engaged in agricultural occupations in England and Wales in 1851 were hired hands, a larger proportion probably than ever before or since. Everywhere rural populations tended to grow in absolute terms, and the real exodus did not begin until later in the century with the rationalization of agriculture which followed the spread of improved farming techniques, more efficient land uses, and the expansion of railroads and transoceanic transportation. Even in the United States, industrial manufacture was able to recruit most of its labor before mid-century from the native-born population, especially younger females and youths, in spite of the "pull" of the frontier or the statistical fact that seventy per cent of the entire labor force was still either self-employed (farmers, mechanics, small tradesmen) or slaves.

THE LEVEL OF LIVING

Whether the structural shift from relatively undifferentiated labor forces of pre-industrial times to the more specialized non-agricultural labor forces represented a marked improvement in the average level of living of the working classes is another question. It is certain, as the tragic hunger in Ireland and some of the German states was to demonstrate in the 1840's, that if output and employment opportunities had not risen, there would have been a drastic fall in living standards that could not simply be attributed to a breakdown of communications or the gross incompetence of the ruling class. The structural shifts did enlarge national output. Over the century after 1750 national product per capita in Britain rose almost two and a half times, bringing with it the probability that average levels of living could roughly have doubled. A slow incremental output commenced just prior to 1750 and by the decades 1780-1800 had become a surge, growing at an annual rate of 1.8 per cent, or roughly twice the rate of half a century earlier. At the 1780-1800 rate, national output would have been likely to double in seventy or eighty years but population was meanwhile increasing so rapidly that the deflated per-capita rate was not above 0.9 per cent annually, or a doubling every century and a half. It is probable that the quarter-century war against the French Revolution and Napoleon (1792-1815) retarded the rate of growth of British output more than it did the rate of natural population increase, and the growth rate itself did not pick up much again before the 1820's and 1830's. However, from the first decade of the century to the 1850's as a whole, national output was rising at an annual rate of 2.9 per cent (a doubling almost every twenty-five years) and, when deflated by interim population growth, output per capita rose at about 1.5 per cent (which implies a doubling about every half century.)

The old upper and new industrial middle-classes received relatively more of the increment than did the enlarged working classes. More of the increase in national income appears to have been paid out proportionately as interest, dividends, and rents, than as wages; but, allowing for the more unequal distribution of incomes that seems to have accompanied economic growth during the Industrial Revolution, a majority of the population was beginning to enjoy a moderate rise in its level of living before mid-century.

The improvement in living conditions that was celebrated in the Great Exhibition in London during 1851 was not merely a "windfall" that had enabled workmen and their masters to sit back and consume the superior productivity of the power technology and factory organization. Some of that increment was already going into the improvement of men and machines, to which the rise in productivity was itself eloquent testimony. Moreover, so long as specialization and differentiation continued, this kind of improvement was likely to become more significant. It is precisely the tendency to involve ever more expert operatives in novel working relationships to more complex and sophisticated technologies that has negated the fear of Adam Smith that the simplification and monotony of tasks in division of labor, for all its economic promise, would render a growing proportion of the town populations "as stupid and ignorant as it is possible for a human creature to become."

The census of occupations in 1851, however, revealed how small a share of the growing labor force was yet, for better or worse, involved with the new factory-organized industries. Far more people were still toiling away in their own homes as self-employed craftsmen or outworkers, or in small-scale workshops, than in all the large-scale factories and mills put together. Nevertheless, regardless of the type of organization in which they might work, most of the 1851 labor force was specialized and full-time; and it is perhaps to this development even more than to the celebrated factory system that productivity increments of the early 19th century were due.

In 1850 much the same can be said for a majority of the non-slave population in the United States, although statistical measures for the years before 1840 are still in many respects unsatisfactory. Before the end of the century, of course, the expectation of improved levels of living was becoming widespread, even though the gap between conditions in high-income and low-income countries was already growing fast. While much of the amelioration in high-income countries was a direct consequence of structural shifts in economy and society out of the traditional-rural into the more specialized industrial-urban sectors, part of it, in the statistical sense, at least, was a consequence of the "safety valve" of emigration. Between 1840 and 1900 well over 20 million emigrants left ports of Europe mostly destined for the Americas and Australasia.

UNEMPLOYMENT AND WAGES

The periodic surges or long swings of growth that followed upon the Industrial Revolution were overlaid with a disturbing business cycle in the form of prosperity alternating regularly with slump. If anything, these industrial business cycles became more regular in their incidence and pattern—those, for example, which peaked in 1825, 1836, and 1845—than the older pre-industrial cycles that had been closely associated with harvest fluctuations and wars. The principal cycles in British business activity that have been identified for the period 1790-1850 seem to have had their origins in: (1) fluctuations in overseas demand for exports (especially textiles), and (2) related fluctuations in domestic investment. Moreover, already well before mid-century, a decline in capital investment opportunities at home was beginning to coincide with the discovery of, presumably, more profitable opportunities for capital export abroad.

For the first generations of industrial workers, the incidence of unemployment connected with these fluctuations in levels of business was perhaps even more insupportable than the hardships involved in migration and structural change. They were at least spared the necessity of stating their choice. For the workers, of course, a job was a living and when it was no longer there, it mattered little whether the proper explanation was structural, cyclical, or merely seasonal. Nor did they debate whether their frustration and distress were related to the old-style instability connected with harvests and wars or the new-style linked to exports, investment, and the adjustment of inventories. During the French Revolutionary and Napoleonic Wars (1792-1815), many workers in Britain seem to have had the worst of both worlds. They experienced all or most of these afflictions at once; but they suffered hunger, taxes, inflation, and the likelihood of falling real wages in place of their peace-time fears of unemployment and "short-time."

In recent years optimistic historians have made much of the evidence for rising real wages. But questions persist concerning whose wages, in what parts of the country, and for what years? Money wages in modernizing industries were higher than they were in agriculture and most traditional occupations. Agricultural wages were similarly higher in counties adjacent to industrial districts than in counties where wages were supplemented under "the Old Poor Law." But the recourse to "outdoor relief" payments to help the underpaid or unemployed rural laborer was itself a symptom of rural stagnation, structural lag, and underemployment rather than, as the economist Nassau Senior and other Poor Law critics maintained, their principal cause. But, in any case, for the masters and journeymen displaced by machines, or for the newly expanding class of hand-loom weavers called forth to take up the output of mechanized spinning "jennies," only to find themselves put out of work by the power loom, another person's higher wages did not compensate. Moreover, as Phyllis Deane suggests, it is only after 1820 that a more comprehensive price history and the measurement of a countrywide set of price relationships becomes possible. Hence, although the evidence on rising real wages is substantial by the 1840's, there is still a period after 1820 when not enough is known about the alleged evidence of "abnormal" unemployment to project the trend back that far with any certainty.

In the United States, there were many of the same conflicts and differentials between old and new sectors as in England, although sabotage and riots by displaced workers were not nearly as severe. Thus when Samuel Slater, who had learned the ways of textile manufacture in Jedediah Strutt's celebrated Derbyshire mill, migrated to Rhode Island in 1790, he brought with him his memory of English factory organization as well as its machinery. His original plans for labor included seven boys and two girls aged seven to twelve and, according to the *McLane Report* on manufactures in 1833, at least half of the Slater Steam Cotton Manufacturing Company's workers were children.

There is also evidence that money wages in the new American textile manufactures were better than those in most gainful occupations outside the factories. The precise differentials for male workers, nevertheless, are difficult to determine since the sources are not only fragmentary and incomplete but nonfactory wages were already affected by factory wages for, as Stanley Lebergott has shown, "both derive from the same labor market." Weekly wages for female factory operatives, still few in number in the early 1830's, were considerably above those of, for example, domestic servants, while the latter worked seven 14-hour work days a week compared with only six 12- or 12½-hour work days a week for most factory women. This differential of about one-third in real terms between factory and domestic service pay sufficed to attract a considerable female force into the mills notwithstanding what Lebergott calls "the confines, intensity, and discipline of factory work." Finally, there is also some evidence both for women and men that, once the factory system had become established and a market for factory labor can be said to have existed in an area, a decline in wage rates might follow before either the onset of a cyclical recession or any sizable body of immigrants had entered factory employment.

PROBLEMS OF TOWN LIFE

But the growth of industrial-urban communities raised many other questions besides those of the relations of men to masters and the "cash nexus" that tied their respective interests together. There were, for example, the old but intensified problems of town life even when the towns themselves were quite new—matters of adequate housing, health, authorities, boundaries, morals, and law and order itself. Large industrial centers with populations ranging up to 100,000 and more often had to cope with those grave features of "the social question" with the powers and finances inherited from a manorial village.

While the balance sheets of firms and whole industries might benefit from the external economies of agglomeration and urbanization, there can be little doubt that property taxes and special assessments on business did not cover the full costs of the external diseconomies that their operations had created. Probably the greatest part of the social cost of industrialization fell to the local body politic in the form of festering "city problems." There was a period in most countries when the quality of the industrial-urban environment rapidly deteriorated. Fire, epidemic, ignorance, and corruption, compounded in a city milieu, all took their toll on life and property before the passion of social reformers combined with the proceeds of economic progress to make some sections of the city a more civilized place in which to work and live. Meanwhile, on numerous occasions in different cities and countries, only the army or militia stood between the disaffected populace on the one side and the representatives of its ruling classes on the other.

THE INDUSTRIAL REVOLUTION AND PEOPLE

Clearly, the Industrial Revolution compounded political and economic problems even as it began to provide the material means for their solution. But its greatest and most lasting impact was on people—most severe of all upon newcomers to the cities who had grown up amid the still lingering ways of the traditional order but had not yet found a place in the new. Many, such as the displaced German artisans of the 1840's, felt themselves to be "invisible," virtually excluded from society and the state, unable to participate in the new order to which they were required to conform. Not only did they often lack education and understanding of the world forming about them, but lacked, too, the legal right of association with their fellows, not to mention the political right to make themselves heard.

The generalizations and impressions offered in this chapter on the social impact of the Industrial Revolution go far beyond the usual considerations offered in the historical debate concerning the workers' standard of living. They nevertheless stop short of a general theory of social change. For the moment, it must suffice to recall the principal structural movements within a population—the occupational, territorial, and social mobilities—which transform traditional social orders along the characteristic industrial-urban lines of specialization, differentiation, and growing interdependence of functions. What we have called "the trend toward rationality" involved a continuous interchange among different facets of social life—the economic, demographic, political, ideological, and so on—in order to bring about sustained economic growth and material well-being. In course of this unfolding, the economic criterion of "efficiency" was extended into many non-economic dimensions of life and, while the benefits of this convergence were duly accounted in rising per-capita product and income, the costs were largely excluded from the reckoning and the side effects ignored. Meanwhile, the emergence of so many new modes of personal and collective behavior within the short span of a few generations entailed far-reaching adjustments in society. These in turn involve profound alterations in the psychological and moral dimensions of human experience.

Whether or not *homo consumens* turns out to be the pathetic alienated figure that is often depicted in the literature of the 20th century is surely too large a question to be considered within a framework of the costs and benefits of the Industrial Revolution. It is certain that in recent years most students of economic growth have tended to ignore the transition costs or, in Martin Bronfenbrenner's words, have assumed them to be no greater than transition benefits. If in the present chapter we have sought to rescue the first generations of industrial workers from "the enormous condescension of posterity," we nevertheless owe it to them and their counterparts in today's developing countries not to romanticize the material and moral conditions of life in the pre-industrial world. For, regardless of whether one is inclined to be a pessimist or an optimist about the outcome, the social impact of industrial revolution is ultimately to be understood and evaluated in terms of what it does to the quality of mens' lives.

Part **IV**
The Age of Steam and Iron, 1830-1880

19 / The Invention of Invention

JOHN B. RAE

The great expansion of technology which came with the Industrial Revolution focused attention on the causes of technological advance and the individuals responsible for it. Through most of human history, recorded and unrecorded, the process of invention had been either ignored or taken for granted. For example, the inventor of the wheel and axle is forever lost in prehistoric obscurity; so, in much later years, are the inventors of the horseshoe, gunpowder, the spinning wheel, and the mariner's compass. The idea that a new technological discovery is a valuable contribution to society, for which the discoverer should be rewarded by receiving an exclusive right to his accomplishment, is strictly a Western concept dating from the Renaissance. It was then that the first known patent was issued by the Republic of Florence in 1421, and the first general law was enacted by Venice in 1474. The protection of invention by patents then spread throughout Europe.

PATENTS

The English Statute of Monopolies of 1623, restricting the issue of patents to the first inventor of a new technique, became the model for most subsequent patent systems. Oddly enough, in this action Parliament did not consider that it was doing anything new. The purpose of the act was less to encourage invention than to discourage the Crown's propensity for awarding lucrative monopolies to royal favorites, and Parliament believed it was only restating existing common-law doctrine.

When the Constitution of the United States was written in 1787, the desirability of encouraging and protecting invention was fully accepted, as shown in the authorization given to Congress in Article I, Section 8 "To promote the progress of science and useful arts by securing for limited times to authors and inventors the exclusive right to their respective writings and discoveries."

This recognition of the desirability of invention did not extend to defining exactly what it was. There is a difference between invention and technological change, but the borderline between the two can be extremely vague. A true invention requires novelty, which may consist of creating something new or combining existing mechanisms or techniques to produce a novel result. Improvements based on what is generally known as the existing "state of the art" are not, as a rule, considered to be inventions. For example, the initial creation of a four-cycle internal-combustion engine was an invention, but developing the original one-cylinder engine to eight was not. However, the existence of a voluminous body of patent law and a specialized corps of patent attorneys is clear evidence that the difference between invention and improvement is frequently fine. An invention, in fact, can almost be defined as an engineering achievement which happens to be patentable.

INVENTIONS

The nature of the inventive process has been the subject of considerable discussion. In general, theories of invention tend to polarize about two schools of thought: one, the determinist, holds that invention occurs when conditions are ripe for it and the identity of the individual inventor is an unimportant historical accident; the other holds that invention requires an act of individual inspiration. The history of technology suggests that both theses have some validity and that they are not necessarily mutually exclusive. The determinist can argue that the Western world of the late 18th century was ready for the steam engine, both in terms of economic need and the level of technology, so that if James Watt had not invented it someone else would have. On the other hand, just a little later Sir George Cayley could arrive at an accurate conception of the airplane, but the actual invention had to wait until a suitable power plant came into existence. In addition, numerous situations can be cited—telegraph, telephone, internal-combustion engine, Bessemer process, and so on—in which several individuals were working independently along the same lines and arrived at similar solutions almost simultaneously.

The rejoinder is that the solutions as a rule were similar but not identical. Although one can immediately cite exceptions—both Bessemer and Kelly arrived at the same solution for producing steel, and Hall and Héroult devised the same method for producing aluminum—the story of invention cannot be written off in terms that at the appropriate moment someone was going to come along and do it anyway. Most of the major inventions carry the distinctive imprint of their inventor, some novel idea for which there was no guide in established practice, such as the decision of Bessemer and Kelly to use an air blast for purifying molten iron, or the Wright brothers' wing-warping technique. While it can be readily conceded that the great epochal inventions are rare and that the great preponderance of invention consists of step-by-step additions to existing knowledge, the nature of these steps and the direction they will take depend upon the individual inventor. We can accept as a certainty that if James Watt had not invented the steam engine someone else would have, but it would not have been the same engine and it would not have come into use under precisely the same conditions. The differences might have been minor, but minor differences can have major consequences.

INVENTORS

Who, then, were the inventors? They did not run to type. The 19th-century inventors came from all levels of the social scale, ranging in education from university graduates to men with little or no formal schooling. Indeed, the only characteristic they had in common was that they did not conform to the popular image of the inventor as the untutored genius who solved his problem by inspiration and intuition. These were qualities the successful inventor had to have, but the record is quite clear that all of them also took pains to acquire a thorough technical background in their areas of interest, either through formal academic work or on-the-job training or self-education. They also sought the aid of scientists and engineers when circumstances called for it and such assistance was available. As a group, they exemplified the philosopher Alfred North Whitehead's statement that the most important invention of the 19th century was the invention of the method of invention.

A sampling of mid-19th-century inventors shows that Bessemer was a successful manufacturer with considerable experience in working with metals; Elias Howe (sewing machine) was a machinist who had worked in the textile industry; Ottmar Mergenthaler (the Linotype) was a watch-maker; Alexander Graham Bell had made an intensive study of speech and hearing as a teacher of the deaf; and Nicholas Otto (four-cycle internal-combustion engine) was a salesman with a good educational background in basic science. George Westinghouse's father was a mechanic and inventor, and Westinghouse himself was an assistant engineer in the United States Navy during the Civil War. Cyrus McCormick and Thomas A. Edison appear to have begun their careers with no direct training. McCormick, however, grew up on a farm where his father operated a sawmill, a grist mill, and a smelting furnace, and during Cyrus's boyhood the elder McCormick was unsuccessfully trying to build a mechanical reaper of his own. Edison cannot be fitted into any mould, but his amazing capacity for self-education enabled him to get a thorough grasp of the fundamentals of whatever problem he was working on.

The successful inventor has been the individual who not only had the creative insight to envisage a novel solution to a technological problem, but who also possessed or could acquire the necessary scientific and technical knowledge to make his insight a reality. The history of invention is a consistent record of moving from the known to the unknown by systematic, persistent trial and experiment. Invention, indeed, can be given the definition sometimes ascribed to genius: 10 per cent inspiration and 90 per cent perspiration.

What then distinguishes the inventor from the engineer, since both are concerned with technological change? There is no simple, clear-cut answer. Some inventors have been engineers and some engineers have been inventors. The two functions are not identical, but the distinction between them is often indeterminate. One is that the engineer is not primarily concerned with novelty. If existing techniques and methods are adequate for what he is supposed to do, he will use them, with such refinements and improvements as circumstances may suggest. Since most technological change comes in small accretions to existing practice, the engineer can properly claim a more important role in technological change than the inventor.

THE ORIGINS OF ENGINEERING

The engineer is more clearly identifiable in history than is the inventor. In the river-civilizations at the dawn of history the engineer was usually a member of the priesthood—logically so because this was the educated class—concerned with irrigation and flood control and the building of temples, palaces, and pyramids. We know that an engineer named Imhotep built the first great pyramid in Egypt and that he later was proclaimed a god.

In these early days the engineer and the architect were identical, since most engineering was concerned with structures. In Greece he was the *architecton* and in Rome the *architectus,* trained by apprenticeship as a master craftsman, but also something more. Our modern term originated about 200 A.D. when the term *ingenium,* from which we get the word "ingenious," was applied to certain military machines, and the man who could design such ingenious devices became known as the *ingeniator.* Engineering, in fact, had a strong military context throughout the Middle Ages and the Renaissance. When Leonardo da Vinci applied to the Duke of Milan for a job, he stated that he was thoroughly versed in the fifteen branches of military engineering.

It was not until the time of the Industrial Revolution that the engineer began to acquire status as a designer and builder of non-military works and a distinction began to be drawn between "military" and "civil" engineering, the latter meaning simply all engineering performed for non-military purposes. The great 18th-century English engineer John Smeaton appears to have been the first to call himself a civil engineer.

THE PROFESSIONALIZATION OF ENGINEERING

The first recognition of the need for more systematic training of engineers came in France with the founding of the École des Ponts et Chaussées (School of Bridges and Roads) in 1747, the first professional engineering college in the modern world. It still had a strong military flavor, being an offshoot of a military corps of engineers founded some thirty years previously, but it owed its creation to the need of France, as a major land power, for good internal communications. However, the school taught civil as well as military engineering, and one of its students, the distinguished mathematician Gaspard Monge, founded the next major college of engineering in 1794, the École Polytechnique, where students had two years of education in the basic sciences as part of the three-year curriculum. Among its graduates were Sadi Carnot, one of the pioneers in the study of thermodynamics (his father, Lazare Carnot, the "Organizer of Victory" during the French Revolution, was a product of the École des Ponts et Chaussées), and Gustave Eiffel (builder of the Eiffel Tower). The French example was widely followed; during the first quarter of the 19th century, polytechnic institutes were established in most of the principal countries of Europe as well as in the United States.

The outstanding exception was Great Britain, the home of the Industrial Revolution. There, perhaps characteristically and perhaps because of Britain's success in the early stages of industrialization, which seemed to make changes unnecessary, the traditional methods of training engineers persisted. Until well into the 19th century, British engineers continued to be predominantly the products of apprenticeship. The nearest equivalent to the technical schools being established elsewhere were the "mechanics' institutes" of early 19th-century Britain, and these were essentially trade schools giving their courses after working hours. Not until the 1840's was engineering offered at the universities of London and Glasgow, and it would have required some time for the subject to be generally accepted as an academic discipline.

Yet British engineers were as conscious of their professional identity as any others. If France pioneered in establishing institutions for the formal education of engineers, Britain led the way in organizing societies for the purpose of securing recognition of engineering as a profession. It is, in fact, an interesting commentary on the relationship of technology to its social environment that in France the government took the primary responsibility for the advancement of engineering while in Britain it was done by voluntary associations.

The first such organization to achieve stature and permanence was the Institute of Civil Engineers, which began to hold regular meetings in 1820 and was chartered in 1828. Its charter contains a definition of civil engineering written by Thomas Tredgold, "Civil Engineering is the art of directing the great sources of power in Nature for the use and convenience of man—The most important object of Civil Engineering is to improve the means of production and traffic in states, both for external and internal trade." It was an all-inclusive definition and was so intended. Tredgold himself was a well-known civil engineer in the present-day sense of the term, but was also the author of a number of treatises in other fields, including what was then the standard work on steam engineering.

Technical change, however, was moving too rapidly in the 19th century for engineering to remain unspecialized. The first step in Britain came with the founding of the Institute of Mechanical Engineers in 1846, with George Stephenson, the railroad engineer, as president. The separation would undoubtedly have occurred eventually, but it was precipitated when the older Institute of Civil Engineers refused to admit Stephenson unless he wrote an essay demonstrating his qualifications as an engineer. Other specialized groups followed in due course as the complexities of engineering multiplied: the Iron and Steel Institute in 1869, and in 1871 the Society of Telegraph Engineers and Electricians, which became the Institution of Electrical Engineers in 1889.

The mid-19th century was the great era of the British engineer. Names like Telford and Stephenson, Brunel and Whitworth were literally household words, and an author like Samuel Smiles could write a series of best-sellers on the lives of the great engineers. They were the peak products of the system of master-to-pupil training of craftsmen, having learned their art from predecessors like Bramah, Maudslay, and Smeaton; and they had the additional advantage of working in the world's most advanced industrial society. Their very brilliance, indeed, was probably a handicap to the progress of engineering education in the United Kingdom, since the apprentice method appeared to be providing an adequate supply of engineering talent. There was insufficient appreciation of the fact that the growing complexity of technology would require a new breed of engineer, with more systematic training in science and mathematics than could be imparted under the old system.

This is not to say that the quality of British engineering deteriorated in the twilight of the Victorian era. The careers of Sir Charles Parsons (turbines) and F. W. Lanchester (automobiles and aerodynamics) offer ample evidence to the contrary. Rather, British technology was overtaken by the products of the institutes of technology, the polytechnics, and *Technische Hochschule* in the rest of the industrial world. Yet if the craft tradition hung on too long, it also proved to be adaptable when the nature of the problem was recognized. The Livery Companies of London, the direct descendants of the medieval guilds, founded the Institute for the Advancement of Technical Education in 1876 when it became evident that Britain was lagging in this respect. One of the offshoots of this body was the City and Guilds College, which in due time grew into the present Imperial College of Science and Technology.

INVENTORS AND ENGINEERS OF THE UNITED STATES

In its broad outlines, the American experience conformed to the European pattern. The American patent system was based on the British, and the education of engineers followed both British and French models. Nevertheless, the American situation was sufficiently different to produce some interesting variations. At the start of the 19th century the United States was a new country, engaged in a vast physical expansion into an empty continent with a tremendous wealth of natural resources waiting to be developed. There was not enough capital or labor for all the opportunities that awaited, and the shortage of trained technical talent was still more acute.

American conditions therefore put a premium on getting things done as expeditiously and economically as possible and preferably with a minimum of labor. The ability to make devices that worked was important; knowing why they worked was not. The folk-hero of American technology was the ingenious gadgeteer with little or no formal training—such men as Samuel F. B. Morse, Charles Goodyear, and above all Thomas A. Edison. In other words, the inventor outranked the engineer in public esteem. The glamor attached to the successful inventor made it easy to overlook the invaluable assistance both Morse and Alexander Graham Bell received from the distinguished American physicist Joseph Henry, or the skilled technical staff which Edison assembled.

However, even if there was an excessive optimism about the ability of natural genius to solve the nation's technical problems, there was also an ample awareness of the need for trained engineers. During the first half of the 19th century the United States procured its engineers from three main sources. The first was Europe. European engineers in considerable numbers supervised the building and operation of American canals and railroads. Some remained to achieve distinction as American engineers, like German-born John A. Roebling, noted for his suspension bridges and the introduction of the manufacture of wire rope, or English-born James B. Francis, who did outstanding hydraulic engineering to provide power for New England textile mills and also initiated the manufacture of railroad locomotives in New England.

The second source was the United States Military Academy. This institution got its start in 1802, when Congress authorized the Corps of Engineers to train a limited number of cadets at West Point, then being used as a depot for the Corps. After the War of 1812 the school was reconstituted as a full-fledged military academy, with a continuing strong emphasis on engineering in its curriculum. The engineering curriculum was strongly influenced, as has been mentioned, by the École Polytechnique; in fact one of the first professors of engineering at the Military Academy was Claudius Crozet, a graduate of the École Polytechnique. For an entire generation, the Military Academy was virtually the only institution in the United States where academic education in engineering was available. Rensselaer Polytechnic Institute of Troy, New York, founded in 1824, and Norwich University in Vermont, founded in 1820 but not chartered until 1834, were such small-scale operations at this time that their contributions to the number of engineers were minor.

The roster of West Point engineers is long and distinguished. Many of the young men who entered the Academy during this period, perhaps most of them, did so to get an engineering education and not because they wanted to become army officers. This function, indeed, probably saved the Military Academy from extinction during the heyday of Jacksonian Democracy. It was then repeatedly under attack on the ground that the creation of a professional officer class was undemocratic, and the defense invariably pointed to the great national service the Academy was performing by providing an urgently needed body of trained engineers. The Corps of Engineers was the elite corps of the Army; it was the one normally chosen by the top-ranking graduates of the Military Academy, including Robert E. Lee and George B. McClellan of Civil War fame.

Most of the graduates left the Army for civilian careers after a short period of service. Among them were George Washington Whistler and his brother-in-law William Gibbs McNeill, both of whom made their mark as railroad builders, and Herman Haupt, who started the Hoosac Tunnel and built the first long-distance pipeline. Whistler's son, James Abbott McNeill Whistler, followed the family tradition by entering West Point, but his talents lay elsewhere and his stay was brief. In later years the great artist explained the termination of his cadet days and the consequent change in his career thus, "If silicon had been a gas, I would now be a major-general."

The third, and by far the most prolific source, of American engineers (as also of doctors and lawyers) until well after the mid-point of the 19th century was self-education, or more frequently on-the-job training. It was an astonishingly successful method in its day. American history invariably hails De Witt Clinton as the promoter of the Erie Canal, but it seldom gives credit to the men who actually built it: James Geddes, Canvass White, and Benjamin Wright. All three were local landowners with some training and experience in surveying. They not only became competent civil engineers in their own right, but they made the Erie Canal into a school from which numerous talented engineers went out to build canals and railroads elsewhere. A conspicuous example was John B. Jervis, whose career as builder and superintendent of railroads extended from the Delaware and Hudson to the Rock Island line and who also designed the first locomotive with a swivel truck at the forward end.

Almost as influential as the Erie Canal group were the three generations of Baldwins of Massachusetts: Loammi, Sr., Loammi, Jr., and James. Apart from their various achievements, primarily in canal and harbor works, the Baldwins established a standard of professionalism for engineers by insisting on being treated as expert consultants and not just as hired technicians. James Baldwin ran a sort of engineering school of his own. Young men paid him 200 dollars for the privilege of working in his office, and while they were studying under him earned their way by working for him as field assistants. When Baldwin was satisfied with a candidate's proficiency, he issued a certificate testifying that the recipient was qualified to practice as an engineer. Matthias Baldwin (no relative of the Massachusetts family) moved from jewelry- and watch-making to become one of the world's leading designers and builders of railway locomotives, and George H. Corliss, a self-taught genius from Rhode Island, was hailed in Britain as the greatest mechanical engineer of his day for his improvements on the steam engine.

The general state of American engineering in the early part of the 19th century was summed up by the Scottish engineer David M. Stevenson, uncle of Robert Louis Stevenson. He toured the United States in the 1830's and commented admiringly on how American engineers adapted to the conditions under which they worked. "At the first view," he observed, "one is struck with the temporary and apparently unfinished state of many of the American works, and is very apt, before inquiring into the subject, to impute to want of ability what turns out, on investigation, to be a judicious and ingenious arrangement to suit the circumstances of a new country, of which the climate is severe—a country where stone is scarce and wood is plentiful, and where manual labor is very expensive."

Nevertheless, as the American economy expanded and technology became more complex, it became clear that the existing methods of training engineers were no longer sufficient. The generation after 1850 saw a striking advance in formal engineering education in the United States; engineering was introduced at Harvard (1847), Yale (1850), and in several state universities during the 1850's. New technological institutions were founded, including the Polytechnic Institute of Brooklyn (1854), Cooper Union (1859), Massachusetts Institute of Technology (1861), Stevens Institute of Technology (1871), and the Case School of Applied Science, later Case Institute of Technology (1880). Meanwhile Congress provided a powerful stimulus to engineering education in the Morrill Land-Grant College Act of 1862, granting land to the states for the support of "colleges of agriculture and the mechanic arts." From this act a total of sixty-seven land-grant colleges eventually came into existence.

At the same time, American engineering was attempting to secure its professional status by organizing on the British model. The American Society of Civil Engineers was founded in 1852. Whereas the mechanical engineers were the first to split off in Britain, the mining engineers were the first to form a separate group in the United States—a reflection of the importance of the extractive industries in the American economy. The American Institute of Mining Engineers was established in 1871, followed by the American Society of Mechanical Engineers in 1880.

Despite this rapid advance, the tradition of the engineer as the self-taught, versatile handyman persisted. As late as the early 20th century the college-educated engineer in the United States was likely to be viewed with suspicion as a man who might know a great deal of theory but lacked the ability to get things done.

SCIENCE AND TECHNOLOGY

The eventual triumph of the academically trained engineer was inevitable in view of the accelerating pace of technological change in the latter part of the 19th century and the increasing interdependence between science and engineering. The technologies themselves are described in other chapters; the point to be made here is that while engineering remained an art, the engineer needed more and more to be able to utilize scientific knowledge and principles if he was to perform his functions efficiently and economically. There was still room for intuition, provided the intuition was based on a solid grasp of fundamentals, but the "cut-and-try" technique was far too wasteful for technologies requiring the utmost possible accuracy and precision.

This is not to say that engineering became completely dependent on science or that modern technology is entirely a product of scientific research. There have been situations in which the obverse has been true; namely, that scientific investigation has been a response to an advancing technology. For instance, the development of thermodynamics in the mid-19th century came largely because of the need engineers had to find accurate ways to calculate the efficiency of a steam engine, and the 19th-century experiments with flight created the science of aeronautics. Nevertheless, as existing technologies became more elaborate and new technologies began to proliferate, the rule-of-thumb methods which had served well enough in the past ceased to be adequate; engineering henceforth would have to utilize the growing body of scientific and mathematical knowledge. Electrical engineering, for example, was just being born in the 1870's and 1880's; with due allowance for the work of an individual genius like Edison, it was manifest from the start that the electrical engineer who worked by rule-of-thumb and knew nothing of Maxwell's equations would be a poor prospect.

The benefits to be derived from a systematic co-ordination of science, technology, and industry were vividly demonstrated by the industrial history of the latter half of the 19th century, most conspicuously in the rise of Germany to predominance in the manufacture of chemical products, optical instruments, and electrical equipment. These, it may be noted, are fields in which scientific knowledge and technical skill are the components most essential to success. Germany pioneered the establishment of governmental research laboratories; three were in existence shortly after 1870, in Berlin, Leipzig, and Bonn. Industrial laboratories were also developed in Germany at this time on a more elaborate scale than elsewhere, and effective co-operation in both basic and applied research was worked out among government, industry, and the universities.

The United States organized the National Academy of Sciences in 1863, initially as an advisory body to assist the government in utilizing the nation's technical resources in the Civil War. However, until the end of the century there were only a few isolated examples where the application of science to technology achieved significant results. Thus John Wesley Hyatt had the aid of a chemist, Frank Vanderpoel, when he developed celluloid during the 1870's in an attempt to find a substitute for ivory as the material for making billiard balls. The Standard Oil Company called on Herman Frasch to find methods of processing the highly sulfurated crude oil of the Lima-Indiana field, an operation cited as the first major achievement of chemical engineering in the United States. Charles Martin Hall, the inventor, along with Paul L. T. Heroult, of the electrolytic process for reducing aluminum, was a student of chemistry at Oberlin College when he did his first experimenting.

Similarly, the great British physicist William Thomson (Lord Kelvin) was a consultant on the laying of the Atlantic Cable, and two young chemists, Percy Gilchrist and his cousin Sidney Gilchrist Thomas, made a vital contribution to the steel industry by their discovery (patented in 1878) that lining converters of furnaces with dolomite would permit the use of high-phosphorus iron ores. It was not until the 20th century, however, that either Great Britain or the United States began to match Germany in organized research for the application of science.

THE DAY OF THE SPECIALIST

By 1880 two major developments were taking place in engineering, although neither was as yet approaching completion. First was the process just described, whereby engineering emerged as an academic discipline, closely integrated with science. The second, closely related to the first, was a growing specialization in engineering. A generation earlier the engineer was expected to be able to design and build with equal proficiency canals, railways, lighthouses, bridges, and locomotives—and many of them did. But the term civil engineer presently ceased to be all-inclusive, not so much from conscious choice as because the branches of engineering made it progressively more difficult for any individual to work effectively outside his own area of specialization.

The differentiation in the field of engineering accelerated slowly until the last quarter of the 19th century, when it perceptibly picked up speed. Electrical engineering became a separate branch in the 1880's, followed by the heating and ventilating engineers a decade later, the automotive, radio, and industrial engineers in the opening years of the 20th century, and from there to continuing elaborate ramification.

There was much more involved in this process than just the formation of new engineering societies. What was happening was that the engineer as he had existed through all previous history, the individual who could apply his talents to virtually any technical problem, was now becoming extinct. There were too many problems and they were far too complex for any one individual to master. The expert on soil mechanics was not likely to be qualified to design electrical circuits, and the individual who knew all about steam turbines would probably not be the man selected to build an oil refinery.

One effect of this trend was to add an additional dimension to the question of how and by whom technological progress was achieved. Until the latter part of the 19th century this function was shared almost exclusively by the engineer and the inventor, with a considerable overlap between the two. Then, as technology moved into a closer relationship with science, it became difficult in some areas to distinguish between the engineer and the applied scientist—between, for instance, the chemical engineer and the chemist, or the electrical engineer and the physicist. There are, of course, situations not merely difficult to classify, but impossible. Ethyl gasoline was discovered by Charles F. Kettering, an electrical engineer, William H. Midgley, a physicist, and T. A. Boyd, a chemist.

Through all the growing complexity of its various fields, the engineering profession retained a remarkable sense of unity. While American engineers were organizing specialized professional societies, they were simultaneously joining forces to found in 1893 the Society for the Promotion of Engineering Education (now the American Society for Engineering Education). Its purpose was to establish and maintain rigorous standards for the education of *all* engineers, and it has pursued a consistent policy of recommending that engineering education at the undergraduate level emphasize fundamental principles rather than specialized techniques. Specialization would come later in due course. Essentially the profession has adhered to the position that, while the modern engineer must of necessity concentrate on his specialty, there is a common body of engineering knowledge and skill which all its members ought to share, and above all a common approach to the engineer's function. This function remains as it was defined by Thomas Tredgold, "the art of directing the great sources of power in Nature for the use and convenience of man."

20 / Energy Conversion

JOHN B. RAE

The preceding chapters have made it clear that what historians have termed the Industrial Revolution was a complicated assortment of interactions between technological development and the economic and social structure of the Western world. One of the primary elements which distinguished the Industrial Revolution from earlier phases of the history of technology was the innovation and continuing development of new processes for the conversion of energy into forms producing useful work.

Initially these energy-conversion processes were arrived at by strictly cut-and-try methods, rather than by derivation from scientific principles. When usable sources of energy had been those of water, wind, human, or animal power, the amount of energy capable of being put to work was so limited that it was not really necessary to know or understand the principles of energy conversion. Besides, it took time for the implications of the steam engine to make themselves felt; even at the end of the 18th century the use of steam was so far from common that Edmund Cartwright used a cow to operate his power loom, and the first two power looms installed in Scotland (1793) were powered by a Newfoundland dog on a treadmill arrangement.

Cows and dogs were exceptional, of course, but even the usual mechanisms of the 18th century were very limited in their power output. The water wheels of the period at their best delivered about five horsepower. Newcomen's engine had an efficiency of only 1 or 2 per cent, and Watt's of about 5 per cent. (This was by subsequent calculation, for neither Newcomen nor Watt knew how to measure the efficiency of their engines—that is, the ratio of the output of work to the input of energy; what they did know was that their inventions produced power more effectively than anything that had existed before.) This level of performance was good enough to start the process of industrialization, but it had to be improved if progress was to continue.

The most obvious and direct method of improvement was to refine the machinery already in use. During the 18th century the possibility of designing water wheels so as to utilize more of the potential power in the head of water attracted the attention of some distinguished engineers, notably John Smeaton in England and Oliver Evans in the United States. As a result, by the middle of the 19th century it was possible to build more efficient overshot and breast wheels—types which derive their power from the weight of the water in the buckets rather than the force of the fall—capable of delivering as much as 50 horsepower.

THE WATER TURBINE

Of greater significance for the future of technology was the introduction of the reaction turbine as a practical operating device. This was a wheel propelled by the force of water coming out (rather than in) through blades or vanes on its circumference—in effect, the jet principle.

This concept had existed for centuries, and experimentation goes back to Heron of Alexandria in the 3rd century A.D. The Swiss mathematician Leonhard Euler had studied the problem in the 18th century and concluded theoretically that a reaction turbine could be made to convert the entire energy of the water into useful work. Another investigator, Claude Burdine (1790-1873) of France, coined the word "turbine" from the Latin *turbo-turbinis* meaning "that which spins"; more important, he was the teacher of Bénoit Fourneyron (1802-67), the man who finally translated theory into practice.

The reaction turbine, indeed, provides an excellent example of the frequent gap in technology between knowing how to do something and actually doing it. It took years of experiment for Fourneyron to work out the correct setting of blades, the proper dimensions for the wheels, and so on. However, by 1837 he had a water-driven turbine in operation capable of 2300 revolutions per minute, 80 per cent efficiency, and 60 horsepower, all with a wheel a foot in diameter and weighing 40 pounds.

Besides its superiority over the conventional water wheel in power output and efficiency, the turbine had the additional advantage that it could be installed as a horizontal wheel with a vertical shaft if desired, an arrangement virtually impossible for a wheel which depended on the weight or velocity of the water. Consequently, water turbines came into widespread use in Europe and the eastern United States, although not in Britain where coal was plentiful and cheap, and water power limited. Uriah A. Boyden (1814-79) introduced the Fourneyron water turbine, with improvements of his own, to the New England textile industry in the 1840's, delaying for decades the dominance of steam in that industry.

Boyden's work was continued and extended by James Bicheno Francis (1815-92), a British engineer who migrated to the United States and became distinguished in the field of hydraulics (the application of the mechanical properties—flow, weight, etc.—of water to engineering). He eventually turned from the Fourneyron-Boyden type of turbine to the Jonval turbine, also of French origin, in which the water flowed downward rather than outward. The Jonval wheel was first modified by an American, Samuel B. Howd, so that the water flow was inward and downward, and Howd's work in turn was improved on by Francis so that this kind of water turbine became known as the Francis turbine. In a report entitled *Lowell Hydraulic Experiments,* published in 1855, Francis calculated that in the textile center of Lowell, Massachusetts, alone, almost 9000 horsepower was produced by water wheels of various kinds.

As it turned out, the future of the water-driven turbine was not to lie in functioning as a direct source of power for factories but as a means of driving electric generators. For this purpose turbine wheels were developed of much greater complexity and power than the pioneering efforts of Fourneyron and Boyden, but without any essential change in principle. The engineering work done on the water turbine also contributed to the development of the steam turbine near the close of the 19th century.

THE COMPOUND ENGINE

As with the water wheel, so with the steam engine. With a growing demand for power in industry and transportation, there was a strong incentive to produce greater efficiency and economy. It is worthwhile pointing out, however, that efficiency is not necessarily the governing consideration in a power plant; economic and other factors may make the less efficient machine more acceptable. For example, the reciprocating steam engine at its best never approached the efficiency of the water turbine as a converter of potential energy into work, but the steam engine could be conveniently and economically used in many situations where water wheels of any kind were impractical.

Watt's engine was a major technological achievement, but in its initial form it had a thermal efficiency of only 5 per cent. With a work output of only 5 per cent of the heat capacity of its fuel, there was ample scope for improvement in Watt's invention, and one of his contemporaries, Jonathan Hornblower, saw the possibility of using the same steam in more than one cylinder—the principle of compounding—and patented such an engine in 1781. However, the five to seven pound pressures used in these early engines were too low to permit operation of the two cylinders at two different pressures, the most efficient method of compounding. Moreover, Hornblower also ran into patent difficulties with Watt. The latter was quite aware of the desirability of using the tendency of the steam to expand, and he developed a cut-off for his own engines to stop the flow of steam into the cylinder at the proper point. The old method had been to fill the entire evacuated cylinder with steam. By shutting off the supply of steam before the cylinder was completely full, say three-quarters of the way through the stroke, it was believed that one-quarter of the steam was saved. The remaining one-quarter of the cylinder was filled through the expansion of the steam already introduced. The idea was sound enough, and when higher steam pressures came into use early in the 19th century, the compound engine came into its own.

The first practical application of compounding seems to have been made by an Englishman, Arthur Woolf, who between 1804 and 1819 built several compound pumping engines. A Scottish engineer named McNaught worked out an adaptation which was more successful commercially because it required less expensive installation. "McNaughting," as it came to be called, was a technique for adding a low-pressure cylinder to an existing engine.

The compound engine found its greatest use in ocean-going steamships, where economy in fuel consumption meant increased cargo capacity. These engines materially hastened the triumph of the steamship over the sailing vessel. By the 1860's, for instance, before the opening of the Suez Canal, steamers equipped with compound engines were able to compete with sailing ships in

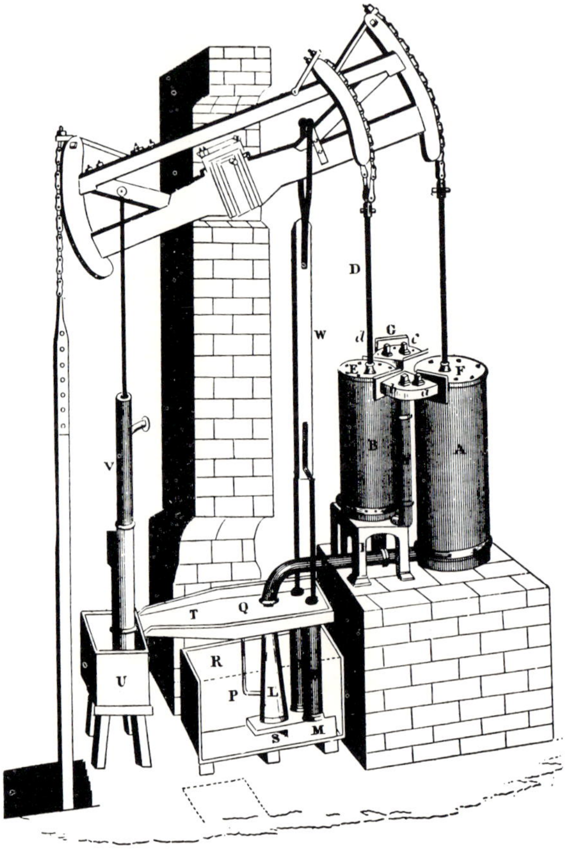

Fig. 20-1. Compound steam engine, Hornblower type. (From *Edinburgh Encyclopedia*)

the Australian trade, whereas previously the amount of fuel needed for the long run between South Africa and Australia had been prohibitive. By the 1880's, marine engines built on this principle were capable of delivering one horsepower for each two and one half pounds of coal, a figure not appreciably inferior to the performance of modern steam turbines. The triple-expansion engine, which appeared in 1906, had an efficiency of 23 per cent; in this engine the steam exerted its power in three stages before exhausting into the condenser.

By the close of the 19th century the reciprocating steam engine was gradually being superseded by more efficient power plants such as the steam turbine and the internal-combustion engine. Despite the appearance of the triple expansion engine, the heyday of the compound engine was over. Yet in its time it served an important technological purpose. It not only facilitated the adoption of steam power, it also stimulated much empirical work on the problem of the utilization of energy. The other method of getting more power from a given quantity of steam, namely super-heating, was essentially a 20th-century development, stemming from the growing body of knowledge about thermodynamics. Superheating was used with good results on some large railway locomotives, but its greatest advantages were realized in stationary power plants using steam turbines. By the mid-20th century temperatures of over 1000 degrees Fahrenheit and pressures of over 3000 pounds per square inch were being employed in such plants, giving thermal efficiencies between 30 and 40 per cent.

THERMODYNAMICS: THE SCIENCE OF HEAT-ENGINES

It has been said with considerable justice that science owes more to the steam engine than the steam engine does to science, and in Chapter 19 the point was made that thermodynamics grew out of the need to find a method of calculating the efficiency of a steam engine. It is true that the properties of heat had previously received some scientific attention from such men as Joseph Black. It was not, however, until the effective harnessing of steam power that intensive study was given to the relation of heat to energy.

The dominant theory of heat in the late 18th and early 19th centuries held that it was contained in a substance called "caloric," which warmed or cooled other substances as it flowed into or out of them. This theory was severely shaken in 1798 by the American-born scientist Benjamin Thompson (1753-1814). Because he was a Loyalist who fought for the British during the American Revolutionary War, he took refuge in Europe, where his services to the Holy Roman Empire won him the title of Count Rumford. While superintending the boring of cannon in Munich, Rumford observed that a dull tool generated more heat than a sharp one, although the sharp instrument, by cutting more metal away more quickly, should theoretically have released more caloric from the metal. After further observation and experimentation, Rumford concluded that heat was a form of mechanical energy.

However, since it adequately explained many observed phenomena, caloric theory took a long time to disappear, and as a matter of fact even today the concept of a "flow" of heat provides a convenient and useful method of looking at the subject. As late as 1852 much publicity was given to a ship built by John Ericsson (the Swedish-American inventor) and powered by a "caloric engine." It was a piston engine propelled by heated air. The unique feature of the design was that the exhaust air was passed through a "regenerator"—actually a complex of wire mesh—where Ericsson believed it would deposit its caloric, to be picked up by the incoming air. The heat would thus be constantly reused, and the inventor asserted that the engine would need only enough fuel to replace the leakage of heat by radiation. Ericsson unwittingly was groping with the idea of the heat exchanger, but he was misled by the notion of caloric, and scientific opinion of his day was unable to offer a clear explanation for the failure of his caloric engine to come up to expectations.

That explanation was in the process of being formulated. It is an interesting commentary on the development of engineering that while British, and to some extent American engineers, predominantly bench-trained rather than college-educated, concerned themselves with the practical improvement of the steam engine, the first theoretical approach came from France's academically trained engineers. As was pointed out in the preceding chapter, Sadi Carnot (1796-1832), a graduate of the École Polytechnique, was the founder of engineering thermodynamics. His essay, *Reflections on the Motive Force of Fire,* published in 1824, was the first systematic attempt to work out a theory of heat engines.

Carnot's ideas were fragmentary and frequently incorrect. He accepted the caloric theory and regarded steam merely as the means for carrying the caloric through the engine. In fact he drew a parallel between the steam engine and the water wheel: one powered by the flow of caloric from boiler through cylinder to condenser, and the other by a flow of water from source through wheel to outlet. He did not believe that heat was lost as the caloric passed through the engine. Nevertheless Carnot showed brilliant insight in concluding that the efficiency of a heat engine was measured by the difference in temperature between the heat source, the point at which the heat entered the engine, and the "heat sink," the point at which it left. Shortly before he died in a cholera epidemic in 1832, Carnot abandoned the caloric theory and was approaching the concept of the mutual convertibility of heat and energy.

This same problem was attracting attention in various fields of science and technology, but the complete history of thermodynamics is too complex a subject for extensive treatment here. In its application to 19th-century technology the most important figures are British. First came James Prescott Joule (1818-89), a businessman of Manchester, England, who made an avocation of science. In the 1840's Joule provided the experimental proof for the principle of the conservation of energy: energy can neither be created nor destroyed; this was later formulated as the First Law of Thermodynamics. He also worked out the mechanical equivalent of heat: 778 foot-pounds of work equivalent to the heat required to raise one pound of water one degree Fahrenheit. The great British physicist William Thomson, Lord Kelvin (1824-1907), was the first to appreciate the significance of Joule's work, although he had misgivings for a while because Joule's results contradicted the accepted concept of caloric. At any rate, Kelvin, along with the German physicist Rudolph Clausius (1822-88), completed the theory of heat engines by defining entropy (in very simplified form, heat divided by temperature; or, the amount of energy, for example, steam, in a heat-engine which cannot be transformed into mechanical work) and showing that it had a natural tendency to increase. This finding, the basis for the Second Law of Thermodynamics, accounted for the observed loss of heat in engines. The principle can be considered to have been established by 1865, so that after forty years, the flow of heat initially recognized by Carnot and appropriately named the "Carnot cycle" was satisfactorily accounted for.

The practical application of thermodynamics to the design of steam engines was promoted by William Rankine (1820-72), Professor of Engineering at Glasgow University, whose *Manual of the Steam Engine and Other Prime Movers* (1859) was for long the standard work on the subject. As it happened, most of the improvements on the reciprocating engine, which thermodynamic theory might have suggested, already had been made pragmatically by this time: specifically, high pressures and maximum use of the expansive power of the steam. The growing body of theory contributed somewhat more to the development of the steam turbine.

By the time intensive development began on the internal-combustion engine, most engineers had some grasp of the basic principles, although it is difficult to find a conscious application of thermodynamic knowledge on the part of any one man. Before 1880 the most direct use of this growing body of knowledge about the properties and behavior of heat was the introduction of mechanical refrigeration, which essentially involved using mechanical power to reverse the flow of heat. What is important is that by 1880 the fundamental principles of the relationship of heat to energy and the conversion of energy through heat-engines had been established and were now part of the technological equipment of industrial society.

THE USES OF COAL GAS

By 1830 coal, either coked or in its natural form, had been in common use as a fuel for over a century. It came to be discovered, however, that coal yields a variety of economically valuable by-products. The first of these to be observed and put to use were the combustible gases, which can be disastrously explosive and which were sometimes given off under natural conditions (such as "firedamp" in coal mines). In 1730 an Englishman named James Lowther piped firedamp to the surface and ignited it, but he appears to have done so only as a safety measure; there is nothing to suggest that he thought of using the gas for either heat or light.

If he did not, others thought of it fairly soon afterward. Active experimentation with coal gas began about 1760, and a decade later Archibald Cochran, Earl of Dundonald, an amateur scientist of some note, used it as fuel for tar ovens in Kilross, Scotland. Before the end of the century two demonstrations of lighting by coal gas were made in Germany: Johann Georg Pickel, a pharmacologist, illuminated his laboratory by gas in 1786, and in 1799 Wilhelm Lampadius did experimental gas lighting at Dresden, in the palace of the Elector of Saxony.

The first public illumination of a building by gas was done at the Hotel Seignelay in Paris in 1801 by a French engineer named Philippe Lebon (1767-1804). Lebon did not use coal gas, however, but made his fuel by distilling gas from wood in "thermo-lamps" of his own design. As this record of experimentation indicates, the utilization of gas for light and heat shares with most major technological advances the characteristic that it was easier to conceive the idea than to make it work. Lebon could pass beyond the stage of experimentation because he had some previous experience which proved useful. He was an engineer who had lived among charcoal-burners in his youth and therefore knew something about the behavior of wood under heat. As it turned out, Lebon's technique was almost immediately superseded, because it proved more efficient and more economical to distill gas from coal than from wood.

In this area, also, previous experience was important. The utilization of coal gas was achieved by men who had worked with coal, especially with converting

Fig. 20-2. Lebon's first gas-making plant. The sheet-iron retort (AA) is encircled by the flue (FF) from the furnace (EE). The whole is set in fire-brick. The gas, conducted from the retort through G, is to be passed through water and the by-products are to be collected. From Lebon's patent specifications, 1799. (Reprinted from Singer *et al., A History of Technology,* Oxford, 1954-58)

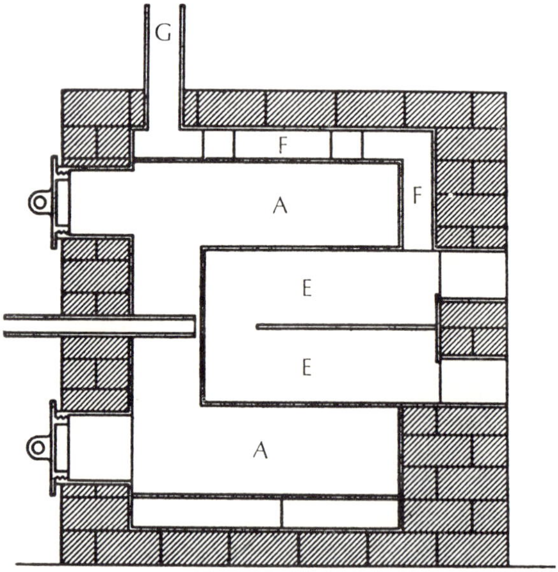

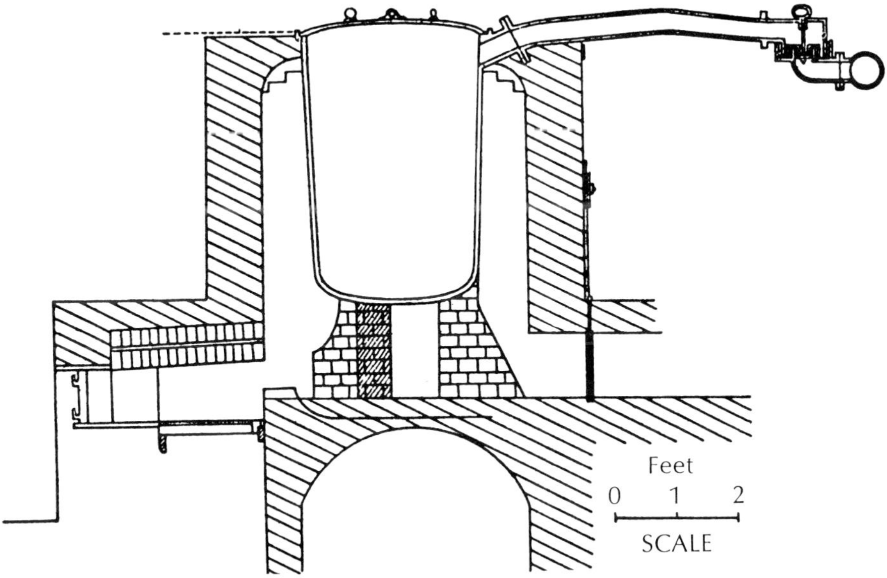

Fig. 20-3. Section of Murdock's first retort for Phillips and Lee, 1806. (From Singer *et al.*, *A History of Technology*)

it into heat and power. William Murdock (1754-1839) and Samuel Clegg (1781-1861) both worked as engineers with Boulton and Watt. Murdock was in fact a major figure in the development of the Boulton and Watt engine. He is credited with designing the sun-and-planet gear that Watt adopted when patent conflicts blocked his use of the crank. Clegg, in the long run a more important figure in gas lighting than in steam-engine design, had also been a pupil of the great chemist, John Dalton.

Murdock began his work with gas in the 1790's. He lighted a room in Redruth, Cornwall, with coal gas in 1792—a great improvement upon contemporary experiments in that he first washed the fuel by passing it through water, thereby removing some of the impurities which affected its burning qualities. Several years later he illuminated the engine house at the Boulton and Watt Soho Works, in honor of the short-lived Peace of Amiens between Britain and Napoleon in 1801, and in 1806 provided lights for the textile factory of Phillips and Lee at Salford, then one of the largest factories in England. At the same time Clegg was doing the same thing in another textile mill in Halifax. At this point both men were primarily interested in meeting the demand for some kind of lighting system that would permit factories to operate with reasonable efficiency at night. Gas solved the problem until it was superseded by electricity at a later date.

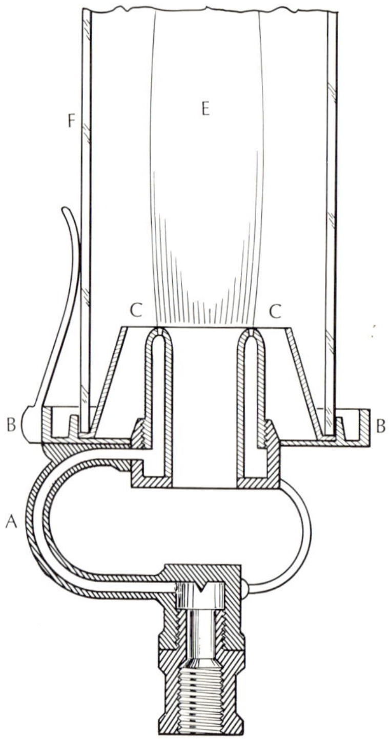

Fig. 20-4. Argand burner, adopted as the British government standard burner in 1869. Supply-tube to chamber of combustion (A); support for chimney (B); cone, outer air supply (C); flame (E); chimney (F). (From Singer *et al., A History of Technology*)

In the meantime there were still problems to be worked out in connection with gas illumination. At Murdock's installation at Salford, the retorts in which the gas was generated carbonized badly. In addition, it was necessary to find out how to burn the gas for optimum results. Murdock and Clegg used both the Argand burner, initially designed for oil lamps, and the "cockspur" burner, so named because of the shape of the burner and the flame it emitted. Clegg went a step beyond Murdock in purifying his gas by adding lime to the washing process. In 1825 Thomas Drummond (1797-1840) produced a brilliant light by using a stick of lime in a gas flame, a technique from which we have derived the term "limelight."

The idea of using gas to illuminate whole areas—including streets—rather than single buildings must have occurred to several of these experimenters. The first steps in this direction, however, were made by a man who was a promoter rather than an engineer. He was a German named Friedrich Albrecht Winzer, who had been impressed by Lebon's demonstration in Paris and realized that gas lighting should be organized on a broad scale rather than simply building by building. He moved to England in 1803, became Frederick Albert Winsor, and in 1806 formed the world's first gas company, the National Light and Heat Company, later called the Gas Light and Coke Company.

Gas lighting was installed on Pall Mall in London in 1807, but Winsor's technical skill did not match his promotional abilities, so that the project floundered until Clegg joined the company in 1812. A great deal of skilled work was required to solve the problems encountered in providing acceptable service. Coal gas is toxic, it has a strong odor, and is explosive. Provision had to be made for producing, distributing, and using it both safely and economically. In addition, fumes and smoke from insufficiently purified gas and poorly designed burners created a strong distrust of gas lighting for many years. Efficient burners, with air introduced into the flow of gas just before combustion, appeared in the 1840's, and the well-known Bunsen burner was introduced by R. W. Bunsen (1811-99) for use in his laboratory at Heidelberg in 1855.

There were business problems also. In the early days of the industry, competing gas companies frequently were formed in the same community, with an inevitable waste of much effort and frequent financial disaster. Charging for the service proved a difficult problem until efficient meters were introduced in the 1850's. Nevertheless, the use of gas in urban communities spread steadily. The first American city to have gas lighting was Baltimore, where installation began in 1816. Although the United States had a special advantage in having supplies of natural gas, the major gas deposits were in the west, so that although Fredonia, New York, used it on a limited scale in the 1820's, natural gas came into general use only in the late 19th century. By the mid-20th century natural gas had virtually supplanted coal gas in the United States.

With the major technical and organizational problems satisfactorily resolved by about 1880, the gas industry enjoyed a boom for some twenty years. Then electricity appeared as a superior competitor in illumination. Gas lighting was preserved, although on a diminishing scale, by the invention in 1885 of the incandescent, fabric mantle by Carl von Welsbach (1858-1928), now widely used in camp lanterns, which gave a more brilliant light. Although electricity finally came to dominate the lighting field, gas retained a secure place as a fuel for both industrial and domestic heating. Compared with coal and to a lesser extent with oil, gas has advantages as a fuel in giving clean and even heat and in being noiseless.

The economic and social implications of illuminating with gas during the 19th century were considerable. Permitting industrial plants to operate at night—the purpose for which it was first developed—gas contributed markedly to the expansion of industrial activity in the Victorian era. It has been claimed that gas lighting had an adverse effect on labor by enabling factory owners to extend the length of the working day, but this claim is not supported by evidence; the working day of twelve hours and more existed long before illuminating gas came into use, and the great era of gas lighting, approximately 1840-90, saw most of the Western industrial world moving toward amelioration of the condition of its industrial workers. It can be argued, indeed, that good illumination encouraged the limitation rather than the extension of the working day. When factory hours were limited to the period between dawn and dusk, a single day-long shift was the only practical labor arrangement. Once it became possible to keep open at night, then a two-shift system became feasible.

The social effects were incalculable. Given good artificial light it was possible for people to engage in a variety of group activities after working hours that would have been out of the question previously. Adult education, for example, was one of the great enthusiasms of the 19th century, but its progress would have been severely retarded if evening classes had had to be held by candlelight. Yet there was one inherent limitation to the use of gas which made it inevitable that it would give way to electricity as an illuminant. Its advantages were restricted to urban areas because it simply was not economical to extend gas mains into rural areas where consumption would be slight.

Still, gas paved the way. The technology of the central generating station and the distributing system was first worked out by the gas companies, and one of the major reasons for the success of Thomas A. Edison and other pioneers in the commercial utilization of electric power and light was that they could follow the trail already blazed by the gas industry.

SUMMARY

In this chapter we have attempted to show how the growth of industrialism, and especially the development of power machinery, created in the Western industrial civilization a need to understand the nature of the energy and the methods of putting it to work as power, heat, or light. This problem had attracted scientific interest earlier, but the great upsurge in the study of the conversion of energy that took place during the 19th century must be attributed primarily to the pressures of accelerating technological change. At the same time it is important to remember that inventors did not wait for scientists to discover new principles, but continued to work with the ones already available, inadequate as some of them were.

Until close to the end of the 19th century, the principal mechanisms for harnessing energy were the water wheel and the steam engine. In terms of the problem of energy conversion, and in spite of Sadi Carnot's perceptive and useful analogy between the two devices, it must be pointed out that they represent quite different processes for putting energy to work. The water wheel, whether a conventional wheel or a turbine, uses the kinetic energy of falling or running water. The steam engine, on the other hand, brought with it the more complex and more significant technique of utilizing energy in the form of heat. The key realization was that heat could be converted into power, as in the steam and other heat-engines, or into light, as in the gas technology.

From the study of heat-engines came the science of thermodynamics, a distinctively 19th-century achievement. It developed from a combination: first, practical experimentation with steam engines to achieve better performance; second, the formulation of scientific principles not only for the technology of heat-engines but for other phenomena in physics, chemistry, and even in fields so apparently remote from the steam engine as biology and human society. It is understandable that Ernst Mach (1838-1916) could have become so enthusiastic as to maintain that the entire universe could be explained in terms of thermodynamics, and that the American historian, Henry Adams (1838-1918), could predict pessimistically that just as some of the steam engine's energy could not be transformed into mechanical work in a thermodynamic system, civilization was tending toward complete entropy, with energy gradually being made unavailable.

It is not necessary for us to become so pessimistic. For the advance of technology it was essential that the fundamental principles of the conversion of energy should be discovered and applied, and the process by which this was done represents a fascinating study in the development of an increasingly close tie between science and technology. The essential feature is that by the close of the 19th century both the theory and the practice of energy conversion were well understood, and man was equipped with an effective and powerful tool for extending his control over the forces of Nature.

21 / Mining and Metallurgical Production, 1800-1880

CYRIL STANLEY SMITH

As has been shown in previous chapters, one of the fundamental changes wrought by the Industrial Revolution was the transition from an industrial establishment based on wood-working to one based on metalworking. Indeed, it is impossible to think of the 19th century apart from the metals which were the very fabric of its structure—iron and steel in particular were available in quantities and at a price never before possible. This difference in quantity, however, soon amounted to a qualitative difference: somehow during these years, the total influence of the industry became more than just the sum of the various technical innovations which so increased the production of metals and the range of their usefulness.

MINING AND ORE-DRESSING

The larger scale of smelting operation in the 19th century was accompanied by an increased scale of mining and ore-dressing operations. Though gunpowder had been used for blasting in mining by mid-17th century, the introduction of guncotton (1846) and especially explosives based on nitroglycerin, such as dynamite and blasting gelatin (1875), made breaking up the rocky ore at the mine face much easier. Working in mines became safer as more attention was paid to mine ventilation (the use of fires and chimney shafts to produce drafts, although known in the 16th century, became widely applied in the 19th century), and when the Davy miner's safety lamp (1815) minimized the danger of explosion in coal mines.

Power hoisting and pumping allowed ever deeper mines, with improvements even in the carrying of men up and down the shaft. An early elevator was the "man engine" in which steps were attached to reciprocating rods extending down in the shaft, driven by a water wheel or beam-engine, with the rods often doubling as pump rods. Matching these was a series of small stationary platforms onto which the men could jump to await the next stroke which carried them either up or down a distance of 6 to 9 feet, 4 to 6 times per minute. This practice began in the Harz Mountains in 1833, but it is particularly associated with Cornish tin mines, where several such elevators operated over a 1600-foot depth.

In ore-dressing, the use of jaw crushers and rotary ball mills (where the ore was broken into small pieces by revolving in a drum with heavy balls of hard material) cheapened the production of finely ground ore. The Wiffley table (1895) greatly improved the washing operation. This was a mechanically shaken, inclined table with a stepped series of parallel grooves across which the water and finely ground ore fell, separating the heavier mineral particles from their rocky companions.

A radical change occurred with the introduction, by the Bessel brothers in 1877, of a totally different principle of mineral separation, that of flotation. This new method did not depend on different rates of settling of particles in water as in the previous panning or washing operations (used so successfully, for example, in the California gold fields during the Gold Rush). Instead, the new process was based upon a gas method: because of difference in surface energy, when water wets particles of rocky matter, bubbles of gas will stick preferentially to some mineral particles; the bubbles then literally lift out the mineral particles. This process was a radically new invention, though an analogous process, depending upon the difference between the adhesion of grease and water to minerals, had been exploited in the 14th century in preparing ultra-marine pigment from ground lapis lazuli, and later in recovering diamonds. The Bessel patent covered the production of gas by chemical reaction. It was not at first applied on a very large scale, but after the introduction of froths mechanically produced with air in 1901-10, flotation rapidly spread to the treatment of most non-ferrous ores. The use of chemical additives later improved the process, and flotation is of immense economic significance today.

Magnetic separation of iron minerals, experimented with in the 19th century (among others by Thomas A. Edison on New Jersey ores around 1890), did not become economically important until after the depletion of the high-grade ores.

IMPROVEMENTS IN IRON PRODUCTION

Ferrous (iron and steel) metallurgy in the first quarter of the 19th century is notable for the consolidation of the theoretical gains of the late 18th century, and the general increase in the scale of operation of both blast furnaces and the puddling process for making wrought iron. Damascus steel, with its texture of crystalline origin, had been successfully duplicated in 1821 by Breant in France, partly as a result of the application of new chemical knowledge, but its importance by this time was slight.

In the smelting of ore to pig iron, blast furnaces continued to increase in size, and coke was replacing charcoal throughout the world, except in the Pennsylvania area, where anthracite coal was used for many more years, until finally displaced by cheaper coke. The use of anthracite was facilitated and the economy of operation in any furnace was greatly increased by the introduction of the hot-blast process by Neilson in 1828. Upon discovering that production and quality could be improved if the air blown into the furnace was heated beforehand, ironmasters employed a separate furnace with cast-iron pipes heated with a coal fire to produce hot air for the blast. However, before long, waste gas from the blast furnace itself was being used as fuel, once the furnace top had been properly modified to allow for the collection of gas. The regenerative Cowper stove, in which an open stack of brickwork is first heated by a gas flame and then used to heat a through-flowing blast of air, dates from 1857. Blast-furnace gas was also burned in boilers to drive blowing engines for the blast and, after methods of removing dust from the gas had been developed, was used directly in huge internal-combustion engines.

Henry Cort's puddling process (Chapter 16) was itself improved by the introduction, in 1816, of an air-cooled cast-iron hearth bottom which reduced the erosion of the silica refractory lining of the furnace. In 1839 this was radically improved by Joseph Hall who began to use the "bull dog," an oxidized iron cinder that was not only refractory and resistant to the silicate slag but which

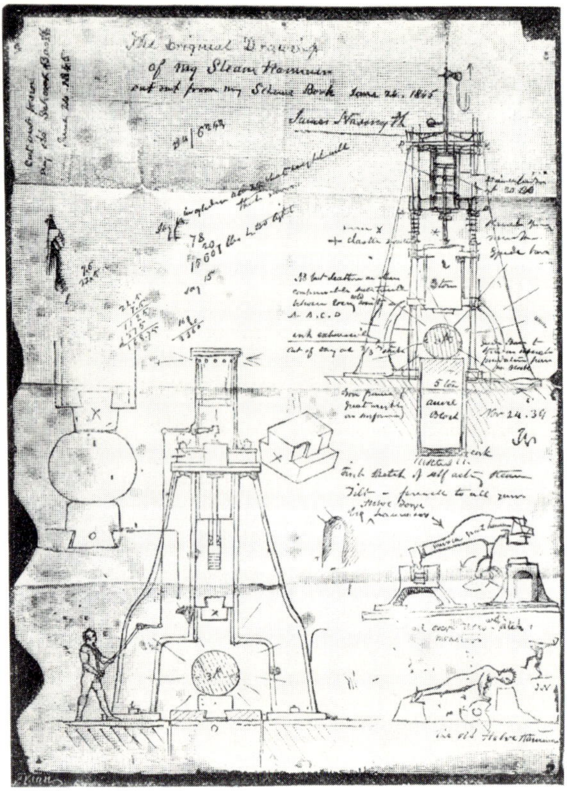

Fig. 21-1. A page from John Nasmyth's notebook showing his first sketches of a steam hammer. (From his *Autobiography,* 1885)

also contributed chemically to the refining operation. This was known as the "wet process" (from the fluidity of the oxide slag used), or as "pig boiling" from the apparent ebullition as the carbon-monoxide bubbles rose to the surface to burn as "puddler's candles." Very frequently, success in metallurgical operations lies in the proper refractory, the non-combustible lining of the furnace; this is true in the puddling process, as well as in the basic Bessemer and open-hearth furnaces.

The production and shaping of wrought iron was improved when the water-driven hammer, falling by gravity, was replaced by Nasmyth's steam hammer (invented in 1838), which was both more powerful and more controllable. This in turn was replaced by an eccentric rotary squeezer for the squeezing of puddler's balls.

BESSEMER STEEL

Steel remained essentially a special-purpose material during the first half of the 19th century, with only small amounts being produced. In 1850, on the eve of Bessemer's invention, only 60,000 tons of steel were made annually in England, compared with 2½ million tons of wrought iron. In 1898, the year of Bessemer's death, 5 million tons of steel were produced.

Before Bessemer, the best steel was still being produced by the slow cementation process, in which iron (usually Swedish) was remelted and poured into ingots to make cast or crucible steel for those tools, cutlery, and springs which needed the most homogeneous slag-free metal. Nevertheless, hearth-produced "natural" steel still found considerable use in cruder applications, and the cementation process was much less popular on the Continent than in Britain, where it was necessary for the growing tool industry. In 1843 a French author reported that cementation steel exceeded the production of natural steel only in Great Britain and Russia, while in France the amounts were about equal, and in Austria—the largest continental steel producer—no cementation steel whatever was made.

Much of the success of British steel was based on the use of Swedish wrought iron, which was very low in phosphorus content. At least some of the delay in the development of the French steel industry can be explained by the unwillingness of the government to allow free importation of the Swedish iron. Soon, however, one of the greatest technological revolutions in metallurgical history was to change the very meaning of the name "steel," and to make wrought iron little more than a curiosity.

Up to this point, the story of iron had been mainly the story of wrought iron. During the preceding centuries wrought iron had been growing in production and decreasing in cost, but there had been no essential change in its quality. The softest iron was regarded as the best, although plenty of harder, less refined material was made. The word "steel" was reserved for high-carbon steel, a special material used for tools, springs, knives, and weapons which needed the extra hardness and strength acquired by heat treatment. When Bessemer found a process by which low-carbon "steel," free of slag, could be produced cheaply in the molten state, an attempt was made to call it "homogeneous iron," but commercialism resulted in the debasement of the word "steel" previously associated with the quality product.

The strong personality of Henry Bessemer (1813-98; knighted in 1879) has tended to overshadow the essential contributions of others to the development of the process which bears his name. His process involved no new chemical or physical discoveries: it consists simply in blowing air through molten cast iron, simultaneously removing the silicon and carbon by oxidation, the heat of the reaction being enough to raise the temperature above the melting point of the fully decarburized metal. (Cast iron melts at about 1200°C, but pure iron is liquid only above 1530°C.)

Both elements of Bessemer's process had been known for centuries. The heating of a lump of iron sponge in the bellow's blast, doubtless known earlier to practical men, was recorded in scientific literature in the 17th century and several times thereafter. The chemical effect accompanying the conversion of cast iron to wrought iron was the removal of carbon and silicon by the oxidizing effect of the blast in the refinery hearth, or of the air and ore in the puddling furnace. However, the full explanation of this refining process could not be given until 1786, shortly after the discovery of oxygen. In China, the partial improvement of molten cast iron by running it through the open air had been practised at least since the 17th century.

In 1847, William Kelly, an ingenious ironmaster in Kentucky carried out experiments aimed at making wrought iron by a fuel-less pneumatic process, but it seems that he neither sought nor produced a molten product. He operated on a fitful production scale between 1851 and 1856. When, in 1857, he heard of Bessemer's work in Britain, he hastily filed application on his similar process with the United States Patent Office. The Patent Office ruled in favor of Kelly's priority, and a complex commercial struggle followed, which was eventually settled (in the United States but not elsewhere) by a merger of the Kelly and Bessemer interests in 1866. It can be seen, therefore, that the Bessemer process had ample precedents as well as an actual predecessor; yet this does not detract from the significance of Bessemer's work.

Unlike Kelly, Bessemer was a complete outsider to the iron industry, though he had been working on ingenious and profitable inventions since his youth. He had developed a method of making "gold" flake for paint by a cheap rolling process which he succeeded in keeping secret for forty years and which provided the income to support his other projects, which included the development of a successful rotary press and other machinery for sugar refining. His first experiments involving high temperatures were with the melting of glass in a reverberatory furnace and the rolling of plate glass. The trail to his invention of the steel process began with the invention in 1854 of an aerodynamically spun projectile to be used in smooth-bore cannon; tests of this in France impressed him with the weakness of the cast-iron guns at the time and led him to attempt to improve cast iron by melting it with steel additions in a reverberatory furnace into which a channel had been built for introducing secondary air into the flame.

A chance observation of an unmelted shell on a pig of iron exposed to the draft showed that air was a powerful decarburizer, and prompted Bessemer to experiment with air blown through molten iron in a crucible in a furnace. He soon discovered that external heat was not necessary if the operation was carried out fast enough, and he then devised large containers arranged for bottom blowing; these were later replaced by the tilting vessels that became the symbol of steel for a century.

Early in August 1856, Bessemer demonstrated the process on an 800-pound charge of pig iron before the famed engineer, George Rennie (1791-1866), who urged him to present a paper at the meeting of the British Association for the Advancement of Science. The paper immediately created a great sensation, and five licenses for use of Bessemer's patent were sold. At first the licensees could produce nothing but brittle metal: Bessemer's process did not seem to work successfully, and it was some years before the nature of the difficulties could be found and the process made to work properly. What had happened was that Bessemer, totally ignorant of the complexities of the iron business, had by sheer luck acquired from a local merchant a grade of pig iron unusually low in sulfur and phosphorus. It was this iron which he had used in his experiments; when his licensees tried the process with other pig iron it would not work. Bessemer sought the advice of metallurgical chemists, and at first settled on the use of a pure pig iron from Sweden, although suitable deposits of hematite (ferric oxide, Fe_2O_3) ores were later found in England. His troubles were by no means over, however, for something had to be done to counteract the harmful effects of traces of oxygen and sulphur. This was accomplished by the addition of manganese. The use of manganese in cast crucible steel had been patented by J. M. Heath in 1839, and Robert Mushet had applied for and received a patent on its addition to air-carburized iron immediately after Bessemer's public announcement of his process. Hearing of this, Bessemer adopted the use of manganese in the form of *Spiegeleisen*, a manganese-rich cast iron which also restored needed carbon; his troubles were almost over by 1859. While Bessemer admitted that he got the idea of using manganese from Mushet, he refused to pay royalties, thereby creating much bitterness, even though Mushet's patent was of doubtful validity.

The first commercial success of the Bessemer process was actually in Sweden, where G. F. Goransson found that by stopping the blowing at the right moment to produce the desired carbon content, the metal was less oxidized and retained the necessary manganese. Shortly afterward, Goransson adopted the routine use of the Eggertz method of carbon analysis, based on measuring and comparing the colors of solutions of steel in nitric acid; this method was very rapid and provided the basis for accurate control of the blowing procedure.

An important role in industrializing the Bessemer process was played by Arthur L. Holley in Troy, New York, who added a removable bottom to the converter and improved the general plant layout. Bessemer steel was made in huge quantities in America, beginning in the 1860's, with almost all of it going into the production of rails for the expanding transcontinental railroads. With

Fig. 21-2. Plant for making Bessemer steel, *c.* 1860. While the converter on the right is being poured into the ladle for casting, the one on the left is in full blow. The converters could be hydraulically rotated on their axes, and the central crane was a hydraulic one which could rotate to carry the ladle successively over the various moulds. (Courtesy *The Iron and Steel Institute*, London)

vastly increased production, it soon became apparent that the metallurgist's job was quite as much to devise efficient methods for handling large quantities of material as it was to modify their chemistry.

Once production and deoxidation difficulties were overcome, the Bessemer steel quickly replaced wrought iron both in old applications and in new ones. The first steel rails were laid in 1862, and proved greatly superior to those made of wrought iron, which were softer and more likely to split because of internal stringers of slag. Bessemer steel was first used in ship construction in 1863 and in building construction in 1888.

OPEN-HEARTH STEEL PROCESS

Although a charge of steel in the Bessemer converter could be made in about fifteen minutes and needed no fuel beyond that required to melt the cast iron, this very speed militated against accurate control. Often the product was deficient in ductility, especially at low temperatures, because of inclusions of oxides and nitrides (the latter only recently recognized) originating in the air blast. Moreover, it was difficult to use any large amount of scrap iron in the converter, which would have made the process more economical, because of the difficulty in maintaining the proper amount of heat to keep the scrap molten. The way remained open therefore for the development of a steel of better quality.

The open-hearth process—the reverberatory furnace with regenerative heating—provided the answer. By mid-19th century, the value of a hot blast in the blast furnace was well known, and the inefficiencies of the puddling furnace were notorious: boilers were often placed in the stacks to utilize their waste heat. In 1856, William Siemens, who knew something of the developing thermodynamic theory, conceived the idea of regenerative action (that is, making use of heat which would otherwise be wasted) by heating a mass of loose brickwork with the waste gases leaving the furnace, and by reversing the direction of flow periodically, using it in turn to heat the incoming air. This scheme was first used in coal-fired furnaces for melting glass and for crucible-melting of steel. In 1857 it was applied by Cowper, one of Siemens's associates, to the heating of air for the blast furnace, using blast-furnace gas itself as a fuel. By 1861 Siemens had developed an efficient coal-gas producer, and built furnaces in which both gas and air were heated regeneratively.

A true open-hearth furnace for steel-melting followed, and the metallurgy of the pig-and-ore and pig-and-scrap methods was quickly developed, the latter especially by Emile and Pierre Martin (father and son), in Sireuil, France, who first successfully used the open-hearth process in 1863. The use of pig iron and scrap steel to produce steel in an open-hearth furnace is thus rightly known as the Siemens-Martin process.

Steel-makers now had to learn how to use the cheap, controllable, high-temperature heat available to them. The first process involved simply the melting of wrought-iron, scrap, and pig iron in appropriate proportions to give the desired carbon content without refining (much as in the American version of the crucible process, though on a much larger scale). The wide hearth was adapted particularly for reactions between the metal and the slag, and Siemens soon developed a process by which iron ore was used to react with the carbon and silicon in molten pig to give steel directly, duplicating the chemistry of the puddling process at a higher temperature of operation. The resulting steel was cast into ingots of fairly large size and subsequently rolled into bar sheet rails or other sections for use. The open-hearth furnace, though initially of only about 4 tons' capacity, could eventually be scaled up to handle batches of about 50 to 100 tons, and today 500 tons is not uncommon. This capacity was much larger than that of the Bessemer converter, and the slowness of the open hearth—about 10 hours per heat (batch)—was actually an advantage in that it made possible better control.

Primarily because of better quality production, the open-hearth furnace, for all its slowness and thermal inefficiency, increased output while the cheaper Bessemer process declined in use. World open-hearth furnace production passed that of the Bessemer process in 1907. It should be noted, however, that in the 1950's the principle of the Bessemer process was revived in a vastly improved form, using top-blowing (as in Bessemer's first experiment) and employing oxygen in place of atmospheric air. It seems likely that this oxygen process, first applied in Austria, will eventually displace the open hearth in the production of unalloyed steels, just as the open-hearth furnace has already been largely replaced by the electric furnace in the manufacture of high quality alloy steels.

The refractory lining adopted for both Bessemer and open-hearth furnaces at first was silica, usually a ground sandstone, tamped in place or premoulded into bricks with a little lime for bonding. To prevent undue deterioration of the lining, it was necessary to keep the slags silica-rich. Chemically, this method prevented the elimination of phosphorus from the metal; the sulfur, however, could be removed by simple oxidation.

Finally, two Englishmen who were cousins, Sidney G. Thomas and Percy C. Gilchrist, showed that a lime-rich slag would remove phosphorus; more important, they found in widely available magnesite rock an appropriate basic refractory to withstand corrosion. This development in 1875 was of immense importance. Using the Thomas-Gilchrist discovery, both basic open-hearth and basic Bessemer processes developed very rapidly after 1880, especially in France, Germany, and Belgium, since the Thomas-Gilchrist process made available to steelmakers previously useless but extensive deposits of phosphorus-bearing iron ore. Even the slag, which now contained rich phosphate materials, became useful as a fertilizer.

Curiously, the English ignored their own phosphorus-bearing ores until the end of the century, preferring to use the acid process on iron smelted from low-phosphorus ores imported from Sweden and Spain. In the United States, the vast amount of good ore in the Mesabi Range of upper Michigan and Minnesota enabled American steel-makers to use the acid Bessemer process during the period of great growth, but the basic open-hearth process eventually outstripped it.

SHAPING OF METALS

The mechanical shaping of metals improved as fast as or faster than the methods of their reduction and refining. Foremost among these developments was the rolling mill.

Lead pipe, which initially had been made either by soldering or, in the 18th century, by the rolling of cast pipe in grooved rolls over a mandrel (removable core), after 1797 was made by extrusion (that is, by being forced or pushed out)

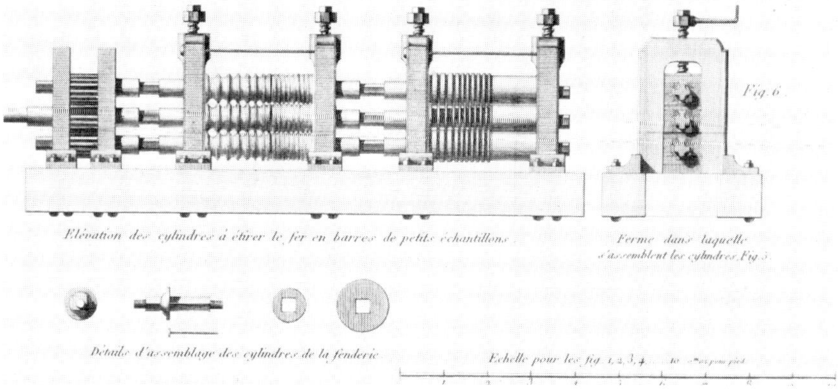

Fig. 21-3. An English three-high mill for rolling iron bars, 1827. By using three rolls, the metal could be worked while travelling in both directions through the mill, almost doubling the output. Rolls with differently shaped grooves were later used to produce rails, I-beams, angle irons, and other shapes. (From Du-Frenoy and de Beaumont, *Voyages métallurgiques*, Paris, 1827)

in a hydraulic press. This technique was later used for applying a protective lead sheath to electric cables and as better die steels became available, for the shaping of brass, aluminum, and even steel.

Henry Cort's development of grooved rolls for consolidating wrought iron and producing simple rod sections was followed early in the 19th century by the use of specially shaped grooves to make rails and later the T, L, and I beams (so named because their shapes resembled those letters of the alphabet), needed in ever greater quantities by engineers and builders.

Rolled-steel products everywhere came to displace not only iron forgings but also castings of iron or non-ferrous metals. Larger and faster rolling mills were developed, and great ingenuity was shown in the design of rolls to produce rails and achitectural shapes.

For faster rolling of heavy ingots, two kinds of mills were developed: first, the reversing mill in which the direction of rotation of the rolls could be reversed to compress the ingot as it went back and forth; second, a three-high mill (which could be driven with a smaller engine having a large flywheel to equalize the load), the ingot passing alternately above and below the middle roll as it was carried down. Both of these devices had been used on lead before 1700 and had found some previous use in iron-rolling, but their development on an appropriate scale for the new steel industry is generally attributed to Lauth in England in 1862.

Steel wire, no longer used in mail armor, was needed for wire-rope cable (invented in 1834) for mine hoists, suspension bridges, wire nails, and for the telegraph (which absorbed large quantities of galvanized wrought iron and later, mild steel wire of about 0.18-inch diameter). The wire fence, greatly improved by the addition of barbs after about 1870, was an important factor in the expansion of the western United States; furthermore, it added greatly to the demand for wire, which could be met only by the development of high-speed mills for rolling small rods and the associated machinery for drawing them into wire. By the time copper wire was needed in quantity for the electric power industry, the methods had already been developed for steel; these methods were far removed from the simple techniques that had been used in making the first precious metal wire for jewelry.

Wire is the strongest form of a metal of a given composition, for the mere act of drawing increases its strength. The strongest steel wire was heat-treated by running the wire continuously through (in sequence) a heating furnace, a quenching bath, and a lead tempering bath, followed by still further cold drawing. The "patented" wire used in cables for suspension bridges was made by this process which was invented in 1854.

ENGINEERING AND THE TESTING OF MATERIALS

In the 19th century the increasing application of science, which resulted in the better understanding and control of the old types of metals and alloys, did not immediately result in the development of many new materials that were attractive to the engineer. Innumerable short-lived proprietary (patented) compositions—usually modifications of the standard alloys—were developed which were mostly of interest to the compilers of recipe books. Repeated failures in iron railroads, bridges, and cannon led, by the middle of the 19th century, to intensive programs of testing in order to compare available materials and to obtain numerical data on which to design. Bit by bit the variability of standard materials was exposed and, at least to some extent, removed.

A particularly impressive series of mechanical tests of both iron and nonferrous metals for cannon was carried out for the United States Army in 1840-60. In 1872 the government, after appropriate urging, appointed the United States Board on the Testing of Iron, Steel, and Other Metals, composed of eminent engineers, both military and civilian, to undertake a comprehensive study of the strength of materials. Their first report, published in 1880, contains a summary of the knowledge of the mechanical properties of materials at that time, and was followed by many detailed stress-strain curves made with the testing machine that had been designed and constructed at the Board's direction at the Watertown Arsenal. Many private testing laboratories were established at about the same time.

After about 1865, strength and ductility requirements were included in the purchase specifications for materials for most major construction in the United

States. These requirements were at first resented by the producers of materials, but they soon found that the tests were useful in process control and led to a deeper understanding of materials and processes. On the whole, the engineer was satisfied to adapt his design to what was available, and only occasionally realized that the few general-purpose materials developed many centuries earlier could be supplemented by much better materials designed to optimize the combination of properties needed for any specific application.

Several studies of alloy properties were undertaken; most notable were those done about 1880 by R. L. Thurston for the United States Board, covering in an impressively systematic way the whole ternary series of copper-zinc-tin alloys. Commercial brass was improved by being made with metallic zinc instead of by the calamine process, which ceased after about the middle of the century. The cold-drawn brass condenser tube grew up with the steam turbine. A high-zinc brass called Muntz metal with about 40 per cent zinc could be hot rolled and found application for ships' bottoms after 1832. Similar compositions with additions of iron and manganese, misleadingly called "manganese bronze," gave mechanical service under mildly corrosive conditions. Aluminum bronze was known after 1856 but found no engineering use for a long time. The older tin bronzes were greatly improved by the addition of phosphorus as a deoxidizer after 1870.

Most heavy-duty bearings were still made of bronze as they had been for centuries. Tin in the form of pewter had been used in lathe bearings in the 18th century and was improved by Babbitt in 1839 by hardening with antimony and copper. The greatest changes, however, happened in steel, at first slowly and then with gathering momentum at the end of the century (see Chapter 36).

ELECTROMETALLURGY

The discovery of electricity had as profound an effect on metallurgy as on most other human activities. Its first application to metallurgy was in plating one metal with a surface layer of another for improving the appearance or corrosion resistance, and to duplicate detail, especially for printing and decorative use. The electric generator had its first application in plating.

Coating materials with precious metals was an old art. Gilding was best done with an amalgam of gold, spread on the surface and heated to drive off the mercury. One of the most successful forms of silver plating was that used in producing Sheffield plate, objects made from silver-clad copper sheets produced by rolling a soldered composite ingot. This had reached high popularity in England after its development by Thomas Bolsover in Sheffield in 1743, but the electrodeposition of silver in the production of relatively inexpensive but glamorous tableware, started in 1841 by the Elkington Brothers in Birmingham, quickly replaced the more expensive Sheffield plate. Electroplating of

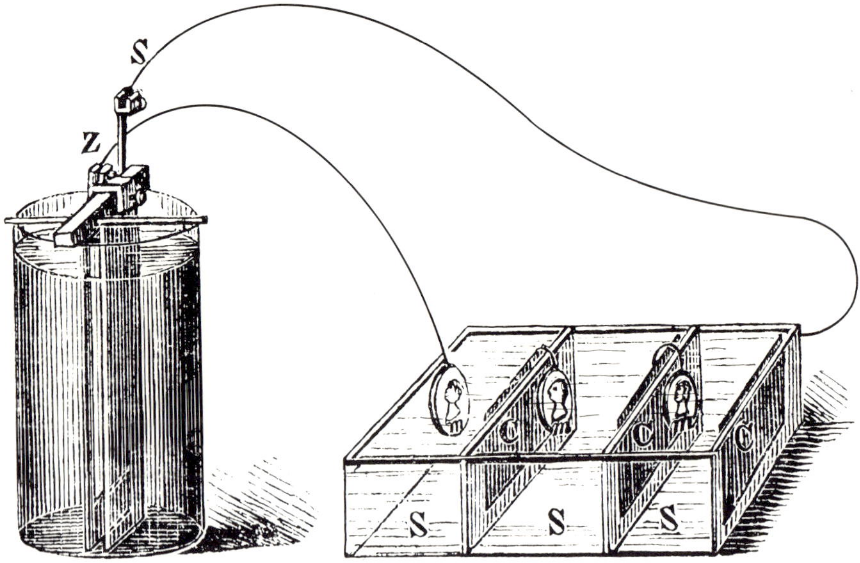

Fig. 21-4. Battery and electrolytic cell for making electrotypes. The making of electrotype plates for printing followed immediately after the telegraph as the second important practical use of electricity. (From Smee, *Elements of Electro-metallurgy*, London, 1841)

silver was especially successful when the newly developed white alloy called "German silver" (see below) was used as a base for the plating so that worn spots were not conspicuous. The corrosion-resistant qualities of electrodeposited pure nickel were extensively used after about 1870 and had a virtual monopoly as a decorative corrosion-resistant metallic finish on iron and steel objects until the development of chromium plating and solid stainless steels in the 20th century. When appearance was not important, iron was coated with tin or zinc, applied by dipping the iron into a molten bath of the non-ferrous metal. Tin was so used in the Middle Ages, and reached its apogee in the tin can for food in the 19th century; zinc-coated ("galvanized") iron sheet was introduced about 1840.

THE NON-FERROUS METALS

The unrecognized presence of nickel had caused trouble in smelting many copper ores for a long time—its name is supposed to have originated as *Kup-fernickel* (Old Nick's copper)—but the element was not isolated and recognized until the work of Cronstedt in 1751. Copper alloys with about 20 per cent nickel had been used early in the 2nd century B.C. in Bactria, and a copper-nickel-zinc alloy known as *paktong* ("white copper") was known in China and had actually been imported into Europe in some quantities from the 17th century on. This Chinese metal was analyzed by Von Engestrom in 1776 and found to be a nickel alloy, but it was not successfully duplicated in Europe until 1823, and did not become commercially significant until, under the name "German silver" or "nickel silver," it was joined with silver plating to give a popular product.

Nickel is easily reduced from its oxide by heating with carbon, an operation that used to be done below the melting point to produce porous metallic "rondelles." The first ores commercially worked were the nickel-cobalt-bismuth-silver ores of the Saxon *Erzgebirge*. Then in 1863 came the discovery of the easily treated nickel-magnesium-silicate deposits in New Caledonia, a French colony in the Southwest Pacific, which for a time supplied most of the small needs for the metal.

The modern and much larger nickel industry essentially had its start in the discovery of large deposits in Ontario, Canada. But these ores, containing both nickel and copper, posed a severe metallurgical problem until solved by R. M. Thompson's "Orford" process, developed in 1893, which depends on the fact that alkali sulfides and the sulfides of most heavy metals except nickel will not mix, and the interesting Mond process (1889) which utilizes the volatility of a compound of carbon monoxide with nickel and its easy decomposition to re-form the pure metal.

The growth of the electric telegraph had an important impact on metallurgy. The building of the Atlantic Cable required the investigation of the electric conductivity of different materials and disclosed that different kinds of commercial copper differed greatly in conductivity. It was not until late in 1857, when the cable was 70 per cent complete, that the conductivity of the wire to be purchased was specified. It was found that Lake Superior copper had a conductivity 92.6 per cent that of the purest electrolytic copper available, while some Spanish copper, containing 0.2 per cent arsenic as an impurity, had as little as 14 per cent conductivity. These observations sparked scientific research on the relationship of electrical conductivity to alloy constitution.

Copper was commercially produced by an electrolytic process using batteries in 1865, but it was not a large-scale operation until after the development of large direct-current generators in the 1870's. It should be noted that pure high-conductivity copper not only enabled the development of the electrical light and power distribution system, but was itself made possible when the resulting cheaper power was applied to the electrolytic bath in which this had all started. Electrolytic copper would, however, hardly have been commercially attractive had it not been that the process yielded as a by-product enough precious metal (present as "impurity" in most copper ores) to pay for the operation. The ingenious and ancient pyro-metallurgical procedures for de-silverizing copper were displaced.

Fig. 21-5. Generators and cells for the electrolytic production of pure copper from copper matte. The plant, with 20 water-driven generators, was capable of producing two tons of copper per day. It was established about 1880 at Sestri-Levante in the Italian Alps. (Engraving from *La Lumière Electrique*, 1884)

The development of the Bessemer converter for removing carbon from iron suggested the use of a similar pneumatic process for the removal of iron and sulfur from copper matte—this was first used on a large scale at the Anaconda Refinery in Montana in 1880. There had been experiments on the use of reverberatory furnaces fired with long-flamed coal for the smelting of both copper and lead about 1700, and the old blast-furnace and refining-hearth procedures of the 16th century had been largely superseded early in the 19th century by the "Welsh" process in which alternate roasting and reverberatory-furnace smelting greatly decreased the number of operations and enabled larger castings of refined copper to be made. The blast furnace continued for the primary smelting of matte, especially for "pyritic" smelting which, after 1880, utilized the heat of oxidation of pyrites in the furnace. Beginning in 1850, multi-stage roasting furnaces with mechanically operated rotary rakes for handling fine ores were introduced. The MacDougal roaster of 1873, for example, had six hearths with water-cooled rotating rakes alternately carrying the ore from the inner to the outer part of the hearth and dropping it to the one below.

The metallurgical principles involved in the fabrication of platinum, which were at first unique to it, later served to make available a new class of materials, today's powder metallurgy products. Platinum had been used in Ecuador before the Spanish conquest to make jewelry by an ingenious process in which small piles of selected fine grains of native platinum were soldered together with gold and then worked to shape. The precious metal was first studied in Europe in 1741, after an Englishman residing in Jamaica had sent samples for study. As an exciting new metal, platinum was subject to systematic studies of its alloying behavior, but no satisfactory way of working it was found for forty years. The melting point of platinum (nearly 1800°C) was beyond the reach of early furnaces, but like iron, platinum may be welded by mechanical compression when hot. In 1786, Janety in Paris used a fusible alloy of platinum and arsenic which was cast to shape and then heated at a low temperature to evaporate the arsenic, leaving a porous mass of platinum behind which was heated until very hot and then forged to solidity. At almost the same time, Charbaneau in Spain developed a better process in which a chemically pre-

Fig. 21-6. The Parks mechanically raked ore-roasting furnace, *c.* 1850. The introduction of mechanical raking greatly improved the uniformity of ore roasting as well as enabling higher output. (From Kerl, *Grundriss der Metallhüttenkunde*, 1875)

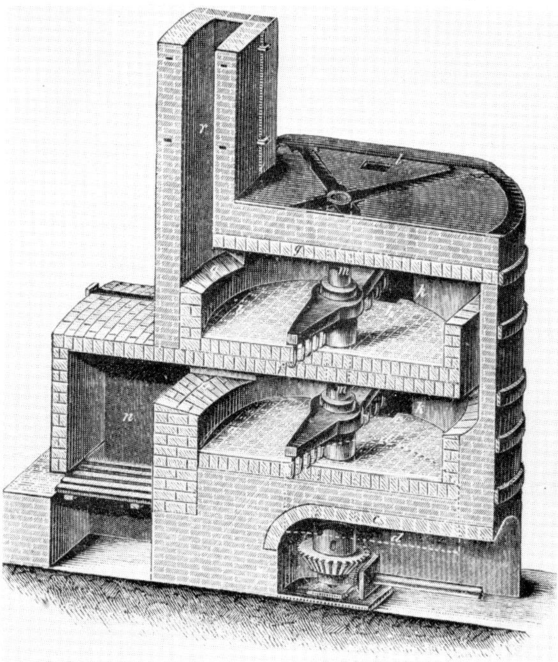

cipitated powder was consolidated by pressing, then heated and forged into pieces for further working. By 1787 he produced ingots weighing 63 pounds. A large chalice made from his platinum in 1789 is still preserved in the Vatican.

Although platinum found some application for jewelry, coinage, and for gun touchholes, its greatest significance has undoubtedly been its value in the scientific laboratory: as a material for inert crucibles, for high-temperature resistance elements, for conductors fusible into glass, as a catalyst, and as a material for high-temperature measurement, it became indispensable in the laboratory and eventually in industrial operations. The platinum industry developed especially in England and in Russia. An important new use of the metal was as coinage (in Russia only), and a spectacular if not large use was in the new metric standards of weight and measure made by Janety in 1795-1802 for France.

The principles of modern powder metallurgy stem from 18th-century platinum metallurgy, but it is interesting that the powder process was abandoned for the making of platinum itself when the oxyhydrogen blowpipe was developed, thus making it possible for platinum to be melted and cast in 1857. This was the work of Deville and Debray, and in five years ingots of 100 kilograms were produced. A new standard meter was made in 1878-79 of a platinum-iridium alloy.

Gold, which retains an economic importance quite independent of its minor industrial uses, continued to be extracted from its ores by the old washing process, greatly improved by new crushing mills and by the use of amalgamation tables to collect the gold. A revolution followed the introduction of chemical extraction of gold and silver by the use of dilute solutions of potassium or sodium cyanide, the metal being subsequently recovered by replacement with zinc. The process, patented by R. W. and W. Forrest in 1887, is operated cold and can be profitably used with as little as five pennyweights of gold per ton of ore (0.0008 per cent). Cyanides would not have been available in quantity had it not been for previous developments of electrolytic methods of making sodium, at first intended for the aluminum producers, but later directed to the manufacture of alkali and the production of cyanamide for fertilizers.

SUMMARY

The period from 1800 to 1880 witnessed a major transformation in man's basic materials—from wood to iron and steel. Our modern industrial civilization is thus founded on the new processes—Bessemer, Siemens-Martin, and Thomas-Gilchrist—which greatly increased the production of steel while reducing its cost. At the same time, developments were occurring in non-ferrous metals which were to lay the groundwork for further metallurgical change in the 20th century.

Although (as we shall see in more detail in Chapter 36) the rise of the metals industry during the years before 1880 was not unrelated to the growing ability of scientists to explain and predict the behavior of the physical world in general, it remains true that most advance still came from practical men. It is fruitless to debate whether the chicken of industrial growth during these years hatched the egg of a vigorous metallurgical industry, or whether the former grew out of the latter. What is clear is that the two developed together—a growing demand for a wide range of metals to meet new needs and the growing ability of the industry to provide them.

If, as some scholars have said, the history of technology is the history of man's growing ability to make use of the materials provided by nature, then the metallurgical history of the first three quarters of the 19th century represents one of man's greatest technological advances. In increasing measure men found how to extract metallic ores from beneath the earth's surface, to reduce these ores to useful metals, and to fabricate useful articles from them. The great advances in the metals industries in the 19th century are truly an indication of man's technical ingenuity in fashioning the raw materials of nature to human use.

22 / Buildings and Construction

CARL W. CONDIT

Building technology from 1830 to 1880 was characterized by revolutionary developments in structural techniques and architectural forms. Of primary importance were the continuing progress of iron framing and suspension-bridge construction, the introduction of the iron-truss bridge, and the widening use of concrete as a basic structural material. Among traditional techniques that go back to the origins of the building arts, stone and brick masonry continued an active life up to the mid-century, but the day of fundamental innovation in the use of these materials had long passed. Timber construction flourished vigorously, mainly in the United States, where several inventions greatly broadened the range of its structural possibilities. Because the history of building techniques in the middle period of the 19th century is extremely complex, it will be necessary to divide our account into separate sections according to technique and type of structure, and to examine briefly the background of this period.

BACKGROUND

The most revolutionary event in more than five centuries of building was the introduction of iron as a primary structural material. This achievement coincided with the Industrial Revolution of the 18th century and had two enormous consequences: one was a radical change in the form and size of buildings and bridges, and the other was the transformation of building from a pragmatic craft to a scientific technology. Indeed, without the latter change, the revolution in the structures themselves could never have occurred.

The first systematic use of iron for building in place of masonry came in the decade of the 1770's largely through the influence of the great British engineer John Smeaton. For nearly seventy-five years after the introduction of iron for building purposes, cast iron was used exclusively because it could be cast in shapes large enough for structural purposes. Although wrought iron had higher tensile strength—that is, it could stand more stress tending to stretch it out along its length—its use had to wait for the development of mills capable of rolling beams of adequate size.

The evolution of iron frames for multi-story buildings begins with the construction of fireproof textile mills in the Midlands and the North of England around 1800. The first was the six-story cotton mill built by William Strutt at Derby in 1792-93; there the floors rested on brick arches supported by timber beams spanning between the masonry exterior walls and two interior rows of cast-iron columns. Four years later Charles Bage designed a five-story mill at Shrewsbury (1796-97), in which the beams as well as the columns were cast iron. By the turn of the century Matthew Boulton and James Watt were building mills with complete interior frames of cast iron. Their chief contribution was the introduction of the inverted T-beam, a step on the way to the modern I-beam. A forerunner of the iron-truss frame came in 1786, when Victor Louis designed a structure of this kind to carry the domed roof of the Théâtre-Français in Paris. By 1830 the internal iron frame was common among industrial and public buildings in England and France, although it was not to appear in the United States until 1848.

The timber frame had reached a high stage of development in medieval Europe, but its continued refinement and enlargement during the early industrial period came chiefly in America. There were good reasons for this: wood was so plentiful as to seem inexhaustible; it is easy to cut, shape, and join; it is durable if it can be protected from fire; and it is an elastic material, strong both in tension and compression. The braced frame of the medieval carpenter was imported into the American colonies by the first English settlers in the 17th century. When the construction of American textile mills began around 1800, timber frames were used for three-story buildings carrying heavy loads of machinery. The basic structure was simple: interior rows of massive columns supported stout girders which in turn carried the lighter joists under the floor planking; the connections were originally mortise-and-tenon joints secured with wooden pins but were later fixed with iron spikes. In the United States, wood remained the dominant material for industrial and railroad structures until the mid-century.

More important for the subsequent evolution of the building arts, however, was the use of wood for bridges. The timber-truss bridge reached its greatest development in the United States, where the high cost of iron or masonry was often prohibitive. A truss is an assemblage of separate members arranged in such a way that they form a rigid structure and function together as a single element, like a beam. The properly designed truss was invented by the Italian architect Andrea Palladio in the late 16th century but was not put to practical use until 1792, when the American carpenter-builder Timothy Palmer built a long timber bridge over the Merrimack River near Merrimac, Massachusetts. The channel spans of the bridge rested on arch ribs stiffened by trusses with posts set on radial lines and single diagonals fixed between adjacent posts. At the turn of the century Palmer was the leading builder of timber bridges, but his reputation was soon eclipsed by that of Lewis Wernwag, who adapted the arched Palladian truss to bridges of longer, clear span.

The first builder to strike out in a new direction was Theodore Burr, whose bridge over the Hudson River at Waterford, New York (1803-04), consisted of flat or parallel-chord trusses combined with timber arches that formed separate structural elements outside the planes of the trusses. Bridges like Burr's were being constructed in Europe at the same time by Karl Wiebeking of Bavaria. In 1820 the American architect Ithiel Town took the decisive step of freeing the truss bridge from its dependence on the arch through his invention of the lattice truss. In this form the horizontal timbers along the top and bottom (the chords) held between them a large number of diagonal members arranged in a tight lattice-like pattern. Since Town's invention acted exclusively as a truss it exerted only a vertical load on the abutments rather than an outward thrust, as in the case of the arch. The subsequent evolution of the truss was marked by a progressive simplification of form that was eventually adaptable to construction in iron as well as wood.

The early history of the iron bridge was primarily an English achievement. The first such structure was built by the iron-founder Abraham Darby III over the River Severn at Coalbrookdale in 1775-79. Darby adopted the shape and proportions of the Roman masonry arch in the semicircular profile of the ribs. The rapid progress of iron-arch construction closely paralleled the growth of the iron frame. The bridge over the River Wear at Sunderland, England (1795-96), designed by Rowland Burdon, was marked by a great increase in length of clear span, a much flatter arch, and a general lightness and openness of construction that revealed how far the builder had departed from masonry precedents. The rigid iron bridge was restricted to the arch form until 1840, when the iron truss made its appearance.

The iron suspension bridge was another product of late 18th century technology, and in its pioneer form it was chiefly the work of an American builder, James Finley. In 1792 he invented a bridge with a timber deck that was maintained rigidly in a level position by means of stiffening trusses on each side and supported by a pair of wrought-iron chains carried between masonry anchors on timber-framed towers. It was a prophetic structure, for the elements of cable, anchor, tower, and stiffening truss remained as essentials throughout the subsequent history of the form. The greatest work in the pioneer phase of suspension construction is Thomas Telford's Menai Straits Bridge in Wales (1820-26), the 580-foot main span of which originally hung from wrought-iron chains carried on masonry towers. The first attempt to substitute wire cables for chains came in 1816, when Josiah White and Erskine Hazard built a suspension bridge over the Schuykill River at Philadelphia, but this structure had no immediate influence. Further progress in wire-cable construction came mainly after 1830 through the work of Marc Seguin and Louis Vicat in France.

TIMBER FRAMING

When we move into the period 1830-80 we must first consider new developments in the traditional systems of timber framing, which played a significant role in American building. The timber frame was composed of a relatively small number of massive pieces which could be enlarged and multiplied up to a point for the increasing loads of multi-story mills and warehouses. For smaller structures such as houses, barns, and stores, however, the heavy members were awkward to handle and were often unnecessarily large in section for the modest floor and roof loads.

A radical improvement in the construction of timber framing came with the invention of the balloon frame, the first of Chicago's revolutionary contributions to building. The most reliable evidence suggests that the inventor was Augustine D. Taylor, who used the new system in the first St. Mary's Church in Chicago (1833).

In Taylor's invention the massive timbers of the medieval braced frame were replaced by a large number of light boards nailed together into a rigid cage of sills, floor joists, studs (posts), and roof rafters. In addition to the rectangular system of vertical and horizontal members, there were usually diagonal pieces in the lower corners of the wall frame to provide rigidity against wind loads. The walls of a balloon-frame house ordinarily consisted of an under layer of wood sheathing nailed directly to the studs, covered in turn by an external layer of clapboard siding. Offering obvious economic advantages in a rapidly expanding nation like the United States, balloon framing quickly spread throughout the whole domain of small-scale building and remains nearly universal today for single-family houses.

Other innovations in timber framing arose from the problem of carrying roofs over extensive interior areas where intermediate supports are undesirable. The earliest examples were the truss frames designed to carry ceilings and roofs

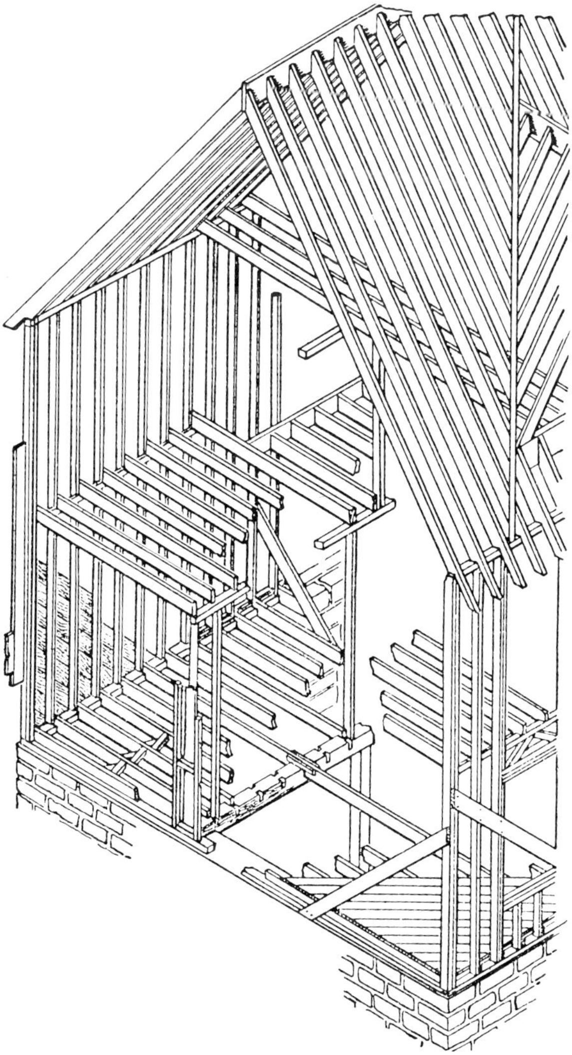

Fig. 22-1. Isometric drawing of a typical balloon frame. (From Singer *et al.*, *A History of Technology*, Oxford, 1954-58)

over the naves of churches. The greater economic and technical resources of Europe made it possible to solve the problem in the early 19th century by means of iron and masonry construction, but in the United States the builder usually had to rely on timber frames up to the Civil War. As in the case of the timber bridge, he exhibited astonishing ingenuity in meeting unusual structural

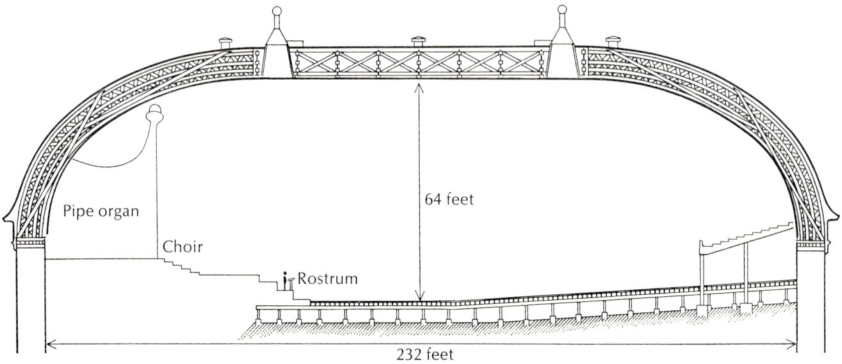

Fig. 22-2. Tabernacle, Church of Latter-Day Saints, Salt Lake City, Utah, 1863-68. Longitudinal section shows the semi-arches at the domed ends. William H. Folsom and Henry Grow, architects. (From *Scientific American, Inc.*)

problems. Roof trusses of advanced form appeared in large churches, a particularly sophisticated example being the system of transverse and longitudinal trusses that carry the roof over the nave vault in the Old Cathedral at St. Louis, Missouri (1831-34), designed by George Morton and Joseph Laveille.

By the mid-century decade the railroad station had supplanted the church as the chief test of the builder's ingenuity in creating wide-span roof frames. Although iron had replaced wood for the European train shed as early as 1835, the older material retained its dominant position in the United States until the end of the Civil War. As the size of the station grew, the old gable roof had to give way to vaulted forms, which necessitated the replacement of the familiar triangular truss by arched forms. The Philadelphia terminal of the Philadelphia, Wilmington, and Baltimore Railroad (1851-52) was the pioneer example of the vaulted shed on timber-arched trusses. These were reinforced against the horizontal thrust of the arch by wrought-iron tie rods extending between the springing points of the trusses.

The masterpiece in this tradition is the Tabernacle of the Church of Latter-Day Saints (Mormons) in Salt Lake City, Utah (1863-68). The architect in charge of the whole project was William H. Folsom, but the designer of the timber structure was the bridge-builder Henry Grow. The roof of this extraordinary essay in building consists of a central vaulted area closed at the ends by semi-domes, the maximum span of the vault being 150 feet. The plank roof rests on an elaborate system of arched lattice trusses with a uniform depth of nine feet. The central vault is supported by full arches extending transversely between opposite pairs of the sandstone buttresses, while the semi-domes at the ends are carried by semi-arches set in radial planes. The Tabernacle is the greatest work of timber framing still surviving, and it represents the culmination of the technique, for by the mid-1860's iron had already become the dominant material for wide span enclosures of this kind.

TIMBER BRIDGES

Even more impressive examples of the carpenter's art appeared in the timber-truss bridge, which in the United States served the railroad almost exclusively for the first two decades of its existence. Wood retained its traditional importance in the building arts because of economic factors peculiar to the United States. High labor costs meant that the fabrication and erection of iron members and the quarrying of stone were prohibitively expensive; the skills of the carpenter, on the other hand, were nearly universal among able-bodied men who worked at any kind of craft. Wherever wood was plentiful and industrial techniques less advanced than in Western Europe, timber construction was bound to be the natural choice. The same conditions existed in 19th-century Russia, where building in timber persisted as widely and for as long a time as it did in North America. Moreover, American bridges were generally built for enlargement and replacement rather than permanence, as they were in England and on the Continent. But the vulnerability of wood to fire and its inability to sustain heavy rail loads eventually forced the builder to abandon it for all but the shortest spans. Nevertheless, his solutions in wood to the problems posed by the novel form of transportation provided the basis for the later evolution of the iron truss.

The chief innovator in this period was Stephen H. Long, but the immediate history of his inventions suggests that they were too advanced to be appreciated by practical builders, including himself. Long was granted a patent in 1830 for a timber truss in which the arrangement of the members and their exact proportions suggested a mature understanding of truss action. Between the parallel top and bottom chords there were two diagonals in each of the panels formed by the posts, and the various members were designed for the most efficient distribution of stress: the top chord and the posts were in compression, while the lower chord and the diagonals acted in tension. In the horizontal planes between the chords of parallel trusses Long introduced sway bracing, in which the diagonals of each panel defined by the transverse beams were arranged like the letter K, which is why these are known today as K-trusses. Although Long's invention was a perfectly valid design, it was embodied in very few bridges of wood, and his later patents were so far inferior to it as to suggest that he himself failed to understand the soundness of his original achievement.

The truss which quickly came to supersede all others in wood was patented by William Howe in 1841. It was nearly universal among railroad bridges as long as timber construction survived, and remained in use for highway spans until

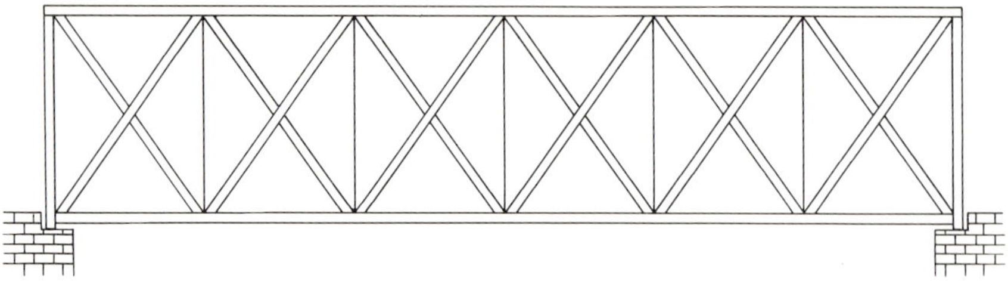

Fig. 22-3. The Howe truss, 1841. (Drawing by author)

1948. The arrangement of members was exactly like that of the Long truss except that the diagonals were designed for compression and the posts for tension. In this respect the Howe truss was inferior to its predecessor, since the long diagonals were more likely to buckle under compression than were the shorter posts. But it was a strong and rigid form which was made even sturdier by the later practice of adding arches outside the truss planes to carry part of the load.

Among structures of pure truss form, one of the most spectacular was the bridge built by the Buffalo and New York City Railroad over the Genesee River at Portage, New York. Erected in 1851-52 under the direction of Silas Seymour, the structure stood at a maximum height of 234 feet above the stream bed. The bridge was destroyed by fire in 1875. That was the fate of most timber spans, and whenever they burned down they were usually replaced by iron bridges.

IRON FRAMING

Internal frames of iron combined with external bearing walls of masonry were, as we have noted, fairly common in England and Western Europe around 1830. The next stage of development was to make the frame the entire supporting structure, so that the masonry wall might be reduced to a non-bearing curtain or dispensed with altogether in favor of glass.

The earliest free-standing frames were those of market halls, the first of which was probably the Madeleine in Paris (1824). In this structure the broad central hall was covered by a gable roof carried on triangular trusses, the ends of which rested on thin columns, with the whole structural system composed of cast iron. More advanced was Charles Fowler's design for the Hungerford Fish Market in London (1835), a long shed-like building without the usual enclosing walls. The most important feature of this cast-iron structure was the wind-bracing in the form of diagonal brackets located at the connections between the outer columns and the primary girders.

Market halls were always single-story buildings, but this limitation was surmounted when James Bogardus built his foundry in New York City in 1848-49. One account of this building describes it as having timber floor-beams, but otherwise the structural system consisted entirely of cast iron. This included the walls, which were reduced to areas of glass set into the rectangular bays enframed by adjacent columns and girders. The framing described in Bogardus's patent of 1850 was wholly iron, with flooring composed of cast-iron plates.

Bogardus's influence as a builder was so great that by 1860 commercial buildings with interior iron frames and iron fronts had multiplied by the hundreds in the major cities of the eastern United States. The great advantage of this technique was the ease and economy with which the iron members could be cast, assembled, and bolted together. But it possessed certain defects, especially the vulnerability of exposed cast iron to fire, the inadequate rigidity of the unbraced bolted connections, and the low tensile strength of the metal.

Fig. 22-4. Foundry of James Bogardus, New York City, 1848-49. (*The Smithsonian Institution*)

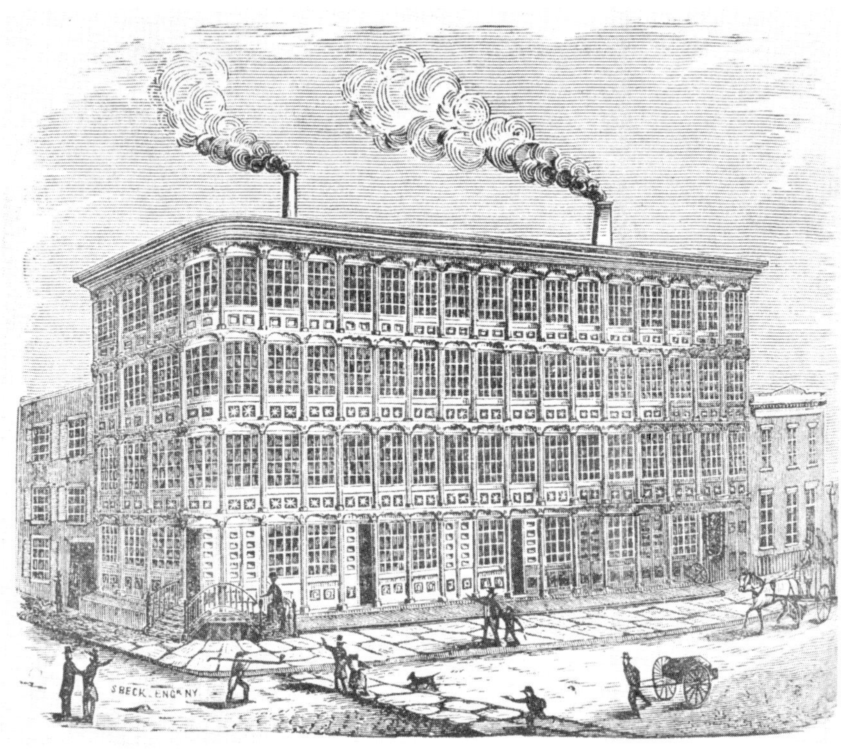

SCIENTIFIC BUILDING TECHNOLOGY AND THE I-BEAM

In spite of its widespread use, iron was still an unfamiliar material, and the design of the iron framework was often the product of rule-of-thumb experimenting. A scientific attack on the whole problem was a matter of necessity if the builder was to make further progress without wasting money and sacrificing lives. The introduction of theoretical and experimental science into the building arts had come in the 18th century, continuing in a piecemeal way up to 1820.

Systematic experimental investigations into the behavior of iron beams was initiated in 1826 at Manchester, England, by Eaton Hodgkinson, a professor of civil engineering at University College, London, and William Fairbairn, a practical builder. The results of their collaboration were published at Manchester in 1831 and serialized during the following year in the American *Journal of the Franklin Institute*. Hodgkinson and Fairbairn next turned their attention to the behavior of wrought-iron bridge girders, publishing the results at London in 1849. Fairbairn's experiences as a builder, along with the results of scientific investigations, were brought together in his *Application of Cast and Wrought Iron to Building Purposes* (1854), which quickly became the gospel in the British construction industry. Hodgkinson's work was paralleled by the development in France of a theoretical science of structure and materials, initiated in a systematic way by Emiland Gauthey and Louis Marie Navier early in the century and extended by the brilliant group of mathematicians at the École Polytechnique in Paris.

Perhaps the chief practical consequence of these inquiries was the development of the wrought-iron I-beam, which is now the standard shape in steel-framing systems. The English ironmaster Henry Cort had begun to roll wrought-iron rails in 1784, and Thomas Tredgold had proposed the I-section as the most efficient shape in 1824, but it took the Hodgkinson-Fairbairn tests to establish the superiority of wrought iron in tension and the structural necessity of the flanged section. Both these discoveries appear to have been put to practical use for the first time in 1841, when Robert Stephenson adopted the built-up plate girder of wrought iron for railroad bridges.

Wrought-iron angles and tees had previously been used in roof trusses (e.g., in the gable train shed of Euston Station, London, 1835-39), but the application of the I-beam to building frames was held up by the inability of the mills to roll a beam of sufficient depth. This limitation was surmounted in 1847 when Ferdinand Zorés of Paris began to roll beams with a depth of at least six inches. During the decade of the 'forties the wrought-iron I-beam steadily replaced its cast-iron counterparts; by 1850 it had wholly superseded the latter in Europe, but it was another ten years before the change occurred in the United States.

Fig. 22-5. Crystal Palace of the London Exhibition, 1851. Joseph Paxton, designer. Interior of the central vault.

THE CRYSTAL PALACE

The stage was now set for a dramatic exploitation of the new techniques, if only a builder of sufficient daring and imagination was willing to try his hand. The English gardener Joseph Paxton, in collaboration with the engineers Charles

Fox, C. H. Wild, and William Cubitt, was the first to show the way in his design for the Crystal Palace of the London Exhibition of 1851.

The Crystal Palace was a prophetic work in many respects: it proved to be the first great iron-framed building, the first in which the outer walls were reduced entirely to glass curtains, and the first for which the structural units were prefabricated at the factory and erected on the site. The major part of this vast building (408 x 1850 feet in plan) was a flat-roofed enclosure in which the primary structural members were hollow octagonal columns of cast iron and trussed girders of both cast and wrought iron with a maximum span of 72 feet. The vaulted central court was carried on arched ribs springing from the two rows of columns at the sides of the vault. The most advanced feature of the supporting structure was the system of wind-bracing known as portal bracing: the main girders were rigidly fixed at their top and bottom chords to ring plates bolted to the column flanges.

The glass-and-iron construction of the Crystal Palace was soon repeated with suitable variations for the famous market building known as Les Halles Centrales in Paris (1854-57). The work of the architect Victor Baltard, it was designed as the first produce market for a city of more than one million inhabitants and was an important feature of Georges-Eugène Haussmann's plan for the rebuilding of Paris, executed under the authority of Emperor Napoleon III. Les Halles is a rectangular enclosure consisting essentially of parallel rows of market stalls divided by a central covered street. The skylighted gable roofs covering the street, walkways, and market stalls are carried on light wrought-iron trusses with lower chords in the form of arched ribs. These ribs spring from rows of columns located on the walkways in front of the shops, the closely spaced columns joined by arches along the longitudinal line. The extreme attentuation of the columns and the lightness of the wrought-iron angles reveal the extraordinary delicacy and buoyancy possible with iron framing.

DEVELOPMENT OF THE IRON-FRAME BUILDING

The next step was to translate the iron construction of these glass-walled enclosures into a true skeletal system for multi-story buildings. In the decade following 1855 the groundwork for this achievement was substantially laid in a series of remarkable works. The first was a curiosity, the importance of which was largely overlooked at the time. In 1855 James Bogardus built a shot tower for the McCullough Shot and Lead Company of New York in which the brick panels of the inward-sloping walls were carried on eight rings of beams supported in turn by eight stands of columns. The columns rose one above the other to the full height of 175 feet, high enough to make the tower a skyscraper for its day. A similar structure was used in the conventional rectangular building for the Boat Store of the Royal Dockyard at Sheerness, England (1858-60),

Fig. 22-6. Warehouse, St. Ouen Docks, Paris, 1864-65. Hippolyte Fontaine, architect. A portion of the main elevation showing the exposed iron frame. (From *The Guilds Engineer*)

designed by Godfrey T. Greene. The presence of timber beams spanning between the main girders belongs to an earlier tradition, but the wrought-iron roof trusses and girders, the cast-iron columns of the now standard H-section, and

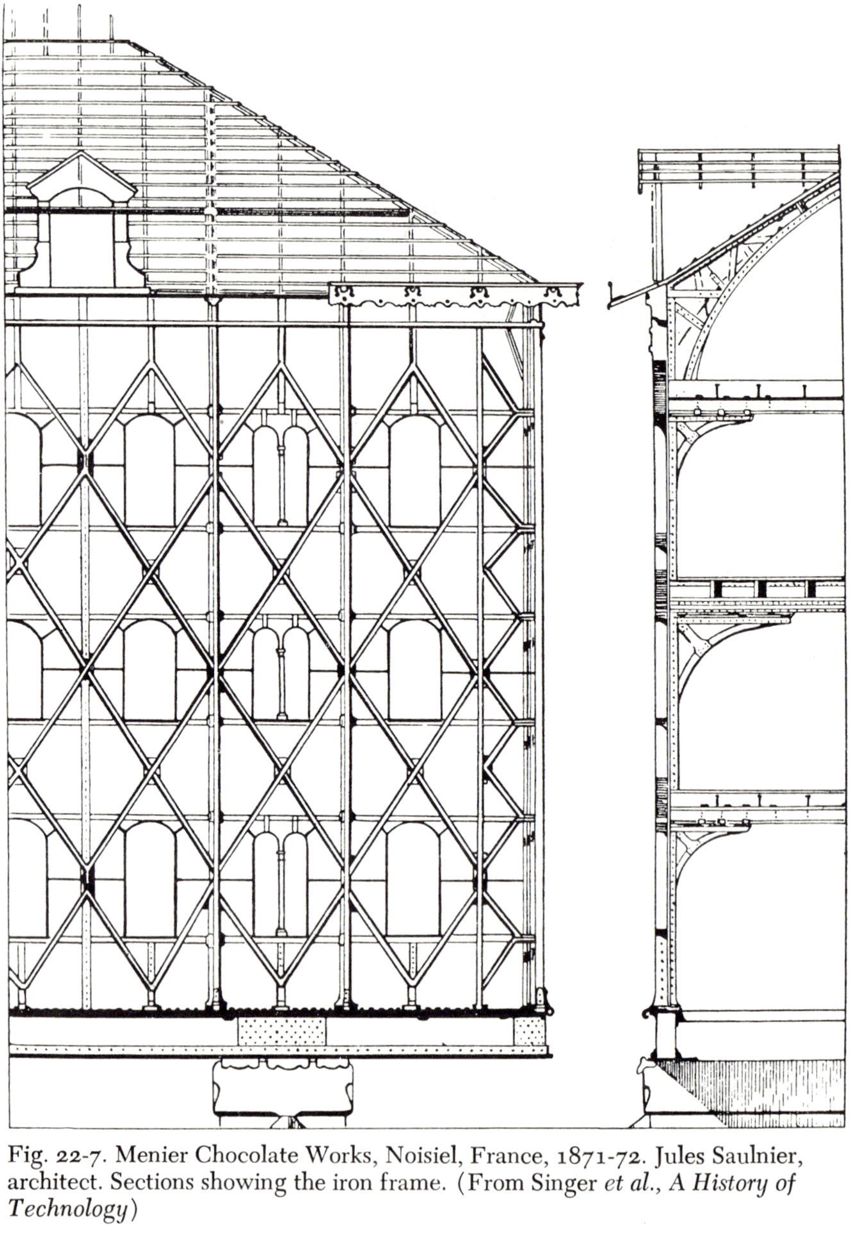

Fig. 22-7. Menier Chocolate Works, Noisiel, France, 1871-72. Jules Saulnier, architect. Sections showing the iron frame. (From Singer *et al.*, *A History of Technology*)

the portal bracing place this long four-story building in the front rank of structural evolution.

The problem was now that of combining the brick curtain walls of the

McCullough shot tower with the braced frame of the Sheerness Boat Store. The solution came with the construction of the warehouse for the St. Ouen Docks near Paris (1864-65), the work of Hippolyte Fontaine. Here the brick wall panels are supported by cast-iron girders spanning between the cast-iron columns, the whole framing system clearly visible in the outer wall surfaces of the building. The floor structure is more advanced: the concrete flooring is carried on hollow brick arches set between wrought-iron plate girders which span 27 feet between adjacent rows of columns. No masonry performs a wall-bearing or load-bearing function in the St. Ouen warehouse. In this work the multi-story, iron-framed, fireproof building was established.

The French builders were to add one further innovation to round out the progress that led to the iron- and steel-framed skyscraper. Jules Saulnier successfully solved a peculiar problem in his design of the Menier Chocolate Works at Noisiel, France (1871-72). In order to carry the five-story building over the Marne River on a few widely spaced piers, he virtually transformed the frame into a braced iron bridge. The hollow-brick panels of the curtain walls are set into the openings enframed by the cast-iron columns, the wrought-iron beams, and the wrought-iron diagonal bracing that extends through the bays up the entire height of the building. The wall frame is thus a lattice truss which is further braced by curved knee brackets and by roof trusses with arched bottom chords. The Noisiel factory is something of a curiosity, but large-scale truss framing to carry floors over extensive open interiors was to be a striking feature of the first mature skyscrapers, which were created by the Chicago architects in the next decade (see Chapter 37).

ARCHED TRUSSES

The special problem posed by the train shed of the big railroad terminal led to the most impressive feats of framing in arched trusses, and these in turn provided the antecedents for such wide-span structures of the 20th century as aircraft hangers and covered stadiums. As the metropolitan terminal continued to grow in size, the all-covering vault over the track area eventually had to be built with iron supporting structures. Vaulted sheds on iron ribs appeared in England as early as 1840, but the difficulty of rolling and placing heavy ribs soon led to their replacement by arched trusses, the relatively small members of which could be readily handled from platforms on scaffolding.

The first shed to embody the new principle is that of the Gare de l'Est in Paris (1847-52), designed by the architect François Duquesney. Covering six tracks and the associated platforms, the shed is a segmental vault with a longitudinal light monitor extending along the crown. The primary supports are shallow, arched trusses of cast iron whose profiles exactly match the cross section of the vault. Under each arch is a curious truss-like system of wrought-iron tie rods and cast-iron struts arranged in a form somewhat like that of the truss invented about 1845 by Antoine Polonceau. This framework absorbs the horizontal thrust of the arch by means of the horizontal and diagonal ties, which are kept in a rigid position by the central hanger and the two intermediate struts.

The rather homely ingenuity that marks the Gare de l'Est shed soon gave way to a more sophisticated design. The emancipation from the pragmatic approach is strikingly revealed by the great train shed of St. Pancras Station, London (1863-76), the work of the architect George Gilbert Scott and the engineers W. H. Barlow and R. M. Ordish. Spanning 240 feet clear, the vault is supported by wrought-iron arched trusses springing from cast-iron shoes. The horizontal thrust of the arch is taken by a wrought-iron I-beam set in a waterproof enclosure extending under the track and platform level. Longitudinal trusses tie the arched members together and thus act both as roof purlins and wind-bracing. The later introduction of the hinged arch made it possible to calculate

Fig. 22-8. Gare de l'Est, Paris, 1847-52. François Duquesney, architect. A cross section of the train shed. (From C. L. V. Meeks, *The Railroad Station*, Yale University Press, 1956)

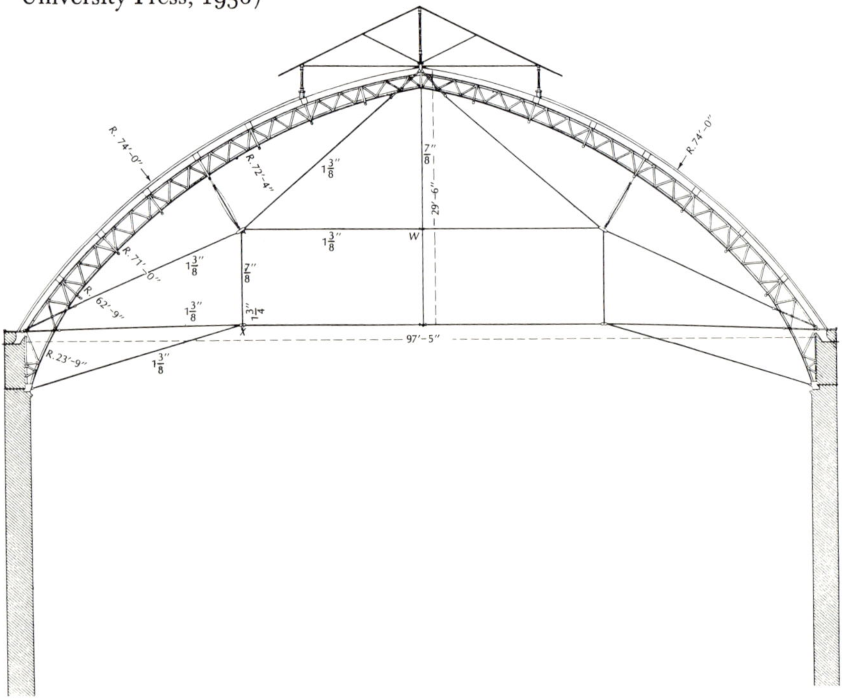

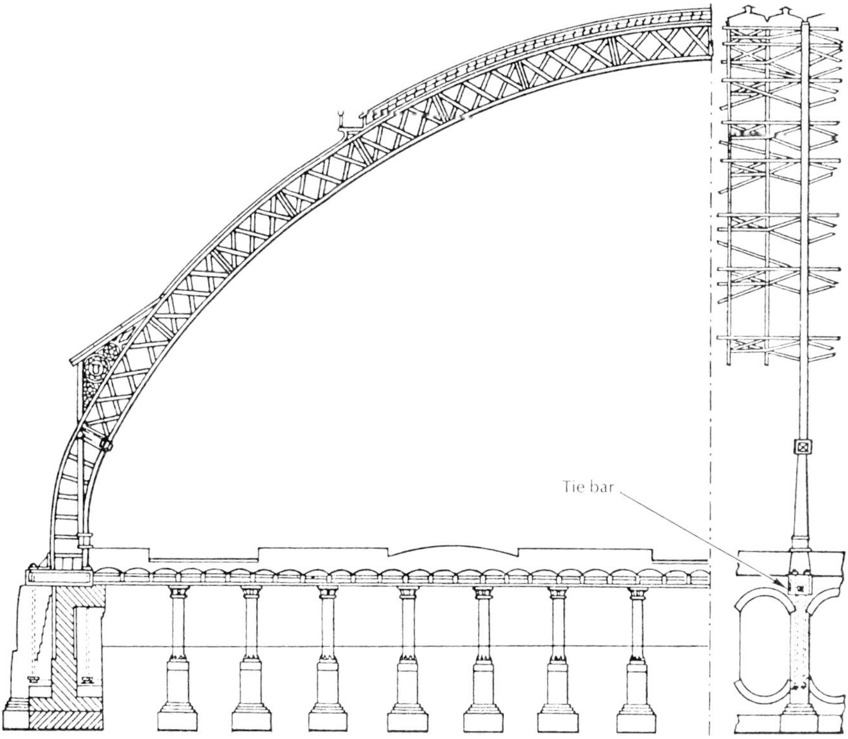

Fig. 22-9. St. Pancras Station, London, 1863-76. George Gilbert Scott, architect; W. H. Barlow and R. M. Ordish, engineers. Half-cross section and detail showing truss framing. (From Singer *et al.*, *A History of Technology*)

exactly the size and proportions of the truss and to increase the span to more than 300 feet.

IRON-TRUSS BRIDGES

The iron-truss bridge was another creation of the prolific decade of the 1840's, and the thorough experience of the American builders with the timber truss placed them in the front rank of its development. Earl Trumbull built the first truss composed entirely of iron in 1840 to carry a highway over the Erie Canal at Frankford, New York. Extending 77 feet between abutments, the deck was carried by a pair of parallel Howe trusses of cast iron throughout, the two were joined and braced by means of a deck frame between the lower chords that almost exactly duplicated the form of the primary trusses.

A long forward step came in 1842-44, when Caleb Pratt and his son Thomas received patents for a truss that differed in a small but significant way from

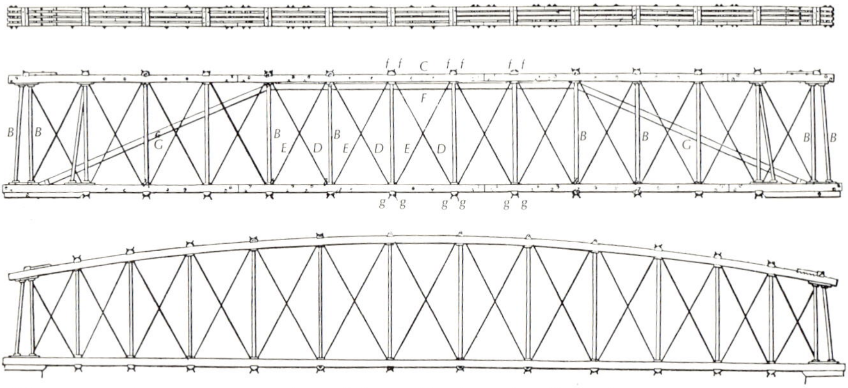

Fig. 22-10. Pratt trusses, 1844. (From *Engineering News*)

Howe's invention. The Pratt truss was designed for construction in iron and timber or in iron alone, but in either case the posts were treated as compression members and the double diagonals as tension, which reversed the action in the Howe system. The iron Pratt truss revealed the usual division between cast and wrought iron: the top chord and the posts (under compression) were cast iron, while the diagonals and the bottom chord (under tension) were wrought iron. The original truss contained two diagonals in all the panels, but the second diagonal was later seen to be unnecessary. By 1860 the diagonals had been reduced to single members in all but the two center panels, and in the next decade the paired diagonals were eliminated in favor of a single system throughout.

The Pratt truss was accepted rather slowly because of competing forms that appeared to offer greater strength and rigidity, but eventually it came to be one of the two most common truss forms in bridge construction. The classic iron bridge of Pratt trusses is the high span built in 1875 by the New York and Erie Railroad to replace the wooden structure at Portage, New York. The engineer of the iron span, George S. Morison, provided an impressive demonstration of the new structural science in his exact calculation of the size and distribution of the members to take account of the wind stresses as well as traffic loads.

The chief reason for the slow acceptance of Pratt's valuable invention was

Fig. 22-11. The Whipple truss, 1847. (Drawing by author)

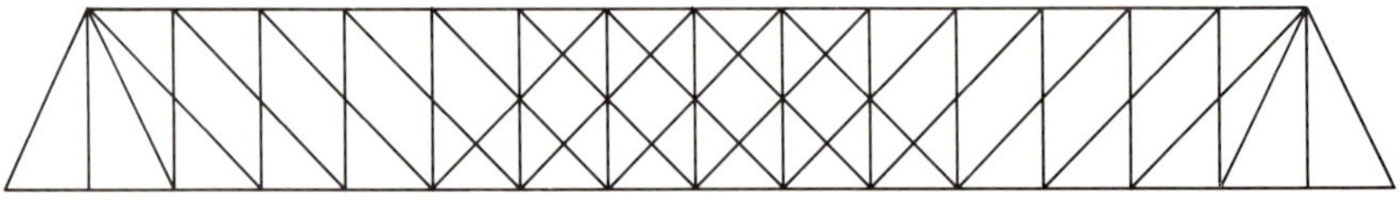

the truss patented by Squire Whipple in 1847. This celebrated engineer and theorist was not only the leading authority in American bridge design until his death in 1888 but was highly respected among European engineers as well. Whipple's reputation rested in good part on his treatise, *A Work on Bridge Building* (1847), one of the most important pioneer texts on the calculation of stresses in the members of a truss. The distinguishing features of his truss were the close spacing of the posts and the unusual pattern of the diagonals, each crossing two panels and all of them in one half of the truss sloping downward toward the center. The Whipple system provided for a considerable increase in the depth of the truss with a corresponding increase in the strength, while the elaborate sway bracing in the top and bottom frames added further rigidity to the whole structure. The distribution of metal followed the usual division of stress: top chord and posts were cast iron, and bottom chord and diagonals were wrought iron.

The Whipple truss remained the standard for long-span railroad bridges until about 1890. Most important of these was the bridge built in 1876-77 to carry

Fig. 22-12. Forms of the Warren truss: (A) Original patent design, 1848; (B) Single-diagonal system with posts; (C) Double-diagonal system. (Drawing by author)

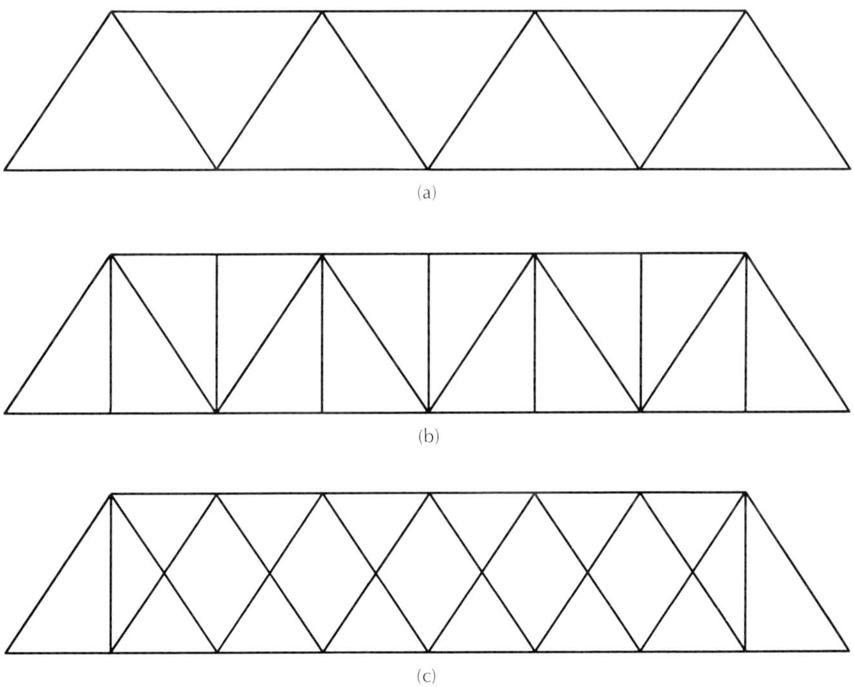

the line of the Cincinnati Southern Railway over the Ohio River at Cincinnati. Designed by Jacob Linville and Louis F. G. Bouscaren, the bridge was the first to include a trussed span exceeding 50 feet in length and the first to embody the entire modern program of bridge design. Two years after its completion Whipple trusses were used in the first steel-truss bridge, built by the Chicago and Alton Railroad over the Missouri River at Glasgow, Missouri (1878-79), with William Sooy Smith as chief engineer.

Destined to share with Pratt's invention a nearly universal role in 20th-century bridges was the truss form patented in England by James Warren and Willoughby Monzani in 1848. The original design was unique in its absence of posts, the compressive force in these members being taken by alternate diagonals, but most contemporary Warren trusses include the usual vertical members. More thorough investigation into the action of the Warren system later revealed that all the diagonals may be subjected alternately to both tension and compression as the load moves over the span, so that engineers adopted the practice of designing all web members for both kinds of stress. The belief that the Warren truss lacked rigidity for rail loads prevented a ready acceptance of the British invention. A more active role in bridge construction was assured when Albert Fink developed a subdivided form of the Warren truss for the 400-foot channel span of the bridge built by the Pittsburgh, Cincinnati and St. Louis Railroad over the Ohio River at Louisville, Kentucky (1868-70).

GIRDER BRIDGES

There were alternatives to the truss among iron structures, and the railroad builders explored them all in their search for strength and economy. For short spans the simple girder bridge proved most efficient, and it came into use along with the first iron trusses.

The wrought-iron plate girder was introduced by Robert Stephenson in 1841 for a little railroad bridge that gave no indication of the extraordinary work that the British engineer was to initiate five years later: the Britannia Bridge. Built in 1846-50 to carry the line of the Chester and Holyhead Railroad over the Menai Straits in Wales, the Britannia Bridge was in many respects the greatest girder bridge ever constructed. Although Stephenson called it a tubular bridge, it was what we now call a box girder, large enough for trains to pass through. The two parallel girders of the Menai span, joined at the top and bottom by a cellular framework of beams, extended continuously over four spans, two of them a record length of 459 feet. Before the bridge was built a variety of tubular forms were tested by William Fairbairn and Eaton Hodgkinson at Stephenson's request, with the result that the rectangular section was shown to have the greatest strength. But the problem of placing the enormous girders on the piers proved so difficult and the quantity of metal in the solid-web girders so redundant, that is, excessive for the job to be performed, that except for the Conway Bridge in England the form was never used again.

The box girder of welded steel or concrete is now a common form, but its ancestor was the simple plate girder of short span that has multiplied by the thousands for rail and highway bridges all over the world. The standard form of the girder, which was patented by Fairbairn in 1846, is in effect an I-beam built up of riveted and later welded plates.

IRON ARCH BRIDGES

For long spans where the characteristics of the stream made it impossible to erect trusses from falsework, the iron arch offers the only alternative capable of sustaining modern railroad loads. The iron arch had been established as a valid form in the late 18th century, and it was again the prolific Stephenson who first demonstrated its possibilities for rail service. In his double-deck rail and highway bridge over the Tyne estuary at Newcastle-on-Tyne (1846-49), the primary members of the six spans are solid cast-iron arch ribs which are supplemented by an iron frame of posts and beams designed to transmit both deck loads to the arches. The horizontal thrust of the arch is sustained by wrought-iron tie rods rather than by the piers and abutments.

THE EADS BRIDGE

The hollow tube rather than the solid rib is a structurally more efficient form which offers the additional advantage of enabling the builder to increase the span length of the arch by bolting together short rolled segments like barrel staves. The masterpiece of this kind is the Eads Bridge over the Mississippi River at St. Louis, constructed in 1868-74 under the direction of James B. Eads as the chief engineer and Charles Shaler Smith as the chief designing consultant.

A double-deck rail and highway structure, the Eads Bridge established a record for length of span outside the suspension category. Of the three spans in the river crossing, the one at the center extends 520 feet and the two flanking it 502 feet—extreme lengths demanded by the influential steamboat interests that still dominated the Mississippi Valley freight traffic. The structural system of each of the river spans consists primarily of two ribs on each side, one above the other, which are joined and braced by a Warren truss set in the vertical plane between the paired ribs on one side and by a braced framework set transversely between the lower ribs on either side. The load of the two decks is carried to the arches by a rectangular framework of deck beams and posts.

Eads originally specified cast steel for the tubular segments but later demanded the expensive and hitherto little-used chromium steel when samples of

Fig. 22-13. Eads Bridge, St. Louis, Missouri, 1868-74. James B. Eads, chief engineer. *Top*: a sketch of the completed bridge. *Bottom*: elevation of bridge and section through river bed. (From Henry G. Tyrrell, *History of Bridge Engineering*, Chicago, 1911)

the cast metal failed in the testing machines of the manufacturer. Steel had been introduced for structural purposes in 1828, when Ignaz von Mitis used it in the eye-bar chains of a suspension bridge over the Danube Canal at Vienna, Austria, but the Eads span represents the first use of the metal in a rigid structure. The masonry piers of the St. Louis bridge were built up from bedrock inside of pneumatic caissons, which had been developed mainly by Thomas Cochrane and John Wright in England between 1830 and 1845.

ROEBLING AND SUSPENSION BRIDGES

The suspension bridge passed through a more systematic evolution than the arch, although it reached its maturity about the same time. The wire cable began to replace the iron chain around 1840, chiefly through the practical demonstrations provided by Marc Seguin in France and Charles Ellet in the United States. Ellet's bridge over the Ohio River at Wheeling, West Virginia (1846-49), was the first with a span greater than 1000 feet in length. The partial destruction of this daring bridge by storm in 1854 coincided with the end of Ellet's career as a builder, while its reconstruction in the same year brought John Augustus Roebling to prominence.

Born in Germany and trained in a German technical school, Roebling had established himself as a practical builder of the first rank by means of a number of suspension aqueducts constructed by the Delaware and Hudson Canal (1845-50) and the double-deck Niagara River span at Niagara Falls (1851-55), the first successful railroad suspension bridge. The Wheeling and Niagara projects launched Roebling on the most productive bridge-building career of the century. His mechanical ingenuity as well as his structural imagination appeared in the construction of the Sixth Street Bridge over the Allegheny River at Pittsburgh, Pennsylvania (1857-60), where he first used the method of spinning the cable on the site by means of travelling sheaves that moved on temporary cables over the towers from one anchor to the other. (This technique had been independently invented by Vicat in France.)

This spinning device was used on a far greater scale for the construction of the Ohio River Bridge at Cincinnati, which dragged on through eleven bitter years of financial failure, Civil War, ice, flood, and storm. Begun in 1856 and finally opened to traffic in 1867, it was one of the most spectacular bridges of its time, with a record span length of 1057 feet. All the essential features of the characteristic Roebling bridge are present in the Cincinnati span—the bound

Fig. 22-14. Suspension bridge, Ohio River, Cincinnati, 1856-67. John A. Roebling, chief engineer. (Courtesy of *Keller Studio*)

cables of parallel wires, the stiffening trusses that extend along the length of the deck, the wire-rope suspenders, and the auxiliary stays that radiate downward and outward from the tops of the towers to the deck beams to provide aerodynamic stability.

Roebling's greatest work is the Brooklyn Bridge (1869-83). Like the Cincinnati span, it stands somewhat modified but in full use today. The Brooklyn project established a new record, which it was to retain until well into the following century: the main span extends 1595 feet 6 inches between the towers, and the two anchor spans add another 930 feet at the shore ends. Each of the four cables contains 5434 steel wires, representing the initial use of the stronger metal for wire cable. The towers of the Cincinnati structure were founded on a compact bed of gravel not far below the surface of the ground along the river banks, but those at New York had to be carried to bedrock far below the alluvial sediments of the East River. As a consequence, the masonry had to be laid up in pneumatic caissons like those that Eads introduced at St. Louis. Unfortunately, John Roebling saw none of the actual construction work on the Brooklyn Bridge: he died of a tetanus infection in July 1869 following an injury suffered while directing the final survey. His son was to carry on and transfer his father's design into actuality.

CONCRETE CONSTRUCTION

The period from 1830 to 1880 saw the rise of iron to a dominant position as a structural material, but at the same time concrete was finding a widening range of uses in the building arts. The great advantage offered by the newer material is that since it is prepared in a plastic state, it can be readily poured into any shape for which the carpenter can make a form, and it will harden into a durable fireproof substance that gains strength with age.

Structural concrete is a mixture of cement, sand, water, and gravel (the aggregate), which enter into a chemical union during setting. Cement is made from lime and a small proportion of clayey materials that give it its hydraulic property, that is, its capacity to set under water. Although hydraulic concrete had been used by the ancient Romans, knowledge of the material was lost from the fall of the Roman Empire until the mid-18th century. Its revival as a structurally useful material depended on the development of hydraulic cement, which began in an unsystematic way around 1760. James Parker of England in 1796 invented a process of manufacturing hydraulic lime from shale-bearing limestone found on the Isle of Sheppey, but the production of such natural cements, as they are called, is seriously restricted by the limited quantities of clayey limestones. Artificial cements made from carefully proportioned mixtures of chalk and clay were invented simultaneously around 1811 by Louis J. Vicat in France and James Frost in England. The decisive step came in 1824, when

Joseph Aspdin of Leeds, England, began the manufacture of Portland cement (so called because of the similarity of its appearance to that of Portland limestone). His process involved burning the limestone and clay at a high enough temperature to produce a glassy clinker, which when ground into cement powder yielded a stronger and more reliable product than that of his predecessors.

Since plain concrete, like stone, has only a compressive strength, it must be confined to simple compression structures like footings, walls, and small arches. In order to use it in structural members subject to tension, the material must be reinforced with iron or steel bars. This technique was developed mainly in the last third of the century and will be treated in Chapter 37.

Building in concrete was inaugurated in 1760 by John Smeaton, who first used it for the cores of lock walls in the River Calder. By 1830 the material had become fairly common in the foundations of bridges and in harbor works. One of the earliest examples of complete concrete construction was a house at Marssac, France (1832), designed by François-Martin Lebrun, who decided on this novel technique through the influence of Vicat's *Experimental Researches into Building Limes* (1818).

By mid-century the leading French pioneer of concrete construction, François Coignet, had embarked on his highly productive career. His first major work was a chemical factory at St. Denis (1852-53) in which the vaulting, stairs, and lintels as well as the walls were poured as monolithic units. More ambitious because of its complex structure was the parish church at Le Vésinet (1864), a monumental work of Gothic Revival design with iron vault ribs and concrete walls and buttresses. By this date, however, Coignet was experimenting with reinforcing, thus helping to give the French builders an early lead in this valuable technique, a pre-eminence that they have held to this day.

Work in England and America closely paralleled the French development. Obadiah Parker built a Greek Revival house in New York City in 1835 in which the walls, columns, and main horizontal members were poured concrete monoliths. He used either Portland cement imported from England or a mixture of his own, possibly derived from Vicat's prescription. The chief American practice until 1870, however, was to pour the concrete into precast blocks and to lay them up in simulation of stone masonry. The English, on the other hand, tended to follow the French technique of pouring walls and other structural members as monolithic elements. They were building large houses and multi-story tenements in this way by the decade of the 1860's, their primary intention being to secure complete fireproofing in economical and durable structures.

Two ambitious works in the United States brought the pioneer phase of concrete construction to its culmination at the very time that reinforced concrete was beginning to grow toward maturity. The South Pass Jetties at the mouth of the Mississippi River (1875-79), built under the direction of James B. Eads, are long channel-protecting walls poured like a dam in enormous concrete blocks. Eads chose concrete to withstand Gulf Coast hurricanes, since stone blocks of the required size would have been impossible to ship and raise into place. The Ponce de Leon Hotel at St. Augustine, Florida (1886-88), designed by the New York architects Carrère and Hastings, sprang from the same purpose. In this remarkable building, still functioning in its original form, the footings, foundations, exterior walls, and some interior partitions are monolithic concrete with an aggregate of marine shells. This rather primitive material was first used two hundred years earlier in the same city for part of the flooring of the Castillo de San Marcos, built by the Spanish governor of Florida as America's first work of military architecture in masonry.

SUMMARY

The middle years of the 19th century were, then, times of development and increased application, rather than a period of fundamental innovation. In England and on the continent of Europe, the structural properties of iron for both buildings and bridges were studied with new scientific interest, and both cast and wrought iron were widely applied to the building arts in novel ways. At the same time, hydraulic concrete (based on Portland cement) came to be used, though still un-reinforced, for a wide range of structural purposes.

In the United States, and in other nations where wood was cheap and plentiful, the use of these new materials had been delayed. During the period from 1830 to 1880, however, here too the advantages of strength and durability—together with a dwindling supply of virgin timber—won the day for iron, steel, and concrete. The sophisticated constructions of timber framing developed by American builders were to stand them in good stead when they developed an appreciation of the new materials which allowed them to build wondrous bridges and, within a few years, to create that hallmark of American technical achievement, the skyscraper.

23 / Machines and Machine Tools, 1830-1880

CARROLL W. PURSELL, JR.

The half-century between 1830 and 1880 defines the very heart of the Machine Age in the Western world. Previous decades had witnessed the birth of the Industrial Revolution, and subsequent generations were to live with the logic of mechanization. But it was during this middle period that most people came into contact with the machine in its full glory, with consequent changes in their lives. Gradually, during these fifty years, machines were brought into one industry after another, as well as eventually onto the farm, into the office, and even the home. At the same time they were spread more widely about the world—triumphant capitalism carried its allied technology into such diverse and unlikely climes as the recently opened Japan, the American West, and the interior of Czarist Russia.

It was a period of great ferment and of system building in Europe. During these years Victoria came to the British throne (1837-1901), France accepted its second Napoleon to restore the greatness of empire (1852), the modern United States grew out of civil war (1861-65), Canada was unified with Dominion status (1867), Italy was unified (1859-70), the Dual Monarchy of Austria-Hungary was established (1867), and the German Empire formed (1871). After being forced open by the American navy, Japan was modernized and westernized (1854-68), and the emancipation of the serfs in Russia (1861) marked the beginning of a slow and painful reorganization of that sprawling empire into a modern state.

It was, in short, the period during which Western Europe first perfected her systems and then, by 1880, began to impose them upon the remainder of the world: in economics, capitalism; in politics, nationalism; in philosophy, materialism. Underlying this golden age of European culture was an increasing reliance upon machinery. It was this ability to make and use machines which ultimately enabled Western Europe to flourish and to dominate the world. It was a time, as the German Chancellor Otto von Bismarck said, of "blood and iron."

THE SPREAD OF MACHINE-BUILDING

The Industrial Revolution had started in Great Britain, and in 1830 she was still the innovator and exporter of machine technology. It was here that new machinery, and new processes based upon it, had revolutionized such basic industries as those of iron and textiles. Perhaps more importantly, it was through the agency of such inspired mechanics as Henry Maudslay, James Nasmyth, and Joseph Whitworth that a machine-tool industry was created and, within their lifetimes, raised to great heights of perfection.

In the earliest years of the Industrial Revolution, the British government had passed laws forbidding the export of certain machinery, drawings, plans, models, workmen, and so forth, in an attempt to maintain a monopoly on the new technology, especially in the iron, steam, and textile industries. To a certain degree such laws seem to have been successful in delaying the spread of these technologies to other countries. Nevertheless, such restrictions, at least in a day when restriction itself was rudimentary, could delay but not prevent altogether the escape of such knowledge. As mechanization progressed, too many machines were in operation, and too many men knew about them, to keep their details forever secret. Before these laws were repealed, British machine builders had collaborated in the wholesale transgression of them and had in fact aided the establishment of machine-building both in the United States and on the continent of Europe.

By 1829, according to one observer, "in respect to the construction of machinery, all the latest and most approved machines of every description used in England have been introduced and are now in actual operation in the best mills in the United States." In 1841 a select committee of the British Parliament looked into the charges that the restrictive legislation still on the books was hurting rather than protecting the machine trade of England. They discovered, for example, that a flat surcharge of 7.5 per cent was levelled in Liverpool to cover the costs of smuggling machines to America.

It was further revealed that Joseph C. Dyer, an American by birth who had long resided in Manchester, had very extensive foreign interests. In the first place, he acted as an agent for American inventors who wanted their devices perfected, patented, and marketed in Great Britain. Based on these and other machines, Dyer built up a considerable machine-building establishment in Manchester. At the same time, he set up his sons in France under the firm name of Dyer Frères, as machine-builders and cotton-spinners. It was the purpose of this firm to build in France those machines which could not be exported from England. Besides this establishment, Dyer maintained business connections with the house of Messrs. Escher, Wyss & Co., in Zurich, Switzerland, and with firms in Austria, Italy, and other nations, through which he worked to introduce his machines into Europe.

The British monopoly was being broken down through other avenues as well. Testifying before Parliament in 1841, Matthew Curtis, successor to Dyer, pointed out that the very success of the British in perfecting machine tools undermined markets for other machines: "The foreign machine-maker is especially benefited," he stated, "inasmuch as those tools stand in the place of a skilled artisan, who must have served a seven years' apprenticeship; but you may now put an ordinary labourer to attend to the machine, and he will do it as well as any skilled artisan of some years' standing."

Not infrequently, the machinist as well as the machine was British. In 1850 a Scottish chemist touring the United States was being shown about New York City by a rather patriotic American who was praising his country's mechanical genius. "I went into some of the machine-shops where the materials for the new line of steamers were in process of manufacture," he reported, "and I heard almost every workman talking with either an English or a Scottish tongue. I remarked this to my New York friend who accompanied me. 'Yes,' he observed, 'but the head-man is an American.' " In describing his experience later, the Scotsman concluded, a bit tartly, that "workshops filled with British workmen are British workshops, on whichever side of the Atlantic they may be."

Nor did Americans limit themselves to producing machines for their own use once representative examples had been smuggled from England, and British workmen hired. The Parliamentary committee of 1841 was told that for the past decade the United States had been exporting an estimated £20,000 worth of machines each year, primarily to markets in Russia, Austria, Prussia, Spain, the West Indies (and Cuba), to Mexico, and the nations of South America. A merchant from Liverpool, travelling through the United States in 1843, wrote angrily to the *Liverpool Times* that Russian interests, prevented from placing orders in England, had just purchased half a million dollars' worth of machinery from the United States, that the two governments were on the best of terms, and that more orders were expected. In 1850 a Baltimore newspaper announced that two complete flour mills had just been loaded for transport to the west coast of South America, accompanied by three millers from Delaware who were to set them up and get them running.

This is not to say, of course, that all American machines were merely derivative. Curtis testified to Parliament that "the chief part, or a majority, at all events, of the really new inventions, that is, of new ideas altogether, in the carrying out of a certain process by new machinery, or in a new mode, have originated abroad, especially in America." Such a statement is difficult to prove, of course, but the United States does appear to have been a leader in developing some types of machines, particularly those designed for wood-working. In a country still blessed with an abundance of wood, the working of this material reached a very high degree of development.

Joseph Whitworth and George Wallis, reporting to Parliament in 1853, claimed that "in no branch of manufacture does the application of labour-saving machinery produce by simple means more important results than in the working of wood. Wood being obtained in America in any quantity, it is there applied to every possible purpose; and its manufacture has received that attention which its importance deserves." A Liverpool firm, they stated, was importing some of the best wood-working machines with a view toward introducing them in England. By and large, such machines were similar to the machine tools used to work iron (planers, lathes, and so on), but those used for wood were of lighter construction and often more intricate. The circular saw, although developed first in England, was introduced into America about 1814 and found extensive use, especially in cutting veneers and small boards. The large band saw, used to saw logs, was patented in 1869 by an American, Jacob R. Hoffman, and became standard for heavy work.

MACHINE-TOOL INDUSTRY IN ENGLAND

When Henry Maudslay died in 1831, England was still the great center for the invention and production of machine tools. The work of such giants as Joseph

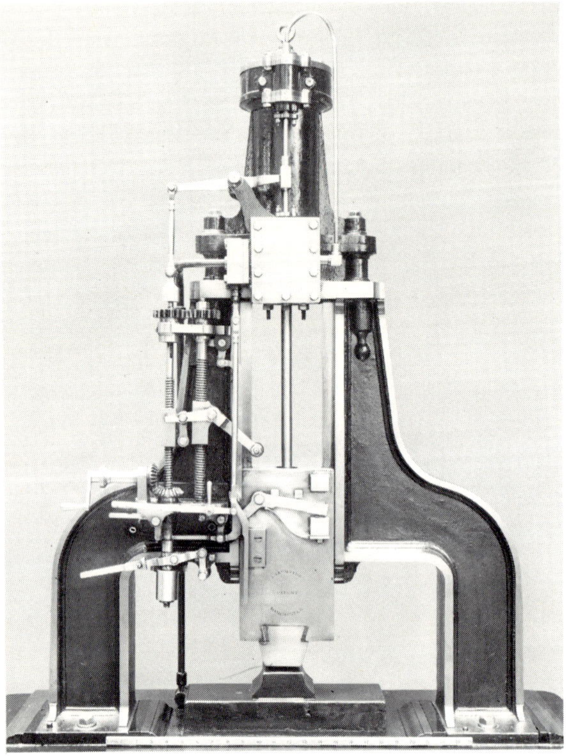

Fig. 23-1. Model of Nasmyth's original steam hammer. (British Crown Copyright. *Science Museum*, London)

Bramah (1748-1814) and Maudslay was ably carried on by Joseph Whitworth (1803-87), Richard Roberts (1789-1864), and James Nasmyth (1808-90), all of whom had worked in Maudslay's shops. Before 1850, however, the leadership even in this field had begun to pass to the United States, where it still rests.

Although many of the basic machine tools were already invented, if not perfected, by 1831, the industry was yet in its infancy. The growing amount of precision manufacture of interchangeable parts, especially in the United States, and the birth of the railroad industry made new and increasing demands upon the machine-tool makers who were required to come up with machines to do ever more precise and specialized tasks. Roberts, for example, although perhaps best known for his development of the self-acting mule, also produced, in 1847, a punching machine to place holes at precise intervals in steel plate—a task made absolutely necessary by the demand for strong boiler plate and the increasing use of iron and steel in boats, bridges, and other structures.

Nasmyth, in 1836, invented the shaper which, though improved, is still in use. This machine, in which the workpiece is clamped horizontally on a table and worked on by a cutter with a reciprocating motion, was able to plane small surfaces, cut keyways, or produce any other surface made up only of straight lines. In 1839 Nasmyth came up with his most famous invention, the steam hammer, developed to handle the larger forgings required by industry. First using the steam only to raise the hammer, it was soon made double-acting by allowing the steam to aid gravity in driving the hammer downward onto the piece to be forged.

Whitworth, later knighted for his accomplishments, began his career as an independent tool-maker in 1833. It was in his shops that machine tools, formerly made primarily to build other machines, were first produced on a scale considerable enough for manufacturers who produced consumer as well as other goods. At the 1851 Crystal Palace Exhibition, Whitworth had no less than twenty-three exhibits, displaying a range of machine tools, each of which was specifically adapted to a certain purpose: punching, drilling, shearing, and so forth. So completely did Whitworth dominate the field that, at the International Exhibition of 1862, his firm's showings took up a quarter of all the space devoted to the display of machine tools.

Among his many improvements were an improved planer able to cut in both directions (1842), and the development of a standard screw-thread in 1841. This standard remained in use until 1951, although widely replaced, especially in the United States, by the standard thread proposed in 1864 by the American tool manufacturer William Sellers. In fact, Sellers's more successful standard was symptomatic of a shift in the geographical center of the industry itself.

MACHINE-TOOL INDUSTRY IN AMERICA

In 1849 the British House of Commons established a select committee to investigate expenditures on army and ordnance. There was at the time no national armory where guns were made, and both the initial price of weapons and the expense of repairs were considered too costly. This matter attracted further attention with the appearance, at the Crystal Palace Exhibition, of six American rifles manufactured with interchangeable parts. When the United States held its own Crystal Palace Exhibition in 1853, Parliament sent a commission including Joseph Whitworth and George Wallis to investigate what progress Americans had made in this and other lines of manufacture. The next year the British decided to establish their own armory, at Enfield, for the manufacture of small arms, and another commission was selected to equip such a manufactory with American machines. On this tour 131 American machines were purchased at a total cost of $105,000.

Among this initial order of 131 machines were 51 milling machines, one of the first important American developments in the machine-tool trade. The milling machine replaced rigid single cutting edges (really like chisels held firm) with rotary cutters, that is with wheels having a series of cutting edges at the circumference, rather like a cogwheel with sharpened teeth. The building of such a machine was only justified if there was sufficient identical work to be done, and if the problem of keeping the greatly increased number of cutting edges sharp could be solved. The advantage was that work was done much more quickly and with less heating of the tool and of the material being cut.

The first milling machine manufactured for sale rather than private use was designed in 1848 by Frederick W. Howe (1822-91) for the Robbins & Lawrence Company of Windsor, Vermont, makers of the famous Sharpe's rifle. By adding more and more cutting wheels, milling machines have since been developed which perform marvels of workmanship in cutting and shaping metals. Robertson's milling machine of 1852, although not the most important economically, was the first to receive an American patent.

The milling machine was almost entirely an American development, at least during the 19th century. It quickly found extensive use in the manufacture of small arms, clocks, sewing machines, and the other light metal products mass-produced at this time. By 1860 it was firmly established in American shops but it was not until nearly 1870 that continental European manufacturers began to use the machine. At first those built in America were used, but by 1873 European-built examples were available. British shops were even slower in adopting milling machines, and it was not until after 1890 that they were found in numbers.

Another new machine tool to make its appearance was the grinding machine, developed about 1830. Previously, grinding had been used merely to polish or sharpen metal rather than to shape it. Grinding machines began to appear early in the century near the armories and textile mills of New England. In the early 1830's Alfred Krupp in Germany, and a little later Nasmyth and others in England, also began to build such machines. By mid-century power-driven, heavy-duty grinding machines were found in many shops in America, Great Britain, and on the Continent. After 1860 work concentrated on two problems: first, achieving that precision which had already been built into such machine tools as the lathe; and second, hitting upon the proper speed of rotation and composition of abrasive for the grinding wheels. The need for precision was met through the work of the American machinist Joseph R. Brown, and demonstrated in the universal grinding machines exhibited by Brown and Sharpe in Paris in 1876. For the wheels themselves, natural emery, sandstone, and corundum were all used before 1830. After that time, artificial grinding wheels were developed and experimented with. Solid-bond emery wheels were on the market in England by 1837, in France by 1843, and in Germany by 1850. A typical

Fig. 23-2. Milling machine designed by F. A. Pratt and manufactured after 1855 by George S. Lincoln & Co., Hartford, Conn. These machine tools were made in large numbers, from interchangeable parts, and many remained in use for decades. (C. H. Fitch, "Interchangeable Mechanism," *United States Census: Report on the Manufactures of the United States.* Tenth Census, 1880, II, Washington, D. C., 1883)

wheel, made of emery, clay, and feldspar, mixed in proper proportion and fired in a kiln, was developed by Swen Pulson and patented in the United States by F. B. Norton in 1877. By the turn of the century, wheels made of emery or corundum, using vulcanized rubber, glue, cement, silicate, shellac, or frit as a bond, were in common use.

Other older types of machine tools, such as lathes, shapers, planers, drill presses, and so forth, were also improved during these same years. The gear-cutting machine for example, although known for many years, was not common in shops until after 1850. The turret lathe, designed to bring several tools to play upon the workpiece as needed, was also a member of what has been called the second generation of machine tools. The improvement of the older, and the development of a new generation, was largely called forth by the spread of mass production from small arms and clocks to other, often totally new, devices such as the sewing machine, the typewriter, and the bicycle in the early 1880's. In other words, the principle of precision tooling, based in large measure upon the spread of mass production and interchangeability, put a higher premium on the speed and accuracy of machine production than upon the elegance of hand manufacture.

THE AMERICAN SYSTEM

The concept of making large numbers of identical parts, which could be used interchangeably for machines or devices, has obscure beginnings, but the fact that it came to be called universally the "American system" shows that it was developed in the United States. Although legend attributes invention of this system to Eli Whitney (1765-1825), inventor of the cotton gin, we know that the Swedish engineer Christopher Polhem was using elements of it in the 1720's and that the Frenchman Blanc, whose work was known to Thomas Jefferson and others, had a very clear idea of the process as early as 1790.

The attribution of interchangeability to Whitney stems from the fact that in 1798 he contracted with the United States government to make 10,000 rifles in two years, and hoped to accomplish this astonishing feat by some new process involving machines run by water. No clear description of Whitney's method survives, but it is clear that any interchangeability he may have achieved was only a by-product of his attempt to cut labor costs by using unskilled labor to manufacture small arms. All the evidence, however, including Whitney's failure to fulfill his contract, points to the fact that he was not in fact the first to apply successfully the American system of manufacture.

Where then did the American system come from? Although the question cannot be answered with certainty, the system appears to have been the result of complex social, economic, and technical forces which were operating in the United States during the early 19th century. The technical requirements of the system—mass manufacture, use of power machinery especially designed for the work to be done, use of gauges to ensure uniformity, and some degree of interchangeability—were at first met in the small-arms industry in this country, especially as it was carried on at the national armory at Springfield, Massachusetts, and by John H. Hall at the Harpers Ferry National Armory in West Virginia. Here machines powered by water performed various steps in the manufacture of small arms and accuracy was checked against standard measures. By the 1830's, concepts, machines, and workmen from these centers had spread, bringing the system to other manufactories, especially those producing light metal products such as clocks and locks, particularly in the northeastern part of the United States. The results of this system were presented dramatically to the British at the Crystal Palace Exhibition in 1851.

TEXTILE MACHINERY

The Industrial Revolution had first transformed the textile industry, particularly the cotton industry of Great Britain. Most of the great developments of the industry were made during those early years, but in all four stages of textile production—preparation, spinning, weaving, and finishing—important improvements continued after 1830. In the area of preparation, for example, Americans, between 1830 and 1840, developed a device to crush burrs in raw wool, thus eliminating a source of considerable hand labor and waste material. The Goulding condenser, first used in the wool industry about 1826, fed wool automatically from one carding machine to the next and finally at the end of the process formed a rove, or small twisted sliver of wool which could be transferred to the spinning equipment. This device eliminated the need for small children to stand between the carding machines, passing wool from one to the other. With the process now automatic, speeds were raised from 75 to 100 rpm and the machines could be placed closer together. In 1853 two Americans invented another device which allowed carding machines to rid themselves of dirt and waste automatically, thus eliminating another hand operation. Cleaning, carding, and combing became entirely mechanical operations during these years, and these processes were most advanced in Great Britain.

One of the most important developments in spinning was the great improvement in the mule which took place in England. Richard Roberts, the machine-builder at Manchester, began to develop a self-acting mule when the mule operators, skilled workmen who already received premium wages, went on strike in England in 1825. Roberts patented two such self-acting mules, in 1825 and 1830, the latter machine, after considerable development, proving ultimately successful. (The first self-acting mules reached the United States by the late 1830's. Their introduction into this country was facilitated by a special Act of Congress, supported by one representative from Kentucky who declared his willingness to do anything in his power to help improve the breed!) Many other attempts were made during the 1830's to speed up the process of spinning, the most successful being the American invention of 1828 known as ring-spinning, brought to England in 1834 but not in wide use until the 1850's. The ring-spinning frame substituted a flanged ring and a light traveller for the flyer; the traveller held the thread under the proper tension and in the best position for winding upon the spool, which could now attain speeds of several thousand revolutions per minute.

After the successful invention of the power loom by Cartwright (1785), and its successful application to the factory system in Waltham, Massachusetts, in 1814, four main problems remained to be solved: first, the quality of power-woven cloth had to be improved; second, output had to be increased; third, the looms had to be adapted to weave wool as successfully as cotton; and fourth, looms had to be applied to fancy fabrics. All four problems were solved during the next few decades.

Among several looms developed specifically to weave fancy fabrics, was that patented in 1837 by William Crompton, an English weaver living in Massachusetts. He later adapted his looms to handle woolens, a task commissioned by manufacturers in Lowell, and an accomplishment which greatly affected the entire wool industry. The single most important loom for fancy weaving was that of J. M. Jacquard (1752-1834), invented in France in 1801 for weaving silk and long kept secret. By about 1830 the Jacquard loom was brought to England, and in 1833 power was applied to it for the purpose of weaving woolens. After 1850 a simplified Jacquard loom, called a "dobby," was widely used for cotton fabrics.

In 1837 Erasmus Bigelow, of Massachusetts, patented a loom for making coach lace, and soon thereafter turned to the problem of weaving carpets. By 1841 his machine-woven carpets were judged equal in quality to hand-woven products. By the mid-1840's, several major developments had made power-weaving more economical and more widely acceptable. One of these, developed in England between 1834 and 1842, was a device to stop the loom when a thread was broken, thus making it possible for one workman to watch many looms since none of them needed close attention. As early as 1840 it was said that while England was ahead in the processes of carding and spinning, the United States led in the application of power to weaving. Nevertheless, even in this country the skill of hand-weaving died hard, and such activity was commercially important in Philadelphia as late as the Civil War.

In the finishing process, wool was again somewhat behind cotton and its problems lasted well into the period 1830-80. Yarn-dyed woolens were unchallenged before the perfection, in the United States, of color printing on pure woolens and cotton-wool mixtures, sometime after 1840. The problem of raising the nap on woven cloth, then shearing it off so as to present an even finish, had been mechanized to some extent in early years but the use of gig mills for this purpose was much resisted by workmen and led, in 1812, to the famous Luddite riots in Yorkshire, when workingmen smashed the machinery they believed was depriving them of their jobs. The mill was introduced into France in 1802 and spread quickly in that country; however, its acceptance in England varied from region to region, being widely used in some by 1830 but in Yorkshire, for example, not universally adopted until after 1850.

Although the mechanization of textile manufacture had its greatest impact on the industry prior to 1830, it continued past that date with unabated and even increasing changes. It is said, for example, that the great Lowell mills were completely reconstructed in the decade of the 1850's, so numerous were improvements in machinery. Furthermore, textile companies were no longer de-

Fig. 23-3. Singer sewing machine, 1854. (British Crown Copyright. *Science Museum*, London)

pendent upon their own machine shops for their machinery. The making of textile machinery had become a specialty trade, carried on in separate shops located near the mills, and concentrated, in the United States, in the New England towns of Taunton, Providence, Worcester, and Lowell, and in the Middle Atlantic states in Paterson and Philadelphia. Innovation in these shops continued to aid the whole textile industry.

SEWING MACHINES

A prime example of the way in which machines replaced even the most complex handwork during these years was the development of the sewing machine. Credit for inventing the sewing machine, or at least for solving the most vexing problems associated with its development—such as placing the thread-hole at the point rather than at the top of the needle—must go to Elias Howe (1819-67), a Massachusetts machine- and instrument-maker. After taking out a patent in 1846, Howe went to England to market his machine with the aid of William

Fig. 23-4. View of the main machine shop of the Wheeler & Wilcox sewing machine manufactory, Bridgeport, Conn., in 1880. (C. H. Fitch, "Interchangeable Mechanism," *United States Census: Report on the Manufactures of the United States.* Tenth Census, 1880, II, Washington, D. C. 1883)

Thomas, a London corset manufacturer. Upon returning home, he discovered that sewing machines had found a ready market and that many competitors were already in the field, in several cases with machines that infringed upon his own patent.

One competitor was Isaac M. Singer (1811-75), who had, in 1850, marketed the first practical sewing machine to appear in this country. Although Singer moved his factory from Boston to New York City in the early 1850's, the mechanical nature of the sewing machine made it natural that its manufacture be centered in New England. Made in new factories requiring special power tools, and made of small precision parts forming identical machines, the sewing machine drew heavily upon techniques already developed for the small-arms and clock industries. In fact, men moved freely from one industry to another.

Although an inconspicuous article of trade in 1850, the 111,000 sewing machines produced in the United States in 1860 were worth nearly as much as all the textile machinery made in the country that year. Sold by an aggressive force of some 3000 salesmen, sewing machines began to make their appearance in the garment districts of cities such as New York, and in thousands of homes across the nation and throughout the world. In Troy, New York, alone, in 1858, there were 3000 sewing machines making shirts and collars. The value of ready-made clothing produced in this country rose from $40,000,000 in 1850 to over $70,000,000 in 1860.

The Singer machine was sufficiently sturdy to sew through leather and was soon adopted to make the upper parts of shoes. In 1858 Lyman R. Blake (1835-83) patented a machine capable of sewing the tops of shoes to the bottoms, and this in turn was improved by Gordon McKay (1821-1903). In the United States, the area around Boston had since colonial times dominated the field of commercial boot and shoe manufacture, and it was here that new improvements were made. Early sewn shoes tended to lose their outer soles if exposed to dampness, but improvements in waxing the thread and controlling the tension, introduced in 1867, largely answered this problem. By 1880 the mechanization and division of labor in the shoe industry in the United States was virtually complete, and further developments took the form of perfecting existing machinery. The McKay machine reduced the cost of sewing shoes from seventy-five cents (when done by hand) to about three cents a pair. Between 1864 and 1870, the number of shoes worked on McKay machines increased from 5 million pairs per year to more than 25 million. By 1895 this total had reached 120 million.

MACHINES IN THE HOME

The Machine Age had an impact on homes as well as upon places of work. For one thing, the home itself and its furnishings were now likely to be the result of machine manufacture. On their tour of the United States in 1853, Whitworth and Wallis noticed that home furnishings, particularly draperies and carpets, were made inexpensively by machine "to look well for [2 or 3 years] . . . and then others are brought in to supply their places. And this habit of almost constant change," they pointed out, "is said to run through every class of society, and has, of course, a great influence upon the character of goods generally in demand." At the same time, house construction was changing under the influence of machinery. One social critic, casting an eye on the "Carpenter Gothic" style of the age, asked rhetorically, "to whom can this period of decadence in household art and architecture be attributed, if not to the pestilent inventor of the buzz-saw?"

It was the period also when the ubiquitous "appliance" began to make an appearance. Just as the factory system was taking certain machines, like the spinning wheel, out of the home, inventors were introducing new labor-saving devices to take their place at the domestic hearth. In 1815, ex-President Thomas Jefferson expressed the hope that a small portable steam engine could be made which could be used to "raise from an adjacent well the water necessary for daily use; to wash the linen, knead the bread, beat the hominy, churn the butter, turn the spit, and do all other household offices which require only a regular mechanical motion." It took longer than he probably expected for this dream of cheap, divisible, domestic power to materialize, but eventually it did. In 1869, for example, the American I. W. McGaffey patented a suction-type vacuum cleaner, and in 1876 Melville R. Bissell invented a practical carpet sweeper. The first motor-driven vacuum cleaner was patented in 1899 by John Thurman.

The telephone, patented by Alexander Graham Bell in 1876, was first installed in a home in 1877; by 1884 people could call from their homes in New York to others as far away as Boston. Whether the telephone saved or created labor is a moot point, but the habit of having a mechanical and electrical marvel in the home was easy to acquire and hard to break. The domestic market for mechanical contrivances, at least in the United States where servants were always a "problem," seemed inexhaustible. One recent study has discovered nearly 600 American patents issued through 1847 for "machines and implements for domestic purposes," of which no less than 228 were for washing machines alone. There can be no doubt that the trend continued throughout the 19th century.

MACHINES IN THE OFFICE

The great triumph of capitalism in the 19th century, that is, the weaving together of the entire world into one very complex and delicate market for the production, distribution, and consumption of goods, was marked by innovations in the office as well as in the mill or factory. As business enterprises became more complex and extensive, machines were increasingly called upon to help control the gathering, recording, and dissemination of data required for sound operation. The telegraph, for example, was soon applied to the needs of quick intelligence by commercial firms, and many corporate, brokerage, and banking offices had instruments installed for their own use.

The typewriter was introduced in the 1870's after the American C. L. Sholes began developing ideas first conceived in Europe; he and his associates made it commercially possible in 1873. In that year the Remington arms-makers were chosen to manufacture the machine. Thus began the production of a new type of precision instrument, put together from mass-produced interchangeable parts. By 1890 there were thirty factories in the United States alone making such machines.

Calculating machines were also developed during these years. Among the more important were the first successful recording adding machines, developed by William S. Burroughs in 1888, and the cash register, developed during the

Fig. 23-5. Sholes's typewriter. Woodcut, *c.* 1873. (*The Bettmann Archive*)

1870's and 1880's in Dayton, Ohio, by the National Cash Register Company. Although not yet applied to private industry and commerce by 1880, very complex calculators had been developed in the 1820's by Charles Babbage with money from the British Parliament. They were brought to practical fruition in the United States when Herman Hollerith, after some years of work, developed a machine which used punch cards to sort, tabulate, and analyze data. Hollerith applied his system with spectacular success to the United States Census of 1890, and processed the data in about one-fourth the time required for that of 1880. In 1896 Hollerith set up the Tabulating Machine Co. to develop his machine for commercial use. Although one may say that the modern generation of computers dates only from Mark I, developed by International Business Machines and Harvard University in 1943, the ancestors of this were developed and successfully applied during the late 19th century.

CONCLUSION

The years 1830 to 1880 marked the maturing of the Machine Age born during the Industrial Revolution. One by one, older technologies gave way to new ones, based on power-driven machinery. Machinery was enlisted to help the housewife, the office worker, and the factory worker. Entirely new devices, such as the sewing machine and the typewriter, were invented, developed, mass-produced with specialized machine tools, and then marketed. They soon became indispensable.

During these same years Europe reorganized its economic system to accommodate the new industrialism, and its political system to make the economic effective. Before the end of the century the great nation-states of Europe, supported by machine technology, were able to gain economic control of the rest of the world with a systematic thoroughness never before dreamed of. In a world in which both economic and political power were becoming rationalized and centralized, more precision and mass production were demanded to fill the channels of domestic and foreign commerce. The American system, as it came to be called —the mass production of devices put together from precisely-made interchangeable parts—characterized the making of such products as small arms, clocks, textile machinery, and sewing machines, among a host of others. In short, machinery, and machine tools, came to characterize the major technologies of the age, and therefore its fundamental nature.

24 / Food and Agriculture in the Nineteenth Century

GEORG BORGSTROM

The world of the farmer was turned upside down in the 19th century. He began the century, in all parts of the world, as a small-scale operator growing crops for his own consumption, for limited distribution to local markets, or on occasion, for export in comparatively small quantities to specific areas. Crops were won from the soil primarily by the sweat of human beings, free, serf, or slave. Despite the interest of gentleman farmers and the "enlightened" absolute rulers during the late 17th and the 18th century, neither science nor improved technology had done much to change the age-old conditions of farming for most people.

By 1900 most of this had changed for many farmers. A power revolution—from human to horse and the beginnings of machine power (as in steam-powered tractors)—and the introduction of some helpful machinery had lightened the farmer's burden and increased his productivity. At the same time, political unification of such areas as Italy (1859-70), Canada (1867), and Imperial Germany (1871), the freeing of serfs in much of Europe in 1848 and in Russia in 1861, the opening of vast new areas to cultivation in Argentina, Australia, Russia, and the United States, all had a profound impact upon the market position of farmers everywhere. Railroads, steamships, and the telegraph made the fluctuations of the world market for agricultural products felt on the farms of North Dakota. The primary technological improvements which made farming easier, more productive, and more expensive to engage in, as well as the secondary improvements which shrank the world and put all farmers into competition with one another, increased both the potential dangers and rewards of agriculture. Farming became as much an industry as a way of life.

Although these changes were worldwide, the farmers of the United States were among the first to feel their impact. American agriculture at the close of the American Revolution, like that of most nations in the 18th century, was still very primitive by modern standards. Nevertheless, by the end of the War of 1812, the land along the eastern seaboard was under cultivation and many new emigrants were settling between the Appalachians and the Mississippi. The drive continued with increasing force until the entire area along the Mississippi River from the Great Lakes to the Gulf was under cultivation. It was the fertile prairies, however, already encountered in Illinois, that encouraged the beginnings of large-scale farming operations which have become characteristic of our own time. This movement was stimulated in the second half of the century, by emerging urbanization, swelling immigration, strong population growth, lively industrial expansion, intense railroad construction, and the final settlement of the western frontier.

BEGINNINGS OF MECHANIZATION

Agricultural technology during the first half of the 19th century shifted only slowly from manual to animal labor. By 1840 many crude contraptions were commonplace on American farms. The invention of the horse-drawn hay rake (1856) increased the productivity of this important cash crop and started the technological march of new implements and machines. As cheap cast iron became available, the supply from a monopoly of local carpenters and blacksmiths was supplanted by deliveries from large-scale manufacturers of machinery.

One of the major mechanical improvements in agriculture during the 19th century was the reaper invented by the American Cyrus McCormick (1809-84). Harvesting consumed more labor than any other agricultural task, and it is not surprising that many men had turned their attention to the problem of mechanizing the reaping operation, or that another American, Obed Hussey (1792-1860), invented a reaper similar to McCormick's almost simultaneously and quite independently. McCormick's reaper, patented in 1834, utilized the knife and cutter-bar principle, but unlike some earlier reapers which employed the same system, McCormick's reaper was pulled rather than pushed by the horses. It took some years after the invention for McCormick to find sufficient capital to begin commercial manufacture of his reaper, but this problem was resolved about 1840, and soon his reaper dominated the field.

Plowing also underwent a major technical advance during the 19th century, foreshadowed by the development of the cast-iron plowshare by Robert Ransome of Ipswich, England, in 1789. Ransome introduced the chilled cast-iron share, so called because the under surface of the plow blade had been cooled more quickly than the upper surface, with the result that the blade sharpened itself through use. A great advance occurred in 1837 when the American John Deere (1804-86) introduced the steel plow. The advantage of the steel mouldboard (that part of the plow which turns the soil over) was that sticky soil would not cling to the steel; this made the steel plow of great importance in turning the rich soil of the American Midwest.

Other important developments included the clod-crusher, the grain drill, and, topping them all, the combine harvester drawn by teams of horses, capable of transferring 30 acres of standing grain per day into sacks for bulk storage. Agricultural production increased rapidly with the aid of these new improved machines, which were still drawn by animals. Wheat and corn farmers could plant more acres. Many other tasks were also lightened as more horse and steam power

Fig. 24-1. The McCormick reaper of 1847.

Fig. 24-2. Large-scale plowing in the American Midwest, *c.* 1880. (From *Farm and Home*, 1882)

made their way onto the farm. Farmers on the prairie found these new machines indispensable for their livelihood.

POWER FOR THE FARM

The power revolution in agriculture during the 19th century was based upon the use of horses (and in some cases oxen, which still remain the beasts of burden in many parts of the world) rather than upon any of the mechanical prime movers associated with industry. As long as farmers depended upon their own muscle power machines were unthinkable. The early hay-mowers and rakes, the harvesters, and other labor-saving devices introduced by mid-century were horse-powered either directly as with the devices above, or through the intermediacy of "horse-powers," that is, inclined treadmills set up near the work to be done and used to operate saws, blowers, and other machines.

The use of horses for field work, added to the demand for horses for personal transportation and, in the growing cities, for such new uses as horse-drawn streetcars and fire engines, made the raising of horses a major agricultural enterprise. Despite the increased use of steam engines for agricultural purposes in the United States after the 1840's, the horse was not really displaced on American farms until the successful advent of gasoline tractors and trucks in the first half of the 20th century.

A total of 1860 stationary steam engines were in operation in the United States by 1838. Primarily they served industrial purposes in mines and mills, but they were also found in considerable numbers on the sugar plantations of Louisiana. The steam engine was indispensable on these plantations: the raw cane was of little use unless converted into sugar products by grinding. By 1859 three-fourths of the southern planters had resorted to steam power for their sugar mills. However, the adoption of steam for more general purposes, and on farms in other sections of the country, was slower. The cost of steam engines prohibited more general use until after mid-century, and up to that point they provided only a small portion of the power on American farms.

The importance and success of railroad engines encouraged the mechanics to harness the "iron horse" for field work. The first steam traction engines appeared in the mid-1880's. They were a welcome sight to the prairie farmer with hundreds of acres under cultivation. Farming operations attained a gigantic scale on these prairies with the dramatic appearance of the largest agricultural engines ever built rolling from one farm to another. Some of the machines were a towering twelve feet tall and cost several thousand dollars, far beyond the economic resources of an average farmer. They could be purchased only by the largest bonanza farms or by private and itinerant harvesters who moved through the grain belt with their engines doing custom work in season.

FARM PRODUCTION

With the development of the steel plow, the reaper, the portable steam engine and other farm machinery, production jumped ahead. Modern power-farming slowly replaced subsistence farming. John Deere's steel plow, the reaper, and the self-binder increased cereal crop production ten- to fourteen-fold, primarily by allowing the cultivation of more land. Man-hours for harvesting were reduced to one-sixth of the time previously required.

Corn was the leading crop in the East for the early settlers, but it gradually lost its importance. Wheat acreage increased fastest of all the plant crops. Between 1858 and 1866, wheat acreage doubled, and wheat became the chief crop of the Great Plains.

The northeastern part of the country raised over 70 per cent of the potato crop in 1870. In addition to being cooked, some potatoes were fed to livestock and some were converted into sugar, syrup, and starch. A new disease-resistant seedling revolutionized the sugar cane industry in Louisiana and added millions to the wealth of this state. Production of hay on the plains increased tenfold between 1880 and 1900, facilitating the feeding of cattle during the severe winters in these new lands.

IMPROVEMENTS OF FARMLAND

Clearing timber was a laborious task for the homesteaders in the American wilderness. It took several years to clear a 100-acre farm. But by 1840 the westward-moving population had emerged from the timberland—almost continuous from Maine to Kentucky—and reached the prairie region. Despite the fact that it took over one month to clear an acre of farmland, by 1850, when the plains were reached, some 100,000,000 acres were ready for the plow.

American farmers had been draining their land since the first settlers arrived. Pipe tiles for soil drainage were a revolutionizing innovation brought to the United States in 1840 from Scotland. Drainage added 50 to 60 million acres of fertile land and increased production on an additional 75 to 100 million acres.

Irrigation has been practised in the Western hemisphere since prehistoric times. But modern irrigation began in 1847, when the Mormon pioneers in Utah started diverting water from City Creek near present-day Salt Lake City. The creation of irrigation on the high plains was closely associated with the shift of the livestock industry from the open range to the ranch.

The fencing of land goes far back in history, but for ranchers on the Great Plains it was a novelty—desperately fought—and yet this was one of the most significant technological breakthroughs both in the United States prairies and the South American pampas. The costs for fencing were prohibitive on the treeless plains until the technical innovation of barbed wire was introduced, which enabled cattle raisers to secure and hold range land, to introduce central feeding, and initiate organized breeding. With fences, land values rose or were stabilized; the new fences also hindered raids by the Indians.

Commercial fertilizers in the United States date back to 1830 when Chilean nitrate was first imported. The first chemical fertilizer plant was established in 1850 in Baltimore, and other plants were soon built along the Atlantic Coast. The four most common commercial fertilizers were Peruvian guano (bird dung), fish guano, gypsum, and superphosphate. The use of fertilizer developed at a slow rate because large areas of fertile virgin lands were still available for production. As time passed, the nutrients in the soil, even in the new regions, became depleted and the use of fertilizers rose rapidly. In the United States, by-products from meat-packing plants, especially bones, residues from plant oil industries, and scrap from fish canneries were some of the earliest fertilizers. Potash from Germany and basic slag, a by-product of the steel industry, were also important. In 1903 atmospheric nitrogen was "fixed" in a form so that it could be used in the United States as a fertilizer for the first time.

A technical advance of singular importance in the 19th century was the discovery of several potent chemicals to eliminate insect pests and crop-killing diseases. A major event was the discovery of the Bordeaux mixture, released in 1885 by the French scientist Millarde, and still a key weapon against fungal disease. New chemicals were provided to control such prevalent diseases as the potato blight and apple scab.

THE DIET

The diet at the start of the 19th century differed considerably from that of the 20th. Without home refrigeration, families could not depend on fresh meat, poultry, or dairy products. In order to conserve butter during the long winter months it had to be treated with brine or salt. To protect against vitamin C deficiencies people drank spruce beer and chewed spruce twigs. Meat was abundant but the diet was dominated by salt pork. The early settlers, milling their own flour, used whole meal rather than refined flour for their bread.

The early farm wife in the United States preserved most of her foods by drying. Canning was not yet known, so dried products such as apples, corn, squash, and beans were part of every meal in the winter. Very few farmers or inns employed ice for preservation. Cool springs and wells were used to keep perishables, such as milk and butter, from spoiling. Most farmers slaughtered their own hogs and cattle and then salted, pickled, smoked, and potted their own meats. Venison was also popular. A farm that was not self-sufficient in respect to meat was considered poorly managed.

The diet of the city-dweller, particularly among the working class, was less varied than that of the rural population. Bread formed the core of daily food. In the cities food spoilage, along with sewage disposal and inadequate water supplies, was a critical issue.

TRANSPORTATION

A new era for United States agriculture opened up in 1807 when Robert Fulton's steamboat made its first trip up the Hudson. Steamboats could deliver produce to the markets along internal waterways faster than any previous means of transportation. When railroads came upon the scene they gradually replaced steamboats for internal transportation, although the boats retained some significance locally.

Perishable food supplies delivered to the cities greatly increased as railroads were built, for they hauled produce faster and over greater distances than could canal barges or wagons. Commercial dairying, market gardening, and horticulture began to spread out from urban centers. Railroad mileage tripled within twenty years with the construction of transcontinental lines. With new lines and new freight equipment, more goods could be carried safely to market. Another effect of the railroads was the ready arrival of foreign emigrants, recruited in Europe by the rail companies, who provided cheap labor for agriculture.

From 1870 on, shipments of cattle and hogs from Wyoming and other states on the high plains were important trade for the railroads. At this time, freight charges were based on the carload, and the tendency was to crowd as many

Fig. 24-3. In this scene, cattle are being loaded upon a Mississippi River steamer at New Madrid, Missouri, for shipment to New Orleans, along with a cargo of corn. (*The Bettmann Archive*)

animals as possible into each car. The hardships of the journey caused the animals to lose much weight before they got to market. The development of the refrigerated car made it possible to send dressed quarters to Chicago. After 1850 the development of steamships and railways widened the transoceanic markets, particularly to the growing industrial areas of Europe. Canadian exports to Europe also increased substantially.

The great cattle drives from the lower tip of Texas up the Chisholm Trail became an epic of the West. The peak of these came in 1880, when more than four million Texas Longhorns were sent north. Sheep were also driven for miles to slaughter in the East, South, and West. The development of refrigeration and of transportation by trucks and railroads ended this picturesque phase of man's conquests of the North American continent.

MARKETING OF PERISHABLES

Early in the 19th century methods of preserving food were limited, and perishables could be eaten only in season. Rural people got their fresh meat from the

Fig. 24-4. Shipment of American cattle at New York for Great Britain. (*The Bettmann Archive*)

cattle driven through the streets to urban markets. They grew their own vegetables and fruits, but without methods of preservation they could not hold them through the long winters. In winter they were limited to a monotonous diet of turnips and potatoes. Although larger cities like New York, Boston, and Philadelphia had more diversified markets, they were still far behind large European cities in the availability of fresh produce.

Milk was carried in tin cans on yokes by boys and women or delivered by horse cart to individual customers in town. Swill milk (watered milk) constituted half of the supply of New York City in 1853. The situation changed gradually as the railroad system was extended, allowing milk to be shipped greater distances. The 19th century also introduced a second potent factor in the worldwide distribution of foodstuffs, namely, refrigeration.

REFRIGERATION

Large-scale refrigeration started in the United States near the beginning of the 19th century with a unique worldwide trade in natural ice. In 1820, storage houses for ice were built in New Orleans, and American ice shipments extended as far as Calcutta, Bombay (India), Manila (Philippines), Singapore, Yokohama (Japan), and Australia. The shipments went around Cape Horn, for the Panama Canal had not yet been built. The natural ice trade rose from 180 tons in 1803 to over 3 million tons in 1880. Soon after this, artificial ice entered the scene and its natural counterpart started to vanish.

The first United States refrigerator, patented in 1803, was a double-walled cabinet with a partition for ice. The spread of refrigeration was retarded due to the high cost of ice, although an ice-cutter devised in 1827 simplified harvesting from ponds and streams, and an icebox patented in 1858 made cold storage through ice more satisfactory.

The ice cabinets were soon enlarged for use as railcars; but they were not very effective until adequate ventilation, insulation, and better ice storage were developed. Such improvements were patented in 1867 and gradually introduced for the carrying of meat, fish, fruits, and vegetables. Fruit steamers similarly equipped were put into traffic on the high seas. Expanded markets for fish were developed in steps after 1830 with the direct icing of fish. Cross-continental deliveries began in 1884. All year round the curing of bacon was practised by resorting to ice-cooled cellars.

Patents for producing cold and making ice, either by expansion of compressed air or by the evaporation of volatile liquids, were applied for from about the middle of the 19th century. The most important was Carl Linde's ammonia compressor of 1876; it and other types were installed in warehouses and food plants. Refrigeration became particularly valuable to the fish and dairy industries, but the meat, poultry, and baking industries were also among those early resorting to refrigeration in the handling, processing, and distribution of food.

The success of mechanical refrigeration marked the beginning of a new era for cold storage. The United States, Great Britain, Australia, Argentina, and Germany pioneered the erection of cold-storage buildings mechanically equipped in the last two decades of the 19th century.

By 1870, United States shipments of chilled beef and other foodstuffs were made successfully to England and other European ports. Refrigerated vessels were soon bringing to Europe an endless flow of meat, dairy, fruit, and poultry products from transoceanic sources.

Freezing foods for preservation had been well known to the ancients, and for centuries the Eskimos and Siberian people preserved game, fish, and milk by freezing. Ice-salt mixtures were used in the freezing of ice cream in the 17th century, and the treatment of fish in this way was done in New England in 1850. Methods were constantly improved—in 1882, by enclosing the fish in metal containers, then dipping the container into the freezing medium. The greatest development and most extensive application of this type of freezing seems to have been in the fisheries around the Great Lakes. The frozen-food industry thus made a start even prior to the development of a system of mechanical refrigeration.

A number of freezing-houses were built along both the New England and Pacific coasts during the 1880's and 1890's. Herring on the East coast, salmon, sturgeon, halibut, and smelt on the West coast, were frozen and then distributed throughout the United States, and some even sent to Europe.

Other items were soon frozen by a variety of methods. Meat, liquid eggs, fruits, and vegetables—all on a minor scale before the 20th century—were preserved by freezing. By the end of the 19th century, railroad cars and ships could handle frozen products, although defrosting still caused difficulties.

Australia was one of the first countries to install freezing-plants for its meat industry. Sheep, beef, poultry, and fish could be frozen for both the local market and for transoceanic delivery (started in 1882). Plants for freezing beef, mutton, and other meats started in Europe.

The first successful shipment of frozen mutton from Argentina to France was in 1877; the meat, carried at $-17°C$, arrived in excellent condition. By 1892 transoceanic trade in frozen meats amounted to two million carcasses, and ten years later to four million carcasses. The foundation for a flourishing trade in frozen beef and mutton was well laid.

FRUITGROWING

The rapid development of commercial fruit industry was a remarkable feature of world agriculture in the 19th century. Up to this time vineyards and olive groves had dominated fruit production for centuries. Under the influence of improved shipping on both sea and land the markets for fresh, perishable fruits broadened rapidly. Apples, pears, peaches, apricots, and many others became common cargo loads. The improved land transportation system in the United States led to drastic overproduction, especially in the South where the incentive of high prices for early fruits in northern markets induced large plantings of perishable fruits. Refrigeration soon aided in alleviating these gluts. Canning and drying also allowed consumption to be spread more evenly throughout the year.

PASTEURIZATION

The great French bacteriologist, Louis Pasteur (1822-95), at mid-century, developed a method for destroying most of the harmful microorganisms in wine without deactivating the useful ones. This was done by a low-level heat treatment that killed the detrimental organisms that were less heat-resistant. The application of this process to the preservation of milk greatly improved its shelf-life and immortalized Pasteur's name, although he had not developed the process with milk in mind. The process also eliminated several milk-borne diseases conveyed to man from cows, particularly a type of tuberculosis and undulant fever. Pasteurization became an essential prerequisite for the marketing of milk in the 20th century.

FOOD TECHNOLOGY

Converting raw materials from agriculture and fisheries into finished products involves several fundamental technologies, such as cleaning, processing, grading, and blending. In such processes as baking, further steps of applying heat, cooling, and packaging are important. All these operations moved ahead under the influence of steam and steel in the 19th century.

It is noteworthy that there had been no major innovations in the field of food preservation until around 1800. Until then the ancient techniques of salting, drying, and smoking prevailed. The first real innovation in food preservation to reach global significance was the canning process invented and developed by the French confectioner, Nicolas Appert, in 1810 after thirty-one years of painstaking effort.

Canning, the hermetic sealing of food in combination with heat processing, quickly passed from the experimental stage and became a leading industry. Several improvements were made on Appert's method during the 1800's. These included reducing cooking time by adding calcium chloride to the water bath, thus raising the boiling temperatures from 100°C (212°F) to 115°C (240°F). This was later accomplished by closing the cooking kettles and converting them into high-pressure cookers (retorts). Appert had used glass containers sealed by cork stoppers; subsequently tin-plated cans were introduced.

Several other concomitant inventions are part of the remarkable progress stimulated by canning, such as machines for cutting corn off the cob, pitting machines, fruit-peeling and paring machines, pineapple cutters, fillers, and so on. The development of canning made a greater variety of safe, wholesome foods available to all people. For the first time in history, men could begin to control nutritional diseases, such as scurvy. Canning placed fruits, vegetables, and fish on the shelves in many homes; it was almost a century later that controlled refrigeration took the second important step by prolonging the shelf-life of perishable foods, and the freezing of foods for extended periods of storage came even later.

In the United States, canning did not become popular until the 1840's; even then, America lagged behind Europe and Australia. From 1840 to 1850, however, the industry grew rapidly, as the Gold Rush gave an impetus to canned foods by guaranteeing subsistence for the long journey to California. Corn, tomatoes, green peas, and fish were then the leading packs.

The Civil War stimulated the canning industry to answer the urgent needs of the armies. Indeed, it was chiefly military needs that created the stimulus for the growth of canning industries in England, France, Germany, and Japan. Sailors and soldiers became accustomed to eating canned products, thereby removing some of the prevailing prejudices against canned foods. The industry began to expand, and several canneries were built in the Midwest. After the Civil War the Libby firm was founded in 1868, and the meat canning industry started in 1872. Olives, salmon, and tomatoes were canned in California, and as early as 1880 that state began shipping canned goods to England and other foreign ports.

Sun- and air-drying of foods had been practised for centuries, but not until the end of the 18th century were technical devices such as precooking, air-heating, and piece-cutting introduced. Artificial dehydration, which originated in France, called for the food to be compressed by hydraulic pressure after being dried. This procedure removed air and facilitated preservation, storage, and transportation. California's weather conditions favored drying in the open air, and many fruits such as figs, prunes, and grapes (raisins) were processed in this manner.

Some entirely new food products appeared on the market during the 19th century. Most important was the first evaporated condensed milk. At first only skim milk was condensed, but by 1902 whole milk powder was also made available. Appert had attempted to can milk, but the most famous name to Americans is that of Gail Borden (1801-74), who also worked with instant coffee.

In Europe fruit and berries had been preserved since the 17th century with sugar taking the place of honey, which had previously been used for this purpose. Jam was made in British homes by 1730. These traditions were carried to the New World, but sugar was not always available in abundance. In searching for new markets, British farmers created the jam industry in the 1870's and jams became an indispensable bread spread. The American jam industry started twenty years later.

Embedding meat products in melted butter or lard for preservation was still practised only one generation ago. This process excluded air and hindered access of contaminating organisms and water. This was the simple device behind pemmican—the survival diet of the American pioneer.

Several chemical preservatives were also in use: salicylic and benzoic acids for jam and fruit syrups, sulfurous acids for the bulk preservation of fruit pulp. The meat rations of the American troops in the Spanish-American War of 1898 were preserved with formaline. However, salt was unquestionably the prime preservative in common use for fish, meat, butter, and cheese.

NEW DAIRYING METHODS

Up to the Civil War there was little progress in dairying. Most families owned their own cows to supply the household with milk and butter. In American towns of colonial times, like Boston, these animals were allowed to graze on the town common. Increasingly though, villagers began to depend on farmers in the vicinity for their dairy products.

The average daily per-capita consumption of milk in the cities during the 19th century was small; less than ⅛ pint in Washington, and a ½ pint in New York. The growing cities were nonetheless forced to extend their milk sheds considerably and bring in fresh milk from pasture-fed cows miles away. Eventually the expanding urban population, improved transportation, greater recognition of the nutritive merits of milk, and more effective sanitary regulations were responsible for a quick build-up of the fluid milk trade and for a rapid growth of its domestic market.

After the Civil War the traditional practices of the dairy industry underwent profound changes. Cheese factories multiplied, and creameries and condensed-milk plants were built. Better transportation and refrigeration also intervened. Particularly significant to dairy progress in the 19th century were five major technical advances: commercial refrigeration, the condensed-milk industry, pasteurization, the continuous milk separator, and tests for rapid determination of fat content. The centrifugal cream separator (by the Swedish engineer Gustav de Laval) was the prime invention and is still in worldwide use. The separator enabled dairies to economize on the time and labor required to skim the cream and on the space for storing the milk while the cream rose to the top.

The first major efforts to preserve milk had been undertaken by Europeans. These included a British patent for condensing milk (in 1835), and one for dried-milk powder (1855) obtained by adding sodium carbonate prior to evaporation. Twenty years later J. R. Meyenberg in the United Kingdom sterilized condensed milk in pans and without adding sugar. After mid-century, Americans began to make great strides in milk preservation, the most notable development being the successful operation of the condensed-milk industry established by Gail Borden in 1856-59. The milk was concentrated in a vacuum and sugar added for good keeping properties. Other new products introduced in this century were "solidified" or dried milk, milk tablets, and spray-dried milk.

Until 1900 the making of butter was largely a home operation; however, the switch to creameries began about 1861. The benefits of large-scale operations became increasingly evident, and with the advent of the fugal separator in the 1890's the trend became more marked. Butter was usually packed in wooden tubs or kegs. Since butter is perishable, even if heavily salted, it had to be marketed within a reasonable length of time. Therefore, until refrigeration and

Fig. 24-5. Meat-dressing room in an American packing house, *c.* 1880. Note the conveyor method. (*The Bettmann Archive*)

rapid transportation were developed, large-scale manufacture of butter was delayed.

MEAT-PACKING

Meat-packing used to be a local business until cities grew in size. Packing-houses increased in number in New England and along the frontiers of the western states. Settlers in the prairies soon found it more profitable to feed grain to cattle and then drive them over the Alleghenies to the packing-houses than to ship feed-grain. Commercial meat-packing in the Midwest started in Cincinnati and quickly spread to Chicago and St. Louis. By the time of the Civil War, Chicago ranked first.

Meat-canning expanded rapidly after mid-century. Canned meats, especially corned beef, became popular items with homemakers eager to try these convenient, economical, and nutritious products. The Morgan machine for compressing meat into cans displaced hand packing in 1871 and is still an important device.

Refrigeration was particularly important to the meat industry. By 1880, refrigerator cars could transport meat from Omaha to New York without spoilage. Instead of 1000 pounds of live steer, 500 pounds of dressed beef could be sent to consumers. Other advantages included using less salt, offering more appealing smoked meats, and the possibility of utilizing a portion of the animal previously considered waste product.

MILLING AND BAKING

Large merchant mills which bought grain and milled it to flour were producing more than nine-tenths of America's flour by 1900; a century earlier they had accounted for only a little more than half. Pioneer America had been dotted with a host of small grist mills, grinding for local customers, and only the cities had merchant mills.

Flour-milling was, for a long time, the most important industry in the nation. The mills gradually grew in size and efficiency through new mechanized devices. Most revolutionary, after Oliver Evans's automatic mill of the 1780's, was the supplanting of grindstones by rollers. This change, initiated in Hungary (1840), produced a finer grind and better grading. When this type of milling was introduced to the American Midwest in 1870, it resulted in a white flour—more attractive and keeping better—which soon gained worldwide acceptance. Furthermore, it permitted the planting of hard-shelled wheat strains which were particularly suitable to the plains states.

Commercial baked products started to supplant household use of flour. This was characteristic of a whole series of technical changes which transferred the processing of food from hearth and home to large-scale commercial establishments. Along with bread, jam and beer, previously produced in the home, started to be made on a significant scale by commercial manufacture. Home-brewing in Europe had been on the decline since the Middle Ages. Porter-brewing pioneered this trend in England, since it was a product with particularly good keeping properties that could be widely distributed. By 1873 there were some 4000 breweries in the United States with an output of 10 million barrels; however, the next step of consolidation into bigger units had not yet started.

FOOD CHEMISTRY

An expanding knowledge of chemistry, and a concomitant willingness to apply it to food products brought many innovations, some of which have already been mentioned. The rapid development of the oleomargarine industry stemmed from the discovery of ways to convert beef suet into edible butter-like fat, based on methods invented in 1860 by the French chemist H. Mege-Mouries. Gradually, animal fats were replaced by vegetable fats such as cottonseed oil, corn oil, and coconut oil. Improvements were also made in the hydrogenation process—adding hydrogen to oil in order to produce a solid fat—which made it feasible to upgrade low-quality fats and oils. By the end of the century, millions of people in Europe and America depended on this inexpensive fat-substitute made from vegetable oils, since dairy cattle and hogs no longer provided enough fats.

The cane plantations of the West Indies supplied the world with most of its sugar throughout the 19th century. Several attempts to produce beet sugar in France and in parts of the United States in the early 1800's failed, but later efforts in Germany, Utah, and California were more successful. Some of the original plants for processing the beets into sugar are still in operation.

Salt was indispensable as a food preservative and was used also in the preparation of hides for tanning. Salt springs in Salina and Syracuse, New York, were the prime sources for a long time. Two million bushels of salt were boiled down in every year (around 1850) in the evaporating plants, packed in barrels, and transported via canals and lakes to Canada, Michigan, Chicago, and the Far West. The wood fuel requirements for this evaporation soon devastated adjacent forests. Salt production in Michigan started during the Civil War, and that state soon became the leading salt producer. Frequently, auxiliary enterprises and byproducts eventually became more profitable than the original operation.

SUMMARY

The Industrial Revolution profoundly affected the productivity of agriculture and gave birth in the 19th century to a diversified food industry richly endowed with new methods. These methods saved food otherwise wasted and made feasible major increases in the production of perishables, so essential to health as sources of key nutrients.

At the same time, problems of food supply of a dimension unprecedented in its entire history were posed for the Western world. In the course of the 19th century the population of Europe grew by no less than 200 million, not counting the large overflow into the Americas. Europe's domestic agricultural production no longer sufficed to feed its teeming millions in cities and industries, and new sources of food were needed.

By the close of the century, the gravity point of world agriculture shifted decisively from Europe to the United States: the significance of no less than 400 million acres being brought under cultivation in the United States, in the period 1860-1900, was tremendous.

Technical advances, such as improved urban water works, large-scale vaccination schemes, improved availability of more diversified food, and industrial preservation of food on a major scale, were all factors which contributed to creating a serious imbalance between birth and death rates. Population growth accelerated sharply toward the end of the century with potentially grave consequences to Europe and the world. Yet these grave consequences did not materialize by the end of the 19th century: progress in agricultural and food technologies not only kept Western man from starvation but made him better fed with a wider variety of foods than ever before in history.

25 / Locomotives, Railways, and Steamships

ROGER BURLINGAME

Without adequate transportation, the Industrial Revolution could never have prospered. Previous chapters have chronicled this awareness on the part of such leaders of that revolution as Josiah Wedgwood in England and Oliver Evans in the United States. In the half-century before 1830 much had been done to meet this need for transportation, and for the first time in centuries improved travel was made possible. Most of this improvement, however, came in the form of improved roads—often turnpikes built for private gain—and new canals, most often the result of government aid and encouragement. On land, very little improvement was made in the vehicles which traversed these new avenues of commerce, although "steam carriages for common roads" were often spoken of and sometimes tried. On the water, of course, the steamboat had revolutionized water transport on the lakes and rivers of the United States, and to a lesser extent in England and on the European continent.

During the next fifty years—that is, the middle half of the 19th century—two more critical developments gave the world, and particularly the major industrial nations of the West, a transportation system which was at last of sufficient size and efficiency to meet the new demands of an age of steam and steel. During these years, great land masses were spanned by railroads, which in turn were interconnected by steam-powered vessels of iron (and then steel) which truly made the oceans highways of commerce. In these developments Great Britain and the United States led because of their great industrial activity and extensive merchant marines. At the same time, the island and the continent had differing needs which prompted different developments.

THE BEGINNING OF THE RAILWAY ERA

By 1830, what is known to historians as the "Railway Era" had already opened in England. The year before, the locomotive *Rocket*, designed by George Stephenson and constructed by his son Robert, had come out ahead of the two other contenders in the difficult trials at Rainhill, near Liverpool, and had won for the Stephensons a prize of £500. The *Rocket*, which had travelled 12 miles in 53 minutes, was adopted by the Liverpool and Manchester Railway which opened for business in 1830. It then drew an average gross payload of 40 tons behind the fuel-tender and attained a speed of 13⅓ miles per hour. The Com-

Fig. 25-1. The *Rocket* and a later British locomotive. (*The Bettmann Archive*)

pany was so impressed with this performance that it gave the Stephensons an order for seven more locomotives. These had the essential elements of the modern steam locomotive.

These events in 1829 and 1830 put England ahead of the rest of the world in railway construction and operation. But there were still many problems. Conservatives were not convinced that locomotives could master any but the gentlest slopes, and stationary engines were still used to pull trains up hills. There was a persistent belief that smooth wheels must inevitably slip on smooth rails, and various devices were tried, such as roughening the tracks or laying a track in the middle with which a cogged wheel on the locomotive should engage. However, as the locomotive engines became more and more powerful, the doubters were persuaded that steeper gradients were possible. The success of the Liverpool and Manchester spurred many new projects, and by 1840 some 1331 miles of track had been laid.

NATIONAL DIFFERENCES IN RAILWAY DEVELOPMENT

Most of the railways on the continent of Europe had a later start and a slower development. Although France and Austria opened their first lines at about the same time as England, their operations remained experimental for at least a decade. Belgium, Germany, Russia, and Italy began ten years later, Switzerland and Denmark in 1844, Spain in 1848, Sweden in 1851, Norway in 1853, and Portugal in 1854. There was no railway in Turkey until 1860 and none in Greece until 1869. In Canada and South America, there was no activity until the decade of the 1850's. In the United States, however, the railway era began at about the same time as in England, but its subsequent development far outran every other nation so that, even by 1850, its mileage was the greatest in the world.

There were reasons for these national differences in the rate of rail progress. England's Industrial Revolution began in the mid-18th century: by 1800, the steam engine had become commercially practical; ten years later, steam navigation had begun. In the early years of the 19th century, the technologies involved in coal mining, iron manufacture, ship-building, canal construction, and many kinds of metalwork were brought to a high degree of excellence.

The fact is, however, that the United States, far behind Great Britain in the technologies that had brought Britain to such a high industrial level, was able to build and operate steam railroads and to experiment with steam navigation even before the turn of the century. Yet at one time, when factories all over England were being operated by steam, there were only three steam engines in the whole of the United States, and these three were below the standards set by Boulton and Watt in London. Furthermore, the new nation had not, even by 1830, developed its economy to the point where there was sufficient capital for novel and uncertain enterprises.

There are two possible explanations of this rapid American development. One is the presence of several truly great inventors or engineers in the eastern United States: such men as John Fitch, Henry Voight, Oliver Evans, John and Robert Stevens, Peter Cooper, Benjamin Dearborn, and Nicholas Roosevelt; of persistent promoters such as Robert Livingston, Robert Fulton, and Horatio Allen. Such men gave rise to the term "Yankee ingenuity"; they had knowledge ahead of their time and little else to work with except ideas, in a country which, apart from its vast wilderness, was almost wholly agricultural.

The second explanation has to do with these very wildernesses. To explore and settle them, to consolidate an immense continent into a single nation set a crying need for transportation that invention had, somehow, to satisfy. As soon as knowledge of the steam engine reached America, the inventive prophets knew that it need not be used to turn the wheels of mills which could, for a long time, operate with water power, but rather that it should go primarily into steamboats and what were at first known as "steam carriages."

BRITISH AND AMERICAN RAILWAY ENGINEERING

From the first it was evident that railway construction and operation in England and the United States would present wholly different national patterns. In

Fig. 25-2. Early locomotives of the Great Western Railroad, England, 1837-55. (British Crown Copyright. *Science Museum*, London)

insular Britain, the lines would inevitably be short, and the terrain was not likely to pose any colossal engineering problems. In the United States, on the other hand, the distances which would eventually have to be traversed were enormous, and the mountain ranges between the Atlantic and Pacific coasts would set obstacles that would tax the skill and ingenuity of the best civil engineers the nation could produce.

In England, the pattern of the "way" was set by the construction of the first locomotives. The English locomotives were massive and rigid. Both the driving wheels and the front supporting wheels turned on axles that were rigidly attached to the engine's frame. A glance at the accompanying illustration will reveal the locomotive's inflexibility, its exclusive adaptibility to straight-line travel. Furthermore, its lack of equalizing devices made necessary completely even rails always precisely on the same level, since any roughness in the roadbed would be damaging to the locomotive. On a track that was straight or whose curves were very slight, whose ties were rigidly set, and whose rails were without defect, the English locomotives performed magnificently.

To avoid spiral hill-climbing, tunnels or cuts were necessary, and viaducts had to be built over valleys to keep the way as level as possible. Construction which met these needs was naturally expensive. But for the short distances of the planned lines, it was not prohibitive, especially in view of the heavy traffic promised by the mines and the factories already established. The result was a network of beautifully constructed railways which promised both speed and safety and, for the passenger traffic, comfort.

When railroad operation began in the United States, it was obvious that the locomotives imported from England could never be adjusted to the roadbed inequalities and curves of long-distance American transportation. And it was equally obvious that railroads built on the British plan would be far too expensive for the vastly increased mileage involved. Since it was impossible, therefore, to make the roads fit the locomotives, it was necessary to design a locomotive that would fit the roads. Such a locomotive had to follow a quickly built track with curving ascents instead of costly tunnels, cuts, and viaducts; it should not be derailed by a rough roadbed laid at minimum expense, and, finally, it had to be light yet powerful enough to pull a long train up far steeper gradients than could be found on any English railway.

To do away with what was known as the "hammer blow" caused by wheels pounding on uneven roadbeds, Joseph Harrison, a Philadelphia mechanical engineer, invented the equalizing lever. This was a flat bar, fastened at its center to the locomotive frame, its ends resting on the journal boxes of the driving wheels. With this device shocks were evenly distributed between the wheels.

At the very start of American railroad construction in 1831, John Jervis, a locomotive builder from upstate New York, devised the swivel or "bogie" truck which soon became famous all over the world. Having two or four wheels, it was placed under the front end of the boiler and could follow the curves of the track as the front wheels of a road vehicle follow the curves of a common road. With the bogie truck the locomotive could be made longer, and there was also a better distribution of weight. When Jervis built his "American locomotive No. 1, Second Series," he equipped it with a boiler the unprecedented length of five feet. Finally, came the cow-catcher, devised by Isaac Dripps to remove the obstructions which fell or wandered on the track as the train wound its way through open fields having none of the safety precautions that distinguished the English "way." This completed the characteristic American locomotive.

The "basket" construction, as it came to be called, of the American locomotive combined with the T-rail track, laid on wooden ties (sleepers) and set in loose gravel, to give maximum flexibility—the American ideal. All these factors plus the extraordinary ingenuity of the mid-century American engineers in solving the problems (of culverts, truss bridges, snow sheds, loops, switchbacks, embankments, and so on) presented by the effort to traverse the mountain ranges, resulted in the transcontinental system which opened in 1869—certainly one of the greatest engineering achievements of all time.

It is true that American railroads had many disadvantages which the British and continental European roads did not share. Safety and comfort to passengers were sacrificed to speed; the wood-burning locomotives which persisted until after the Civil War caused disastrous fires, especially where the roads ran directly through towns; and there were countless accidents at grade crossings. Nevertheless, the American railroads did much to advance the settlement of the

Fig. 25-3. The *Experiment* built by John Jervis. (*Science Museum*, London)

Fig. 25-4. American locomotive. (*Science Museum*, London)

country, and developed meanwhile a whole technology which was eagerly adopted in Europe and Asia. They also paved the way for America's industrial leadership of the world.

It should be noted that with railroads, as with machine tools and many other devices, the American requirement of flexibility meshed nicely with the abundance of wood as a construction material. While English engineers built with stone, Americans built with wood. It was the use of wood, as much as anything else, which gave the American system its native character—one which often offended the sensibilities of English engineers but as often caught the imagination of those from other nations, notably Russia, which met with similar problems and advantages.

Accuracy of schedule was, of course, greatly increased in Britain, the United States and, indeed everywhere else, by the application of the electric telegraph to railway operation: during the 1840's and 1850's in Britain, it was the needle telegraph of Cooke and Wheatstone, and in America, the Morse magnetic telegraph. Even so, in the United States time was kept on a local standard until 1883, when over fifty local times were abolished in favor of the time-zone system now used. This standardization of time, which now seems so necessary and obvious, was a direct result of the need of the railroads to help prevent accidents and make possible more efficient train scheduling.

BRITISH IRON SHIPS

As wrought iron came to be more widely used in Britain late in the 18th century, it was proposed that this material be used to replace, or at least supplement, the stout oak of the British fleet. Despite the growing shortage of suitable timber for ships, such a move was vigorously opposed. Some objected that iron was too heavy, and that an iron ship would not therefore float; others believed that a hull of iron would destroy the accuracy of the mariner's compass; or that an iron bottom would foul easily and might not easily be cleaned of weeds and barnacles. But the most curious prejudice was that the traditional national boast —that England was protected by its "wooden walls" (of ships)—would no longer be true.

Ship-builders began by using iron only in the internal structure of wooden ships for added strength. Various experiments with the use of iron plates on the hull of canal- and riverboats in the late years of the 18th century were followed by the first true iron steamship, the *Aaron Manby,* built in 1820 for direct service between London and Paris. She was constructed inland in sections which were then shipped to London for assembly.

In 1832, came the *Alburkha* and *Aurora,* all-iron steamships built by the Lairds of Birkenhead, and the *Garryowen* which had water-tight bulkheads. Yet even after these ships had answered conservative objections, some of the largest steamships, destined for transatlantic passage, continued to be built of wood. This was true of the celebrated *Great Western* which displaced the unprecedented tonnage of 2300. Launched in July 1837, she made her maiden trip from Bristol to New York in fifteen days in April 1838. The smaller *Sirius* at the same time made the trip from London to New York in sixteen days. Both ships, which made their entire voyage by steam, were propelled by side paddle-wheels.

By 1838 it was evident that the steamship was to move away from both wooden hulls and paddle-wheels. Isambard Kingdom Brunel (1806-59), one of the greatest engineers of the century, was the mastermind of the change. While still in his twenties Brunel had been appointed engineer of the projected Great Western Railway, for which he designed tunnels, bridges, and viaducts. In 1835, he suggested to the directors of the company that they "make the railway longer"—extend it to New York by way of transatlantic steamers. The first of these was the *Great Western,* but already his imagination was reaching beyond wooden hulls and side paddle-wheels. Soon after the *Great Western* made her maiden voyage, he made plans for an even larger ship to be built of iron and moved by a screw propeller. This ship, the *Great Britain,* measured 3443 tons and displaced 2984. Her first transatlantic voyage was in August 1845, making the run from Liverpool to New York in 14 days, 21 hours.

By this time the screw propeller, a device suggested by Archimedes' screw, had had a long history, although on the *Great Britain* it emerged definitely from the experimental stage. Its first use is ascribed to John Fitch who is said to have used it to propel a small skiff in New York in 1797. In 1804 the American John Stevens built a vessel which was equipped with twin screws. In England a

Fig. 25-5. Brunel's *Great Eastern*, launched in 1858, employed both paddle-wheels and screw propellers. The ship was later used for laying the Atlantic Cable. (*Science Museum*, London)

patent for a screw propeller was taken out by Thomas Pettit Smith in 1839 and was tried in that year on a large steamer, the *Archimedes*. But the man who patented and made this method of propulsion a success was the Swedish-born American citizen, John Ericsson (1803-89), who later designed the Union warship *Monitor* which won one of the famous naval battles of the American Civil War. Brunel's largest ship, the *Great Eastern*, launched in 1858, employed both paddle-wheels and screw propellers. She was later used in the experiments for the laying of the first Atlantic cable.

British steam navigation and railway progress in Great Britain not only moved along parallel lines but were clearly associated. The Great Western Steamship Company was an offshoot of the Great Western Railway Company, and Brunel used his talents in both directions. Both stimulated coal and iron production. In the public mind, also, both were closely associated.

AMERICAN IRON SHIPS

By 1830, when the first American railway was complete in South Carolina, steamboats were already moving in very considerable numbers to help settle the banks of the Ohio, Mississippi, and Missouri rivers. On the other hand, far less attention was given to ocean-going steamships in America than in England. Always a seafaring nation, Britain, with its growing colonial empire, was more than ever in need of rapid ocean transportation. The American-built and owned *Savannah*, which made the first transatlantic crossing by a steamship in 1819, was an exception to this general rule, but since she operated by steam for only 80 hours out of the 29½ days of the voyage, the rest of the time operating under sail, she can hardly be considered a full-fledged ocean steamship.

In the meantime, American marine builders and steam engineers concentrated on steamboats for the inland waters, and these great, wood-burning Mississippi vessels established one of the romantic American traditions. Simultaneously, however, the clipper ship era in America saw ocean-going sailing ships reach their peak of speed and efficiency. The great speed, efficiency, and beauty of these graceful ships no doubt helped retard the development of a fleet of steam-powered iron ships.

The rapid rise which characterized the railroad story in America found no parallel in the building of iron boats. While Great Britain was applying the iron boat with screw propeller to marine service, American ship-builders were more conservative. Part of the reason probably lies in the fact that the high point of the American merchant marine in the 19th century came in the 1850's, coinciding with the era of the clipper ships which brought profits to their owners and glory to American ship-builders. These clippers were the very acme of wooden sailing ships.

Probably originating from the two-masted Yankee schooners of the 18th century, which skimmed over the water, the clipper was distinguished by its sharp, long bow, its hollow lines, and its raked masts. The emphasis on speed, required for long-distance trade with China or travel around Cape Horn to the California Gold Rush, meant that the skippers put on as much sail as possible, rather than "snug down" for the night by reducing the sails during the evening hours. Some of the clippers were very large, exceeding 2000 tons, and could be driven through any kind of weather; their speed was phenomenal compared with the older ships, being the only sailing ships known to have made 400 sea miles in a single day. They marked the climax of the development of the sailing ship, for even as the clippers were being designed and built, steam was already beginning to play a large role in transport along inland waterways, and it was but a matter of time before it could compete effectively even on long oceanic shipping routes.

American builders, of course, were perfectly capable of building steamboats for marine service. This industry was concentrated in New York City, much as the clipper-building was concentrated in New England. Charles H. Crump, after the Civil War, a leader among American iron ship-builders, thought little of New York's dominance, claiming that the marine engineers concentrated in New York were "adherents of the paddle-wheel, walking-beam type of engine, and nothing would do but that type of engine. . . . [They] spoke of propeller engines with the most profound contempt." Although Philadelphia builders made propeller engines, their New York competitors were able to dominate the field with their conservative ideas because New York dominated the steam fleet—and indeed all the marine commerce—of the nation.

While ship-builders continued to work in plentiful, workable, and familiar wood, the first iron vessels in the country were being built in Philadelphia by steam engineers and boiler-makers accustomed to working in iron. The first iron boat in this country was the *Codorus*, 60 feet long, built in 1825 by John Elgar, a Pennsylvania boiler-maker. By 1840 it is estimated that some 100 of these vessels had been constructed in the United States, most at Pittsburgh for use on the canals of Pennsylvania. In addition, John Laird of Birkenhead, sent over at least seven from England. In 1841 the United States government ordered its first iron boat, the *Michigan*, which was to have a long career on the Great Lakes. All of these boats, however, were considered to be more experimental than compelling instances; wood and sail continued to dominate, especially on the high seas. Nevertheless, experience was gained by the new industry which was clearly established along the Delaware River: between 1845 and 1857 at least 60 vessels were built, of which 35 were made at Wilmington, Delaware, 13 at Philadelphia, and only 4 at New York, the center of wooden ship-building.

Even when the American merchant marine was at its height, during the 1850's, it was becoming technologically obsolete. The British fleet was rapidly being taken over by iron vessels, although the significance of this development was lost on many even in England. In 1860 Sir John Pakington, First Lord of the Admiralty and president of the newly formed Institution of Naval Architects, warned that his nation was falling behind the Americans, who were building clipper ships and who had won the America Cup for racing with the famous yacht *America*.

The formation, in England, of a new and separate professional society of naval architects pointed up one of the major weaknesses of the American industry. It had been the habit, among wooden ship-builders, to work only from a model whittled from solid wood. As in most ancient crafts, the artist's sense of fitness and proportion was the critical factor in producing excellent work. This system, which seemed to work well enough in wood, was totally inadequate for working with iron. There it was the practice to design as one went along and it was not uncommon for up to one-third of the iron plates used to be wasted by cutting and punching holes in the wrong places. As late as 1860, when the United States Navy asked for bids on iron vessels, only one out of thirty proposals submitted was accompanied by detailed plans. The need of the consumers (navies, shipping lines, and private yachtsmen) to know what they were paying for finally forced iron ship-builders to make plans and work from them.

The importance of iron for ships was finally called forcefully to the attention of Americans when the Southern Confederacy converted the burned-out *Merrimac* into an ironclad ship during the Civil War. Faced with the large-scale destruction of her wooden ships by the *Merrimac*, the Union government promptly ordered a large number of iron-built *Monitors* and purchased all the iron boats available in the North. The resulting American iron ship-building industry, however, did not survive the accelerated demands of the Civil War, and immediately after the war it fell into the doldrums.

A revival ensued in the late 1860's, and the Delaware River came to be known as the "American Clyde," after the British river which was the world-wide center of iron ship-building. The honorary title referred only to quality, however, not to quantity. The introduction of labor-saving machinery, especially in the handling of materials, helped bring the costs of American construction closer to a competitive level, but total volume remained small. A revival of naval building by the United States government in the 1880's helped more, and brought further advances in technique—the first naval vessels in the country to be made of steel. In 1887 the Navy also began to order quantities of the new steel armor plates—a demand which forced an increase of capacity and new knowledge upon the American steel industry. Steel naval vessels were, however, still relatively exotic, and it is a reflection of the general state of American society that the iron and steel ship-building industry depended frequently upon orders for yachts from the new breed of millionaires which the free-for-all economic system was spawning. As with iron, the use of steel by American ship-builders lagged behind British practice.

CONCLUSION

In the half-century from 1830 to 1880, the railroad and the iron ship were developed to become—what they have largely remained since—the workhorses of transportation. In the process they served two vital functions: first, they provided the transportation needed to complete and to make effective the Industrial Revolution; second, they in turn greatly stimulated that revolution by their nearly insatiable demand for the very products of that revolution, particularly coal and iron. In the United States, for example, it has been estimated that in 1856 some 38 per cent of the wrought iron consumed in the nation went into rails. During the railroad boom of the early 1880's an incredible 90 per cent of the rolled steel in this country went into rails. When rolling stock, steam engines, and sea-going vessels are added, it is obvious that the iron and steel production of the 19th century was intimately tied up with the expanding transportation network.

The social and political consequences of the railroad and steamship matched their technological impact. Quickened transportation served to tie nations, and the entire world, more closely together. Indeed, the completion of the trans-continental railroad in the United States had an incalculable effect in making possible the westward movement and tying the country together into truly united states; the railroads provided the technological sinews for an organically integrated nation. Similarly, the railroad system was to help the new German Empire, created in 1871, become a unified nation economically as well as politically.

The basic transportation patterns and modes of the Western world were revolutionized by the railroad and the iron steamship. In 1880 it no doubt seemed that the "modern world" had arrived; that progress, which was universally worshipped as the source of this miracle, could hardly be expected to overturn it within an equally short time. Yet this is, in effect, what was to happen. In the half-century following 1880 the internal-combustion engine in automobiles, trucks, and airplanes, created its own network of transportation, sometimes supplementing, often conflicting with, those of the railroad and merchant marine. But the complex transportation needs of an industrialized world assured that the railroad and the ship would retain their significance.

26 / The Beginning of Electricity

BERN DIBNER

Electricity is one of the first, most welcome, and most gifted children of science. While most areas of technology—metalworking, agriculture, warfare—were improved during the 19th century by the growing ability of science to organize and explain physical phenomena, the electrical industry, from the beginning, was especially dependent upon the increase of scientific knowledge and the research of scholars. From Gilbert's investigation of magnets in the 16th century down to the latest researches of our own day, science and technology in this field have been virtually inseparable. If science owed anything to the steam engine, it paid its debt to technology with electricity.

THE HERITAGE OF GILBERT

While it is true that some ancient Greeks had noticed the power of attraction of rubbed amber (Greek, *elektron*), no advances were made in amassing knowledge of the properties of rubbed substances until the year 1600, when a modest but important book was published in London. This was the *De magnete* of William Gilbert (1544-1603), physician to Queen Elizabeth I, and one of a company of men whose restless nature helped carry the flag of England around the world and who, at home, worked in science, philosophy, and letters on a scale comparable only to the age of Pericles in ancient Athens. Behind Gilbert stretched more than two millennia of indifference and superstition with regard to the emission of sparks or the crackling noises heard when certain garments were rustled. For example, sailors had viewed with superstitious awe "St. Elmo's fire" (fireballs) at the tip of a ship's mast or at the ends of the rigging, but no one could come forward with a rational explanation of this natural phenomenon. Gilbert's approach was fresh and, for his time, thorough, although the chapter in his book devoted to electrical phenomena was primarily intended to show the distinction between electrical attraction and magnetic attraction, his primary interest. Yet this was the introduction of literature on electricity and the beginning of proper analysis of electrical behavior.

In his book, Gilbert not only presented the most comprehensive study that had yet been made in magnetics, but also used for the first time the procedure of thoroughly analyzing one area of physical phenomena by a series of experiments. It was this procedure—inquiry through the "scientific method"—that helped establish his century as the threshold of modern times. At the expense of most of his fortune and eighteen years of work, Gilbert transformed the mystical and often contradictory notions regarding "electrics" and "magnetics" held by the natural philosophers and navigators of his time into a body of data proved by his repeated experiments, confirming by many trials the results of his tests. A sound body of knowledge thus began to replace the age-old accumulation of mystery, fancy, and faith regarding the magnetism of the lodestone and the attraction of light bodies when rubbed by amber.

The next major advance in electrical knowledge was brought about by Robert Boyle (1627-91), who experimented in many new fields of science. To the work of Gilbert, he added new electrical relationships which he published in 1675 in the first electrical tract in English. Boyle, one of the earliest experimenters with the vacuum pump, performed several electrical experiments in a vacuum which he created.

After the first electrical indicator devised by Gilbert, the earliest operating mechanism is credited to Otto von Guericke (1602-86), long-time burgomaster of Magdeburg and experimenter in pneumatics. The electrical machine that Guericke built was described in a book he published in Amsterdam in 1672. Guericke cast a sulfur globe on a shaft that could be turned on its bearings. When so turned, and with a dry hand held on the sulfur globe, an electrical charge would gather on the globe's surface. If the globe on its shaft were then lifted from its supports and carried about the room, Guericke observed that it would attract a feather floating in the air. He also noticed that on its touching the globe, the feather would be immediately repelled. Most important, Guericke must be credited with having noted small sparks when the globe discharged and having heard their crackling sound. (These observations are the first credited to man-made electricity.) He also observed that an electrical charge travelled out to the end of a linen thread and that bodies became charged when brought near to an already charged sphere. We thus have instances of both electrical conduction and induction, fields investigated by later electricians. These were small steps indeed but the evolution of technological expansion in electricity began with such trifling phenomena, puzzling to an inquiring mind.

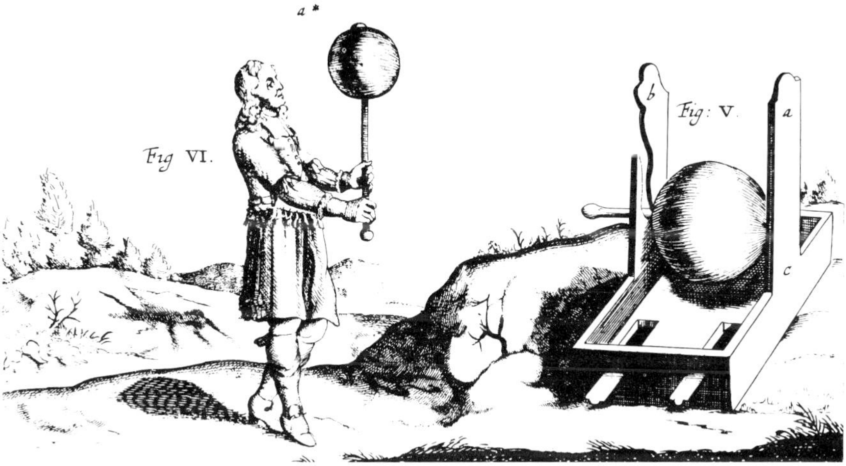

Fig. 26-1. In his studies in attraction and gravitation, Guericke devised the first electrical generator. When a hand was held on a sulfur ball revolving in its frame, the ball attracted paper, feathers, chaff, and other light objects. The ball could be carried about the room to attract or repel a feather (a). (From Otto von Guericke, *Experimenta nova,* 1672. Courtesy of *Burndy Library*)

Much of the earliest experimenting with electricity was done in England by men like Stephen Gray and Francis Hauksbee; but science knows no geographical borders, and experimenters were soon active in France and Germany. When the physiological effects of a shock were first noted, the application of electrical charges to the human body for therapeutic treatment became inevitable. Not only were muscles contracted by the charge but there seemed to be an increase in the pulse beat and an acceleration of the circulation of the blood. Congestive ailments, fevers, and lameness were electrically treated. Some quacks even claimed that the electric charge was a panacea for all of man's ills. Yet it was to be some time before great progress was made in understanding the nature of electricity or in getting it to perform work.

A major advance on the extended road of electrical evolution was made in Leyden in 1746 when the genial professor Pieter van Musschenbroek (1692-1761) announced that an electrically charged glass jar partly filled with water and held in one hand drew a charge from a gun barrel and delivered it suddenly to the professor with "such violence that my whole body was shaken as by a lightning stroke." This was the Leyden jar, which consists of a glass jar coated inside and out with tin foil and having a metallic rod attached to the inner foil lining and passing through the lid; it serves as a condenser for static electricity. With further experiments and resulting improvements in the shape and lining of the jar, heavier charges could be contained. Leyden jars were

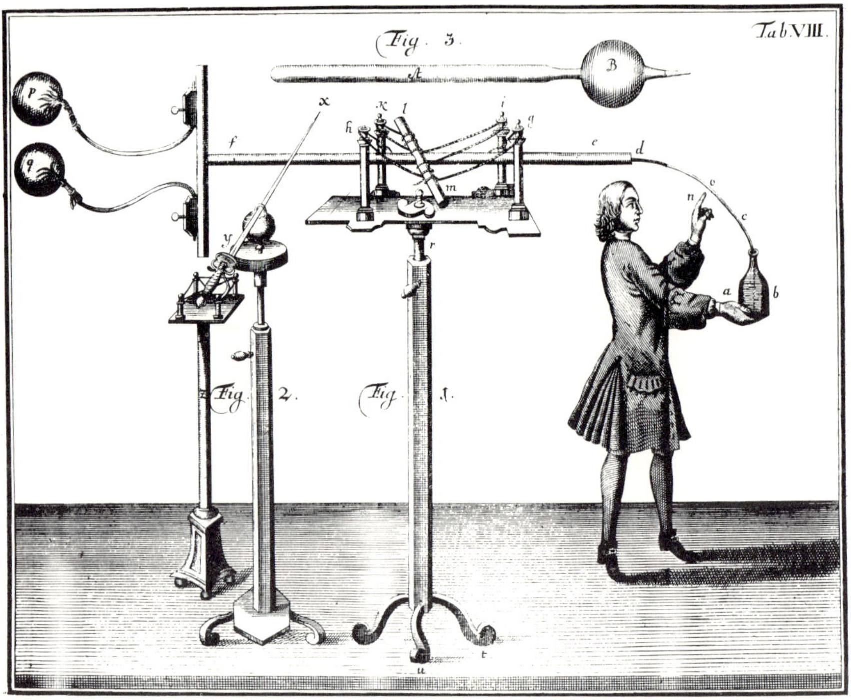

Fig. 26-2. The earliest illustration of a Leyden jar for retaining electric charges. The charges developed from the rotating glass globes, left, were transmitted through the central conductor, and led down a wire to the bottle, partly filled with water. Held in one hand, the other hand equalized the circuit—with a resulting shock. (From J. H. Winkler, *Staerke der Elektrischen Kraft,* 1746. Courtesy of *Burndy Library*)

gathered into groups (called batteries), set on metallic bases, and arranged with multiple connections, thereby further improving the discharge power. With higher potentials, longer distances were tried over which a discharge would register. Dr. William Watson (1715-87) traversed a range of 1200 feet across the Thames River in England in 1747 and later successfully tried an 8000-foot circuit using the earth as a return. At such distances Watson was tempted to determine the velocity of electrical conduction and concluded that it was "nearly instantaneous" over a four-mile circuit.

With the Leyden jar as a new storage facility, the experimenters on both sides of the English Channel turned to improving further the electrostatic generator by which the jar was charged with electricity, and which consisted basically of a device to rub a glass sphere with a non-conductor such as leather to develop a static charge. The spherical body gave way to cylindrical shapes, and glass and heavy cardboard discs sometimes replaced these. The leather rubbing pads were treated with amalgams of tin and mercury, and Joseph Priestley in 1775 visualized a great machine of 20 or 30 globes charging heavy electric batteries, sufficient to give "a shock as great as a single common flash of lightning."

FRANKLIN AND LIGHTNING

The Thor who was destined to "snatch lightning from the sky and the scepter from the tyrant" was the Philadelphia printer-philosopher and electrical experimenter, Benjamin Franklin (1706-90), whose electrical interests began when he attended a lecture and demonstration in 1744 in Boston. So immersed did Franklin become in his electrical experiments that he retired from his lucrative printing business and in the late 1740's devoted himself entirely to his scientific studies. Moving from the simpler apparatus such as rubbed glass tubes to the more complex spherical and cylindrical revolving machines, Franklin found his keenest attention drawn to the performance of the Leyden jar. He became convinced that in a charged jar it was the glass of the jar rather than the outer foil or the inner liquid that held the charge. Further analyzing these charges, Franklin was led to the conclusion that the old notion of two kinds of electric fluid could be better explained as a state of imbalance, positive or negative, from an otherwise normally neutral electrical state. Franklin's theory that electricity consisted of two conditions of a single fluid replaced the idea of two different kinds of fluid, answered many questions otherwise explained with difficulty, and brought his name to many science circles. His ideas were made public in England in 1748 and served as the beginning of much fruitful correspondence with his English and French colleagues.

Following his "one fluid" theory, Franklin's most important scientific achievement was the realization and experimental proof that lightning was an electrical manifestation. Franklin first mentioned this hypothesis in 1749 in one of his many letters to England; it culminated in the famous kite experiment in June 1752. His letters were gathered into book form and printed in London, and French, German, and Italian editions soon appeared and spread his fame. Franklin's dissertation on the lightning rod as a means for effectively draining the electrical charge from clouds into the earth was confirmed by significant experiments in May 1752 in a field near Paris. In a thunderstorm an attendant drew a blue spark from a 40-foot rod; the spark, by color, smell, and sound, was identified as of true electrical nature. This experiment was followed by erecting a 99-foot rod at a Paris home. Thus, from Franklin, colonial American, originated the practice of using the rods which in untold numbers have everywhere protected structures against destructive lightning.

Once it was shown that knowledge of electricity could be turned to practical use, electricity became the subject of wider experimentation and of correspond-

Fig. 26-3. Franklin and a section of the lightning rod erected in his Philadelphia home in September 1752. Shown, left, are two bells and the silk-suspended clapper between them, also the two-ball electroscope attached to the right bell. Divergence of the balls indicated a charged cloud overhead. With this apparatus Franklin discovered that most clouds were negatively charged and that " 'tis the earth that strikes into the clouds, and not the clouds that strike into the earth." (From a painting by Mason Chamberlain in 1762, engraved by C. Turner. Courtesy of *Burndy Library*)

ence between experimenters. Larger and more powerful machines were built and larger batteries were assembled to hold the heavier charges. This culminated in a giant machine having glass discs 65 inches in diameter and mounted on an impressive insulated stand with very large brass standards to receive the electrostatic charges and batteries totalling 135 Leyden jars, sufficient to generate a spark two feet long. This monumental apparatus was constructed by Dr. Martinus van Marum in 1784 and may still be seen at the Teyler Museum in

Haarlem, Holland. With this powerful generator, metal wires were fused, wood blocks were sundered, and iron wires were magnetized or had their magnetism reversed.

VOLTA'S CONTINUOUS CURRENT PILE

As America and the nations of Europe moved toward the close of the 18th century in a blaze of revolutionary thought and experimental attainments, the work of the electricians of the 1600's and 1700's closed neatly on a new stage of electrical development. The precise event was the reading on June 26, 1800 of the first portion of a paper in French sent to the Royal Society of London from Como, Italy. The author was Alessandro Volta, a member of the Royal Society and professor of physics at the University of Pavia. This paper announced the results of several years' investigation by Volta of a phenomenon first noticed by his compatriot, Professor Luigi Galvani, a physician of Bologna. Volta's announcement contained the seed that was to blossom into the electrical age.

The new branch of scientific knowledge began quite unexpectedly in the laboratory of Galvani who, as an anatomist, was dissecting a frog when he noticed a violent twitching in the severed lower half of the frog. This was a most unexpected motion, since the frog had been dead for some time. On retracing the steps that caused this strange phenomenon, Galvani observed that a nearby electrical machine had been in action but was in no way connected with the frog. Further investigation showed that merely using the scalpel under certain conditions induced twitching even without the use of the electrical machine. More experimenting prompted Galvani to conclude that any electrical action in the vicinity, such as lightning, could also affect the frog's legs. There followed a comprehensive series of experiments with frogs that broadened into the use of warm-blooded animals, all of which responded to electrical stimuli in various ways. The careful and retiring Galvani realized he had come upon an important discovery which resulted from experiments that stretched over a half-dozen years and culminated in the publication of his findings and theory in the *Transactions of the Science Institute of Bologna* in 1791. An offprint of this important paper was sent by Galvani to Professor Volta and this paper stimulated an entirely new and revolutionary trend in electrical discovery.

Volta, on reading the Galvani paper, at first agreed with Galvani that the manifestation of muscular action in a sectioned frog might indeed indicate what Galvani termed "animal electricity." However Volta, a physicist rather than an anatomist, soon became doubtful of Galvani's conclusions and thereupon began a series of experiments of his own. Turning his attention from the frog to the other elements that stimulated the action, Volta concentrated on the possibility that the metal elements that touched the frog's nerves might be the source of the action. After two years of experimenting and testing, he became convinced

Fig. 26-4. Volta's crown of cups. The "pile" of pairs of zinc and silver discs that formed Volta's first electric battery was modified to form a series of cups filled with weak acid. These were connected by metallic strips terminating in alternate plates of silver (A) and zinc (Z). Reduced evaporation extended the battery's usefulness. (From *Transactions of the Royal Society*, June, 1800. Courtesy of *Burndy Library*)

that "animal electricity" should more correctly be expressed as "metallic electricity." To Volta it became evident that it was the contact of two dissimilar metals, moist and touching at one end, and in contact with the frog's nerves and muscles at the other ends, that caused the twitch. Experimenting further, he concluded that certain pairs of metals were more effective in their stimulus than other pairs and after this investigation he arranged the metals in a series of relative effectivenes that evolved into the modern electrochemical series.

This difference of interpretation between the natural electrical property of animal tissue as held by Galvani and the purely metallic nature of Volta's electrical source brought on a controversy in which whole schools of scientists took part with considerable vehemence and which ended with the retirement of Galvani and the announcement by Volta of his electrochemical pile.

As described by Volta in his classic paper, the most effective combination of elements in his pile consisted of pairs of zinc and silver discs separated by brine-soaked cloth or paper. An assembly of 30 such pairs was sufficient to cause a sensation of continuous flow of electric current through a person placing a hand at each end of the pile of discs. Because the current tended to decrease with the drying of the wetted cloth separators, Volta devised his "crown of cups," which consisted of a ring of cups filled with brine into which were placed alternating strips of silver and zinc. This form gradually evolved into various forms of the wet battery, which provided a continuous flow of current rather than the single instantaneous discharge of the earlier electrostatic generators.

FIRST USES OF THE WET BATTERY

Volta's masterful invention was immediately put to experimental use in England and on the Continent. Nicholson and Carlisle assembled a voltaic pile immediately after the necessary elements were announced and while experimenting with it they observed that a gas formed when a drop of water was placed where a wire joined one of the battery terminals. They thereupon carried brass wires into each end of a glass tube filled with water and after several hours noticed that a gas had gathered at one of the terminals. This gas (hydrogen) was mixed with an equal volume of air and exploded by a waxed taper. Refinement of the apparatus showed that oxygen collected at one terminal and hydrogen at the other, thereby proving that the electric current was decomposing water into its elements.

Later the effectiveness of the pile was improved and its active life extended. With the improved battery, copper and silver were electrodeposited on baser metals, and an important industry was born. Humphry Davy at the Royal Institution of London built a 2000-plate voltaic battery and with it extracted a new metal from caustic potash to which he gave the name *potassium*; he followed this discovery by isolating *sodium* from caustic soda. These discoveries were followed by the reduction of the metals calcium, magnesium, barium, and strontium from the alkaline earths. In 1809 and 1810 Davy opened a new chapter in electrical engineering by drawing an electric arc with charcoal electrodes which created a "light which was so intense as to resemble that of the sun, [and] produced a discharge through the heated air of nearly three inches in length, and of a dazzling splendour." This was the introduction of the first practical electrical illumination that was improved and remained in use for nearly a century. It also provided an electric furnace that reached heats of hitherto unattained intensity.

The threshold that separated the electrical practices of the 1700's and those of the 1800's was indicative of the trend that science and technology were to follow in the succeeding century and a half. The giant electrostatic machines with their revolving glass cylinders or plates creating sparks, halos, and pinwheels were impressive and promising but proved quite useless beyond introducing the new natural force. The electric battery on the other hand became a new research and industrial tool, more prosaic and corrosive, but providing a *continuous flow* of electric current that was soon to be put to a series of remarkable uses.

The ability of an electrostatic charge to travel along a damp string or a metallic wire led investigators to try to determine the velocity of flow of an electric charge. The general conclusion was that it was "instantaneous." With the coming of the new source of continuous-flow electricity the effort to measure its speed became equally challenging. Giovanni Aldini (1762-1834), nephew and champion of Galvani, stretched a wire across the Bay of Calais with underwater return. He concluded that the speed of electric effect moved with astonishing rapidity. Such a unique property of speed of travel lent itself to several technical problems immediately.

In the first two decades of the 1800's, the new electrical force was applied to the ignition of gunpowder by the British, and a Russian diplomatic officer exploded mines under the River Neva near St. Petersburg using an insulated copper wire that was placed underground and underwater. The military potential of such an application was of prime importance. The use of electric currents for some form of telegraph communications also seemed evident, and inventive minds devised several approaches, but a practical system was not hit upon immediately.

OERSTED AND ELECTROMAGNETISM

Incredible as it now seems, during the first twenty years of the experimental and applied use of electric batteries no one, so far as we know, had observed a phenomenon which first attracted the attention of a professor of physics while lecturing at the University of Copenhagen. During those two decades many instances were recorded of attempts to understand more fully the chemical and metallurgical characteristics of the voltaic pile, and to apply its electric current to ever-widening practical uses and philosophical investigations. While demonstrating the properties of the electric battery before his class, Professor Hans Christian Oersted (1777-1851) was distracted by something that had eluded his many colleagues.

Oersted had done no more than to place a conducting wire over and parallel to a pivoted magnetic needle when he noticed that the needle swung perpendicular to the wire. The professor made no mention of this strange behavior at the time, but later he repeated the conditions using a more powerful battery and a larger conductor. The results were the same, and this prompted him to organize the problem. The confirmation was then issued in one of the most famous announcements in the literature of science; on July 21, 1820, a four-page tract appeared in Latin announcing that an electric current in a conductor created a circular magnetic field around the conductor.

The results of Oersted's discovery spread quickly to the important scientific centers in Europe and were confirmed in every laboratory that possessed a voltaic pile. Public demonstrations of the new discovery were held, and at one in Paris in September 1820 an interested member of the audience was the mathematics professor André-Marie Ampère (1775-1836). Ampère saw its significance and within the week presented a paper on the mutual action of two electric currents, the first of his many studies in electrodynamics. He also showed that an electric current both influences a magnet and itself produces magnetism, and that two solenoids carrying currents behaved like true magnets in polar attraction and repulsion. In addition, electromagnets were much more powerful magnetically than lodestones.

With the new knowledge now made available to electrical experimenters, two major paths of activity were opened. The first led to the discoveries by Michael Faraday (1791-1867) of the generation of electric voltages by mechanical rotative and motive power. The second branch of activity led via the electromagnet through the complex system of telegraphy and telephony and into the vast network of communication and broadcasting.

It had taken two centuries to progress from Gilbert's first description of electrical phenomena to Volta's announcement of continuous-current electric generation; and it required two decades (a remarkably long period) for Oersted to discover the link between electricity and magnetism. The next major step in electrical development occurred only eleven years later.

FARADAY'S ELECTROMAGNETIC INDUCTION

With Oersted's announcement that a circular magnetic field surrounded a wire carrying an electric current, the scientists and mathematicians began to plot the extent and relationships of this new tool of investigation. William Sturgeon in England in 1825 wound 16 turns of an electric wire around a soft iron core and produced an electromagnet. In Albany, New York, the teacher and experimenter Joseph Henry (1797-1878) applied a silk winding around the conductor and closely packed many turns about the iron core to produce electromagnets capable of holding 3600 pounds. While using an electric battery no larger than that with which Oersted first discovered electromagnetism, Henry's electromagnet supported a ton of iron.

This very impressive production of magnetic effects by use of electricity prompted many investigators to seek some method of reversing the conversion, producing electric effects by use of magnetic effects. It remained for the humble laboratory assistant at the Royal Institution in London, Michael Faraday, to show the way to this epochal attainment. This was no easy step because the best experimental minds had been directed for more than ten years on this effort. Faraday, doggedly seeking means of converting magnetic energy into electrical energy, succeeded only in arranging two glass vessels nearly filled with mercury in which a pivoted, movable magnet was made to rotate about a fixed electrical conductor dipped into the mercury. At the same time a movable conductor was able to revolve about a fixed magnet resulting in what Faraday called "electromagnetic rotations." However, ten years after this demonstration, in the summer of 1831, the fifth of yet another series of investigations began in the effort to produce electric effects by use of magnetic effects.

In October 1831, using an iron ring upon which two coils of wire were found, one connected to a voltaic battery and the other to a galvanometer (an instrument, named after the famous Galvani, for determining the direction and intensity of an electric current), Faraday noticed that when the battery circuit was closed, the galvanometer needle swung out and returned to the zero position; when the circuit was broken, the needle swung in the opposite direction and returned. Pursuing this lead and mounting a copper disc between the pole pieces of a powerful magnet, Faraday showed that a steady electric voltage was generated between the outer rim and the axle of the disc when it was rotated. A new link in the lengthening electric chain was therewith forged, for Faraday showed that when an electrical conductor continued "cutting the lines of magnetic force," an electric voltage was generated in the conductor. In feverish haste and completely exhausted by ten intensive days and nights of effort to complete and record a series of confirming experiments, Faraday, on November 24, 1831, read a paper before the Royal Society describing his discovery.

Having shown that electric voltages were induced when either the magnet or the conductor moved in relation to the other, Faraday and others followed with a series of improvements to the initial apparatus. More powerful electromagnets replaced the original bar magnets and wire loops replaced the copper disc; the loop ends were terminated with rings which provided improved current collection. It was observed that the current in the loop appeared as alternating pulses as the wires alternately passed the north and south poles of the magnet. Faraday countered this by splitting the terminating rings into a true commutator and thus provided all the elements of a practical direct-current modern generator. From here on, design moved into the refinements of improved materials, narrower magnetic gaps, carbon instead of metallic collecting brushes, multi-polar yokes, and better revolving armatures with copper electrical

Fig. 26-5. Joseph Henry's electromagnet. Direct current from the voltaic pile was applied to a coil wound around an iron horseshoe core to produce a powerful electromagnet. It held over a top of suspended iron. (From Joseph Henry, *Galvanic Multiplier*, 1831. Courtesy of *Burndy Library*)

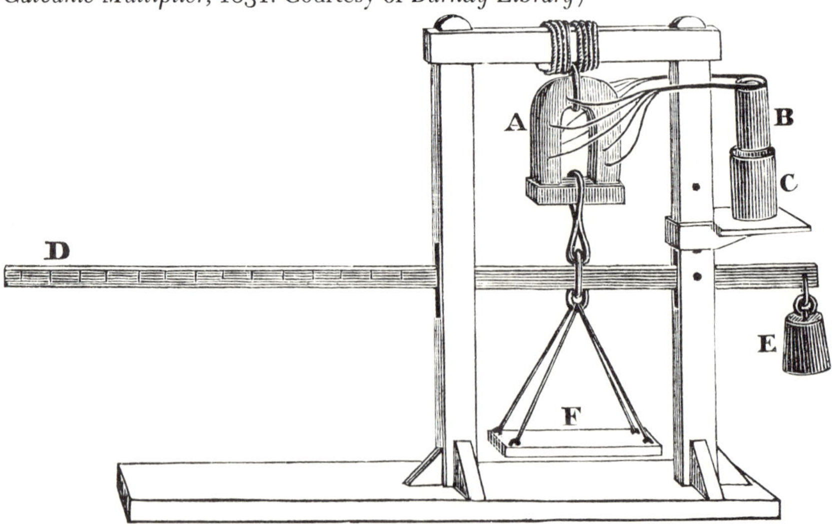

Fig. 26-6. Faraday's electric generator. Faraday first devised the means for generating an electric current by moving an electric conductor in a magnetic field. This simple generator evolved into the great alternators of today. (Courtesy of *Burndy Library*)

and iron magnetic circuits. Acting under his own maxim of "work, finish, publish," Faraday published the first of twenty-nine reports on his researches that covered the period 1832 to 1852; this constitutes one of the richest sources in scientific literature.

There now existed two prime sources of electrical energy, the electrochemical cell that had evolved from the voltaic pile, and the electromagnetic generator that evolved from Faraday's rotating disc. The further evolution of each of these devices depended in large measure on the practical use which was made of them, and, of the two, the success of the cell and the battery of cells came earlier because they were used in the simpler electrical telegraphy system. It took, on the other hand, half a century to turn Faraday's laboratory generator into a functioning central power-station device.

THE BATTERY

As a philosophic instrument, Volta's pile was impressive enough but its practical limitations soon became apparent. The electrolyte into which the metal plates were dipped tended to dry out and the acid corroded the metals. To avoid the constant corrosive action, some electricians devised the "plunge" battery, which permitted the plates to be withdrawn from the electrolyte when not in use.

Impurities in the metals set up local electrochemical action that corroded the metal away; in some types of battery mercury was applied to the zinc electrode in order to extend its life. It was also found that a gaseous or solid film tended to cling to some electrodes, acting as insulation and thereby constricting current passage. A step to prevent polarization by contaminants was to introduce a membrane or unglazed porcelain cylinder that held two distinct electrolytes. This concept was modified into the "gravity" cell in which two fluids of different densities reduced polarization. An improved form of cell, the use of which continued into the 20th century, was one in which a charcoal rod was placed in nitric acid and a zinc plate in dilute sulfuric acid. Gaston Planté introduced the first storage battery in 1859. In this improved device, a layer of lead oxide formed in the lead electrodes that were immersed in dilute sulfuric acid; the battery performance was further improved by applying a paste of red oxide of lead onto the plate surfaces.

The essential parts of an electrolytic cell (called a "primary" cell) consisted of two dissimilar electrodes immersed in an electrolyte of some acid, alkali, or salt dissolved in water. In this form of "wet" cell, electric current was generated by the chemical action of the elements. A "dry" version of a primary cell was one designed for use until exhausted at which time it was discarded; a wet cell could have elements replaced. The introduction of storage cells (called "secondary") was made possible by the reversible chemical reactions of its elements permitting the cell to be recharged and used again. The recharging could be done from primary batteries or, later, by generators. Their main use was in telegraphy, for short trolley runs, train lighting, and telephony. After 1894 the storage battery was introduced in automobiles.

Thomas A. Edison (1847-1931) added a novel battery to his many inventions. After thousands of trials, he developed a battery having nickel hydroxide as a positive plate separated by an alkaline electrolyte from an iron oxide negative. This highly successful device was further improved and applied widely throughout the world in industrial trucks, stand-by power supply-systems, automatic instrumentation, mines, and portable instruments. As battery efficiency increased and size decreased, new applications were found. Today's nickel-cadmium batteries using potassium hydroxide as an electrolyte find application in military, aircraft, and space use. They can be recharged hundreds of times, can operate in temperatures down to $-40°$ F and can be permanently sealed. They are used in hundreds of kinds of portable equipment as a source of power, of control, and for communications.

DYNAMOS AND MOTORS

The commanding importance of the telegraph in the early years of electrical application led to a concentration of research on the battery as a power source, but the discoveries of Faraday and Henry were by no means ignored. With the increase of interest in electric illumination (Chapter 34), work pressed forward more rapidly on the generator.

Modifications in Faraday's electric generating device improved one of the two weak elements of the illuminating system then being worked on, for by 1853 the first attempts had been made to feed an arc lamp by a magneto generator (a dynamo in which a permanent magnet produces the magnetic field) rather than battery. The earliest application of the new arrangement was for lighthouse illumination on the Straits of Dover in 1862. By a series of improvements to the generator, such as replacing permanent magnets with electromagnets, improving armature design, narrowing the gaps between pole-pieces and armature, and shortening the magnetic circuits, more reliable equipment was produced, especially in France. By 1882 England had five lighthouses and France had four equipped with electric arcs fed by rotating generators, driven by steam engines.

Self-excitation of the magnetic poles further improved the generators, and the substitution by Siemens in 1856 of the drum armature for the earlier disc further raised the generator efficiency and reliability. Z. T. Gramme's ring armature of 1867 greatly advanced the generator art and proved his ability as a designer and his skill as a promoter. The improved self-excited generator was now termed a "dynamo-electric machine," later shortened into "dynamo," a term that has persisted for a century.

In the evolution of more efficient dynamos it was discovered in 1873 that when a larger dynamo was electrically connected with a smaller one and supplied with voltage from it, the latter would revolve and act as an electric motor. By extending the wires of the interconnection, the motor, fed from a dynamo, could be installed at a considerable distance; this provided for the electric transmission of power which was rapidly applied to traction, machine tools, pumps, and for general utility. This also led to the development of the electric elevator, that element essential to the skyscraper. Such broad application of power in turn signalled the need for more and better generators.

GROWTH OF THE ELECTRICAL INDUSTRY

Older industries such as agriculture, forestry, mining, and manufacturing felt the quickening touch of the Industrial Revolution and surged forward to new levels of activity under the stimulus of improved technology, but one industry, the electrical, was entirely new and fully developed in only one century. The industry began with the simple battery of Volta in 1800 and built thereon with the inventions and discoveries of the electrical pioneers; by the close of the last century investments in the electrical equipment of 1899 totalled four billion dollars. This phenomenal growth has continued at the same rate into the present century.

While each move forward consumed the genius and energy of individual experimenters (and later, teams in research laboratories and entire organizations), greater honor must yet be accorded the work of the pioneers—Gilbert, Franklin, Volta, and Faraday. Their careful experiments were the seeds that sprouted into the intricate, multibillion-dollar electrical industry, still young, but indispensable to our entire society.

27 / Communications

BERN DIBNER

The ability of man to communicate with ease helps to distinguish him from all the other creatures that inhabit this planet. The pattern of development of methods of communication can well form an index of his individual and social progress. The extent to which he has been able to communicate, store, and retrieve information is the bold and measurable mark of his progress in recent centuries.

Signal codes were developed early in man's history and these helped lead to systems of graphic communications and counting. Beacons and crude semaphores were highly developed in Roman times; Saracen and Crusader transmitted intelligence during their military campaigns. Ships at sea and armies on land maintained elaborate systems of communication and command. Each system, however, had its limitations: mist, darkness, and terrain interfered with and disrupted communications; and in the case of the military signal systems, the enemy could easily intercept the visual signals and devise counter-measures.

THE ELECTRIC TELEGRAPH

Telegraphy had begun with visual signals by day and the light of fires at night, usually for the transmission of military messages. In 1792 Claude Chappe had introduced his optical telegraph in France, and his network of visual semaphores spread with the advances of the French Revolutionary armies. It was the Chappe semaphore telegraph that alerted Napoleon to the disaster at Munich and caused him to leave for Bavaria in time to command the city's liberation. The Chappe semaphores had to be placed at high points for maximum visibility; over such a system messages could be transmitted for 150 miles in a quarter of an hour. The semaphores were spaced at about fifteen-mile intervals, and similar devices were used in New England harbors to signal the arrival of fresh cargoes.

The Chappe semaphore telegraph, the most advanced in its time, was nevertheless limited by storm, mist, distance, and its relative slowness of transmission. In the mid-1800's the electric telegraph was developed; its value was so quickly

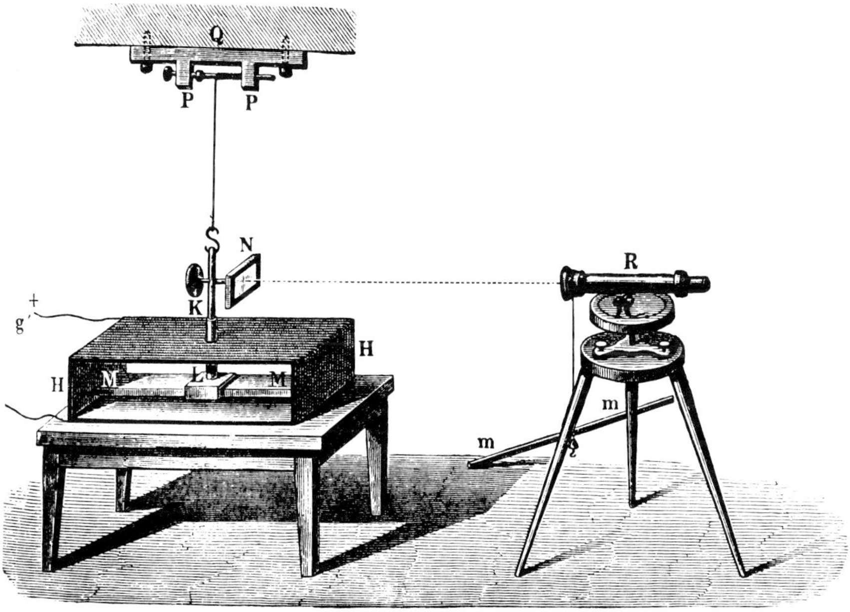

Fig. 27-1. The Gauss-Weber telegraph. Two experimenters in Göttingen connected a magnetic laboratory and an astronomical observatory by an electric telegraph in order to communicate more readily. An electromagnetic impulse on a suspended coil and mirror was observed through an optical magnifier. (From G. B. Prescott, *Electricity and the Electric Telegraph*, 6th edn., 1885. Courtesy of *Burndy Library*)

realized that it became universal in only a few years. The electric telegraph was the first great application of the growing body of electrical research of which it was a direct outgrowth.

Again, it was the mechanism rather than the idea which proved difficult. Early applications of electricity to ringing bells by electrostatic discharge and by complex systems of decomposition of water in pre-Oersted days proved ineffective. Causing magnetic needles to move by systems of wires required so many wires that it proved equally impractical. Joseph Henry at Princeton successfully constructed a telegraphic device in which an electromagnet drew an armature which rang a bell, but the code required to make it practical had not yet been developed. Gauss and Weber in Göttingen, Germany, in 1833 erected a telegraph system between a laboratory and an astronomical observatory, a distance of 8000 feet, over which signals were transmitted. After a considerable amount of introductory work in the development of a practical telegraph system by Schilling and Jacobi in Russia and by Cooke and Wheatstone in England, the result was a five-needle system which was tried in 1837 on the London and

Birmingham Railway. Later this was simplified to two needles which proved practical for railway signalling. A single needle two-wire system which Cooke and Wheatstone devised in 1845 remained operative in England until the beginning of this century.

Cooke's and Wheatstone's device was a direct application of Oersted's discovery that a compass needle oriented itself perpendicularly to a wire carrying a current, and that reversal of the current's direction in the wire reversed the needle's direction. They set up a diamond-shaped board with five magnetic needles spaced across the widest part of the diamond; a separate wire under each needle excited the needle to move to the left or right when a current was passed through it from the transmitter, depending on the direction of the current. When current was sent through two wires at a time, two needles moved, and the point where these two needles intersected indicated the proper letter or number written on the diamond-shaped board. Although this device was simple to operate, for the operator need only know how to read the alphabet rather than be trained to understand a code, it had the disadvantage of requiring six wires (the five wires to the needles plus a return wire) between the transmitter and receiver.

In America, efforts were made to parallel the telegraph developments of Europe by using the many combinations of mechanical and chemical writing devices that ticked, scratched, perforated, and inked electromagnetically operated keys and magnets. The simple equipment of the American painter-inventor, Samuel F. B. Morse, finally triumphed. The essence of Morse's invention was his code. The transmitter emitted an electric current which could be of short or long duration (dot-dash), which in turn energized an electromagnet to push a pencil to make a mark on a moving paper tape. Although more complicated to operate than the Cooke-Wheatstone device, for it required training operators in the code, the Morse telegraph was more rugged and much cheaper to construct. Basically, the Morse telegraph was a device for transmitting information by electrical means.

Morse's equipment evolved into an instrument which could, by November 1837, transmit signals for a distance of ten miles. By 1843 Congress had appropriated $30,000 for a line between Baltimore and Washington, over which the now-famous message, "What hath God wrought," was sent. However, this line was rarely used afterward. In 1845 Morse organized the Magnetic Telegraph Co., and the line built between Washington and New York proved profitable. This firm and its early competitors, however, were troubled more by business problems than technological ones.

Within a short time telegraph lines were extended over hundreds of miles, and by September 1846 joined Boston to Washington. The following year saw telegraph lines extended westward to Pittsburgh, Cincinnati, and Louisville. During 1848 every state east of the Mississippi except Florida was connected to

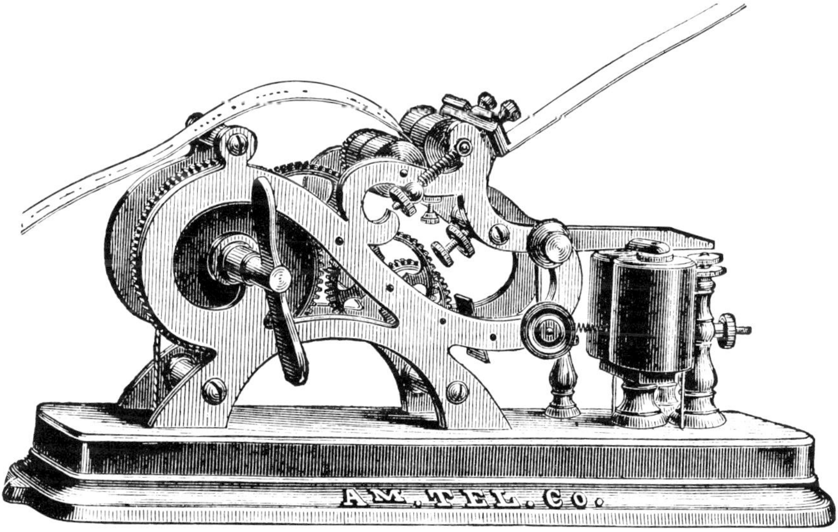

Fig. 27-2. An early Morse telegraph for recording the code of dots and dashes on a moving tape. The electric telegraph in this form was the first important application of electromagnetism and heralded the conquest of space by electrical communications. (From Prescott, *Electricity and the Electric Telegraph*, 6th edn., 1885. Courtesy of *Burndy Library*)

the telegraph network. In December of that year President Polk's message to Congress was directly telegraphed as far west as St. Louis where the newspapers carried it within twenty-four hours of its presentation. The telegraph became essential to efficient railroad operation, and weather reports gathered and communicated by it helped agriculture and commerce.

Political interest in the telegraph grew, and the newspapers used it for quickly gathering and publishing events of the day. The war with Mexico in 1847 turned everyone's attention to news of the campaign as it was flashed over the constantly extending telegraph wires. Wire-stringing techniques improved, as did the setting of poles; the application of insulators and an improved conductor further helped establish the reliability of telegraph lines. The organization of the Western Union Telegraph Company in 1856 lent stability to the system of many individually owned lines with varying standards and practices. The westward movement, the construction of railroads, and the outbreak of the Civil War, each contributed greatly to expanding the telegraph network.

Constant improvements in synchronizing the sending and receiving portions of the several telegraph systems combined to produce apparatus that sent and printed up to 60 words per minute by 1860. President Lincoln's call for troops to repress the rebellious South, on April 15, 1861, was made to all the state governors by telegraph. During the Civil War over six million messages were transmitted by some 1500 telegraphers, and by the end of 1865 more than 200,000 miles of telegraph line were in service. With continued crowding of messages over the wires, duplex and multiplex systems were developed for sending several messages simultaneously over the same conductor. Extension of telegraph lines for long distances became feasible with the introduction of the relay, proposed by Joseph Henry in the early 1830's.

No picture of Wall Street, which grew enormously during the third quarter of the 19th century, is complete without the glass-domed stock ticker. This was a special type of telegraph, invented in 1870 by Thomas Edison, that printed 285 characters a minute. It was Edison's first major invention and helped to finance his later work on the electric light.

THE TELEPHONE

Multiple telegraphy improvement suggested the introduction of transmitting systems employing alternating currents of different frequency with corresponding systems of selection of individual frequencies at the receiving station. Inevitably coarse sounds were produced by some of the frequencies tried, which pointed to the possibility of transmitting the human voice over electric wires.

One experimenter in this line was Philipp Reis, who in 1860 in Frankfurt devised a telephone based on the principle of an imperfect contact made by a vibrating diaphragm intermittently pressing on a metal point in a circuit. The Reis receiver consisted of an electromagnet mounted on a resonator. In spite of its theoretical possibilities the Reis instrument was not successful.

Another experimenter in the transmission of sound was Elisha Gray of Western Electric, who patented several sound-transmitting and reproducing instruments. On February 14, 1876, when he filed a caveat for one of his most important patents, he was astonished to learn that on the same day a patent for a similar telephone mechanism had been applied for by Alexander Graham Bell (1847-1922); this led to long litigation by the inventors and their backers. The Bell patent finally prevailed and has been considered the most valuable patent ever issued by the United States.

By October 1876 Bell and his assistant, Thomas A. Watson, had introduced a sufficient number of improvements in the diaphragm and circuits so as to transmit human speech for several miles. Bell's success stimulated many inventors, including Emile Berliner and Thomas Edison, who in 1877 invented a speaker which consisted of a vibrating diaphragm pressing against a mass of carbon. The vibrating diaphragm changed the carbon resistance and so modulated the current. Protracted patent litigation was inevitable where such a valuable device was involved and so many interests. The original telephones had been set up in pairs for local communication; a single iron wire connected the two phones

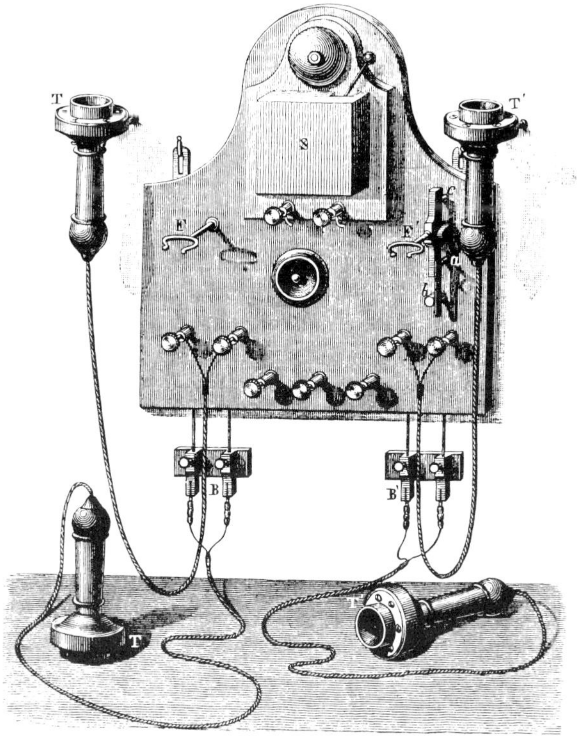

Fig. 27-3. Bell telephone. In the original Bell system, the speaker and receiver were similar. Continued improvements in call systems, exchanges, batteries, and lines made the telephone the most used invention of modern times. (From Du Moncel, *Le Téléphone*, 1882. Courtesy of *Burndy Library*)

with a grounded return circuit, and there were no switchboards to permit communication among a number of users. As the telephone became more popular, a central switching system became necessary, and the Bell Telephone Company expanded the service to interconnect individuals and cities in switching exchanges at an ever-growing pace. Boston was connected to New York and Philadelphia in the middle 1880's and Bell inaugurated the line to Chicago in 1892. By 1900, over 300 phone patents had been granted and over 1,500,000 phones were in service.

THE ATLANTIC CABLE

The failure of overseas communications to get through to their destinations in time has been the subject of more than melodrama. One such tragic example is the Battle of New Orleans (1815), in which more than 2000 combatants were killed, not knowing that the United States and England had signed a peace treaty fifteen days earlier in Ghent, Belgium. Even the fastest ships carrying messages from England to the port cities of the new United States could not have carried the news of the treaty in less than three or four weeks, either directly by sail to New Orleans, or through Boston, New York, or Philadelphia and thence overland by military messenger. Such incidents—and the development of telegraphy—made men dream of the day when the continents could be linked by instantaneous communication.

One of the most brilliant chapters in the history of electrical technology involved a combination of the astute vision and perseverance of a financial promoter, the co-operation of men from two often competing nations, the ingenuity of electrical inventors, and the skill and courage of manufacturers of electrical equipment. The resulting triumph was an operating transatlantic cable that spanned the vast ocean and linked two surging world communities.

As more and more cities and towns along America's eastern seaboard joined the network of telegraph systems, it was inevitable that the terminal reaching farthest eastward would look to its equivalent in Western Europe with the hope that somehow the wide gap would be spanned. Underwater telegraph wires had already been successfully laid between England and France, between Corsica and Italy, and between Prince Edward Island and New Brunswick, in Canada, the first important submarine installation in North America. The chance meeting of a veteran of this last submarine telegraph installation with a still young but retired industrialist Cyrus W. Field (1819-92) brought about the transatlantic telegraph line.

Convinced of the soundness of plans for a transatlantic underwater cable, Field left for England to seek aid in financing the enterprise. After making two voyages to sound the bottom of the ocean, Field again sailed to England in July 1856, where he organized the Atlantic Telegraph Company. The new company ordered 2500 miles of cable to be manufactured under very loose specifications. With little installing or operating experience, the cable fabricators drew and insulated 1200 lengths, each about two miles long, which were then joined. Half of the finished cable was loaded on the American frigate *Niagara*; the other half of the cable was to be carried by the British man-of-war *Agamemnon*. On August 5, 1857, the cable ships drew anchor at the eastern terminal and moved westward paying out their precious line. However, 330 miles out the cable became separated and the ships returned.

For their second attempt the ships were fitted with improved cable and a better means of braking the cable drum to keep it from reeling out too quickly; the two cable ships met in mid-Atlantic in June 1858. Their cable ends were spliced and the ships now moved in opposite directions, the *Niagara* toward America and the *Agamemnon* to the European terminal at Valentia, Ireland.

Fig. 27-4. Laying the Atlantic Cable from the *Great Eastern.* (From Figuier, *L'Elettricità*, 1886. Courtesy of *Burndy Library*)

Again the cable broke, and the ships were reunited to try again. Further breaks occurred, and a third effort was begun in a few days. Several hundred miles of cable were now lowered, and again a break disrupted the operation. A fresh expedition was therefore organized, and by mid-July the squadrons left Cork for their fourth attempt to lay the cable that year. Although plagued by a series of insulation and circuit breaks which were repaired only under great difficulties, and storms that threatened to snap the cable, the *Niagara* on August 5, 1858, connected her cable end at Trinity Bay, Newfoundland, and the *Agamemnon* joined her cable at Valentia. There was celebration in England and America, with Queen Victoria and President James Buchanan exchanging felicitations by direct telegraph. However, after about a month of slow deterioration, the cable ceased to function and the high hopes of the initial success were shattered. Nevertheless, enough of the sweetness of success had been tasted so that new funds and new enthusiasm (most of it generated by Field) led to a reorganized enterprise.

By 1865 a committee of experts had studied the cable and its laying facilities and had rendered a comprehensive report. A new cable was to be built in a single length of 2300 nautical miles; it was fortunate that the ship *Great Eastern* was available to bear such an enormous burden. Launched in 1858, this steamer was five times the size of the largest ship then afloat. For all her size a concentrated burden of 5000 tons of cable remained a major problem; this was solved by building three tanks within her hold in which 2000 tons of water kept the cable immersed. On July 23, 1865, the great ship left Valentia, paying the newer, heavier cable out over her stern. Several interruptions caused by "shorts" in the cable delayed the effort, and with the ship only 600 miles from Newfoundland the cable broke and the end dropped into the sea. Efforts were made to grapple for the cable; it was found and dragged up part way from the two and a half miles of ocean depth. After four failures of the lifting tackle, the effort was abandoned and the ship returned to port.

Again improvements were made on the machinery for letting out and hauling in the cable for the next trial, and better insulation and armor were specified for the new cable. Twenty-four hundred miles of the new cable were stored aboard ship when she left port at the end of June 1866 and in mid-July the fifth and most successful trial was made in laying the transatlantic cable. With only one interruption due to cable fouling, the voyage was uneventful, and on July 28, 1866, anchor was dropped in Trinity Bay. When the cable was connected with the telegraph system, the link between Europe and America was finally established. To add to this triumph of organization and technology, the lost cable of 1865 was grappled and joined to a new section. It was brought to Trinity Bay on September 7, thereby providing a double success for the persevering staff. To help celebrate that momentous event the mayor of Vancouver, in western Canada, cabled the mayor of London informing him that by the existence of the Atlantic Cable the colonial western outpost was only seconds away from Mother England, though separated by over 8000 miles.

PHOTOGRAPHY

The expansion of communication through the medium of the telegraphed word was paralleled by the discovery of practical forms of photography in the 1830's. The forming of pictures by light entering a *camera obscura*, a darkened chamber with a small hole in one wall through which an image is projected onto the opposite wall, had been known for centuries. With these, solar eclipses were observed, and the processes of copying, drawing, and painting were aided by the image (inverted) cast upon a screen. Then the observation that silver salts darkened under exposure to light prompted the first steps toward practical photography. In a series of evolutionary steps—mostly the work of Niepce, Daguerre, and Fox Talbot—combining optics and chemistry with patience and perseverance, light images were permanently fixed, and the process of photography was made public in 1839.

Shortly after 1816 Joseph Nicephore Niepce in Paris had placed a sheet of paper sensitized with chloride of silver in a *camera obscura* and exposed it to a view from his attic window. After exposure of more than an hour the impression was fixed with nitric acid (which in time bleached the picture). After further experimenting with bitumen on glass and attempts at optical photo-engraving on zinc and pewter, Niepce succeeded in fixing a photograph of a view from the dormer window of his home using light-sensitive bitumen on pewter that required an exposure of eight hours of summer daylight. This photograph was probably made in 1826, and in the following year Niepce sent a notice of his success in a short communication to the Royal Society in London. However, Niepce's visit to England to generate interest in his process was not successful. In 1829 he formed a partnership with L. J. M. Daguerre that was to last four years; working with an improved camera fitted with an achromatic lens, they made successful exposures on glass.

When Niepce died in 1833, Daguerre, continuing the work of his former partnership, discovered that the latent image on a plate could be exposed and subsequently developed by mercury vapor. The exposure time was reduced and, in 1837, he developed the technique of fixing the image by washing the plate with hot brine. The first "daguerreotype" was now a reality.

Daguerre's attempts to interest financiers in his invention were not successful until 1839, when Arago introduced the subject to the Academy of Sciences in Paris. Scientists were joined by politicians in acclaiming the marvels of the daguerreotype; painters and engravers saw the doom of their art and craft. Opticians' and chemists' shops were raided by amateurs buying cameras, plates, iodine, and mercury with which to try their hand at capturing Nature. In the first two years, 1839 and 1840, a pamphlet describing the process of Daguerre was printed in over thirty editions in eight languages.

Daguerre personally demonstrated his process of cleaning and polishing a silver-copper plate and sensitizing it by the vapor of iodine, thereby forming a thin layer of silver iodide on the surface to be exposed. On exposure, a latent image was formed and developed by the vapor of mercury, the mercury affecting only those parts of the silver iodide that had been exposed to light. The picture was then fixed with salt (after 1839 with hyposulfite of soda). The plate was finally cleansed in distilled water. Clumsy as the process seems in the light of today's advances in instantaneous photography, it swept the world of affairs. However, while still-life and panorama photographs lent themselves to the lengthy exposures, portraits requiring a sitting of 20 minutes in strong sunlight proved impractical. Inventors therefore applied themselves to shortening the time and improving the quality of the pictures.

We owe to W. H. Fox Talbot of England the introduction of the process of exposing a negative from which an unlimited number of positive prints could be made. In late 1840 Fox Talbot found that by treating exposed photogenic paper with gallic acid the sensitivity was greatly increased and the exposure time could be cut down to very few minutes. By this calotype process, paper was coated with solutions forming silver iodide and brushed with gallic acid and silver nitrate. The treated paper was then exposed and fixed with potassium bromide solution, dissolving the unexposed silver.

Fig. 27-5. This photo portrait was made in 1867 by Julia Cameron; it shows Sir John Herschel, a contributor to photographic progress, at the age of 79. (Courtesy of *Burndy Library*)

In the early 1840's portrait studios were opened in the major cities, for exposure time had been cut from two minutes to ten seconds, and daguerreotypes became practical as well as fashionable. A red light was added to the dark room as it was found not to affect the sensitized plate. Cameras were enlarged and lenses improved for both speed and clarity.

Fox Talbot's introduction of the calotype process using sensitized paper developed into many variations, one of which was the "blueprint." Its simplicity in handling and its fine registry of dark and light lines made it the medium of communication in the construction and machinery worlds. At the suggestion of Sir John Herschel in 1842, paper was sensitized by being coated with cyanogen and iron; its low sensitivity and low price made it ideal for contact prints to reproduce drawings and writing on a semi-transparent positive. The print was fixed simply by washing in cold water. The blueprint and its variations continue as major means of the transmission of all kinds of technical information.

The introduction of the emulsion spread on glass produced the form of negative that made photography such a success in the second half of the 1800's. In spite of the disadvantages of weight and fragility, glass was inexpensive compared to metal, and its transparency was far superior to that of paper with its inherent grain. By 1848 the process had moved out of its experimental stage, and the delicate details possible with a glass negative enhanced this process even further by the unlimited number of positives that were possible from a single negative. The glass plate was coated with albumen (white of egg) containing a few drops of potassium iodide solution, dried and washed with silver nitrate, exposed, and developed with gallic acid.

In 1851 the wet collodion process was introduced which remained the main method of plate production for thirty years. F. S. Archer, a London sculptor, was responsible for this process which involved coating a glass plate with potassium iodide and collodion (nitrocellulose dissolved in ether) and sensitizing it immediately before use by dipping it into a silver nitrate solution, using it while still wet. The result was an amber colored photograph, which was much used for inexpensive portraits.

The simplified collodion process popularized photography in the 1850's and 1860's. Photography became a subject taught at technical institutes, at military and naval academies, in courses for architects and engineers, in medicine and in exploration. Manuals appeared, exhibits were held, and societies were organized to promote the new art. Professional photographers grew in number, and camera manufacturers as well as plate and chemical producers evolved into a new and flourishing industry. By 1861 some 33,000 persons earned their living from photography in Paris alone, and photography became an art for the millions.

The news value of the photograph was first realized in the Crimean War, and the coverage of the American Civil War by Mathew Brady not only brought views of the battles on land and sea back for the news media but has preserved for posterity the most reliable record of events and personalities of the war. Before the war Brady had gathered a considerable fortune from his New York and Washington studios where in 1861 he had taken over 30,000 portraits. He resolved to record the Civil War and covered it with a staff of twenty photographers and photographic equipment that cost him over $100,000. The effort cost him his health and fortune, but also immortalized him.

The change from the wet-plate process to gelatin dry plates began in the early 1880's. It provided means for truly instantaneous photography; exposures required less than a second, and plates and processes became simplified and standardized. It was the introduction of photography on film that provided the most revolutionary step forward in the whole history of photography. The light weight and flexibility of celluloid film as a base for photographic emulsion was introduced by Hannibal Goodwin of Newark in 1888 and was furthered by the energetic promotion of the new photographic film by the Eastman Company which exploited the advantages of the flexible film by making it in film pack form and roll films. This step made possible the reduction of the size and weight of the camera, making it universally popular, and led to its later miniaturized form. George Eastman popularized the handy "Kodak" by simplifying the ten usual exposure steps to only three, further shrinking the camera weight and size, and permitting 100 exposures without reloading.

The extension of the photograph for multiple reproduction on the printed page was accomplished through the development of the photo-engraving. After various pioneering attempts at etching the daguerreotype plate, true photo-engraving was accomplished by Paul Pretsch of Vienna in 1855. By 1879 the photogravure process using a copper plate grained by fine resin dust and treated by a carbon tissue printed under a diapositive proved highly successful, especially for long runs of printing. For rapid printing a cross-line screen was substituted for the aquatint grain. Photogravure and photo-lithography were developed, each improving the quality and reducing the cost of multiplying the photograph by using printer's ink.

POSTAL SERVICE

Like administrative records, private correspondence has an ancient and useful history. Throughout the ages letters have been made to serve many diverse purposes. Epistles from the Apostle Paul were used to exhort the faithful in early Christian times. During the Renaissance business houses like that of the Fuggers in southern Germany received long letters from their agents abroad describing market conditions and the probable effects of political trends. These are, in fact, newsletters of the type which served as one of the antecedents of the newspaper.

Letters continued to be carried privately, but the beginning of an authorized postal service seems to have come in the mid-1500's, when government officials were permitted to transmit private letters in the extending empires of Spain and Germany. Special containers were carried in relays by men and horses operating between established stations. In addition to those under state control, some universities and merchant guilds also tried to establish and maintain postal service. A weekly mail was established between London, Antwerp, and Brussels that required four or five days for delivery. This was improved by 1635 to two deliveries requiring two days. In the mid-1600's patents were issued to Count Thurn establishing him as hereditary postmaster for the Holy Roman Empire. In 1680 a penny post, so called because the basic postal charge was a penny, was established in London in which letters, weighing up to one pound and of up to £10 in value, were delivered in a system comprising some 400 to 500 stations and wall boxes. Collections were hourly, and ten daily deliveries were made in central London and six in the suburbs. This monopoly, held by the

Duke of York, was later extended and absorbed by the general postal service.

Postal service in America was first established in Massachusetts in 1639, operating out of Boston. The New York service began in 1672 with a monthly service between that town and Boston; the first colonial Postmaster General for America was designated in 1692. The waylaying of postmen by robbers was considered the normal risk on any road, and this prompted banks and money houses to resort to the exchange of checks and money orders rather than cash. In England in 1784 the security of mail coaches was improved by increasing their speed, limiting the distance travelled by each horse, and providing an armed guard to each coach; in the subsequent twenty years the mail revenue multiplied fivefold.

The last and best Postmaster General in colonial America was Benjamin Franklin, who had been associated with the postal service for forty years. He had been appointed Postmaster of Philadelphia in 1737, and as Postmaster General in 1753 had visited personally all the post offices in New England, New York, New Jersey, and Pennsylvania according to the terms of his agreement with the Crown. Franklin lost £900 in his first four years in office, but he later made the service profitable to both himself and the Crown partly by using the mails to carry copies of his newspaper.

A great surge forward in the efficiency of postal service came with reforms proposed in Great Britain by Rowland Hill in 1837. After a careful analysis of the postal traffic he recommended that a uniform rate of one penny for each half-ounce letter be instituted for delivery to any other post-town in the British Isles. He also advocated prestamped envelopes and the first of the billions of gummed stamps that have made the universal postal service practical. The first postage stamps were introduced in May 1840, and brought about the doubling of the number of letters delivered in the next two years. The service was later extended to include inland newspapers, books, and merchandise. Legislation protected the sanctity of mail in the postal service under the Hill reforms of 1840, and the number of letters handled multiplied sixfold in the following twenty years to more than half a billion. Over 600 million books, circulars, and samples were delivered annually in the United Kingdom by the end of the 1800's.

The number of United States post offices increased from 75 in 1789 to over 900 in 1800, and then multiplied rapidly to reach 77,000 in 1900. In the 1840's the rates were based on distance of mail delivery, the postage being three cents up to 300 miles and ten cents beyond that. Gradually the rates were reduced and distances increased to the rate of ten cents for over 3000 miles. In 1863 first-class postage was fixed at three cents irrespective of distance, and twenty years later the postage rate was reduced to two cents. By 1838 there had been constructed approximately 100,000 miles of post roads, with new roads being opened by surplus funds of the Post Office Department. In that year a department agent first handled the mail on trains operating between New York and Philadelphia. Two years later clerks began to sort mail on moving trains, a service that grew in time and replaced delivery by stagecoach and steamboat. Further to speed the handling of train mail, night service on the railroads was instituted.

Although the increased cheapness and efficiency of postal service did not bring about the millennium of international goodwill and understanding which the more sanguine reformers predicted would come about through better communication, the improvements in postal service throughout the world have been valuable on their own terms.

NEWSPAPERS

The single aspect of the communications revolution which had the most direct access to and impact on the new Common Man who was coming into his own during these years was the wide availability and low cost of newspapers. The newspaper as a form of mass communication began in the 17th century and soon became popular in England. In the American colonies, the first effort to establish a paper, the *Public Occurrences* of Boston, was suppressed by the government after one issue in 1690 for criticizing the official conduct of a war. The Boston *News-Letter* (1704) became the first newspaper in America to publish on a regular basis. By 1790 there were 92 newspapers being published in various parts of the new United States. During the 19th century technological developments radically changed the nature of American and foreign newspapers. With these changes the industry flourished: the approximately 150 American newspapers in 1800 grew to 863 in 1830, to 2800 in 1850, and to the astonishing total of 3725 (of which 387 were dailies) in 1860.

The most fundamental technological development of these years was in the press itself. The old hand press, such as that used by Benjamin Franklin and now preserved in the Smithsonian Institution in Washington, was improved to make its action faster and easier: the press came to be made of iron rather than wood, the platen was lowered onto the flatbed by a lever rather than a screw, and springs and rollers made it easier to push the bed back and forth under the platen. In 1827 Daniel Treadwell of Boston introduced a steam press capable of producing 500 impressions an hour. The adoption of this machine was slow, however, and in 1837 Harper and Brothers, New York book publishers, still used 37 hand presses and only one press powered by a mule. The real breakthrough in press design came with the cylinder press, used by the London *Times* as early as 1814. American printers were slow to adopt this new design.

Between 1828 and 1847 Robert Hoe and his son Richard developed the Hoe revolving press which eliminated completely the flatbed of type and substituted for it a revolving cylinder of type which made impressions on paper

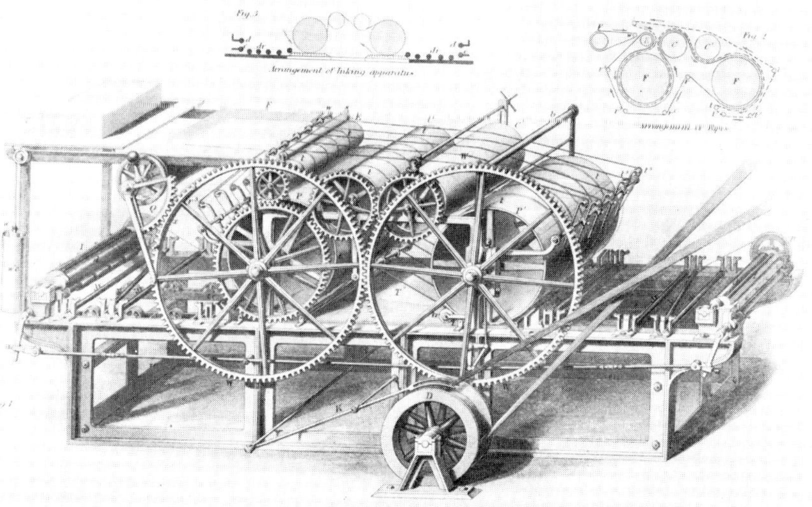

Fig. 27-6. The wide dissemination of news awaited the development of rapid printing by the double cylinder rotary press. This advanced machine enabled both sides of the paper to be printed simultaneously and the ink to be evenly and quickly spread. (Courtesy of *Burndy Library*)

carried by another cylinder. Driven by steam, some of these "Lightning" machines, as they were called, contained ten cylinders, were as large as a two-story house, weighed over 60,000 pounds, and could turn out 20,000 impressions an hour.

As so often happens, mass production and mass consumption went hand in hand. The ability of the newspaper to gather news of all kinds and from all places by telegraph, and to print it quickly in large editions, was paralleled by the growth of a literate and responsive reading public. This demand for newspapers was met after 1833 in the United States by the rise of the penny press. In that year Benjamin H. Day (1810-89) established the New York *Sun* as a one-cent newspaper and by 1840 it boasted a daily sale of 40,000 copies. A special edition on June 1, 1839, commemorating the arrival in New York of the steamship *Great Western* sold a staggering total of 57,000 copies. Innovations such as street-corner newsboys, circulation-building hoaxes, crime and society reporting were also added during these years. The telegraphic reporting of sports events helped turn the playing of games into an industry. Other newspapers followed the lead of the New York *Sun,* including the *Philadelphia Public Ledger* and the Baltimore *Sun.* The Census of 1840 showed that although the population of the country had risen 32 per cent during the preceding decade, newspaper circulation had increased 187 per cent.

Increases in both mechanism and market continued after the Civil War. William Bullock's web press, first used by the *Philadelphia Public Ledger* in 1865, drew paper from a continuous roll and printed both sides. In 1871 the *New-York Tribune* installed an improved model in which all operations, save folding, were fully automatic. In 1886 the *Tribune* also pioneered in the use of the Linotype, developed by Ottmar Mergenthaler, an employee of the Federal Patent Office. Operated from a keyboard similar to that of the new typewriters, it produced lines of type on matrices rather than letters on a paper. Again the increased ability to produce was matched by a growing market; the November 8, 1876 edition of the New York *Sun*, marking the disputed presidential election between Rutherford B. Hayes and Samuel J. Tilden, sold an unprecedented 220,000 copies.

CONCLUSION

Communication is a skill basic to all forms of social intercourse. The improvement of communications during the 19th century was basic to the development of a worldwide commercial, political, and intellectual arena in which were played out the great dramas of the century. Farmers in Kansas discovered that sales of Argentine wheat in London were immediately made known in Chicago, and prices were fixed accordingly. Miles of telegraph line were strung out behind advancing Union armies during the Civil War. Explorers in Africa were found and interviewed to the relief and gratification of the readers of the penny press. John Quincy Adams, ex-President of the United States and son of the lawyer who defended the British soldiers involved in the Boston Massacre, sat for a portrait by a photographer. It was hoped by many reformers, and especially those who championed the penny post and the International Postal Union, that communication would promote understanding and thereby world peace. Unfortunately, like much other technology, it has also served other ends.

28 / Industrial Chemistry in the Nineteenth Century

ROBERT P. MULTHAUF

THE EIGHTEENTH-CENTURY BACKGROUND

Nowadays when science and technology are so closely related, it is difficult to imagine a time when the two were quite isolated. Yet, as we have seen, through most of history, science and technology followed separate paths and until the time of the Industrial Revolution only occasionally did these paths cross. The present close ties between the two developed largely during the 19th century, and first in the field of chemistry. Yet the roots of this development can be traced farther back—to the relations between chemical science and industry in the 18th century and earlier.

The possibility of improving the "useful arts" through science was evident enough to induce a few governments from the mid-17th century to give official sanction, and in some cases even material support, to the abstruse activities of scientists. The scientists in turn made a concerted effort to accomplish such improvements. For example, the German chemist, J. H. Pott, made about 30,000 experiments in behalf of the King of Prussia in an effort to discover the composition of the famous Saxon pottery. Most chemists, however, found the most profitable occupation to be the production of "chemically prepared" medicines. A few carried on instruction in this art, and were remarkably successful, considering the general predilection for secrecy, in disseminating information on this subject.

By the end of the 18th century we are confronted with a large literature from the chemists of Western Europe which interestingly reveals the national characteristics of the most technologically advanced nations. In France the Royal Academy of Sciences produced such splendid and expensive monuments as the famous Diderot *Encyclopédie* on the arts and sciences, and an elegant ten-volume description of machines. At the same time German professors were founding many journals and writing ponderous bibliographies of a technological literature which was already surprisingly large and conveyed through a multitude of publications of small circulation. Both the French and the Germans were studying the practitioner of the arts rather than helping him. It was England which was to pioneer industrial chemistry as it developed in the 19th century, and only in England do we find simply written books of modest size on chemical technology.

One of those who published such books was Peter Shaw, a physician, whose *Chemical Lectures*, delivered in 1731 and on several subsequent occasions, aimed at improving the art of chemistry. Most chemical processes involved heating, and Shaw's book, like most others, began with a discussion of furnaces, which were then nearly as various as the processes to which they were applied: there were furnaces for metallurgy, glass and pottery production, sugar-making, and above all, for the distillation of beverages, perfumes, the mineral acids, and a variety of drugs. These furnaces reveal that complex chemical machinery is not new. The variety of distilling apparatus known in 1800, for example, was perhaps greater than that used in the distilling industry today.

Shaw was concerned with useful chemical processes of any kind and interested in convincing his readers that the circle of industries which would be called "chemical" was larger than they realized. Characteristic of modern "industrial chemistry" is the convention of restricting the term to chemically prepared raw materials as separate from consumer chemical goods whose inclusion would make the field almost limitless. Between the time of Shaw and the end of the

18th century the development of the "heavy chemical" industry made that separation necessary.

NEW HEAVY CHEMICALS

Certain substances had been produced in large quantities since antiquity: common salt and sulfur, which were purely extractive industries; (anhydrous) gypsum and quicklime, which were produced through the simple heating of minerals. The alkali, vitriol, and alum industries were more complex in that they involved the differentiation of several members of a chemical family, and production processes of somewhat greater complexity. In addition, there were in the 18th century a few heavy chemicals which had not been known in antiquity, notably saltpeter, borax, and sal ammoniac (ammonium chloride), all of which were imported to Europe from eastern regions. There was an evident

Fig. 28-1. Sulfuric acid. Production of "oil of vitriol" (sulfuric acid) in galley furnaces, as practised in Central Europe from the 16th century. (From L. Figuier, *Les Merveilles de l'industrie*, Paris, 1873. *The Smithsonian Institution*)

Fig. 28-2. Production of potash by burning wood (right), leaching, and concentrating (left), as practised in the Middle Ages. Soda was similarly produced from seashore plants. (From M. B. Valentine, *Museum museorum*, Frankfurt, 1704. *The Smithsonian Institution*)

concern for European self-sufficiency in these substances, and one of them, sal ammoniac, was successfully produced artificially in the last half of the 18th century. Its production, moreover, was accomplished through development of a variety of processes, and effected almost simultaneously in Germany, France, and England.

The most direct process for the production of sal ammoniac is through the reaction of hydrogen chloride and ammonia, and the problem of reacting these gases in large lead-lined chambers was successfully confronted by one of the proprietors of the sal ammoniac industry. Others tried to produce sal ammoniac through large-scale reactions in the liquid phase—producing ammonium sulfate by filtering ammonium carbonate solution through gypsum, followed by the addition of salt and differential crystallization of sodium sulfate and ammonium chloride, again on a large scale. Factories existed for these and other processes, a fact which has been forgotten because they all failed. They did not fail because they could not produce sal ammoniac, but because the market for sal ammoniac failed. Its principal uses had been for the cleansing of metal surfaces and as a source of ammonia; it was supplanted by other chemicals in the 19th century and thus reduced to what we now call a "fine" chemical.

A longer history was in store for another heavy chemical, sulfuric acid, the industrialization of which also came about in the last half of the 18th century. This substance, which remains today the keystone of the chemical industry, was scarcely more important than sal ammoniac in 1750, but it had been used for several centuries as an agent in the preparation of medicinal ingredients. It had been produced in some quantity since the 15th century, from the destructive distillation of vitriol or alum. By the 17th century it was also produced by burning sulfur and absorbing the effluent gas in water, and it had been discovered that this process could be improved by adding what we now call a catalyst, a small quantity of saltpeter. The producers in general used laboratory apparatus, either multiplied in number or enlarged in size, until in 1746 John Roebuck thought of replacing the enormous glass flasks then in use with a brickwork chamber lined with lead. Thus was the lead-chamber process for sulfuric acid evolved out of greatly enlarged laboratory apparatus.

The modern history of industrial chemistry is in large part the history of by-products, for the profitable utilization of waste materials characteristically determines the success of a process. In most of the sal ammoniac factories sodium sulfate was a by-product. This substance had for a century been marketed as a medicine, the "miraculous salt" of Johann Glauber, but the commercial success of "miraculous" medicines depends upon their scarcity, and the sal ammoniac manufacturers ruined the business in "Glauber's salt" at no profit to themselves. Their attempts to find uses for it are prominent in histories at the end of the 18th century, the most common idea being to convert it into soda (sodium carbonate) in one of a variety of processes.

One of the most important industrial chemicals, soda was used in making soap, glass, and saltpeter; and as saltpeter, an ingredient of gunpowder, became increasingly a "critical material" so consequently did soda. Soda was largely obtained by leaching (that is, separating from insoluble material by dissolving in water) the ashes of sea plants and was traditionally an industry of the Spanish coast. The analogous material, potash (potassium carbonate), obtained by leaching wood ash, was equally useful for the same purposes and had long been imported from forested Eastern Europe. However, the drive for new sources of alkali in 18th-century Europe concentrated on soda, perhaps because of an inevitable decrease in potash availability due to deforestation.

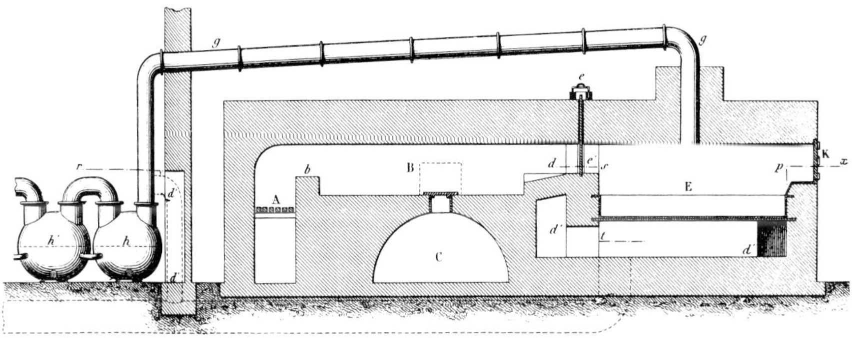

Fig. 28-3. Leblanc soda. "Salt-cake" furnace for the first stage of soda production. Common salt and sulfuric acid are reacted on a lead tray (E). The effluent hydrogen chloride is dissolved in the carboys (h). The reaction is completed to obtain acid-free sodium sulfate by moving the mass to another part of the furnace (B) by means of rakes. (From A. Payen, *Précis de chimie industrielle*, Paris, 1859. *The Smithsonian Institution*)

It remains unclear who first succeeded in producing an artificial soda. It was a goal toward which many were striving in the last quarter of the 18th century, including most of the sal ammoniac manufacturers, who were trying to make it from sodium sulfate. The offer of a prize by the French government in 1783 gave an additional incentive which was further increased when, during the French Revolution, France was cut off from its customary sources of both soda and potash. In the successful process sodium sulfate is reduced by charcoal to sodium sulfide, following which further heating with calcium carbonate (chalk) induces a change in chemical partners which yields the desired product, sodium carbonate, and a new by-product, calcium sulfide. This process is credited to, and named after, Nicolas Leblanc, who claimed the prize (unsuccessfully) for a factory set in operation at St. Denis, near Paris, sometime between 1791 and 1794.

Military necessity did not suffice to produce a commercially viable artificial soda. Instead, it appeared only a decade after the end of the Napoleonic Wars, after Europe had passed into one of its most peaceful eras. It was the requirements of a world of rising living standards, of houses with glass windows, of more frequent bathing and changes of clothes that inspired an increased demand making profitable the artificial production of soda, and it was accomplished in England, then the "workshop of the world." One should not suppose that this was simply a matter of transferring a laboratory process to large apparatus. The peculiar problems of chemical engineering became apparent in the soda industry; for example, the problem of extracting the soda from the furnace product, which was known as "black ash," is indicated in the following passage, written in 1901 (from F. H. Thorp, *Outlines of Industrial Chemistry*):

> If the black ash is put directly into cold water, it often agglomerates in hard lumps, which dissolve exceedingly slowly; the free lime present forms calcium hydroxide, which reacts with the sodium carbonate solution, forming some caustic soda; the solution of sodium carbonate, especially if hot and dilute, reacts on any calcium sulfide present, forming some sodium sulphide; moreover, moist calcium sulphide oxidizes rapidly to sulphate in the air, and this reacts with the sodium carbonate. Hence the lixiviation must be done as rapidly as possible, at a low temperature, and without exposing the wet black ash to the air.

BY-PRODUCTS

The soda industry also revealed the full complexity of the problem of by-products. The sal ammoniac industry seems to have merged with the soda industry when its by-product, sodium sulfate, became an essential intermediate in the Leblanc process. The fact that sodium sulfate was made through the action of sulfuric acid on common salt created a partnership between the sulfuric acid and soda industries which extended through the 19th century and brought into existence the complex of processes and products which has come to characterize the chemical industry.

Fig. 28-4. Sketch of a row of salt-cake furnaces. (From *Pictorial Gallery of Arts*, London, 1845-47. *The Smithsonian Institution*)

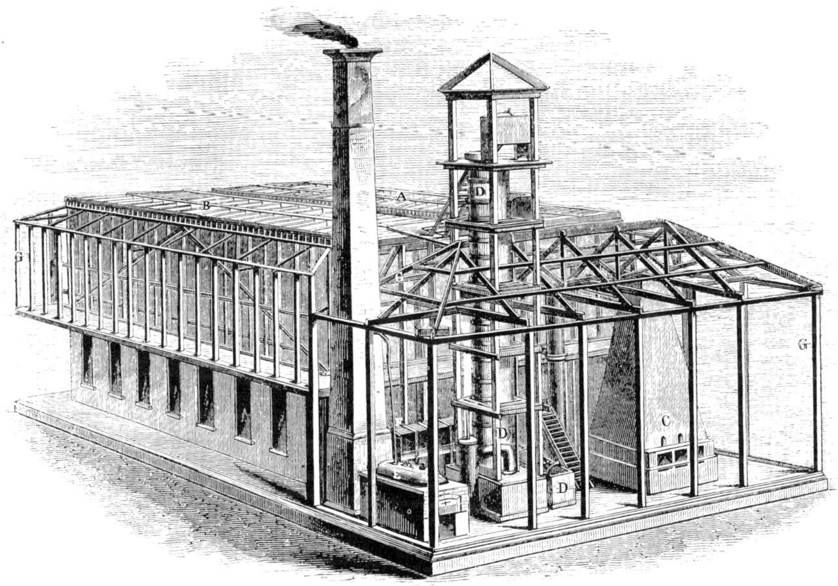

Fig. 28-5. Sulfuric acid: chamber process. A typical English works of 1863. Note the sulfur burner (C) and Gay-Lussac tower (DD) for absorbing "nitrous gas," and hence recovering the nitric acid. (From Richardson and Watts, *Chemical Technology*, London, 1863. *The Smithsonian Institution*)

The sulfuric acid-soda industrial complex produced two new by-products, hydrochloric acid and calcium sulfide, posing a challenge to the chemical manufacturer which he cannot be said to have met very successfully. In the 1830's whole areas of the English countryside were laid waste by the ejected fumes of hydrochloric acid. Within about a decade both prudence and economy dictated the condensation of these fumes and their conversion into chlorine, which had come into increasing demand as a textile bleaching agent since the discovery of that property in France, in the 1780's. Until the 1880's the calcium sulfide remained "a great nuisance," as Thorp called it, for

> The air is contaminated by the hydrogen sulfide and sulphur dioxide liberated, and the soluble polysulfides of calcium and sodium formed are dissolved by rainwater making the objectionable "yellow liquors" which run into streams and sewers.

By the 1880's Alexander Chance had devised an economical process for converting this "tank waste" into lime and sulfur which could be reused in the sulfuric acid works. By this time, ironically, the formerly wasted hydrochloric acid had become the principal economic support of the Leblanc soda industry, for as a source of soda it had been largely superseded by the Solvay process

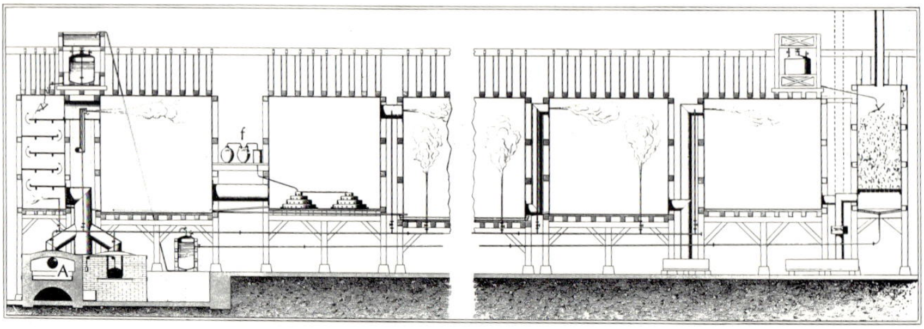

Fig. 28-6. Sulfuric acid: chamber process. A French works of 1859 (cross section). It features steam jets in various chambers and hence was more complicated than the English system. The sulfur burner is at A and the nitric acid generator at f. (From Payen, *Précis de chimie industrielle. The Smithsonian Institution*)

(see below). The British soda firms had reorganized into a gigantic combine, the United Alkali Company, but the development of more economical processes for hydrochloric acid after 1890 sealed the final doom of the enterprise, and the United Alkali Company was dissolved in 1926.

ELECTROCHEMISTRY

We generally think of electrochemistry in terms of the passage of an electric current through a material in which it induces a chemical change (electrolysis). However, current electricity was itself, through most of the 19th century, produced chemically, in the "battery" introduced by Volta in 1800. Hence the "perfection" of an economical battery was a primary concern.

A multitude of different types of batteries were invented in the 19th century, none of which proved economically useful for other than highly specialized purposes, for the value of the ingredients consumed in a chemical battery exceeded that of the substances produced in the electrochemical processes. As chemists came to realize this a sense of discouragement set in which caused them to overlook the solution of this problem for some time after it had been found, in the form of the mechanical generator of an electric current—the dynamo.

The first substances produced in electrochemical processes were at that time of interest only to "pure" science, being new materials as sodium and potassium. These substances revealed themselves as deposits on the cathode in electrolysis, and it was soon found that such useful metals as copper could be similarly deposited. The use of a metal object as cathode in the electrolysis of a copper sulfate solution proved an ingenious method of copper plating that object, and led in the 1840's to the establishment of electroplating as an industry. The use of electroplating to cover cheap metalware with the nobler metals delighted both producers and consumers, and a substantial industry arose based on this process. It was further discovered that non-conducting objects could be endowed with metallic surfaces by giving them a conductive coating of powdered carbon, a form of electroplating which became known as "galvanoplasty."

Electroplating was economically practical only because the quantity of metal deposited, and hence the current required and energy consumed, was really very small. If one wished to deposit large quantities of a metal through this process the large current required and the large amount of electrical energy used made it impractical. Before mid-century the dynamo had become an alternate source of electricity, and although its efficiency was so low that it was scarcely superior to the voltaic battery, it was good enough for electroplating.

ALUMINUM

The principal incentive to the development of a large-scale electrochemical industry was the mystery of aluminum, a metal which was known to exist in enormous quantities in the crust of the earth, but which was difficult to reduce to the pure metallic state. Although the Danish chemist H. C. Oersted had produced minute amounts of it in 1825, he did nothing further about it. In 1827 the celebrated German chemist, Friedrich Wöhler, succeeded in reducing anhydrous aluminum chloride with potassium, a chemically active metal which Humphry Davy had discovered in some of the first electrochemical experiments seventeen years before.

The years between 1827 and 1854 were filled with experiments, finally successful, in preparing aluminum in a pure state and in a quantity sufficient for the chemical study of its properties. From 1854, H. St. Claire Deville's efforts to produce aluminum in commercial quantities were supported by the French government, primarily because someone had conceived the idea of making aluminum helmets for soldiers. It was by then common knowledge that metallic ores could be "reduced" by electrolysis, and Deville first tried electrolysis of aluminum salts, using the voltaic battery as a source of current. This failing, he tried substituting the cheaper sodium for potassium in the process used by Wöhler. This process, in which the sodium used represented about 50 per cent of the cost, was "commercialized" in 1874 to produce an aluminum which was a semi-precious metal, used in jewelry and for other special purposes (among them the fabrication of a cap to top the Washington Monument).

Deville's process was improved in 1886 when Hamilton Castner of New York succeeded in reducing the cost of sodium by three-fourths by manufacturing it through the electrolysis of common salt dissolved in fused caustic soda. In testimony to his ultimate objective, the company he established in England to exploit this process was named the "Aluminum Company." But it was too late for this type of process, for the direct electrolysis of aluminum salts had already been accomplished by improved dynamos.

There had been a widespread belief that the dynamo was subject to the same unavoidable limitations as the voltaic battery, since the dynamo also depended ultimately upon a chemical reaction—the combustion of the coal used to fire the boilers of the steam engine which drove it. Unlike the ingredients of the chemical battery, however, coal was not expensive. The real difficulty with the dynamo was its inefficient design, and this was rectified by Edison and others in the early 1880's. Any lingering doubts the chemists may have had of its utility were banished by the simultaneous development of hydroelectric power, which was not dependent upon a chemical energy source and yet was ideally situated for the convenience of chemical factories. The rise of electrochemistry as a great industry coincided almost precisely with the development of hydroelectric power.

Castner, himself, had used dynamos, not voltaic batteries, in his sodium process. Others were simultaneously working toward the direct electrolysis of aluminum salts, notably R. Gratzel in Germany, P. L. T. Héroult in France, E. H. and A. H. Cowles and C. M. Hall in the United States. Gratzel's process of 1883 was apparently the first, but Hall's, which involved the electrolysis of a molten mixture of aluminum oxide (alumina) and aluminum fluoride (cryolite), proved the most successful. This process, on which the Pittsburgh Reduction Company (subsequently the Aluminum Company of America) was founded in 1888, was developed by Hall when he was a twenty-two-year-old student at Oberlin College.

A by-product of the search for aluminum was the discovery of a commercial process for making calcium carbide. T. L. Willson, a Canadian engineer working at a hydroelectric power site in Spray, North Carolina, was somewhat behind the times in attempting in 1892 to substitute calcium for sodium in the old-fashioned aluminum process. In the course of his experiments in the reduction of lime to metallic calcium by heating it with coal in an electric furnace, he discovered, apparently by accident, that the product was instead of calcium, calcium carbide. The French scientist and electric furnace expert, Henri Moissan, made the same discovery at nearly the same time, if not earlier, but failed to exploit it commercially. Its commercial importance was to be immense, for "carbide," familiar to us as a source of fuel in portable lamps, was to become one of the most important raw materials—as the source of acetylene—in the industry of synthetic organic chemicals.

Thanks to the immense energy sources of hydroelectric power the electric furnace simultaneously joined the electrolytic cell as a factor in the chemical industry. Through the use of the electric furnace Edward Acheson, an assistant to Thomas Edison, founded an industry in 1891 for the production of the artificial abrasive material, carborundum. A few years later in a similar process he developed an artificial graphite. In the field of electrolysis, Castner in 1894 developed an electrolytic process for caustic soda. The explosive growth which has characterized the chemical industry in the 20th century was already fully manifest in electrochemistry in the last decade of the 19th.

RUBBER

In 1736 the Academy of Sciences in Paris received from C. M. de la Condamine, who with other academicians had been dispatched on a scientific expedition—primarily geodetic—to South America, a report on a peculiar Brazilian tree, the milky sap of which hardened into an "elastic resin" called caoutchouc. This material, which is our rubber, had in fact been noticed long before, by Europeans on the second voyage of Columbus (1493-96) who saw the "Indians" playing with rubber balls. However, it did not attract attention until Condamine's report, after which time the chemists of the Academy undertook its systematic examination.

Practices for systematic chemical examination were just then being developed by the chemists of the Paris Academy, and included the investigation of the texture of a substance, its solubility in various solvents (water, alcohol, ether, etc.), its behavior when heated, and its action on acids, alkalis, and other familiar active materials. The most useful properties of rubber, flexibility and resistance to water, appear to have been noticed already in Brazil where the explorers experimented in impregnating cloth with rubber and forming the cloth into waterproof garments or shoes.

Although the milky juice (latex) of the rubber tree quickly hardened into its familiar solid form, by the end of the 18th century it was discovered that the juice could be transported in liquid form in air-tight containers. This opened the way to a European rubber industry, which began in 1824 with the establishment of a raincoat factory by Charles Macintosh, an enterprising Scot who had earlier been prominent among the sal ammoniac manufacturers.

Macintosh founded an industry which became widely disseminated in Europe and America, producing many of the rubber products which are still familiar. Unfortunately, when untreated, rubber tends to become brittle in winter and sticky in summer humidity and the industry was much troubled by the instability of its product. Rubber had been treated as though it were a kind of leather (it was thought of as a suitable material for bookbinding, among other things), and the problem was in its "tanning." Among those who set themselves to the solution of this problem was Charles Goodyear, an American whose principal qualification was persistence. Goodyear's story is legendary. Beginning about 1831, at the age of thirty-one, he set about random experimentation in the chemical treatment of rubber. Repeated failures depleted his resources, impoverished his family, and sent him to debtor's prison. He persevered, reputedly even selling his children's school books, and finally discovered, by accident in 1839, the "cure" for the instability of rubber: treating it with sulfur, in the heating process known as vulcanization. Henceforth rubber was approximately as useful as it is today.

OTHER PLASTICS

Rubber was not without rivals. A vegetable gum which softened when boiled and could be moulded like clay was known in the East Indies, and in 1832 a British military surgeon named Montgomerie thought of using this material, which the natives called "gutta percha," as a substitute for rubber. It was flexible and waterproof, and more resistant than rubber to summer humidity although it tended to melt when brought near a fire.

In 1840 another vegetable gum was found in Central America and marketed as a rubber substitute under the name "balata." Although an American manufacturer advertised balata as the "best rubber in the world," neither this plastic nor gutta percha proved to be competitive with vulcanized rubber. They remained commercially viable through the discovery of particular uses. Balata was to become, under the name "chicle," the base of chewing gum. Gutta percha had a more dignified future, for it was found ideal as an insulator in the new electric industries. As a covering for underseas telegraph wires gutta percha was a critical factor in the success of the transatlantic telegraph cable.

The shape of the future plastics industry was indicated at the London Great Exhibition of 1851, where, according to Andrew Ure, the West Ham Gutta Percha Company exhibited "a beautiful group representing a boar hunt, covered with a metallic coating in imitation of bronze." This example of galvanoplasty was only one way in which the plastics inventors were busily engaged in "product improvement." They also mixed rubber, gutta percha, and balata with each other and with other resinous materials such as asphalt and pitch, and tried to devise "tanning" treatments analogous to vulcanization for the lot of them. The next important development in plastics, however, was to be a by-product of the soberer occupations of the organic chemist.

INDUSTRIAL ORGANIC CHEMISTRY

Between 1833 and 1845 a number of studies were made in France and Germany concerning the action of acids on such materials as starch and cellulose. They culminated in 1845 in the discovery by the German chemist, C. F. Schönbein, of nitrocellulose, or as he called it, "explosive cotton wool." It was a product of the action of nitric acid on cellulose. Schönbein predicted that it would soon replace gunpowder, and experiments were immediately initiated, only to be abandoned in the 1850's after a series of disastrous accidental explosions. The methods of safely handling this "smokeless powder," which did result in its replacement of gunpowder after about 1890, were largely worked out in connection with its development as a plastic.

In experimenting with solvents for the new "powder," Schönbein found that it dissolves in a mixture of ether and alcohol, and that the evaporation of the solvent leaves a kind of liquid glue. This glue, when painted on a surface, leaves a skin-like plastic film, the use of which was advocated in surgery, as a temporary substitute for skin, as early as 1848. In 1851 the film was brought into use in photography, at which time F. S. Archer gave it the name "collodion" (from an allegedly alchemical name meaning glue-like). The first attempt to apply it to uses requiring a larger bulk of the material seems to have been made about the same time by Alexander Parkes, who treated the nitrocellulose with a variety of mixed solvents designed to give it a mouldable form like rubber or gutta percha. Objects of "Parksine," as the resulting plastic was called, were shown at the Exhibitions of London (1862) and Paris (1867). Although the material was awarded medals on both occasions, the numerous combs and similar articles which Parkes had sold began to come back from irate purchasers in wrinkled and shrivelled condition, and Parkes' firm suspended operation in the same year in which he received the medal at Paris.

In chemical terms cellulose is a long-chain (aliphatic) hydrocarbon containing a number of "hydroxyl groups" (—OH) which may react with nitric or some other acid. The number of hydroxyl groups which react with the acid depends upon the conditions under which the reaction occurs. In the mid-19th century it was only known that there were different "kinds" of nitrocellulose, and this justified additional experiments, patents, and commercial operations directed toward the solution of the problem of obtaining from nitrocellulose a plastic competitive with rubber. The particular problem which had defeated Parkes was that of finding a third substance to add to the nitrocellulose and its solvent to give bulk and stability to the resulting plastic. It was solved about 1870 through the use of camphor, to form "celluloid," the first commercially successful nitrocellulose plastic.

Celluloid was the commercial name of a plastic developed by John W. Hyatt, an Albany printer, and his brother, Isaiah. John had become an experimenter in plastics in 1863, when he read of a $10,000 prize for a substitute for solid ivory billiard balls offered by the Phelan & Collender Co. of New York. He tried various mixtures suggested to him by the scientific literature, taking up collodion about 1868, and finally mixing it with camphor, which was a commonplace material among plastics experimenters. Although he made use of the work of earlier experimenters he was apparently unaware of Parkes and other earlier manufacturers who had failed. He owed his success more to his method of moulding the plastic under heat and pressure than to any unique properties of the mixture he used. He also owed something to the fact that American dentists just at this time were receptive to plastic substitutes for rubber in the fabrication of dental plates, because of the allegedly exorbitant prices charged by the "rubber monopoly." In 1870 Hyatt organized the Albany Dental Plate Co., which became the Celluloid Manufacturing Co. two years later.

Experimentation in the action on cellulose of acids other than nitric was a matter of particular interest to practical chemists who sought a plastic form of cellulose which would be free of the inflammability of nitrocellulose. During the two decades after 1869 a number of investigators succeeded in reacting acetic anhydride with cellulose to produce cellulose acetate. It was not inflammable, but being, like nitrocellulose, a powder it involved the same problem of dissolving and reducing it to a plastic form. This was not solved until the 20th century, when the first important use of cellulose acetate plastic was the coating of airplane wings, where it was painted on much as collodion had been.

The action of alkalis on cellulose was also investigated, and as early as 1844 John Mercer had observed that caustic soda (sodium hydroxide) changed the texture of cotton and gave it a silk-like transparency which textile manufacturers exploited under the name "Mercerization." Attempts to imitate silk with threads of cellulose plastic were the most popular form of experimentation in plastics toward the end of the century. The quest which triumphed in the 1930's with the invention of "nylon" (which is not made of cellulose) began in 1885, when St. Hillaire de Chardonnet formed threads of nitrocellulose by forcing collodion into a coagulating solution through a disc pierced with fine holes. A highly inflammable artificial silk was of limited use, and in 1890 threads of "regenerated" cellulose were made by extrusion of a solution, discovered by M. E. Schweizer in 1857, in which cellulose is dissolved in ammoniacal copper hydroxide. Two years later C. F. Cross, E. J. Bevan, and Clayton Beadle patented a process for forming regenerated cellulose thread in a similar process using a solution of cellulose reached by successive treatment with caustic soda and carbon disulfide. Both in this case (the "viscose" process) and in the "cuprammonium" process deriving from Schweizer, the coagulation was accomplished by a bath of sulfuric acid. The product, in both cases, was what we know as rayon.

COAL GAS

It had been known since the early 18th century that coal emitted inflammable vapors when strongly heated in a closed space, and the possibility of using this as a source of gas for artificial illumination had been noticed. Its successful accomplishment was a substantial engineering problem, and is generally credited to William Murdoch, an engineer with the steam-engine manufacturers Boulton and Watt. In 1792 Murdoch used such illumination in his home, and ten years later he brought it impressively before the public in an extensive illumination of the Soho factory of Boulton and Watt, timed to celebrate the Peace of Amiens (a temporary truce in the war between France and England). By 1822 London had, according to Ure, nearly 17,000 lamps (10,660 private lamps, 2248 street lamps, and 3894 theater lamps), and by 1834 the total number had increased to 168,000.

The gas industry was probably equal in importance to the soda-sulfuric acid industry in the development of what we now call chemical engineering processes. There were similar problems in achieving a sufficient yield of the desired product to make the process economically sound, and there was also the problem of by-products. It was such a by-product, the ammonia from gas works, which finally sealed the doom of the sal ammoniac industry. The coal distillation process also yielded other gases, low-boiling liquids, high-boiling liquids, and a viscous residue called "coal tar." Experiments in the fractional distillation of coal tar showed that it contained a great many ingredients. In Friedrich Knapp's *Chemical Technology* (1847) we find the author declaring that the process by which coal gas, coal tar, and coke were produced

> . . . admits of the production of such an innumerable series of bodies as will never be exhausted by science, nearly as many as there are mathematical combinations, binary and ternary, depending upon the temperatures. Most of these are of constant occurrence, and some are of importance, and deserve notice.

COAL-TAR DYES

Considerable notice had already been taken of the by-products of coal distillation, principally in the laboratory which Justus von Liebig had headed since 1825 at the University of Giessen. Liebig had been trained in France but he returned to Germany to found what is probably the most famous laboratory in the history of chemistry. He trained many of the most distinguished chemists of the next two generations, including A. W. Hofmann, who was to be primarily responsible for the spectacular consequences of the study of coal tar.

From 1825 to 1840 Liebig and his students were especially concerned with organic analysis, a field to which they gave a new foundation. Hofmann, who was the son of the architect of Liebig's laboratory, entered as a student in 1837 and made coal tar his special province. He was not the first to do so. Numerous more or less distinct substances had been obtained from it, chiefly by fractional distillation, and some of the same substances had been obtained from sources other than coal tar. The whole subject had inevitably fallen into a confusion of names and properties. Hofmann's first publication (1843), "On the Volatile Organic Bases in Coal Tar," was primarily concerned with the resolution of this confusion.

Hofmann became and remained the "court of last resort" on coal tar. In 1845 he accepted a temporary appointment to head the new Royal College of Chemistry which had been founded in London by Liebig's English admirers. His "temporary" appointment lasted for twenty years, after which he returned to Germany and continued, in a private laboratory at the University of Berlin, to play his role as the authority on coal-tar chemistry. This move was to play a decisive part in the subsequent development of industrial chemistry in both England and Germany.

In 1856 Hofmann's eighteen-year-old assistant at the Royal College of Chemistry, W. H. Perkin, patented as a dye a purple residue he got as the result of an unsuccessful attempt to synthesize quinine through the action of potassium dichromate on one of the coal-tar chemicals, aniline. This was not the first discovery that a dyeing substance could be obtained by the chemical manipulation of coal tar, but it was the first attempt to exploit that fact, and it represents the beginning of the coal-tar dye industry.

Perkin's "mauve" was scarcely on the market before a red dye of similar origin (which was obtained by reacting aniline with carbon tetrachloride) had been patented in France. This dye, called "fuchsin" (after the flower fuchsia), was even more popular than mauve, and its introduction was followed by an explosive development of the dye industry. By 1862 there were twenty-nine synthetic dye companies in Western Europe. At the London Exposition in that year five dyes were exhibited, mauve, fuchsin, and three called "aniline blue," "aniline yellow," and "imperial purple." Hofmann predicted on that occasion that synthetic dyes would replace many of the time-honored natural dyes, as was indeed to be the case, beginning in 1869 with the synthesis of the vegetable dye derived from madder, alizarin, and culminating in Adolf Baeyer's analysis of indigo in 1883 and its synthesis about a decade later.

SOLVAY SODA AND CONTACT SULFURIC ACID

In the course of the 19th century the industrial chemist earned the right to call himself an engineer. Not only did he adopt materials and machinery to the conduct of traditional chemical processes, but he made a very significant beginning in the development of industrial processes which were entirely new and which depended on novel techniques. Among the consequences were processes for the production of soda and sulfuric acid which by the end of the century had brought near to collapse the Leblanc soda and chamber sulfuric-acid business which had previously been the backbone of the chemical industry.

This revolution began in 1863, when the Belgian Ernest Solvay introduced a process for the production of soda by the carefully controlled reaction of concentrated brine (sodium chloride) first with ammonia and then with carbon dioxide, both in the gaseous state. The reaction,

$$NaCl + NH_3 + H_2O + CO_2 \longrightarrow NH_4Cl + NaHCO_3$$

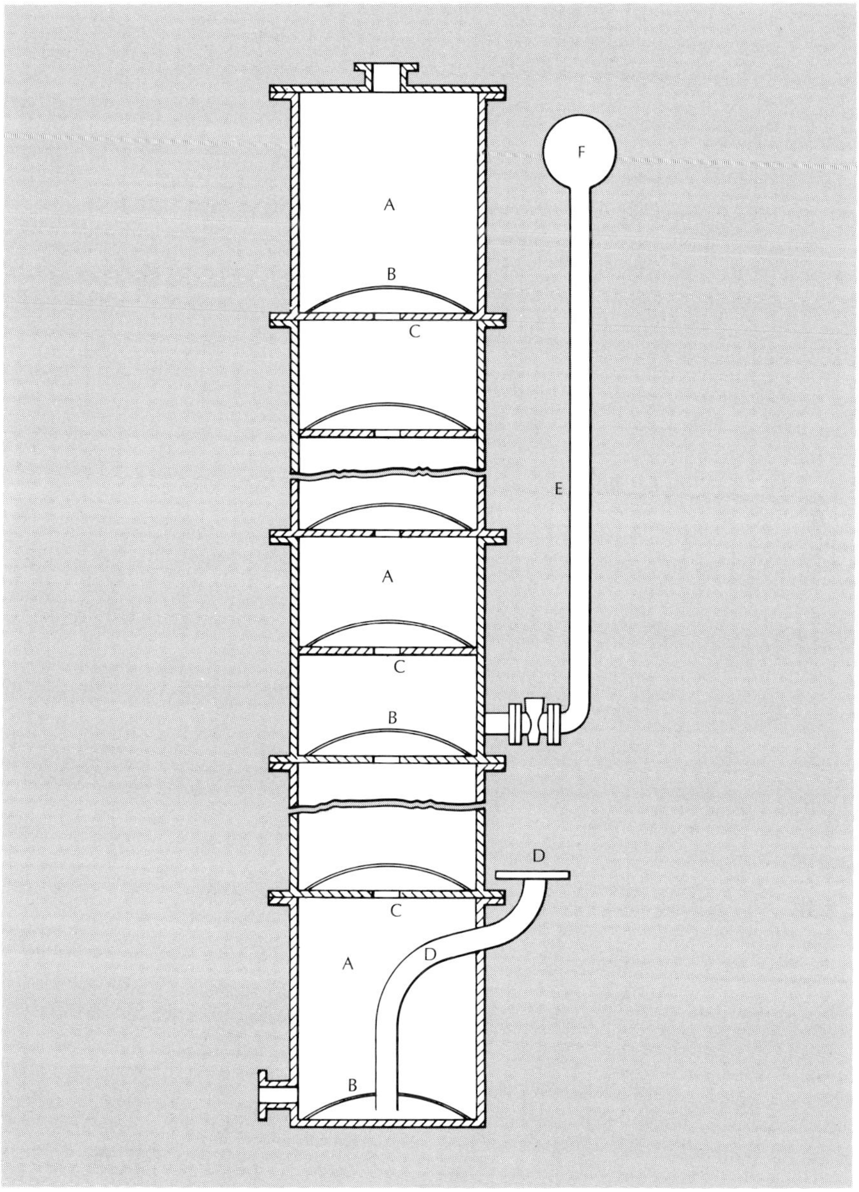

Fig. 28-7. The Solvay reaction tower, which is the heart of the process for the production of soda. The ammoniacal salt solution enters under pressure at E, and the lime-kiln gas (carbon dioxide) at D. The gases are atomized as they rise, by fine holes in the passage through each tray. The liquid with sodium bicarbonate leaves at the bottom. (From Solvay's U.S. patent of 1873. *The Smithsonian Institution*)

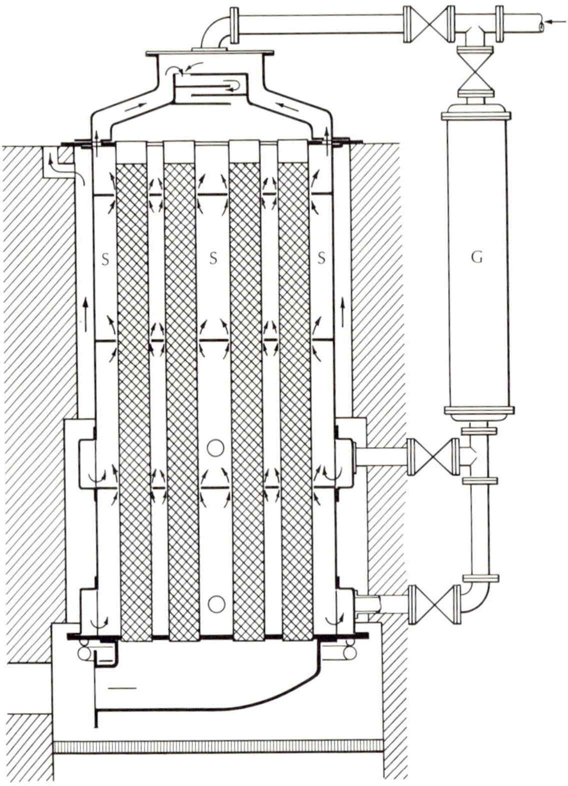

Fig. 28-8. Sulfuric acid: contact process. "Pipe stove" in the process patented by the Badische Anilin und Soda Fabrik (Germany) in the 1890's. The crosshatched contact pipes are filled with clay balls impregnated with platinum black. A mixture of oxygen and sulfur dioxide is passed through these pipes and converted into sulfur trioxide. The gaseous reaction products are recirculated (through G and S) to maintain the "stove" at the proper reaction temperature. (German patent of 1898. *The Smithsonian Institution*)

was curiously similar to those used just a century earlier to produce sal ammoniac. When the objective was to recover sal ammoniac, the 18th-century chemists had resorted to the introduction of compounds of calcium or magnesium which yielded relatively insoluble carbonates. Solvay's objective, however, was to recover the sodium bicarbonate (which further heating reduced to the carbonate) and he found it possible to cause the bicarbonate to precipitate by taking advantage of the fact that it is only slightly soluble in a cold ammoniacal solution of common salt. As is characteristic of industrial chemistry, the implementation of this experimental discovery was in fact very difficult to accomplish. The heart of the Solvay-process plant is in a "carbonating tower" down which flows the ammonia-saturated brine, meeting along the way an upward current of carbon dioxide gas. The flow is controlled by a series of perforated "shelves" within the tower and by the maintenance of a certain liquid level on each of these shelves. The temperatures are controlled by cooling coils, and the concentrations of the gases are carefully maintained through pressure regulation. The carbon dioxide is obtained from a lime kiln which also provides calcium oxide (quicklime) which decomposes the sal ammoniac to recover the ammonia, leaving calcium chloride as the only by-product.

The Solvay process is continuous; that is, the ingredients are continuously fed into the carbonating tower and the product is continuously recovered from it. It presented intricate problems—in the synchronization of flow of materials, temperatures, and pressures—which required about a decade to solve, but by the end of the 19th century Leblanc soda was no longer in a competitive position. Since fortunes had been invested in immense chemical works using the Leblanc process it remained in use into the 20th century, until they were worn out. Leblanc soda was for several decades kept economically viable through the sale of the once despised by-product, hydrochloric acid.

The chamber process for sulfuric acid lasted somewhat longer, and was only replaced by the "contact" process after the development of the synthetic dye

Fig. 28-9. Synopsis of contact-process operations, as shown in the U.S. patent of 1902, of the Badische Company. The contact chamber (H) is simpler than that shown in Fig. 28-8. E, F, I, and J are vessels for washing or absorbing gases. G and K are testing apparatus. (From George Lunge, *Sulphuric Acid and Alkali*, London, 1903. *The Smithsonian Institution*)

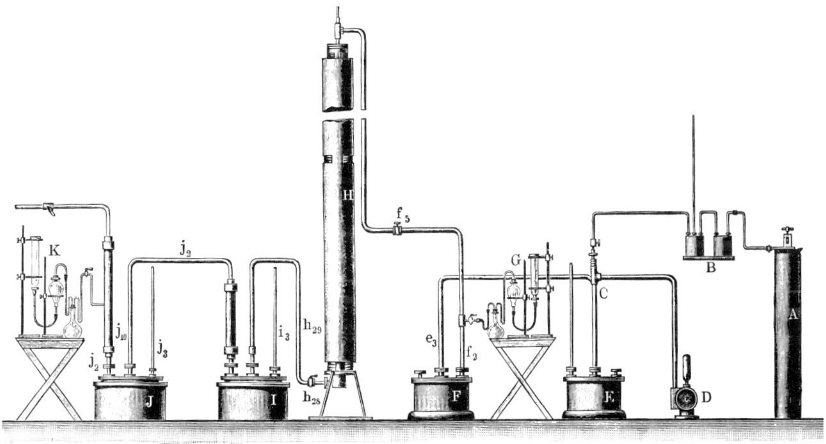

industry in Germany had generated a demand for sulfuric acid of a concentration which the chamber process could not accomplish. The contact process in fact had many advantages: whereas the chamber process required enormous spaces (the chambers) for the leisurely conversion of sulfur dioxide into sulfurous acid, and then into an unstable combination with nitric acid (known as nitroso sulfuric acid), and finally into sulfuric acid, the contact process proceeded quickly and simply through the direct combustion of sulfur dioxide with oxygen through contact with heated platinum to yield sulfur trioxide which was converted to sulfuric acid by solution. The platinum functions as a catalyst, a substance required only in small quantity which is regenerated on completion of the process. (It is supposed that it forms in the course of the reaction some intermediate compound, and thus plays a role corresponding to that of nitric acid in the chamber process.)

A contact process was patented in England as early as 1831, and others appeared at frequent intervals thereafter, but attempts to make it work in practice were not successful. It seems very likely that this was partially due to the fact that there was little demand for it, but it was also a consequence of the frequent failure of the platinum, after a short period, to function as a catalyst. "Poisoning" of the catalyst, as this was called, called forth much research, not only on the platinum contact process but on similar processes using other catalysts. Although most of this work was done in England, the final success of the contact process was the result of a patent taken out in Germany in 1875 by Clemens Winkler.

The success of Winkler's process was assured by the support of the emerging dye industry, which required sulfuric acid of a peculiar kind, known as oleum. This acid is technically stronger than 100 per cent since it consists of sulfuric acid containing dissolved sulfur trioxide. It could not be obtained from the chamber process, which produced a weak acid which could be concentrated only to about 94 per cent. Like the Solvay process for soda, the contact process requires careful control of the flow of gases and the maintenance of steady conditions of temperature and pressure in a continuous process. The lessons gained in developing a catalytic process of this kind were used by German chemists in the early 20th century to develop catalytic processes for ammonia and nitric acid, processes both more difficult and more important, for they made Germany self-sufficient in nitrates and saltpeter, the only traditional heavy chemicals for which Europe was still dependent upon outside sources.

The suppression of the Leblanc and chamber processes by the Solvay and contact processes brought forcibly to the attention of the proprietors of the chemical industry the ascendancy of physico-chemical science over the older empirical methods. The state of the electrochemical and plastics industries in 1900 showed that the potentialities of a relatively unalloyed empiricism were not exhausted, but in the case of the coal-tar dyes the power of the scientific approach to create wholly new industries had been demonstrated. In the 20th century, empiricism was to give up further ground and science to make further conquests. In chemical technology, however, these conquests owed much to the "chemical engineer," trained in science, but no less indebted to empiricism. Whether his advent be regarded as a sign of the conquest of technology by science or as an indication of the technologist's mastery of science depends upon one's point of view.

29 / Military Technology

THOMAS A. PALMER

Modern warfare begins with the French Revolution. In 1789 the French people had risen against the absolute rule of their Bourbon monarch. King Louis XVI had reluctantly given in to their demands for reform, coupled with revolutionary violence, and had agreed to the establishment of a limited monarchy. But this experiment in constitutional monarchy proved short-lived, for the Bourbon ruler was not content to reign but not rule, and the princes of the neighboring states invaded France in order to force a return to absolute rule by the divine-right king. The result was the overthrow of the monarchy and the proclamation of the First Republic under the guidance of advanced revolutionary leaders who controlled the National Convention, the governing body of the Republic. Establishment of the Republic was met by further aggression from the rulers of the European states who wanted to destroy the French Republic before the revolutionary virus spread to their own subjects and brought about their own downfall.

To meet this military aggression from the crown heads of Europe, the French Republic called for all citizens to support the defense effort. In 1793 a strident decree issued by the National Convention directed the French people to defend their beleaguered homeland—

> From this moment until that in which our enemies shall have been driven from the territory of the Republic, all Frenchmen are permanently requisitioned for service in the armies.
>
> Young men will go forth to battle; married men will forge weapons and transport munitions; women will make tents and clothing . . . children will make lint . . . old men will be brought to the public squares to arouse the courage of the soldiers, while preaching the unity of the Republic and hatred against kings.
>
> The public buildings shall be turned into barracks, the public squares into munition factories . . . the interior shall be policed with shot

> guns and with cold steel and all saddle horses shall be seized for the cavalry.

The revolutionary spirit of the radical revolutionary leader, Robespierre, despising moderation on the battlefield as well as at home, thus called an entire nation to arms. The mass armies and economic mobilization that followed saw the beginning of modern war.

A REVOLUTION IN ARMS

Up to this time, 18th-century warfare had been regulated by a system of universally understood rules. Men, equipment, and money were scarce: therefore mutually destructive battles of annihilation were not sought. The martial ardor of sovereigns and their military commanders was usually satisfied by skillful maneuvers and the attainment of limited objectives, such as the siege and capture of a fortified city. The civil populace, while exposed to the normal depredations of armies on the move, were by and large expected to be nothing more than spectators. Marshall Saxe, a leading strategist of the 18th century, expressed the dominant military theory in these terms, "I do not favor pitched battles . . . and I am convinced that a skillful general could make war all his life without being forced into one."

The intensity and emotion of the French Revolution swept away this pastoral view of war. Conscription, economic mobilization, propaganda, and internal security measures transformed warfare from a limited to an unrestrained affair. The restricted and rather formalistic operations of early 18th-century armies were rarely to be seen again. The French Revolutionaries called for a *levée en masse,* the rising of the entire population, not just a professional soldiery, in order to carry on war. Henceforth, war would call increasingly for "total" response of a nation.

The development of nationalism not only created larger armies but made battles more frequent and intense. Even during the wars of Frederick II (Frederick the Great of Prussia), which had been exceptional in the 18th century for their size and vigor, the armies had averaged only 47,000 men, and battles had been fought on an average of one a month. A few decades later, during the Napoleonic Wars, which continued those of the French Revolution, the average size of the armies had jumped to 84,000, and battles were fought at the then amazing rate of eight per month. Whereas Frederick's wars saw only 12 battles with over 100,000 men engaged, 37 such battles were fought during the Napoleonic Wars, including the Battle of Leipzig (1813), in which half a million men participated, ranking it as the greatest of Europe's battles until the outbreak of World War I in 1914. The new scope of military operations can be further gauged by the fact that Napoleon gathered and equipped an army of 612,000 men for the invasion of Russia (1812).

This expansion of the scope and intensity of warfare was prompted by social and political factors rather than by advances in weapons technology. With the exception of modest improvements in the performance of artillery, the armies and navies of the Napoleonic era went into battle with essentially the same equipment as in the previous century. However, if technology lagged, strategy did not; maneuver, mass, and shock action were well, sometimes brilliantly used. It was, in fact, one of those rare but recurring periods of history where generals often proved better than the weapons at their disposal.

This fleeting situation was reversed by the great technological advances of the 19th century. One hundred years after Waterloo (1815) the sophistication and lethal nature of the instruments of war had outstripped the ability of the strategists to employ them decisively. Wars of maneuver gave way to the linear stalemate, that is, to warfare where millions of men faced one another in opposite trenches without either side being able to gain a quick and decisive victory.

Although weapons technology did not advance greatly during the French Revolutionary and Napoleonic Wars, the wars did have a profound effect on other technological elements. For one thing, the vastly increased scope of warfare—the mere task of feeding, equipping, and supplying such large numbers of men under arms—required a heightening of manufacturing effort. Mechanization was increasingly applied in order to meet the demands for an outpouring of war materials. In addition to the stimulus given to production, the large-scale warfare resulted in a search for substitute materials (to make up for the deficiency of imports formerly obtained from the enemy territory) or for new techniques and devices to carry on total warfare. Thus, for example, French chemists sought better and cheaper means to manufacture gunpowder, and Napoleon's efforts to supply his vast armies with sufficient food led directly to the invention of canned food. On the other side, the British efforts to finance and equip the armies of their continental allies in the struggle against Napoleon led to an acceleration of the industrialization which had already commenced in Britain a half-century earlier. Of course, it is impossible to say how much of this industrialization would have taken place without the impetus given by war production. The fact is, however, that industrialization proceeded apace, and the demands of war provided the stimulus of large markets for industrial production.

SMALL ARMS AND ARTILLERY

After the Napoleonic Wars, political expediency and the memory of war's bloodshed and horrors contributed to several decades of peace for the major European powers. Nations turned their energies to domestic growth and industrial expansion. Military establishments declined in both size and effectiveness, while for the troops it was a period of grinding monotony and attention to minutiae. Officers, notably in the cavalry, turned to brilliant uniforms, corseted waists, and social preoccupations to dispel boredom. Military efficiency became far more apparent than real. It was, as one French general caustically said, "a halt in the mud."

In 1854 hostilities in the Crimea brought war again to England and France, as they allied to defend the Turkish empire against the aggression of the Russian tsar. The military stagnation of the previous forty years soon came to light as the Crimean War turned into an epic of military mismanagement, symbolized by the ill-fated "Charge of the Light Brigade."

The military historian, J. F. C. Fuller, in a summary of the Crimean operations stated, "as regards the act of war there is little to learn. . . . Generalship was beneath contempt, leadership was of a low order, and tactics were prehistoric." To this evaluation might be added that it was the last 19th-century campaign to be fought with 18th-century weapons. But the years from Waterloo to the siege of Sebastopol, the major military effort of the Crimean War, cannot be passed over quite so lightly in all areas of military importance. Although generally obscured by the lag-time between invention and adoption, new technological developments, particularly in metallurgy, chemicals, and precision tooling, made this a significant, if unplanned, period of research and development in weaponry.

Certainly, some of the greatest technical achievements of the 19th century were in the improvement of small arms. Two in particular stand on an almost equal footing with the invention of gunpowder as the most important developments in the history of military firearms: one was the percussion cap; the other, the innovation of breech-loading.

Invented in 1807 by the Scottish clergyman, Alexander Forsyth, the percussion principle, in which a chemical compound detonates and gives a fire which ignites the propellant gunpowder, replaced the unreliable flint-and-steel spark and remains the basis for all modern propellants. As is usually the case with great

Fig. 29-1. An English muzzle-loading flintlock, used in the United States during the American Revolution. (*The State Historical Society of Wisconsin*)

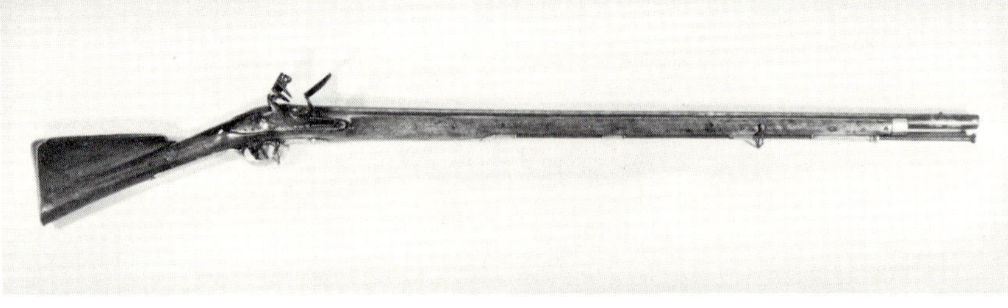

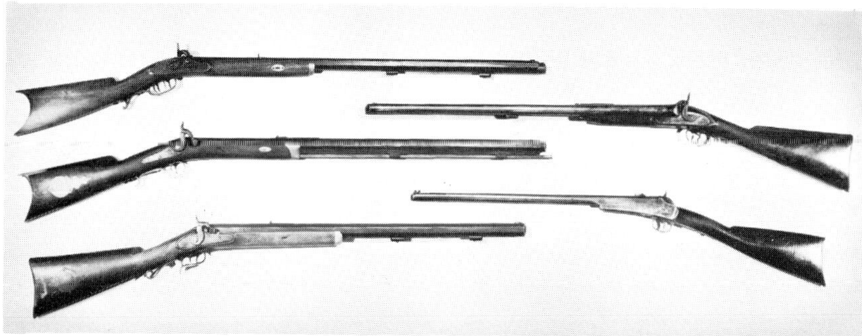

Fig. 29-2. (Right top) Combination percussion rifle and shotgun, *c.* 1850. The right side is a .41 caliber rifle and the left side is a 12-gauge shotgun. (Right bottom) Breech-loading cartridge carbine: rim fire, single shot. This model was submitted by the inventor to the United States government during the Civil War. (Left) Three percussion rifles made by gunsmiths in Wisconsin, 1851-65. (*The State Historical Society of Wisconsin*)

technical accomplishments, further development was required for practical use. In this case, Joshua Shaw, an English portrait painter, created (1816) a copper percussion cap using fulminate of mercury as the detonator, which led to the general adoption of Forsyth's idea.

The muskets used previously had been smoothbore, that is, the inside of the gun barrel had been perfectly smooth. They fired rifle balls, small ball-shaped pellets, and were notoriously inaccurate unless the enemy was close enough to "see the whites of their eyes." To improve the accuracy, a conical bullet was developed and the principle of rifling was employed. Rifling involved the cutting of spiral grooves into the inner surface of the barrel to make the bullet spin through the air when fired, thus giving it greater accuracy and distance.

Muskets had another disadvantage: they were muzzle-loading, that is, the powder and shot had to be loaded through the front end of the barrel of the firearm. Not only did this take time, involving ramming the shot down to the bottom of the barrel, but the loader had to stand up to do the job properly, thus exposing himself to enemy fire.

The percussion cap, development of the conical bullet, and refinement of the principle of rifling converted the erratic smoothbore musket into the military rifle which was effective at 600 or 700 yards and dependable in any type of weather. Yet even before these developments could be fully absorbed, the Prussians in 1841 issued a workable breech-loading rifle to selected army units. The needle gun, as it was known, derived its name from a slender firing pin which a pull of the trigger sent forward to detonate the bullet. This revolu-

Fig. 29-3. Gatling gun, famous ancestor of the modern machine gun.

tionized tactics because it allowed soldiers to load and fire from a prone position or on the run. The needle gun proved a decisive advantage in the Austro-Prussian War of 1866 in which, on the field of battle, it was estimated that one Prussian soldier armed with the needle gun was the tactical equal of three Austrians armed with the muzzle-loader. The lesson was not lost; only four years later the Franco-Prussian War (1870-71) was almost entirely a breech-loading affair.

Rapid-fire guns also made their battlefield appearance in some quantity. Most famous of these ancestors of the modern machine gun was that invented by the American, R. J. Gatling (1818-1903). His gun consisted of a number of separate gun barrels clustered around a circular axis; when rotated by a crank the barrels were successively discharged, thereby giving rapid firing. The American Civil War had seen no more than sporadic use of the famous Gatling gun, but the French took to the field in the Franco-Prussian War in 1870 with several units of horse-drawn, twenty-five-barrelled "grapeshooters" of considerable fire-power.

The percussion cap had given dependability to the small arm, and the breech-loader introduced flexibility in its employment. Metallic cartridges and smokeless power in the 1870's added an even greater dimension of utility. By 1890, the transitional period in small-arms development was over. The small bore, breech-loading, magazine rifle and the single-barrel rapid-fire gun with automatic cartridge loading and ejecting mechanisms were in common use.

Artillery was, during the same period, rescued from subordination to the improved infantry weapons. Rifling, breech-loading, and smokeless powder were advantageously adapted for larger weapons, and a series of brilliant metallurgical developments set in motion a progression from cast iron to forged iron and ultimately steel in the manufacture of cannon. The new-found increases in metal strength provided a challenge to the chemical industry to produce more powerful explosives. Scientists such as Alfred Nobel responded with nitroglycerin, guncotton, and a host of other fulminate innovations to provide additional range and destructive power for the artillery.

THE RAILROAD AND TELEGRAPH

In addition to creating new weapons, the unprecedented speed of industrial developments transformed the military concepts of time and space. The instruments for this transformation were the railroads, which allowed great armies to be mobilized, transported, and maintained; and the electric telegraph, which allowed generals to control their armies once deployed.

Generals Helmuth von Moltke (1800-1891) and Ulysses S. Grant (1822-85), two of history's most diverse military contemporaries, were responsible for converting rail systems into strategic weapons. Moltke, as Prussian chief of staff during the Franco-Prussian War, was the first to make strategic plans for the use of railroads, while Grant was the first to achieve strategic wartime victory from their use. Grant improvised during the course of the American Civil War (1861-64) what Moltke so meticulously prearranged before the start of hostilities with France in 1870. The end result was the same; both sustained vast armies in combat for indefinite periods with an uninterrupted flow of fresh manpower and equipment.

While Grant's armies derived increasing strength from the efficiently functioning Northern railroads, the Confederacy was correspondingly weakened as its rail centers were captured or destroyed. Transportation between the limited Southern industry and its source of raw materials was blocked, and the strategic movement of Southern armies rendered almost impossible. In the Franco-Prussian War, Moltke's plans sent 1,183,000 regular soldiers and reservists through the processing centers and into the wartime army within eighteen days after the declaration of war. During this same period almost half a million were transported to the French frontier to open the campaign. The Prussian railroads, planned with strategic purposes in mind, as opposed to the

French system based on regional monopoly, provided an overwhelming margin of victory.

Larger armies, increased firepower, and enhanced mobility strained to the limit the ability of generals to exercise the personal command relations of the past. Growing operational complexity was in part resolved by utilization of the telegraph for first, strategic, and later, tactical control measures. Co-ordination of the vast theaters of the American Civil War proved the greatest 19th-century military achievement of the telegraph.

The handling of local command problems in this conflict was also facilitated by the laying of insulted copper telegraph line from commanders to lower units, even in the midst of combat. The telegraph cart and men were often seen well up to the front of battle and exposed to all its consequences. In Europe it was the same. Moltke and his staff controlled Prussian forces in 1870 with a telegraph network that eventually totaled almost 7000 miles of line and provided armies, corps, and divisions with rapid communications to each other's and higher headquarters.

In addition to widening the span of military control, the telegraph injected public opinion and close political supervision into command relationships. This was demonstrated in the first wartime use of the telegraph in the Crimean War (1854-56), when Paris and London were brought, within hours, news of the distant battlefield, and advice to military commanders flowed in a never-ending stream. Napoleon III was so fascinated by use of the telegraph that the French Commander Niel complained bitterly about being placed at the "paralyzing end of an electric cable."

The rapid transmission of information from the battlefield produced a new awareness of military matters among the citizenry. While telegraphed news could produce national heroes overnight, at the same time, military shortcomings could be exposed to immediate public scrutiny. These facts were not lost on officials, particularly elected ones, and the lot of the soldier became a matter of national concern. National clamor over the inadequate sanitation and medical facilities for British forces in the Crimea led to the selfless efforts of Florence Nightingale (1820-1910) and the birth of modern military nursing. Of even greater significance, the telegraph, allied with the printed word, was responsible for the conversion of the soldier from a stranger, who might be a dangerous brute, to a fellow citizen in the eyes of his countrymen.

TOTAL WARFARE

Wide-ranging improvements in weapons, transport, and communications coincided with the development of American-inspired mass-production techniques. As we have seen, the handiwork of the individual craftsman who produced by "knack" was superseded by automatic machinery and the use of interchangeable mechanisms. These innovations were particularly adaptable to the production of weapons—indeed, interchangeable parts had first been widely used for musket manufacture—and by mid-century over one hundred different automatic power tools were used to shape the parts of U.S. Army rifles manufactured at the government's Springfield Arsenal.

The American Civil War demonstrated the immense diversification and potential of machine production. During the four years of conflict, Northern industry provided four million muskets and a billion and a quarter cartridges for the Union forces. Adoption of the sewing machine in factories allowed these armies to be outfitted with standard, ready-to-wear uniforms and shoes. Simultaneously, the mass-produced McCormick reaper and the steel plow converted the prairies of the Midwest into an abundant granary for the North.

The Civil War proved to be the first occasion when the achievements of the Industrial and Scientific Revolutions were put to large-scale military use—a war in which the artisan, farmer, and mechanic as well as the soldier played essential roles in determining the final outcome of the conflict. From 1860 on, warfare became irretrievably committed to the machine, which gave an increasing advantage to the rich, educated, and scientific combatant nation.

The battlefield lost its singular position as the only projection of a nation's wartime effort. "Home front," a 20th century phrase, was nonetheless an established concept of the 19th century. More competent weapons and production techniques to meet greater military needs proved to be only the most obvious demands of the industrialized society at war; there arose further the requirement for homogeneity of outlook and purpose among the populace; the need for civilian armies to be gathered and fed without living off the land; the need to avoid the debilitating ravages of epidemics in the fields and at home; and the demand for the development of domestic self-sufficiency which was curiously paralleled by a need for international expansion to seek markets, raw materials, and prestige. All of these requirements were met by 19th-century technology, which provided the bridge between the ideological birth of the nation-in-arms and its growth to maturity in the great World War of 1914.

During the French Revolution the National Convention had been forced to depend on elders in the public square "to preach the unity of the Republic and hatred against . . . [the enemy]." The 19th century replaced orators with the rotary printing press to achieve the same end of arousing patriotic enthusiasm. The mass printing of newspapers brought into being the power to unify nations by reaching a larger audience than could orators in public squares. News correspondents and photographers became respectable military camp followers, and no campaign, however small, escaped detailed coverage. The printed word, supported by photographs, dispelled the former sense of remoteness of the battlefield and helped every individual during wartime to feel himself involved in a mighty communal effort.

Nor was medical technology ignored in the mobilization of national resources. The last half of the 19th century marked a turning point in the epidemic history of the Western world. Communicable diseases were, of course, still plentiful, but the widespread pestilences of earlier days showed a drastic decline. The introduction of vaccination by the English physician Edward Jenner (1749-1823) had conquered smallpox, and the value of rail transport in the distribution of food and the development of intensive agriculture provided safeguards against famine with its attendant diseases. Concurrently, the rise of modern medicine and the organization of local, national, and military sanitation supervision made their distinct contributions. Disease still took its toll, but after mid-century the conquest of widespread epidemics greatly reinforced the strategic strength of nations and armies.

As national armies increased in size and the threat of blockade assumed new importance for industrialized states, the requirements grew for a larger and more dependable food supply. The early 19th-century beginnings in agricultural science expanded rapidly after mid-century to meet these needs. Developments in chemical fertilizers and farm machinery allowed much greater acreage-return with the expenditure of less manpower.

Not only was more food grown, but methods were found to prevent spoilage. In France, Nicolas Appert successfully preserved food in hermetically sealed bottles and later cans. Acceptance of his products by the French navy for shipboard consumption marked the first military use of the then revolutionary concept of food preservation.

In America, the Civil War spurred the use of canned goods. The great commercial food houses of Borden and Armour got their start by supplying tinned milk and meat to the Union armies. The development of canning methods in volume gave birth in turn to the modern military food-supply system. Troops were no longer required to spend as much energy in foraging for food in the surrounding countryside. Just as in the case of railroads, canned food allowed war to be geographically expanded, for armies could subsist adequately, if monotonously, on tinned rations in areas that provided no adequate sources of native food supplies for large numbers of outsiders. The late 19th-century campaigns of imperial conquest in Africa owed much of their success to the availability of canned food supplies which met the otherwise almost impossible provisioning problems encountered in tropical or semi-arid lands.

INDUSTRIALISM AND TACTICAL LAG

The American Civil War and the European wars in 1866 and 1870 stimulated the spread of industrialism and, in turn, nationalism. Nations sought to become industrially self-sufficient as a basic strategic goal, and they imposed severe tariffs on external manufactures in order to safeguard and develop domestic production. Rapidly expanding industrialism, however, required outside help in the form of raw materials for industrial production and of markets where excess manufactured goods could be sold. The major states found themselves pursuing two courses of action: protectionism against industrial equals and expansionism against technological inferiors.

The United States was able to resolve this paradox happily by a nation-building, westward expansion which required only a small professional army to fight against the technically inferior Indians. The European powers, and Japan, emerging in the Far East, turned to imperialism. Great Britain, France, Germany, Italy, and Russia engaged in an expansionist rivalry, and technological attention was showered on their respective armies and navies as instruments of this drive.

As the major powers pushed their way into underdeveloped areas, it became obvious that no primitive people could withstand the onslaught of even a small force with modern arms. Some theorists felt it reasonable to assume that products of machine production would likewise replace men in battles between technological peers. Certainly, it was argued, wars could now be determined by small professional armies if they were provided with superior firepower and mobility. The reverse of this hypothesis came to pass.

The technological increases in the lethal power of weapons had in almost all cases created advantages which accrued to the defensive. Yet for the military strategists it was axiomatic, and correctly so, that only the offensive could win victories. The defensive thus became technologically stronger, while the offensive was forced to seek its strength in the psychological domain of military *élan* (enthusiastic spirit). In the final analysis, the only choice offered by technology as the 19th century ended was to combat superior defensive firepower with superior offensive manpower. To this end, in the era of machine guns, magazine rifles, and quick-firing artillery, paragraphs 346-348 of the German infantry drill regulations of 1899 are of interest:

> When the decision to assault originates from the commanders in the rear, notice thereof is given by sounding the signal "fix bayonet" . . . .
>
> As soon as the leading line is to form for the assault, all the trumpeters sound the signal "forward, double time," all the drummers beat their drums, and all parts of the force throw themselves with the greatest determination upon the enemy. It should be a point of honor with skirmishers not to allow the supports to overtake them earlier than the moment of penetrating the enemy's position. When immediately in front of the enemy, the men should charge with bayonet and, with a cheer, penetrate the position.

By the turn of the century, technology had created a vast incompatability between tactics and weapons; not for forty years was the balance to return, and then only through the imaginative use of the internal-combustion engine both on the ground and in the air.

CHANGES IN NAVAL WARFARE

Naval warfare was influenced by 19th-century technology to an even greater extent than land warfare. In the thirty years after 1860, sails, wooden hulls, and smoothbore guns were replaced—in short, everything important about a ship of the line changed. The transformation in the means of propulsion, construction, and armament of ships demanded a concomitant switch in naval tactics.

The great transition from wood and canvas was made possible by the introduction of malleable iron, and later, steel to ship construction and the even more important development of compound engines capable of driving a vessel for sustained distances. Commencing with the first steam warship, a harbor defense vessel built by Robert Fulton in 1814, the use of steam propulsion steadily advanced during the 19th century. Early steamships, however, lacked range and power. In 1830, an average of nine pounds of coal was required per horsepower hour—a consumption rate which used up 40 per cent of the ship's total cargo capacity for carrying fuel. By 1870, improved engines required only two pounds of coal to produce the same amount of energy. This more efficient conversion of coal to power made the steamship a truly global weapon, greatly enhanced the potential military power of industrial states, and for the first time made naval fuel a strategic problem. Coaling stations became important goals of the new imperialism; they were built and seized throughout the world, and in addition to supplying military needs, spurred the development of commercial steamers and international trade.

The use of wooden ships clad with iron plates had been a tactical innovation of no little importance to naval warfare. But the second great 19th-century strategic naval development was the construction of iron ships which, together with armor plate, made the modern battleship possible. In 1860, the British launched the 9000-ton ship *H. M. S. Warrior.* Far larger than any wooden ship could possibly be, she boasted 4½ inch armor plate and was the first capital ship built throughout with iron. Two years later at Newport News, the success of the Confederate ironclad *Merrimac* against conventional ships stimulated a technological spiral of mighty proportions. Only another ironclad ship, the *Monitor,* could halt the depredations of the *Merrimac.*

The construction of sailing men-of-war had been a lengthy and complex process, consuming ten, sometimes as much as twenty years. It required specialized craftsmen, properly selected and seasoned wood, and extensive government shipyards. These factors were acceptable since the sailing ship might reasonably be expected to stay in commission a half-century or more.

Iron construction completely outmoded the concepts and techniques of traditional ship-building. Warships could now be built in private shipyards in two

Fig. 29-4. The great fight between two ironclad ships, the *Merrimac* (left) and the *Monitor* (right), March 9, 1862. Note the revolving gun turret on the *Monitor*. (*The Bettmann Archive*)

years or so, using relatively unskilled labor. Improvements in ordnance, design, and propulsion could be readily assimilated. Rapid obsolescence was the natural result, and far from being shunned it was accepted by governments as a necessary technological by-product. Naval ordnance was a case in point. In the twenty-five year period following 1860, the maximum weight of the shipboard gun changed from the 4¾ tons of the smoothbore, cast-iron, 68-pounder to the 111 tons of the British steel, 16-inch, naval rifle.

By 1895, the warship had attained its modern form—powerful in speed and destruction, protected by armor two feet thick in places, she was a stirring instrument of national power. The traditionalists, who had decried the descent of the sailor from the rigging to the boiler-room, were left unheard. Technology gripped naval science and held it fast. Offense was pitted against defense, and one could gain technological ascendancy over the other for only a brief period. As the century passed, even land-oriented nations could not resist the lure of the grand fleet, and the naval race between Britain and Germany that followed was to bring Europe to the brink of war on more than one occasion. By the end of the 19th century the submarine had passed from the experimental to the practical stage. Its future role in naval warfare was assured by the concurrent development of the Whitehead torpedo (named after its English inventor). This marine projectile, stabilized by gyroscopic devices and powered by compressed air, had before 1900 attained a speed of 28 knots and a possible range of 1000 yards. The first victim of the modern torpedo was claimed during the Chilean Civil War of 1891, when torpedo boats sank the warship *Blanco Encalada* from a distance of about 50 yards. Merging the submarine and torpedo created a deadly weapons system which would completely alter naval concepts.

MILITARY TECHNOLOGY AND PRIVATE INDUSTRY

Broad military and naval technological changes brought new industries into being and also created closer ties between government and business. The increase of military contracting to private companies and the growth of the joint-stock principle started a modest but growing trend toward tying government expenditure to private dividends. By the 1880's some English firms engaged in naval construction were granting returns of 15 per cent to their stockholders. The expectation of profits spurred the technological race even further by creating and stressing competitive research and development facilities among industries.

Technological rivalry also increased among nations, and new developments with possible military application became jealously guarded state secrets. Even the venerable art of spying was affected as technological espionage assumed greater importance than procurement of more traditional military information.

DEATH-DEALING TECHNOLOGY

In the 19th century, particularly from 1850 on, the range, scope, and destructiveness of war increased to an unprecedented degree. Weapons shared generously in the technological abundance of the period, so much so that many thought the limits of lethality had been reached. Others, in some cases those charged with sending men into battle, chose to downgrade the human spirit and ignore the effects of technology in resolving the little-understood dilemma of man against machine.

Thus an unnamed Prussian general, at the end of the 19th century, could say of a generation that was shortly to endure with dogged courage the greatest blood-letting in history:

> Our modern personnel have become much more susceptible to the impressions of battle. The steadily improving standards of living tend to increase the instinct of self-preservation and to diminish the spirit of self-sacrifice. The spirit of the times looks upon war as an avoidable evil, and this militates directly against that courage which has a contempt for death. The fast manner of living at the present day undermines the nervous system, the fanaticism and the religious and national enthusiasm of a bygone age are lacking, and, finally, the physical powers of the human species are also partly diminishing.

30 / The Spread of the Industrial Revolution

HERBERT HEATON

Great Britain was the birthplace and cradle of the Industrial Revolution, and Britons were the pioneers on many technological frontiers. They mechanized the processes for producing cloth, harnessed steam power, and by drawing on their rich deposits of coal broke the metallurgical bottleneck—charcoal fuel—which had condemned iron to be a rare metal and steel almost a precious one. They increased the efficiency of their machines and engines by developing machine tools which made cylinders that were cylindrical, parts that were precise in size and shape, and cogwheels that meshed snugly together. They transformed transportation by putting engines on wheels and in ships. They gave some thought to applying the growing body of scientific knowledge to industrial problems. To exploit this new technology they developed the factory organization of production and sank large sums of fixed capital into their enterprises, with little or no direct aid from their government or foreign investors. It was almost a classical case of "free enterprise"—or of do-it-yourself.

From the first, British observers were quick to note that the occupant of this economic cradle was a rather remarkable babe; they felt that "a revolution is making," and forecast that the early spinning machines, coke-fed blast furnaces, and steam engines would "produce great changes in the appearance of the civilized world." Foreign observers were no less impressed, though their admiration might be tinged with fear, envy, or covetousness. A Swedish visitor to a Welsh iron works in 1767 wondered if what he saw there might not do dreadful injury to his country's export of iron. Prussian bureaucrats who managed state-owned mines and manufactures came to peep at (or purchase) Watt's engines.

When peace reopened lines of communication across the North Atlantic, Americans learned with surprise that the British had not been industrially idle since 1775. "Strange as it may appear," wrote Tenche Coxe, a federal employee who surveyed the American economy shortly after the foundation of the United States, "they also card, spin, and even weave, it is said, by water in the European manufactories." It therefore followed, in the opinion of Coxe, Alexander Hamilton, and many others, that the newly won freedom of the United States must be supplemented by economic independence, that Americans must develop their own manufactures, and begin by procuring "all such machines as are known in any part of Europe." France, long famous for her state-fostered production of luxury goods for kings and courtiers, had by 1750 realized there was a growing market for necessities as well; that Britain was well ahead of her in supplying some of this market; and that she would be wise to copy "the English model." And so it went on throughout the first half of the 19th century as Britain became known as "the workshop of the world."

THE TRANSIT OF TECHNOLOGY FROM BRITAIN TO CONTINENTAL EUROPE

From admiration to imitation seemed a desirable, even an essential step, and not an impossible one. For history abounds in illustrations of the transit of technology, despite the efforts of innovators to ward off imitators and of governments to protect any monopolies, resources, and technical superiority enjoyed by their country's economy. Instances range from ancient China's attempt to maintain its stranglehold on the world's silk supply by banning the export of silkworms to the United States' effort after 1945 to prevent the spread of nuclear know-how. Yet virtually every government has been eager to welcome skilled immigrants and to import materials or equipment that will add variety and strength to its industrial capacity.

This two-edged policy of "beggar my neighbor" is well illustrated in British history. For example, Belgian Protestant refugees were welcomed since they would introduce the making of "new draperies" to Tudor England, as were French Huguenot (Protestant) silk-workers when Catholic France became intolerant toward their faith in 1685. In 1718 Parliament granted first a patent, then a gift amounting to $70,000, to an Englishman who had constructed a silk-processing machine based on patterns and drawings he had smuggled out of Italy. On the other hand, Britain forbade the export of wool from 1660 to 1825, while a series of 18th-century laws banned the emigration of skilled artisans and manufacturers and the export—except by special license—of tools, utensils, and machines, or models or drawings thereof, used in making cloth, hardware, paper, glassware, and cannon.

These prohibitions fell far short of achieving their objectives. For one thing, enforcement was administratively impossible in a land which had a long coastline. Smuggling was an ancient art, and countless artisans undoubtedly escaped by giving false names or by falsifying their occupations. Meanwhile foreigners snooped around factories, iron works, and mines, and frequented taverns in search of artisans who might give them information, smuggle them into industrial plants, or be willing to emigrate. Eventually Parliament recognized these realities when it lifted the ban on emigration in 1824 and on machinery exports in 1843.

By that time the exodus of men and machines had been going on for at least a century, increasing in volume to continental Europe after 1750 and to North America after the American Revolution. Only a few outstanding instances need be cited. John Holker, a Lancashire cotton manufacturer, escaped from jail and from probable execution for treason, landed in France in 1750, and quickly won the French government's approval of his plan for modernizing that country's textile industries by introducing English machines, artisans, and methods. By 1754 he was running a factory thus equipped and staffed. From 1755 to 1786 he also served as government inspector-general of factories, toured the country stimulating the transformation of cloth-making, and influenced to some extent the methods of producing iron, arms, and chemicals. To Holker and the emigrants who joined him or went to other countries goes much of the credit for introducing mechanical spinning in France, Belgium, Germany, Switzerland, and the province of Alsace by about 1800.

The Cockerill family took the continental Industrial Revolution at least two stages further. William Cockerill, a Lancashire carpenter skilled in making spinning jennies, found a home and a job in 1799 in Belgium, a land at that time occupied by French troops. He and his sons established workshops in Verviers, Liège, and Rheims, in which they built machines for at least a score of Belgian and French textile firms. In 1817 the youngest son, John, purchased the chateau and estate of Seraing, near Liège, and developed there a vast engineering complex which pushed forward from making textile equipment to producing engines, steamboats, locomotives, rolling stock, bridges, cannon, and other heavy iron products, while at the same time proceeding to mine coal, smelt ore in coke-fed blast furnaces, puddle and roll it into plates, rails, and other shapes. His industrial empire was thus integrated from raw materials to finished products; he controlled or owned sixty separate units scattered over Western Europe, and boasted that he had all the new inventions at Seraing ten days after they came out of England. The firm, a joint-stock company bearing his name, still survives as one of Belgium's leading iron and steel enterprises, a memorial to the family which enabled Belgium to follow most closely in Britain's footsteps and provided the gateway through which the Industrial Revolution entered the continent of Europe.

The early stages of the new-style industrialization in other countries benefitted from the entry of British machines, artisans, managers, entrepreneurs, and investors, as well as from the visits, fleeting or prolonged, of continental observers. Alfred Krupp of Essen spent a year under a false name learning how steel was made in Sheffield and various other processes. Every later mechanical innovation led contintental manufacturers to buy the new models and engage British artisans to install them, keep them running, and teach the natives how to operate them. The dawn of the Railroad Age in 1825 opened up a vast field for British investors, construction firms with their gangs of workers, engineers, locomotive builders, and producers of railroad iron. The capital, engines, and drivers for the first important French railroad—Paris to Rouen (1843)—were all British. Brassey, builder of that line, spent the next two decades doing similar jobs elsewhere, and at one time was busy simultaneously on four continents. Even the standard width of track (4′ 8½″) was a British export.

THE TRANSIT OF TECHNOLOGY TO THE UNITED STATES

The transit of technology to the United States has long been exemplified in the history textbooks by narrating how young Samuel Slater slipped out of England in 1789. Since he had been an apprentice, then overseer, in a mill filled with Arkwright's inventions, the details of their construction and operation were all tucked away in his head. He made contact with a Providence, Rhode Island mercantile firm that was struggling to develop a spinning mill. By the end of 1790 he had built three machines "all on the Arkwright principle," rebuilt and improved others, and was driving them by water power in an old fulling mill. The firm's output doubled within a year; the British ban had been bypassed; and Slater thus embarked on four prosperous decades of textile manufacturing in addition to winning repute as the "father of the American cotton industry."

Slater was one of thousands of British industrial emigrants entering the United States between 1783 and 1812. While many of them knew only the old ways of work, a goodly number came from counties where new equipment and know-how had made headway. Some had been "seduced" or "decoyed" by the offer of a well-paid, responsible job by American labor recruiters; these words being used in the British laws which threatened "divers ill-disposed persons" who induced artisans to emigrate with imprisonment and a fine amounting to $2500 for each person decoyed. This recruiting became vigorous in the mid-1780's when a heavy surplus of imports in the United States provoked an economic and patriotic clamor for reduction. Societies "for Establishing Useful Manufactures" popped up all the way from Boston to Richmond. Bounties, premiums, and prizes were offered—it was news of such rewards that drew Slater to America. The crusade to lure British technicians became so loudly active that President Washington had to decline further direct participation in it because "it certainly would not carry an aspect very favorable to the dignity of the United States for the President in clandestine manner to entice the subjects of another nation to violate its laws."

This zest for "instant industrialization" weakened when some grandiose schemes ran out of funds and when the outbreak of the French Revolutionary Wars in 1792-93 attracted capital and enterprise into the profitable pastures of neutral shipping and trade. Only when difficulties with Britain after 1805 led the United States into an economic "cold war"—with embargo, non-intercourse, and non-importation laws—then into the "hot" War of 1812, did some merchants once more transfer their capital and attention to producing the goods they could not import. When that war came, all unnaturalized British males over fourteen years old were ordered to register as "alien enemies." The register, far from complete, gives the occupations of about 7500 men. Of this number at least 3000 were industrial workers; about 1000 of them were making cloth or constructing textile equipment, and since over half of the latter group had come since 1808 many of them were familiar with the latest textile techniques. One recent arrival was managing a new firm established at Ballston, near Albany, New York, to produce the superfine woolen broadcloths for which his native West of England was famous. Of him his employer wrote,

> He is a mechanic, artist, and in many respects a man of science. His knowledge in these fields has induced me, together with one or two others, to put our children apprentices to him in order to learn the business in the most perfect and extensive manner he is capable of teaching it.

THE ADAPTATION OF TECHNOLOGY

Industrialization involved much more than merely acquiring men, machines, and methods. Once these were obtained, what followed depended on many other factors: the availability of natural resources and raw materials; awareness of further technological improvements; the supply, cost, and efficient use of capital and labor; facilities for transportation; the size and range of the market; and the political and social climate, stormy or sunny, in which entrepreneurs operated. Lack of space makes it impossible to examine all these factors, and we will concentrate on the supply of raw materials, the effect of technological changes on their possible uses, and the role of transportation improvements, with special reference to the key industries of iron and coal.

THE BASIC MINERALS

The marriage of coal mining and iron-working to make the modern complex of heavy industry could not take place unless both coal and iron were present. Only in England had they approached this stage by 1800. On the Continent they had scarcely met, since iron-workers were wed to charcoal and concentrated their activities, often small-scale and even a peasant's by-occupation, in regions which had ore deposits, forests to provide charcoal, and streams that supplied power to operate the furnace bellows and forge hammers. This juxtaposition of ore, wood, and water accounts for the distribution of iron works in the late 18th century over a broad belt of hills, plateaus, and low mountains covering much of France, eastern Belgium, and the southern half of Germany as far as Silesia.

The map (Fig. 30-2) indicates the coal resources known to the 19th century. The "exposed" deposits outcropped on the sides of hills or the high banks of rivers, and had long been worked to provide heat for brewing, distilling, evaporating brine, and kindred industries or for domestic warmth in parts of northern France and the Low Countries. But wherever charcoal and firewood were abundant and cheap, coal received little attention. As a French writer put it (1839), coal was "no more than an object of mineralogical curiosity" in his country until the late 18th century, but since then had been transformed into "a combustile of great importance . . . the raw material of all raw materials."

The date and the success of the "marriage" depended on four favorable conditions. The first was adequate supplies of coal, especially of good coking coal. These rarely appeared in the exposed outcrops, but were discovered during the 1830's and 1840's more than a thousand feet underground in northern France, Belgium, Lorraine, the Saar, and the Ruhr region. The second condition was ample deposits of ore. Few of the older workings combined high quality with great quantity and petered out if vigorously exploited. What proved to be one of the world's largest ore beds outcropped or lay near the surface under much of Lorraine, with extensions into Luxembourg and southern Belgium. Since this *minette* contained phosphorus it could be made into castings or poor puddled

Fig. 30-1. Iron works, late 18th century. Each dot represents a single works; no distinction is made between furnaces and refineries. (Adapted from Pounds and Parker, *Coal and Steel in Western Europe*, Indiana University Press, 1957)

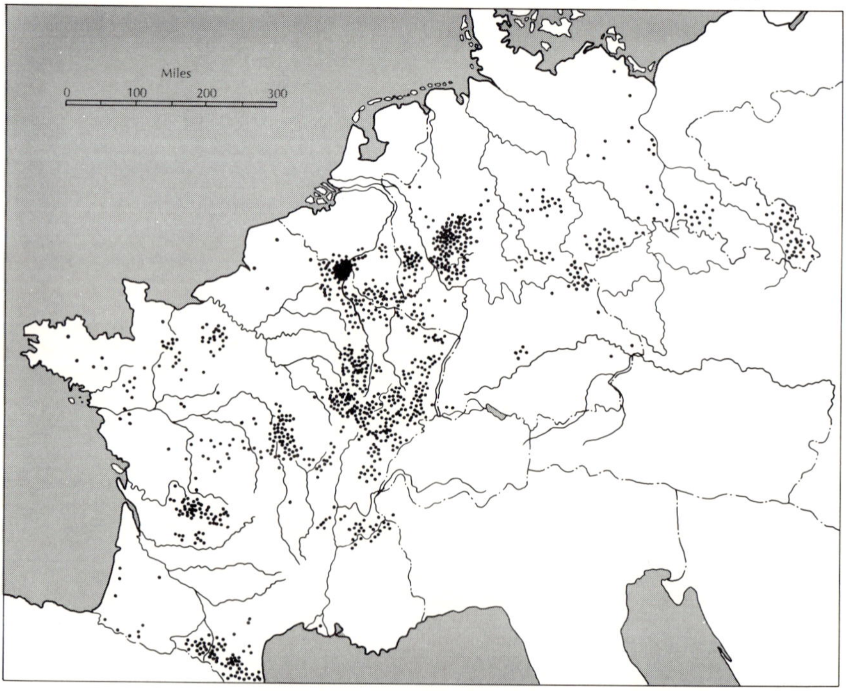

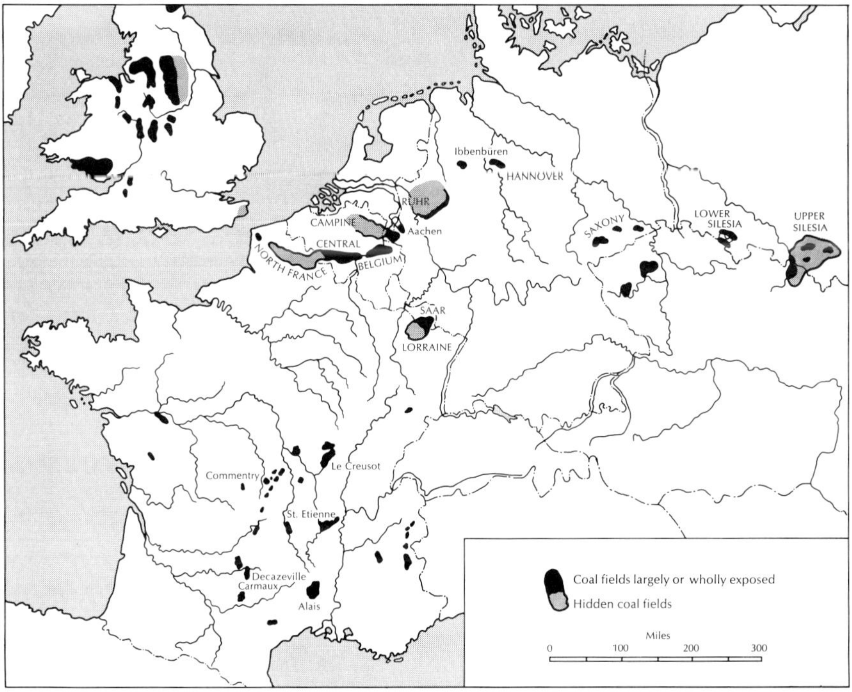

Fig. 30-2. Coal fields of Europe known to the 19th century. (Adapted from Pounds and Parker)

wrought iron, but not into steel until the Thomas process (1878) offered a way to remove the phosphorus. The third favorable, or even essential, condition was cheap transportation of the heavy raw materials and their products to consumers. The fourth was a supply of capital for building the much larger coke-burning blast furnaces, for reaching and working the deep coal seams, and for improving transportation facilities.

BELGIUM

Belgium was the first continental nation to enjoy these conditions. Substantial coal and iron deposits lay close together on or near rivers and canals, of which the small country had by the 1830's one navigable mile for every fifteen square miles of its area, and the government was already laying out trunk railroad lines. In 1823 the state financed Cockerill's first coke-burning blast furnace, by 1842 there were 45 coke-burners against 75 charcoal users, and by 1864 the former were producing an individual average of 10,000 tons per annum while the five charcoal survivors averaged less than 1000 tons apiece. Pig-iron output more than trebled in three decades (1836-66). Puddling, machine-making, and the use of steam power expanded rapidly, while coal mines increased their output ninefold between 1830 and 1870. By the latter date Belgium's heavy industry was well established, integrated, and concentrated on the coal fields. But this rapid rate of growth could not continue indefinitely. Ore supplies ran out, and by the 20th century, imports had to take their place. Deep coal seams, one of them more than 3000 feet down, became more difficult, costly, and dangerous to mine.

FRANCE

Since one-eighth of France was covered with forests, the charcoal furnaces were slow in yielding ground to coke. In the mid-1820's over four hundred of them were producing virtually all the country's pig iron. They doubled their output during the second quarter-century, but their share of the total supply had dropped to just over half by 1850, and fell to a third in 1860, and to a mere 8 per cent in 1870.

By that date total production had quintupled since 1820, but coke had become the normal fuel, first for puddling and later for smelting, only after a hard struggle, till at least mid-century, with problems of capital supply, quality of product, and most vital of all, transportation. For France, unlike Britain and Belgium, was a large political unit—200,000 square miles—but geographically and economically divided into many distinct regions and markets. Her coal and iron deposits lay far apart, especially in the central region, and were remote from customers and seaports. Her roads had fallen into deplorable disrepair during the Napoleonic Wars, and their improvement was intermittent, difficult, and costly. Not until the 1830's did it become obvious that the canals dug since 1815 to link the upper reaches of rivers in the central and northern regions, like the short railroads built by 1830 to give the coal- and iron-workers of Le Creusot and St. Etienne access to waterways, were of limited value because river navigation was possible only part of the year and in the case of the turbulent Rhone, impossible upstream. Coal shipped by canal and river from St. Etienne to Paris, 360 miles away, sold at ten times its pithead price, while that despatched from Mons in October might not end its 150 miles journey to Paris until the following March. Hence, as A. L. Dunham puts it,

> The distances were so great, the watersheds so high and numerous, and the volume of traffic was so slight and grew so slowly that even the wisdom of Solomon could not have provided France with an adequate system of waterways before the advent of the railroads.

The master plan for a national railroad system was approved in 1842, with most of the nine main lines radiating from Paris to ports, industrial areas, and frontier points. Construction progressed well for a few years, but the boom in railroad stock speculation burst in 1847 and the Revolution of 1848 upset the country; also, defects in the plan delayed the completion of the trunk lines till nearly 1860.

But the mileage rose from 500 in 1842 to 2000 in 1850 and to 11,000 in 1870; the economy enjoyed greater mobility at lower costs. Coal prices dropped from ten times to only twice the pithead price, and the journey from Mons to Paris was cut from months to days. The textile and other light industries benefitted from lower freight charges on their raw materials, and the heavy industries found their transportation barriers removed. France shared in the general wave of prosperity which swept the Western world during the 1850's and 1860's. It set up the Crédit Mobilier in 1852 as an investment bank to raise capital by selling bonds to small- and middle-sized savers; the capital was used to finance railroads, mines, and other heavy industries. Coal and iron production mounted to meet the growing demand, and the heavy industries became firmly established.

Unfortunately, France proved to be poor in coal reserves, and continued to be dependent on imports for a third of her total supply. Some of her ores gave out, and she lost a large fraction of her mining capacity to Germany in 1871 as spoils of war. Nevertheless, much ore remained on the French side of the border, and the Thomas process enabled France to turn this ore into steel. In 1913 France's steel output was 5 million tons—Germany's was 17 million tons. But in that year she exported half of the 22 million tons of ore she mined, most of it to Germany, and imported 45 per cent of her coke needs in return. This dependence on imported coke and exported ore assumed sinister importance when the First World War broke out a year later.

GERMANY

Germany, that is, the more than 200,000 square miles of Central Europe that was merged into the newly created German Empire in 1871, was somewhat slower to exploit its mineral wealth than were its two western neighbors. Its charcoal-burning iron industry, like its farming and handicrafts, was widely dispersed and operated on a small scale. Its large area created much the same diseconomies we have noted in France, but in addition there were diseconomies caused by the political fragmentation of Germany: the Napoleonic Wars had reduced the number of German states from hundreds to less than twoscore, but each was surrounded by its own tariff walls and burdened by heavy tolls on rivers, bridges, and roads. These hobbled trade and tended to limit the market to the state in which products originated. Further, there survived traces of serfdom tying the peasant to the soil, of medieval guilds which bound craftsmen to the job for which they had been apprenticed, and of state decrees which said "thou shalt" or "thou shalt not."

The transition from this old economy and patchwork polity to the new was facilitated by three outstanding developments. The first was the territorial expansion of the Kingdom of Prussia from its Berlin base in the heart of the German plain. In the mid-18th century Silesia was annexed, and Prussian officials strove to develop the mineral deposits of this southeastern frontier by introducing British techniques and technicians. Then the peace settlement of 1815, which brought to a close the long series of French Revolutionary and Napoleonic Wars, chose Prussia to keep "the watch on the Rhine" against France by giving her about 160 miles of the Rhine River, with about 10,000 square miles of the Rhine Province on the left bank and 8000 of Westphalia on the right. This proved good pay in advance for the watchman, since it included the Ruhr and Saar coal fields, some rich ore deposits, some good roads made by Napoleon, and a river system capable of great improvement, including steam navigation.

To the development of its expanded domain Prussia devoted serious efforts. Its officials visited Britain and Belgium or sent representatives there to look at the latest innovations, established an Industrial Institute in Berlin to teach the new ways of work, encouraged road-building, and controlled in every detail the opening up of the Ruhr coal field by private firms. The state had earlier taken steps to abolish serfdom and curtailed the power of the guilds in 1848.

The second constructive development was the movement toward a German "common market." Prussia led the way in 1818 by sweeping away its interprovincial customs duties and replacing them with a moderate duty on goods entering the kingdom. Later on she used bribes or bludgeons on other states, and on the first day of 1834 the customs barriers surrounding seventeen states and 23 million inhabitants gave place to one tariff wall around the new Prussian-dominated Customs Union (*Zollverein*). Other states joined sooner or later, thereby producing an economic bloc almost identical with the German Empire created in 1871.

The third useful development was the coming of the railroads as "carriers and customers." The first short lines, laid in Saxony between 1835 and 1840, immediately proved so useful and profitable that other German states quickly followed. These railroads were built by state governments or by companies to which the state contributed some capital and/or guaranteed interest payments on the remainder. The railroad mileage jumped from 300 in 1840 to 3700 in 1850 and 11,000 in 1871. By that year the newborn German Empire was bound together with rail tracks, and well on its way to becoming knit into the economy and polity of continental Europe.

Thus aided, Germany threw off what one historian has called its "economic stagnation" and the "inferiority complex" which had made German ironmasters lament that they could never hope to compete with the richer coal fields, more abundant iron ore, and cheap bar iron of Belgium and Britain. Other things helped: the coal industry discovered large deposits, often deep down, concentrated in a few regions, in Silesia, the Saar, but most of all in the Ruhr Valley and the Emscher Valley a few miles to the north.

The first deep Ruhr mine was opened near Essen in 1841, and further prospecting revealed the wide extent of the field. Soon the railroads came, as did the river barges, and capital flowed in, especially after 1851 when the government gave owners and managers freedom to run their own businesses and halved the royalty on gross output payable to the state. A veritable boom in 1854-57 saw German and foreign investors increase the number of Ruhr mines from 200 to 300, including two financed by an Irish group and christened "Hibernia" and "Shamrock." Germany's total coal production trebled between 1830 and 1850, then quintupled during the next two hectic decades. The figure for 1870—nearly 30 million tons of hard coal—exceeded the joint French and Belgian output and was just over a quarter of the British.

Meanwhile, the ironmasters had discovered that the deep-dug Ruhr coal made good coke, and that some of its seams were separated by layers of iron ore. Since the 1820's they had been using coal for puddling pig iron bought from German charcoal furnaces and British or Belgian smelters. The find of local ore in the 1840's led to the erection of the first coke-fed blast furnace in the Ruhr in 1849. By 1857 there were thirty of them, after joint-stock companies, supported by new investment banks, poured capital into the industry. The Ruhr set the pace in iron as in coal production, and contributed mightily to the leap forward which increased Germany's pig-iron output sevenfold between 1850 and 1871. In the latter year Germany's 1.6 million tons surpassed the output of France and was a quarter that of Britain. Germany was now definitely in the Industrial Big League.

The German "leap forward" fell flat on its face in the world economic crash of 1873, and by 1876 half its blast furnaces were idle. A tariff on imports, and the Thomas process for making steel got the wheels moving again, with the Lorraine ores, now exploitable by that process, carrying Germany into the Age of Steel. Between 1880 and 1900 steel output increased tenfold, and Germany's 7.4 million tons placed it well ahead of Britain, made it second only to the United States, and accounted for a quarter of the world's production. The German steel industry was providing material for trebling the Empire's railroad mileage; for building a merchant marine that by 1914 ranked second, albeit a poor second, only to the British; for expanding the making and exporting of a vast variety of machines and machine products which introduced the label "Made in Germany" to the world; and for the creation of a navy which proved formidable in World War I.

Meanwhile, the electrical and chemical discoveries described in Chapters 26 and 28 above were made by or were quickly attended to by German scientists, of whom the nation had many, as well as by industrialists and bankers who saw the chance of developing two new "growth industries." By 1913 Germany was producing 85 per cent of the world's dyes and housed a quarter of its chemical industry. In the electrical field Germany's progress after 1880 in electric lighting, central power stations, long-distance transmission, streetcar services, production and export of equipment, and discovery of new uses for electric power went ahead at a pace often equalling and sometimes exceeding that of the United States. Steel, electricity, and chemicals had made Germany a highly industrialized nation.

SUMMARY

Against the dominant theme of British economic and technological superiority which characterized the 19th century, there arose a counter-theme of foreign competition which grew steadily through the decades of the Pax Brittanica (that period from 1815 to 1914 when the British navy ruled the waves and when there was no general European war). As we have seen, Britain had been the birthplace of most of the technological innovations of the Industrial Revolution, typified by those in the iron and coal industries, and therefore had a head start in the economic race which ensued. Despite firm and serious national policy, however, Britain's efforts to maintain exclusive possession of the keys to change were thwarted, often through the enterprise of her own people, who wished to make profits through the introduction abroad of British machines and techniques.

By the end of the 19th century, such countries as Belgium, France, the German Empire, the United States, and Japan were using the tools and methods of the Industrial Revolution and were competing successfully against British industry. To the extent that world power is based upon technological superiority operating through the medium of economic and military institutions, it was slowly moving from the exclusive hands of Britain. Technological advances, by their very nature, cannot be monopolized by any single nation. The diffusion of technology and the proliferation of industrial potential among more and more nations of the earth is a process which has not yet reached its end; it continues in the 20th century as the so-called underdeveloped nations seek to borrow the technology of the industrially advanced countries.

31 / The Economic Consequences of Technological Change, 1830-1880

NATHAN ROSENBERG

Nineteenth-century industrialization seems to have wrought certain very similar economic consequences upon countries which were fully caught up in it. Hence a detailed, country-by-country examination of the economic consequences of technical change in the period 1830-80 is unnecessary. Instead, we will focus attention specifically on the American case for illustrative purposes—a case where, essentially, the emerging European technology was transferred wholesale and planted in an environment vastly richer in its natural-resource endowment.

MAN AND NATURAL RESOURCES

Technology commonly refers to all those available means which may be used by men to convert scarce natural resources into forms which satisfy human needs. In his attempt to satisfy his voracious appetite for goods and services, man is everywhere confronted with resource limitations. If we are to appreciate fully the decisive economic consequences of technical change, we must explore the meaning of this relationship.

What is meant by "natural resources"? It might seem as if this question is answerable by an exhaustive inventory of natural surroundings. One might catalogue, in as great detail as one wishes, the extent of mineral deposits, soil content and topography, and rainfall of the physical environment. But what would be the *economic* meaning of such a catalogue? How deep beneath the earth's surface should oil deposits be counted? What minimum ore content of iron deposits should be included in such an inventory of natural resources?

These questions, as posed, are essentially unanswerable. For Paleolithic men neither oil nor ore was at all usable, and the answer given today would differ drastically from the answer given one or two hundred years ago. Clearly, it is not even possible to talk about natural resources in a meaningful way until the level of technology is specified. For example, parts of our natural environment which were useless two hundred years ago are highly valued today—for example, petroleum or uranium deposits.

Economic history is, in large measure, a study of the continually shifting relationship between man and nature resulting from man's growing capacity to exploit his natural surroundings. The interplay in this relationship, however, is a subtle and shifting one which continually alters the relative economic importance of different resources and different geographic regions. Thus, the improvements in transportation which provided low freight rates for the agricultural products of the American Midwest made inevitable the decline of New England farming. The diffusion of the steam engine as a source of power gradually eroded the locational advantages of firms which possessed sites endowed with sources of water power. The Thomas-Gilchrist technique of steel-making uniquely favored German industry since, by making it possible to produce steel from phosphoric ores, it opened up for exploitation the immense iron-ore deposits of Lorraine.

TECHNOLOGY AND ECONOMIC GROWTH

Changes in technology, then, are the critical instruments enabling man to utilize his resource environment more efficiently and to generate the increase in productivity which is at the heart of the process of economic growth. An economy may increase its output in a variety of ways. This may be accomplished first of all, by using more inputs in the productive process: by mobilizing a larger fraction of the population into the labor force, by increasing the number of working hours per laborer, by withholding a larger fraction of resources from current consumption and devoting them to a more rapid accumulation of capital goods (capital formation), and so on. Alternatively, output may be increased by technical innovations—whether developed domestically or borrowed from abroad—which increase the productivity of the existing supply of resources.

Several attempts have been made to measure the relative contribution to economic growth resulting from using more resources, and raising the productivity of resources. Although such studies have involved the use of statistical data of imperfect reliability and heroic simplifying assumptions, they nevertheless all point forcefully and conclusively in the same direction: that is, that the great bulk of the historical increase in output per capita, which was already well under way during the period 1830-80, was due to raising the productivity of resources rather than to using more of them. Thus, if American economic growth had consisted only in the addition of resource inputs of *unchanged productivity*, the resulting increase in output per capita would have been only a rather small fraction of the amount which actually took place.

Nevertheless, it would be erroneous to conclude, for example, that capital formation was therefore relatively insignificant during these years. In fact, capital formation is essential for technical change, since technical innovations acquire economic significance only when they are embodied in new productive equipment. Furthermore, technical change, as it becomes embodied, frequently requires adaptations and adjustments on the part of both capital and labor, as well as shifts in factor proportions. The responsiveness of the factor inputs to the requirements of a changing technology is one of the most important determinants of the rate at which successful technical change will occur. Finally, since technical change involves qualitative as well as quantitative adjustments in factor inputs, it is often extraordinarily difficult to separate the relative contributions of technical change from alterations in factor inputs.

THE INSTITUTIONAL FRAMEWORK OF TECHNOLOGICAL GROWTH

Other forces than technology contribute to economic changes, and although there is a strong presumption that technological change is the predominating force, no satisfactory method has yet been devised which will enable us to assign appropriate weights to the relative contributions of all the forces making for economic growth. Furthermore, we must distinguish between technical change as an addition to knowledge on the one hand, and the actual application and utilization of this knowledge on the other. We know remarkably little about the determinants of technical change, but there is considerable evidence to suggest that many technical changes originate as a response to certain kinds of economic problems, and that an economy's capacity to produce such changes is strongly influenced by the general level of education, and especially by the mastery of technical skills by the working population. Economic systems are complex collections of institutions which hang together and interact in subtle and often unpredictable ways. To disentangle the cause-effect relationships and to weigh costs and benefits in such a way as to arrive at a properly balanced judgment on the role of particular institutions is enormously difficult.

Many people argue, for example, that the American patent system played a vital role in encouraging technical change by holding out to the potential inventor the incentive of high profits during the period of protection offered by the patent law. Yet a priori reasoning does not establish a clear case, and the historical record can hardly be said to pronounce an unambiguous verdict. It is certainly apparent that some people were influenced in undertaking inventive activity by the apparent protection offered by the patent laws. But it is also apparent that, in the burgeoning economy of 19th-century America, powerful and pervasive incentives to undertake particular kinds of inventive activity already existed even in the absence of these laws. What the "incremental" influence of our patent system was it is hard to say. There is no known way of determining how many of the total number of inventions we owe specifically to the particular inducements of the patent system. On the other hand, neither do we know how to appraise the costs of the system. For society may be said to have "purchased" these incremental inventions by granting to all patent holders the right to exclude others from the use of their inventions for a period of seventeen years. This is, of course, one of the central paradoxes of our patent system. Its ultimate justification is that it is supposed to increase the number of inventions available to society. But the system by which this is supposed to be accomplished is one which allows the patent holder to restrict the diffusion of new inventions until the inventor has reaped his reward by exploiting his short-term monopolistic position.

To evaluate the effects of the patent system we would really have to compare the *increase* in output which society owes to that class of inventions which would not have been made at all in the absence of the patent system (plus the increase in output owing to the earlier introduction of inventions which would have come anyway, but later) with the *restriction* in output which the patent laws have been responsible for by allowing all holders of patented inventions to exclude others from using their inventions during the period of patent protection. For, to the extent that the patent system offers protection, it slows down the diffusion of patented inventions. We should also have to add to the costs of the patent system the many well-known abuses of it which have made it possible for powerful firms to exploit the protection afforded by patent rights in such a way as to perpetuate positions of monopoly power. Thus, the patent system has increased output in some ways (benefits) and restricted it in others (costs). What the net balance of the system has been it would be impossible to say in the absence of reliable quantitative information on these two magnitudes.

POPULATION AND PRODUCTIVITY

The most important single overall economic consequence of technical change is a rising per-capita output, that is, a greater quantity of goods produced per worker. But such scanty data as are available do not provide any solid basis for estimating the rate of growth in overall output per capita, since a conspicuous feature of economic growth is that different sectors of the economy grow at appreciably different rates and that therefore the relative importance of these sectors is also shifting over time. Indeed, these shifts constitute, as we will see, some of the most important economic consequences of technical change. For the overall rate of growth of the American economy we can derive some sense of the order of magnitude from an estimate made by Professor Raymond Goldsmith for the period 1839-79. Goldsmith has estimated that, for this period, gross national product per person in real terms (that is, corrected for price changes) grew at an average rate (compounded) of 1.55 per cent per year. The available evidence suggests, moreover, that the period 1830-80 was one during which the rate of growth was accelerating. This rapid rate of growth of output per capita, furthermore, occurred during a period when the growth of total population was extremely rapid, actually exceeding 30 per cent per decade. According to U.S. Census figures, the total population of the United States rose in the following fashion over the period:

Population of the United States

1830	12,901,000
1840	17,120,000
1850	23,261,000
1860	31,513,000
1870	39,905,000
1880	50,262,000

Within the American context, then, technological change must be seen as a force which made it possible for a population, growing with enormous rapidity (quadrupling over the fifty-year period), to attain very substantial improvements in its material well-being. These improvements, to be sure, occurred within a vast geographic environment which contained a rich abundance of natural resources, in sharp contrast to the British or French cases, where industrialization occurred within an environment much more severely constrained by resource limitations. This profusion of resources influenced the pace of American growth and left a distinctive imprint upon its character as well. However, it was the mastery of a sophisticated technology which made it possible to exploit the great potential afforded by this rich resource environment. After all, the million or so Indians who roamed the American continent for centuries before the arrival of European settlers enjoyed the same environment but lacked the technology to exploit it fully for human material wants.

Closely associated with the rise in output per capita during this period were major shifts in the composition of total output. These shifts reflect the changing pattern of expenditures by consumers as their incomes mounted. The most outstanding of these changes was a relative decline in the amount of money spent on food. Although absolute dollar expenditures on food may have increased with rising incomes, they failed to increase as rapidly as income, and therefore the percentage of incomes spent on food tended to decline. Unfortunately, we have no reliable estimates of consumer expenditure patterns for the period 1830-80, but we have estimates of sector shares in commodity output which go back as far as 1839 and which reflect the changing patterns of consumer expenditures. These show that in 1839 some 72 per cent of commodity output originated in agriculture, whereas 17 per cent originated in manufacturing, 10 per cent in construction, and 1 per cent in mining. Agriculture's share of commodity output had declined to 49 per cent in 1879, 41 per cent in 1884, and by the end of the century (1899) had fallen to 33 per cent. On the other hand the manufacturing share rose from its 17 per cent level in 1839 to 37 per cent in 1879, 44 per cent in 1884, and 53 per cent in 1899.

These shifts in the composition of output have their counterpart, of course, in the changing industrial composition of the labor force. For, although many of the most dramatic changes associated with industrialization became most conspicuous in the decades just following the 1880's, it is nevertheless true that by 1880 the United States was no longer a predominantly agricultural society of widespread and loosely connected units, as it had been in 1830. Whereas in 1830 some 70 per cent of the total labor force was engaged in agriculture, by 1880 just under one-half (49.5 per cent) of the labor force was so engaged (by 1900 this figure was to decline to 37 per cent). On the other hand, a category including construction, manufacturing, and independent hand trades, which comprised just over 12 per cent of the labor force in 1820 (when agriculture accounted for 72 per cent), rose to include 23 per cent of the labor force by 1880. Although data on the service industries (including transportation and other public utilities, trade, finance and real estate, education, professional, domestic and personal services, and government) are fragmentary until later in the 19th century, this sector obviously grew substantially, and it is estimated that 24 per cent of the labor force in 1880 was engaged in the provision of services.

Underlying these figures are pervasive and interrelated technical changes in agriculture and transportation. For what they reveal is a new form of social life and organization increasingly freed of its agricultural and rural ties. In 1830 it was still necessary for most of the working population, employing primitive hand implements, such as the sickle or cradle, to engage in food production. This was transformed by the development of cheap iron and steel and by a series of mechanical inventions drawn by animal power which included seed drills, steel plows, reaping and threshing machines, mowers, rakes, and cultivators. The westward movement and heavy investment in farm machinery and implements, and the increasing application of the fruits of scientific research were further encouraged by the passage, in 1862, of the Homestead Act and the Morrill Land-Grant College Act. In the South, the westward movement of cotton culture into the Gulf states was initially a response to the rapidly expanding demands of the British textile mills and had been given a fillip by the invention of Eli Whitney's cotton gin in 1793. Indeed, on the eve of the Civil War cotton alone accounted for over one-half of the total value of American exports.

THE TRANSPORTATION REVOLUTION

The westward movement of agriculture, its growing regional specialization, and the increasingly commercial orientation of agricultural activity were all dependent upon improvements in transportation. Indeed, improvements in the nation's transportation network were critical to the massive geographic redistribution of people and industry throughout the period. Without them the United States would have remained a collection of largely self-subsisting regional units, with interregional trade confined to those few commodities of small bulk and high value which could bear the cost of long-distance transport. The great grain and livestock producing resources of the Midwest would have been of limited national consequence if the cost of transporting their bulky output to eastern metropolitan markets remained prohibitively expensive. The point is general and fundamental: a truly national and integrated economy, where each region specializes in those outputs in which its comparative advantage is greatest, and acquires other goods on more favorable terms through interregional trade, was the creation of what has aptly been called "the transportation revolution."

The "transportation revolution" was a revolution of steam and iron. For the railroad was, above all, the union of steam power with cheap iron and steel. Moreover, steam and iron technology was also transforming water transport in forms such as steamboats on the Mississippi and Ohio river systems, steam-driven ore barges on the Great Lakes, and coastal and transoceanic steamships.

The spectacular nature of the "railroadization" of the United States should not obscure the contributions of the earlier forms of water-based transport. Canals were already providing cheap freight charges to areas they serviced. The Erie Canal, for example, that remarkable achievement of public enterprise, first made possible the agricultural specialization of the trans-Appalachian region. By connecting Buffalo on Lake Erie with the Hudson River, the canal provided a very cheap form of all-water transport between the Old Northwest, bordering on the Great Lakes, and New York City. Indeed, the opening of the Erie Canal reduced the cost of transporting a ton of freight between Buffalo and New York City from $100 to an average of less than $9 during the period 1830-50. New York's unquestioned commercial leadership in fact asserted itself with the superior freight-rate position provided by the Erie Canal.

The emergence of the railroad network coincides with the beginning of our chronological period and continues beyond its close. From its modest beginning with the Baltimore and Ohio in 1830, the railroads grew at an accelerated pace, with well over 20,000 miles of road laid down in the 1850's alone. By 1860 the major cities of the northwest quadrant of the country were joined, albeit imperfectly, to one another and to the eastern seaboard. Railroad building rose to a peak in the 1880's when almost 70,000 miles of rails were laid down, including the completion of a transcontinental system. The technical innovations in transport and the resulting decline in transport costs were, then, a critical link in the creation of a unified national economy and rapidly growing mass market in a country which was already experiencing rapid growth both of total population and per-capita income.

The railroad embodied important accomplishments in engineering in the development of a high-pressure steam engine, in metallurgy (including the later substitution of steel for cast-iron rails), and in civil-engineering techniques in the construction of the railroad right of way. Here, as in so many other areas of iron and steel technology, there were important interdependencies in the process of technical change. On the one hand, the building of railroads gave a strong stimulus to the designing of iron bridges. This experience in the use of iron and in such problems as the calculation of stresses led to further improvements in the construction of steam engines, which fed back into locomotive design. As another example we may cite the improvements in iron-rolling methods from the mid-1840's onward in response to the need of the railroads for increasingly heavy and durable rails. The resulting advances in rolling-mill techniques and in understanding the composition of metals were quickly transferred to other users of iron. In numerous ways the huge demands of the railroads for steam power, metals, and machinery generated improvements which spilled over into other sectors of the economy which relied on steam and iron technology.

The frequency with which, during this period, techniques developed in one industry found useful applications in other industries, should occasion no surprise. For underlying the apparent proliferation of new techniques and new products was a small clustering of technical problems shared by a growing range of industries. These problems principally involved (1) the production of power, (2) the transmission of power, and (3) techniques for shaping metals. The solution to a particular problem in any of these categories, regardless of the industry in which it originated, usually had applications elsewhere.

ECONOMIC ORGANIZATION

The reciprocal relationship between technical change and economic growth must be stressed. We have so far emphasized the impact of technical change in generating a rapid rate of economic growth. It is equally true that a rapid rate of economic growth, with its consequent expansion of markets, is one of the strongest possible encouragements to technical innovations. This is a particularly important factor when, as was increasingly the case during this period, the new technology involved substantial capital expenditures on large-scale plants and equipment. Such commitments are more readily undertaken by entrepreneurs when rapidly expanding markets create hopes for large profits. Furthermore, adaptation to technical change by the labor force, involving as it did industrial, occupational, and locational shifts, encountered less resistance than would have been the case in a less burgeoning economic environment.

The economic consequences to the growing industrial sector of the economy presented by technical change and rapidly expanding markets defy simple summary and description. For what was involved was nothing less than a fundamental change in the whole structure of economic organization. For one thing the technology of steam and iron, with its concomitants of cheap transportation and specialized machinery employing steam power, meant a great shift toward the centralization of productive activities, that is, a smaller number of units. In part this was reflected in the decline in household production, as the factory took over the production of goods which previously were made within the home. This included clothing and textiles generally, household furnishings and implements, and increasingly after the Civil War, the processing and preparation of food products. In addition it meant the decline of numerous industries which previously existed in extremely small workshop units which were widely scattered through the small towns and villages of the country.

The small-scale craftsman catering to a local market, whether he produced shoes, hardware, firearms, clocks, or agricultural implements, was swept away by the new technology. For a distinctive feature of the new technology was a significant increase in the optimum size of the producing unit. In order to exploit the economic advantages of the new techniques, production frequently had to be undertaken on a scale far beyond either the limited markets or meager resources of localized producers. Much of the new technology required a sizable investment in fixed plants and machinery and achieved low per-unit costs only when the volume of output was very large. This was conspicuous, for example, in iron and steel, where technical progress produced continual increases in blast-furnace size. One scholar has found that from 1860 to 1875 the daily output of the largest blast furnaces increased from 40 to 100 tons of pig iron. By 1900 the record was above 800 tons, and the national average was 500 tons per furnace. The large Pittsburgh furnaces were smelting as much daily as charcoal furnaces had made yearly around 1850. In the production of so prosaic an item as barbed-wire fencing (which played an important role in the economic development of the western regions of the United States) specialized machinery made low production costs possible, but only when the volume of output was extremely large. For many kinds of durable goods, although small units were *technically* feasible they were economically inefficient; considerations such as fuel economy or disproportionately large construction costs rendered very small blast furnaces or steam engines inefficient economic units.

The rapid growth of markets, then, made it possible to realize the growing benefits to be achieved from economies of large-scale production, and brought in its wake an increasing centralization of productive activity. By the last decade of the 19th century industrial bigness and a growing tendency, in many industries, toward the concentration of a large fraction of output in the hands of a few firms, had become a matter of widespread national concern. Out of this concern emerged the Sherman Antitrust Act and a host of regulatory legislation and agencies to prevent monopolies by large industrial corporations.

It is perhaps partly because of the increasing public preoccupation with the economic, political, and social problems associated with a relatively small number of industrial giants that much less attention has been devoted to accompanying changes in industrial structure which did not necessarily involve very large size or high degrees of industrial concentration. For, if we think of productive activity in its vertical dimension as a sequence of stages from the extraction of raw materials through the various refining and processing stages until eventual conversion into a final product, and if we consider furthermore the range of products produced by an individual firm (that is, its horizontal dimension), we encounter significant changes which have been comparatively neglected.

One important characteristic of the age of steam and iron is the elongation of the vertical structure of industry in the sense of the springing up of forms of activity and specialization at the intermediate stages which had no counterpart in the earlier, more primitive division of labor. The hand-loom weaver was not only replaced by the power loom but also by the growth of a new textile machinery industry and, further, by the emergence of yet another specialized industry which produced the machines which, in turn, made the textile machinery. Similarly, small farm households producing their own food supply were replaced not only by factory workers producing agricultural machinery but also by a transportation and distribution network which made agricultural products available to city-dwellers far from the sources of food production.

It is apparent that we cannot encompass the economic consequences of technical change by concentrating on the activities of an individual firm or even an individual industry. We will, therefore, concentrate on one industry which was, in a sense, at the heart of the entire process of industrial development: the machine-tool industry. A brief discussion of this industry will not only serve as an example of the elongation of the productive process; it is additionally important because machine tools constitute perhaps the most significant example of this process. It was the industry which produced the machines which in turn produced the other machines which embodied the new technology.

THE MACHINE-TOOL INDUSTRY

At the beginning of our period (1830) there was no separately identifiable machine-tool or machinery-producing sector in the American economy. Although machines of varying degrees of complexity were, of course, being used, the production of the machines had not yet become a specialized function of individual firms. Machines were, by and large, produced by their ultimate users on an ad hoc basis, that is, as they were needed for particular purposes. Machinery production was not only unspecialized in the sense that each producer typically undertook a wide range of output, it was also a highly localized operation, where each producer restricted his work to a limited geographic radius. For many years, the most intractable problems associated with the introduction of techniques of "machinofacture" lay in the inability to produce machines which would perform according to the special and exacting requirements and specifications of the machine user. The prolonged difficulties which Eli Whitney encountered in attempting to fulfill his government musket contract is a well-known example. A major episode, then, in the process of industrialization lay in the emergence of a specialized collection of firms devoted to solving the unique technical problems and mastering the specialized skills and knowledge requisite to machine production.

The growth of independent machinery-producing firms occurred in a continuing sequence of stages roughly between the years 1840 and 1880. These stages reflect both the growth in the size of the market for such machines and the accretion of technical skills and knowledge (and growth in the number of individuals possessing them), which eventually created a pattern of product specialization by machine-producing firms which was closely geared to accommodating the requirements of machine users.

In the earliest stages, machinery-producing establishments made their first appearance as adjuncts to factories specializing in the production of a final product. Thus the first machine-producing shops appeared in New England attached directly to such textile firms as the Amoskeag Manufacturing Company in Manchester, New Hampshire, and the Lowell Mills in Lowell, Massachusetts. As such shops achieved success as producers of textile machinery, they gradually undertook, not only to sell textile machinery to other firms but to produce a diverse range of other types of machinery—steam engines, turbines, mill machinery, and (most important) machine tools—as well. In the early stages, then, skill acquired in the production of one type of machine was transmitted to the production of other types of machines by this very simple expedient whereby a successful producer of one type of machinery expanded and diversified his operations. Thus, with the introduction of the railroads in the 1830's, the Lowell Machine Shop (which became an independent establishment in 1845) became one of the foremost producers of locomotives. Similarly, locomotives were produced by the Amoskeag Manufacturing Company; and the locomotive works in Paterson, New Jersey, grew out of the early cotton textiles industry in that city. The most successful of all American locomotive builders, the Baldwin Locomotive Works in Philadelphia, grew out of a firm previously devoted, among other things, to textile-printing machinery. An official British observer who visited the United States in 1851 wrote:

> The practice which prevails of combining various branches of manufacture in the same establishment, would . . . render separate descriptions of each somewhat complicated. In some cases the manufacture of locomotives is combined with that of mill-gearing, engine-tools, spinning, and other machinery. In others, marine engines, hydraulic presses, forge-hammers, and large cannon are all made in the same establishments. The policy of thus mixing together the various branches arises, in addition to other causes, from the fact that the demand is not always sufficient to occupy large works in a single manufacture.

Whereas the production of heavier, general-purpose machine tools—lathes, planers, boring machines—was initially undertaken by the early textile machine shops in response to the internal requirements of their own industry and of the railroad industry, the lighter, more specialized high-speed machine tools—turret lathes, milling machines, precision grinders—grew initially out of the production requirements of the arms-makers. Somewhat later, the same role was played by the manufacturers of sewing machines and, around the turn of the century, by the demands of manufacturers of bicycles and automobiles.

The machine-tool industry, then, grew out of a response to the machinery requirements of a succession of particular industries. While still attached to their industries of origin, these establishments undertook to produce machines for diverse other industries, because the technical skills acquired in the industry of origin had direct application to production problems in other industries. Finally, with the continued growth in demand for an increasing array of specialized machines, machine-tool production emerged as a separate industry consisting of a large number of firms most of which confined their operations to a narrow range of products—frequently to a single type of machine tool, with minor modifications with respect to size, auxiliary attachments, or components. In the last decades of the 19th century the industry had also acquired a high degree of geographic centralization. In casting a backward look, the 1900 Census pointed out:

> In late years . . . manufacturers starting in this branch of industry (metal-working machinery) have very generally limited their operations to the production of a single type of machine, or at the most to one class embracing tools of similar types. For example, there are large establishments in which nothing is manufactured but engine lathes, other works are devoted exclusively to planers, while in others milling machines are the specialty.
>
> This tendency has prevailed in Cincinnati perhaps more than in any other city, and has been one of the characteristic features of the rapid expansion of the machine-tool industry in that city during the past ten years. During the census year there were in Cincinnati 30 establishments devoted to the manufacture of metal-working machinery, almost exclusively of the classes generally designated as machine tools, and their aggregate product amounted to $3,375,436. In seven shops engine lathes only were made, two were devoted exclusively to planers, two made milling machines only, drilling machines formed the sole products of five establishments, and only shapers were made in three shops.

The Census also reported that there were ninety metalworking machinery firms in the five leading centers of Cincinnati, Philadelphia, Providence, Hartford, and Worcester.

Thus the machine-tool industry itself was generated as a result of the specific production requirements of a sequence of industries which adopted techniques of machine production in the last sixty years or so of the 19th century. In each case, the introduction of a new process or a new product required an adaptation and adjustment in the industries which produced capital goods (such as machine tools) to new technical requirements and specifications which did not initially exist. There took place, as it were, a period of technical gestation at the intermediate stages of production, during which time the appropriate accommodations were made to the specific technical needs of the new process or product. As the demand for particular kinds of machines became sufficiently great, reflecting the fact that the same machines came to be employed in a progressively increasing number of industries, the production of that machine itself came to constitute a specialized operation on the part of individual establishments. The end result was the creation of an entirely new industry.

URBANIZATION

Some of the further economic consequences of technical change have been implicit in our earlier discussion but must now be brought to the forefront. Perhaps the most significant in terms of its far-reaching implications to a whole society's way of life was the growing proportion of the population which came to reside in urban areas. The technical revolutions in agriculture and transportation sharply reduced the proportion of the population engaged in agricultural pursuits. At the same time technical considerations in manufacturing brought about a vast rise in the optimum scale of productive activity and resulted in a concentration of factory workers in a relatively small number of locations. This concentration was further reinforced by growing specialization and the rise of ancillary services, not only in industry itself but in finance, advertising, and distribution, which heightened the advantages of a location close to them.

The fact that raw materials are themselves geographically concentrated and not randomly distributed, and the unique advantages in terms of transport costs which are afforded by geography to certain sites, all favored a limited number of locations. Thus, the growing concentration of America's iron and steel complex in a few urban areas, stretching along the southern portion of the Great Lakes from Chicago to Pittsburgh, was primarily the result of Mesabi iron ores and Appalachian coal deposits and a magnificent natural system of internal waterways which made it possible to bring these materials together cheaply. Finally, urban concentrations, whatever the reasons for their initial growth, developed certain self-reinforcing advantages. For large population densities tend to be inviting to firms because of the markets which they offer to producers for the sale of their products and for the purchase of their labor inputs.

The period 1830-80 was, then, one of accelerated urbanization. Whereas in 1800 6.1 per cent of the population lived in urban areas (over 2500 persons) and in 1830, 8.8 per cent, the number rose sharply to 19.8 per cent in 1860. By 1880 the figure had risen to 28.2 per cent and at the turn of the century it was 39.7 per cent. In 1830 there was only one city in the United States with a population over 100,000; in 1880 there were twenty. The changing patterns of both production and consumption associated with the growing urban concentration of the population were, of course, intimately connected with the growth in the proportion of the labor force engaged in the service industries. Moreover, although satisfactory data are not available, the whole complex of technical and economic changes which have been referred to brought with them important alterations in the occupational categories and "skill-mix" of the working population.

NEW INSTITUTIONS

Increases in scale and changes in industrial structure and geographic location required new techniques for the organization and management of economic activity and for mobilizing the economy's resources so that they could be directed to their most efficient uses. From this point of view the period 1830-80 was one which saw the evolution of institutions which were to place a decisive stamp upon the character of the 20th-century American economy. The management of technically complex, large-scale enterprises, the increasing importance of decisions involving long-term planning, and above all the huge growth in financial requirements, all rendered single proprietorship or partnership increasingly inadequate. Although the corporation as a legal device has a history long antedating the period with which we are concerned, the development of its modern form as a way of organizing large-scale industrial units occurred in the middle decades of the 19th century.

By issuing equity stock, the corporation draws upon the savings of large numbers of individuals, and by pooling their resources, achieves a scale of operation far beyond that of the family-owned firm or partnership. Moreover, its shareholders are limited in their liability to the size of their investments; that is, if the corporation went bankrupt, they lost only the amount of their investment in its stock. The corporation had been employed in the Massachusetts textile mills in the 1820's, and became common in transportation and financial institutions before the Civil War. The widespread adoption of the corporate form in manufacturing, however, took place only after the Civil War, coinciding closely in time with the introduction of technical innovations which increased substantially the scale of operations of the individual firm.

Closely associated with the rise of the corporation were the growing importance and increasing sophistication of financial institutions and capital markets. The natural caution of the individual investor in the face of his own ignorance and desire for safety typically lead him to channel his savings into forms of economic activity with which he is personally familiar. Such behavior on the part of savers, unless it can be altered, is likely to have a seriously retarding effect upon an economy, for technical change and economic growth require a continual shifting of resources from one sector of the economy to another and from one geographic region to another.

This mobility was gradually improved by the growth of financial intermediaries (primarily savings banks and life insurance companies) which were able to make investment decisions based upon a knowledge of alternatives and degree of expertise far beyond the capacity of the small saver, and by appropriately diversifying their investment portfolios, could offer the small saver a degree of safety beyond what he could achieve by his own efforts. The securities markets also provided the meeting place between the suppliers of funds for investment purposes and those demanding such funds. These markets, primarily in Boston, New York, and Philadelphia, catered initially to local needs but gradually assumed the proportions of a national market. This growing and increasingly unified network of financial institutions and markets matured in the decades following the Civil War. Indeed, the massive sale of treasury bonds during the war itself played a very important role in the development of these institutions and in accustoming the public to holding a portion of its wealth in a form which was still comparatively strange. The degree of capital mobility which was achieved through these new institutions and markets was sufficiently flexible to accommodate the basic needs of a rapidly growing and changing economy.

By 1880 the American economy and society had been transformed in numerous significant ways and already displayed many of the distinctive features of the 20th century. Total output was rising with such speed as to permit a rapid rise in output per capita in spite of an extremely fast growth of total population. The age-old predominance of the agricultural sector was being eroded by improvements in agricultural productivity and shifts in the composition of expenditures associated with rising incomes. American society was becoming increasingly urbanized, with a growing proportion of the labor force engaged in manufacturing and services. Within the industrial sector there were changing patterns of specialization as new industries arose and old ones declined, and economic activity was being organized in larger industrial units. The labor force, which was acquiring a wage-earner and salaried status, undertook activities involving a new sort of work discipline as well as changes in skill, industry, occupation, location, and educational background. The economy had grown to continental scale involving a high degree of regional specialization. It was becoming a unified and integrated system, not only as a result of new forms of transportation, but also because of improvements in the functioning of markets and changes in economic organization which made possible a high degree of flexibility in the factors, and which provided more effective means of mobilizing resources. Underlying all of this were the pervasive technological changes brought forward by the new era of steam and iron.

EUROPEAN DEVELOPMENTS

The countries of Western Europe offer not only interesting contrasts with the United States but, equally, with one another. In none of the "old" countries, to begin with, did population grow at anything approaching the prodigious rate of the United States. During the 19th century, the population of the United States increased by more than 1300 per cent, that of Great Britain almost 250 per cent, and Germany over 200 per cent. France, on the other hand, remained a virtual outsider to the explosive growth in human numbers experienced by the other major Western nations. Her population in 1900 was a comparatively scant 51 per cent over her 1800 level. One result of the slow rate of population growth in France which should be borne in mind is that French economic growth in the 19th century is much more impressive when expressed in per-capita rather than aggregate terms.

It was in Great Britain, of course, that the Industrial Revolution began. In Great Britain, therefore, many of the economic changes which have been discussed with respect to the United States were apparent much earlier than was the case elsewhere. Not only is the acceleration of the rate of growth of per-capita output clearly visible by the last two decades of the 18th century, but the transformation of the economy away from its agricultural center had already been carried very far by the beginning of the 19th century.

As early as 1801 the proportion of the British labor force engaged in agriculture, forestry, and fishing, was estimated at about 36 per cent, a figure doubtless much lower than any other European country, and something like one-half of the corresponding American figure. This figure continued to decline throughout the 19th century. By 1831 it had fallen to less than 25 per cent. The estimate for 1851 is 21.7 per cent, for 1881 12.6 per cent, and for 1901 a mere 8.7 per cent. Great Britain, therefore, not only began her large-scale transfer away from agriculture earlier than other nations; it has also carried the process farther than any other major economy of the world.

The period of most rapid growth of the British industrial labor force occurred in the first thirty years of the 19th century. After 1830, although the industrial sector grew slowly, the most rapidly growing sectors were trade and transport, particularly the latter. Great Britain, of course, was very favorably endowed with respect to water transport. In 1830 the country was already equipped with nearly 2000 miles of canals which, together with her river systems, provided an effective internal network of waterways. The decades after 1830 saw the application of steam power to surface transport, in the railroad, and water transport, in the steamship.

These innovations brought in their wake, as in America, growing markets and changing patterns of specialization. One consequence was to accelerate the decline in the agricultural sector. For the steamship and the railroad opened up the vast agricultural resources of the Americas and later Australia and New Zealand, and made their food and raw material products available in Great Britain at prices with which British agriculture could not possibly compete. Great Britain, now becoming the "workshop of the world," was also becoming a country which acquired an increasing proportion of its food supply through foreign trade, in exchange for the products of its vast industrial establishment. Indeed, foreign trade generally played a far more important role in Great Britain than in other industrializing countries, particularly after Britain adopted a policy of free trade in the 1840's. In the five-year period 1880-84 imports reached their all-time peak of almost 36 per cent of the national income.

The extent of the decline in the agricultural sector in Britain was unequalled in America or on the European mainland. In France, for example, over half of the labor force was engaged in agriculture in 1866, and as late as 1901 the figure was still 43 per cent—higher than it had been in Great Britain a century earlier. The causes behind this relatively limited decline in French agriculture are numerous and complex, reflecting the French social structure and legal forces as well as, no doubt, the well-known Gallic enjoyment of food (and wine) which has caused Frenchmen to allocate a larger fraction of their incomes to expenditures on food than other people at comparable income levels. But there is a much more important point. Given her resource endowment and the peculiar technology which emerged in the 19th century, France could not as advantageously as some other countries shift her resources out of agriculture. The apparent failure of France to shift from agriculture to industry may not therefore have been a failure at all, but merely a reflection of where—knowing the resource requirements of 19th century industrial technology—her greatest comparative advantage lay.

As we have seen, 19th-century Western technology was primarily a technology based upon the use of coal and iron. France, which is rather meagerly endowed with industrial minerals, was particularly handicapped by the relative poverty of her coal deposits. Since the production of a ton of malleable iron or steel required an input of two or three tons of ore but somewhere between seven to ten tons of coal, the development of iron and steel industries were critically dependent upon the location of coal deposits. France was seriously handicapped not only by the inferiority of her coal deposits but by their inopportune location with respect to iron ore as well. Most French coal and iron ore were relatively far apart, as compared with the much greater proximity of these resources in Great Britain and Germany. The development of the French iron industry was therefore largely delayed until the coming of the railroad which reduced the transport costs in iron production from levels which were almost prohibitively high. But the railroads by themselves could only reduce the handicap of France's inadequate resource base, not eliminate it.

As might be expected from the much greater persistence of agricultural activities in France, the social and economic changes associated with industrialization were much slower and more limited in making themselves felt. Perhaps the most significant index of these changes is the extent to which the population shifted from a rural to an urban location. In 1801 the proportion of the French population living in cities was estimated at 21 per cent. Half a century later, in 1851, it had risen only to 26 per cent, and in 1891 it was 37 per cent. For the same years the figures for England and Wales, where urbanization had been carried the furthest, were 26, 45, and 68 per cent respectively. Thus France presents a picture, relative to other Western industrial countries, of a slower rate of growth of total population and a much more modest place in the decline of the agricultural sector and the urbanization of the population. Closely associated was a much greater persistence, in France, of the small independent producer or retailer; similarly, the family as a producing unit has shown a remarkable durability in France. Much of the French agriculture remained wedded to small-scale peasant proprietorship, and in industry and commerce family ownership has remained, to this day, a more common phenomenon than in most other high income, industrial countries.

While technical change therefore imposed the basic direction of the economic changes experienced by industrializating economies, differences in social and legal institutions and in natural resource endowments shaped both the pace and the extent to which individual economies were transformed by these changes.

32 / Population Movements and Urbanization

SAM B. WARNER, JR.

The commonplace generalizations of Western history—population explosions, industrial, scientific, and technological revolutions—obscure many of the basic events which underlie the recent transformation of Europe and world areas of European settlement. Such terms falsely suggest that a long, static past preceded the dynamic history of more recent times.

Actually, the pace of pre-18th-century historical change is better understood as a slowly moving mean about which there occurred wide and often violent alternations. This perception of history as the story of ceaseless fluctuations is especially important to an understanding of 18th- and 19th-century populations, migrations, and urbanization because the modern history of European populations has been the history of ancient demographic responses to new opportunities. Specifically, Europeans responded to abundant new land, to modern urbanization, and to modern industrialization with waves of increases in births.

WORLD POPULATIONS AND MIGRATIONS

Because the poor are and always have been so numerous, the history of populations is the history of the responses of the common people to slight gains and losses in their opportunities and environment. The sheer numbers of the poor magnify these slight adjustments into changes of huge dimensions and often violent outcome. It is in this sense that ancient and modern history, American, European, African, and Asian experiences resemble one another. In American terms our history at least until 1880 was largely the history of the cabin, shack, slum, and farmhouse shelter and the cereal, pork, and milk diet. In European and Asian terms, their history was the history of the cottage, hut, and village shelter and the potato, cereal, pork, milk, and fish diet. At such levels of living any slight change in the rate of births, the amount of land for cultivation, or the movement of farmers and peasants has an enormous impact upon the size of an area's population, its living conditions, its urban patterns, and even its industrial structure.

The novelty of the 18th and 19th centuries lay in the scale and speed with which population interactions took place within Europe, and the world areas of European settlement. The abundant cultivable land in the Americas and the technological capability of moving large quantities of food, goods, and people in ever shorter intervals of time sent quickening reverberations of change throughout America and Europe. In the sense that the masses of poor people responded to slight changes in their opportunities and environment, recent history has thus repeated events as ancient as history itself. However, the uniqueness of these centuries came from the speeding up of these reverberations which successively added to the power, intensity, and scope of these interactions between population and the environment.

Many of the conventional subjects of history can be interpreted in the terms of population interactions. Thus demography (the science of population, including vital statistics) can shed new light on some of the more traditional aspects of history. Furthermore, it can easily be seen how technological changes and demographic changes interacted. For example, abundant new land fed a rapidly growing rural population on the American continent. It also yielded a surplus for American and European cities. The paths of Atlantic commerce carried agricultural staples to purchase manufactures, they also carried migrants. An outmigration of starving peasants or discontented farmers to urban areas created new jobs and markets in the large cities; the same migration created new agricultural opportunities for those who stayed behind. Canals, steamboats, and railroads made the growth of large commercial cities possible, and the new scale and pace of business of these cities forced the modern organization of finance, marketing, and manufacture upon city-dwellers.

WORLD POPULATION:	IN MILLIONS				GROWTH BY PER CENT		
	1750	1800	1850	1900	1750-1800	1800-1850	1850-1900
Europe & U.S.S.R.	144	192	274	423	33.3	42.7	54.4
U.S. and Canada	1.3	5.7	26	81	338.5	356.1	211.5
Latin America	11.1	18.9	33	63	70.3	74.6	90.9
Oceania	2	2	2	6	0	0	200.0
Sub Total: Areas of Heavy European Settlement	158.4	218.6	335	573	38.0	53.2	71.0
Asia	475	597	741	915	25.7	24.1	23.5
Africa	95	90	95	120	—5.3	5.6	26.3
Total World Population	728.4	905.6	1,171	1,608	24.3	29.3	37.3

Note: All figures from A. M. Carr-Saunders, *World Population* (Oxford, 1936), p. 42; in Walter F. Willcox, *International Migrations* (National Bureau of Economic Research, *Publications #18*, New York, 1931) II, 78, estimates are up to 17.5 per cent higher for Africa, and sometimes 15.4 per cent lower for Asia. Population has been allocated to Asiatic Russia according to the method of Warren S. Thompson, *Population and Progress in the Far East* (Chicago, 1959), p. 12.

Every city and country in Europe and America became flooded with newcomers, while at the same time changes in the ways of working, homemaking, learning, and socializing came on so rapidly and so pervasively that traditional institutions and inherited constraints could no longer contain many kinds of individual or group conflicts. In America the 18th century closed with the Revolution, and during the next hundred years there followed the nation's heaviest rural and urban violence and the bloodiest war of the entire 19th century. In Europe, the 18th century closed with a prolonged world war—the French Revolutionary and Napoleonic struggles—and was followed by continuous urban violence and a succession of national wars and revolutions.

INCREASE OF POPULATION

A brief comparison of Chinese, American, and European population changes gives some sense of the blend of old and new which underlies modern demographic history. The Chinese repeated an ancient alternation: rural prosperity led to a surging population growth; this in turn was followed by overpopulation, drastically lowered rural standards of living, famine, pestilence, and wars. At least until 1860, before extensive urbanization and industrialization, the American case resembled the Chinese: extended rural prosperity was accompanied by a very high rate of rural population growth. The European case blended old and new. In the 18th and 19th centuries marginal improvements in peasant and village conditions brought surges of rural population increase. These local increases were in some instances followed by the ancient scythe of falling living standards, famine, and disease; in other instances they were relieved by massive migrations to European cities and to America.

From the 14th century to the 17th century China's population grew in an interrupted fashion, now booming, now halted by wars and famines. With the Manchu Conquest and the beginning of the Ch'ing dynasty (1644-1911) there commenced a 150-year period of peace, order, low taxes, and prosperity. During the 18th century the population of China more than doubled. No agricultural or industrial revolutions supported this growth; the only innovations were government encouragement of sweet potato and maize culture, the clearing of new upland forest lands, and the draining of swamps for cultivation. By the time of the American Revolution, Chinese authors were commenting on the spendthrift habits of the Chinese peasant and the lavish excesses of wealthy city-dwellers. The major explanation for this expansion of the Chinese population seems to lie in the confidence of the common people which led to a heightened birth rate, and perhaps a reduction in infanticide. Once the population had doubled, however, even a sharp reduction in the rate of births and family formation could not stop the further multiplication of the population. By 1800 the optimum matching of China's agricultural population to its resources seems to have passed, and by 1850 the pressure of population upon resources created the modern impoverished and war-torn China. Since the mid-19th century, wars, emigration to North and South America and Southeast Asia, floods, famines, and disease have held the Chinese population to a lower rate of growth and a lower level of living than in the 18th century. The post-World War II forced-draft Communist industrialization program is now attempting to head off the further massive disasters which have in the past sharply reduced the population of China.

The American case repeated the 17th- and 18th-century Chinese experience under even more favorable conditions. The abundance of cheap, albeit often poor quality, farm land kept the great mass of marginal farmers sufficiently prosperous to support large families. Until 1840 almost all the American population growth was supplied by natural increase. Although the birth rate fell steadily from 1830 to 1860, it exceeded any recorded contemporary European rate. After the Civil War, American birth rates continued downward in the general pattern of the urbanized and industrialized European world. At the same time occasional immigration peaks of 800,000 to 1,200,000 new persons per year often held native increases to less than two-thirds of the total national population growth. Even in the 20th century, Appalachia, the rural South, and the mountain states—the nation's poorest, least urbanized, and least industrialized areas—have maintained the nation's highest birth rates. This lingering pattern stands as a reminder of the ancient response of the world's rural peoples to marginal human prosperity. It was this massive native rural population increase which filled the American continent and which created America's 19th- and 20th-century cities.

The novelty of European population experience in the 18th and 19th centuries lay in the availability of transportation to the New World. By the mid-18th century the basic lines of Atlantic commerce had been knit together. Frequent sailings linked British and European ports so that immigrants could reach major Atlantic ports from fringe areas like southwestern Germany. A heavy volume of staples, sugar, flour, tobacco, and lumber flowed eastbound from the New World thereby creating spaces in ships for westbound passengers. Subsequent to the Napoleonic Wars, the growth of a vast cotton acreage in America and cotton manufacturing in England and Europe added a heavy volume to Atlantic shipping. Thus, from the mid-18th century on, cheap westbound passage to America, and to a lesser extent South America, became a constant factor in European population growth and migrations.

EUROPEAN MIGRATION TO AMERICA

European migration to America fluctuated widely during the 19th century, and these variations did not reflect any slight changes in the cheap price of Atlantic passage. Rather, the variations in flows of European migrants seem to have been caused by the interaction of European and American job opportunities and living conditions.

The European birth rate and migrations responded both to the lure of opportunities and the pressure of poverty, famines, and political conflict. For instance, Irish migration to America had been quite heavy prior to the American Revolution. Northern Irish, especially, moved to America as English textile competition and high farm rents drove down living standards at home. After the Napoleonic Wars (1815) Dublin's cotton and silk industry collapsed from similar causes, driving another wave of immigrants to America. Between 1819 and the famine of 1845-49 Irish tenant farmers and village artisans came to America in increasing numbers, not so much driven out by a sudden depression as seeking new opportunities to practise their trade or buy cheap farmland in the United States. The heavy migrations of Scottish Highlanders, rural Englishmen, and Germans in the 18th and 19th centuries similarly reflected a patchwork of specific American attractions and local European expulsions. Indeed, after 1870 a substantial number of skilled workers moved back and forth between Europe and America depending on the job opportunities in their trades.

The contribution of natural increase and net arrivals to the decennial growth of U.S. population, 1810-1910.

	NATURAL INCREASE (PER CENT)	NET ARRIVALS (PER CENT)	TOTAL GROWTH
1810-1820	97.0	3.0	2,399,000
1820-1830	96.2	3.8	3,228,000
1830-1840	88.3	11.7	4,203,000
1840-1850	76.8	23.2	6,122,000
1850-1860	68.0	32.0	8,251,000
1860-1870	75.1	24.9	8,375,000
1870-1880	74.2	25.8	10,337,000
1880-1890	58.8	41.2	12,792,000
1890-1900	71.6	28.4	13,047,000
1900-1910	60.4	39.6	15,978,000
1910-1920	83.6	16.4	13,738,000

Adapted from Conrad and Irene B. Taeuber, *The Changing Population of the U. S.* (S.S.R.C., Census *Monograph Series* I, New York, 1958), p. 294.

The most dramatic of the many European population waves was the Irish expulsion of the mid-19th century. Since the end of the 17th century, protected by peace and feeding off potatoes, the Irish population had grown rapidly; just before the great famine of 1845-49 it was estimated to be 8,500,000. Even before 1845, about 2,250,000 Irish lived every winter unemployed and in semi-starvation. The four-year failure of the potato crop brought disaster to Ireland. Because of the extent of the crop failure, and the conscious policy of the British government, only a pitiful fraction of the available world grains were imported to the island, yet Irish grains were exported as usual. As a result, about 1,500,000 Irish died of starvation and disease on the island during the years 1846-49.

Half a million Irishmen, some by selling property and household goods or using savings, some assisted by relief committees, some aided by American relatives, others shipped out by landlords, attempted to reach the New World. Perhaps 200,000 died in the attempt, either starving or dying of typhus at Irish and English ports, or stricken by disease in the crowded "fever ships" that sailed to Canada and America. Another half million fled to Great Britain where many died, with the rest filling the slums of England and Scotland's new industrial cities.

By 1849, when good potato harvests returned, the population of Ireland numbered 6,500,000, a drop about equal to the semi-starved group of 1841. Massive immigration to America and a reduced birth rate followed these years of horror, so that by 1890 when emigration began to wane, the 18th-century peasant standards of living had been fully recovered. Peasant confidence, however, did not return, and today the population of Ireland, north and south, stands at little more than 4,000,000, only half its 1840 peak.

In general, the waves of immigration to America depended upon a slow tide of marginal rural prosperity which spread eastward across Europe during the 18th and 19th centuries. Slight improvements in transportation and agriculture, migrations to nearby cities, employment in new industries all seem to have contributed to this marginal increase in European rural prosperity. This marginal prosperity created times of peasant confidence, rising birth rates, and ultimately new pools of excess rural population.

The peaks of European excesses of births over deaths occurred in 1825, 1840-45, 1860-65, 1885-90. These heavy increases meant a large number of rural poor and unemployed when the children grew up. Hence nineteen to twenty-eight years after each of these birth peaks came the heaviest migrations to America and presumably also to the nearer booming European cities. The heaviest migration years were 1844-54, 1863-73, 1880-84, and 1902-15. None of these migration peaks, of course, came during American depressions. Indeed, these waves of births, migrations, and business conditions are evidence of a mutually dependent social and economic system which embraced both sides of the Atlantic.

EUROPEAN URBANIZATION, 1800-1900

Just as European world population growth and migration mixed ancient responses with new possibilities, so too the urbanization of the European world blended the steady multiplication of small cities of the ancient commercial type with the modern novelty of numerous huge industrial metropolises. A brief list of the world's major cities reveals the uniqueness of the 19th-century European world. Despite substantial variations in the pace and patterns of industrialization, all the European areas, North and South America, and Australia experienced the very rapid growth and multiplication of industrial metropolises. Asia, on the other hand, remained a continent of many villages and a few large port cities. New York, Paris, and London dwarfed the largest Asian cities. By the same token, Japan, the first Asian nation to industrialize, held that continent's largest cities, Tokyo and Osaka.

The expansion of European commerce since the Middle Ages had, by 1800, created a number of reasonably large cities. Indeed, by 1800, London, a commercial giant and center of manufacturing, could be said to have reached the scale of a modern industrial metropolis. Had there been no further technological change it seems likely that the European world would have been dominated in 1900 by cities of the 200,000 to 300,000 size, with perhaps New York grow-

Recorded immigration by countries and continents 1819-1930
Thousands

	GREAT BRITAIN	IRELAND	GERMANY	SCANDINAVIA	AUSTRO-HUNGARY	RUSSIA	ITALY	ASIA	AMERICAS	OTHER	TOTAL
1819-20	2	4	1	—	—	—	—	—	—	1	8
1821-30	25	51	7	—	—	—	—	—	12	48	143
1831-40	76	207	152	2	—	—	2	—	33	127	599
1841-50	267	781	435	14	—	—	2	—	62	152	1,713
1851-60	424	914	952	25	—	—	9	41	75	158	2,598
1861-70	607	436	787	126	8	3	12	65	167	104	2,315
1871-80	548	437	718	243	73	39	56	124	404	170	2,812
1881-90	807	655	1,453	656	354	213	307	68	427	307	5,247
1890-1900	272	388	505	372	593	505	652	71	39	291	3,688
1901-10	526	339	341	505	2,145	1,597	2,046	244	362	690	8,795
1911-20	341	146	144	203	896	921	1,110	193	1,144	638	5,736
1921-30	330	221	412	198	64	62	455	97	1,517	751	4,107

Adapted from Conrad and Irene B. Taeuber, ***The Changing Population** of the U. S.* (S.S.R.C., Census *Monograph Series* I, New York, 1958), pp. 53, 56.

Some major world cities 1800-1900

	POPULATION IN THOUSANDS				PER CENT GROWTH			
	1800	1850	1880	1900	1800-50	1850-80	1880-1900	1800-1900
United States								
New York (without suburbs)	68	696	1,912	3,437	924	184	80	4,954
Chicago	—	30	503	1,699	—	1,577	238	—
Philadelphia	41	121	847	1,294	195	600	53	3,056
St. Louis	—	78	351	575	—	350	64	—
Boston	25	137	363	561	448	165	55	2,144
Baltimore	27	169	332	509	526	96	53	1,785
Pittsburgh	2	68	235	452	3,300	246	92	22,500
San Francisco	—	35	234	343	—	569	47	—
Latin America								
Mexico City	130	N.A.	N.A.	345	—	—	—	165
Rio de Janiero	43	266	275	811	519	3	195	1,786
Buenos Aires	40	76	236	821	90	211	248	1,953
Europe								
London (without suburbs)	959	2,363	3,830	4,537	146	62	18	373
Paris	547	1,053	2,269	2,714	93	115	20	396
Berlin	172	419	1,122	1,889	144	163	68	998
Vienna	247	444	726	1,675	80	64	131	578
Leningrad	220	485	877	1,133	120	81	29	415
Moscow	250	365	612	989	46	68	62	296
Warsaw	100	160	252	638	60	58	153	538
Rome	153	175	300	423	14	71	41	176
Africa								
Cairo	300	N.A.	375	570	—	—	52	90
Asia								
Tokyo	800	N.A.	N.A.	1,819	—	—	—	127
Peiping	700	N.A.	N.A.	1,000	—	—	—	43
Osaka	350	N.A.	N.A.	996	—	—	—	185
Shanghai	300	N.A.	N.A.	870	—	—	—	190
Calcutta	600	N.A.	612	848	—	—	39	41
Bombay	200	N.A.	773	776	—	—	.4	288
Oceania								
Melbourne	—	—	283	496	—	—	75	—
Sydney	—	—	225	482	—	—	114	—

Adapted from Wladimir S. and Emma Woytinsky, *World Population and Production* (New York, 1953), pp. 121-2.

ing to the size of the Paris of 1800. Instead, the new 19th-century technology of transport and manufacture created the possibility of numerous metropolises. Cities of half a million and more became commonplace elements in the European world.

AMERICAN URBANIZATION, 1830-80

America's experience during its years of transition from a rural to an urban nation demonstrated many of the important interactions which accompanied the recent urbanization of the European world. At the outset one must distinguish between two kinds of urbanization: first, the multiplication of small towns and cities which had always accompanied rural prosperity and easy conditions of commerce; second, the growth of very large metropolises which followed upon 19th-century transport, business, and industrial innovations.

At least until 1880 the United States was a prosperous agricultural nation. The great mass of Americans, native and immigrant alike, lived in scattered farms beyond the limits of even small towns. During these years urban America was small-town America. In 1830 only New York, the nation's commercial capi-

Number of places by size and per cent of total U.S. population living in places of various sizes:

	1800		1830		1860		1880		1900	
PLACES	NO.	PER CENT U.S. POP.	NO.	PER CENT U.S. POP.	NO.	PER CENT U.S. POP.	NO.	PER CENT U.S. POP.	NO.	PER CENT U.S. POP.
1,000,000 and more	—	—	—	—	—	—	1	2.4	3	8.4
500,000-999,999	—	—	—	—	2	4.4	3	3.8	3	2.2
250,000-499,999	—	—	—	—	1	.8	4	2.6	9	3.8
100,000-249,999	—	—	1	1.6	6	3.2	12	3.6	23	4.3
50,000- 99,999	1	1.1	3	1.7	7	1.5	15	1.9	40	3.6
25,000- 49,999	2	1.3	3	.8	19	2.1	42	2.9	82	3.7
10,000- 24,999	3	1.0	16	1.8	58	2.8	146	4.4	280	5.7
5,000- 9,999	15	1.8	33	1.8	136	3.1	249	3.4	465	4.2
2,500- 4,999	12	.8	34	1.0	163	1.9	467	3.2	832	3.8
TOTAL URBAN	33	6.0	90	8.7	392	19.8	939	28.2	1,737	39.7
Places 1,000-2,499	—		—		—		—		2,128	4.3
Places under 999	—		—		—		—		6,803	4.0
TOTAL RURAL		94.0		91.3		80.2		71.8	8,931	60.3
TOTAL U.S. POPULATION IN THOUSANDS		5,297		12,901		31,513		50,262		76,094

Adapted from U. S. Bureau of the Census, *Historical Statistics of the U. S., Colonial Times to 1957* (Washington, D.C., 1960), p. 14.

tal, exceeded 100,000 inhabitants. Most of the balance of urban America lived in small towns and cities which served the people and products of a rural hinterland.

High costs of transport kept millers near the wheat, pork-packers near the hogs, miners and founders near the ore and forests. Also, until stationary steam engines became commonplace in the 1840's those factories using large quantities of power had to stay close to waterfalls. As a result, the early industrialization of America, especially its textile and machine-tool industries, grew up in small cities and towns in the northeast, not in the large coastal commercial cities. The cities of Boston, New York, Philadelphia, and Baltimore did do a great deal of manufacturing in the years before the introduction of big steam mills, but this was mainly carried on in many small shops, not in large factories, and it often lagged in the introduction of new machinery.

By 1880 Americans had filled their continent with farms and small cities, but urban America, now 28.2 per cent of the nation's population, had split into two parts—small cities and towns, and large industrial metropolises. In that year twenty of the nation's cities exceeded 100,000 inhabitants. These metropolises accounted for 12.4 per cent of the population of the United States. Since the middle decades of the 19th century, America's largest cities had become places of enormous economic potential, places whose common standard of living exceeded the amenities of much of rural America, places where a few grew fabulously rich and the many were able to live a little easier than before. Rural America and rural Europe poured into this new environment.

The growth of these large industrial metropolises imposed a totally new set of conditions upon the history of the nation and indeed upon the history of the modern European world. Among the many changes which took place within American cities three ought to be singled out even in the briefest survey of the subject: first, the substantial reorganization of work and business life; second, the development of large-scale segregated settlement patterns; third, the breakdown of local government. None of these events was planned, none anticipated by contemporaries; all were, like the waves of population and migration, the outcome of millions of individual anonymous decisions.

REORGANIZATION OF WORK IN CITIES

The succession of transportation innovations from 1800 to 1880—turnpikes, canals, steamboats, railroads, telegraphs, clipper ships, and steamships—vastly increased the volume of inter-city traffic; it also cut communication and shipping from weeks to days, and from days to hours. The immediate and enduring result of these innovations was an enlargement in the scale and quickening of the pace of business in the large cities of America. As the supply and sales areas of businesses increased, handsome profits accrued to those who could simplify their financing, production, and marketing by specializing, rather than continuing in the old, undifferentiated ways. Business specialization totally reorganized the work patterns of the city in the years 1820-60. It was a reorganization that generally took place prior to the introduction of much new machinery, and which probably would have taken place had there been no technological changes save those in transport.

At the time of the American Revolution the nation's trade moved without banks or insurance companies. In each city general merchants gathered in coffee-houses where they pooled their personal resources by day-to-day bargaining to provide the services of banking and insurance. By the Civil War this undifferentiated activity of merchants had developed into the whole modern set of business institutions; commercial and savings banks, marine, fire, and life insurance companies, stock and commodity exchange, and stock and commodity brokerage companies. In addition, a whole nest of wholesale, freight handling, and shipping companies had grown up to improve the unspecialized middleman services of the colonial merchant.

The process of differentiation and specialization did not restrict itself to businessmen; it reached down into almost every trade and job in the city. Every conceivable activity from shoe-making to house-building was similarly reorganized during the years 1820-60. Petty capitalists, often onetime artisans and shopkeepers themselves, organized small groups of workmen for production for specialized wholesale and retail markets. In Philadelphia, shoe-makers, who formerly had made the entire shoe, now often worked on lasts and soles imported from New York or Boston, and their final assembly was sold in large city retail outlets or to the western and southern wholesale market. Houses were no longer constructed by a mason and a carpenter, but were built by men of many trades, each having its own contractor. Some of these trades worked in turns on the site, others like the fixture-, moulding-, sash-, and door-makers manufactured their products in small shops, and their work was delivered to the building for final placement by others.

The heavy flow of Irish and German immigrants to the nation's largest cities accelerated this process of specialization. The immigrants created a pool of cheap unskilled labor in the metropolises; this pool could be tapped profitably by any who could break down a skilled craft into discrete semi-skilled tasks. Many trades were reorganized in this way in the mid-19th-century city, the classic case being garment manufacture, which was taken out of the hands of the skilled tailors and placed out to step-by-step tenement work for poor native and immigrant labor.

By the time of the Civil War, after a generation of turmoil, the large cities of America had, by this process of task reorganization, trained a mass of workers to long hours of repetitive work, the wage system, and shop discipline. This largely pre-machine process laid the groundwork for the swift growth of the industrial metropolis in the late 19th century.

Because the process of reorganizing old crafts and shopkeepers' ways affected so many, and because it touched the daily rhythms and habits of family life, it proved extremely disruptive to the working class and middle class alike. The increase of pace and competition, loss of traditional ways, community status, and control over tools and product created an enormous pool of tension in all the major cities of America. Some of this tension flowed constructively into labor union or benefit association activities, some into ethnic, social, fraternal, and church group activities. Much tension, however, remained diffuse and undissipated. It found its expression in sporadic violence, especially against Catholic Irish immigrants and Negroes. Race and ethnic riots broke out intermittently in all major cities from 1830 to the Civil War. The most violent were the Philadelphia Riots of 1844 against the Irish, and the New York Draft Riots of 1863 when the Irish themselves went on the rampage. In addition to tacit approval of such behavior, middle-class and working-class city-dwellers turned to bigoted and xenophobic politics. Various forms of nativist parties played a major role in city politics from 1840 to the Civil War.

In the absence of well-developed residential segregation to isolate groups in conflict and without modern police forces to contain the violent, all the large cities of the nation suffered from constant gang warfare and periodic rioting. Professional big-city police date from the 1840's and 1850's when they were created to restore order to America's large cities. Within a decade or two the police succeeded in confining most illegal business, crime, and violence to one cordoned area of each metropolis. Also, during these same years the mass of city-dwellers became more accustomed to the new patterns of modern business and industry.

SEGREGATED SETTLEMENT PATTERN

Equally important to this task of pacifying the large metropolis was the development of the modern segregated settlement pattern. The growth of factories and railroads at the edge of large metropolises, where land was cheap and plentiful, created self-contained and self-policed mill towns in sectors of all large cities. These mill sectors provided socially stable, if economically unsatisfactory, elements in the metropolis—places of disciplined workers and their families. Elsewhere in the city the old patterns of jumbling occupations, classes, and ethnic groups came to be replaced by income-segregated residential tracts.

By 1850 the popular taste in housing for separation of home from work, the rising demands for amenities, and the new municipal fire codes combined to separate cheap alley shacks and cottages from middle-income housing. Hereafter new construction could only be rented or purchased by the upper third or half of the population. New construction thus became middle-class building. By the operation of economic forces and social taste, large income-segregated tracts grew up in all large American cities. The introduction of the horse-drawn street railway in the 1850's and its associated fashion of suburban living accelerated this change.

By 1880 the two-part settlement pattern of the industrial metropolis of 1880-1930 came into being—within, a core of shopping, offices, small manufactures, luxury housing, and old slum buildings; without, sectors of manufacturing and vast tracts of new middle-class homes and recent second-hand working-class houses. Except at the margins where one type of settlement met another, segregation damped tensions among city-dwellers by reducing contact with hostile and unfamiliar groups. The poor and the recent arrivals formed ghettos for mutual help and comfort, and the working class and middle class likewise found comfort and relief in being surrounded by those of similar background.

MUNICIPAL GOVERNMENT

Although this pattern of income segregation helped to pacify the metropolis, it had the unfortunate consequence of also making the metropolis politically weak. To be sure, as cities grew, their very large size rendered it impossible for most citizens to know more than a fragment of their city. Separation of work from home and segregation by income intensified this growing ignorance, so that in most voters' minds the city became a mixture of the fragment they knew and their fantasies of unknown lands of rich or poor, Irish, German, Jew, or Italian. At the same time the new regional and national scope of much of the city's business drew the old merchant leadership away from local concerns toward state and national politics.

Local politics, like retailing and real estate, became a specialty of its own. In the middle-class suburbs, as well as the ethnic enclaves, politics fell into the hands of local tradesmen, petty lawyers, leaders of social and church groups, and those with business with the municipal corporation—civil servants, transit, real-estate, and utility interests. Since the 1830's when this process began with the native American Jacksonian leaders, minor ward bosses and firehouse politicians began to move to the foreground. After the Civil War, when the merchant leadership abandoned the city and the metropolis assumed its segregated form, boss politics became the dominant feature of metropolitan government. The Gas House Gang of Philadelphia or Boss Tweed of New York were symbols of the new order.

The bosses ruled, some more effectively than others, some more corruptly than others, until the automobile and shifts in the national economy altered the spatial and industrial patterns of the metropolis. For the modern observer, the important fact of boss rule lies in its being a weak form of government. The bosses' power rested on ever-shifting coalitions of minority groups within the segregated districts of the city. The lack of powerful common interests among all groups in the city made this sort of coalition government possible, but it also ended the bold leadership which had characterized the earlier era of merchant-led city government and which had carried through difficult and expensive projects. In the merchant era before the Civil War every major municipal function performed today, except zoning and housing, had been introduced to America. Since that time the political weakness of the American metropolis has kept it from being a place of political innovation or major reform. Instead, the large industrial metropolis has become an enormous passive space—socially, remarkably static; economically, wonderfully productive.

DEMOGRAPHY AND TECHNOLOGICAL CHANGE

From the beginnings of the Industrial Revolution in the mid-18th century to the end of the 19th, the demography of the Atlantic community had undergone major changes. Industrialization had been accompanied by urbanization and widespread migration from those areas in Europe which suffered economic distress to new opportunities in America. At the same time the development of industrial metropolises changed the patterns of work, living, and local government.

In the 20th century, as we will see, further demographic changes were to take place. The metropolis was to grow into the "megalopolis," and the ills of the city, aggravated by the automobile and air and water pollution, were to become major problems of modern life.

33 / Effects of Technology on Domestic Life, 1830-1880

ANTHONY N. B. GARVAN

Despite such evidences of public concern as international exhibitions and numerous popular articles, sermons, and cartoons, the impact upon domestic life of the great industrial and technological changes that separated 1750 from 1880 still remains largely speculative. The cell walls of the home seem, on first inspection, to have been impervious to the forceful virus of industrialization and mechanization.

Naturally, the new products of mass manufacture, the new foods, the medical discoveries, the new methods of communication were not kept out of the home. Quite the contrary. Wherever cost permitted, they were embraced. But also wherever possible they were warped into a crude implementation of remembered, nostalgic customs. On the surface, domestic life continued apparently unchanged by the steam engine, the factory, or the railroad.

THE HOUSE AND ITS FURNISHINGS

The house is, of course, the primary artifact of domestic life. Semi-permanent, expensive, and fixed in location, it withstood all but the most determined efforts to alter its basic design. Only those countries with the most rapid population growth and the largest population migration developed significant house design changes, and the effect of these changes on family life may only be surmised. The balloon frame, the use of cast iron, central heat and plumbing, the elevator before 1880, the steel frame after 1880, had their chief impact upon the United States and other areas of European domination where rapid population growth overlay technological competence; one without the other (as in Brazil and France) had little effect.

Even in the United States it is hard to assess the direct consequence of technological change for life within the home. The ability to enclose space more cheaply in terms of labor and materials certainly made possible more specialized rooms for bathing, washing, cooking, eating, sleeping, reading, and entertainment. For the first three of these chambers, devices of a new sort were designed—toilets, bathtubs, washing machines, and iron coal-stoves—which either improved the quality of the chore or shortened the time necessary for it. In each instance, new combinations of devices well known for their industrial application were modified for domestic use. In every case there was at least a potential saving of woman's time. The toilet for the pot, the coal stove for the wood stove, the washing machine for the tub, the radiator for the fireplace, the icebox for the market; all substitutions made, in part, before 1880, gave every woman who could afford them a little more time and freedom.

More spectacular, but far less significant, were the "household inventions": folding beds, semi-automatic vegetable peelers, knife sharpeners, and similar devices of apparent, if not real, utility. Towering in importance over them all was the sewing machine. Widely manufactured and easily marketed, it permitted speedy and cheap fabrication of clothing for family and sale. Its convenience, efficiency, and low cost kept many lower-income women as domestic producers long after canned and prepared foods had penetrated their kitchens.

Meanwhile, the homes of the middle class and the wealthy also underwent profound changes. Rooms became increasingly specialized and designated as dining room, library, parlor, bedroom, or, after 1840, bathroom. With each of these divisions personal privacy and isolation increased to a marked degree and enhanced the "mystery" of bodily functions. Bathing, elimination, and sexual intercourse could now be carried on behind closed doors and with a degree of isolation unknown even to the wealthy in the 18th century.

Furnishings of all sorts became cheaper and available to a larger number of people. The application of the power saw, factory techniques, and wide distribu-

Fig. 33-1. An urban middle-class family in 1871. Oil painting, "The Hatch Family," by Eastman Johnson. (*The Metropolitan Museum of Art*. Gift of Frederick H. Hatch, 1926)

tion made elaborate furniture cheap and available. Ceramic tablewares continued to drop in price and increase in quantity. Imported and domestic rugs left the tables and covered the floor. Lighting devices improved and multiplied rapidly. In short, the consequence of technical improvements was to enrich the middle- and upper-class home with the confusion of material goods of the Victorian era.

Such clutter did not offend the Victorian eye. Varied in its design source, incredible in its symbolisms and analogies, the furnishings of the Victorian home were an early reflection of the increased role of woman as buyer and ruler over the home. The delicacies of the "genteel female" and the masculine vigor of the woman reformer both owed much to new wealth and the new freedom from constant domestic chores. Restless, literate, and self-conscious of herself, the middle-class woman in industrial society spent her energy upon conspicuous consumption, the arts, and political reform.

SELECTIVE IMPACT

Other direct consequences of new domestic devices upon society as a whole must, at the present, remain conjectural. First of all, they were unevenly enjoyed. The poor, because of price, and the wealthy, because of the domestic servant, failed to take advantage of many improvements. The middle classes, where their use presumably was greatest, also felt most critically other forces for change, so that the influence of technological development is obscured by related changes in the economy, class roles, religion, and education. It is not at all clear that the new household machines directly changed major family rituals. It does seem certain that the immigrant and, indeed the native American or European, preserved a surprising amount of his family traditions, changed perhaps in appearance or taste, but little changed in substance.

Moreover, the influence of these innovations was infinitely greater in America than in Europe because the rapid growth of the American population constantly required new construction. In Europe, where alteration was expensive and slow and new building seldom needed, few persons, except the transient traveller, enjoyed such luxuries as the complete bathroom, central heating, or refrigerated food. In Africa and Asia the entry of these devices, except again for the use of the foreign visitor, awaited the end of World War II and the tardy recognition of new foreign markets.

This is not to say that technological change had no effect upon domestic life. To the contrary, it was felt more powerfully in its indirect consequences than in its direct. The rise of the city, the increase of industry, the speed of transportation, and the ease of communication, all affected the family and domestic life. New sources of power, the mechanization of the farm, and the creation of the suburb were important, even dominant, factors in the lives of many families. All these developments were the cause or consequence of other technological changes, and each of them caused profound alteration of domestic life.

INDUSTRIALIZATION

Commercial, religious, fortified, and administrative cities had been common phenomena during five centuries of European civilization. Their growth and size had been limited by geographical and regional characteristics. However, the new industrial city, after 1750 in Europe and after 1820 in America, had seemingly no demographic limitations. Its population increased geometrically according to the sale of its manufactured products. Since new techniques and competition generally depressed prices, the market usually widened with the passage of years. As machinery improved and increased in cost, its geographical distribution became concentrated, so more and more persons were attracted to manufacturing centers distant from their home.

Division of labor and the simplification of tasks attracted less skilled persons to factories; for the first time single women and children were employed, often at a distance from their parents' residence. Abuse of the apprentice system brought the child only a very narrow acquaintance with the process of manufacturing in which he assisted. Indeed, his knowledge was more of a handicap than an advantage, for he emerged from his youth with only enough skill to guide a few machines and seldom had the ability to rise to the status of a mechanic. His survival thus depended on the success of a narrow range of factories, and his absence of generalized skills made him more dependent upon his employer than had been the serf upon the lord of the manor.

Often in the search for a suitable fall of water, which might be far from existing towns, factory- and mill-owners had to build housing for their workers, further increasing their dependence upon the mill-owner. The church which the mill-owner patronized prospered; his taste governed the town's appearance; his demands determined its rents; and the wages he paid fixed its income. Only in similar but rival establishments could the factory worker escape his particular employer.

THE FAMILY AND THE FACTORY

For the woman, the same conditions applied but with several important added factors. As a mill worker, she now had a possible alternative to early marriage. Mill work gave her independence of family and rural community and the possibility at least of saving toward her dowry. Thus both men and women of the mill-working classes, for different reasons, tended to delay marriage and to postpone the establishment of homes. Once married, the urban couple postponed in some measure the birth of children. The roles of disease, of age, of crude birth control, and of undernourishment in this delay have not been made clear. Each played a part which substantially and continuously depressed the size of families in cities generally and in manufacturing centers particularly.

Rural districts felt the consequences of these changes. Youthful labor tended to migrate to industrial centers at home and abroad, and although women often returned to marry and raise a family, the men often did not. Conditions of life which had affected the American colonists' life for two centuries became, after 1800, conditions of the lives of a far broader segment of the population. Contact with older generations became remote. The individual, or the nuclear family of a married couple and their offspring, became a steadily more common order of domestic life.

Their habitation, the small row house, the urban flat, the rented room, seemed to emphasize their isolation from their extended family, their dependence upon the neighborhood for amusement, and upon the large-scale employer for livelihood. Moreover, a subtle change in family ritual was caused by industry's alteration of the daily routine. Work from dawn to dusk had almost always been customary for the farmer, but now no possibility remained that the male worker would return at midday. Instead he was called to work at dawn by the factory bell and returned home at dusk in the company of his fellow workers. Rain, snow, or severe cold affected his schedule little. Tardiness and absence carried severe penalties; his holidays became fewer and more fixed. In short, the father's apparent control over his own routine became weakened.

The increased role of the factory was not always harmful. The autobiography of Samuel Gompers, the great American labor unionist, describes the cigar factory as an educational experience which gave men of a common craft an introduction to letters, sociology, and economic theory which they later applied to labor organization:

> The craftsmanship of the cigarmaker was shown in his ability to utilize wrappers to the best advantage to shave off the unusable to a hairbreadth, to roll so as to cover holes in the leaf and to use both hands so as to make a perfectly shaped and rolled product. These things a good cigarmaker learned to do more or less mechanically, which left us free to think, talk, listen or sing. I loved the freedom of that work, for I had earned the mind-freedom that accompanied skill as a craftsman. I was eager to learn from discussion and reading or to pour out my feeling in song. Often we chose someone to read to us who was a particularly good reader, and in payment the rest of us gave him sufficient of our cigars so he was not the loser. The reading was always followed by discussion, so we learned to know each other pretty thoroughly.

Thus for Gompers the cigar factory becomes the dominant method of socialization and education.

In the management and owner classes, the impact of the new methods of manufacture and of living had an equal but quite different effect. The manager who determined policy, the mechanic who repaired and improved machinery, and the owner who financed the factory system within their respective spheres of primary activity controlled the new technology. Moreover, the process they directed was so much more powerful and so much more productive than its agricultural predecessor that it seemed to embody differences in kind, not degree. Its fantastic potential seemed unlimited and no doubt encouraged the archetype mill-owner, aggressive, self-confident, even brutal, but above all, successful.

At the same time, the new devices required close supervision and constant improvement. If not so maintained, the mill quickly fell behind its rivals and failed. Caught in a spiral of narrowing profit margins and increased production, the mill-owner, despite his personal power, had little security, much anxiety, and great responsibility.

The career of Patrick Tracy Jackson, a founder of the great textile mills at Lowell, Massachusetts, well illustrates the compulsive hold of industry upon its leaders:

> Having succeeded in establishing the cotton manufacture on a permanent basis, and possessed of a fortune, the result of his own exer-

> tions, quite adequate to his wants, Mr. Jackson now thought of retiring from the labor and responsibility of business. He resigned the agency of the factory at Waltham, still remaining a director both in that company and the new one at Lowell, and personally consulted on every occasion of doubt or difficulty. This life of comparative leisure was not of long duration. His spirit was too active to allow him to be happy in retirement. He was made for a working-man, and had long been accustomed to plan and conduct great enterprises; the excitement was necessary for his well-being. His spirits flagged, his health failed; till, satisfied at last that he had mistaken his vocation, he plunged once more into the cares and perplexities of business.

Thus, even when successful, the mill and the economic system of which it was a part controlled domestic life. For owner, superintendent, and mechanic no less than for worker, the power and potential power of the factory became the dominant force of life.

Life lived away from the home had little in common with domestic life. Instead, work and relaxation became separate; the home grew into a haven secure from outside pressure to which the father returned daily. Within the home the woman's role increased, encompassing almost sole responsibility for religious and moral teaching, and for instruction in the arts, reading, and manners. To learn the masculine world, the boy fled the house to the street to learn from other boys what was left unsaid at home. From the street came his knowledge of masculine bearing, physical prowess, and personal courage.

Henry Adams, the child of the wealthy and politically prominent Massachusetts family, described the "virtues" he learned on the Boston Common:

> One of the commonest boy-games of winter, inherited directly from the eighteenth-century, was a game of war on Boston Common. In old days the two hostile forces were called North-Enders and South-Enders. In 1850 the North-Enders still survived as a legend, but in practice it was a battle of the Latin School against all comers, and the Latin School, for snowball, included all the boys on the West End. Whenever on a half-holiday, the weather was soft enough to soften the snow, the Common was apt to be the scene of a fight, which began in daylight with the Latin School in force, rushing their opponents down to Tremont Street, and which generally ended at dark by the Latin School dwindling in numbers and disappearing. As the Latin School grew weak, the roughs and young blackguards grew strong. As long as snowballs were the only weapon, no one was much hurt, but a stone may be put in a snowball, and in the dark a stick or a slungshot in the hands of a boy is as effective as a knife. . . . the boy Henry had passed through as much terror as though he were Turenne or Henry IV, and ten or twelve years afterwards when these same boys were fighting and falling on all the battlefields of Virginia and Maryland, he wondered whether their education on Boston Common had taught Savage and Marvin how to die.

result was often more disastrous than the introduction of new machinery. Starvation, poor shelter, and disease became common in backward districts.

The chief measure taken by both the prosperous and the more backward landholder was to increase his holdings. Large holdings permitted a more scientific use of land and a more economical use of labor and machines. The separation between the farmer and his laborers grew. Spinning and weaving declined in the rural cottage because of factory competition. In greater numbers than ever before, emigration and military service drained away the young European man. The factory, the sea, and frontier did much the same for American youth. Domestic service and the factory took many of the young women in both America and Europe. The rural laborer and the landowner, like their urban contemporaries, became specialists in increasing degrees. The middle class of farmer, increasing in both numbers and prosperity, was able to introduce rural sports, education, and a measure of the world's goods into his family's life. The lot of the laborer in Europe worsened; in America it improved in a straightforward reflection of the market for his services. In general, domestic rituals in the country changed far less than in the city as a consequence of technological innovation, because hired laborers and larger farm units preserved customs of an earlier time, but even in rural areas the family became dispersed earlier and formed later in life.

HEALTH AND SANITATION

Despite these negative aspects of adjustment to change it does seem clear that technology most profoundly affected domestic life in the extension of the life-span and improvement of life expectancy. Although the chief factors have not been separated, some of them can be listed. First of all, agricultural improvements enriched the diet, particularly with meat and other protein sources. Though not equally priced or distributed, this improvement came first to countries with rich virgin soil or advanced agriculture. Certainly it directly benefitted men and women and enabled them to survive the physical hazards of life.

Secondly, the improvement of midwifery, added to a richer improved diet, steadily increased the chance of infant survival. Such increased chances of life had profound domestic consequences. Reducing, as it did, the constant presence of infant mortality, it permitted greater attachment between parent and child, less fear of imminent loss. First noted in America by the French observer, Alexis deTocqueville, these attitudes are similarly found in mid-Victorian households of the middle class in Europe.

A most useful by-product of the growth of cities was the necessity of sanitation. Practices which had been tolerable in a village or small town became intolerable in large cities. The new concentrations of manufacturing and the general increase of town size demanded pure water supplies and adequate waste disposal. With the improvement of each, the townsman's health, despite plagues and epidemics, improved. A partial corollary of these developments was the substitution in small part of tea and coffee for alcoholic beverages, especially at home and during working hours. Before 1880, diet, medicine, and sanitation had more direct effect upon domestic life than all the other changes in industry and agriculture. These laid the foundation for the reduction of family size, a longer and easier life—especially for women—and sponsored more affectionate relationships within the family.

At the same time, even slight improvements in general medicine and the few major breakthroughs of vaccination, anesthesia, and antiseptic techniques seemed to promise more than they could fulfill. Individuals lost much of the religious resignation with which they had faced death and in its place substituted deep sentimental and romantic attachments which made unexpected death, such as that of Prince Albert, consort to Queen Victoria, lifetime burdens.

THE DECORATIVE ARTS

Technology's impact upon the furnishing of the home, a major aspect of the decorative arts, is not very clear. Although mass-production techniques had long been employed in ceramic and metal manufacturing processes, the introduction of high-speed and high-output power machinery immensely cheapened and extended the market for fashionable furniture, prints, and wallpaper.

Whether a broader market or new techniques of manufacture had any direct impact upon design before 1880 is uncertain. The same machine which could produce the sharp lines and geometric surfaces of Neo-Classicism could be altered to produce the rich variety of later Victorian periods. Although both styles had equally gifted apologists ready to theorize and attribute change to the new technology, in all probability change in design only reflected cyclical changes between extremes of simplicity and complexity, speeded but not significantly altered by new machinery.

However, the lowered relative cost of furniture, textiles, and ceramics gradually did have an impact upon domestic interiors. The uncluttered room became rare, and in its place richly colored walls, elaborately carpeted floors, and well-stuffed furniture attested to the wide interests and varied taste of the owner. In the interior, at least, the open spaces, plain surfaces, and precise mouldings of architectural design became subordinated to the eye-catching mixture of framed pictures, rich wallpapers, elaborate curtains, and swelling furniture contours. The quantity and distribution of such materials, if not their design, seems directly dependent on mass manufacture for their lower costs; on speeded transportation for wide distribution; and on mass communication for advertisement and sale.

A second equally significant consequence was the destruction of rural and lower-class decorative arts. The rapid change of style and the availability of great amounts of discarded, high-style furniture, on the one hand, and the dramatic drop in cost of textiles and ceramics, on the other, tended to make the labor of hand-craftsmen, especially unsophisticated ones, unnecessary and too costly even for personal consumption. Moreover, since most isolated and rural communities tended to remain static or actually decline in their population, the rural craftsman found his market constantly shrinking and, because of the availability of cast-off urban models, found his own and his customers' tastes pruned of independence. Finally, at the end of the 19th century the folk skills which had been the direct consequences of isolation and rural life were embraced by the urban artist and incorporated into the decorative arts of the elite and wealthy as self-conscious efforts to attain purity and simplicity of design, becoming, thereby, in many instances, patterns for more mass manufacture.

DIRECTIONS OF CHANGE

Quite certainly 19th-century changes in technology reinforced already established directions of change in domestic life. The increase of privacy, the decline of wide social contact, emphasis upon distinctions between child and adult life, the reduction of the size of both households and families, the increased rigidity of class barriers—all were abetted by the Industrial Revolution. But at the same time all these trends were present in West European and colonial societies as early as the mid-17th century.

As yet, the impact of technological innovation upon domestic life has not invited precise investigation. The changes noted in this chapter in fact may prove to be coincident alterations which have a common cause still undescribed. In any event, family life changed most rapidly in those European and American nations which experienced the rise of the city, the introduction of the factory system, the speeding of transportation, and use of new sources of power and light in the years between 1830 and 1880. Why this should be so is far less certain but invites speculation.

The problem is a very complex one since few of the principal causes were coincident and fewer occurred in all countries. No technological change had a more direct effect upon the family than the introduction of power machinery. The cost, limited flexibility, and great capacity of such machinery forced the development of the factory and the withdrawal of the worker, both man and woman, from the home. But these changes occurred in different years in different countries and bore different roles in each national economy. At most, one can demonstrate a vaguely parallel force which moulded historical and cultural conditions in each of the countries affected. The great rapidity and dynamism of industrial change between 1790 and 1850 in Great Britain and the United States and the comparatively slower adaptation of new techniques on the European continent in part may depend upon the existence of a strong and wealthy mercantile class who were more easily attracted to the expansion of industry than a wealthy but landed aristocracy.

On the other hand, the example of industrial nations and the values which their technological changes nurtured became forces in the less rapidly changing societies of France, Germany, and Holland, even though the technological change itself was not initially developed. Moreover, the subsequent implications of the rise of the factory and changes in military and transportation technology could be ignored only at the risk of national disgrace. What was not apparent in the experience of the masses had to be compensated for by state education, and so nations with lagging economies often had more elaborate and competent technical and scientific schools than those who led economic advances.

In any event, directly or indirectly as a consequence of industry, warfare, or education, the individual was forced to turn away from his fireside, from his church, and from his land for values. He was at once partially freed from his family and at the same time enslaved to new large-scale organizations: the factory, the army, and the state school. His loosened dependency demanded privacy at home, separation from his children, and mobility of his household.

At the same time, the function of the family changed. Less and less a productive unit, it enjoyed, in differing degrees, a steady increase of leisure. The year's routine was changed, and vacations for the middle class and wealthy became commonplace. Industry and transportation determined the hours of rising, breakfasting, and dining. Midday meals were seldom taken at home; the man arrived home from work later and later. In short, tangible evidence of the economic value of the family and the role of the man in it became more and more rare.

In its place was substituted the leisure of first, Sunday, then the weekend, better health for women and children, and greater isolation of the family from society. In short, there was every encouragement of unreal sentimentality, deep but questionable affections within the family, and great personal self-awareness. The separate roles of child, father, and mother became obscure, and the maintenance of the family despite the inevitable inroads of the passing years and death became a major concern of each member, often to his deep psychological and biological hurt.

Only rural and agricultural societies defied these changes—Ireland, Italy, and part of France perhaps best of all. Elsewhere as a consequence of actual technological changes, warfare, or threatened national security, these changes came about and were chronicled in the novels of Dickens, Tolstoy, James, and Hamlin Garland. But any psychological insecurity was obscured by the vast improvement of health, diet, clothing, and shelter which made Victorian life so comfortable, at least for the middle classes. Blessed in large part with only minor wars, the Victorian could indeed easily overlook the domestic consequences of technological change in his time.

In a curious fashion technological change in industry and agriculture both strengthened and weakened the family between 1830 and 1880. First of all the family unit shrank in number of persons in households. Because of strong barriers, some social, some moral, some psychological, members of the family were separated. The aged seldom lived with their adult offspring in small city homes; the father's work, especially in winter, absented him from the household. Technical schooling and discipline became a goal of most nations, purchased at the expense of the child's absence from the home. The woman after marriage rarely had many contacts beyond her household. The rural women no longer worked in the fields; the urban middle-class woman saw only her own servants and social equals. Since she lived apart from her parents, the poor married woman, when she worked, neglected her children. In short, at every stage of life from childhood to old age the individual faced a separation of family from social life. The home became a refuge dominated by women, secure from the outside world of factory and warfare, ancestors and descendants. Its unchallenged role and its isolation encouraged deep sentimental attachments which, as Freud claimed, often bordered on the incestuous.

At the same time that the internal bonds of the family nucleus were strengthened, the household lost prestige in terms of its external social power, and its role declined in education, production, religious worship, decorative arts, games, and, above all, value formation. The factory produced clothing, processed food, and building materials. The school dominated the child's education. Religion became confined to Sunday worship and the Sunday school. Folk arts, though often revived artificially, largely disappeared as natural phenomena. Games were played by youthful peers for larger and larger audiences. New values like those of Social Darwinism and pragmatism arose from industrial experience and scientific observation.

Thus even by 1880 the fundamental conflict between family and society, almost unknown in the 18th century, had become a central concern of industrial society.

Part V
The Promise of Technological Fulfillment, 1880-1900

34 / Applications of Electricity

HAROLD I. SHARLIN

Electrical energy can be converted into heat, light, and mechanical energy (motion), all three of which are now utilized for the convenience of man. Nature itself exhibits the first two conversions in lightning, but the conversion of electricity to mechanical energy was unknown until produced artificially in the 19th century. The earliest electric machinery, the static electricity machines of the 18th century, were used to make these same first two conversions, as when a spark was made to jump with a flash of light or when Benjamin Franklin cooked a turkey with an electric discharge.

It was to be expected, then, that when the battery was invented in 1800, the continuous current made available would be used to produce the familiar electric spark. Further research on this spark led to the development of the arc light in which electricity was converted into light. A second source of illumination developed later, the incandescent lamp, used electricity to produce heat which, in turn, caused a filament to incandesce, that is, to give off light. Finally, in the electric motor, electricity was converted directly into mechanical energy. All three of these types of conversion were developed during the 19th century.

ELECTRIC LIGHTING

During the first two-thirds of the 19th century, gas seemed to be doing an increasingly good job of providing illumination for domestic, municipal, and industrial purposes. Far from seeming a crude and burdensome expedient to be replaced as quickly as possible, gas lighting was one of the marvels of the age and a primary hallmark of progress. The ability of electricity to produce light was well known, but not much considered. Humphry Davy (1778-1829), for example, while experimenting with the voltaic pile (battery) as a source of energy for chemical analysis, produced a carbon arc light by bringing two pieces of charcoal together and causing an arc to jump between them. He did not follow this up nor did anyone else.

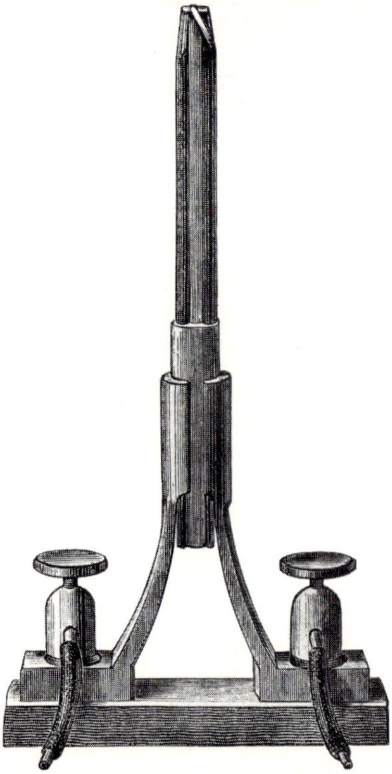

Fig. 34-1. An early form of electric illumination was provided by the "candle" of Paul Jablochkoff. Two carbon rods were separated by a volatilizing insulator and operated on alternating current. A 10-inch candle lasted about 1½ hours. (Courtesy of *Burndy Library*)

The discouraging effect of a flourishing gas-light industry on the infant study of arc lights was complemented by a second handicap: the lack of a suitable source of power. The voltaic batteries available were adequate for the commercial success of the telegraph, which used comparatively little power, but hardly sufficient for the needs of illumination. After the development of better batteries in the 1830's and 1840's, some interest was rekindled. Many ingenious devices were worked out, notably to advance the carbon rods of the lamps toward each other as they burned away on the end, but these did not really help solve the central problem of power supply. The economic disadvantages (strong competition from gas and dependence on very expensive power) under which the arc-light system suffered discouraged inventors to the extent that no further improvements were patented between 1860 and 1870.

The commercial stage of the arc light began with the invention, by a Russian living in Paris, Paul Jablochkoff (1847-94), of the Jablochkoff "candle"; this required no mechanism for moving the carbon rods, between which the electric arc was formed, and seemed to be the ultimate in economy. For uniform consumption of the carbons, the Jablochkoff candle required alternating current. Zénobe Théophile Gramme designed a generator especially for this arc light which had to be of very high voltage. Previous generators, which were used for electroplating, produced relatively low voltage. The system—the Jablochkoff light and the Gramme generator—was widely used in Europe until about 1881, but four years of experience showed that the system consumed carbons at too great a rate. An important step had been taken, however. The use of the arc light with a generator seemed to promise an economical electrical light.

In 1878 the Ohio inventor Charles F. Brush (1849-1929) invented an arc lighting system with important improvements in the high-tension direct-current machine, and with a simple form of arc lamp. This system met with immediate success for lighting streets and other large spaces. A year later, in 1879, two Philadelphia high-school teachers, Elihu Thomson and Edwin J. Houston, also got their start in the electrical industry by designing an arc lighting system. Both the Brush and the Thomson-Houston generators had magnetic fields of increased strength to deliver a large current for arc lights. In 1886 three-quarters of the world's arc lights were run by either the Brush or the Thomson-Houston generators. This commercial success of arc lighting led inventors to consider the possibility of using electricity to light homes.

The search for a suitable domestic light was taken up by many men in Europe and the United States, and quickly resolved itself into the problem of "dividing the light." This meant that a lamp had to be devised which would give less glare than the arc light so that it could be placed in the rooms of a house. The immediate competition for house lighting, as for street lighting, was gas. The search centered on finding a material which would heat to incandescence without burning when an electric current passed through it. The light source, in order to be cheap enough, could not be consumed as was the carbon in arc lights, for then the cost of the lamp would be too great in relation to the small amount of light which it would produce.

Humphry Davy, in his early experiments with the battery, had produced incandescence by passing a current through a platinum wire and a carbon rod. However, the glow was short-lived, because these substances were quickly oxidized in the atmosphere. In the following years, many others had experimented with the idea of incandescence. The favorite materials for their experiments were platinum, iridium, and carbon, because they were the only materials known with the desired characteristics of a high melting point and ability to conduct electricity.

To prevent the filaments from being consumed by the atmosphere it became standard practice to put them in an evacuated glass globe. By 1860 the requirements for a successful incandescent lamp were known and had been tried. Many inventors were using carbon filaments of one shape or another and most experimenters were using evacuated globes, exactly the elements used in the successful bulbs of the 1870's. Yet in 1860 the problem was abandoned by most inventors, because they could not produce a bulb cheaply enough that would also last a reasonable length of time. The trouble was that the kinds of carbon used did not have the necessary qualities for a filament, and the equipment of the time did not produce enough of a vacuum to protect the filament. Nevertheless, some inventors continued to work on the problem, and in the process, came close enough to success to claim to be "the" inventor of the electric light.

In 1865, Hermann Sprengel invented an exceptionally good mercury vacuum pump, which Sir William Crookes used for evacuating glass globes for his radiometer in 1875. Joseph W. Swan (1828-1914), a British chemist, had designed a low-resistance carbon lamp in 1860 but found the results unsatisfactory. He returned to research around 1877 after hearing of Crookes's success with the new vacuum pump. Swan demonstrated a carbon rod lamp in 1878 but it had too low a resistance and deteriorated quicky. Paying no attention to the other components of a complete incandescent lighting system, Swan made the mistake of working on the lamp as a separate unit. His results had some serious drawbacks: for example, he wrongly believed that the best system was to connect the lamps in series (that is, with the parts connected end to end—positive pole to negative pole to positive pole, and so on—so that the current flows from part to part in succession), as with the arc lamps.

William E. Sawyer, another experimenter in this field, came to incandescent lighting by way of telegraphy. While working as a telegraph operator, he picked up enough of an interest in, and knowledge of, electricity to begin concentrated efforts to solve the incandescent problem in 1877. He was joined by Albion Man, a Brooklyn lawyer who gave financial support and some useful technical help. Sawyer and Man did all of their work with carbon and at one time experimented with, and then rejected, using it in a filament shape. They settled on a short, pencil-shaped carbon in an atmosphere of nitrogen. They rightly concluded that the nitrogen would reduce the oxidation of the carbon at high temperatures, but they did not foresee that the nitrogen would permit destructive vaporization of the carbon. Nevertheless, their method of "flashing" the carbon in a hydrocarbon atmosphere was a lasting contribution to electric lamp technology.

Great sums of money were to be gained by the man who would successfully stake a patent claim to the incandescent lamp. The inventor was certain to reap the added benefit of fame. The certainty of these rewards, promised but not guaranteed, came from the knowledge that artificial lighting had a large and growing market. At the very time experimenters were struggling with the design of a commercially usable incandescent lamp, the gas manufacturing companies were expanding at a prodigious rate. In 1850 there were thirty companies manufacturing gas in the United States and by 1870 there were 390. The capital invested in these enterprises in 1850 was over $6.5 million and had increased to almost $72 million by 1870. In addition, the discovery of oil in 1859 had started a boom in kerosene lighting.

With stakes involving money and glory, it was not surprising that there should be several claimants to the laurels of priority. Sometimes the supporters of the claimants in their zeal overlooked essential factors. One supporter of the Sawyer-Man claim wrote, "The essential element which differentiates the successful modern lamp from its unsuccessful predecessors is the arch shaped illuminant of cellular organic material." Although not incorrect, this claim is not completely correct. The author then made the mistake of assuming that the object was to produce a successful laboratory experiment, rather than a commerically usable lamp. He wrote that the other characteristics of the lamp, "such as high resistance, small radiating surface, high vacuum, hermetically sealed one-piece glass globe, and platinum lead-in wires, have one and all proved to be mere incidents and not of essence of the modern incandescent lamp." A high-resistance filament is *not* incidental to the financial success of the electric lamp, nor are the other features unimportant in the manufacture of lamps for mass distribution. All of these problems were to be successfully met by Thomas A. Edison.

THOMAS A. EDISON

Thomas A. Edison (1847-1931) came late to the search for a practical incandescent lamp. He had already acquired reputation and money through his inventions of a stock ticker, the multiplex telegraph, and other devices. Edison began the manufacture of telegraphic instruments and, eager to continue his inventive efforts, built a laboratory at Menlo Park, New Jersey. In 1878 he was casting about for the next problem whose solution promised a profitable return. Visiting a factory which was making generators and arc lamps, he was so impressed with the progress made in electric power, as exhibited by the generator there, that he decided that electric illumination was the most promising field for him.

Edison was no scientific trail blazer. His success was both a technical and commercial achievement. His idea was to invent a serviceable lamp and also to build a system for marketing this low-cost light, which meant competing with gas lighting for a mass market.

In keeping with his mass-distribution concept, Edison wanted widespread publicity for his work and was happy to give newspaper interviews. In one of the first he outlined his scheme:

> I have an idea that I can make the electric light available for all common uses, and supply it at a trifling cost, compared with that of

> gas. There is no difficulty about dividing up the electric currents and using small quantities at different points. The trouble is in finding a candle that will give a pleasant light, not too intense, which can be turned on or off as easily as gas.

How consistent he was in his approach can be seen by the way he finally marketed the electric light. The Edison light was sixteen candlepower, the same as a gas jet. Like the gas companies, the electric company used monthly billing, and the electric lights were referred to as burners. To emphasize to the customer that he was buying an old familiar product—light—and not a new, unfamiliar product—electricity—the bills were for light-hours rather than kilowatts.

However, Edison began on the wrong track by trying platinum as the incandescent material; nevertheless, he was bound to succeed because he stuck with the correct principle: the lamps were to be at the customer's control and within his reach. The street arc lights were well beyond reach so that the very high voltage necessary for series operation of those lights was not a hazard to the public. The handy location of the incandescent lamp left no alternative but parallel connections, or so Edison believed.

A parallel combination is one in which the lamps are strung between the lines instead of in the lines themselves. Each lamp will thus have the same voltage applied, and the turning on and off of any number of the lamps does not affect the voltage applied to the others. Since all of the lamps are at the same voltage in parallel, the current in the line is the sum of the currents in all of the lamps. (In a series arrangement, on the other hand, the current in the lamps and the lines is the same.) In order to avoid excessively large and expensive line conductors, the current in each of the lamps in a parallel connection must be kept small. The only possible answer was to have a lamp whose filament had a high resistance. Edison, therefore, was not only looking for a material which would remain incandescent for a long time when a current was passed through it, but in addition something that would have a high resistance—a commercial rather than a technical requirement.

In his idea that an electric-light system required a high-resistance lamp connected in parallel, Edison was putting engineering planning, or what was the same thing, economic reasoning, ahead of some of the best thinking in science of that day. In this he might be compared with Marconi, whose stubborn attempts to span the Atlantic by wireless were also contrary to the best scientific advice of his time. Several eminent scientists and engineers stated that parallel operation was impossible. Among them was Sir William H. Preece, an eminent English electrical engineer. Preece, talking about parallel operation, said in 1879, "it is, however, easily shown (and that is by the application of perfectly definite and well-known scientific laws) that . . . a subdivision of the electric light is an absolute *ignis fatuus*."

Edison left the calculations of lamp resistance, voltage, and conductor size to

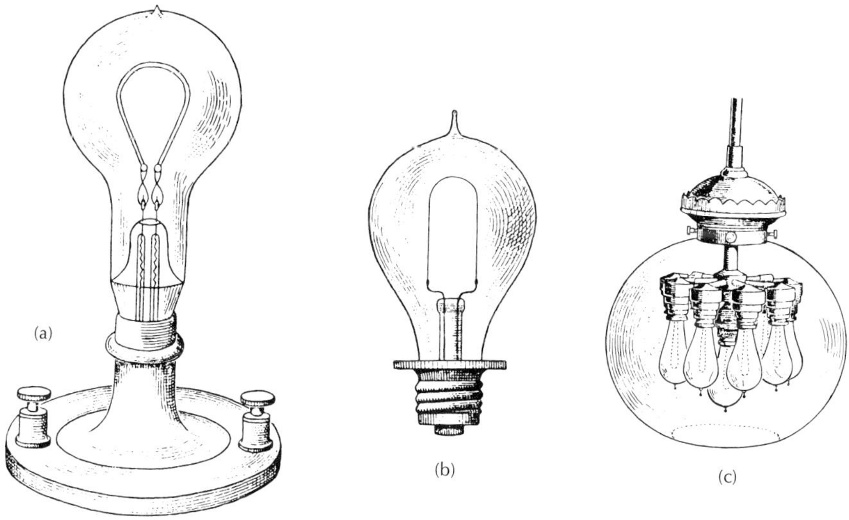

Fig. 34-2. (a) Edison carbon-filament lamp, *c.* 1881; (b) Edison carbon-filament lamp, *c.* 1082; (c) Edison lamp cluster, *c.* 1804. (Reprinted from Singer *et al., A History of Technology,* Oxford, 1954-58)

his hired mathematician because he recognized his own limitations in the field. Francis R. Upton, the mathematician, had to be re-educated by Edison in the simple application of Ohm's law to parallel operation. Upton was prone to agree with the other scientists of the time that such an operation was impossible and, as he said later, "I cannot imagine why I could not see the elementary facts in 1878 and 1879 more clearly than I did. I came to Mr. Edison a trained man, a post-graduate at Princeton; with a year's experience in Helmholtz's laboratory; with a working knowledge of calculus and a mathematical turn of mind."

The fact that Edison placed two not necessarily complementary requirements on the material for his lamp explains his worldwide and seemingly disorganized search for an "ideal" filament material. He eventually found the material by lucky accident in the form of a carbonized thread. Other materials which could be turned into a thin high-resistance thread were substituted later. The incandescent lamp which Edison invented in 1879 was part of a larger technological development. His design of a whole system was the key to commercial success. Others could, and did, make incandescent lamps which would burn for a considerable length of time. But these lamps were only laboratory models. Edison's engineering sense and commercial acumen produced a technological innovation. His invention is an excellent example of what is meant by technology as contrasted with science or invention.

Particularly noteworthy in this connection were the largeness of Edison's con-

Fig. 34-3. The dynamo room of the first Edison electric lighting station in New York. (Courtesy of *Consolidated Edison Company*)

ception and the attention he paid to commercial and financial factors. The incandescent lamp, which was demonstrated at a much-publicized exhibit at Menlo Park, gave a soft light usable in the home and it lasted an appreciable time. But it was only the start of the Edison "system." Edison designed the base for the lamps, the wiring for houses, the underground cable system for the streets, a meter for measuring the amount of electricity used by the customer, and the generator.

Edison was able to undertake the development of the complete system, and see it through profitably, because he had adequate financial backing. The advantage that he had over the other inventors in this field was his reputation, gained from successful inventions in telegraphy. Grosvenor P. Lowrey, general counsel for Western Union Telegraph Company, had become interested in Edison's work on the incandescent lamp, and he interested about a dozen other men in underwriting Edison's plans for a central electric lighting system. The Edison Electric Light Company was incorporated in October 1878 with capital of $300,000. Among the twelve subscribers of the stock were eight men who had had some contact with the telegraph industry, including Dr. Norvin Green, President of Western Union, and Egisto P. Fabbri, a partner of the financier J. P. Morgan. Before the Edison system began to pay a return, these men had invested half a million dollars.

The investment paid very well. Since the Edison Company was not only first in the field but was also aggressive, its growth was extremely rapid. There were two plans for marketing electric power: the customer could get electric power from an Edison-operated central station or he could have his own individual power plant. Two different companies were formed. One, an operating company, began business in 1882 and was immediately supplying power to 2323 lamps from a central power station. By 1884 this same Pearl Street Station in New York City was supplying 11,272 lamps in 500 homes. The other company, for the consumer's individual power plant, had installed 10,424 lamps in the first six months of its operation in 1882, and 59,173 lamps by 1884. Edison also manufactured his own lamps. By 1885 the Edison Company still had a virtual monopoly in the field, selling 200,000 of the 250,000 lamps in use. The company has had an unbroken record of dividends dating from that year.

THE STREETCAR

The second great use for electricity, after illumination and outside of communications, was the electric motor. If the arc light had a vigorous competitor in the gas-light industry, the electric motor seemed at first hopelessly outclassed by the magnificent steam engines of the 19th century. Indeed, as with the early gas engines, many of the early electric motors betrayed their object and inspiration by being clear imitations of steam engines, complete with walking beam, pistons, crank and flywheel, and so forth. Both Michael Faraday (1781-1867) and Joseph Henry (1797-1878), the co-discoverers of electromagnetic induction, built small electric motors as a part of their investigations. Henry's motor featured an electromagnet, pivoted in the middle, with a permanent magnet under each end. As an electric current was passed through it, the electromagnetic beam rocked to and fro.

A fair number of experimenters and inventors on both sides of the Atlantic worked to improve this basic motor. Basically their approach was the same: an electromagnet was made by winding wire on a soft iron core; this magnet was then made to attract or was attracted by other permanent magnets. In order to produce continuous motion, the poles had to be constantly changed to disrupt the equilibrium, and it was to this end that most inventors directed their ingenuity. Many clever devices were produced, but the motor as a whole remained something less than a commercial success. For one thing, the soft iron core of the electromagnet proved to be a hindrance rather than a help, since the coil of wire itself was sufficient, and the soft core took long enough to magnetize so that the action of the motor was noticeably retarded.

An even greater obstacle, however, was the same reliance on batteries that had hindered the development of successful illumination. Although several spectacular experiments in electrical transportation were made before mid-century, all were brought down by the limitations of the batteries available. In Russia, for example, M. H. Jacobi operated a small boat on the River Neva in 1838 using an electric motor powered by battery. In England Robert Davidson achieved a speed of four miles per hour with his electric-powered locomotive, the *Galvani,* also using batteries. In the United States, a blacksmith, Thomas Davenport, applied electric motors using magnetic induction and powered by batteries to operate such shop machinery as lathes and drill presses. All of these came to naught for roughly the same reasons: a lack of both sufficient economic incentive and an economical source of electric current.

In 1886, S. P. Thompson saw an "immense field" for industrial application of motors. All that was required was "regular town supplies" (central stations) to make possible the economic use of motors to drive lathes and sewing machines and to operate elevators. But the electric motor was developed *before* such "town supplies" were available. The impetus to its development came from electric traction.

Electric traction moved from experimental efforts to practical success in eight years. The first experimental streetcar line was built by Siemens and Halske for the Berlin Industrial Exhibition of 1879. Edison toyed with the idea of electric traction at Menlo Park for a few years beginning in 1880. Frank J. Sprague (1857-1934), a onetime employee of Edison's, was the successful pioneer in this field. Sprague went to the Crystal Palace Electrical Exhibition in London in 1882 as a representative of the United States Navy. His report to the Secretary of the Navy, dated March 12, 1883, includes a detailed discussion of electric machinery. Sprague's interest in electric machinery had begun before the Exhibition, and he had already been actively experimenting with the dynamo in connection with a lighting system.

In 1884 the Sprague Electric Railway and Motor Company was incorporated; its chief asset was a motor designed by Sprague. But there was more to a successful street railway system than a well-designed motor, as Sprague found out when his company was awarded a contract, in 1887, to install a twelve-mile-long street railway in Richmond, Virginia. No other company appeared to be daring enough, or foolish enough, to undertake this contract with its almost impossible provisions. Although the largest streetcar system up to that time had only twelve cars, the Richmond contractor was to supply forty cars, of which thirty had to operate simultaneously. The contract added the further insuperable burden of a ninety-day completion date. Even Sprague did not know just how difficult a job he had undertaken. He had never seen Richmond and therefore did not know of its many hills, which were a particular obstacle to the satisfactory operation of his electric motors.

The project sounds like the kind of impossible task which would be assigned to engineers in wartime. Apparently it also did to Sprague, the former navy officer. For key positions he hired a former army engineer and another ex-navy man. The type of man he needed, Sprague said, had to have "nerve, grit, and coolness, one who, if he gets on the front end of a car going seventy-five miles an hour and there is danger ahead, will stay there."

The roadbed, constructed in a great hurry, had been built in a shoddy manner, and the contractor had thoughtlessly built many sharp curves in the track. The continuous changes in speed which the twisting course required were to strain the endurance of the streetcar motors to the point of breaking. The powerhouse had to be large enough to handle so many cars at once, but in this case the previous work on generating stations for arc and incandescent lights was of considerable help. However, bringing the power to a moving car presented a unique and much more difficult problem; before that was solved, more than fifty different trolley poles had been tried.

The motor, in practice, was found to have fallen short of requirements. It had to be fixed to a constantly bucking mounting, and then geared to the drive wheels. There was no past experience to help in designing a lubricating system for such an arrangement. The brushes, which brought the current to the rotating part of the direct current motor, had to supply an unusually large amount for uphill pulls, and had to be rugged enough to withstand the jolting trip. The brass brush which Sprague settled on was an expensive solution. In operation, the trolley cars left behind a shower of brass dust which alone cost the company nine dollars a day.

Furthermore, some system of motor control had to be devised that would be able to stop and start a car often, one that could also withstand a heavy power surge as the car went uphill. The system of control finally worked out was to serve as a model for power installations of less exacting demands. Each of these challenges Sprague met in turn, while being driven by the impossible time limit.

Sprague improvised a controller for the cars but was not able to overcome the basic trouble. When the controller was moved a notch it was equivalent to opening and closing a switch, and with that amount of current a large arc was formed. The arc wasted power and, even more important, was destructive to the contacts of the controller. In 1892, too late for the first Richmond installation, Elihu Thomson thought of the idea of placing a magnet near the controller contacts so that the magnetic field could "blow out" the spark. This approach not only solved the problem of controllers but has since been used in opening much larger circuits in power stations.

The same year, 1892, that Thomson devised the magnetic controller, Sprague developed the multiple controller for operating several motors at one time, making the subway and elevated multiple-car electric train economic competitors of steam. Although steam locomotives were impossible in an enclosed subway, they were being used on elevated tracks within the city. The multiple controller gave the motorman in the lead car control of all motors in the train. This device added versatility to the advantages of economy and cleanliness which the electric railway had over steam. Rapid urban transit, therefore, came to be electrically powered.

The Richmond line was successfuly operating in 1888 to the specifications of the contract. Because it was not completed in time, Sprague did not collect the full price but was paid $90,000. His cost was $160,000. The contracting company got a bargain, but Sprague had also gained. His company was well established after that, and the concept of an electric street railway system itself was successfully demonstrated. The high cost was due to the fact that so little preliminary work had been done for the project. No engineer today would undertake a project of that magnitude and uncertainty without first making a careful study of its feasibility.

Yet Sprague's daring and resourcefulness had paid off. In 1887 there were no testing laboratories and no engineering experience that would have enabled any company to go into such a project any better prepared than Sprague. The electric street railway could not be tested anywhere but on the streets. Like the laying of the Atlantic Cable, the experimenting for the street railway had to be done as the project was being completed. Later engineering practice has shown how expensive this method can be, but it is a tribute to the men who successfully completed these pioneering projects that they were done at all. Once Sprague showed its feasibility, other companies moved quickly into the street railway business. Some of the older lighting companies bought up or merged with the pioneer railway companies. Edison's company finally selected the Sprague Company for purchase on the strength of Sprague's position in the field.

The horsecar thus became obsolete in 1888. One electric motor took the place of two to four horses, for four horses were regularly used in snow. A team wore itself out in four years, and every year a horse ate its original cost in feed. The electric street railway eliminated the huge stables, the army of stablemen, the relay station, and the danger from costly horse diseases. Fourteen years after Sprague's Richmond line—the first complete electric street railway system—the country had 22,576 miles of track. All but one per cent of the old horse railways had been converted to electricity. In 1882 the investment in street railways, predominantly horse-drawn, was $150,000,000. By 1902 the investment had risen to over two billion dollars.

Yet a surprisingly short time later, in 1919, the secretaries of Commerce and Labor addressed a letter to President Woodrow Wilson in which they called attention to the electric railway problem which had "recently assumed such serious national proportions as to warrant the prompt attention of the Federal Government. Already fifty or more urban systems," they warned, "representing a considerable percentage of the total electric railway mileage of the country, are in the hands of receivers. . . . Other large systems are on the verge of insolvency, for the industry as a whole is virtually bankrupt."

What had gone wrong? The American Electric Railway Association complained of the high rate of obsolescence resulting from rapid improvements in technology. The cars of 1888 had been sound in principle, but improvements, if not of a fundamental nature, apparently were significant enough to warrant change. The association also complained of rapidly rising labor costs, in the face of which the state regulating commissions had stubbornly refused to allow fare increases.

These complaints sound very much like those of the commuter railroads of a later date. The electric street railway, like the commuter trains, had much in common with other industries which were faced with rising labor costs and prices that were not keeping pace. The street railways rose to a crest of prosperity during the time when urbanization was moving too quickly for the limited range of the plodding horsecar. But as urban sprawl increased, the population of cities spread out too thin for any type of mass-transit system to be able to maintain the revenue which their large investment required. The mass-produced, cheap automobile came in at the very time when more individual transportation was needed. The electric railway therefore declined.

ELECTRIC MOTORS

The rise and decline of the electric railway is important to the history of electric power for other reasons than its role in the history of transportation. In the development of this system, means for generating, distributing, and using large amounts of electric power were successfully worked out. The rugged d.c. (direct current) motor and the complementary control devices were perfected on the streets, and the lessons learned were of great importance to the electrification of manufacturing industries. The electric railway motor with its controller was a progenitor of the electric motors used in steel-rolling mills, where thousands of horsepower are applied, with speeds rapidly increased, stopped, and reversed. The first stage of development of the steel-rolling mill motor was on the streets of Richmond, Virginia.

The d.c. motor has the characteristic of readily adjustable speed and is uniquely suited for transportation. Because of its qualities, and a long established precedent of using d.c. for most power applications, no other type of motor was developed up to 1887. In spite of the subsequent development of a variety of other motors, the d.c. motor has remained the first choice for some applications where speed control is of prime importance. But the d.c. motor must have a commutator (a revolving part that distributes the current to the brushes), a factor that weighs heavily against the d.c. machine because of the maintenance it requires.

An entirely new type of motor was proposed by Nikola Tesla (1857-1943) in a paper before the American Institute of Electrical Engineers in 1888. In the course of the paper Tesla criticized the illogical construction of the d.c. motor:

> In our dynamo machines, it is well known, we generate alternate currents which we direct by means of a commutator, a complicated device and, it may be justly said, the source of most of the troubles experienced in the operation of the machines. Now, the currents so directed cannot be utilized in the motor, but must—again by means of a similar unreliable device—be reconverted into their original state of alternate currents. The function of the commutator is entirely external, and in no way does it affect the internal workings of the machines. In reality, therefore, all machines are alternate current machines, the currents appearing as continuous only in the external circuit during their transfer from generator to motor. In view simply of this fact, alternate currents would demand themselves as a more direct application of electrical energy, and the employment of continuous currents would only be justified if we had dynamos which would primarily generate, and motors which would be directly actuated by, such currents.

If there was to be no switching device like the commutator, how could the effect of a rotating magnet be produced? The answer was to use two or more alternating currents. This discovery of how to produce a rotating magnetic field by means of two currents was made independently by several persons. Galileo Ferraris published a paper in Italy entitled "Electrodynamic Rotations Produced by Means of Alternate Currents" in the same year (1888) as Tesla's. Another motor was invented by M. von Dolivo-Dobrowolsky of Berlin, who took out patents in 1889. He used three different currents to produce the rotating magnetic field and he gave credit to Ferraris for discovery of the principle. The motors which Tesla and Dolivo-Dobrowolsky invented were known as induction motors, because a voltage was induced in the rotating member, or rotor, by the rotating magnetic field. This same action occurs in transformers, where the primary coil induces a voltage in the secondary coil.

The induction motor reached a high state of development in a much shorter period because engineers and designers were able to apply what they had already learned from improving the d.c. machine. The induction motor was barely out of the experimental stage when it was shown by several companies at the Frankfurt Exhibition of 1891 and the Chicago Columbian Exposition of 1893. Yet Gisbert Kapp, an electrical engineer who reviewed the state of a.c. (alternating current) motors in 1910, said that "the manufacture had become perfectly standardized, and no startling improvement can be expected in the ordinary induction motor."

The a.c. induction motor and the d.c. motor met the electric power demands, both great and minute, of countries which were rapidly urbanizing and industrializing. The induction motor was rugged, and turned at nearly constant speeds regardless of the load. It was built in all sizes, speeds, voltages, and could be made sparkless (or explosion proof), and water-tight. One failing the induction motor had was that it ran only at the speed for which it was built and it was very difficult to adjust. But the d.c. motor met just exactly that need; its speed could easily be changed over a wide range merely by turning a rheostat.

The first factory in the United States to be completely electrified was a cotton mill. The system was completely a.c. and went into operation in 1894. The electric motor replaced the cumbersome central steam engine then being used in factories. One steam engine supplied the power needs for a whole building, the power being distributed by means of innumerable shafts, pulleys, and belts. With steam power, the mills were best built several stories high, but with electric motors each machine could have its own motor drive and a single-story structure could be used with a resulting improvement in materials handling. In addition, it was no longer necessary to locate near a water supply (for the steam boilers) since it was possible to transmit electric power over great distances; one requirement of plant sites was thus eliminated. Steel mills especially became major users of electric motor drives. Direct current motors of capacities in the thousands of horsepower were used in rolling operations.

Beginning in the 20th century, all new factories used electric motor drive. The decade from 1899 to 1909 was the turning point in electrification of industry. At the beginning of the decade, of a total of 160,000 motors manufactured, 36,000 were industrial motors. In 1909, 243,000 motors were manufactured for industrial purposes from a total of 504,000. The number of industrial motors produced increased 584 per cent while the number of all motors increased 216 per cent. The a.c. motor became a major factor in industrial power during this decade. Whereas one-fifth of the industrial motors made in 1899 were a.c., by 1909 over half of the motors were a.c. In 1899 there were only 16,891 motors with a capacity of less than half a million horsepower. By 1909 a total of 388,854 electric motors were installed in factories with a total capacity of 4,817,140 horsepower.

At the beginning of the 20th century, the number of electric motors rose, as they began to replace other means of power; in the streets the motor replaced the horse, and in the factory it replaced water-wheel drive and the steam engine. Since that period, the use of electric motors has continued to rise, but in recent periods the increase has been due to the total growth in the use of power. Whenever power is needed today, the usual choice is one of the many types of electric motors.

SUMMARY

The electric light was the first personal introduction to the public of the marvels which electricity could perform. The mystery attached to such a feat, and the pleasing glow of the lamp, probably sold more lamps in the first months than did any real evidence of the electric light's economy. Light was electricity's first commercial success, but today lighting consumes only a minor portion of the electric power generated. Electric communications—the telegraph, the telephone, and the radio—are a more glamorous consumer of electricity but they use an even smaller amount of the total power presently generated.

Less glamorous than radio and television, and not so well known as lighting, is electricity's role as a producer of mechanical power through the electric motor. Electricity for mechanical power had competition from the start. Inanimate power had been supplied for centuries by the water wheel and the windmill. The steam engine was reaching its peak in performance and application when the electric motor was being introduced commercially. The fact that an electric motor can exert a sizable, continuous torque to do useful work would simply have added one more item to the list of available power sources. What finally separated electric power from all others was its unmatched versatility. Great electric motors can move mountains, and minuscule motors can make fine adjustments on precision machinery, or shave whiskers. Large or small, an electric power source is always tightly reined; that is, it can be more quickly and accurately controlled than any other power source. The electric motor is the muscle of the age of electricity. The electric light and electric communications give spirit to the age but the electric motor animates it.

35 / Electrical Generation and Transmission

HAROLD I. SHARLIN

The population of the United States was migrating to urban centers in the latter part of the 19th century. In 1860 only 21 per cent of the population lived in cities with 2,500 people or more. By 1880 the percentage had risen to 28 per cent and by 1900 almost 40 per cent of the population of the United States lived in cities of 2500 or more. New York, Chicago, and Philadelphia had populations of over one million by 1890. This urbanization of the United States was a part of its growing industrialism. America after the Civil War rapidly moved from a primarily agricultural economy to an industrial one. Manufacturing became America's biggest business, and by 1894 the United States was the leading manufacturing nation of the world. The growth and concentration of industry placed electricity in the most favored class of energy. For the electric-power industry is a subsidiary industry which depends on the growth of other industries, manufacturing and transportation, for its expansion.

The 19th century was the age of steam power, but even at the end of the century a good deal of water power was still in use. As to electricity, the Census

Fig. 35-1. Confusion of wires due to the development of electrical communications, Broadway and John Street, New York, in 1890. (Courtesy of *American Telephone and Telegraph*)

of 1880 had no report of its being used in manufacturing. However, the 1900 Census reported 300,000 electrical horsepower in use, and by 1914 the horsepower had increased to 8,847,622, the same as was produced by steam in 1900.

The development of the United States as an urban industrial nation during the latter part of the 19th century accounts for the increased demand for energy, for lighting, transportation, and industrial power. But the fact that the energy was to be used in the form of electricity was due to technological developments which demonstrated electricity's natural advantages over all other forms of energy.

ELECTRICAL EXPERTS

As electricity became the basis for the important new industry dealing in power and light, there arose a need for men technically competent in the new technology and for others able to manage the business end. The ideal would have been men equipped for both aspects of the industry.

The first and probably most versatile of the new breed of men in the electrical industry were the arc light salesmen. These men had to be a combination of missionary and promoter because they were selling something—electricity—which was completely new to some communities. A salesman was expected to win over the men of means to get them to finance the new company locally, find property sites for the plant, get the town government to grant franchises (whose terms he frequently drew up), assist in the organization of the company, sell it the necessary equipment, and draw up the contract of sale. He did not leave until the construction men arrived to make the actual installation.

Unable to find technically trained men, the Thomson-Houston Company, one of the largest manufacturers of electrical equipment, trained its own, who were called "experts" after finishing a short course of study. The company recruited machinists, steam fitters, carpenters, school-teachers, and anybody with mechanical skill. "We took agents and peddlers, prize fighters and preachers," President Charles A. Coffin of Thomson-Houston said. Coffin himself had been a shoe salesman before entering the electrical business; in spite of very little knowledge of electricity, he became one of the most successful electrical company executives because of his organizational ability. He realized early that there was a shortage of good managers in the new industry, and when a manager complained that he did not even know what electricity was, Coffin said: "That's just our trouble. We've too many men who know what electricity is, or think they do. What we want now is somebody to care for the commercial side. Perhaps the less you know about what electricity is, the better."

Men like Edison and his major competitor, George Westinghouse (1846-1914), were self-taught and had acquired considerable competence in electricity. Both men had a good sense of what was commercially feasible. At the same time, both hired highly competent professionally trained men. Edison had F. F. Upton, a university-trained mathematician, and Frank J. Sprague, an electrical engineer. When General Electric decided late in the game to enter the field of alternating current (a.c.) it promptly used Charles Steinmetz (1865-1923), one of the few men in the field who had been trained in science.

For the most part, the early technical men in the field of electricity were either those who had gained experience in telegraphy and/or telephony or were self-taught. Primarily the theoreticians were mathematicians or physicists. However, it was but a short time before the needs of the electrical power industry influenced the engineering colleges, which began to introduce curricula in electrical engineering. The colleges did not seem to be farsighted about the needs. In 1884 the president of Stevens Institute of Technology said: "Ten thousand mechanical engineers are wanted to every ten electricians; and it would be a mistake for a very great number of young men to determine to devote themselves to electrical enterprises."

ELECTRICAL SYSTEMS

In the early stages of the development of electrical power, systems were designed for specific purposes. The arc light, electric motor, and incandescent lamp were each marketed as a separate system with a generator specifically designed for the installation. For example, Edison's central power station supplied electricity to many customers but the power was used only for incandescent lighting. There was no need for this segregation by use and customer except that, until the end of the 19th century, there were not enough users of electricity to warrant grouping them together and developing a flexible system suitable for all. One important stipulation was implicit in any such merger of uses and users into a single system, and that was that they use the same type of power, either alternating current or direct current.

Some applications of electricity, like the electrotype process and the early d.c. railway motor, would operate only on d.c. Some, like Jablochkoff's arc candle and Nicola Tesla's induction motor, required alternating current. The incandescent lamp worked as well on either a.c. or d.c. The invention of the a.c. induction motor not only made it possible to use motors in a.c. systems, but the motor itself had such useful characteristics that it was an important factor in favor of a.c. use. The practical transformer which was developed in 1885 was the strongest argument in favor of a.c. It was the last one of the very important electric power apparatuses to be developed, although the principle upon which it is based had been discovered by Michael Faraday and Joseph Henry as far back as 1831.

In Faraday's first successful experiment in electromagnetic induction, he used two coils of wire wound on an iron core and noted that the stopping or starting of a current in one coil induced a voltage in the other. This experimental device was the prototype of the transformer, and although the usual use for the transformer is to change the voltage, Faraday was interested only in demonstrating a principle. In 1838 Joseph Henry was able to change an "intensity" current, as he called a relatively low current associated with high voltage, to a "quantity" current, by which he meant a relatively high current associated with low voltage. He used a primary coil of many turns of wire and a secondary coil of a few turns of wire: what we call today a step-down transformer because with the high voltage in the primary a low voltage appears in the secondary.

Nothing was done with these voltage changers at first, except to use them in laboratory demonstrations. In 1882 Gaulard and Gibbs took out English patents on a transformer which they designed for isolating the customer from the main power line and reducing the main line's voltage to a safe value. Since the transformer works by means of a magnetic flux connection between circuits, the two circuits can be electrically isolated. This patent was later invalidated because it was found that the system did not operate satisfactorily. The Gaulard-

Gibbs system had the primaries of the transformers in series in the line, and as the number of lamps connected to each transformer changed, the voltage in the line was affected.

The difficulty was removed when three Hungarian inventors, Zipernowsky, Beri, and Blathy, devised the parallel connection for transformers. From that year, 1885, the transformer was an established electrical device, and its uses multiplied. The most useful application was for raising and lowering voltages, which was accomplished with practically no power loss. The transformer was used to raise the voltage of the generator to a very high value, the energy was transmitted at this high voltage, and at the consumer's end, the voltage was reduced again to a safe value. The higher voltage of transmission was desirable because there was less energy loss on the line than when electrical energy was transmitted at lower voltages.

In 1891 at the Electrical Exhibition at Frankfurt, Germany, a transmission line 109 miles long was erected. Transformers were used to raise the voltage of transmission to 25,000 volts. The test successfully demonstrated that electric power could be transmitted long distances by means of transformers and that the transmission could be accomplished efficiently, with a relatively small loss of energy. Electrical energy could be transmitted long distances by using alternating current systems and transformers; the question was, *should* it be?

Several things had to be determined before that question could be answered. The engineers had to know what was to be the major use for the electricity. In 1890 the major use was lighting, but would it remain so in the future? Other things had to be determined, such as where the electricity was to be used, how fast the demand would grow, and what adjustments could be made for a power demand like lighting, which fluctuated sharply over a twenty-four-hour period. A group of promoters who sought to harness Niagara Falls for electric power purposes needed to answer the question, a.c. or d.c.?

HARNESSING NIAGARA FALLS

In 1895, the tremendous hydro-energy of the falling water at Niagara Falls was successfully tapped and converted into electrical energy for commercial use. The achievement was spectacular in its own right, but of more lasting importance was the use made of electrical engineering to solve commercial problems and the consequent decision to build a large central station which would distribute power over a wide area. Electrical theory was used by engineers to demonstrate the superior economy of the alternating-current power distribution system over that of the more limited direct-current system. The project at Niagara was a pioneering move in the development of the modern systems of generating and distributing electricity. The challenge of harnessing Niagara was intimidating in its magnitude, and the methods to be used were unproved; yet

American capitalists were willing to undertake the risk and to wait five years for returns on their money.

The development of plans for the Niagara Falls power project follows the historical development of the use of power from water wheels directly coupled to mechanical transmission, thence to direct current transmission of electricity, and finally to alternating current transmission. When the first scheme for utilizing the Falls was suggested in 1886, water wheels were proposed to produce the mechanical power to generate the electricity. Between that date and 1893, when the decision was made in favor of alternating current, the capitalists who organized the power project made use of the most advanced thinking of the world's leading scientists and engineers. On the basis of this thinking, they were able to make decisions which have since proved absolutely correct.

No one can adequately describe Niagara Falls. Since they were first discovered, men have been awed by the spectacle of grandeur and power. The total head, or drop, in the immediate vicinity of the Falls is about 200 feet, and an estimated 210,000 cubic feet of water per second flows over it. If all of this could be converted to usable energy, six million horsepower would be developed. Not only is there great power available but the constancy of flow of the Niagara River makes it a very dependable and steady source of power. A great reservoir of four of the five Great Lakes maintains the level of the river, and hence the potential power of the Falls remains almost unchanged from year to year. At an early date mills, which derived power from water wheels, had been built along the banks of the Niagara River, but by 1882 only 7000 horsepower of the vast potential was being used.

The harnessing of the Falls to produce electrical power came as the direct result of a rapidly increasing demand for that product. Since 1882 when Edison's Pearl Street Station was built, the idea of a central station for supplying direct current to incandescent lamps had spread rapidly. Other uses of electricity also increased; the use of electricity to power street railways was pioneered by Frank J. Sprague. His successful installation of an electric streetcar system in Richmond, Virginia, in 1887, and the electrification of elevated lines gave "electricity such an advantage over all other forms of motive power that it soon was recognized as the only motive power suitable for urban rapid transit." The United States in 1890 had 1261 miles of electrified street railway, compared to over 5000 miles of railway which used animal power. By 1902, electric street railways covered 21,920 miles, while animal, cable, and steam-power railways together amounted to only 669 miles. In 1886, however, the possibilities of exploiting electricity were just beginning to be realized. The demand for lighting and traction was to require large investments which were to pay handsomely. Electric power was to find even more extensive use in manufacturing, but in 1886 this application was only in the experimental stages, and electrification of factories did not begin in earnest until 1900. The electrolytic refining of aluminum ore was patented in 1886. This and other electrochemical processes were to be large consumers of Niagara power.

The citizens of Buffalo, New York, a promising industrial location, took a forward-looking view of the application of power. The first alternating-current central station in the United States was installed there in 1886. At that time, Buffalo's industries were powered by approximately 30,000 horsepower, little of which was electricity. In 1887 the businessmen of Buffalo pledged $100,000 to be offered as a prize "to the inventors of the World" who would design an appliance which would utilize the power of the Niagara River "at or near Buffalo, so that such power may be made practically available for various purposes throughout the city. . . ."

At about this time, New York State acquired land along the Niagara River both above and below the Falls in an attempt to preserve the natural beauty of the site. This move placed serious, but as it turned out fortunate, limitations on any proposal to harness Niagara. Thomas Evershed, a division engineer on the Erie Canal, in a letter to a Lockport, New York, newspaper in February 1886, had suggested a tunnel with openings beyond the borders of the state reservation. Water was to be taken from the river above the Falls at several different spots. It was then to be fed through the water wheels of 238 mills (maximum) and pass through the tunnel out into the river below the Falls.

In the summer of 1889, a group of bankers, including an associate of the capitalist Cornelius Vanderbilt, met to make plans for carrying through the Evershed scheme and formed the Cataract Construction Company. The Cataract Company purchased the entire capital stock of a previously formed company, the Niagara Falls Power Company; Cataract became the financial agent for the Niagara company and was to undertake all construction. Edward D. Adams, a partner in Winslow, Lanier, and Co., a firm of New York bankers who took a half-interest in the Cataract Company, and a member of the Board of Directors of Edison General Electric, was delegated to conduct an investigation to determine the merits of the proposal. Adams became the capitalist most responsible for the success of the Niagara Falls project; he was connected with the enterprise for the next thirty-seven years, until 1926.

The Evershed scheme, as conceived in 1890, was solely for the use of water wheels within individual factories. Yet the list of subscribers to the first construction fund contains many people who were important in Thomas Edison's electric enterprises. Three men on the list of subscribers, J. P. Morgan, Darius Ogden Mills, and M. K. Twombly, comprised a committee of three which had represented the Edison General Electric Company in the merger with Thomson-Houston which later formed the General Electric Company.

It was estimated that a fund of $2,630,000 would be enough for financing a complete plant which would produce 20,000 horsepower by hydraulic means. At $10 per horsepower per annum, the income would be enough to pay 5 per cent interest on $3,0000,000 worth of bonds and leave $50,000 for a sinking fund and to pay operating expenses. The raising of capital was accomplished quickly, being completed early in 1890, and in February the Cataract Company was authorized to proceed with construction. It is interesting to note that the financial backing for the Niagara project came largely from American sources, though most of the 19th-century American industrial enterprises had been largely dependent on foreign capital, for the most part British. By 1890 American capital was well on its way to independence from foreign sources. Upon Adams fell the responsibility of turning Evershed's basically sound scheme into a profitable venture. Engineers were agreed that the plan for producing power at the Falls by means of water wheels or water turbines, was feasible, for the Swiss had been successfully manufacturing water wheels and turbines of the size needed. The village of Niagara Falls, however, hardly seemed the place to market 20,000 horsepower. It had a population of only 5000 in 1890. The power would somehow have to be transmitted to Buffalo. But how?

A.C. OR D.C.?

The distance from Buffalo to Niagara Falls was twenty miles. Never before had such a large amount of energy been transmitted so far. Several methods of transmitting power were in use in the United States and Europe. They were by wire or manila rope, by water under pressure, by compressed air (Westinghouse in 1889 seriously urged this system because he believed it was the only one which was perfected enough to handle the magnitude of the Niagara power), and by electricity. Although mechanical means were not ruled out until the end of 1891, the scientific and engineering world believed that electrical transmission would surpass the others in efficiency. The real battle was over the question of whether the transmission system was to be alternating or direct current. A decisive encounter in this so-called "battle of the systems" took place at Niagara.

The main advantage of a.c. over d.c. is the efficiency of transmission. Alternating currents can easily be changed to high voltage, which reduces the loss of energy on the transmission line beyond anything that is possible with d.c. But much of the argument of a.c. versus d.c. in 1890 centered on the economics of the two systems and turned on the question of what the electricity was to be used for.

If a.c. generators and transformers were used at full capacity, they were unquestionably more efficient than a comparable d.c. system. One pound of coal burned to produce steam which ran turbines used to drive generators would deliver more energy at the end of a 20-mile transmission line if a.c. were used. But the trouble was that the demand for electricity in 1890 was mainly for lighting. That meant that during the daylight and late night hours, when little electricity was used, the generating equipment was not employed to capacity. Under conditions of a varying load over a twenty-four-hour period, d.c. was much more efficient than a.c. One engineer estimated that under such conditions 20 pounds of coal would be needed to deliver a unit of energy in the a.c. system whereas only 10 would be needed for the d.c. system.

The solution to the problem of a varying load had already been worked out for the d.c. system. Small generators were used in the central stations in conjunction with a group of batteries. When the load was very light one or two generators were used. As the load increased, generator after generator was added to the system. During the short periods of peak demand the batteries would always be working at maximum efficiency, and therefore the cost of power was kept down.

Using more than one generator on a transmission line at the same time is called paralleling. Although the theoretical problem had been solved as early as 1884, the building and operating of actual machines was another matter. So little was known about the characteristics of a.c. generators in 1890 that many engineers were not convinced that parallel operation was feasible in that system.

The Cataract Construction Company decided to consult several authorities in the field. Among the first to be contacted was Thomas A. Edison, whose advice was sought partly because of his experience and partly because many of the financiers associated with the Niagara project had also been associated with Edison in his electric light ventures. Edison was enthusiastic. When asked via cable to Europe: "Has power transmission reached development that in your judgment seems practicable?" He replied, "No difficulty transferring unlimited power. Will assist. Sailing today." After a survey, Edison recommended generation of d.c. at Niagara Falls for transmission to Buffalo. But because of the interest in alternating-curent power, the Cataract Company decided to delay a final decision until that method had been fully investigated. Edison apparently was piqued by what he considered a rebuff and he refused an offer of $10,000 for his work. He said that he would prefer to keep his information.

Frank J. Sprague, who had just installed his electric trolley system in Richmond, was also consulted. He was cautious about using a.c. at Niagara:

> My own feeling is simply this: with ample means, and with an assured demand for the power, I would not hesitate to transmit any amount of power from Niagara Falls to Buffalo, but, although I would feel capable of doing this, if I were at the same time asked if I would invest any money in the enterprise, I would decline to do it. . . .

Professor Henry A. Rowland of Johns Hopkins University, one of the country's leading physicists, was asked, has the remarkably expanding field of electrical transmission reached the commercial stage? He replied:

> That power can be transmitted to a great distance by electricity and with reasonable certainty is a matter determined today [1889]. But the

> practical and commercial question is of a different nature from the scientific one and may be stated thus: At what distance from cheap water-power can such power, transmitted electrically, compete with steam in cost and certainty of operation?

Several groups involved in the Niagara project travelled through Europe hoping to gain some help. These men travelled through England, France, Italy, Germany, Hungary, and Switzerland. The tour gave Adams an idea for an international contest to develop plans for the Niagara project; the plans were to be judged by an International Niagara Commission with Sir William Thomson (Lord Kelvin), famous British physicist, as its chairman.

Letters of invitation to compete went out in June 1890. Twenty-eight invitations were sent, twenty-three to Europeans and only five to Americans. There were three different categories with three prizes in each. The largest award was to be $3000. Commissions ranging from 2.5 per cent to 5 per cent of the manufacturing costs were to be paid to the winners if they were not also chosen to build the equipment.

There was some very strong criticisms of the Commission's competition principle. Westinghouse was quoted as saying, "Those people are trying to get $100,000 worth of information for a prize of $3000." At a meeting of the British Institution of Electrical Engineers one engineer said the undertaking at Niagara would be carried out on the "unrecompensed work of the whole electrical world." The remark was greeted by applause from the other English electrical engineers.

Adams, in defense of the Commission idea, cited the fact that only one entry was considered worthy of first prize and that one was for the hydraulic turbines which were designed according to well-established principles. No prize was awarded for a system of distribution. The fees paid to foreign engineers and to the Commission, a total of $75,872.91, were far in excess of Adams's original estimate. An additional sum of $430,000 was paid to foreigners for engineering fees, machinery, and material. These sums make the charge of "unrecompensed work" difficult to sustain, but it is hard to estimate what a "fair" price would have been.

The International Niagara Commission's report marked progress toward acceptance of the principle of electrical transmission. It was the general opinion of the Commission that electrical methods be adopted as the chief, although perhaps not the sole, means of distributing power. However, "in the selection of electrical methods they are not convinced of the advisability of departing from the older and better understood methods of continuous currents in favor of the adoption of alternating currents."

By the end of 1891, Evershed's original scheme for diverting the Niagara River's waters through a tunnel which was to pass under mills, had been considerably modified. Adams wrote: "I came to the conclusion that our true way possibly might be to build this tunnel and develop the whole power in this one central station, transmitting the power to different places." He reached this conclusion in the summer of 1890.

The final decision in favor of electricity was made by the Board of Directors in December 1891, on the advice of the company's technical advisers. One of these advisers was Professor George Forbes, who was a Fellow of the Royal Society and had been president of the British Electric Light Company. His competence in the design of electrical and mechanical machines was to prove essential to the success of the Niagara project. The technical advisers had been swayed by successful transmission of electrical power over considerable distances from Tivoli to Rome (16 miles), in Portland, Oregon, and at Telluride, Colorado.

Once the major decision had been made in favor of electricity, the Cataract Construction Company sent letters of invitation to a select list of companies which had "a scientific staff competent to design such novel machines as were required, and of ample facilities for their construction." The only American companies to receive invitations were the Thomson-Houston Company, the Edison General Electric Company, and the Westinghouse Company. Nothing was heard from these companies for a year. Meanwhile a Swiss manufacturing company had submitted a likely proposal for an a.c. system, but negotiations with this firm were reluctantly stopped when it was decided that the disadvantage of a 40 per cent freight charge was too great.

In March 1892, Westinghouse notified the Cataract Construction Company that it was ready to submit plans for the generators at Niagara. Representatives of the Cataract Company visited Westinghouse and were impressed by its staff of mechanical and electrical engineers. Not only did Westinghouse have competent theoretical men, but, as Adams noted, it had manufacturing facilities capable of handling the unique problem of the Niagara generators. Adams later wrote, "Moreover, the fact that the Westinghouse enterprise developed from a machine shop as a foundation into a great engineering structure, including important electrical works as its latest production, was a promise of favorable results. . . ."

The important electrical works produced by Westinghouse were some 300 central power stations supplying electricity for an estimated 500,000 sixteen-candlepower incandescent lamps. This growth had been accomplished in a short span of four years, from 1886 to 1890, and according to the engineering judgment of the day, the work was technologically superior.

All of the stations built by Westinghouse were for alternating current, for the company at an early date had committed itself to this system. Not only had it bought the American rights to a system designed in Europe, but it had Tesla's patent on an a.c. motor. At the Chicago Exposition of 1893, Westinghouse exhibited a complete polyphase a.c. system, including an a.c. generator, transformers for raising the voltage for transmission, a short transmission line, trans-

Fig. 35-2. Generators in the first Niagara Falls power station, Photo 1896. (Courtesy of the *Niagara Mohawk Power Corporation*)

formers for lowering the voltage, induction motors, synchronous motors, and a rotary converter which supplied d.c. for a railway motor.

Finally, in March 1893, eighteen months after the letter of invitation had been sent, both Westinghouse and General Electric submitted plans for a.c. systems consisting of generators, transformers, and transmission lines for Niagara. The Cataract Company's consultants noted that both firms had submitted almost identical plans except for a difference in the number of phases. The Cataract Company decided, however, that there was no advantage in selecting either proposal and rejected both. Instead, in the spring of 1893 the Cataract directors decided the company would design its own generators and placed Professor Forbes, who favored the use of a.c., in charge of the work. By this time the tide of scientific opinion seemed to favor this system. Accordingly, the directors approved the use of alternating current for transmitting power from Niagara Falls.

Practical evidence of the feasibility of a.c. transmission was not lacking. The International Electrical Exhibition at Frankfurt, Germany, in 1891 displayed the latest advances in electrical technology. One of the main features of the exhibit was an a.c. line for carrying sizable quantities of power from Frankfurt to Lauffen, a distance of 110 miles, the farthest electric power had been sent to date. Tests upon the line showed an efficiency of transmission of 77 per cent; that is, for every 100 watts fed in at one end of the line only 23 were wasted in the line and the other 77 were delivered as useful energy. The success of this line was particularly gratifying to the proponents of alternating current. The directors of the Cataract Company anxiously awaited the results of the Frankfurt tests and were told of their success in January 1892.

In a paper delivered before the Institution of Electrical Engineers in 1893, Forbes explained the unique generator design which he had created for Niagara and the reasons for it. He prefaced his remarks by saying that, if he had given the paper a few years before, he would have had to devote considerable time to discussing the relative merits of a.c. and d.c. and to justifying the Cataract Company's choice of the former. What had changed since the company's first inquiries in 1889? What made Forbes and the company so sure they had made the right decision? As a matter of fact, the decision at Niagara was truly a trail-blazing one.

There were, of course, some very strong voices of disapproval. Many experienced and capable electrical engineers severely criticized Forbes at the Institution's meeting, where the discussion was long and heated. Indeed, less than a week before the company made its final decision, Lord Kelvin, who had served as chairman of the International Niagara Commission, cabled Adams, ". . . trust you avoid gigantic mistake of adoption of alternate current."

Different arguments were employed on either side of the Atlantic. The English debated the scientific and economic questions, while the Americans seemed more concerned over the safety of a.c. Edison, for example, campaigned against a.c., maintaining that it was dangerous. His warnings seemed to be underscored when the New York legislature chose electrical shock as a means of carrying out the death penalty and chose Westinghouse to furnish the a.c. generators. Forbes and all of the people connected with the Cataract Company and the Westinghouse Company were correct in seeing that a.c. was the best form for electrical power. The question was both technical and economic: could a.c. be generated, transmitted, and used so that it gave dependable service? Events were to prove that it could.

What the critics of a.c. did not foresee was that electricity would have many more uses than lighting. One of the largest and growing uses of electricity in the United States was for motors on street railway cars. Electric motors were also being used for driving sewing machines and elevators. These diverse uses meant that demand for electricity over a twenty-four-hour period would not vary as much as it had when used only for lighting. Forbes reported to the Institution of Electrical Engineers that "the officers of the company and myself . . . looking at the purposes for which our machinery is being set up, felt sure that the proportion of electricity which would be used for lighting purposes would not be large, and that we must look upon our whole plant as a *power* producing and distributing plant, and that our object must be to distribute *power* in the most efficient and economical manner."

Time proved Forbes and the directors of the Cataract Company to be wiser than they knew. The beginning of the 20th century marked a sharp increase in the use of electricity for many things other than illumination. The constant flow of water down the Niagara River meant that power could be produced continuously for a twenty-four-hour period with hardly any extra cost. The availability of this constant power attracted electric furnace (chemical) industries producing such commodities as aluminum, abrasives, silicon, and graphite. For example, the cheapness of Niagara power made the manufacture of aluminum profitable. For the first time such electrochemical industries were economically feasible, and these industries in turn provided the constant demand for electricity which was to make the Niagara project even more profitable than anticipated.

THE NIAGARA LESSON

After Forbes submitted his plans for the Niagara generators, the Westinghouse Company was asked to construct them. The Westinghouse engineers balked, since Forbes's design was a departure from the standard machine. They would give no guarantee of performance, basing their objection on the fact that they had had more experience in the construction of a.c. generators than the Englishman and hence were more cautious. As a result, Forbes's revolutionary ideas were tempered with the manufacturing experience of the Westinghouse men to produce generators which performed even better than anticipated.

The generators were finished and installed in 1895. Power was being supplied locally in the summer of 1895, and the first transmission of power to Buffalo took place in November 1896. Years later, when more was known about a.c. generators, a Westinghouse engineer reviewed Forbes's original plans. He found that he would have made no changes, even in the light of advanced knowledge.

The Cataract Construction Company was voluntarily liquidated in 1899, ten years after its incorporation. At the close of its construction operations, the Niagara Falls power plant was well established, with eight 5000-horsepower units in operation. These were providing enough revenue to pay interest charges on $9,000,000 worth of bonds, operating expenses, and still have $100,000 a year surplus. During the ten years of operation, the stockholders of the company had paid in a total of $7,044,500. Their faith in the Niagara project was well rewarded, for on liquidation they received from the Niagara Falls Power Company $8,832,000 in first mortgage 5 per cent bonds, $3,974,000 in capital stock, and $289,750 in cash.

The plan to produce large quantities of power and distribute it over a distance of twenty miles proved a success. The engineering world hailed the achievement as representing the coming of age of electrical engineering and a final demonstration of the superiority of the alternating current system over the direct current system. Niagara had pioneered the way in the development of large centralized electric utilities which supplied the power for the 20th-century expansion of the American economy.

36 / Metallurgy: Science and Practice Before 1900

CYRIL STANLEY SMITH

SCIENCE COMES TO THE STEEL INDUSTRY

The great advances in physics and chemistry in the 17th century had no immediate effect upon the metallurgical industries, for the processes had reached a stage of empirical development far beyond the theoretical understanding of the day. By the end of the 18th century, however, things had changed. Great advances in methods of chemical analysis (largely by Swedish chemists) led to the discovery of new metals and also provided techniques for revealing the minor amounts of impurity that had been responsible for the vastly different qualities of iron from different ores and for the differing effects of various methods of processing. In addition, they were able to determine the nature of various alloys.

The identification of tin as the significant alloy in bronze was obvious from the admixture necessary in its manufacture. In brass the former mystery was dispelled when, early in the 17th century, it was found that metallic zinc (collected as an accidental condensate in furnace crevices) added to copper in the right amounts, gave a metal identical with the ancient calamine-made material.

Sulfur had long been associated with hot shortness (hot brittleness) in iron since it could sometimes be identified by its smell, but Bergman's identification of phosphorus in cold short iron in 1781 solved a great mystery.

Far more important, however, was the definite identification of carbon as the substance in iron which was responsible for its change first into hardenable steel and then into fusible cast iron; the fact that carbon was used as fuel in turning iron into cast iron and steel had completely disguised its chemical role, especially since such very small amounts produced such profound effects—about 1 per cent carbon in hard steel and 3 per cent in cast iron. Aristotle's idea that steel was a more refined form of iron was natural enough because steel resulted from a more prolonged treatment in the fire, and fire, in common experience, does indeed purify. The French scientist René Réaumur (1683-1757), on the basis of experiments done in the manner of the best modern applied scientist, had concluded in 1722 that some material substance went from the cementation compound into the iron during the making of blister steel. That this "something" was the charcoal itself was not proved until the 1780's after some rather fruitless speculations on the role of phlogiston in steel.

The discovery of carbon in steel occurred in Sweden, where studies of Damascus steel had shown that acids would attack iron cleanly but would leave a dark deposit on the surface of steel and cast iron. Rinman in 1774 studied the residue left on dissolving cast iron in nitric acid and saw its similarity to "black lead" (plumbago or graphite). Torbern Bergman confirmed this with quantitative measurements in 1781, and in 1786 the essentially modern theory of steel, free from phlogistonic overtones, was expressed independently in an encyclopedia article by Guyton de Morveau and in a paper before the Académie des Sciences by Vandermonde, Berthollet, and Monge, all working in France. However, the theory of the latter savants—two outstanding mathematicians and a chemist second only to Lavoisier—misinterpreted the nature of white cast iron, which they attributed to the presence of both carbon and oxygen.

By no means all metallurgists accepted the carbon theory for several decades, but it did explain the differences between the forms of iron and helped in understanding the processes of their manufacture. The operations and the phenomena of puddling took on new meaning, although it should be noted that no theory whatever had been involved in the invention of the puddling process. In fact, England, which had led in the industrial development of iron and steel in the 18th century, lagged far behind Sweden and France in metallurgical science.

After this, the chemist came to play a more and more obvious role, and a few of the larger iron manufacturers began to employ chemists to examine their raw materials and products. The close association of the scientifically trained chemist with the practical problems of production helped to accelerate the change in character of the industry from pure empiricism to a more scientific approach. By the end of the 19th century both buying and selling to chemical specification were common.

THE MICROCRYSTALLINE STRUCTURE OF STEEL

The mechanical engineer, testing products both for reliability and to obtain values of strength for use in design, had an equally profound impact on the understanding of metals. At the very beginning, metallurgists must have employed a crude mechanical test of their product simply by hammering and breaking a sample. The appearance of the fractured surface was properly regarded as an index to the character of materials of all kinds and thus related to their suitability for different sorts of service. Réaumur based his scientific studies of iron and

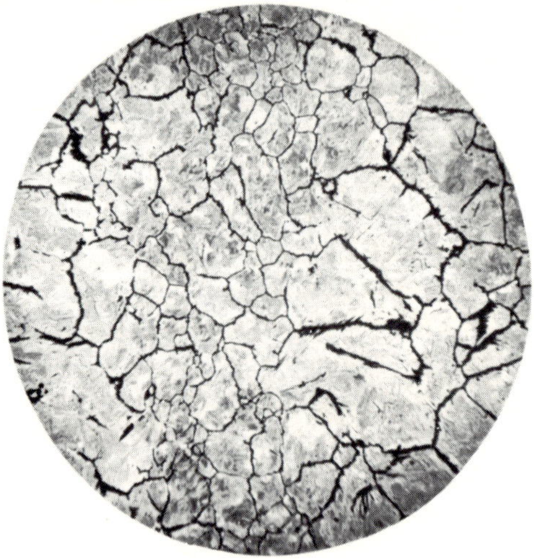

Fig. 36-1. The structure of high-carbon steel, photographed under the microscope at a linear magnification of 9 after polishing and etching. This historic photograph was made by H. C. Sorby in 1864. (From *Journal of the Iron and Steel Institute,* 1887)

steel on the change of texture accompanying the conversion of iron into steel.

Not, however, until late in the 19th century was it definitely shown that the appearance of the fracture depended upon the size and distribution of metallic microcrystals, an idea so common to us that it is hard to realize how recent it is. For example, the fracture of a piece of tough copper has a smooth silky appearance, while that of wrought iron shows elongated fibers: since fatigue fractures of the same metal often show crystalline facets, it was once assumed that vibration caused crystals to grow in a material that was previously amorphous. Moreover, malleability was rarely associated with crystallinity, for the commonest natural crystals are brittle minerals.

Although there had been previously hints as to the true nature of metallic structures, the work of H. C. Sorby in 1863-64 in the microscopic study of metals led to the most important discoveries. A non-professional scientist working at his own home in Sheffield, England, Sorby had ready access to the products of the local steel-makers and to the polishing techniques used for the manufacture of steel plates for engravers. Prior to his work, microscopists had seen only burnished or broken features; he produced unstrained surfaces on the metal and revealed, by the application of chemical etching reagents, the microconstituents in their true shapes and arrangement. He was able to identify seven different chemical phases and to relate the various amounts of iron carbide and graphite in the samples to the old grades of steel and cast iron. Sorby showed how the little crystals in iron were packed together, how they could be deformed by working and would recrystallize on heating. Steel was seen to undergo a complete change in structure when it was hardened.

For the first time, it was possible to explain the diverse properties of metals in realistic terms based on observable features, quite different from the "molecular" explanation of the chemist and physicist that was current at the time. Nevertheless, Sorby's techniques were slow in being widely adopted. Indeed, it was not until after similar independent studies in Russia, Germany, and France from 1878 to 1885 that their full significance was realized in his native country.

In 1887 Floris Osmond in France combined the microscopic observations with a study of heat evolution by thermal analysis to show the temperatures at which structural changes occur in iron and steel, and metallurgical science began to take on its modern appearance. The next decade saw the collection of a perplexing mass of observations on the structures of a wide variety of alloys as they

Fig. 36-2. The great testing machine in the Watertown Arsenal, built in 1879. This machine had a capacity of 800,000 pounds in tension, 1 million pounds in compression, and could test specimens up to 30 feet in length. It was designed by A. H. Emery and uses a hydraulic load-balancing system. This illustration formed the frontispiece of the *Report to the U.S. Board for Testing Iron, Steel,. and Other Metals* (Washington, 1881).

were affected by changes in composition and heat treatment. In 1899 all this mass of complicated data was put in order by the application of the "phase rule" of the American scientist Willard Gibbs and by the drawing of diagrams depicting the dependence of microstructural constitution on temperature and composition. By the early 20th century, metallurgists had systematically collected data on the constitution of alloys, had related microstructural details to service failures, and learned to use structure as a key to the control of reproducible quality in production operations.

The unravelling of the microstructures of metals was paralleled by the development of the mathematical theory of elasticity, often misnamed "strength of materials." Testing methods for mechanical strength, ductility, hardness, and fatigue resistance provided a quantitative check on the effectiveness of the compositional and structural changes wrought by the metallurgists and gave a numerical basis for engineering design which permitted lowering the factors of safety as confidence increased.

The next stage in scientific understanding would involve the determination of the actual arrangement of atoms within the crystals of metals, possible only after the introduction of X-ray diffraction in 1912. Though this period is beyond our concern, it should be emphasized that the modern success in developing materials for strenuous service in nuclear and space applications stems directly from a long period of concern with structure, and that the textured metals of the Orient provided the initial incentive for the metallurgical studies by European scientists.

DEVELOPMENT OF ALLOY STEELS FOR TOOLS

No application of iron and steel before the last years of the 19th century, whether for general structural purposes or in specialized tools or weapons, seemed to need anything beyond the range of properties that could be obtained by the adjustment of the carbon content and by heat treatment. Carbon itself is the most effective alloying element for iron, and none but the most recent alloy steels are stronger than plain carbon steels when properly heat-treated, provided that the sections are sufficiently small so that the center cools rapidly enough during quenching. The critical thickness in this process is not over about half an inch, and therefore when engineers began to think of heavy sections bearing high unit stress, the effect of alloying on steels was intensively studied. The stories told of alloy steels in antiquity are mythical, for the superiority of steel from one source or another depends more on the absence of impurities than on the presence of alloying elements, and even more on the skill of the artisan.

The effect of some alloying elements on iron was investigated in the 18th century, partly in connection with magnetic studies and partly for general interest, but no useful compositions emerged. In 1819-21 a famed London cutler named James Stodart enlisted the aid of Michael Faraday in an attempt to duplicate Oriental wootz (a high-carbon crucible steel, which probably originated in southern India in antiquity, and which possessed superior qualities), and they published studies of tool steels alloyed with many different metals. Following their work some cutlery steels containing silver were made, and a Swiss metallurgist, John Fischer, who had pioneered the crucible process on the Continent, marketed nickel tool steels from about 1830 on; both, however, were of trivial economic significance.

Tungsten tool steels were first made commercially in Austria in 1855, and Robert Mushet in England, beginning in 1868, marketed large quantities of his "air hardening" tool steel. This contained about 9 per cent tungsten and 1.7 per cent carbon, and the dispersion of numerous microscopic particles of tungsten carbide enabled the tool to last longer without resharpening. This alloy in turn was improved spectacularly by a high temperature heat treatment devised by the Americans F. W. Taylor (later the originator of scientific management) and M. White in 1898-1900. After adjustment in composition, this became the steel for modern high-speed tools that will not soften at a red heat, and can cut metal at enormously increased speeds. The proper utilization of these tools required the development of an entirely new line of much sturdier machine tools with more powerful driving mechanisms. It was Taylor's general concern with machine-shop efficiency that had focused his attention upon the need for better cutting tools, but his discovery of the high-temperature heat treatment was an astute following up of an accidental observation, owing little to science.

ALLOY STEELS FOR STRUCTURES

By the end of the 19th century, steel had ceased to be a material only for tools and was everywhere replacing iron in machine parts and building materials. Herein lay the main opportunity for the use of alloys. Chromium steel (first made by Berthier in France in 1821) was the first alloy steel to be used in structures, as distinct from tools, in the bridge at St. Louis built by James Eads in 1867-74. Crucible-cast ingots of the steel were forged to shape and supplied without heat treatment, but they varied in composition and often did not meet the proof-stress specification: in the last parts of the bridge to be constructed, the steel members were replaced by heavier sections of wrought iron. Recent analyses have shown the presence of almost no chromium in the metal.

Naval armament gave the first real demonstration of the advantages of alloy steels in large structural applications. The heavy wrought-iron plates first used in ironclad ships were improved either by welding on a facing of carbon steel, or by a deep surface carburizing process ("harveyizing") which had the same effect.

A French metallurgist, Marbeau (who had played an important role in the development of methods of smelting nickel from New Caledonian ores) experimented with nickel steel armor plate in 1885. Some of Marbeau's steels were examined by an English metallurgist, James Riley, who published in 1889 a comprehensive study of the mechanical properties of a range of nickel steels which showed the exceptionally high strength and ductility conferred by the addition of about 3 per cent nickel. He emphasized that the steel could easily be made in the open-hearth furnace without change of practice. Trials in England and the United States in 1890 and 1892 left no doubt of the superiority of the new material, and it shortly became standard in the principal navies of the world. Subsequently, nickel steels were used extensively in case-carburized automobile gears. Nickel was a component in the later important nickel-chromium steels and in many more recent alloys, especially the austenitic stainless steels.

Manganese, it will be remembered, had an important role as a minor addition—less than 1 per cent—to deoxidize and counter the effects of sulfur in crucible and Bessemer steels. With 5 per cent manganese, however, the steels become brittle, and in 1882 Robert Hadfield made the spectacular discovery that over 10 per cent manganese produced a quite different type of material—known today from its structure as an austenitic steel—which is softened, not hardened, by quenching. Steel castings containing 12 per cent manganese and 1.2 per cent carbon found use in rail joints, steam-shovel teeth, and other parts where extreme wear resistance was needed. But Hadfield's steel was perhaps even more important in publicizing the effects of alloying elements in steel and making it obvious that their study would be profitable. Shortly after 1900 Hadfield had magnetic measurements made as a part of systematic studies of iron alloys, and uncovered the improved permeability and low magnetic losses in silicon iron alloys, which have been used in electric power transformers ever since, with vast savings of power.

The automobile gave the principal incentive for the development of deep-hardening alloy steels. This began with vanadium steels in 1904, a French discovery, improved in England but applied most widely (by the Ford Motor Company) in the United States. Since that time, growth of the alloy steel industry has been as closely allied with the internal-combustion engine as with armaments. With the new techniques of metallography, the effects of alloying on the microstructure of steels could be followed closely. Alloying was found to effect changes in three quite distinct ways: by delaying or suppressing the high-temperature transformation of steel to soft products; by refining the grain sizes; and by causing mild hardening of the iron itself. Not until 1930 were the effects properly understood, and even today parts of the theory are in dispute.

In all the older alloy steels, the elements were added mainly to ensure that the carbon could have its full strengthening effect. However, after the introduction of the austenitic stainless steel following World War I, a different class of iron-like material became available which depended more on the non-ferrous metallurgist's principles of solid-solution hardening and precipitation hardening. In many of these "steels," carbon, far from being considered a desirable strengthener, is regarded as an undesirable impurity. The recognition of the similarities between all metals, indeed all materials, marks today's exciting period in the growth of knowledge of materials.

ALUMINUM AND TUNGSTEN

While the story of iron and steel during the 19th century shows the growing influence of science on metallurgy, the history of the development of commercially useful aluminum is perhaps tied even closer to the parallel development of scientific theory and laboratory practice. A large proportion of the men who advanced the technology of aluminum themselves had scientific training. The cheapness of the metal was dependent upon advances in the new electrical industry, and its wide application required the development of a new series of alloys based on a new principle (precipitation hardening), dependent, in turn, on a growing scientific understanding of the intricacies of metal structure.

The production of aluminum from its chloride was the first large-scale metallurgical process which involved reactions other than those of the oxides and sulfides known to antiquity. It was a precursor of the use of halides in the preparation of the more reactive metals today and, probably, will be of much wider use in the future. The alkali metals, sodium and potassium, had been made electrolytically by Humphry Davy in 1807, but he was unable to produce any metal from aluminum compounds.

Aluminum is a strongly electro-negative metal, but the main difficulty lay not so much in its reduction as in the oxidation of the surface of the reduced metal, which prevented its consolidation. The Danish chemist, H. C. Oersted, produced in 1825 minute amounts of a metal "in color and lustre like tin" by reacting potassium amalgam with anhydrous aluminum chloride and then distilling; but he did not follow this up, and for a time it was believed that F. Wöhler's production of a metallic gray powder in 1827 was the first aluminum. In 1845 Wöhler produced tiny globules of malleable metal and measured its density. Development of a usable process continued to interest men in several countries. Commercial production began in France after H. Sainte-Claire Deville modified Wöhler's process by using sodium to react with a fusible double salt of sodium and aluminum and recovered large coalesced drops of metal. By 1855 the metal was offered publicly at $200 per kilogram; thereafter, the story has been one of decreasing cost and increasing scale. However, for a long time its high cost meant that the chief uses of aluminum were in jewelry and trinkets rather than in engineering applications. One exception was the cap of the Washington Monument, put in place in 1884. It is a 100-ounce casting of chemically prepared aluminum.

Although the sodium process had become much cheaper, it was suddenly

Fig. 36-3. Electrolytic cells producing aluminum in the Pittsburgh Reduction Company's plant, *c.* 1890. (Courtesy of the *Aluminum Company of America*)

displaced in 1886 by an electrolytic process developed almost simultaneously by C. M. Hall of Oberlin, Ohio, and P. L. T. Héroult of Gentilly in France, both of whom applied for patents early in 1886 on the electrolysis of alumina dissolved in molten cryolite (sodium aluminum fluoride), which has a density low enough to allow the metal to collect in a pool at the bottom of the cell. The resistance of the electrolyte with the high currents used gave enough heat to keep the electrolyte molten without external heating. The successful development of the process required improvements in the carbon anodes (large quantities of which were consumed) and was critically dependent upon the availability of cheap electrical power. Equally important was the preparation of the alumina raw material. Though aluminum is a major constituent in common clays, only that rich in aluminum bauxite has served as an economical source. The Bayer process, which was developed just in time to feed the expanding electrolytic industry, is based on the selective precipitation of aluminum hydroxide and other hydroxides from solution in caustic soda as the concentration is changed.

Producers of aluminum at first had some difficulty in selling their product. The first extensive market was in domestic cooking-ware. Despite the development of some good casting alloys, aluminum did not become a significant engineering material until after the development of relatively high-strength wrought alloys. The greatest single step was the discovery of age-hardening (precipitation hardening) alloys by a German engineer, A. Wilm in 1906. The theoretical explanation of this discovery by Merica and his colleagues in 1919 provided a great stimulus to metallurgical research in general, for age hardening was the first and even today practically the only useful metallurgical phenomenon which was not exploited in some form by the ancients. The first use of the new alloy, which was called Duralumin after the Durener Metallwerke, A. G., came even before the theory existed, in the construction of the British airship *Mayfly*, built in 1910.

The availability of aluminum enabled the development of the aluminothermic method of reducing other metals from their oxides. In this process, aluminum powder is mixed with oxides such as manganese, chromium, titanium, and vanadium. The reaction, once started, will proceed exothermically producing a high enough temperature to melt not only the metal but the resulting alumina slag. For several decades after about 1890, this found considerable application in producing the new metals needed for alloy steels, and also in reacting with iron to provide a portable means of producing hot steel for welding operations.

Tungsten, the metallic element of highest melting point (3410° C), was first reduced to the metallic state in 1783 and found important use as an addition to tool steels, which require special properties of hardness. The tungsten carbide cutting tool appeared about the late 1920's, and the ductile tungsten electric lamp filament, in 1910. Many other metals, such as vanadium, tantalum, titanium, zirconium, colombium (niobium), molybdenum, and uranium, were identified chemically in the 19th century or before, although they played no significant role in industrial metallurgy until the improved vacuum and analytical techniques of the 20th century permitted their production in a high purity form.

UNITY AND DIVERSITY IN METALLURGY

One of the major problems connected with applying science to the technology of metalworking has been that while the laws of behavior of all materials are characterized by a natural unity, the metals trades themselves have historically been divided, both in their economic structure and in the very techniques employed. The smelter with his blast furnace, the smith with his hammer, and the foundryman with his crucible and mould, each exploits a distinct metallic property and works in virtually separate establishments: the three professions have remained apart to the present day. Even those who used the same techniques on different metals or in making different objects often failed to recognize their kinship: blacksmiths and whitesmiths, for example, went their separate ways.

But the greatest distinction has been between the operations of the numerous and scattered artisans using metals diversely, and the more organized metallurgy of large-scale smelting, which has appealed to both princes and scholars. Between the two, the merchant played an essential role. Nowadays, of course, the *science* of metallurgy covers the whole gamut, for the physics behind the alloying and treatment of metals to develop specific combinations of properties, like the thermodynamics and chemical kinetics of ore-dressing and smelting, are similar for all metals and incapable of proper study without some consideration of all metals and even of all forms of matter. The unifying force of a desire for general understanding, as opposed to the divisive needs of practical application, is nowhere better illustrated than in this area.

The discovery of the range of diverse combinations of useful properties obtainable by combining metals with each other occurred in the Middle East two millennia or more before the rise of the Greek and Roman civilizations. It seems that sheer enjoyment of the marvellous transformations of the qualities of matter that are produced by heating things prompted the empirical experiments that led to the initial discovery of many alloys. Their first use was usually in works of art, and their utility had to be evident before large-scale production could begin. Necessity has rarely been the mother of a truly basic invention, but it is certainly the mother of development. At the present day there is a clear two-way interaction between adjusting needs and the possibility of designing materials to meet them, but the first suggestions of radical improvements still often come by chance from areas outside the industries that will be most affected. Although today metallurgy is truly beginning to be an applied science, until a few decades ago the practical metal-worker's accumulated experience had prompted the scientist's interest in the behavior of materials far more often than the scientist's knowledge guided the artisan.

The craftsman and the assayer applying heat and stress to mixed materials disclosed most of the interesting properties of matter, and their empirical discoveries of chemical fact and physical behavior are as important for modern science as the more widely heralded mathematical regularities found by the practical astronomer. Man's view of his place in the universe depends on his picture of the crystallinity, the cellular, molecular, and atomic natures of the materials around him quite as much as it does on the cosmological concepts that have hitherto attracted more attention from philosophers and historians.

37 / Buildings and Construction, 1880-1900

CARL W. CONDIT

The fundamental inventions from which modern steel-framed construction evolved had all been introduced by the last quarter of the 19th century. The primary achievement during the period from 1880 to 1900 was the development of the internal skeleton to the point where the high tower-like building known as the skyscraper became a practical possibility.

The skyscraper is a peculiarly American phenomenon, and until the end of World War II it did not exist outside the United States. Many of the economic determinants that led to the creation of this unprecedented kind of building appeared in Europe in the late 19th century, but it was only in New York and Chicago that they came together in such a way as to make the skyscraper inevitable. The accelerating expansion of industry, the parallel increase in urban size and urban density, and the need for the centralization of administrative and financial institutions were dominant features of the Western European economy as well as the American, but characteristics peculiar to the United States gave these forces a special intensity.

THE URBAN SETTING

The economic expansion that began in England at the time of the Industrial Revolution was concentrated in the United States mainly in the period following the Civil War. One consequence was an extremely intensive land use in the core areas of the main commercial centers, with the result that speculation in land drove the price of real estate upward at an astronomical rate. The steady outpouring of inventions provided the mechanical utilities—elevator, steam heat, electric lighting—necessary for the large-scale construction of office buildings. The development of mass-transit systems meant that the administrative work force could travel conveniently from the urban periphery to its place of business in the urban core. All these factors operated in a most concentrated way in the city of New York, yet the contemporary steel-framed curtain-walled skyscraper was the achievement of Chicago builders.

Certain characteristics peculiar to the history and the location of Chicago made it the likely center for this accomplishment. The rapid development of mechanized agriculture in the Midwest, the joining of the Great Lakes and the Illinois-Mississippi waterway systems by canal, the completion of railroad lines to the Atlantic, Gulf, and Pacific coasts—these made the city the industrial, financial, and transportation hub of America by 1870. This unparalleled development was suddenly arrested in 1871, when the entire commercial center of Chicago was destroyed by fire. The resulting demand for office buildings reached an irresistible level and quickly attracted the greatest concentration of talent in the construction industry. Following the precedent of such iron-framed structures as the warehouse of the St. Ouen docks, the architects and engineers soon created an entirely new kind of building, the unusual height of which established the word *skyscraper* in common currency by 1895.

EARLY CHICAGO BUILDINGS

The Chicago precedent for the new office tower was the Montauk Building (1882-83), a work of rational and empirical architecture designed by Burnham and Root as a combination of masonry bearing walls and internal iron frame. The great step forward came with the office building of the Home Insurance Company (1884-85), designed by the architect-engineer William Le Baron Jenney. The building represented the initial application of skeletal construction to a large work of careful architectural design. All but a small fraction of the floor, roof, and wall loads were carried by a framework of cast- and wrought-iron columns, wrought-iron girders and floor beams up to the sixth floor, and steel beams in the remaining three stories. Jenney's adoption of steel for the upper floor frames marked the first use of the metal in a building. Possessing high tensile as well as compressive strength and stronger than either cast or wrought iron, steel was the logical choice for the skyscraper, but its high cost discouraged builders from readily accepting it until nearly the end of the century. The curtain walls of the Home Insurance, largely glass, were carried bay by bay on shelf angles fixed to the outermost girders of the frame. Fire protection of the metal was achieved by covering the framing members with fire-resistant tile, which had been invented in 1871.

Certain features prevented the Home Insurance Building from reaching the full maturity of the complete skeletal skyscraper. Part of the load fell on the granite envelopes of the columns at the base of the façade and on a brick party wall along one side. In addition, the bolted frame lacked wind-bracing; this was probably felt to be unnecessary because of the extensive masonry covering of the outer columns and girders.

In spite of these defects, Jenney had clearly pointed the way, and five years were enough for the Chicago builders to bring skyscraper construction to its maturity. The riveted frame came with the Tacoma Building (1887-89), the work of the architects Holabird and Roche and the engineer Carl Seiffert. Although the Auditorium Building of Adler and Sullivan, built at the same time, marked a return to the older combination of masonry bearing wall and internal iron frame, the need to roof a 4000-seat theater inside the building and to provide space for diversified hotel facilities required that the designers exhaust the structural possibilities of the age. Elliptical trusses spanning 117 feet at the maximum, a variety of parallel-chord trusses, and an elaborate system of column-and-girder framing compose the interior structure of this huge block-long building.

The Manhattan Building (1889-90), designed by Jenney and the engineer Louis E. Ritter, may be regarded as the first high office building of which the structural system was the product of scientific design throughout. The supporting skeleton of iron and steel includes a full system of portal bracing supplemented by diagonal bracing in the outer bays of the basement, where the bending forces exerted by the wind reach a maximum.

Full emancipation from a century's dependence on cast and wrought iron came with the Rand McNally Building (1889-90), the work of Burnham and

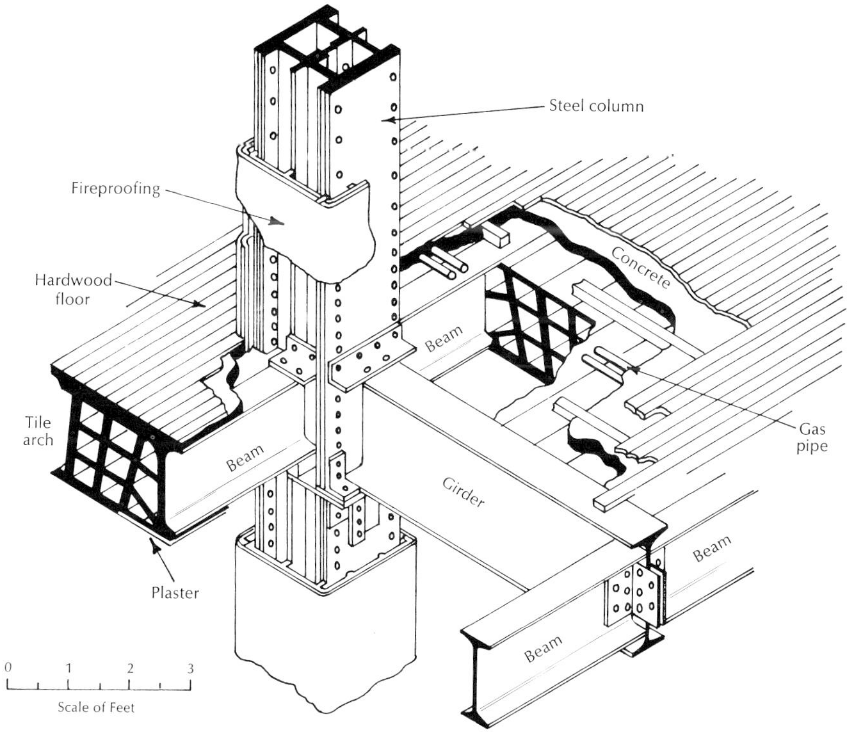

Fig. 37-1. Column and beam connection in the steel frame of the Fair Store, Chicago, Illinois, 1890-91. William Le Baron Jenney, architect. Cut-away drawing showing the elements of the steel frame as developed by Chicago architects, 1885-90. (From *Industrial Chicago*, Goodspeed Publishing Co., 1891)

Root in collaboration with the engineers Wade and Purdy. This was the first building carried on an all-steel frame and the first to be covered with glass and a sheathing of terra-cotta tiles, a material obviously without bearing capacity.

The presence of an interior theater under an office tower in the Garrick Theater Building (1891-92) again required the architects, Adler and Sullivan, to include truss framing in the steel supporting structure. The columns of the upper seven floors in the central portion of the building rested on a series of trusses spanning the theater (occupying the full width of the building) and rising through a height of two stories. The trusses of the Garrick were the first to be designed to carry superimposed floor loads. All the elements of contemporary steel framing appeared in the Chicago work of the 1890's, and the subsequent evolution of the riveted frame was largely a matter of refinement of form and increasing scientific exactitude.

EUROPEAN DEVELOPMENTS: THE EIFFEL TOWER

The skyscraper as a building did not exist in Europe, but as a work of structural design it was more than matched by the Eiffel Tower (1888-89), the ultimate expression for the time of the long-established pre-eminence of French engineering. Gustave Eiffel (1832-1923), the author of this daring work, had been trained in the École Polytechnique and the École Centrale, which had been from the time of their founding the leading technical schools in the world. He was the foremost authority on hinged-arch bridges and on problems of aerodynamic stability when he proposed the tower that was to be the central theme of the Paris Exposition of 1889. With an overall height of exactly 300 meters (984 feet 4 inches) it remained the highest structure in the world until 1930, and the largest steel structure other than a bridge. The tower embodied all of Eiffel's knowledge of the behavior of structures and the effect of wind upon them.

The primary supports are four trussed frames (we would now call them space frames), square in cross section, rising from independent foundations in hyperbolic curves, each pair on any one side gradually approaching a hypothetical central axis (an asymptote, as it is known in analytic geometry). Each of the four frames is made rigid by a double-diagonal system of trusses lying in the horizontal or transverse planes and in the space extending upward between the pairs of curving legs. The whole structure is braced by a complex system of horizontal chord trusses at the first two stages and by diagonal bracing connecting the four inner legs. This naked framework of steel seems bewildering in its intricacy, yet it forms a coherent geometric pattern with all parts carefully calculated to support the vertical loads and to resist the bending forces of the wind. The curved shape of the chief supporting frames further increases this resistance, although their profile was in part dictated by Eiffel's desire to create an aesthetically pleasing form. Architectural traditionalists long regarded the tower as a monstrosity, but it is now universally admired as the greatest structural monument of its age.

The curved trusses of Eiffel's creation are closely related to similar forms in the broad vaulted enclosures of exhibition halls and train sheds. By 1870 French engineers had greatly improved the efficiency and exactitude of arched trusses like those in St. Pancras Station by dividing the fixed arch into separate halves held in place by hinged connections at the crown and abutments. This refinement turned the arch into a determinate structure, that is, one in which the forces exerted by the arch at the ends and the center can be exactly calculated on the basis of the load it carries. Such forces can only be approximated in the case of the fixed arch. The finest example of three-hinged arch construction appeared at the same exposition for which Eiffel built his tower. The Galerie des Machines, designed by Cottancin and Dutert, was a tremendous glass enclosure

Fig. 37-2. Galerie des Machines, Paris Exposition, 1889. Cottancin and Dutert, engineer and architect. Construction view showing the three-hinged arches. (From Sigfried Giedion, *Space, Time and Architecture,* Harvard University Press, 1954)

carried by hinged arches with a record span of 115 meters (377 feet 4 inches). Since these arched trusses sprang from hinges, their depth was contracted nearly to a point at their lower ends. The absence of a fixed and solid support gave them such a precarious appearance as to seem downright frightening at the time, even to trained engineers.

The hinged arch was introduced into the United States by the Philadelphia engineer Joseph M. Wilson in 1869, when he designed a bridge for the Pennsylvania Railroad in that city. Before the end of his career he was to use it as the chief supporting element under the largest permanent roof for its time. The railroad's Broad Street Station in Philadelphia (1892-93) included a train shed with a span of 300 feet 8 inches. The plank roof rested on twenty three-hinged arched trusses of wrought iron, grouped in pairs and tied by steel beams set below the track level. The close identity between the structural systems of the Broad Street shed and the Galerie des Machines and the short interval of time that separated their completion indicate that the new science of engineering construction had reached the international stage, transcending the regional differences that were once so conspicuous in building.

BRIDGES

Many of the same economic factors that tended to make buildings higher, operated to make bridges longer and heavier. The growth of industry and agriculture brought with them an attendant expansion of the rail network and an increasing weight of traffic. The demands of the railroads for massive long-span bridges of the most economical construction seemed insatiable.

The steel truss bridge has always been the most common type, and after 1880 it increasingly appeared in variations on the Pratt and Warren forms. For the big railroad structures the engineers generally relied on subdivided trusses in which half-length diagonals, intermediate posts, and horizontal struts were added in order to provide a more nearly uniform distribution of the heavy rail loads over the bottom chords of long-span trusses. A typical example among simple spans (those in which the truss is supported at its ends) is the bridge built by the Chesapeake and Ohio Railway over the Ohio River at Cincinnati (1886-88). Designed by William H. Burr, the structure is composed of subdivided Pratt trusses of which the pair in the channel span established a short-lived record length of 545 feet. The Pratt and Warren trusses of railroad and highway bridges reached stable forms by the end of the century and have retained them to this day.

The span length of the truss bridge can be considerably increased through a combination of cantilever and simple trusses. A cantilever is a beam or truss that is supported at only one end, in contrast to the simple member, which is supported at both ends, and the continuous member, which is carried on intermediate as well as end supports. A cantilever will not sustain itself unless it is in some way anchored at its support or held in place by means of another member extending in the opposite direction from the fixed end to balance it. By combining two cantilever trusses with a simple truss set between them at their free

Fig. 37-3. Bridge of the Chesapeake and Ohio Railway, Ohio River, Cincinnati, Ohio, 1886-88. William H. Burr, chief engineer. Subdivided Pratt truss of the channel span. (Drawing by author)

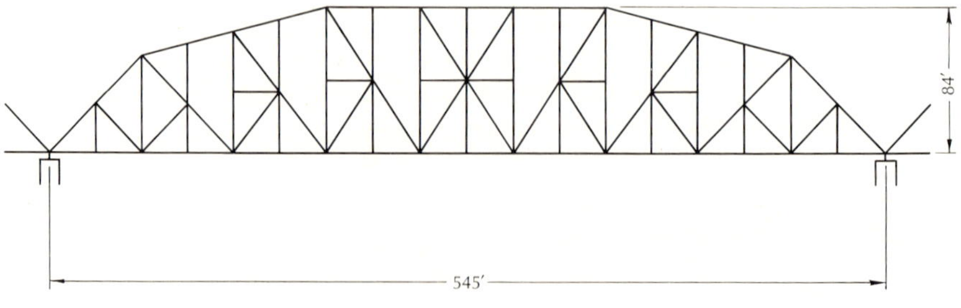

Fig. 37-4. Firth of Forth Bridge, Edinburgh, Scotland, 1883-90. Benjamin Baker and John Fowler, engineers. (From Wilber Watson, *Bridge Architecture*, 1927)

ends, builders were able to increase the length of clear span far beyond the maximum dimension that was the rule around 1880.

The early history of the iron cantilever is obscure, although its primitive forms in timber may be traced back many centuries in China. In 1811 the American carpenter-builder Thomas Pope proposed a bridge constructed of solid timber cantilevers, but there is no evidence that he built such bridges or that he influenced the builders of iron structures. Interest in the form seems to have arisen from scientific inquiry into the behavior of continuous structures, which are more complex in their action than simple forms. Heinrich Gerber designed a continuous-truss bridge built over the Main River at Hassfurt, Germany (1866-67), in which he incorporated a novel device that had no exact precedent. At the points between the piers where the direction of bending changes, he introduced hinges that had the effect of transforming the bridge into a series of cantilever and suspended spans. Gerber's technique was derived from investigations carried on by the engineer and theorist Karl Culmann. The idea was taken up by the American engineer Charles Shaler Smith, who used it in the spectacular bridge built by the Cincinnati Southern Railway over the Kentucky River at Dixville, Kentucky (1876-77). The feasibility of the cantilever was thus established, and in little more than a decade it was to reach a size that dwarfed all other bridges outside the suspension category.

John Fowler and Benjamin Baker astounded the world when they brought to reality the gigantic double-track rail bridge they had designed to cross the Firth of Forth near Edinburgh, Scotland (1883-90). This structure included by far the largest truss span in the world, a record that it was to retain until the opening of the St. Lawrence River bridge at Quebec in 1917. The overall length of more than a mile between abutments was impressive enough at the time, but what seemed incredible is the 1700-foot length of the combined cantilever-suspended trusses. The preparation of working drawings for the bridge was preceded by a series of experiments aimed at determining, as exactly as possible, the effects of wind pressure on a long span. This investigation became a matter of urgent necessity after the collapse in December 1879 of the Firth of Tay bridge, which went down under a seven-car train during a gale that reached a maximum velocity of 80 miles per hour.

The Forth bridge was designed for a wind pressure of 50 pounds per square foot over the side area of the truss, corresponding to a wind velocity of 123 miles per hour. In the completed structure the primary members of the cantilever and anchor spans and of the tower that supports them are riveted steel tubes. The depth of the cantilever truss at the tower is 330 feet, a dimension so extreme as to make the 350-foot suspended span appear to be grotesquely out of scale with the rest of the structure. The main trusses have web members arranged according to the double-diagonal Warren system, with one set in the tubular form to take the compressive load, and the alternate set composed of built-up open-web members to act in tension. Although the system is efficient, the later practice has been to use rolled members throughout to avoid the expense arising from a variety of special shapes.

The final innovation in truss form came in 1896, when the Belgian engineer M. A. Vierendeel invented the rigid-frame truss. In the common forms, such as those invented by Pratt and Warren, rigidity is secured by arranging the individual members in a series of triangles. (The triangle is a rigid figure, that is, one whose shape cannot be altered without deforming one or more of its sides.) In Vierendeel's invention, however, the diagonals are omitted and rigidity is imparted by means of rigid connections between posts and chords. The Vierendeel truss, as a consequence, is composed of a series of rectangles or trapezoids rather than triangles. The inventor's chief purpose was to achieve greater economy by using members of uniform size. The Vierendeel truss has found relatively little use in truss bridges, but in 1929 it was adapted to the steel towers of suspension bridges and later to building frames where long spans are necessary.

The fixed steel arch reached its greatest size in the past century in the Eads Bridge at St. Louis, but as in the case of the train shed and other vaulted structures, the hinged arch was beginning to come into prominence around 1880. The three-hinged arch was found to lack adequate rigidity for long-span bridges, with the result that the two-hinged form was adopted to preserve some of the virtues of the fixed and hinged types. The leading engineer of steel-arch con-

Fig. 37-5. Vierendeel trusses in a street bridge, Los Angeles, California. (Photo by *Los Angeles Department of Public Works*)

struction was Gustave Eiffel, who provided the most impressive demonstration of this structural technique in his Garabit Viaduct over the Thuyère River in France (1880-84). The two-hinged arch truss of this elegant steel bridge spans 541 feet between hinges and carries a railroad track at an elevation of 402 feet above the water surface. Since there can be no bending force at the hinge, the crescent-shaped arch is drawn nearly to a point at its ends, giving the long bridge the same aerial delicacy revealed by the arches of the Galerie des Machines. A little more than a decade after the completion of the Garabit bridge the American engineers L. L. and R. B. Buck carried the steel arch to its greatest span before the end of the century. Their highway bridge over the Niagara River immediately below the Falls (1895-98) extended 840 feet between the hinges of the arch. The bridge served well until January 1938, when an ice jam, squeezed into the narrow gorge, sheared the arch at the hinges causing the mass of steel to drop to the surface of the ice.

After the completion of the Brooklyn Bridge in 1883 the history of the suspension bridge is a matter of the progressive refinement of the main working elements. Towers in the form of three-dimensional frames (space frames) of riveted steel replaced the masonry structures of Roebling's work. The suspension system was reduced to vertical hangers only, the necessary rigidity being secured through deep longitudinal stiffening trusses usually in the Warren form. The number of cables was reduced from four to two as the tensile strength of steel wire steadily increased, and concrete took the place of masonry for an-

Fig. 37-6. Top: Garabit Viaduct, Thuyère River, France, 1880-84. Gustave Eiffel, engineer. Bottom: The hinge at one end of the arch. (From Sigfried Giedion, *Space, Time, and Architecture*)

chors and tower piers. All these elements are present in the last great suspension bridge of the century, Leffert L. Buck's Williamsburg Bridge (1897-1903), the second to cross the East River at New York and the first to exceed the span length of Brooklyn Bridge.

REINFORCED CONCRETE

The history of reinforced concrete is especially complex because of the many parallel developments occurring in Europe and the United States. As we noted in Chapter 22, plain concrete has only a negligible tensile strength and hence cannot be used for structural members subject to bending, twisting, and shearing action. This limitation can be surmounted, however, by placing iron or steel bars in the body of the concrete member and locating them in such a way that the tension and shear are taken by the metal rather than the surrounding material. This is known as reinforced concrete. The background of this idea may be traced to the use of reinforced masonry, which dates from the late 18th century. William B. Wilkinson of England was the first to apply the idea to concrete. In 1854 he was granted a patent for imbedding a grid of wire cables in a floor slab composed of an odd kind of concrete with an aggregate of hay and ashes.

At this point leadership in the development of reinforced concrete passed to the French, who were to retain it with few challenges up to the present time. François Coignet obtained a patent in 1855 for a floor slab poured around a two-way grid of iron rods, but these were introduced as tie rods to take the place of wall buttresses rather than as reinforcing to absorb the tension in the slab. Joseph Monier used the same idea in his patent of 1867, which described a method of strengthening flower pots and tubs by imbedding wire-mesh reinforcing in the mortar shell; again, however, there is no evidence that Monier understood the proper role of the metal. Ten years later he proposed bar reinforcing for concrete bridges, but he still failed to realize that the iron bars were necessary to take the tensile stress. In any member which is deflected downward, like a slab or a beam, the upper half is subject to compression and the lower to tension, and it is in the lower half, accordingly, that the reinforcing bars must be located. When the German engineer G. A. Wayss pointed out this fact, Monier thought Wayss was wrong and insisted that the rods acted solely to increase the general cohesive strength of the concrete.

By the date of Monier's second patent (1877) the idea of reinforcing had spread to other European nations and to the United States. As in the case of iron a half-century earlier, the builder was now faced with the necessity of a scientific investigation of the new technique. The initiator was Thaddeus Hyatt, who carried on his researches so thoroughly that many of his conclusions remain valid to this day. Born in the United States but working in England, Hyatt in-

Fig. 37-7. William E. Ward house, Port Chester, New York, 1871-76. Robert Mook, architect; William E. Ward, engineering designer. (Photo by *Portland Cement Association*)

vestigated most of the modern forms of bar reinforcing for a wide variety of structural members under all conditions of load. The publication of his results at London (*An Account of Some Experiments with Portland-Cement Concrete Combined with Iron*, 1877) established a landmark in the history of reinforced concrete, for it marked the beginning of a scientific technology in what is now the largest domain of construction. Two years later, Wayss bought the rights to the Monier patent for use in Germany, Austria, and Russia, then undertook his own investigations into the proper methods of reinforcing. He published his results under the title *The Monier System in Its Application to Building* (1887), which marked the second major step in the evolution of a scientific understanding of the technique.

Meanwhile, a few builders had been active in putting these ideas into practical form. The leading work in this pioneer stage is the house built by William E. Ward as his own residence at Port Chester, New York (1871-76). In collaboration with the New York architect Robert Mook, Ward created the first structure built entirely of reinforced concrete. The columns were cast as hollow cylinders reinforced with hoops to counteract the tension arising from the buckling tendency of the vertical members. The three-inch floor slabs, poured over a two-way grid of iron rods located in the tension zone, rest on concrete beams reinforced with wrought-iron I-beams. Ward tested the parlor floor with a 26-ton weight left at the center of the 18-foot span throughout the winter and measured a resultant deflection of only 0.01 inch. A massive balcony at the second-floor level rests on concrete cantilevers again reinforced with wrought-iron I-beams. The room is a concrete slab, and the exterior walls and main interior partitions are vertical bearing slabs reinforced against the buckling tendency of vertical members.

Ward's house is a work of structural virtuosity that clearly pointed to the unlimited possibilities of concrete for large commercial and industrial buildings; yet he himself was uninterested in following the path he had laid out, so that it is difficult to determine what his influence might have been. Detailed descriptions of his house were published in various periodicals, including one in France, where it undoubtedly attracted the attention of the French builders. Foremost among those engaged in concrete construction at the time was François Hennebique. He began work in the late 1870's, and in 1880 he discovered the nature of shearing stress in concrete. Shear occurs in any deflected member along the plane where tensile and compressive stresses meet, but is particularly concentrated along diagonal lines near the ends of a simple beam. In order to counteract this stress Hennebique introduced vertical J-shaped bars called stirrups into the ends of the beam to bond the upper and lower layers together. He carried on twelve years of secret research (1880-92) before taking out his first patents. At the turn of the century several engineers and builders in Europe and America, chiefly Wayss, Hennebique, and Ransome, developed various techniques for resisting shear by bending the ends of the tension bars upward at an angle of 45 degrees.

Hennebique established an extremely profitable business as a contractor and consulting engineer in 1892. The durability and economy of concrete construction made it attractive, but its chief appeal was nicely summed up in Hennebique's advertising slogan—*Plus d'Incendies Désastreux!* (No more disastrous fires!) His first building constructed entirely of reinforced concrete—walls and roof as well as framing members—was the sugar refinery known as the Parisian Refinery of St. Ouen (1894-95). Its chief distinction was the skylighted gable roof in which the glass inserts were set into a concrete framework.

Hennebique's most advanced work was the mill building, Le Moulin Idéal at Nort, France (1898). The monolithic frame of concrete embodied all the techniques of scientific reinforcing developed up to that time in ways that are essentially like those used today. The main innovation in the structural system was the use of thin precast slabs for the curtain walls and slender precast posts for the window mullions. Hennebique's reinforced concrete building at the Paris

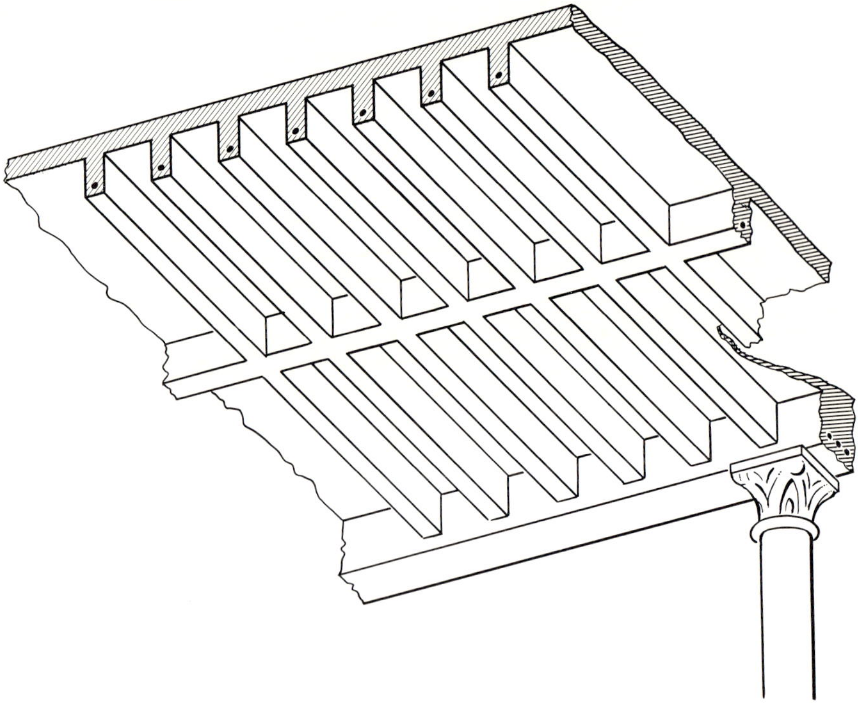

Fig. 37-8. Floor reinforcing system, Bourn and Wise Wine Cellar, St. Helena, California, 1888. Ernest L. Ransome, engineer. (From Ernest L. Ransome and Alexis Saurbrey, *Reinforced Concrete Buildings*, McGraw-Hill Co., 1912)

Exposition of 1900 led to the adoption of the new technique in parts of Europe where it had not previously been used. From there its use soon spread to Latin America and Asia.

Reinforced concrete construction in the United States grew mainly out of the pioneer work of Ernest L. Ransome. His migration from England to California around 1870 and the special problem of designing structures for earthquake resistance gave the coastal state its early lead in reinforced concrete building. Ransome was initially interested in developing floor construction in which the slab and the supporting frame were cast as an integral unit. In the Bourn and Wise Wine Cellar at St. Helena, California (1888), he poured the floor as a series of beams united by shallow arches, the beam reinforced with bars in the tension zone; and in the following year he substituted reinforced T-sections for the arches in the floor construction of the Borax Works at Alameda, California. His first buildings constructed of reinforced concrete throughout were the University Museum and a girls' dormitory at Stanford University (1892). The columns and the main floor beams were cast separately, but the floor sections and the joists formed integral units. By 1900 Ransome was the designing engineer for the construction of large factories of reinforced concrete.

The combined influence of builders such as Hennebique and Wayss in Europe, and Ransome and Julius Kahn in the United States quickly led to daring essays in a form of construction that was still regarded as a novelty by many engineers. Shortly after the turn of the century reinforced concrete was to be used for the entire structural systems of a skyscraper in Cincinnati, Ohio, and a railroad terminal in Atlanta, Georgia.

REINFORCED CONCRETE BRIDGES

The arch bridge of reinforced concrete was late in coming chiefly because builders failed to recognize the value of reinforcing in what had always been regarded as a pure compressive structure. The traditional view, however, was discovered to be inaccurate. If an arch is loaded uniformly along the curve of its axis, the load line follows a curve called an inverted catenary, that is, the inversion of the curve assumed by a flexible cord freely suspended from its two ends. The axes of traditional semicircular, elliptical, and segmental arches deviate from the catenary in varying degrees, and wherever such deviation occurs, especially under moving loads, the arch is subject to tension as well as compression. Tensile stresses may be absorbed by sheer mass of masonry, but under the rigorous standards of design that were becoming the rule around 1880 this came to be regarded as an inefficient and uneconomical practice. By using concrete instead of masonry and by introducing reinforcing at the proper locations, the builder could create a precisely designed arch with a minimum of material.

For at least the first forty years of its history, the concrete arch bridge was built of the plain or unreinforced material. The first such bridge was a little

Fig. 37-9. Floor reinforcing system, Borax Works, Alameda, California, 1889. Ernest L. Ransome, engineer. (From Ernest L. Ransome and Alexis Saurbrey, *Reinforced Concrete Buildings*, McGraw-Hill Co., 1912)

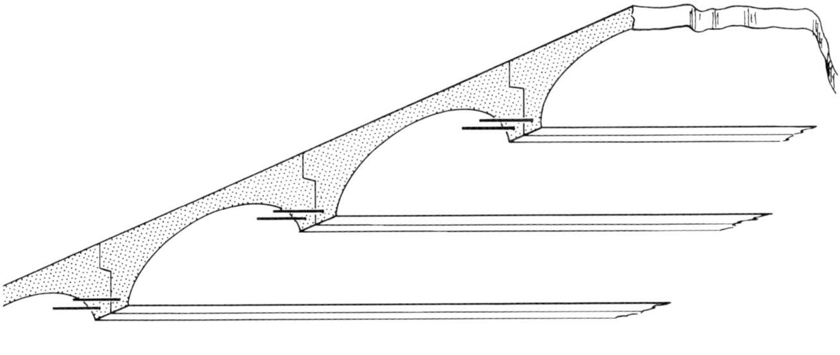

span constructed in 1840 to carry a road over the Garonne Canal in Grisoles, France. The early chronology of reinforced concrete bridges is still in a confused state, but it seems clear that the first such structure could not have been built much before 1885. Monier, in his patents of 1877 and 1879, recognized the utility of reinforcing in concrete bridges and reservoirs, although it is questionable whether he understood its structural role. Shortly after Wayss bought the German rights to the Monier patents in 1879 he began to build bridges and hydraulic works of reinforced concrete which showed a more mature grasp of the technique. Typical of the Wayss arch bridges was the structure built at Wildegg, Switzerland, in 1890. The clear span was a little more than 122 feet for a rise of 11 feet 5 inches, and the arch barrel was only eight inches thick at the crown—unheard-of dimensions that showed how completely the reinforced arch could be emancipated from masonry precedents where the crown would have had to be much thicker. The barrel was heavily reinforced by two-layered grids of bars, one set near the upper and the other near the lower surface.

The first American bridge of reinforced concrete appeared about the same time (1889), and the reinforcing is much like that used by Wayss, although the span length is only 20 feet. Ernest L. Ransome was the author of this pioneer work, which still stands as originally constructed in Golden Gate Park, San Francisco.

The bar reinforcing of Wayss and Ransome proved to be perfectly sound, but the concrete arch was still in an experimental stage in the 1890's, so that other builders were sure that they could find better alternatives. Chief among these was Josef Melan of Vienna, Austria, who obtained a patent in 1894 on a system of reinforcing in which parallel steel I-beams were imbedded in the concrete. The beams were bent into the curve of the barrel and extended the full length of the span. The Melan bridge was introduced into the United States in the same year and for nearly two decades was widely used on both sides of the Atlantic, especially for the heavily loaded railroad spans. The great weight of metal in the massive and closely spaced beams meant that the Melan arches were virtually steel-rib structures with a protective envelope of concrete. This technique of reinforcing apparently provided the precedent for the first concrete girder bridges. F. W. Patterson of Pittsburgh, Pennsylvania, took the initial step in 1898, when he began the practice of pouring concrete around steel plate girders.

But if ribs and girders are regarded as reinforcing rather than primary structural members, the quantity of steel is more than is absolutely necessary for the role assigned to it. Although the Melan system was popular, the more scientifically minded engineers recognized the superiority of bar reinforcing. The earlier form was more economical in two decisive respects: the thin bars could be easily bent and exactly located in such a way that their tensile strength could be made to work precisely where it was needed, and the relatively light

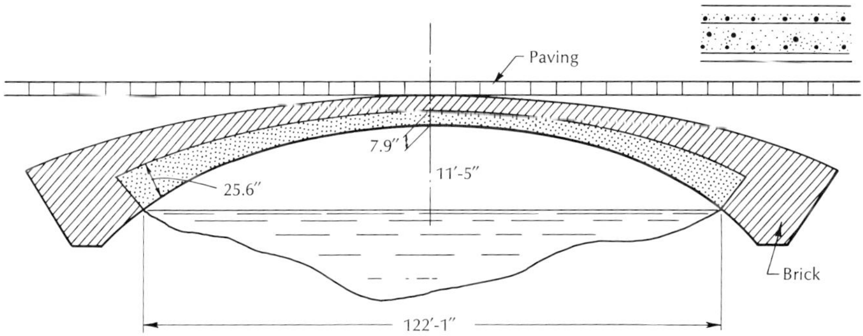

Fig. 37-10. Bridge, Wildegg, Switzerland, 1890. Longitudinal section and detail of reinforcing of a concrete bridge built according to the Wayss system. (From Aly Ahmed Raafat, *Reinforced Concrete in Architecture*, Reinhold Publishing Co., 1958)

weight of the bars meant that they could be readily handled by workers without mechanical aids.

A prevision of the extraordinary possibilities of the technique came with the early work of the Swiss engineer Robert Maillart. In 1900 he began to construct bridges in which the arch was reduced to a pair of thin vertical slabs poured integrally with the bridge deck. In larger structures he kept the traditional barrel and used thin slabs to carry the load of the deck to the arch. Maillart's bridges appeared to be so light and delicate that few engineers had the courage to follow him, but the new structural world of shells, plates, and box girders is plainly prefigured in his work.

SUMMARY

By the turn of the century the innovations of previous decades had been explored and refined, and the future path of the building arts was clearly indicated. Those monumental structures with which the mid-20th century is so familiar—the skyscraper and the suspension bridge—were both brought to near perfection in theory if not yet to maximum size. The combination of new materials, most importantly steel and reinforced concrete, and rigorous scientific analysis of design, made possible feats of construction undreamed of in former years but now made virtual necessities by the economic imperatives of a maturing industrial society.

38 / Machines and Tools

ROBERT S. WOODBURY

Machines and tools in the second half of the 19th century, as always, reflected each other's characteristics. New tools and new forms of old ones made possible important changes in all machines, especially those for mass production. At the same time, the demands of new methods of manufacture and new products gave the incentive for the development of important new tools.

In the preceding half-century a similar relationship had also held. The classical machine tools—the lathe, the boring and the drilling machines, the shaper and the planer—all these were suited to the production of heavy, rather clumsy machines of relatively slow speeds and much noise, designed for the factory conditions of the early Industrial Revolution. These machines were also made to order. Their parts were first rather roughly machined and then laboriously hand-fitted to the necessary tolerances. No two machines, even by the same maker, were exactly alike.

In the second half of the century a profound change, already foreshadowed in some types of machines, became widespread. Machines designed for mass production came into wide use. These machines operated at ever-increasing speeds and required greater precision. Mass manufacture of many devices by means of interchangeable parts increasingly permitted highly specialized machine tools. Lighter, finer, and better designed machines, such as the bicycle, the sewing machine, and the automobile, were built for operation outside factories and by technically unskilled users. All these changes in machines required and permitted corresponding changes in the tools needed to make them.

Precision methods of construction of machines, standardization of parts, and interchangeability of manufacture are all different aspects of the technical means of mass production by machines. All were made possible by technical advances in machine tools.

JOSEPH WHITWORTH

Joseph Whitworth (1803-87), the foremost machine-tool builder of the mid-19th century, recognized the need for precision methods of machine-tool workmanship. In his day these demands were evident in the higher speeds and higher efficiency of the steam engine and other machines, in gears and other mechanical devices for the transmission and utilization of power, and were soon to become even more critical for the internal-combustion engine and the steam turbine. Whitworth foresaw this need for greater precision and standardization, as had Watt, Maudslay, Clement, and others; but he put his ideas into practical widespread use by manufacturing lathes and other machine tools, as well as precision measuring devices, which brought far higher orders of accuracy into the ordinary machine shop. In Whitworth's youth, good workmen could work to 1/64 inch, a margin far too crude for modern machinery. By the time of his death, common machine-shop practice was to work to 1/1000 inch. Whitworth was the founder of practical methods of precision in working metals.

Trained in Maudslay's shop (see Chapter 16) and with very extensive practical mechanical experience, Joseph Whitworth had very early become interested in practical problems of shop precision. He made in his own lodgings three true surface plates by the now familiar method of repeated fitting of one to the other. He used rouge paste to discover the high points, which were then painstakingly worked down with a carbon steel scraper to obtain finally three perfectly flat surfaces which could act as standards for subsequent work.

In 1833 he moved to Manchester and set up in business as a tool-maker. Here he insisted that every one of his machinists have at hand a true plane surface for use in precision lay-out of the parts to be machined and for accurate testing of each flat surface of each part as it was finished. This practice continues in machine shops doing precision work to the present day.

Whitworth also recognized the importance of using exact measurement in building machines, rather than the expensive hand fitting which was the common practice of the day. He constructed for this purpose a bench micrometer screw gauge. On one end of a heavy and rigid base he mounted a back stop. On the other was a pedestal carrying a screw of fine pitch which could be rotated by a large wheel carefully graduated in fine divisions. By 1856 he had such a device theoretically capable of measuring to 0.000001 of an inch. It is doubtful, however, that his screw-thread was of that order of accuracy, but he could probably *compare* two dimensions to about 0.0001 inch. Nonetheless, Whitworth first brought into shop use the method of end measurements, using the sense of touch. He wrote: "We have in this mode of measurement all the accuracy we can desire; and we find in practice in the workshop that it is easier to work to the ten-thousandth of an inch from standards of end measurements, than to one-hundredth of an inch from lines on a two-foot rule. In all cases of fitting, end measure of length should be used, instead of lines." This method of exact measurement is still practised.

His bench micrometer was the standard only for Whitworth himself. He realized that the ordinary workman at his bench could neither be supplied with nor use such a device effectively. Previous bench methods of measurement had been an accurately graduated steel rule used to set either inside or outside calipers. Fair refinement of touch made possible rather good *comparison* of sizes with this method. But there were individual differences of touch from one workman to the next and even greater differences in setting the calipers to the desired dimension from the graduations on the steel scale. The actual *measurement* was therefore often inaccurate; in addition, the whole process was time-consuming.

In the same year (1856) Whitworth therefore brought out his plug and ring gauges. The ring gauge consists of a heavy disc of metal with an accurately bored hole in its center: a shaft of any desired diameter is machined until it just fits into the ring gauge hole. Paired with the ring gauge and kept always with it is a corresponding plug gauge which fits it exactly. Whitworth supplied these gauges in standard sets of various dimensions. They came into wide use and thus made possible, in common workshops, machining to about 0.001 inch.

But more is necessary than just the ability of the machinist to make accurate measurements for precise production of machinery, as Whitworth well knew. The machinist must also have machine tools capable of precision in working pieces of metal of industrial sizes and at production rates of speed. Maudslay and the classical machine-tool builders had shown the way—all metal parts for tools plus refinement of construction. Whitworth's manufactory, producing machine tools, not to special order but for sale in the open market, turned out thousands of machine tools with the necessary strength and rigidity of design, with true surfaces, flat and cylindrical, and with accurate lead screws for thread-cutting.

It was also Whitworth who introduced the box design or hollow-frame construction for the main structure of machine tools. His designs also greatly increased the weight of metal used in certain critical parts of the tools. Both these features produced strong and rigid tools capable of maintaining, under production conditions, the precision built into them and thus able to impart that precision to the work being machined on them. On these machines Whitworth's ideal of precision machining could be put into practice under ordinary shop conditions. And Whitworth machine tools were soon considered standard all over the world —lathes, planers, drilling, slotting, and shaping machines—standard in accuracy and quality of workmanship.

Two other aspects of the new generation of machine tools felt the influence of Whitworth's ideas. He also invented a number of features intended to speed up the production rates of his machine tools, of which the best known are the Whitworth quick-return motion for the shaper, his automatic cross-feed for the lathe, and his lathes designed to use multiple cutting tools.

Although Maudslay and Clement had made a beginning in the standardization of screw-threads, in practice there was chaos, for each shop used its own ideas for a screw-thread. Whitworth collected a number of these and designed a "standard" thread. The Whitworth standard was announced in 1841 and by 1860 became standard in Great Britain, where it is often still found. Difficulties in its manufacture, however, led in 1864 to the Sellers standard thread, which

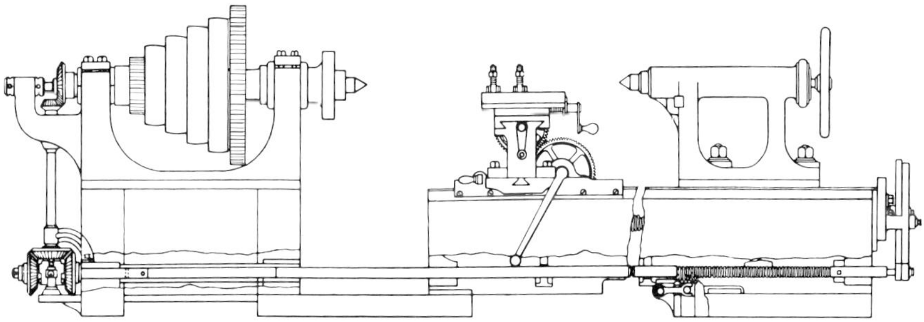

Fig. 38-1. Whitworth's heavy production lathe, 1839.

was adopted officially in the United States in 1872, and in 1898 at the International Congress at Zurich.

The machine tools of Joseph Whitworth were the culmination of the work of the classical machine-tool builders and foreshadowed what was to come later. With tools made in his plant or based on his designs were made automatic textile machinery, steam engines, railroad locomotives, and many other machines which formed the technical basis of the mid-Industrial Revolution. Furthermore, he helped lay the technological foundation for the age of high-speed automated mass production which followed.

Whitworth's machine tools were so successful that his methods dominated British machine-tool practice until Britain lost her leadership in tool-building after 1870. Most of the great advances in the field in the second half of the 19th century were to be made in America.

AMERICAN DEVELOPMENTS

In 1853 Whitworth made an elaborate study of production methods then in use in the United States. He was particularly impressed by the extensive use of standardization and its accompanying automatic machinery. While British machine-tool builders had initiated the age of machine tools and dominated the market in Britain and on the Continent, American tool-builders had developed new machine tools and new methods of using them for mass manufacture. In the second half of the 19th century these important innovations were expanded and added to until the leadership in machine-tool design and manufacture was in American hands. Even French and German machine shops imported the more expensive but vastly superior American machine tools; and in some fields, such as small-arms manufacturing, British shops were using tools based upon American designs, if not actually imported from America.

The American innovations centered around machine tools for mass manufacture largely by means of interchangeable parts. These included more automatic machine tools, more specialized machine tools, improvements in shop precision of measurement coupled with machine tools capable of greater precision. All these advances were made possible by important improvements and modifications of the classical machine-tool designs as well as by the addition of new ones—the turret lathe, the automatic screw machine, the gear-shaper and hobber, the milling machine, and the grinding machine.

PRECISION MEASUREMENT IN THE SHOP

The need for accurate measurement of the work done on machine tools had been apparent since Watt's day, but neither measuring instruments nor machine tools with the necessary precision were available in ordinary shop practice until after Whitworth's time. Whitworth ring and plug gauges were expensive and suitable only for cylindrical work. By the time of the American Civil War, at the Springfield and Harpers Ferry armories and at various other small-arms manufacturers, such as Robbins and Lawrence of Windsor, Vermont, and the Colt Armory at Hartford, Connecticut, a system of receiver gauges specially made for each part or even for each operation had been developed and shown to speed up mass production of interchangeable parts. Sets of these gauges were, however, very expensive and awkward to use. They were therefore gradually replaced, especially in private manufacture of other products, by the limit gauge—often called "Go-No-Go" gauges. These incorporated an entirely new concept of manufacture for interchangeable parts. No longer was the part to be made to "fit" a gauge of a certain size; instead, the limit gauge had two sets of surfaces between which the work was to be measured. One set was calibrated for the largest dimension acceptable for the part being machined, the other for the smallest dimension. The part machined should then "Go" in the larger dimension and "No Go" in the smaller; otherwise it was unacceptable.

These limit gauges were quick and easy to use, much less expensive, and rugged enough for use by the ordinary machinist at his workbench. They came into wide use in American practice by the 1870's, and such gauges are in constant use in shop practice all over the world today wherever accurate parts are mass produced. Of course, both the limit gauges and the Whitworth ring and plug gauges had to be checked against some standard dimension, such as a micrometer screw gauge. This was done from time to time as these gauges wore, by a special workman using precision equipment.

With increasing precision in machine work it was, however, desirable for the machinist to be able to make accurate measurements at his workbench. Watt had made for his own use a portable micrometer screw gauge, now in the Science Museum, London, and of surprisingly good accuracy. In France in 1848, J. R. Palmer had produced a micrometer very much like those in use today, but

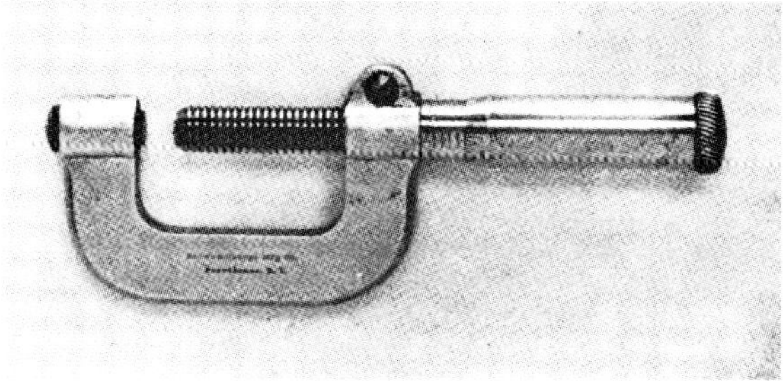

Fig. 38-2. Brown and Sharpe micrometer, 1877.

neither of these devices was to be found in most machine shops of that time. In 1851 Joseph R. Brown introduced the vernier caliper, applying the principle of the vernier to a sliding caliper. Invented by the Frenchman Pierre Vernier (1580-1637), the vernier is a short graduated scale that slides along a larger graduated instrument and can thus be used to indicate fractional parts of divisions with great accuracy. Brown's instrument could read accurately to 0.001 inch and was the first practical precision measuring instrument for ordinary machine-shop use.

Some years later one of the Palmer gauges was seen at the Paris Exhibition of 1867 by Joseph R. Brown and Lucian Sharpe, founders of Brown and Sharpe Co., of Providence, Rhode Island. They recognized its importance and in 1869 put a small one for measuring sheet metal on the market. It was accurate to 0.004 inch. By 1877, Brown and Sharpe were offering a micrometer with one inch between the jaws, and by the 1880's these instruments were in wide use by machinists in shops everywhere in the United States.

However, in order for parts made in one factory to be interchangeable with those made in another in an age of increasing specialization of production, the gauges used by both had to be calibrated to some standard. In addition, precision of 0.001 inch was often inadequate, for the new machines demanded 0.0001 or even 0.00001 inch. These standards came in the form of gauge blocks made up to accuracies of a few millionths of an inch and available in sets by which any dimension could be set up to that order of accuracy. They were the work of the Swedish machinist, C. E. Johansson, in 1896.

Exact methods of shop measurement make it possible to construct machines of greater accuracy and reliability and of higher speeds and production rates. They also make possible greater specialization and mass production of large numbers of interchangeable parts. By means of all these devices, parts can be produced in Detroit, Michigan, interchangeable with those made in Johannesburg, South Africa.

AUTOMATIC, SPECIALIZED, HIGH-PRODUCTION MACHINE TOOLS

Although mass-production methods were used first for small arms, clocks, and a few other devices, a large commercial demand developed for bolts, screws, and nuts, especially after the standardization of screw-threads about 1870. The development of the sewing machine, the typewriter, and the bicycle produced an enormous market for screws and other small parts. Mass-production methods were already known before, but new automatic, specialized, high-production-rate machines were needed to meet these demands. As a result the basic techniques were developed for later use in manufacturing motorcycles, cash registers, small gas and oil engines, telephones, automobiles, and aircraft.

The effects of these techniques on production and labor were enormous. Reduction of manufacturing costs made a mass market for what would hitherto have been only luxury items. Enormous production led to more specialization and the seeking out of worldwide markets. Many monopolies were reduced, but others were created. The problem of training and retraining the individual for a specialized task on a specialized machine brought in 1900 the time studies of Frederick W. Taylor (1865-1915) in the United States. The design of machines was changed to provide for ease and cheapness of production. Not only was the machine designed for the product; the product was designed for the machine. Actual production methods changed, creating for example, the need for inspection of parts on an organized scale—"quality control."

All this was made possible by the invention of automatic and specialized machine tools. Machines of these characteristics had been designed about 1801 by M. I. Brunel and Samuel Bentham for making blocks for the rigging of ships of the Royal Navy. The process was divided into a number of separate operations, and machines were built to perform each special operation. Most of these machines were automatic and acted on a number of pieces of the work at one time. These beautifully constructed machines were thoroughly modern in their conception. Taken together they provided a complete set of tools, each performing a particular part in a correlated series of operations to produce the final product. With them ten unskilled men could do the work of 110 skilled workmen at enormous savings in cost as well as at a high rate of production. This Portsmouth block-making machinery had two important limitations: it worked only on wood, and it was too specialized for general use. The same might be said of Thomas Blanchard's "gun-stocking lathe" built in 1818 for the Springfield Armory.

Yet high production by means of specialized machinery was in the air, as is shown in the work of J. G. Bodmer, described in his remarkable patents of 1839 and 1841. Here are detailed nearly forty specialized machine tools of all types, which were later installed in a systematic plan aimed at an orderly and rapid

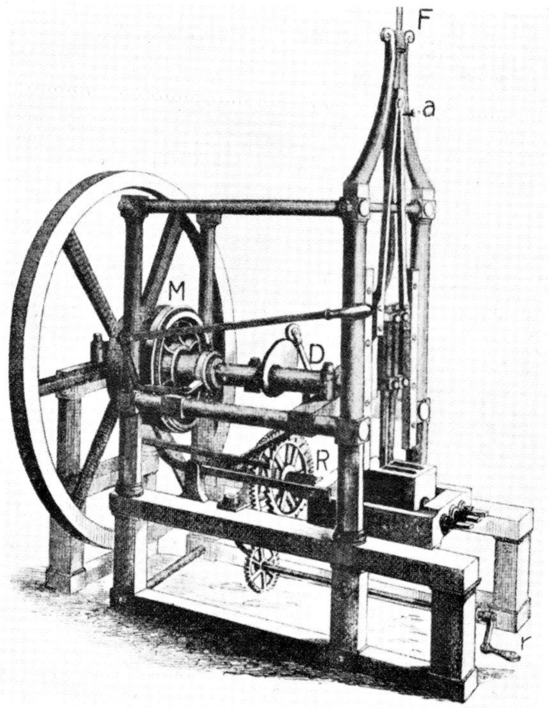

Fig. 38-3. Brunel's mortising machine, *c.* 1801.

series of operations for production of parts in iron and steel. But Bodmer was in advance of his time; more gradual process of development was necessary, and this was brought about by the demands for mass manufacture in the latter half of the 19th century.

This process involved both exploration of further possibilities in the adaptation of the basic machine tools and the invention of new ones, for the materials to be machined were inevitably iron and steel. This development was to take place largely in America. Manufacture by interchangeable metal parts produced on specialized machinery had clearly begun in American arms works, probably in the government arsenals at Springfield and Harpers Ferry. This method of production had passed into the manufacture of clocks, and the interchangeable system reached its highest development in the manufacture of watches. In 1848, A. L. Dennison founded the American Watch Company, later to become the Waltham Watch Company. In this plant watches were produced at high production rates, by semi-skilled labor, and with great precision on a series of highly specialized and fully automatic machine tools. Transfer of the part finished by one machine was even effected by automatic mechanical devices. This was, in effect, the first fully automated process, even though the parts were then assembled by hand. But again this was a very special case, made possible only because the parts were very small.

THE TURRET LATHE

The more general and widespread use of specialized and automatic machine tools first grew out of further development of the possibilities of the lathe. Probably no machine tool has had a greater effect on mass manufacture than the turret lathe, especially as it became automatic.

The turret lathe grew out of the desire to make a lathe capable of producing metal parts at high production rates and requiring an operator much less skilled than the machinist who made these parts on the ordinary lathe. The turret lathe had mounted on it a series of cutting tools so arranged that they could be brought in succession to work on the part merely by operating a simple handle. These tools were "set up" by a skilled workman to do each machining operation with the necessary precision. In this way up to six or eight machining operations could be done rapidly and accurately by a semi-skilled operator, who could then turn out a large number of identical precision parts per day at low cost.

To this end two technical problems had to be solved. First, instead of holding the work between centering pieces or secured to a face plate as in the conventional lathe, the workpiece was held in a chuck mounted on the headstock spindle. Later the spindle was made hollow so that bar or round stock could be fed through it directly into the chuck. Second, the turret head, or means of mounting the tools and bringing them to bear on the workpiece, had to be properly designed. Here two possibilities were tried: swivelling them about on an axis coinciding with the axis of the lathe spindle (horizontal) or on one at right angles to it (vertical).

The turret lathe, although clearly a product of New England tool-builders, was not the invention of one man or even of one date. As early as 1845 a horizontal turret was built by Stephen Fitch at Middlefield, Connecticut. At about the same time the old Gay and Silver Shop at North Chelmsford, Massachusetts, had two turret lathes—one vertical and one horizontal. E. K. Root had his "chucking lathe" at the Colt Armory in 1853. But the most important advance in the turret lathe was its manufacture for sale by Robbins and Lawrence at Windsor, Vermont, based on the designs of R. S. Lawrence, F. W. Howe, and H. D. Stone. All of these turrets except Howe's and one of the Gay and Silver machines seem to have had a horizontal axis instead of the vertical type which later became standard.

However, the turret lathe had not come into widespread use even in America by 1855, for none was included in the large number of machine tools supplied by Robbins and Lawrence to the British Government in that year. But by 1889

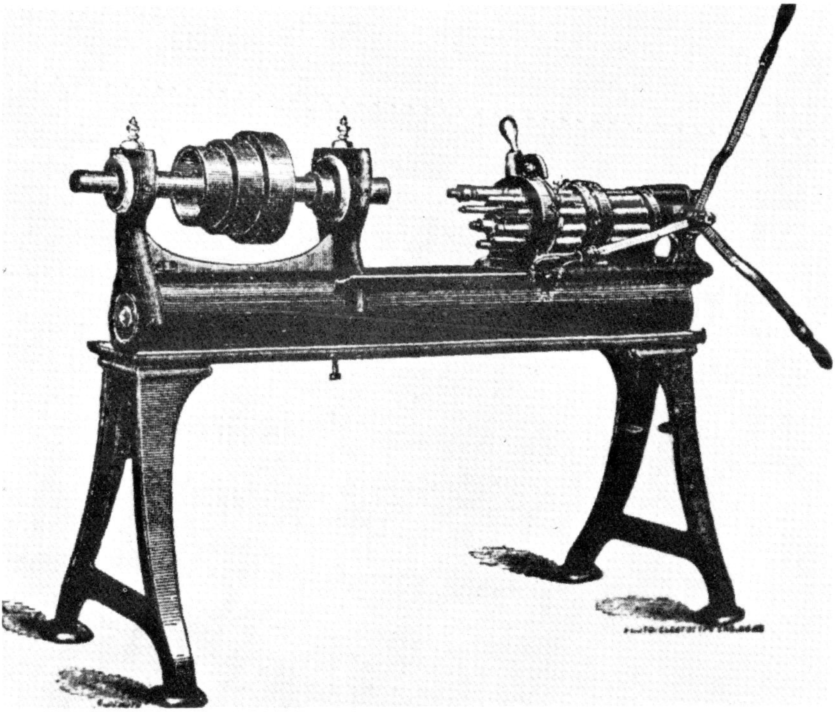

Fig. 38-4. Root's chucking lathe, *c.* 1855.

Jones and Lamson of Springfield, Vermont, were concentrating on the manufacture of turret lathes, which Robbins and Lawrence had been manufacturing since the early 1850's. These machines were of the high-turret, lever-operated type, with power-feed and back gears. Here James Hartness undertook a complete revision of the turret lathe. The result was the "flat-bed" turret lathe and many other improvements. He later discovered company records showing that Howe had anticipated this design forty years earlier.

Meanwhile the turret lathe was being made fully automatic for certain purposes, especially for making screws, which has led to its being often referred to as an automatic screw machine. The first steps were taken in this direction by C. M. Spencer. Having invented a successful machine for turning sewing machine spools, he saw the possibility of making screws automatically. The result was the automatic turret lathe. Spencer's key invention was a blank cylinder on which could be mounted flat strip cams. These could be adjusted for various jobs, and as they engaged in following metal fingers, they caused automatic operation of almost any cycle of tools to engage the workpiece on the turret lathe to which they were fitted. With this device we have an automatic, precision, high-production machine tool, of great flexibility, whose importance cannot be overestimated.

The turret lathe and the automatic screw machine are basic examples of the demand for high production rates at low cost. For industries where a sufficient number of each item was to be made, even lower costs and more rapid manufacture were possible by sacrificing flexibility to specialization. A prime example of this is in the manufacture of automobiles, where today the numbers made are so great that one can make a single enormous machine tool, such as that in the Ford plant at Cleveland, Ohio, just to make the engine blocks for six-cylinder motors. Here a rough casting is turned out every few minutes, milled, drilled, tapped, and so on, by a single fully automatic machine, run by a few men. Of course, the cost of such a machine is enormous, and it cannot make anything else; but if several hundred thousand engine blocks are needed, it is worthwhile. This is just one example of the many specialized and automatic machines that grew out of the work of men like Howe and Spencer.

The demand for high production rates led not only to automatic and to specialized machine tools; it also led to a demand for increased cutting rates of all types of machine tools. In 1880 Taylor and White began twenty-six years of research on cutting metals at the Midvale Steel Plant. The result was a steel for high-speed cutting of tools. They found that by heating tungsten steel to about 1040-1100°C and then allowing it to cool steadily there was a great increase in toughness and hardness. This was the beginning of a new era in tool steel. Later, others used cobalt, molybdenum, and vanadium, and from 1900 to 1925 new tool alloys were developed, some such as Stellite, containing no steel at all. Then in 1925 appeared tungsten carbide, with astonishing cutting properties, permitting much higher cutting speeds than had hitherto been envisioned. Since the 1950's we have had ceramic cutting tools with still greater possibilities. Higher cutting speeds require heavier construction in machine tools and more power to drive them to get maximum production rates and still retain precision.

NEW MACHINE TOOLS

The demand for gears of all types spurred the invention of an entirely new type of machine tool. Gear teeth had been cut on various machines: lathes, shapers, milling machines, and others. But after trying many methods of cutting gears, even on specialized gear-cutting machines, a far better answer was found in the gear-shaper of F. W. Fellows. In 1896, with very little mechanical experience, he invented a machine depending upon great refinement of design, mechanism, and manufacture, yet flexible and capable of turning out rapidly and cheaply almost any kind of gear.

In 1887, G. B. Grant brought out the first successful machine for cutting gear teeth by the hobbing method. He used a special cutter (or hob) fed across the

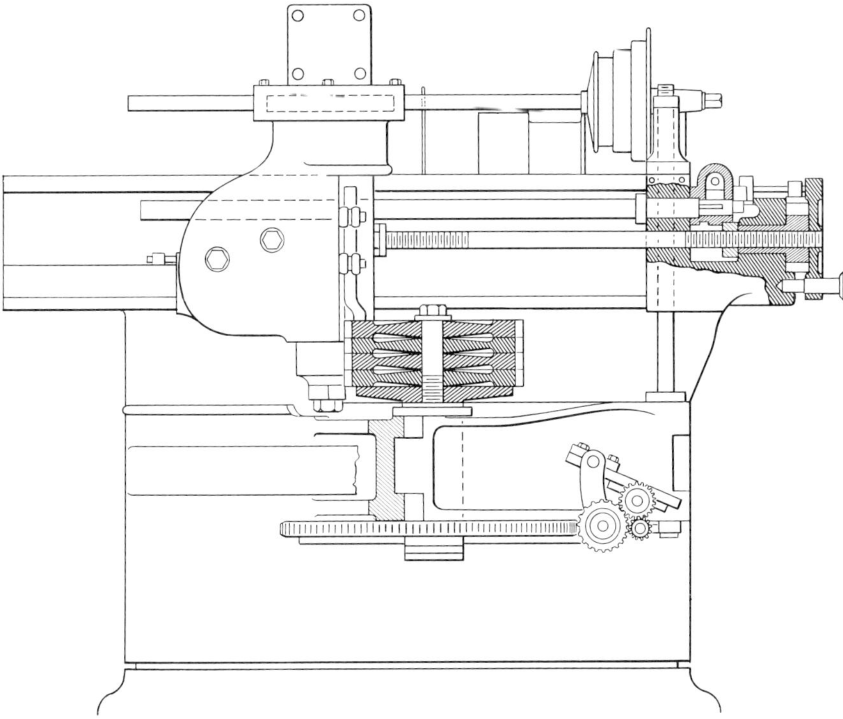

Fig. 38-5. Fellows's gear-shaper, 1896. (U.S. patent)

face of the gear blank while both revolve together. Today only special types of gears are made on machines other than these two.

As we have seen in Chapter 23, the milling machine originated in New England about 1820, principally for use in the manufacture of small arms. More widespread use of milling methods resulted from the "Lincoln" miller designed by F. A. Pratt and and manufactured in quantity by Pratt and Whitney for use in this country and abroad. These machines were most useful for plain and profile milling; but they had limitations, both as to flexibility and size.

In 1861, Joseph R. Brown produced the first universal milling machine, inspired by the need for more rapid machining of the grooves in twist drills required for making parts of muskets for the Civil War. These grooves had hitherto been hand-filed by workmen using a rat-tail file. Brown was asked to find a better and more economical way of doing it. However, instead of designing a machine that would do only this limited and specialized work, he invented the first universal milling machine, which could do all kinds of spiral milling, gear-cutting, and other work previously done by slow and expensive hand methods. The universal milling machine, as manufactured by Brown and Sharpe, came

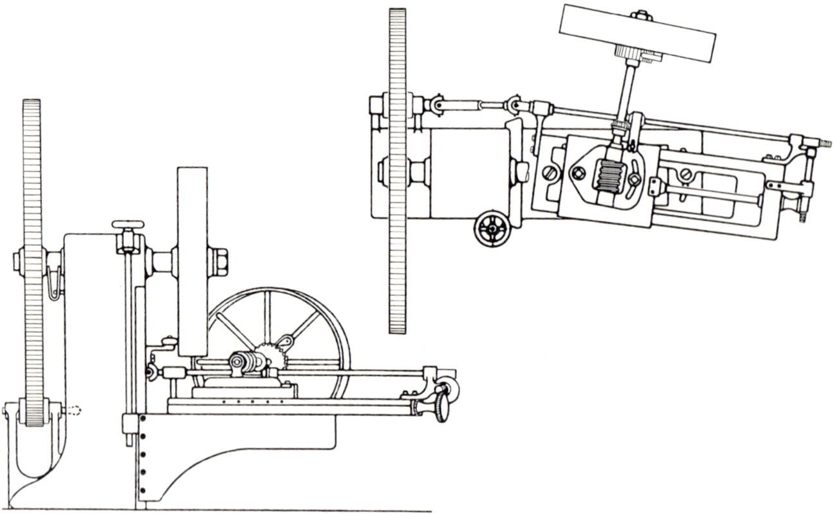

Fig. 38-6. Grant's hobbing machine, 1887. (U.S. patent)

into wide use in America and on the Continent, although English machine shops were slow to adopt it until after 1900.

Between 1890 and 1915 the milling machine revolutionized machine-shop practice everywhere. Since Brown's time it had become a far more sturdy tool capable of heavy work at high speeds. With proper design of the teeth on its cutters, based upon the research begun in 1908 by A. L. DeLeeuw at the Cincinnati Milling Machine Company, it was capable of very heavy cuts, yet left a smooth surface finish. The milling machine has taken over nearly all the work of the planer and shaper. These earlier single-point cutting tools cannot compete with the multiple-point cutting tools of the modern milling machines, especially since the introduction of fluid controls, constant speed, automatic and specialized types from 1900 to the advent of the tracer and electronic machines in the 1950's.

The milling machine proved also to have other uses. Its cutters can be designed for profile milling, that is, they can shape steel or iron into parts having cross sections of intricate curves. More important, the vertical milling machine, first introduced in 1862 by W. B. Bement, made it possible to do cheaply and easily the very expensive work of diesinking—cutting the dies used for die stamping or die forging. Die stamping therefore became a cheap method of producing parts, such as automobile bodies, from sheet metal. Die forging could then be used as an inexpensive and easy method of making parts for many machines, such as the steering knuckles for automobiles.

Perhaps the most spectacular change in high-speed, precision manufacturing came about as a result of the development of the grinding machine in the second half of the 19th century. Grinding as a method of shaping metals was a very old one; however, until the appearance in 1864 of Brown and Sharpe's first commercial grinding machine, the technique was used almost entirely for truing up steel parts which had been distorted by the hardening process. These parts for the sewing machine, the bicycle, and many other devices requiring hardened steel parts of high precision were machined slightly over-size, hardened, and then ground to final size and surface finish with very light cuts taken on the simple grinding machine of that day. The abrasive wheels used for this purpose, even when used on very light cuts, were most unsatisfactory. Their abrasive was emery held in a bond of clay and other substances which were fired into a rule-of-thumb ceramic mix. Nonetheless, skilled and highly paid grinding-machine operators could turn out quite satisfactory work.

The era of grinding as a means of high-speed, precision production of the

Fig. 38-7. Brown's universal milling machine, 1861. (Courtesy of *Brown & Sharpe*, Providence, R.I.)

hardened alloy steel parts increasingly demanded by the turn of the century for locomotives, automobiles, ball bearings, and all kinds of machinery began with new artificial abrasives and an entirely new concept of the possibilities of grinding as a production method.

With the advent of electric power available cheaply and in large quantities, and Moissan's invention of the electric furnace, it was possible to produce new artificial abrasives on a commercial scale. In 1891 E. G. Acheson had heated coke and sand in an electric furnace. Seeking a method of making diamonds artificially he found something which he soon realized was much more valuable —silicon carbide. By 1894 emery—a dark, coarse variety of corundum (aluminum oxide), used for grinding because of its hardness—was being produced artificially. In 1896 wheels using these artificial abrasives were on the market, and by 1898 the Norton Company, at Worcester, Massachusetts, had undertaken systematic research to make possible abrasive wheels vastly improved over F. B. Norton's first vitrified wheels of 1877.

The possibilities of production grinding were first visualized by C. H. Norton in 1897. Norton's idea was to utilize the potentialities of the new abrasive wheels by mounting very large, very wide wheels in a large and heavily constructed grinding machine supplied with great power. These machines were to be made using extremely high precision. On such a machine heavy finish cuts could be taken to remove large amounts of unwanted material from a workpiece of hardened steel quickly, cheaply, easily, and with very high precision and excellent surface finish. The Norton Company built such a machine in 1900, and it proved its value in saving both time and labor cost over the earlier method of finish turning, hardening, and then light grinding. The importance of this new method was noted at once in the automobile industry.

Norton turned his attention to the critical problem of grinding automobile crankshafts. Of necessity these were forged and finally hardened. The existing method required five hours of turning, filing, hardening, grinding, and polishing. The process was one of the most difficult precision operations in making an automobile engine, for relatively large amounts of stock had to be removed from the crankpins, bearing bodies, and side walls, yet demanding standards were required on dimensional accuracy and surface finish. Norton's "plunge grinding" could do the job *better* and in only *fifteen minutes.*

Until 1905 the Norton Company ground and balanced the crankshafts for practically all American automobiles. In 1905 the Locomotive Company was the first to purchase a machine for grinding their own crankshafts. Other companies soon followed suit; by 1911 Ford had ordered thirty-five, and grinding was an established production method in the automobile industry to give the high-precision, high production, low-cost manufacture required. Specialized and automatic grinders of many types soon followed—Landis's automatic-feed camshaft grinder in 1912, the Heald piston-ring grinder of 1904 and their cylinder

Fig. 38-8. Norton's heavy-production grinding machine, 1900. (Courtesy of *Norton Company*, Worcester, Mass.)

grinder of 1905. Automatic features were applied to all these machines, and the method applied even to grinding threads and gear teeth.

The application of the grinding machine in both standardized and specialized form to a number of industries was extended, especially in the automobile industry, which was to become the largest single branch of our metalworking industrial economy. The grinding machine has made work easy and inexpensive which was previously impossible. It has created the modern view of what constitutes first-class production machine work. Like the milling machine, it has taken over some work previously done on other machine tools, but it has far from supplanted them. In fact, the grinding machine has achieved much of its present importance by creating tooling operations and even industries which would not otherwise have been possible.

While much of our industrial economy is not based on principles of mass production, some very important segments of it are. In the automobile industry, for example, interchangeable parts are essential. The old-time machinist would be appalled to see the workmen in a modern assembly line putting together parts assembled from plants all over the country, and calmly assuming that they fit, and fit perfectly too. Underneath all this interchangeability lies precision. The grinding machine showed the way to achieve precision, in both hardened and unhardened parts, quickly and easily. It has allowed the machine designer to do many things that he could not dream of even a generation ago, and in return he has found it valuable to keep the special requirements, as well as the advantages, of the grinding machine firmly in mind in making his design.

The grinding machine is the most significant addition to our production machine tools since Maudslay discovered the idea of the precision machine tool at the end of the 18th century. Not only has it had an important influence on all production metalworking industries, but it has made technically possible our great automotive and aircraft industries which have changed the entire nature of our industrial economy, profoundly changing the way in which we all live.

SUMMARY

In the course of the 19th century a complete revolution took place in the vital area of working and shaping metals. In the first place, hand tools were almost universally replaced by machine tools. This switch brought with it more than just the lessening of cost and introduction of less-skilled workers than were usually associated with machine production. More importantly, the introduction, spread, and specialization of machine tools made possible, perhaps even inevitable, the development of the whole complex and interlocking system of mass production and interchangeability that we now associate with the modern age.

39 / Late Nineteenth-Century Communications: Techniques and Machines

THOMAS M. SMITH

A useful maxim of technological progress that helps to explain what happened in the field of communications toward the end of the 19th century is to be found in the remark, "If it is new, it is still technologically unimportant; when it is important, it is no longer new." The test is that of wide social use. Thus, the first printed book, the first telegraph key, the first newssheet had little social impact, whereas best-sellers, far-flung telegraphic networks, and daily metropolitan newspapers have exerted a profound social influence. The last quarter of the 19th century witnessed a spread and a diversification of communications facilities and techniques that was impressive indeed, compared with the developments of all the centuries before.

The growth of 19th-century communications technology was a symptom and direct consequence of the scientific technology that was emerging, for the first time in world history, during the 18th and 19th centuries in Europe and North America. It came about because of the rapid expansion of the engineering professions during the 1800's and because of the growing realization among scientists and mechanical inventors that the scientific tradition, originating in ancient Greece, and the technical crafts tradition originating in prehistory, could benefit each other mutually on the working level. The scientific area could benefit from the instruments and empirical "know-how" that the technical tradition could provide, and the practical technical side could benefit from the insights into the workings of Nature that theoretical science had to offer. The result of this cross-fertilization of science and technology was the rise of a new scientific technology.

In the closing decades of the 19th century, however, systematic research and development, involving teams of experts working in elaborate industrial or governmental laboratories, was not yet widespread or important. Instead, various promoters, industrial entrepreneurs, scientists, and inventors worked singly or together to provide new or improved communications devices and techniques. Their devices and techniques provided two types of messages—those that were transmitted in recorded form and were relatively permanent and those that were as brief and transitory in transmission as the sound of the human voice.

WRITING MATERIALS AND DEVICES—LEAD PENCILS

A prosaic but highly useful permanent-message device of the 19th century is the so-called lead pencil. It emerged from the craft tradition, and its early manufacture did not require scientific knowledge of the chemistry of graphite and clays. The common, wood "lead pencil"—a slender rod of graphite mixed with clay, fired and enclosed within a wooden cylinder for strength and easy holding—was a device of the 18th century which, although no longer new in the 19th, became technologically important because of its wide use. During the last half of the 19th century, J. von Faber established modern factories all over Europe and in New York and became a leading producer of graphite pencils, watercolor and oil paints, and related materials. His modernization and mechanization of manufacturing techniques typify the movement away from the hand-crafting tradition in order to supply a useful item to a growing mass market at reduced cost.

PAPER

To the old tradition of expensive and often beautiful hand-made paper in relatively small quantities, the early 19th century added machine-made paper in increasing quantities. Pressure generated by volume production during the last half of the century led to the introduction by the 1880's of wood pulp in significant amounts as raw material, replacing rags (except in the finer quality papers) and supplementing esparato grass and other sources of the plant fibers and cellulose essential to paper manufacture. As paper consumption increased toward the end of the 19th and on into the 20th century, wood pulp came into more extensive use at the same time that the varieties, grades, colors, and finishes of paper also multiplied. Newsprint that cost 15 cents a pound in 1867 cost 6 cents a pound just 12 years later. By the end of the 19th century the chemistry of paper-making had become an elaborate applied science.

PENS

Ink pens provide another typical example of the inventive fertility and technical proliferation of new products made possible by the maturing technology of the period. Replacing quills, pens with steel nibs came into wider and wider use between 1800 and 1850, especially after 1820, accompanying the dawn of modern steel-making and metalworking technology. Steel pens could be purchased for $2.00 a gross in 1830 and for 12 cents a gross in 1861. During the last quarter of the century steel pens sold by the millions, as did pencils. They had become technologically important as more people learned to write.

More impressive pens equipped with a reservoir and named the "hydraulic pen" or fancifully, the "fountain pen," appeared in the 1830's, but the smooth-flowing type that fed ink automatically to the pen point by capillary action, and admitted air at the bottom of the reservoir (instead of the top as the early models had done), was not available until the 1880's.

TYPEWRITERS

Machines that could be manipulated to place inked letters and words on paper almost as soon as they came to the writer's mind attracted the interest of inventors long before practical prototypes had been built. Representative of the imperfect but interesting early models are the machines that appeared in 1829 in Detroit and in 1833 in Marseilles. The experimental products of W. Burt and X. Progrin, respectively, were less reliable and practical than the line of machines begun by C. Sholes in the 1860's and available commercially, from the later 1870's on, in the United States.

During the decade of the 1890's hand-powered machines featuring the rapid touch-system keyboard were transforming the typewriter from an occasionally used instrument to a business machine of widening application. Following its successful development in the late 19th century, the typewriter became technologically important in the 20th century as a valuable communications tool. The typewriter had another social effect of subtle, yet profound importance: it helped bring women out of the home into the business world. Women's fingers proved as agile at the typewriter keyboard as with needle and thread. The economic role of women in society at this time was being transformed, and the typewriter was playing a modest part in this transformation.

PRINTING

After the advent of the rapid steam presses (Chapter 27), the most important advances in mass printing were associated with the type itself. Progress in type-manufacturing techniques revolutionized this portion of the printing industry during the later 19th century. The hazards of losing a page of type, easily pied (mixed up) in an accident, especially those accidents caused by the centrifugal force of rotary presses, caused many 19th-century inventors to seek to replace the typeset page with a cast page. Stereotyping was followed by electrotyping, and while neither of these processes reached perfection before the century was out, many improvements were made.

A less conspicuous yet fundamental change was the adoption of standardized type sizes and widths, measured according to the point system (a point equals about 1/72 of an inch) during the 1870's and 1880's in the United States. This done, it became practical to devise machines to produce the type in greater quantities, and from these developments, pioneered by L. B. Benton in the United States in the 1880's, appeared the matrix-cutting machine.

The matrix-cutting machine is a good example of the growing complexity of the communications process that embraces printing, for this machine became important as a contributor to the technique of casting entire lines of type in one piece and thus doing away with old-fashioned, hand setting of type, except where fine custom work was desired. The casting of lines of type was made practical by Ottmar Mergenthaler's Linotype machine, an early version of which was put to use by the *New-York Tribune* in July 1886. From then on, the Linotype machine became an essential item of equipment in any large-scale printing establishment. Its letter matrices provided the patterns from which the linotype slugs were cast, and these matrices were made available in the necessary large quantities by the matrix-cutting machine derived from Benton's punch engraver.

Wood-block line drawings began to be replaced by zinc cuts in the 1880's in the United States. The concept of the halftone picture, in which variations in the density of minute dots provide shading and simulation of contours, was being explored by the middle of the century in England, and from 1880 onward was appearing in United States newspapers. Realistic color pictures were regarded as a possibility before 1870 by du Huron in France, but the three-color process, combining blues, reds, and yellows, faced many technical hurdles before it became commercially practical in the United States during the 1890's. Rotogravure (not a true color process but using rotary presses employing copper cylinders etched from photographic plates) and similar processes also flourished during the last years of the century. Impressive to contemporaries, color printing at the end of the century was only beginning.

In printing, the growth of literacy among the common people and the consequent emergence of a mass market of readers, the evolution of the daily newspaper, and the idea of systematic huckstering in print, which gave rise to advertising as an invaluable adjunct to sales—these and other social and economic factors created a favorable climate in which men of mechanical-design ability could flourish. The result was a burgeoning communications industry built upon

Fig. 39-1. Mergenthaler Linotype machine, *c.* 1890.

the printed word and the printed picture duplicated by the hundreds and the thousands in identical copies for simultaneous, although relatively leisurely, distribution at remarkably cheap cost. By the end of the century the printing trades had become extensively diversified to perform an incredible variety of services for the common man, for only the common man existed in large enough numbers to render such developments economically feasible.

MOTION PICTURES

As 18th-century technical developments had been preliminary to the important spread of the lead pencil in the 19th century, so 19th-century technical developments were preliminary to the 20th-century technological importance of moving pictures, or the cinema. Some of these developments were empirical in nature, and some resulted from scientific understanding of phenomena.

Seventeenth-century investigations of the optics of lenses and their ability to magnify and project images came after the invention of the telescope and microscope. A German mathematician and natural philosopher, A. Kircher, combined a reflector, a light, a lens, a painted glass slide, and a screen, thereby inventing the "magic lantern" in 1645 in Rome, where he was living at the time. All of his pictures were static. It was another 200 years before simulation of motion in projected images was attempted. This lapse of time may be regarded as one indication of the absence of a scientific technology, although the scientific comprehension of the optics of lenses did not by itself provoke the invention of the cinema.

Men had long known, from their own experience, of the inability of the human eye to keep up with rapid motion. Since Greco-Roman antiquity, at least, it was known that a spot of color could become a ring of color on a revolving disc. How long horsemen had argued about whether a galloping horse ever has all four hooves in the air at once is not recorded, but the training of horses and practical knowledge of their gaits were millennia old. Thus, in one form or another there was experience with the phenomenon of persistence of vision. There was not, however, scientific understanding of the cause of this phenomenon by which the eye continues to "see" an object sometimes for as long as a tenth of a second after it has disappeared. Nor did an early 19th-century application of this phenomenon, which produced several intriguing devices, rest upon such scientific understanding.

The gadgets that had appeared in various forms by the middle of the century required new names: wonder turner, magic disc, fantascope, wheel of life, and so on. All of these replaced one picture with one nearly like it in the viewer's line of vision so rapidly that motion could be simulated. The more nearly alike the picture of, say, a dancing figure, and the more pictures in a series, the more gradual were the "motions" of the figure. The fewer the pictures and the more apparent the change in posture, from figure to figure, the faster and more jerky the action. This, of course, is the principle of the animated cartoon.

This principle was applied to the magic lantern during the 1850's, but early applications were not sufficiently practical or refined in their development to qualify as the first movies. Neither were the valuable contributions in stop-motion photography made during the 1870's, although the pioneer, E. Muybridge, is famous for his fast-camera, split-second shots of rapid animal motion. (His photographs, taken at Palo Alto, California, settled the wager in favor of those who argued that all four of a galloping horse's hooves could be off the ground at once.)

The course of events during the latter half of the century is a familiar one in the history of technology: a great many individuals contributed to the emergence of satisfactory photographic film, satisfactory cameras, and satisfactory playback machines—moving-picture projectors. It is an example of the social nature of mechanical invention, where the large number of contributors is more

Fig. 39-2. A cinema house in the 1890's. The motion-picture projector, called the Biograph, is shown on the left.

significant than any particular heroic figure, and where the convergence of scientific inquiry and technical skills from several directions is essential to the development of appropriate instruments and machines, in this case for the movie industry of the 20th century.

The complex inventive process that took place renders it practically meaningless to ask which was the first moving picture. Was it the Austrian Baron Uchatius's moving lantern slides, developed for military instruction in the 1850's, that could produce a sequence of 30-seconds' length, but never saw wide use? Was it Muybridge's high-speed pictures—or was it the elaborate apparatus he used, which was the work of J. Isaacs? For that matter, what about the curiosity regarding horse gaits that impelled the railroad millionaire, Leland Stanford (after whose son a university was named), to set Isaacs and Muybridge in 1872 onto the problem and to provide the money needed? Certainly Stanford's sportsman's curiosity was a significant factor.

If Baron Uchatius's 30-second devices do not betoken the start of the cinema, should Edison's 13-second kinetoscope "peep show" that appeared on the market in the 1890's qualify? A year before Edison demonstrated his first kinetoscope, E. Morey had told the French Académie des Sciences what his "chronophotographic" camera would do. On the other hand, nearly a decade passed before Morey was presumably successfully projecting his movies. And what about Louis and Auguste Lumière who patented their cinematograph and successfully demonstrated it in 1895 in France? Lacking a single inventor standing clearly above the crowd to claim his laurel, we may conclude the motion picture emerged at the end of the century from a complex and vigorous matrix of inventive ability. The emerging scientific technology of the 19th century was thus a social, not a heroic, activity.

TELEPHONE

The telegraph, which had steadily increased its service from the early days of Morse, was still a specialized service, and there were some things it did not do. It did not offer the advantages of voice conversation, of dialogue, even though it provided an expanding, high-speed, electrified-wire, communications network. The theory of the telegraph and that of the telephone had each simulated inventors from the 1830's on, but the theories were different, and the development of an intelligible, practical telephone lagged behind the telegraph by at least a third of a century.

The principle of the telegraph required an electric circuit to be completed and broken in a meaningful, patterned way. The principle of the telephone required the more subtle and complex vibrations of sound to be converted to corresponding variations in an electric current and these to be converted back into identifiable sounds at the other end.

The technical problems of telephonic communication were complicated and challenging. Although their solution depended in part, it is true, upon prior scientific understanding of the phenomenon of electromagnetic induction reached independently by Faraday and Henry, this scientific understanding did not itself automatically indicate or suggest the practical modes of solution. Many methods of transforming sound patterns to electric patterns and back were devised and studied by different investigators during the 19th and 20th centuries. The historical record shows, without detracting from any particular inventor's legitimate claim to originality and fame, that by the 1870's it was no longer a question of "whether" but rather "by whom" the telephone system would be invented and perfected successfully. The selection of famous inventors and their interesting personal achievements for discussion here should not be permitted to obscure the primary fact that many contributions were necessary before any of the communications developments of the 19th century became technologically important.

Among those working on the problem of voice communication by electrified wires in the third quarter of the 19th century, Alexander Graham Bell (1847-1922) is commonly singled out as the inventor who most effectively reduced theory to practice. Like his father and grandfather before him, Bell was an expert in the phonetics and mechanics of speech and hearing. A teacher of the deaf, he also studied the anatomy of the ear; furthermore, Bell was acquainted with electromagnetic induction and with techniques of varying, or modulating, one current by superimposing a variable induced current upon it. Equipped with this knowledge, he deliberately sought a mechanical analogue of the membrane of the human ear to act as a transmitter and as a receiver.

By 1875 his efforts to enlarge the capacity of an experimental telegraph hook-up ("system" would be too elaborate a word to use here) led him to believe that he might make a current of electricity vary in intensity precisely as the air varies in density during the production of sound. This he and his co-worker, T. A. Watson, accomplished successfully for the first time on March 10, 1876. They employed a continuous current, modulated by the sound of the voice and carried by wire from one room to another. They were using a battery to supply the current and an experimental transmitter that featured a wire attached to a diaphragm and dipped in a metal cup of "acidulated water." As the diaphragm vibrated to the sound of the voice, the depth of the wire in the liquid changed and thereby altered the resistance in the circuit, thus modulating the current.

As they began the experiment, Bell accidentally spilled some battery acid on his clothes and called, "Watson, come here! I want you." Watson heard the message from the diaphragm of his receiver, rushed down the hall to Bell's room—and the two young men knew that at last they could send speech over an electric wire in a practical, intelligible way.

The principle of the telephone, as it was subsequently developed, depended upon the deflections of a diaphragm responding to sound waves that actuated an attached steel reed or iron disc in the field of an electric coil. The movements of the piece of iron or steel impressed a pattern upon an existing current flowing through the coil. At the other end, the transmitted electric pattern varied the electromagnetic field surrounding another coil and actuated a diaphragm, setting in motion corresponding sound waves distinguishable to a nearby human ear.

Clearly, a circuit that would work between two telephones in two rooms or over a stretch of 500 feet of wire and be hailed by its inventor as a success was not a circuit that was yet commercially practicable, and as is the case with so many inventions, the practical perfection of telephony posed a host of technical and economic problems that had to be resolved before the telephone *system* became technologically significant. This phase of the development of the telephone lasted the remainder of the 19th century and continued well on into the next.

It is not surprising to find that the busy, expanding telegraph networks could not afford to disrupt their extensive coded-pulse message service in order to convert to the still risky and untested-in-practice telephone. Telephonic net-

Fig. 39-3. Switchboard and operators of the Dane County Telephone Company, Wisconsin. Photo, 1908. (*The State Historical Society of Wisconsin*)

works had yet to be developed. In New Haven, Connecticut, in 1878 twenty-one subscribing parties could use eight grounded lines connected to the first central switchboard. During the rest of the century local telephone exchanges grew up in rural areas especially in Norway and the United States, and in metropolitan centers in the United States and Europe, but the telephone did not compete importantly with the telegraph over long distances until the 20th century. New York was first connected to Chicago in 1892 but not to San Francisco until 1915. By 1888 a telephone cable could contain up to fifty pairs of wires; the single wire, ground-return techniques of the telegraph had proved inadequate even for early commercial telephone service. The possibility of dialling—that is, of automatic switchboards—was investigated during the 1890's by A. Strowger and others, but did not begin to come into important use before 1921.

There were twice as many telephones in the United States (over 155,000) as in Europe in 1885, but between 1892 and 1898 the number of European phones exceeded those in the United States. After that time the United States maintained a widening lead, yet none of these figures can obscure the fact that the telephone was still for many years an interesting oddity of limited use. Between 1880 and 1900 the number of telephones in the United States increased from about 48,000 to 1,350,000; since the population continued to grow in the United States, there were still less than two telephones for every 100 people by 1900. An important beginning had been made under the influence of the emerging scientific technology, but it was still only a beginning. By 20th-century standards, the telephone remained technologically unimportant, although an attractive, promising device. Some inventors even talked enthusiastically of transmitting still and live pictures electrically during the latter part of the 19th century, but little practical work was accomplished.

PHONOGRAPH

Early in the last quarter of the 19th century sound-recorders and playback mechanisms first appeared in the United States as interesting curiosities and engaging novelties worth a passing story in the public press. Before the end of the century they were being advertised and sold to the public, although they did not become technologically important for several decades.

As communication machines they offered so peculiar a hybrid service between the leisurely, permanent message of the printed phrase and the swift transient message of the telegraph that they secured no immediately important social role other than as an entertainment device. Oddly enough, the early devices reproduced music better than speech, so that the major 20th-century use of the phonograph was presaged from the beginning. Not until well into the 20th century did later generations of machines operating on different principles become technologically important as business instruments for dictation and stenography and as accessories to a multimillion-dollar recording industry influencing and catering to public tastes.

Nevertheless, the 19th-century phonograph is another excellent example of the sort of device the new technology could produce, and it demonstrates also the complex modes the inventive process was assuming. Thus, there were inventors of the *idea* of a recording device, such as C. Bourseuil and C. Cros separately in France during the last half of the century; their statements of what might be accomplished did not themselves produce or determine the forms of the prototype machines.

There were also inventors of *instruments* for transcribing sounds, such as L. Scott, whose "phonautograph" of 1857 scratched wavy lines in lampblack-coated paper when sounds caused vibration of a diaphragm to jiggle a stylus. This instrument provided a picture of sound for the scientifically curious, but it could not itself generate sounds. Thomas A. Edison was another inventor who was not satisfied with speculative or with abstract concepts but felt compelled to create the instruments. In 1876, scarcely a year after Bell had designed what is commonly called the first telephone, Edison designed a "telephone repeater" to place a spoken message on the telephone lines (that had not yet been built) without requiring the speaker to be present. Here was an analogy to the telegram, for both could be filed and dispatched in an orderly fashion without delaying the author of the message from going about his affairs.

Fig. 39-4. The first phonograph consisted of three principal parts—the mouthpiece, into which speech was uttered; the spirally grooved cylinder; and the stylus, which, vibrating from the voice impulses, traced by indentations a path in the metal foil that formed the record.

The device that Edison designed recorded and played back for the human ear or for telephonic transmission by impressing a pattern on a moving strip of metal foil. The speaker's voice flexed a diaphragm to which was connected a stylus that made grooves or indentations on the foil sleeve revolving on a horizontal cylinder. Once the pattern was indented, one could at his convenience use it to actuate another stylus attached to a diaphragm. The diaphragm, by flexing in response, could send a telephonic message in the same manner as a Bell transmitter.

However, Emile Berliner in 1887 saw that the up-and-down motions of a scriber on a horizontal cylinder must have included distortions due to gravity. His solution was the "gramophone," which used a plate, as he called it, instead of a cylinder. The plate was a flat disc which spun around a vertical axis, and the stylus moved not up and down but side to side as it responded to the sound being recorded. By 1897 Berliner had a commercially marketable prototype of the flat disc which later acquired the name "record."

All these recording machines were, from the start, voice-powered and mechanical in their principles of recording and playback; electrical circuits in phonograph sound reproduction were to be a 20th-century innovation.

CONCLUSION

In retrospect, machines and techniques of communication of the late 19th century occupied the inventive minds and attracted the developmental efforts of hundreds of men. No longer was an occasional, amateur mechanism contributing an occasional, brilliant new idea. More and more frequently two or more men, not working together, were having the same or challengingly competitive new ideas at about the same time. It was not enough, after a lucky "brainstorm," to sit back and claim a legal right to be the sole producer by right of patent granted, because there were other men eager to investigate, to follow through, and to reduce to practice any promising theoretical insight.

Communications and, more important, the economic and technical capability in Western societies to improve communications, were creating a broader base of diversified skills and abilities persistently dedicated to mastering technical problems and exploring Nature's secrets. Nevertheless, the scientific communications technology that was emerging at this time was itself only a narrow band of the broad spectrum of technological activity that was transforming the United States and Europe and that soon would be transforming the world. Seen in this perspective, the new communications technology was but a small, dynamic segment of a much larger phenomenon.

40 / The Beginnings of the Internal-Combustion Engine

LYNWOOD BRYANT

The thirst for power which characterized the 19th century could hardly be satisfied by the steam engine, even with its many improvements. And although the growing art of handling electricity promised yet another great source of energy, other avenues leading toward greater power resources were explored as well. Besides the general need for more power, these searches were also inspired by such special desiderata as an engine both powerful and light enough to drive carriages on common roads. Again, although steam carriages were always in operation somewhere, the idea of an internal-combustion engine was also pursued. In the last half of the century these hopes were rewarded.

In conception, the internal-combustion engine went back many years. A

French scientist of the 17th century, Denis Papin (1647-*c*.1712), had an idea for an engine driven by the explosion of gunpowder in a cylinder, and for several generations some men followed this mechanical chimera. Others also sought means to produce an explosion within a cylinder, trying alcohol or carbolic engines—but without success. The solution of this problem—the gasoline internal-combustion engine—was to have profound consequences on every aspect of human life.

The story of the development of the internal-combustion engine fits neatly into the 19th century. In 1801, men like the French engineer Philippe Lebon, who was interested in illuminating gas, speculated about a gas-burning engine. By 1860, a hundred such schemes had been proposed, and perhaps a dozen engines had been built. Yet, although the advantages of internal combustion—burning the fuel inside the working cylinder of an engine without any intermediate working fluid like steam—were well understood in the middle of the century, it was a long time before the difficult problems of controlling an explosive fuel like illuminating gas could be solved and a practical engine built.

The decisive breakthrough came in 1876, when N. A. Otto (1832-91), a German self-made engineer, found a good way of compressing the combustible mixture inside the working cylinder before ignition. It took another twenty-five years to work out the details of the automobile engine—to devise a reliable electric ignition, to learn the technique of handling a liquid fuel, and to make an engine light and powerful enough to drive a vehicle. But by the end of the century the spark-ignited piston engine that is now so familiar to us was essentially completely developed, and it is with that development that we shall be chiefly concerned.

NEED AND RESPONSE

To a man brought up on the steam engine and interested in improving its notoriously low efficiency, internal combustion was a very attractive idea. A steam engine requires a furnace to burn the fuel and a boiler to generate the steam, and it loses a good deal of heat up the stack, in the boiler, and in transmission to the engine. The internal-combustion process, which needs no boiler, should make a considerably simpler and cheaper power plant than a steam engine, inventors thought, and it should be more efficient because the heat is generated inside the working cylinder and is converted into work at once, ideally without losses in storage or transmission. The high temperatures and pressures in a gas engine created difficult engineering problems in the early days, but at the same time they offered the possibility of high efficiency because of the wide range of temperature through which the gas engine could operate.

These theoretical advantages of internal combustion were understood by the men who built the first gas engines in the 1860's and 1870's, but it was a long time before engineers came very close to realizing them. The new problems created by internal combustion—problems of fuel handling, ignition, control, and cooling—were very difficult ones, and the early gas engines cost about as much as comparable steam plants, and were considerably more delicate and less reliable. They may have been twice as efficient, but they burned a fuel twice as expensive.

So it was not fuel economy that sold the internal-combustion engine in the early years, but rather convenience of operation, and especially adaptibility to intermittent operation. The steam engine had a long warm-up period and a dangerous boiler that had to be licensed. Someone had to keep the fire going and keep an eye on the boiler whether the engine was working or not. The gas engine, on the other hand, could be started and stopped at will (in theory), it did not have to be fed when it was not working, and it required no fuel storage. Furthermore, it could be installed in a man's shop and operated without extra help and without a license. This sort of flexibility and convenience, and the fact that it was practical in small sizes, made it a sensible power plant for small enterprises. Some of its more idealistic promoters, including Otto and later, Rudolf Diesel (1858-1913), thought of the internal-combustion engine as an instrument of industrial democracy: a steam engine, since it had to be quite large to be economical, required a certain amount of capital and tended to increase the size of industrial units and to squeeze out the little man; but the gas engine would help the small shop to hold out against the large capitalist by giving it the advantage of artificial power on a small scale.

This vision of an engine without a boiler, smaller and cheaper than a steam engine, and perhaps even light enough to drive a carriage, enticed dozens of men before Otto to experiment with ways of controlling combustion within a working cylinder. Scores of their ideas reached the patent stage, and a few model engines may have run a few strokes. But the first engines that were more than philosophical toys were built in France around 1860 by Pierre Hugon and Étienne Lenoir, working independently.

The Lenoir engine of 1860 was the first internal-combustion engine to be sold in some quantity. It was a one-cylinder, double-acting affair like a steam engine that developed one or two horsepower at 100 revolutions per minute. The piston sucked in a mixture of illuminating gas and air for the first half of its stroke. An electric spark then ignited the mixture, and the expanding gases drove the piston through the second half of its stroke. A similar process on the other side then drove the piston back. This engine was not a very good one—it was very inefficient, and ran smoothly, as advertised, only when idling; but it worked, and it was energetically promoted. Distinguished engineers and professors visited the self-taught Lenoir, measured the performance of his engine, and published popular accounts and thermodynamic analyses of it, which were picked up by the popular and technical presses of Europe and America. Lenoir's

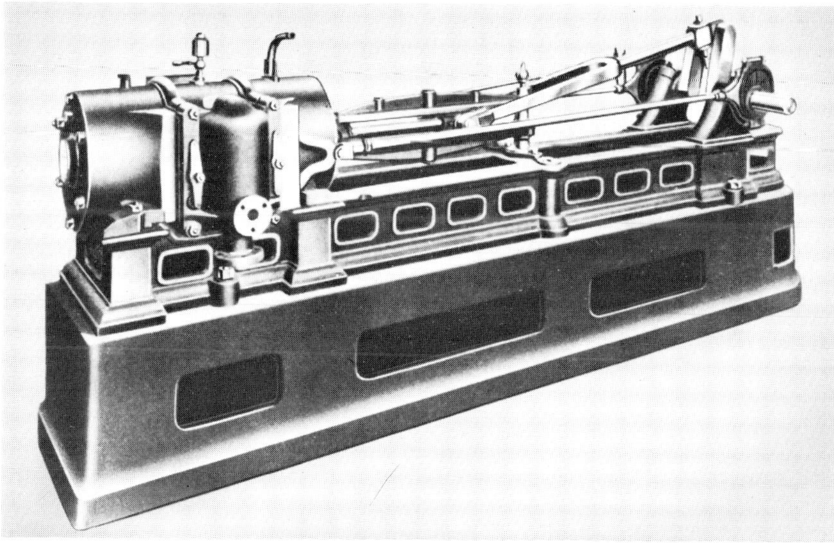

Fig. 40-1. The Lenoir engine, 1860, was very much like a steam engine, double-acting, with a slide valve. It burned illuminating gas, without compression, and ignited it with an electric spark. (*Werkfoto Deutz*)

main achievement was to stimulate a great deal of serious thought about internal combustion as an alternative to steam.

THE OTTO ENGINE

Nicolaus August Otto, a travelling salesman in the Rhineland, read a story about Lenoir in a newspaper in 1860, and had a model engine built, using the newspaper description as a guide. As soon as he started experimenting with this engine in his spare time, he ran head-on into the problem that baffled all workers with gas engines of the day: how to control the severe shocks from the exploding gases. Otto began by varying fuel-air ratio, size of charge, and timing of ignition in an effort to find a combination that would yield a moderate, controllable explosion, but nothing seemed to work. He also tried other approaches to the shock problem common at the time, such as inserting some sort of spring between the explosion and the piston, or injecting water into the cylinder, or using an external burner and feeding the burning gases into the cylinder like steam.

In 1863 he finally gave up the attempt to moderate the explosion, and turned for his solution to the atmospheric principle, which was well known at the time: he used the explosion to drive a free piston up in a cylinder and then let atmospheric pressure and the weight of the piston do the work on the way down.

Fig. 40-2. Otto & Langen atmospheric engine, 1867. The explosion of a charge of illuminating gas, lit from an open standing flame, drove a heavy free piston up. The quickly cooling gases created a vacuum, and the piston was driven down by atmospheric pressure and engaged the load through a special free-wheeling clutch. (*Werkfoto Deutz*)

The suggestion for this solution could have come from the atmospheric steam engine of Newcomen, or from an earlier Italian gas engine, or from his experience with the Lenoir model. It took several years to work out the details of the design, especially to develop a one-way clutch that would allow the piston to rise freely but engage the load securely on the way down, but by 1868 he had a practical engine, the Otto & Langen, which could operate at 80 or 90 explosions per minute, and was able to adapt to an increased load by increasing the number of strokes per minute without increasing the speed of the output shaft. Some 5000 of these engines were built, in sizes up to three horsepower.

Otto who had begun as a solitary amateur inventor, and then run out of money before he had the atmospheric engine working, had formed a lucky partnership with Eugen Langen, who brought capital and sound business judgment to the enterprise, as well as contacts with the larger world of science and business. Together Otto and Langen represent the same sort of fruitful combination of technical and business talent that made Boulton and Watt so significant in the history of the steam engine. After a number of lean development years, the firm prospered and eventually became the largest manufacturer of internal-combustion engines in the world, and a fertile source of new men and ideas in the field.

The Otto & Langen engine laid the foundation for the success of the firm, but it was too inflexible and limited to be a permanent solution to the need for new power. Sales began to decline in 1875 because customers wanted more power, and the engine was not practical in sizes larger than three horsepower. Even in that size it needed twelve feet of headroom, and was as large and noisy as could be tolerated in a shop. Since it relied on atmospheric pressure, it could not be made more powerful without enlarging the cylinders, and two cylinders could not work together on the same shaft. In this critical situation, when the technical men in the firm were trying half a dozen different approaches to the problem of making a more powerful engine, a friend and adviser of Langen's, the distinguished philosopher-engineer Franz Reuleaux, pointed the way to the right answer: internal combustion of a compressed mixture. Under this guidance Otto in 1876 gave up his current work on various ideas for an external-combustion gas engine, and turned back to the direct-acting engine burning gas in the cylinder, the kind that had given him so much trouble with destructive shocks in the early 1860's. This time he came up quickly with the smooth-running Otto Silent Engine, which used the four-stroke cycle to achieve compression within the working cylinder.

Otto's engine of 1876, the earliest recognizable ancestor of today's automobile engine, burned illuminating gas in a single horizontal cylinder, developed about three horsepower at 180 revolutions per minute, and weighed one or two thousand pounds per horsepower. This engine operated on the four-stroke cycle, the one now used by most automobile engines. On the first or *intake* stroke the piston drew in a measured mixture of gas and air, which it compressed on the way back (*compression* stroke) into a volume about 40 per cent of the space it occupied at atmospheric pressure. At the end of this stroke, at the point of maximum compression, a flame ignited the charge, and the expanding gases drove the piston through the third or *power* stroke, and on the last stroke, the *exhaust*, the piston drove the spent gases out the exhaust port. Since this engine had only one cylinder, and only one power stroke for every two revolutions, it had to have a heavy flywheel to carry the engine along between power strokes.

Fig. 40-3. Otto's Silent Engine of 1876, the first to operate on the principle of today's common automobile engine, with a compressed charge and a four-stroke cycle. It used illuminating gas, a slide valve, and flame ignition. (*Werkfoto Deutz*)

This was a very successful engine, about twice as efficient as a comparable steam engine. It embodied three key ideas: internal combustion, the idea of burning the fuel inside the working cylinder; compression of the gas before ignition in order to increase the available energy; and the four-stroke cycle, which is one way of achieving compression within the working cylinder. It was the compression that made this engine such a notable step forward. The idea of compression had been suggested a number of times before 1876, and had been specifically recommended in the published scientific analyses of the Lenoir engine. Otto had also discovered the value of compression himself in his early work on the Lenoir model. Once when he was trying various sizes of charge he drew in a full cylinder load of gas and air, and compressed it on the back stroke. The explosion of this compressed charge startled Otto, and drove the flywheel through several revolutions. This was undoubtedly very interesting, but a very strong explosion was the last thing Otto wanted at the time. His problem was how to get weak explosions, and he was having trouble enough with explosive gases at atmospheric pressures.

But in 1876, when Otto was driven back to the idea of compression as the only way to get a more powerful engine without increasing the size, he had a new idea for a way of controlling the explosion: he thought he could reduce the shock by mixing the gases in the cylinder in such a way that he would have a rich, powerful mixture, easily ignitable, near the point of ignition, and a leaner mixture or even a layer of inert gases near the piston, which he hoped would burn more slowly or not at all, and act as a kind of cushion for the explosion. It was this special way of mixing gases, he thought, that made it possible to use the increased energy made available by compression and accounted for the success of the engine. His theory was wrong, in the opinion of the experts then and now, but nevertheless the engine worked.

The four-stroke cycle, which we see as the key idea in this engine, was something that Otto was forced into for lack of a better way of achieving compression. He never thought of it as the essence of his invention. Everyone at that time approached the gas engine with the steam engine in mind, and in the ordinary double-acting steam engine, every stroke is a power stroke. If the object is to increase the power or smoothness of an engine, to decrease the number of power strokes to one in four would seem to be a movement in exactly the wrong

Fig. 40-4. Indicator diagrams that Maybach took about 1876 showing how the pressure fluctuated within the cylinders of the three engines shown in Fig 1 (Lenoir), Fig. 2 (Otto & Langen), and Fig. 3 (Otto). The vertical axis indicates pressure, measured in atmospheres, and the horizontal axis indicates volume, or piston displacement. This sort of diagram tells the engineer how well the engine is working.

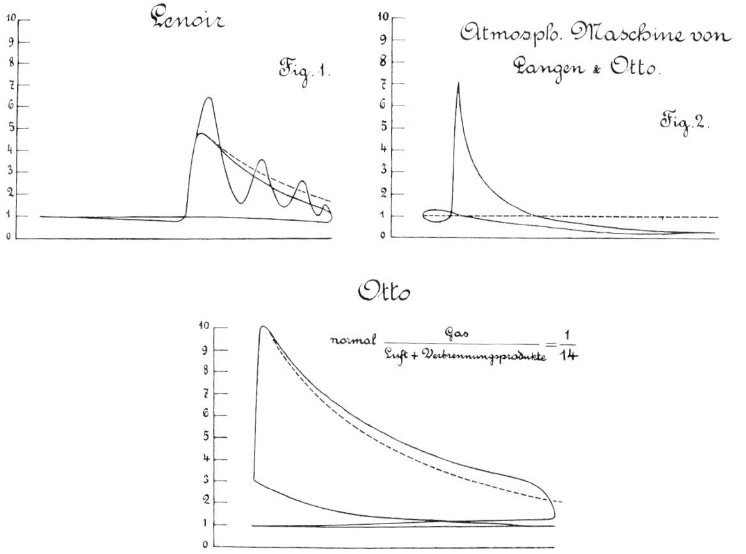

direction. In the experimental stage a man might conceivably begin with one power stroke for every two revolutions, but as soon as he got the engine running he would naturally try to increase the number of power strokes, by working out a two-stroke cycle, or by making the engine double-acting, or both. The four-stroke cycle is one of those ideas that seem simple and obvious in retrospect, but to an engineer in 1876 looking forward into the unknown it was by no means obviously the right answer. It seemed unnatural and awkward at first, even to Otto, and it was some time before engineers fully appreciated that the advantages of compression were so great that they could afford to have only one power stroke in four.

In the years following 1876 a hundred inventors, including Otto, worked on some ingenious schemes for getting more power out of a cylinder by increasing the number of power strokes—usually by performing the compression in a separate cylinder or by using the other side of the working piston for this purpose. But in the early years none of the two-stroke schemes was able to compete in simplicity or economy with the four-stroke cycle.

One curious episode that illustrates how little these inventors were influenced by theoretical considerations is the story of Alphonse Beau de Rochas, an eccentric French engineer who had been working on various projects in the new fields of the telegraph and the railroad, and also reading the early literature of thermodynamics. As early as 1862 he published a peculiar hand-written pamphlet on various ways of improving the efficiency of locomotives, one of which was to combine the steam engine with a gas engine in one system, for example, by using the exhaust heat of the gas engine to generate steam for the steam engine, or by using steam on one side and gas on the other side of the same piston. In the course of this discussion he made a number of interesting suggestions about gas engines that show how far ahead of his time he was. The fuel-air mixture should be compressed as far as possible before ignition, he said, and the limit would be set by the temperature at which the fuel would ignite. So he foresaw compression-ignition, except that it was the fuel-air mixture that was to be compressed, and not the air alone as in the later diesel engine. The compression should take place within the working cylinder, Beau de Rochas said, and it could be achieved by a four-stroke cycle, which he proceeded to describe in detail.

This clear and explicit anticipation of the principle of the four-stroke cycle on purely theoretical grounds is an extraordinary thing, but it hardly seems strange that no one paid any attention to it. The pamphlet was obscurely published, and perhaps no one read it. The four-stroke idea would not perhaps have been an attractive one in 1862 anyway. Apparently Beau de Rochas himself did not take it very seriously as a practical suggestion for an engine, for he did not follow it up or make any attempt to take credit for it later when it was recognized as valuable. He did not build an engine, and he had no ascertainable influence on the history of the internal-combustion engine until patent attorneys discovered his pamphlet in 1883 and were able to use it to invalidate a part of Otto's patent.

Otto's new four-stroke engine of 1876 proved to be a very flexible and versatile type, and a great commercial success. The cylinders could be horizontal or vertical, large or small, and several cylinders could work on the same shaft, in various configurations. It could be adapted to various fuels and to a wide variety of applications. Within fifteen years Otto's engine was being manufactured in a dozen countries, and scores of competitors sprang up to offer the same sort of engine in an expanding market. At the end of the century there were as many as 200,000 such engines in use, mainly in the United States, England, and Germany, providing power for machine shops, hoisting machinery, printing presses, pumping stations, and such applications where power usage was intermittent and the convenience was more important than fuel economy.

After 1880 the electric light offered an important new market. In the days before distribution systems for electric power were well developed, the gas engine was an ideal power source for a small electric light plant serving a unit the size of an opera house or a factory. In the period after 1895, before the potentialities of the steam turbine were fully appreciated, a good many very large gas engines—over 1000 horsepower, fed by individual gas-generating plants or by cheap blast furnace gas—were tried out in electric power stations and even ships in the hope that the superior efficiency of the engine would compensate for the added cost of generating the gas, but they proved economical only where gas was unusually cheap. The gas engine of the heavy stationary type was a more important power source in the range from, say, three to fifty horsepower, but only for perhaps thirty years, until the diesel engine and the electric motor were developed.

FURTHER IMPROVEMENTS

Later engines built by Otto and his competitors weighed several hundred pounds per horsepower and ran at speeds of 50 to 150 revolutions per minute. There was no theoretical reason why such engines could not be built in smaller sizes and used to drive vehicles, but the market for automobiles was not promising enough in the 1880's to attract investors to support the development that would be necessary. The engine would have to be adapted to a liquid fuel, and the weight would have to be drastically reduced. The only way to make the engine light and still powerful enough to drive a car was to make it run much faster, and it could not run much faster without a radical improvement in ignition. These tasks proved to be more difficult than any one had anticipated. The internal-combustion engine proved to be quite sensitive to such things as quality of fuel, fuel-air ratio, and timing of ignition. A steam system could burn almost

Fig. 40-5. Benz motor car, 1888. (British Crown Copyright. *Science Museum*, London)

anything, and it would limp along if some adjustment were not quite right, but a gasoline engine would not run at all unless conditions were controlled within rather narrow limits.

The detailed work necessary to develop a reliable small gasoline engine was carried on by many men in many countries in the last twenty years of the century, often amateurs working on their automobile hobby. But the most significant work was done in Germany in the 1880's, independently by Karl Benz (1844-1929) and Gottlieb Daimler (1834-90), both of whom were experienced professionals, persistent men driven by the impractical ideal of a small fast engine to drive a road vehicle. Daimler had learned about gas engines as production manager for the Otto & Langen engine, and later for Otto's four-stroke engine. In 1882 when he left Otto's firm to devote himself to the light fast engine, he took with him an inspired mechanic named Wilhelm Maybach, who was responsible for some of the notable achievements in weight reduction, carburetion, and cooling that gave the automobile engine essentially its present form by 1901.

The idea of using liquid fuel to make an engine portable was obvious; but in the early days liquid fuels were so difficult to manage that an inventor, after his first exasperating attempts to vaporize or atomize the liquid, mix it in just the right proportions with air, and heat it to just the right temperature, was usually tempted to go back to the convenience of gas, perhaps bottled or produced by a portable gas-generating plant.

For vehicles the only permanent solution was liquid, and for automobiles it had to be gasoline. Everything else was tried first, and gasoline was adopted reluctantly, because although it was easy to vaporize, it was thought to be extremely hazardous, and with an open-flame ignition it certainly was. Carburetors or gasifiers that vaporized petroleum products to supply illuminating gas for locations not reached by urban gas systems were common in the 1860's, and

Fig. 40-6. Daimler motor car, 1895. Rear view.
(British Crown Copyright. *Science Museum*, London)

early inventors tried them to supply fuel for gas engines. They were like a boiler —bigger than the engine—and the mixture of fuel and air that they supplied could not be adjusted easily. Maybach successfully used gasoline for a stationary Otto & Langen engine in 1875; but for an automobile engine, which had to run at fluctuating speeds and loads, the carburetor had to be able to control the volume and the ratio of the mixture precisely. This was a very difficult problem not finally solved until Maybach devised the modern form of carburetor in 1893, a fine jet spraying gasoline into the air-stream being sucked into the cylinder. This small gasoline carburetor was the portable gas-generating plant that made the automobile possible.

The problem of igniting the mixture reliably several times a second was the key problem created by the demand for high speed. The electric spark seemed to be the ideal answer because it was so quick and could be easily controlled. It had been the standard method for igniting gases in the laboratory for half a century, and was proposed for the earliest gas-engine schemes. In 1860 Lenoir used a battery-powered system with the newly developed spark coil and something very much like a modern spark plug. Otto began with such a system himself, but found it expensive and complicated, and sensibly shifted to the much simpler flame ignition. In those days an engine working at atmospheric pressure could easily have a standing flame burning the same fuel as it, and suck the flame into the cylinder at the right time through a touchhole. To ignite a compressed mixture with a flame was more of a trick, but Otto managed it in 1876 through an ingenious system of small passages in his slide valve that enabled him to carry a bit of burning gas away from the external flame and introduce it into the compressed charge. This was a perfectly satisfactory system for low speeds and low compressions, and it did not make sense to consider electrical ignition unless the speeds were much higher or the fuel were a hazardous one like gasoline.

When in 1883 Daimler and Maybach moved at one step up from speeds of 100 to 200 rpm to speeds of 600 to 900 rpm, they needed a quite different type of ignition, and found it in a small tube heated to incandescence by a standing flame, closed on its outside end, but open to the combustion chamber on the inside end so that at the right moment some of the compressed mixture would be forced into the hot tube and ignited. At first Daimler intended to rely on the heat of compression for ignition, and thought of the hot tube as a temporary help to start a cold engine, but it turned out that the hot tube was needed all the time. This was a neat solution to the ignition problem for a fast engine because it dodged the problem of timing by simply providing a steady hot spot in the combustion chamber, just hot enough to fire the charge at the right time, without any moving parts at all. The fire hazard from the open flame was a serious objection, but Daimler simply took this risk. His cars won the early races in the 1890's with an open flame keeping the ignition tube hot.

In the same period Karl Benz, with a slower engine, struggled with electric ignition, which was safer for automobiles, more precisely controllable, and adaptable to fluctuating speeds and loads, but complicated and difficult. It was hard to find materials able to stand the high temperatures of the combustion chamber, and it was very hard to build a system sturdy enough to stand the jolting of hard tires on cobblestones. Benz began with an arrangement of battery, spark coil, and spark plug like Lenoir's, and for twenty years tried many variations to improve reliability. The weight and short life of batteries—Benz had to carry a spare on a ten-mile trip to be sure of getting home—made the engine-driven magneto attractive, but at that time he could not find or build one reliable enough. The fixed spark gap was difficult to maintain and required a high voltage, so that for moderate speeds a low-voltage make-and-break system that produced a hot arc by mechanically separating contacts in the cylinder proved easier and more reliable in the 1890's. The first adequate solution came from Robert Bosch in 1902—a new kind of high-tension magneto able to deliver a hot spark without any battery and without moving parts in the cylinder. This ignition quickly became standard throughout the world.

The success of the light, fast gasoline engine depended on the slow evolution of essential components like the carburetor and the electric ignition, and on a great deal of practical engineering work on the problems of reducing weight and increasing power. In such developments one can pick out a few key ideas or significant steps forward, but it is a matter of continuous refinement of details rather than glamorous breakthroughs. The automobile engine began in the 1880's as a one-horsepower affair weighing 100 to 200 pounds. By 1900 two Daimler engines of sixteen horsepower each drove the first Zeppelin. The following year the first Mercedes, an elegant Maybach design, a modern car in all essentials, with a thirty-five-horsepower engine weighing only 14 pounds per horsepower, achieved a speed of 53 miles per hour.

WHERE THE ENGINE CAME FROM

The evolution of the internal-combustion engine must not be thought of as an isolated strand in the history of technology, but rather as an inseparable part of the evolving complex of 19th-century science, technology, and industry, in which, for example, the gas industry, the steam engine, and the scientific study of heat were intimately related to the gas engine.

The illuminating-gas industry provided an economic incentive for the development of engines, as well as the techniques of handling gases and the beginnings of an interest in combustion. Gas plants became common in European cities by 1850, and wherever they appeared they stimulated the idea of using this remarkably convenient fuel for power. It was the gas industry that worked out the economic techniques and policies for an energy-distributing system of the kind we now have in our electric power systems, with a central generating station and a network of lines reaching into individual houses where appliances can be plugged in at any time to convert energy on the spot into suitable forms of light, heat, and power. The early gas engine was thought of as part of such a system, an appliance that could drive a machine tool, a pump, or a sewing machine, having exactly the same function in an energy-distributing system as an electric motor now has. The concept of such a system, as we look back on it, may seem somewhat less exciting than the parallel development of the telegraph or the railroad, but it was an economically important idea that stimulated a good deal of creative work, some of which led to the internal-combustion engine.

The steam engine was the most important source of ideas and techniques for the gas engine. Early workers on gas engines were able to draw on a century of experience for their cylinders, pistons, cranks, gears, valves, and fabricating techniques. High speeds, temperatures, and pressures presented plenty of new problems, but the basic structures and materials were transferred without essential change from one type of engine to the other. In the early years of the struggle to get a smooth flow of power out of an explosive fuel, inventors relied heavily on their experience with steam, not all of which was helpful. The idea of the four-stroke cycle, for example, which seems so simple and obvious in retrospect, was delayed for some time because of the mental commitment to steam techniques.

The steam engine also provided the background for the developing science of thermodynamics, which began with the obscure publication of Sadi Carnot's theoretical study of heat-engines in 1824. By 1850 the principles governing the conversion of heat into work were beginning to be clarified, and after 1860 the professional journals carried many scientific studies of heat cycles and engines, from which inventors like Lenoir and Otto could have picked up a suggestion or two. The problems that they saw, however, were essentially practical, not theoretical, and the success of the gas engine was a matter of common-sense engineering rather than systematic application of the theory of heat. Still the gas engine shows the same sort of reciprocal relation between theory and practice as the steam engine does. It became the object of study for people interested in the theory of heat like the steam engine earlier, and efforts to improve its performance provided an incentive for further theoretical work. In the early years science revealed the low efficiencies of engines: it told the engineer how much room for improvement there was.

In the 1870's and 1880's engineers knew a little about thermodynamics, and could measure the efficiency of an engine and compare it with the ideal efficiency defined by theory; but they knew very little about combustion and could not tell precisely what was going on inside the cylinder of a gas engine where heat was being invisibly converted into work at the rate of several complete cycles per second. They had one diagnostic tool—the indicator mechanism designed by James Watt, which gave a graphical record of the fluctuating pressures within the cylinder—which they used as a physician uses the pulse as a basis for a guess about what is happening in the body. The indicator diagram gave good information about pressure and volume in slow engines, but one crucial item, temperature, could be known only by inference, and the uncertainties were so large in the time of Otto, and also somewhat later in the time of Diesel, that the most respected experts could disagree about what was going on in the cylinder. This was a matter of considerable importance because in their basic patents both Otto and Diesel had claimed a special mode of combustion as the essence of their inventions. The need for information of this sort for legal purposes provided the incentive for serious studies of combustion in engines, beginning in the 1880's, and it is interesting to note that these studies eventually showed that both Otto and Diesel were mistaken in their original claims.

THE DIESEL ENGINE

The diesel engine, the new variety of internal-combustion engine burning heavy oil, with very high compression and no ignition system, is an excellent and rare example of the application of pure theory to the development of an engine. It was conceived in the 1890's and eventually became a very important source of power for heavy vehicles, but only after a prolonged period of development.

Rudolf Diesel got the original idea when he was listening to a lecture on the Carnot cycle, the ideal cycle for heat-engines, in his thermodynamics course at the technical university at Munich, and he published a book on the theory of his rational heat-engine, as he called it, before he even started to build the engine. The idea was that much higher efficiencies than those attained by the steam engine or the low-pressure internal-combustion engine known at that time could be achieved in an engine that first drew in plain air, compressed to a very high pressure (Diesel was thinking of compression ratios as high as 250 to 1), and then introduced fuel into the air in such a way that it would be burned during the first part of the power stroke at constant temperature (*isothermal* combustion). Expansion would continue through the rest of the power stroke without loss of heat (*adiabatic* expansion) while the temperature dropped ideally to room temperature. This theoretical engine would have in addition to very high efficiencies (Diesel was thinking of 75 per cent) a number of auxiliary advantages: it could use a cheap fuel, because almost any fuel, even coal dust, it was thought, would burst into flame as soon as it met the extremely hot air inside the cylinder; it needed no ignition system and no carburetor at all; and it would not have to be cooled, Diesel originally thought, because so little heat would be wasted.

Diesel finally had an engine in 1897 that would run in exhibition, but it was not very close to the original ideal. It burned heavy oil with compression ignition, had a compression ratio of about 30 to 1, and achieved an efficiency of approximately 26 per cent, about twice as high as the Otto engines of the time. Diesel was able to sell rights to his engine for a handsome fortune, but only after years of the most painful developmental work did he have a practical reliable engine that would sell. One of the most difficult problems was to find a safe and dependable way of injecting fuel into air under such high pressure. Diesel never achieved the isothermal combustion that he claimed.

ACCOMPLISHMENT

By the end of the century, then, the internal-combustion engine had one mature form, the stationary gas engine, of which Otto's engine of 1876 was the prototype. This species flourished for only about one generation after 1880: as a prime mover it could not compete with steam or the diesel engine, and as a small engine using energy distributed to individual customers from a central generating station it had to yield to the electric motor. But it has an honorable place in the history of technology as the ancestor of two forms of the internal-combustion engine that have turned out to be of great importance: the light gasoline engine, which was a vigorous adolescent in 1900 and now accounts for more than 90 per cent of the installed horsepower in the United States; and the diesel engine, which was a sickly infant in 1900 but now supplies most of the power for our heavy trucks, locomotives, and ships.

41 / Expansion of the Petroleum and Chemical Industries, 1880-1900

ARTHUR M. JOHNSON

The last two decades of the 19th century confirmed the place of big business in the American economy and likewise the nation's place as a world industrial power. A leader in this development was the petroleum industry, dating only from 1859 but by 1890 already large and showing the effects of concentration of ownership in refining and pipeline transportation. By comparison, the domestic chemical industry was still in its infancy, located for the most part along the Atlantic coast, and consisting of small firms. Nevertheless, the introduction of electrochemistry, the spread of scientific chemical knowledge, and the rapidly increasing demand for chemicals to feed growing American industries made it clear by 1900 that a new era of chemical progress was already under way.

AMERICAN CHEMICAL INDUSTRY

The American chemical industry was just beginning its modern development at the end of the 19th century. The "heavy" chemicals such as acids, fertilizer, and soda, all of which were basic to intensive agriculture and industries like petroleum refining, could be imported from Britain in many instances more cheaply than they could be produced here. This was so because the more mature British industrial economy had a large demand for chemicals whose by-products could be used to supply the American market at little or no additional cost. The production of synthetic dyestuffs, perfumes, and later, TNT explosives, involved the distillation of coal tar and was based on organic chemistry. Knowledge in this field was most advanced in Europe, especially in Germany.

In short, although not without chemical talent, the United States in 1880 lacked a large group of well-trained chemists and the other resources to support domestic production of chemicals competitive in quantity, quality, and cost with imports from abroad. By the turn of the century, however, this situation was changing rapidly, and by the end of World War I (1914-18) the American chemical industry was to come into its own. This brief chapter deals with these contrasting yet complementary industries, petroleum and chemicals, as they evolved from 1880 to 1900.

RISE OF THE PETROLEUM INDUSTRY

In large measure the dominant element in the petroleum industry, the Standard Oil Trust, organized in 1882, had shown the way for large-scale organization of American enterprises. Structurally this had involved vertical integration, that is, the combination of producing, transporting, refining, and marketing facilities under common ownership. Standard Oil's pre-eminent position as an integrated concern had been achieved by a carefully conceived business strategy, vigorous competitive tactics, and above all careful attention to detail and money-saving techniques. The last of these elements called for full exploitation of technological advances, particularly in the fields of transportation and refining. Although Standard's primary innovating contributions to American industry were more administrative than technical, it was quick to seek and adopt new technology that would reduce costs.

Petroleum refining to produce kerosene (the major product of the oil industry in the late 19th century because of its use in kerosene lamps) was a comparatively unsophisticated operation. Organic chemistry was in its early stages of development when basic oil-refining techniques were developed; yet by empirical processes American refiners arrived at knowledge that later systematic chemical investigation would verify, explain, and build on.

The basic crude-oil refining process remained unchanged in the latter part of the 19th century. Drawing on the earlier technology used in refining coal oil, it involved the application of heat to cylindrical stills containing crude petroleum, resulting in its fractionation into components ranging from light fractions, such as gasoline, naphtha, and benzine, down the scale to tar residue. To obtain higher proportions of burning oil, destructive distillation or "cracking" of the higher boiling-point crude was employed. By such means the kerosene yield was raised from about 50 per cent to 75 per cent of the crude petroleum put into the refining process. Fractional distillates, that is, the products distilled out during the refining process, were treated with sulfuric acid to deodorize them and improve color; caustic soda was used to carry this process further and also to remove the sulfuric acid and compounds that it had formed. Accordingly, the petroleum industry was a leading consumer of sulfuric acid (about 45 per cent of total domestic production in 1884), one of the chief products of the chemical industry.

THE DEVELOPMENT OF PIPELINES

The early development of the American petroleum industry was greatly influenced by the fact that early oil production was concentrated in the western part of Pennsylvania, where oil had been discovered in 1859. But the market for products refined from crude oil lay in metropolitan areas and at the seaboard which lay hundreds of miles away from the oil fields. Therefore, the costs of transportation bulked large in any systematic appraisal of the best refining locations and in the pricing of petroleum products. Here lay the key to Standard Oil's strategy, which sought to bring order out of a chaotic industry by dominating it.

Initially, oil was collected in the fields by wagons for delivery to railroad loading points, or placed aboard barges which were floated out of the producing area on carefully contrived "freshets" for delivery to downstream points such as Pittsburgh. These methods were expensive and unreliable. Wagons became mired in the muck of the oil fields, and barges often could not be floated, or ended up by crashing into one another or into bridges with an attendant loss of oil and a fire menace. In addition, the barrels which held the oil were frequently more valuable than their contents.

To obviate such problems, various efforts were made to pipe oil short distances, but the inability to make tight joints resulted in leakage problems that were not successfully overcome until 1865. In October of that year Samuel Van Syckel, a pioneer refiner, successfully piped oil on a commercial basis from a flush oil field at Pithole, Pa., to a railroad loading point at Miller Farm, a distance

Fig. 41-1. Early oil wells. The oil was pumped into an open vat and from there into wooden barrels for shipment. (From Figuier, *Les Merveilles de la Science*)

of about six miles. This line consisted of lap-weld, two-inch diameter pipe. Three pumps installed along the route moved oil through the line at the rate of about 80 barrels an hour.

The chief investment in oil pipelines is the pipe, acquiring a right of way, and laying the pipe. Since the chief operating obstacle to be overcome is the friction of oil in contact with the sides of the pipe, movement of full loads gives the lowest operating costs. The gradients of the route and the viscosity ("stickiness") of the oil are, of course, factors in the amount of power that must be applied. The addition of pumps increases a line's capacity, which can also be raised by "looping," or the addition of parallel segments at strategic points. Although a line's capacity increases proportionately more rapidly than the investment required to achieve this increase by using larger diameter pipe, the largest pipe in domestic use at the end of the 19th century was eight-inch (compared to thirty-six-inch today). Accordingly, once one line was operating at full capacity, it was customary to add parallel lines to meet additional demands for service.

Van Syckel's line soon proved that it could move oil reliably and at a much lower cost than alternative means of overland transportation. Therefore, it was quickly and widely imitated. Here, as in other cases of technological advance, progress was accompanied by destruction of some existing vested interests. Despite sabotage by displaced teamsters, the gathering pipeline soon displaced the wagon and barge as a means of collecting and moving oil to a railhead for further shipment.

The railroads, and the Pennsylvania Railroad in particular, were still unchallenged as the transportation medium for long-distance oil movements. They moved oil first in barrels, then in large tanks mounted on flat cars, and finally in tank-car cylinders of the type familiar today. By guaranteeing full trainloads of oil, a Cleveland, Ohio, refining firm headed by John D. Rockefeller (1839-1937) in the late 1860's obtained favorable rates that gave it an advantage over smaller-scale competitors. Soon this efficient firm dominated Cleveland refining; from these beginnings grew the Standard Oil Trust.

This Standard Oil-railroad alliance begun in the 1860's led eventually to a transportation bottleneck for others seeking to reach major centers by rail. One objective of Pennsylvania oil producers, whose competitive drilling produced both oil gluts and famines, became access to railroads outside the alliance. To this end they turned to the development of long-distance pipelines. The first so-called "trunk" pipeline, the Columbia Conduit, was built in 1874, and five years later the Tide Water Pipe Line was built over the Alleghenies to a connection with the Reading Railroad at Williamsport, Pennsylvania. This engineering feat inaugurated a new era of oil transportation.

Standard already possessed the leading pipeline gathering system. By using this system and water transportation to play railroads against one another, Standard had proved (1877) that it could call the tune in oil transportation. After the Tide Water line was built, Rockefeller and his associates, who by now consisted of a number of refiners joined in close alliance, could no longer afford to overlook the economies of long-distance oil movement by pipe rather than rail.

In 1881 Standard pipeliners completed a six-inch line from the oil regions to Bayonne, New Jersey. Another line soon paralleled it, and still others were built to Cleveland, Philadelphia, and Baltimore. On the route between Olean, New York, and Bayonne there were eleven pumping stations, utilizing 800-horsepower Worthington pumps operated by steam power. Movement of 24,000 barrels of oil per day through these lines was co-ordinated by telegraph, and "pipeline walkers" patrolled the route on foot to detect and report leaks.

By making the basic decision to refine products in major centers such as Cleveland and the environs of New York City, Rockefeller and his associates had made transportation a key element of their strategy. By requiring large offers of oil for shipment by petroleum producers, and by charging railroad rates for transportation over their trunk pipelines, they kept outsiders off their lines and at the same time satisfied railroad management, which still had Standard's large shipments of refined products as a major source of revenue. Standard Oil, in short, by capitalizing on a technological innovation had achieved transportation savings that competitors and would-be competitors could not match. Accordingly, independent producers became increasingly dependent on Standard's purchases of their crude oil. By the same token, however, the Standard Oil combination was made dependent on the producers for its crude oil supply, a fact that was later to lead to Standard's decision to acquire its own producing properties.

FORMATION OF THE STANDARD OIL TRUST

The success of the early Standard strategy was signalized in the creation of the Standard Oil Trust in 1882. This organizational innovation, soon copied in other industries, involved turning the securities of a number of companies over to a group of trustees for the leading stockholders.

At the time the Trust was formed, its constituent companies held at least twenty patents for processes and apparatus used in general refining. Numerous other patents covered fabrication and filling of cans used to ship refined products, apparatus used in making barrels, and so on. Although otherwise using established techniques and technology, Standard Oil had an advantage over competitors through these patent rights.

The trustees of the Trust guided overall strategy and financial policy. Within the Trust's organization there was a functional distribution of responsibilities between various committees, headed by an executive committee composed of the trustees who were in attendance on any given day. This organizational device provided the leading American oil concern with centralized direction and operational flexibility until one element of it fell prey to an Ohio antitrust action in 1892. The Trust was then voluntarily dissolved and in 1899 replaced by a holding company, the Standard Oil Company (New Jersey).

THE FRASCH PROCESS

Throughout this period the National Transit Company, a Pennsylvania corporation with wide powers, was the Trust's chief pipeline affiliate. It was National Transit that provided the funds and know-how to organize the Buckeye Pipe Line Company, which was formed to serve the new Lima, Ohio, field in the mid-1880's.

In this area of Ohio and in adjoining Indiana came the first substantial oil strikes outside the industry's original oil fields. Lima oil was so heavily laden with sulfur, however, that ordinary refining processes would not render it suitable for use as an illuminant. Pending solution of this problem, a pipeline was built to the vicinity of Chicago to move Lima oil there for use as fuel.

To render the new crude more marketable, Standard Oil in 1886 employed Herman Frasch, a German-born pharmacist whose knowledge of chemistry had helped him to develop, among other things, a process for refining paraffin wax. By 1888 he had solved the Lima crude problem with a commercially feasible process that involved the mixture of kerosene stock with metallic oxides and then treatment of the product with sulfuric acid and caustic soda. The oxides could be recovered and reused by further steps in the process. The success of this applied research justified Standard's major expenditures on Ohio crude oil and construction of a refinery at Whiting, Indiana, near Chicago. Naturally, this refinery was supplied from the Ohio-Indiana fields by pipeline.

In connection with these developments, Standard Oil for the first time began to engage extensively in producing its own oil. The decision to integrate to the producing level of the industry was in part prompted by the pipeline group who wanted the assurance of full loads for their lines, as well as a means of controlling oil production costs.

COMPETITION FOR THE TRUST

The most successful challenge to Standard Oil prior to 1900 rested on pipeline construction by an independent group of producers and refiners who in 1895, in imitation of the dominant concern, organized the Pure Oil Trust. The key to this operation, which involved exporting as well as domestic sale of refined products, lay in the United States Pipe Line Company, which built parallel four-inch pipelines from the Pennsylvania oil regions to the seaboard. After being forced out of New Jersey by court action initiated by a railroad, the lines were rerouted to Marcus Hook, Pennsylvania, near Philadelphia, where the first oil was delivered in May 1901. One of these lines carried kerosene, another innovation in the use of pipelines, because it involved carrying a refined product rather than the crude petroleum.

These developments, as well as a decline in both crude oil and sulfuric acid prices, encouraged independent refiners. As a result, Standard's commanding position in crude oil, refining capacity, and share of major products sold, was significantly challenged between 1880 and 1900. Nevertheless, in all these categories its share of the market—except in lubricants and waxes—remained well over 80 per cent of national totals. In 1898 the combination still produced about one third of the total national output of crude oil and owned nearly 4000 miles of trunk pipelines and over 10,700 miles of gathering lines.

The specialized lube and wax business, based on high-quality Appalachian crudes, offered a particularly attractive niche for relatively small companies to exploit. Standard's share of these markets fell from 75 per cent or more in 1880 to 50 percent or less in 1899.

It was not until after the turn of the century, when the Texas Gulf coast fields with their ready access to water transportation were discovered and developed, that Standard encountered really serious competition in other products. The new competitors, such as Gulf Oil and the Texas Company, also adopted an integrated structure, both as a defensive measure against Standard and as the most economical form of operation. A new era of competition was inaugurated in the oil business, and this time pipelines became weapons in the hands of companies bent on successfully challenging Standard Oil.

In summary, the trunk pipeline was probably the major technological innovation in the petroleum industry during the last two decades of the 19th century. Though refinements in other areas of the industry reduced costs and improved the quality of petroleum products, the reliable, cheap, and flexible carriage of large volumes of oil by pipeline had no counterpart of equal technological and competitive significance in moving the petroleum industry into the forefront of the new industrial age.

THE CHEMICAL INDUSTRY

In the era 1880-1900 the American chemical industry was just beginning to advance toward the size and diversity that were to characterize it in the 20th century. To a large extent the small domestic industry was built around production of sulfuric acid and alkalis such as caustic soda (important for bleaching), soda ash (important in glass-making), and sodium bicarbonate (used in baking powder, beverages, and so on). At the start of the period, however, the demand for these chemicals was being largely met by imports. Freight rates from Great Britain to the United States were low (shippers could usually count on return cargoes of grain), and the tariff was not, at least until 1897, a significant barrier.

British soda and bleaching powder dominated the American market until the late 1890's. British dominance in these alkalis was based primarily on the application of cheap sulfuric acid to salt (the Le Blanc process, discovered in 1789). The acid itself was produced by the lead-chamber process, which involved burning brimstone (sulfur) or Spanish or Portuguese pyrites (iron sulfide, a yellow mineral sometimes known as "fool's gold," which occurs abundantly as a native ore). The resulting sulfuric acid when applied to salt produced salt cake (sodium sulfate) and, as a by-product, hydrochloric acid. The latter, when appropriately blended with lime, produced bleaching powder. The salt cake roasted with carbon and limestone in a rotary furnace produced soda ash, which, with further treatment produced caustic soda.

SULFURIC ACID

The British could afford to import phosphate rock from the United States, treat it with sulfuric acid, and re-export it to America as fertilizer at a price which was remunerative yet highly competitive with the domestic product. American fertilizer and sulfuric acid producers demanded tariff protection, but without success. Even without it, however, domestic fertilizer manufacture grew significantly in the last half of the 19th century. In 1860 there were only 47 establishments manufacturing complete fertilizers; in 1882 there were 278. The amount of fertilizer produced rose from 727,000 tons in 1880 to 1,898,000 tons in 1890 and 2,800,000 in 1900. Since superphosphate fertilizer manufacture involved the combination of phosphate rock and sulfuric acid in equal quantities, the expansion of fertilizer production was also accompanied by increased sulfuric acid manufacture.

The demand for sulfuric acid was also growing elsewhere. In the 1880's, with an annual average output by the petroleum refiners of more than 29 million barrels, the petroleum industry provided significant markets (Standard Oil was itself a major acid manufacturer). To these demands were added those of the textile, tanning, paint, explosives, and soap industries. In 1884 a foreign observer reported that production of concentrated sulfuric acid was running at an annual rate of about 395,000 tons, with fertilizer manufacture and petroleum refining taking about 90 per cent of it.

The sulfuric acid industry was decentralized and small scale, the average plant producing about 4000 tons annually. Compared to European methods, American sulfuric acid production was inefficient and relatively costly (despite a steady decline in the price of the acid). For example, producers abroad had turned to pyrites before 1840, but United States producers continued burning brimstone until the late 1880's. This was all the more surprising in view of the fact that a number of the American manufacturers had brought their chemical knowledge from overseas. On the other hand, the Census Bureau reported in 1900 that there were only twenty-eight trained chemists in the 127 sulfuric acid plants of the country, and this may help to account for the persistence of traditional methods.

The foreign competition generated at least one result of lasting significance to the American chemical industry—the organization of the Manufacturing Chemists' Association of the United States. This early trade association, still in existence, was formed in Philadelphia in 1873 after the tariff had been removed from sulfuric acid. The association's purpose was to attend to the tariff and other legislative matters and to present the industry's case to the general public.

As in other areas of chemical knowledge, the major advance in sulfuric acid production during this period came from Germany, which had the world's best laboratories and most outstanding scientists. The so-called "contact process" involved the reactions of sulfur dioxide and oxygen in the presence of a catalytic mass, which controlled the velocity of the chemical reaction. The patents to one such German process were obtained by August Hecksher of the New Jersey Zinc Company in 1899, and a subsidiary of that company in 1901 manufactured the first contact acid produced in the United States. However, the process was complex and delicate, and the end-product was a highly concentrated acid that did not come into its own until World War I created a large demand for it.

A contact process for producing sulfuric acid had been developed in this country by J. B. F. Herreshoff, a chemist for the Nichols Chemical Company. When German royalty demands for the use of the new contact process proved large, twelve American sulfuric acid manufacturers combined under the leadership of Dr. William H. Nichols to form the General Chemical Company. Although it used the Herreshoff process, this company became involved in a suit with a German concern holding patents on the contact process and was eventually forced to obtain rights from it. Significantly, however, and in sharp contrast to the contemporary wave of mergers in other industries initiated or promoted by investment bankers, General Chemical was one of the few significant combines in the American chemical industry of this era, and it was the product of the initiative of the merged firms.

ALKALIS

Despite the dominance of British alkalis in the American market, there were successful domestic producers who eventually enabled this country to end its dependence on imports. Among these producers was the Pennsylvania Salt Manufacturing Company (Pennsalt). Organized in 1850 to manufacture lye, it later contracted for the exclusive right to import cryolite (sodium-aluminum fluoride) from Greenland to be used in making alkali products. By using the Thomsen process (named after the Danish thermochemist, Julius Thomsen [1826-1909]) to decompose the cryolite, caustic soda, soda alum, and calcium fluoride were produced. In 1867 Pennsalt produced the first commercial-scale output of soda ash and soda alum in this country and was able to sell these products in successful competition with British imports.

A new technique for producing alkalis by an ammonia-soda process was discovered in 1869 by Ernest Solvay (1838-1922), a Belgian. The Solvay process made soda (sodium carbonate) by treating common salt (sodium chloride) with ammonia and carbon dioxide. Under a license from Solvay, the Solvay Process Company was organized in the United States by Rowland Hazard, a Rhode Island textile manufacturer, in 1881. The company began operations near Syracuse, New York, in 1884. At first, it used brine and then solid salt from the immediate vicinity and ammonia purchased from local gas works. This undertaking proved successful, and the company doubled its soda ash output during 1885-87, added bicarbonate of soda to its line in 1888, and caustic soda in 1889. Such a simple device as a bucket-and-cable arrangement to move salt to the works from nearby sources and later the imaginative use of an eighteen-mile pipeline to tap more distant salt beds helped to reduce production costs. By 1896, the Solvay Company had the capacity to produce 500 tons of soda ash per day, and a new plant was soon opened near Detroit.

When the basic Solvay patents expired, the process became available to whoever wished to use it. Captain J. B. Ford, who reputedly established the first plate-glass factory in America in 1869, took advantage of this. He secured land at Wyandotte, Michigan, under which there were extensive rock salt beds. Using the technical knowledge of English and Scottish chemists, the Ford firm, known as the Michigan Alkali Company, began producing soda ash early in the 1890's. By 1895, it had a daily output of 100 tons. A bicarbonate of soda plant was soon added, and quickly reached a daily output of 50 tons. In 1898 a new company was formed by the Ford family to blend and sell alkalis, produced by Michigan Alkali Company, for special customer uses. Eventually these two concerns were merged (1943) to form the Wyandotte Chemicals Company.

ELECTROLYTIC PROCESSES

Meantime, electrolytic production of alkalis began to challenge the older processes. Both the Le Blanc process, used abroad but not here, and the Solvay process were well adapted to producing soda ash and domestic soda. However, electrolysis of salt solutions results in the direct production of caustic soda and chlorine (used for bleaching, water purification, and so on), which were only by-products under the older processes.

The first electrolytic alkali plant in this country was established at Rumford Falls, Maine, in 1892. Other plants, using somewhat different electrolytic processes soon appeared. One of the most important new firms was the Mathieson Alkali Works, organized in 1892 to manufacture caustic soda and soda ash. Again, English technical knowledge was used; the very name of the company shows this, for the Mathiesons had been active in the English alkali trade to America. Their hand-labor production techniques, however, proved too expensive for use in this country, and the company's operations were soon mechanized.

The Mathieson Company used the Castner electrolytic cell at Saltville, Virginia, in 1896 to produce the first bleaching powder made in this country on a commercial scale from electrolytic chlorine. The "cell" was a concrete box containing an anode and cathode with mercury as an intermediary cathode; salt brine and water were brought together here in the presence of an electric current. The need for low-cost power in this process induced the company to move its electrochemical operation to Niagara Falls, where abundant water power had recently (1894) created cheap electricity and attracted other chemical companies.

At this same time, Herbert H. Dow (1866-1930) was also moving into the electrolytic production of chlorine and caustic soda as a by-product of bromine refined from raw brine at Midland, Michigan. This approach also led Dow (the Dow Chemical Company was incorporated in May 1897) into bleaching-powder production, which expanded rapidly. As a result, the price of bleaching powder fell significantly and foreign imports were quickly reduced.

COAL-TAR DYES

The production of coal tar, the basis of many new chemicals, had been neglected in the United States. Although the steel industry might have provided the tar in connection with its coking operations, it had employed beehive ovens which made no provision for recovery of the tar. Seeking more adequate sources of ammonia than local gasworks, the Solvay Company in 1892 built the first by-product coke ovens in America. A few years later, a new concern, the Semet-Solvay Company, was organized to build and operate this type of oven, which soon found acceptance with blast-furnace men and provided the basis for the domestic coal-tar chemical industry.

By 1902, by-product coke ovens were producing 25 million gallons of coal tar annually, and this output was rising rapidly. The Barrett Manufacturing Company discovered that it could recover ammonium sulfate from the by-product coke ovens and found a market for this chemical in fertilizer manufacture. However, the major use for coal tar was in providing synthetic dyes.

The tariff was of critical importance to the infant American dyestuffs industry. Under a protective policy from 1879 to 1883, nine American firms made dyestuffs from coal tar. However, when the tariff was lowered in 1883, the Germans took over, for they were much farther advanced in making better and cheaper coal-tar dyes. As a result, in 1899 only $1.3 million worth of synthetic dyestuffs was made in the United States.

EXPLOSIVES

A new demand for coal tar was created by the explosives industry. A coal tar derivative, toluol, when mixed with nitric acid produced TNT, an explosive just coming into use at the turn of the century. The leading American explosives manufacturer was E. I. duPont de Nemours and Company, established on the banks of Brandywine Creek near Wilmington, Delaware, in 1802 by French refugees. This company had made significant contributions to the Union Army's firepower during the Civil War. After the war the black powder market was demoralized as the result of the wartime influx of a number of small producers

Fig. 41-2. "Atlas" dynamite-mixing machine, which Lammot duPont helped to develop. (*The Hagley Museum*, Wilmington, Delaware)

and a large stock of surplus government powder. To meet this problem, General Henry duPont led the industry into the Gunpowder Trade Association, whose objective was to bring order to the market by fixing prices and using pressure against non-members who threatened to upset the association's arrangements. This device proved quite effective until the black powder industry, like many others, was hard hit by the Panic of 1873.

In the aftermath of this financial panic, duPont acquired complete or partial control of numerous firms. By 1880 the duPonts confronted only one major rival, the Laflin and Rand Powder Company. At the urging of Lammot duPont, the two companies joined forces to organize the Repauno Chemical Company for the manufacture of dynamite, a high-powered explosive discovered by Alfred Nobel (1833-96) in 1866, which was well suited to many construction needs in a raw and expanding country.

The production of dynamite eventually came to combine nitroglycerin and an activating base such as cotton fiber or sawdust impregnated with sodium nitrate solution. Nitroglycerin, the key component of the new explosive, was first made commercially in this country in 1868. It combined sulfuric and nitric acids with glycerin, a by-product of soap-making. To reduce the hazards of working with such a dangerous mixture, Lammot duPont, who later was himself killed in an 1884 explosion at the Repauno plant, helped to develop a power-driven wheel mixer which replaced the rakes and shovels then used in mixing. The demand for dynamite rose rapidly and was met by an expansion of manufacturing facilities by a number of firms. By 1900, there were reportedly thirty-one companies in the business, producing dynamite valued at $8.2 million; twenty years earlier there had been only two firms and sales of less than $750,000.

Another important development in explosives during this era was smokeless powder. Paul Vielle, a French engineer, discovered in 1884 that trinitro-cellulose could be colloided with alcohol and ether and extruded in a macaroni machine. When it hardened, it could be handled safely, and the size and length of the resulting cylindrical rods gave a means of controlling the timing and strength of detonation. DuPont was among the early firms to move into smokeless powder production, building a laboratory and guncotton plant at Carney Point, New Jersey, in the early 1890's. Smokeless powder manufactured there was put on the market in 1894.

The duPont family concern was incorporated in Delaware in 1899, and nearly passed into other hands in 1902 following the death of its president Eugene duPont. However, Pierre, Coleman, and Alfred duPont retained the company and acquired shortly thereafter Laflin and Rand. The old Gunpowder Trade Association was dissolved, and duPont, recognizing the extent of past dependence on European discoveries, began to engage in extensive research of its own in connection with explosives. This gradually, as in the case of cellulose, led them into other fields.

BACKGROUND FOR FUTURE DEVELOPMENT

While the petroleum and chemical industries have become closely linked in the 20th century, this relationship was only an embryonic one in the last two decades of the last century. As independent oilmen, who were otherwise highly critical of Standard Oil, admitted, the oil combination possessed fine laboratories, and it was Rockefeller who hired William Burton, the first chemist to become head of an oil company. However, it was only with the shift of post-1900 demands from kerosene to gasoline, and the accompanying need for new and improved refining techniques, that the chemist's contributions to the petroleum industry began to mount.

In 1900 there were only about 5000 chemists in the country, and as late as 1915 the American Chemical Society had less than 7500 members. The engineering of unit operations to put the fruits of chemical research to industrial use did not formally gain professional status until 1908, when the American Institute of Chemical Engineers was organized; only twenty years before the Massachusetts Institute of Technology had inaugurated the first course in this field.

It is worth noting that these events were contemporaneous with the beginnings of the scientific management movement and professional education for business—further evidence that the United States was entering a new industrial era. The petroleum industry, dominated by a single large combination, and a leader in this movement, was only on the threshold of its greatest development. The chemical industry, destined to become its close partner in this century, was just starting to grow in response to the new industrial needs. Even by 1900 it was beginning to free this country from dependence on some important chemicals. World War I, cutting us off from the German chemical industry and at the same time increasing greatly the demand for chemicals, provided the impetus for large-scale growth of the industry in America. Nevertheless, it would be a mistake to overlook the domestic chemical progress made prior to 1900. In both the petroleum and chemical industries the seeds of rapid and significant change had been planted between 1880 and the turn of the century.

42 / Mass Production For Mass Consumption

HAROLD F. WILLIAMSON

Early in the 20th century the well-known British economist, Alfred Marshall, in his book, *Industry and Trade,* called attention to what he considered the distinguishing, if not unique, features of industrial development in the United States. In a passage contrasting France and the United States, two countries "on the opposite sides of the broad field of industry," he stated, "French instincts enable the hand and eye to make subtle distinction, and ceaseless variations in form and color; and thus to gratify the fancy and artistic taste, [but only] at prices which are generally beyond the reach of the masses of the people. American methods on the other hand make for the production of business plant and of products for immediate consumption in an almost infinite variety of standardized forms."

The first stages of the American system of mass production for mass consumption that caught Marshall's eye can be traced back in American history to the late 18th century. Their subsequent evolution during the 19th century was the result of advances in two major areas of American economic and social development. One was progress in invention and technology which affected the rate at which improved machinery and equipment became available for production. The second was the growth and characteristics of a market that provided an outlet for an expanding volume and variety of standardized products.

INNOVATION AND TECHNOLOGY

Broadly considered, the evolution of mass-production methods in the United States was featured by increasing division of labor and specialization, the advantages of which had been outlined in the famous description of a pin factory by Adam Smith in his classic work, *The Wealth of Nations* (1776). Indeed, because of a relative scarcity of labor in the New World, compared to the situation in Great Britain or on the Continent, American entrepreneurs were quick to appreciate how the close supervision of a labor force under the factory system could reduce costs by improving the efficiency of the worker. Even more attractive in the American environment was the possibility, under the factory form of industrial organization, of reducing costs still further by the use of power-driven machinery.

Increasing mechanization of the labor force, in fact, became a central feature of the subsequent evolution of manufacturing methods in the United States. The characteristics, however, that served most to distinguish American production methods from those used elsewhere in the world, were the early adoption of: interchangeable parts manufacture; the orderly flow of materials through the various stages of the production process; and what has been termed "the culminating achievement of mass production," the combination of these two elements by linking interchangeable parts manufacture with the assembly line.

The major invention and technical innovations contributing to the development of mass production methods in the United States from the late 18th century to the close of the 19th century are described elsewhere in this volume. In brief they included the following:

1. Advances in the production of metals, particularly iron and steel, that increased the availability and improved the quality of materials used both in the construction of machinery and equipment and fabrication of consumer goods.
2. The development of a machine-tool industry which supplied machine tools adapted to the requirements of increasingly complex manufacturing operations.
3. The introduction and improvement in precision instruments necessary to achieve greater accuracy and standardization in manufacturing.
4. The development of more efficient and flexible methods of generating energy, notably the high-pressure steam engine and the complex of electric generation stations, power lines, and electric motor. This development not only relieved manufacturers of the necessity of locating their plants at water-power sites, but made coal the principal source of energy for industrial production.

How these advances affected both the general growth and labor productivity of the manufacturing sector of the American economy between 1859 and 1899 is suggested by the data shown in Table 1. It is noteworthy, for example, that while the number of workers expanded approximately 300 per cent during these

Fig. 42-1. The increase of mechanization is shown in this view of the main floor of a wire insulating factory. (*The Bettmann Archive*)

years, total value added by manufacturers rose over 500 per cent, resulting in a 167 per cent increase in the average value added per worker. Much of the improvement in labor productivity was in turn a result of greater mechanization and the growing use of power-driven machinery and equipment. This trend is revealed in Table 2 by the rise in horsepower installations per worker from 122 in 1859 to 189 in 1899 and the increase in capital (that is, the amount invested in plant and machinery) per worker from $995 in 1879 (when data first became available) to $1699 three decades later.

THE INTRODUCTION OF MASS-PRODUCTION METHODS

As might be expected, there was a considerable variation in the extent to which mass-production methods were utilized by various manufacturing industries during the last two decades of the 19th century. This was largely because in the actual historical process, the conditions that made mass production possible did not appear together and all at once. Thus, their influence was gradual and cumu-

Table 1. U. S. manufactures: establishments, workers, and value added

YEAR	1859	1869	1879	1889	1899
Number of Establishments (in thousands)	140	252	254	354	510
Number of Workers (in thousands)	1,311	2,054	2,733	4,129	5,098
Value Added (in thousands)	$854,257	$1,395,119	$1,972,756	$4,102,301	$5,474,892
Value added per Worker	$652	$679	$722	$994	$1,074

SOURCE: U. S. Bureau of Census, *Historical Statistics of the United States*, p. 409.

Table 2. U. S. manufacturers: capital and horsepower installations, totals and per worker

YEAR	1859	1869	1879	1889	1899
Workers (in thousands)	1,311	2,054	2,733	4,129	5,098
Horsepower[1] (in thousands)	1,600	2,346	3,664	6,308	9,633
Average Horsepower per Worker	122	114	134	153	189
Capital[2] (in millions)	NA	NA	$2,718	$5,697	$8,663
Average Capital per Worker	NA	NA	$995	$1,380	$1,699

[1] Installed horsepower.
[2] Book value, current dollars.

SOURCE: U. S. Bureau of the Census, *Historical Statistics of the United States*, pp. 410, 506.

lative rather than immediate. Moreover, while the progress of mass production tended to generate its own favorable conditions, .it also posed new problems and difficulties as well.

A brief historical sketch of a selected group of consumer-goods industries that had adopted mass-production methods by or during the 1880-1900 period will illustrate the nature of this process. Each, it may be noted, turned out standardized, finished products in quantity by a connecting series of mechanized operations. In some instances, however, production involved interchangeable parts manufacture; in others it did not. Included in the latter group were meat-packing and the manufacturers of ready-to-wear items, such as clothing and shoes.

MEAT-PACKING

Prior to the introduction of refrigeration during the 1860's, packing operations in the United States were largely conditioned by the fact that fresh meat deteriorates rapidly when exposed to temperatures above 40°F. As a result, slaughtering tended to be concentrated during the winter months; only meat that was canned or otherwise preserved by salting or smoking could be stored or shipped any distance to market. Shipping difficulties not only restricted the geographic area within which fresh meat could be distributed, but also kept the size of packing plants relatively small and made it necessary to drive or ship animals to consuming centers for slaughtering.

By equipping their plants with ice-cooled refrigerator storage chambers, beginning early in the 1860's, packers were able to conduct their operations more nearly on a year-round basis. It was not until the 1870's, however, that shipping limitations were removed on fresh meat by the introduction of the refrigerator railroad freight car and the refrigerator ship. The result was a virtual revolution of the industry. It was now feasible to distribute fresh meat throughout the United States and to overseas markets as well. Meat-packing became increasingly concentrated at locations, such as Chicago, close to livestock growing or feeding areas. With expanding sales, it was further possible to achieve economies of scale by increasing the size of the business units within the industry. This trend is illustrated by the reported net worth of Armour & Company which grew from $200,000 in 1870 to $2.5 million in 1880, and to $10.5 million in 1890.

By the 1890's the major packing-houses were highly mechanized and their operations were marked by an extensive division of labor. For example, as beef carcasses were moved through the various processing stages by a power-driven mechanism, each workman engaged in removing the hide cut only a certain portion, and while the amount done by each was quite small, the quickness with which the work was accomplished made the process highly efficient. Hogs were likewise put through one important processing stage at the rate of twenty per minute.

The impressive growth of the slaughtering and meat-packing industry in the United States between 1860 and 1900 is indicated in Table 3. Especially noteworthy was the increase during this period in the capital employed in the industry from approximately $10 million to over $189 million, and the rise in the value of the industry's output from about $29 million to over $785 million.

Fig. 42-2. The cooking-room in a large Chicago packing-house. (*The Bettmann Archive*)

READY-TO-WEAR ITEMS: CLOTHING

The initial effort to mechanize American industries had come in the production of textiles, just as it had in England. As a result, cotton manufacturing was the first industry in the United States to operate under the factory system. Well before 1800, hand-spun cotton yarn had given way to the product of power-driven spindles, and by 1820 practically all the cotton cloth manufactured in the country was woven on power looms. The spinning and weaving of wool were somewhat slower to move out of the household into the factory, but on the eve of the Civil War the great bulk of woolen fabrics were also machine-made.

However, the production of clothing remained almost entirely a hand operation until the introduction of the sewing machine in 1846. During the 1850's close to 100,000 sewing machines were sold in the United States. They not only enabled the housewife to produce extra garments for the entire family, but provided the basis for a sharp expansion in the output of ready-to-wear clothing previously available on a small scale from tailoring shops.

The factory production of clothing, particularly for men, received a major stimulus as the result of the demand for uniforms during the Civil War. But what can best be described as the "really phenomenal growth of the men's clothing industry" began after the close of the war. Soldiers returning from the war were accustomed to overlooking minor defects in the fit of a garment, while the price advantage of the ready-made suit over its tailor-made competition was large.

Table 3. U. S. slaughtering and meat-packing industry, 1860-1900

YEAR	1860	1870	1880	1890	1900
Number of Establishments	259	768	872	1,118	921
Wage Earners[1]	5,058	8,366	27,297	43,975	68,534
Capital[2] (in thousands)	$10,158	$24,225	$49,419	$116,888	$189,198
Cost of Materials (in thousands)	$23,564	$61,674	$267,739	$480,962	$683,584
Value of Product (in thousands)	$29,442	$75,827	$303,562	$561,612	$785,562

[1] Average annual number.
[2] Book value, current dollars.

SOURCE: Twelfth Census of the United States, 1900, *Manufacturers,* Volume IX, Part III, p. 387.

The subsequent introduction of various mechanical improvements, such as improved cutting machines, button-sewing machines, and machines for barring the corners of pockets opened the way for successful large factories.

The general growth in the output of factory-made clothing between 1860 and 1900 is indicated in the following tabulation (in millions of dollars):

YEAR	MEN'S CLOTHING	WOMEN'S CLOTHING	TOTAL
1860	$ 80.8	$ 7.2	$ 88.0
1870	$148.7	$ 12.9	$161.6
1880	$209.6	$ 32.0	$241.6
1890	$251.0	$ 68.2	$319.2
1900	$276.9	$159.3	$436.2

SOURCE: Twelfth Census of the United States, 1900, *Manufacturers,* Volume IX, Part III, pp. 261, 283.

Largely because of greater variations in style, factory production of women's ready-to-wear clothing tended to lag behind the output of men's clothing during this period. But as far as men's wear was concerned, well before the close of the century, machine production had made it possible for the farmer or the small-town merchant to dress as well as the city banker. His suit was of the same pattern, and was cut in the same style, and only the closest scrutiny could reveal the difference in texture or in the refinements of tailoring.

BOOTS AND SHOES

Although still involving a great number of hand operations, the factory production of boots and shoes, like men's clothing, expanded rapidly in response to the demand for footwear during the Civil War. The subsequent adoption of mass-production methods in the industry was made possible by a series of innovations and technological advances that permitted an increasing use of power-driven machinery.

These included the McKay sole-sewing machine, perfected about 1864, which cut the cost of sewing soles to uppers to approximately 3 cents a pair, compared to a hand-sewing cost of 75 cents a pair. The McKay stitcher was applicable, however, only to coarse, heavy shoes that were stiff and unwieldy. This problem was largely overcome by the development of the Goodyear welt machine, which avoided seams inside the shoe by attaching the upper to the sole by a welt or strip of leather, sewed to the upper and the sole respectively by parallel seams, the seam through the sole lying outside the shoe. By the early 1880's some 3 million pairs of shoes annually were being produced by this method, and by 1895, the number had risen to 25 million pairs.

By the latter date, other machines designed for rolling soles, cutting out soles and heels, attaching heels to shoes, trimming and polishing, and lasting had been adopted. As the Census report on the boot and shoe industry for 1900 noted, "The shoe factory of today provides a perfect system of continuous manufacturing involving in some instances more than 100 operations."

These improvements contributed to a growth in the value of the products of the industry from $166 million in 1880 to $221 million in 1890, and to $261 million in 1900. More important from the standpoint of consumers was that well before the turn of the century, good machine-made shoes could be bought for $5 per pair and in some instances as low as $2, compared to custom-made shoes costing $15 to $16 per pair.

ASSEMBLY LINE PRODUCTS

In addition to the growing volume of mass-produced foodstuffs and ready-to-wear items, the closing decades of the 19th century were marked by an increasing availability of consumer goods involving interchangeable parts manufacture. A part of this expansion came from industries which had successfully adopted this method of production prior to 1860, notably the manufacture of firearms, clocks, and sewing machines. In the generation following the Civil War, other products based on the machine-tool technology of interchangeable parts manufacture that achieved prominence included the manufacture of watches and bicycles. Both were relatively complex metal mechanisms and both could be most efficiently produced by assembling uniform parts.

WATCHES

More than a decade of experimentation with automatic machinery preceded the introduction early in the 1860's of what was described as the "American system" of watch manufacturing. This new refinement of interchangeable manufacture presented many difficulties, partly because of the delicacy and accuracy of the machinery required, and partly because watch-makers and repairers were handicraftsmen and unfamiliar with the engineering trades.

As had been true of other products, the Civil War provided a new and seemingly unlimited market for watches, for apparently every soldier in the army wanted one. As a result watch-making was well established on a factory basis by the end of the war. By the mid 1870's, the output of domestic manufacturers had largely supplanted the cheap and middle-priced watches previously imported into the United States from England and Switzerland, and American-made watches were being exported in increasing numbers. The relative efficiency of the factory system of watch-making over the handicraft methods is suggested by one report which indicated that by 1878 the average annual output of watches per worker in the United States was 150 compared with 50 for Swiss workmen.

The most distinctly American enterprise in the industry began in 1879 when a factory was built at Waterbury, Connecticut, to produce an inexpensive watch with only 58 parts and no jewels. This Waterbury watch, the joy of the small boy of the 1880's, was the pioneer of the ultra-democratic timepieces that soon became so characteristic a product of industry in the United States.

BICYCLES

Developed initially in Great Britain, American production of bicycles was started in 1878 by Colonel Albert Pope at a factory established for this purpose at Hartford, Connecticut. In less than two decades, the original high wheel models with solid rubber tires and chain gears had been replaced by the modern type of vehicle, equipped with pneumatic tires, bevelled gears, ball bearings, and coaster brakes. The industry, which had not been regarded as important enough to warrant a separate category in the Census of 1890, was by 1900 employing some 70,000 workers, and in that year turned out over one million bicycles valued at the factories at over $22 million.

Bicycle factories became a special object of interest, both for their machinery and for their high degree of standardization. The bicycle also attracted a large and devoted public estimated at 4 million persons in 1900. It appears that everyone who could afford one bought a bicycle—doctors, lawyers, bankers, clerks, clergymen, college professors, and even the ladies. Indeed, the bicycle, along

Fig. 42-3. High wheel bicycle, *c.* 1880.
(*The State Historical Society of Wisconsin*)

with the typewriter, served to free women from the age-old restrictions of home life.

Bicycle riders began early to agitate for construction of better roads and in 1887 founded the League of American Wheelmen which launched a nation-wide campaign for improved highways. As the Census report on the industry for 1900 stated, "It is safe to say that few articles ever used by man have created so great a revolution in social conditions as the bicycle."

MASS MARKETS AND MASS CONSUMPTION

The growth of the manufacturing sector of the American economy from 1869 to 1899 would have been impossible without a corresponding expansion of the market which, as Adam Smith so explicitly stated, was a necessary requirement for the extension of the principle of division of labor and specialization.

One major factor which contributed to the wider distribution of manufactured products was the increase in the population of the United States from just under

40 million in 1870 to nearly 77 million in 1900, as shown in Table 4. Even more significant than this near doubling of the population was an approximate fourfold growth in national income, which raised average per-capita incomes—in constant dollars—from $237 in 1870 to $480 in 1900.

What proved to be especially congenial to the growth of mass distribution and mass consumption, however, was the emergence of a vast, homogeneous domestic market within the continental United States. As Andrew Carnegie pointed out in 1902 to an audience at St. Andrews, Scotland, "The American [manufacturer] has a constantly expanding home demand, urging him to extensions and justifying costly improvements and the adoption of new processes. He also has a continent under one government. In short, it is free, unrestricted trade in everything under the same conditions, same laws, same flag, and free markets everywhere." This lack of artificial barriers to domestic trade was in large part the result of the provisions in the Constitution of the United States (Article 1, Sections 8 and 9) which gave Congress the sole right to regulate interstate commerce and prohibited the states from imposing import or export duties on such trade.

Table 4. U. S. population, national income and income per capita, 1870-1900

	1870	1880	1890	1900
Population (in millions)	39.9	50.2	63.5	76.9
Per Cent Urban[1]	25%	28.0%	35.5%	39.5%
National Income[2] (in millions)	$8,995	$15,183	$23,675	$36,557
National Income[2] (per capita)	$237	$309	$383	$482

[1] Living in cities with 2500 or more population.
[2] Constant, 1926 dollars.

SOURCE: U. S. Bureau of the Census, *Historical Statistics of the United States,* pp. 7, 14; National Industrial Conference Board, *The Economic Almanac,* 1956, p. 423.

No less important than the law in fostering a large common domestic market were the improvements and extensions of internal transport facilities. Until the mid-19th century, shipments by rivers and canals were the principal means of supplying consumers west of the Alleghenies with products manufactured largely on the East coast. The subsequent expansion of the railroads from about 9000 miles in 1850 to over 200,000 miles in 1900 provided a far-flung transportation network linking consumers and producers hundreds and even thousands of miles apart. Transport barriers to extensive internal trade were further removed beginning in the 1870's by intensive competition for traffic between the major trunk line carriers and a consequent reduction in the rates on long-distance shipments.

The final and perhaps the most distinctive feature of the domestic market to affect the spread of mass production was the willingness of large groups in the United States to purchase identical or similar commodities. The early acceptance of standardized products was partly the result of the scarcity of craft skills in a new country. Throughout the colonial period and beyond, the products of skilled workmen figured prominently among imports, only to fall subsequently an easy victim to the forces of mechanization and mass production which, by reducing costs and prices, also added many new commodities to the expanding list of consumer goods.

Basically, however, standardized mass consumption was compatible with the class structure and concepts of democratic egalitarianism that were typically American in character. The United States did not, of course, develop a "classless society" by any Marxian interpretation of the term. But the main basis for class status in Europe, such as positions with the church, the government, the military, or membership in the nobility, had relatively little impact on the American class structure. As James Bryce, in his book, *The American Commonwealth,* observed,

> One thing . . . may be asserted with confidence. There is no rank in America, that is to say, no external and recognized stamp, marking one man as entitled to any social privileges, or to deference and respect from others. No man is entitled to think himself better than his fellows, or to expect any exceptional consideration to be shown by them to him.

What did emerge in the New World setting was a society marked by no sharp lines of class distinction, with a high degree of social mobility, and a deeply rooted culture that not only accepted, but put a premium on "economic achievement" as a means of acquiring social status.

This emphasis on economic success was almost inevitably reflected in attitudes toward standards of living. Not only was there a wide acceptance of the idea that it was appropriate for individuals, insofar as their purchasing power permitted, to improve their living standards, but consumption patterns themselves became important indications of social status. Lord Bryce, for example, was also impressed with the fact that in the United States,

> The class of persons who are passably well off but not rich, a class corresponding in point of income to the lower middle class of England or France, but superior in manners, is much larger than in the great countries of Europe. Between the houses, the dress, and the way of life of these persons, and those of richer sort, there is less difference than in Europe. The very rich do not (except in a few places) make an ostentatious display of their wealth.

The drive to emulate the consumption standards of the native-born population also served as a powerful instrument of "Americanization." Successive waves of immigrants drawn from widely diversified racial and cultural backgrounds reinforced their common conceptions of equality and opportunity by taking on at least the superficial similarity of common tastes, which could be satisfied by the products of mass production.

Thus a broad division of society according to income groups emerged that in itself provided the basis for expanding markets for a growing variety of standardized products. This development, already well under way by the 1880's, was in turn closely related to the growing significance of the role of marketing and distribution in the American economy and the general trend toward urbanization.

MARKETING AND DISTRIBUTION

Distributing and marketing functions had already begun to take on added significance by mid-century, when the major portion of American manufactures was being sold beyond local markets. In addition to increasing the general importance of the distributing function, the subsequent growth and geographic expansion of markets in the post-Civil War period also tended to encourage greater division of labor on the part of those engaged in areas of marketing and distribution. This was noticeable, for example, in the decline of the jobbers (wholesale distributors) who handled staples and specialties alike and a sharp growth in the number of wholesalers specializing in particular lines.

The trend toward greater specialization of functions was further evident in new types of retail distribution that grew in importance during the post-Civil War decades, specifically department stores, chain stores, and mail-order houses. One feature common to all three of the new forms of retailing was the separation of policy determination and top-level management from clerking, with management increasingly assuming the function of jobbers by purchasing in large quantities from wholesalers, wholesale brokers, or producers, and by either undertaking or arranging for storage, transportation, grading, repackaging in smaller lots, and other functions previously considered the province of the jobber.

Under the circumstances, department stores operating principally in the larger urban centers, could attract customers by offering a wide variety of merchandise in one establishment at relatively low fixed prices and by emphasizing such services as delivery, credit, and in some instances, the convenience of lunch rooms. As the name suggests, this type of retail establishment was organized on the principle of arranging both the store plan and store accounting by departments. This arrangement permitted buyers and sales clerks in each department to concentrate on one line of merchandise while at the same time each line of goods could be marketed under the advantage of specialists' services in such activities as advertising or display.

Chain store organizations, specializing in various lines of merchandise such as groceries, drugs, household supplies, and hardware, also achieved advantages

Fig. 42-4. Mail-order room of Sears, Roebuck & Company, Chicago, *c.* 1910. (*The State Historical Society of Wisconsin*)

of specialization and economies of scale that were passed on to customers in the form of reduced prices and better quality merchandise. Chain stores emerged initially in large urban centers but, to a much greater extent than department stores, became popular in medium and smaller-sized cities and towns.

The advent of the mail-order house was of particular interest to customers living outside urban centers in the United States. By purchasing in large quantities, often directly from manufacturers, and shipping direct to customers, the mail-order houses cut distribution costs and prices substantially below those offered by the small general country stores, long the only retail outlets available to the rural population.

The growth of competitive advertising was a parallel and logical concomitant of large-scale distribution. This type of advertising increased during the latter part of the 19th century as the markets in which manufacturers were selling became progressively impersonal. Along with the growing use of convenient and distinctive types of packages, advertising served to introduce new products and to build up favorable impressions in the minds of consumers for particular brands of trademarks.

The trend toward urbanization—reflected in a shift in the proportion of the American population living in towns of 2500 or over from about 25 per cent in 1870 to almost 40 per cent in 1900—was an additional factor in facilitating large-scale distribution of standardized commodities. It is true that country stores, mail-order catalogues, and newspaper and periodical advertising tended, over a period of time, to stimulate sales of both old and new products in rural areas. But the mere geographic concentration of groups in urban communities with approximately the same tastes (and incomes) was frequently sufficient to warrant the production and marketing of large quantities of differentiated products. Moreover, urban communities provided an especially favorable environment for the operation of a strong demonstration effect which allowed one to see the products purchased by others and tended to promote greater uniformity of consumption expenditures within each of the various social classes.

SUMMARY

The continued interaction of two sets of forces, already discernible by the beginning of the 19th century, brought about what are perhaps the most distinguishing features of the present-day organization of the industrial sector of the American economy. One was the pattern of resources that put a premium on the development of labor-saving machinery and equipment associated with the spread of methods of mass production. The second reflected the influence of a rapidly expanding population with rising per-capita incomes and a strong desire to improve living standards, in a society loosely stratified socially according to economic status but with a considerable degree of uniformity in the consumption patterns acceptable at any particular time to various groups within each social class. Together these forces provided the setting that according to Alfred Marshall enabled giant business in the United States, which sent large consignments to distant middlemen, to obtain "nearly the full advantages of specialized massive production in many varieties."

43 / Imperialism and Technology

RONDO CAMERON

Imperialism, like technology, is as old as civilization. The earliest civilizations of which we have any record, those of Egypt and Sumer, were societies in which native populations were ruled and exploited by alien conquerors. A large part of all recorded history deals with the rise and fall of empires. Vast well-organized empires existed in Asia when half-naked savages still populated most of Europe. Empires existed, after a fashion, in Africa and America before the voyages of Columbus. People of European culture, however, have been among the most expansionist and imperialistically inclined of all peoples, from the time of their Greek and Roman forebears down to the 20th century.

In the Middle Ages the Europeans expanded throughout continental Europe and into the Near East. Between 1500 and 1800 they spread their civilization to North and South America and established beachheads around Africa and the Indian Ocean. The Napoleonic Wars represented the last of the great colonial struggles of that era. The outcome eliminated France as a rival in the traditional areas of imperial conflict, brought some more or less fortuitous additions to the British Empire, and left Britain the undisputed "mistress of the seas." Shortly afterward, the revolt of the Spanish colonies in America and the separation of Brazil from Portugal reduced the formal hegemony of European powers in overseas areas to its smallest extent in 300 years.

For two generations thereafter, colonial and imperial questions receded in public interest. Continental countries focused their attention on problems at home: reform and revolution, national unification, industrialization, and European power politics. With minor exceptions these left little energy for or interest in the creation of colonial empires. In the mid-19th century even Britain went through a period of anti-imperialism. Then rather suddenly, in the last two decades or so of the 19th century, European nations inaugurated an explosive race for overseas empires that resulted in virtually the whole of the inhabited world's being either subjected to European rule or opened to the subtler influence of European cultural penetration.

It is sometimes asserted that the rapid progress of Western technology in the 19th century was a major determinant of the imperialist drive. That steamships and modern weapons, among other elements of Western technology, facilitated colonial conquests cannot be doubted. Western superiority in ships, navigational techniques, and firearms was a fact of long standing, however. It cannot be used to explain the burst of expansion at the end of the 19th century after almost a century during which Europeans showed little interest in overseas dominion. Western technological superiority is far more relevant in explaining the timing of imperialism in the 16th and 17th centuries than in the 19th. In the early 16th century it enabled a handful of Spaniards to bring down the great Inca and Aztec civilizations of Peru and Mexico. In the 17th century it made possible the establishment of an aggressive and expansionist white civilization in North America. The steamships, telegraph, and Gatling guns of the 19th century merely made the technological gap even more fatal to those who lacked the hardware of modern civilization. Among all the peoples subjected to Western imperialistic encroachments, only the Japanese were both willing and able to adopt Western technology, turn it against their tormentors, and become imperialistic themselves. For the rest, the story was quite different.

In short, technology explains *how* Europeans imposed their wills upon the black-, brown-, and yellow-skinned peoples of the world, rather than *why* they thought it necessary to do so at all.

THE OPENING OF ASIA

Although parts of Asia had been opened to European influence and conquest since the beginning of the 16th century, much of it remained in isolation. In the first half of the 19th century Britain controlled India and some of its surrounding territories; the Dutch held most of the islands of the East Indies; and Spain retained the Philippines. The French and Portuguese maintained small trading settlements on the Indian coasts. The vast and ancient empire of China, however, as well as Japan, Korea, and the principalities of Southeast Asia, attempted to remain aloof from Western civilization, which they regarded as inferior to their own. They refused to accept Western diplomatic representatives, tried to exclude or subsequently persecuted Christian missionaries, and allowed only a trickle of commerce with the West.

Russian frontiersmen began to explore the great wilderness of Siberia about the time of the first English settlements in North America. In 1637 they reached the Pacific coast, and by the end of the 17th century they were encroaching on Chinese borders. The next main thrust of Russian expansion in Asia, in the 18th and early 19th centuries, was directed toward the warmer lands to the south, and was closely connected with the desire to secure ice-free ports. This brought Russia into conflict in the Near East with the Ottoman Empire, Persia, and Afghanistan, and created tense diplomatic relations with the British, who were concerned about the Indian "lifeline" and frontiers. Meanwhile the Russians had been pursuing aggressive policies in East Asia as well. In 1850 they established a settlement on the Amur River, on the border of the Chinese Empire, and shortly afterward extracted from the Chinese two treaties which extended Russian territory to the shores of the Sea of Japan. Toward the end of the century their intervention in the Sino-Japanese War and in Korea, their lease of the Port Arthur naval base on the Yellow Sea, and their occupation of Manchuria brought them into conflict with the by now equally aggressive Japanese.

The Chinese Empire in the 19th century was ruled by the Manchu dynasty (1644-1912), the last in a long succession of dynasties which had risen and fallen in China for more than 3000 years. The Manchu dynasty itself was about 200 years old, and had already begun to show signs of decrepitude before the Europeans intervened to hasten its demise. Chinese tea and silks found a ready market in Europe, but Europe in exchange could offer little that appealed to the Chinese until British traders discovered that they had a marked taste for opium. The Chinese government forbade its importation, but the trade flourished by means of smugglers and the corruption of the customs officials. When one honest official at Canton seized and burned a large shipment of opium in 1839, British forces on the spot disregarded instructions from London and took punitive action against the Chinese. After the brief "Opium War" (1839-42) China ceded to Britain the island of Hongkong, agreed to open five more ports to trade under consular supervision, established a uniform 5 per cent import tariff, and paid a substantial indemnity. Moreover, the opium trade continued.

The ease with which the British forced the Chinese to knuckle under encouraged other nations to seek and obtain equally favorable treaties. Such a show of weakness by the Chinese government provoked the Taiping Rebellion (1850-64), which was both anti-government and anti-foreign. Government forces eventually defeated the rebels, but the general lawlessness that prevailed in the meantime gave Western powers another excuse for intervention. In 1857-58 a joint Anglo-French force occupied a number of principal cities and extorted further concessions in which the United States and Russia also participated. China's history for the remainder of the 19th century followed a depressingly similar pattern. Concessions to foreigners led to fresh outbreaks of anti-foreign violences and lawlessness, which led in turn to further reprisals and concessions. In the end, China avoided complete partition by the great powers only by virtue of great power rivalry. Instead of outright partition, Britain, France, Germany, Russia, the United States, and Japan contented themselves with special treaty ports, spheres of influence, and long term "leases" of Chinese territory.

These continued humiliations resulted in a final desperate outburst of anti-foreign violence known as the Boxer Rebellion (1900-1901). The "boxers" were members of a secret society whose aim was to drive all foreigners from China. They attacked Chinese converts to Christianity and murdered hundreds of missionaries, railway workers, businessmen, and other foreigners, including the German minister to Peking. A joint European military expedition took the capital, meted out severe reprisals, and exacted further indemnities and concessions. Thereafter the Chinese government was in a state of almost visible decay. At length it succumbed, in 1912, to a revolution led by a Western-educated physician, Dr. Sun Yat-sen, whose program was "nationalism, democracy, and socialism." The Western powers were not perturbed, and did not attempt to interfere in the revolution. The Republic of China that resulted remained weak and divided, its hopes of reform and regeneration long postponed.

JAPAN EMBARKS ON WESTERNIZATION

Japan maintained its policy against foreign intrusion more effectively than any other Oriental nation. Gradually, however, under continuous Western pressure for diplomatic representation, missionary activity, and commercial intercourse, the resistance weakened. It collapsed in the face of representation backed by the threat of force made by the United States naval commander, Commodore Matthew Perry, in 1853-54. Soon other Western nations obtained privileges similar to those of the United States. As in China, anti-foreign rioting broke out, lives were lost, and property destroyed. The Western navies began to make reprisals. It seemed as if Japan was destined to repeat the experience of the other victims of Western imperialism, perhaps even more rapidly. Then a remarkable change took place.

Prior to 1868 Japan was ruled not by the emperor but by the shogun, a powerful feudal lord. The anti-foreign sentiment provoked by the concessions of the shogun developed into a movement to restore the emperor to his rightful position. This event, the Meiji Restoration, marks the birth of modern Japan. Immediately upon coming to power the new government changed the tone of the anti-foreign movement. Instead of attempting to expel the foreigners, the Japanese now co-operated with them, but kept them at a polite distance. Intelligent young men went abroad to study Western methods—in politics and government, military science, industrial technology, trade and finance—with a view to adopting those that seemed most efficient. The old feudal system was replaced by a highly centralized bureaucratic administration modelled on the French, with a Prussian-type army and a British-type navy. Industrial and financial methods came from many countries, but especially from the United States. The results of this vigorous, forward-looking policy soon became apparent. Foreign commerce made spectacular gains, modern industries arose where none had been before, and Japanese military might won the incredulous respect of the Western world. In 1894-95 Japan quickly defeated its neighbor and former mentor, China, and joined the ranks of the imperialist nations by annexing Chinese territory and staking out a sphere of influence in China proper. Even more surprisingly, Japan decisively defeated Russia just ten years later.

The Russo-Japanese War resulted directly from the imperial rivalry of these two nations in China and Korea. Japan attempted to negotiate a settlement of disputed claims, but the Russians, who regarded the Japanese as no different from the Chinese and Koreans, treated them with disdain. Provoked by this affront to their national pride, the Japanese in February 1904 bottled up the Russian Asiatic fleet in Port Arthur and declared war two days later. To western observers it looked like an Oriental David attacking the Russian Goliath. The Japanese soon showed their mettle by inflicting a series of defeats on Russian armies along the Yalu River and in Manchuria. To regain control of the seas between Japan and the mainland the Russians sent their Baltic fleet on a seven months' cruise around the tip of Africa; no sooner had it arrived in Eastern waters (in May 1905) than the Japanese sank almost the entire fleet. This shattering and humiliating loss, together with the outbreak of domestic revolution, forced the Russians to acknowledge Japanese predominance in Korea, transfer their lease on Port Arthur and the Liaotung peninsula to Japan, and cede the southern half of Sakhalin. Thus the Japanese proved that they could play the game.

SOUTHEAST ASIA

Indo-China is the term frequently used to refer to the vast peninsula of Southeast Asia. It derives from the fact that the culture of the area is in essence an amalgam of classical Indian and Chinese civilizations. Prior to the 19th century the rulers of the various principalities and kingdoms recognized a vague allegiance to the Chinese emperor. In the course of the century the British, operating from India, established control over Burma and the Malay States, and eventually incorporated them in their empire. In 1858 a French expedition occupied the city of Saigon in Cochin China (Vietnam), and four years later annexed Cochin China itself. Once established on the peninsula the French found themselves involved increasingly in conflicts with the natives, obliging the French to extend their "protection" over even larger areas. In the 1880's they organized Cochin China, Cambodia, Annam, and Tonkin into the Union of French Indo-China, to which they added Laos in 1893.

Thailand (or Siam, as it was called by Europeans), between Burma on the west and French Indo-China on the east, was and remained an independent kingdom. It owed its good fortune to two principal factors: a series of able and enlightened kings and its position as a buffer between French and British spheres of influence. Although opened to Western influence by gunboat treaties, like most of the rest of Asia, its rulers reacted with conciliatory gestures and at the same time strove to learn from the West and to modernize their kingdom. Even this might not have saved it from the fate of Burma and French Indo-China, however, had it not been for the Anglo-French rivalry. Few other non-Western nations were so fortunate.

THE PARTITION OF AFRICA

In spite of vast differences in the cultural level of Asians and Africans in the 19th century, their similarities were more important to an understanding of their vulnerability to Western imperialism. They were alike in three main ways. First, they were technologically backward; secondly, they had weak, unstable governments, shifting political allegiances, and warring internal factions. A third similarity had no functional significance, yet it did have an important bearing on Western attitudes toward Asians and Africans: both had colored skins, in contrast to the predominantly fair-skinned Europeans.

THE BRITISH IN SOUTH AFRICA

The Cape of Good Hope had been settled by the Dutch in the mid-17th century as a way station for ships of the Dutch East India Company. The British occupied it during the Napoleonic Wars to prevent it from falling into the hands of the French. Conflicts between the British and the Boers, or Afrikaners (as the Dutch colonists were called), led the Boers in 1835 to make their "great trek" to the north, creating new settlements in the region between the Orange and Vaal rivers (later known as the Orange Free State), north of the Vaal (Transvaal, which became the South African Republic in 1856), and in Natal on the southeast coast. In spite of the Boers' attempt to isolate themselves from the British, conflict continued to mark the history of the colonies throughout the century. In addition to the conflicts between the British and the Boers both peoples had frequent clashes with native African tribes. These always ended, sooner or later, in defeat for the tribes. Many of these were practically exterminated as a result, and most of the remainder were reduced to a state of servitude not far from slavery, especially those under Boer rule.

Hitherto both the Boer and British settlements had been primarily agrarian, but in 1867 the discovery of diamonds in Griqualand, a region formerly set aside for natives, led to a great influx of treasure-seekers from all over the world. This was followed in 1886 by the discovery of gold in Transvaal. These events completely altered the economic basis of the colonies and intensified political rivalries. They also brought to the fore one of the most influential persons in African history, Cecil Rhodes. An Englishman, Rhodes came to Africa in 1870 at the age of seventeen and quickly made a fortune in the diamond fields, adding to it later in gold mining. Not content with mere profits, Rhodes also took an active role in politics and became an ardent spokesman for imperialist expansion. He entered the Cape Colony legislature in 1880 and became prime minister of the colony ten years later. One of his major ambitions was to build a railway "from the Cape to Cairo"—on British territory throughout, of course. The Boers opposed this project, and Rhodes created "incidents" to provide an excuse for war.

The British government genuinely desired to avoid war with the Boers, but extremists on both sides in South Africa itself pushed the issue to its fateful conclusion. In October 1899 the South African, or Boer War began. The British, with only 25,000 soldiers in South Africa against some 75,000 Boers, suffered a number of early defeats. With the arrival of reinforcements, the British rallied and in less than a year completely overran both the Orange Free State and Transvaal. The Boers then resorted to guerilla warfare, which dragged on for another year and a half before they finally yielded (May 31, 1902).

The British had introduced limited representation in the Cape Colony as early as 1825. In 1853 they introduced a representative assembly, and permitted fully responsible government in 1872. The Boer republics had also been self-governing. Soon after the Boer War British policy toward the Boers changed from repression to conciliation. Transvaal regained responsible government in 1906 and the Orange Free State in 1907. The movement for union with Cape

Colony and Natal gathered momentum thereafter, and in 1910 the Union of South Africa joined Canada, Australia, and New Zealand as fully-governing dominions within the British Empire.

EUROPEAN IMPERIALISM IN NORTH AFRICA

Prior to 1880 the only European possession in Africa, apart from British South Africa and a few coastal trading posts, was French Algeria. France undertook the conquest of Algeria in 1830, but not until 1879 did civil government replace the military authorities. By that time the French had also begun to expand from their settlements on the African West Coast. By the end of the century they had conquered and annexed a vast, thinly populated territory (including most of the Sahara desert) which they christened French West Africa. Meanwhile in 1881 they took advantage of border raids on Algeria by tribesmen from Tunisia to invade the latter country and establish a protectorate. The French finally rounded out their North African empire in 1912 with the establishment of a protectorate over the larger part of Morocco (Spain claimed the small northern corner) after lengthy diplomatic struggles, especially with Germany.

The opening of the Suez Canal by a French company in 1869 not only revolutionized world commerce; it also endangered the British "lifeline" to India—or so it seemed to the British. Thenceforth it became a cardinal tenet of British foreign policy to prevent the canal area and its approaches from falling into the hands of an unfriendly great power. In his efforts to build up Egypt to the status of a great power, the khedive (king) of Egypt had incurred enormous debts to Europeans for such purposes as an abortive attempt at industrialization, the construction of the Suez Canal, and an attempted conquest of Sudan. In an effort to bring some order into the finances of the country, British and French financial advisers were appointed and soon constituted the effective government. Egyptian resentment of foreign domination resulted in widespread riots in which many Europeans lost their lives. To restore order and protect the canal the British in 1882 bombarded Alexandria and landed an expeditionary force. Gladstone, the British prime minister, assured the Egyptians and the other great powers that the occupation would be temporary. Once in, however, the British found to their chagrin that they could not easily or gracefully get out. Besides continued nationalist agitation which required the presence of the army to suppress, the British had, by taking over the government of the khedive, inherited his unfinished conquest of Sudan. In pursuing this objective, which seemed to be justified by the importance to the Egyptian economy of controlling the upper Nile, the British ran head-on into conflict with the French expanding eastward from their West African possessions. At Fashoda in 1898 rival French and British forces faced one another with sabers drawn, but hasty negotiations in London and Paris prevented actual hostilities. At length the French withdrew, thus preparing the way for British rule in what became known as Anglo-Egyptian Sudan.

One by one the Turkish sultan's nominal vassals along the North African coast had been plucked away until only Tripoli, a long stretch of barren coastline backed by an even more barren hinterland, remained. Italy was its nearest European neighbor. In 1911 Italy trumped up a quarrel with Turkey, delivered an impossible ultimatum, and promptly invaded Tripoli. The war itself was something of a farce, with neither side vigorous enough to overcome the other. The threat of a new outbreak by Turkey's neighbors in the Balkans, however, persuaded the Turks to make peace in 1912. They ceded Tripoli to Italy, which the Italians renamed Libya.

THE DIVISION OF CENTRAL AFRICA

Central Africa, much of it covered with dense tropical jungles and inhabited by Negroes, was the last area in the "dark continent" to be opened to European penetration. Indeed, the inaccessibility, inhospitable climate, and exotic flora and fauna—all amply embroidered upon by early explorers and traders—were largely responsible for Africa's sobriquet and formidable reputation.

Prior to the 19th century the only European claims in the region—and those not strong ones—were by Portugal: Angola on the west coast, Mozambique on the east, both on the southern perimeter. The activities of explorers such as the Scottish missionary David Livingstone and the Anglo-American journalist H. M. Stanley, in the 1860's and 1870's, did much to arouse popular interest. In 1876 King Leopold of Belgium organized the International Association for the Exploration and Civilization of Central Africa and hired Stanley to establish settlements in the Congo. Agitation for colonial enterprise in Germany reached a high peak with the formation of the German African Society in 1878 and the German Colonial Society in 1882. The discovery of diamonds in South Africa stimulated exploration in the hope of similar discoveries in Central Africa. Finally, the French occupation of Tunis in 1881 and the British occupation of Egypt in 1882 set off a scramble for claims and concessions.

This sudden rush for territories created frictions which might have led to war. To head off this possibility, and incidentally to balk British and Portuguese claims, Germany and France called an international conference on African affairs to meet in Berlin in 1884. Fourteen nations including the United States sent representatives. They recognized the Congo Free State headed by Leopold of the Belgians, an outgrowth of his International Association, and laid the ground rules for further annexations. The most important of the latter provided that a nation must effectively occupy a territory in order to have its claim recognized.

On the east coast of Africa, meanwhile, much the same process was taking place. In 1882 the Italians occupied Assab at the mouth of the Red Sea and used it as a base for expansion into Eritrea and Italian Somaliland. The French and British, not to be outdone, soon afterward established protectorates of their own on parts of the Somali coast. Farther south two chartered East Africa companies, one German and one British, staked large claims reaching inland to the eastern borders of the Congo Free State. The French, with a long-standing but ineffectual claim to the island of Madagascar, finally annexed it in 1896. One of the few regions of Africa now left unclaimed by a European power was the Coptic-Christian kingdom of Ethiopia, also known as Abyssinia. In 1895 the Italians decided to annex it to their coastal territories, and began a slow and difficult march to the interior. Early the following year the Italian army of 20,000 men was annihilated by barefoot Ethiopian tribesmen at Adowa. That humiliating defeat ended, at least temporarily, Italian designs on the "king of kings."

OTHER IMPERIALISM

Asia and Africa were not the only areas subject to imperial exploitation, nor were the nations of Europe the only ones to engage in it. We have already seen that Japan, once it had adopted Western technology, also pursued imperialistic policies not unlike those of Europe. The United States as well, in spite of strong domestic criticism, embarked on a policy of colonialism before the end of the century. In addition to the acquisition of Alaska in 1867 and the Panama Canal Zone in 1903, the United States participated with Britain, France, and Germany in sharing out the islands of the Pacific, taking Hawaii and part of Samoa as its portion. In the war with Spain in 1898 the United States won the Philippines and Guam, as well as Puerto Rico in the Caribbean.

Some of the British dominions were far more aggressively imperialistic than the mother country itself. The expansion of South Africa, for example, took place largely through South African initiative and frequently against the wishes and explicit instructions of the government in London. The British annexation of southeastern New Guinea in 1884 (after the Dutch had claimed the western half and the Germans the northeast) was directly due to the agitation of the Queensland government in Australia.

A distinction is sometimes made between imperialism and colonialism. Thus, while neither Russia nor Austria-Hungary had overseas colonies, both were clearly empires in the sense that they ruled over alien peoples without consent. On the other side, the imperial powers did not as a rule establish colonies in China; yet China was obviously subject to imperial control. The countries of Latin America experienced no new attempts at conquest by outside powers, but it was frequently alleged that they constituted part of the "informal empires" of Britain and the United States as a result of economic dependence and financial control.

THE ROLE OF TECHNOLOGY IN IMPERIALISM—REASONS AND RATIONALIZATION

The "causes" of imperialism are many and complex. No single theory will suffice to explain all cases. Nevertheless, it is worth reviewing some of the more important interpretations; whatever their objective validity, the interpretations themselves are a part of the intellectual history of the epoch. Furthermore, they help reveal some of the complex ways in which technology affects economic and political life, and vice versa.

In the intellectual cross-currents of the late 19th century, confusion was bound to arise concerning the motives for which imperial expansion was undertaken. While some proclaimed such expansion to be a necessity for Western nations, others demanded the expansion on behalf of the colonial areas themselves. Missionary activity, the desire to bring the comforts of Christianity to the "heathen races," and the belief of well-intentioned humanitarians that non-Europeans would benefit from Western legal institutions, technology, and so forth, even if imposed by force, were advanced as reasons for dispatching expeditionary forces to distant lands. Rudyard Kipling, Poet Laureate of the British Empire, wrote eloquently of "the white man's burden," and French apologists for imperialism spoke of their *mission civilisatrice* ("civilizing mission").

The evangelical zeal of Christian missionaries, both Catholic and Protestant, is beyond dispute. For centuries before the 19th they had carried their message far and wide. They preceded—usually by many years—traders, diplomats, armies, and administrators in both Asia and Africa. Normally, however, they did not ask even for military protection from their homelands, much less for territorial annexation. Theirs was a spiritual mission quite unconnected with politics. When they were subjected to persecution, torture, and even death by the people they sought to convert, the demands for retaliation came most often from journalists, military men, and others who had reasons of their own for desiring imperial conquest. In short, in the imperial game as it came to be played in the late 19th century, arguments from religion were simply convenient excuses, missionaries mere pawns in the game.

One of the most popular explanations of imperialism runs in terms of economic necessity. In fact, modern imperialism is often referred to as "economic imperialism," as if earlier forms of imperialism had no economic content. The arguments appear at first sight to be quite logical, and there is just enough empirical evidence in their favor to make them plausible.

One such argument runs as follows: (1) competition in the capitalist world became more intense, resulting in the formation of large-scale enterprises and the elimination of small ones; (2) capital accumulated in the large enterprises more and more rapidly, and since the purchasing power of the masses was insufficient to purchase all the products of large-scale industry, the rate of profit declined; (3) as capital accumulated and the output of capitalist industries went unsold, the capitalists resorted to imperialism in order to gain political control over areas where they could invest their surplus capital and sell surplus products. Such is the essence of the Marxist theory of imperialism; or better, the Leninist theory, for Karl Marx did not foresee the rapid development of imperialism although he lived until 1883. Nikolai Lenin, who was to lead the Bolshevik Revolution in 1917 which established the Communist regime in Russia, built on the foundation of Marxist theory and in some cases modified it in his widely read little book, *Imperialism, the Highest Stage of Capitalism* (1915). Lenin's argument, it can readily be seen, was indirectly based on the technological transformations which had brought about large-scale industrial development and the consequent mass production of goods.

Lenin was by no means the first person to advance an economic explanation of imperialism, however. On the contrary, he borrowed heavily from the liberal British critic of imperialism, John A. Hobson, who in turn adopted in revised form many of the arguments of capitalist advocates of imperialism: not capitalists themselves as a rule, but journalists, professors, politicians, and military and naval officers in capitalist countries. One such person was Captain A. T. Mahan, an American naval officer who strongly influenced America's leading exponent of imperialism, Theodore Roosevelt. Mahan's dictum was "trade follows the flag." Yet another capitalist advocate of imperialism, Jules Ferry, a French journalist and politician, twice became prime minister and was chiefly responsible for the largest colonial acquisitions of France. Interestingly, on both occasions his policy of colonial annexation cost him his premiership, but the French, like the British, once they were committed to a particular conquest or annexation, found it difficult to withdraw. Equally interesting, Ferry did not utilize economic arguments in defending his actions before the French Assembly; instead, he stressed French prestige and military necessity. Only after he had been permanently retired from office did he write books justifying his actions in which, for the first time, he emphasized the economic gains which France would supposedly realize from its colonial empire.

The facts of the matter, however, simply do not bear out the argument that colonies were needed as markets for surplus manufactures, or were so used after they were acquired. For example, at no time prior to 1914 did as much as 10 per cent of French exports go to French colonies. By and large, the colonies were too sparsely populated and too poor to serve as major markets. Moreover, political control was not required to sell goods. India, "the brightest jewel in the British crown," was indeed a large market in terms of numbers; in spite of its poverty India did purchase large quantities of European wares—but not only from Britain. The Germans sold far more in India than in all their own colonies taken together. France, too, sold more to India than to Algeria. On the other hand, as important as India was for British manufactures, Britain sold far more to Australia, having but a small fraction of India's population. In general, and in spite of protective tariffs, the industrialized, imperialistic nations of Europe contined to trade for the most part with one another.

Perhaps the most important argument for imperialism as an economic phenomenon, at least in Marxist theory, concerned the investment of "surplus" capital. Here again the facts do not substantiate the logic. Britain, with the largest empire, also had the largest foreign investments. But over half of Britain's foreign investments were in independent countries, especially the United States, and self-governing territories. In the case of France the facts are even more surprising: less than 10 per cent of all French foreign investments before 1914 went to French colonies. The French invested heavily in other European countries and in Latin America. Russia alone—an imperialistic nation itself—took one fourth of all French capital exports; the French also invested in Germany and Austria-Hungary, with whom they eventually went to war. German investments in German colonies were negligible. Some of the imperialist nations were actually net debtors: in addition to Russia, they included Italy, Spain, Portugal, and the United States.

Thus the idea that imperialism was an economic necessity for the highly developed industrial nations is essentially fallacious, although it does contain some elements of truth and plausibility. Nevertheless, the most crucial test of the validity of the economic argument is, did imperialism pay?

DID IMPERIALISM PAY?

This question has many aspects, and a complete answer would be very complex. Broadly, imperialism did not pay in a strictly pecuniary sense. With few exceptions—India being the most important—taxes collected in the colonies rarely sufficied to cover the costs of routine administration, much less of conquest. Thus, from the treasury point of view, imperialism was a losing proposition. But it was argued that the indirect benefits through increased trade would make the venture worthwhile. Here the statistics are difficult to unravel and interpret. We have noted that colonial trade did not bulk large in the total international trade; in some cases—notably the French and German—the total value of trade with the colonies did not amount to as much as the expenditure incurred in taking and maintaining them. What is indubitable is that some individuals did make enormous fortunes in colonial ventures—Cecil Rhodes is the outstanding example—and many others gained a modest living; but the profits were by no means equally shared. Taxes had to be raised in the imperial nation to pay for the military and naval expeditions and garrisons and the officials who administered the colonies, as well as for whatever public works they constructed. Manpower had to be diverted from other uses to staff the armies, navies, and colonial services. Under the prevailing systems of taxation, the tax money came largely from ordinary workers and farmers, people who had no pecuniary interest, direct or indirect, in the colonies. In effect, income and wealth were redistributed by the process of taxation and expenditure; the masses paid the costs, the profits were garnered by a favored few.

The ultimate costs, however, should be reckoned in terms of the suffering and dislocation of the peoples subject to Western imperialism, and of the rivalries and frustrations generated in the course of the race for colonial supremacy, rivalries which prepared Europe psychologically for war, and which themselves were factors leading to war. Those costs are still being paid.

WHY IMPERIALISM?

If the economic explanation is not sufficient to explain the burst of imperialism in the late 19th century, how can it be explained? Among many other factors, major responsibility must be assigned to sheer political opportunism, combined with the growth of aggressive nationalism. Power politics and military expediency also played an important role. Britain's imperial policy throughout the century was dictated in large part by the supposed necessity of protecting the Indian frontiers and "lifeline." This explains the conquest of Burma, and Malaya, Baluchistan, and Kashmir, as well as British involvement in the Near and Middle East. Other nations emulated the successful British either in the hope of gaining similar advantages, or simply for reasons of national prestige.

The intellectual climate of the late 19th century, strongly colored by Social Darwinism, likewise favored European expansion. Social Darwinism was based on the application, by analogy, of the biological findings of Charles Darwin (1809-82) to the interpretation of society. Thus society became a vast arena in which "the fittest" nation or individual "survived" the inevitable struggle for existence. According to the Social Darwinists, such competition, whether military or economic, weeded out the weak and ensured the virile continuation of the fittest nation, race, individual, or business corporation. Herbert Spencer, the foremost popularizer of Social Darwinism, was an outspoken anti-imperialist, but this did not prevent others from applying his arguments for "survival of the fittest" to the imperial struggle. Even Theodore Roosevelt spoke grandly of "manifest destiny," and Kipling's phrase, "the lesser breeds without the law," reflected the typical European attitude toward the non-white races. But European racism and ethnocentrism have far deeper historical roots than Darwinian biology. Christian missionary activity itself was an expression of long-held beliefs of European, or Western, moral and cultural superiority. Throughout their history—at least until the mid-20th century—Europeans and Christians have been expansionistic and evangelical. In the final analysis, modern imperialism must be regarded as a psychological and cultural phenomenon rather than simply political or economic.

TECHNOLOGY AND IMPERIALISM

As far as technology is concerned, it entered into imperialism in many different ways, some directly and some indirectly. The technological superiority of the West was certainly an important element underlying the general attitude of Western cultural supremacy over the colonial peoples. Furthermore, Western military technology made easy the conquest and submission of the natives who were the victims of European imperialistic ventures; Western methods of transportation and communication were also essential for the administration of the colonial lands. In certain cases, the level of technology served to determine the direction and scope of imperial activity; for example, 19th-century steamships and naval vessels required frequent recoaling, so nations set out to acquire coaling stations along the routes to their far-flung colonial empires.

Imperialism also served to acquaint the inhabitants of the colonial territories with the products of Western technology. The desire of the colonial peoples for the material advantages of Western civilization proved to be an important stimulus in their push for independence in the 20th century and in their desire for industrial development. Although European political rule of imperial territories was ultimately to wither away in the mid-20th century, the attractions of Western technology were to retain dominion over the colonial peoples, with the result that one of the world's major problems in our own time is the technological uplifting of the so-called "underdeveloped" nations.

44 / Expositions of Technology, 1851-1900

EUGENE S. FERGUSON

The latter half of the 19th century witnessed an impressive number of international exhibitions which mirrored the enthusiasm and optimism with which the Western world pursued technical innovation and novelty. In addition, by bringing together the products and people of many nations, these great world's fairs, starting with the 1851 Crystal Palace Exhibition in London, accelerated the development of technology through exchange of ideas and information.

Between 1851 and 1900, nearly one hundred exhibitions were held. Occurring all over the world, in Paris and New York, Melbourne and Nijni Novgorod, most of these exhibitions were concerned with a single subject, such as electricity or cotton, and were international in name only. Nevertheless, a dozen major international exhibitions, held in London, Paris, Philadelphia, Chicago, and Vienna exerted an unmistakable influence upon technologists and other custodians of progress.

While the 19th-century exhibitions were not quite the no-nonsense, regular trade fairs that flourish today in Frankfurt, Milan, Hanover, and other European cities, neither were they the public-relations carnivals that we have come to expect of world's fairs. While it was possible, for example, for a visitor to the 1889 Paris Exhibition—for which the Eiffel Tower was built—to spend days on the exhibition grounds visiting the art exhibits, the national pavilions, and the "Folies Parisiennes" without setting foot in Machinery Hall, the tangible effects of the exhibitions of technology were so evident that we must conclude that their serious promotional purposes were accomplished. New products, ideas, attitudes, and trade opportunities were brought to the attention of thousands of people who otherwise probably would have missed them.

Although many technical and commercial magazines were being published in 1851, and technical societies, mechanics' institutes, and other institutions for the dissemination of information were functioning, the expensive and elaborate exhibitions played a major role in the wider diffusion of technical knowledge. After all, a moment's reflection will tell us that a surprisingly large portion of our most useful knowledge has been gathered in a most haphazard and uncertain fashion. Chance meetings, chance glimpses that trigger ideas, and coincidental but unlikely series of events occur in great profusion—and all of these contribute to our store of knowledge. And when we are concerned with technical innovations, most of us do not know what to look for or where. Add to this the veil of an unknown foreign or technical language, and the search becomes so nearly futile that it is seldom attempted. Furthermore, an actual object, be it a machine or a piece of window glass, exhibits qualities that cannot be learned from books. Because the exhibitions gathered technical objects from all over the world and explained them in terms comprehensible to a technologist, if not always to a layman, they contributed substantially to material progress.

EARLY EXHIBITIONS

Local and national exhibitions of technology were well established before the first truly international exhibition, in 1851. Nearly a century before (1754), the Royal Society of Arts, in London, had been founded for the purposes of encouraging invention and mechanical improvements, and occasional exhibitions, at which prizes were awarded for original ideas, were held by the Society from 1761. A series of French national technical exhibitions had been inaugurated in 1798 by Napoleon I. In England and the United States, mechanics' institutes commenced holding local exhibits of industry just after 1820. A national industrial exhibition was held in 1846 in Washington, D.C., having been hurriedly organized in order to show members of Congress the extent and excellence of American manufacturers. A temporary T-shaped wooden building, whose stem was 300 feet long, 60 feet wide, and 20 feet high, was put up in three weeks' time. The roof was covered with canvas; the walls were left raw on the outside and draped with red, white, and blue muslin inside, where an impressive display of textiles, farm machinery, and industrial and luxury goods was assembled. The price of each item was given, and those purchased were delivered at the end of the two weeks' show. Some 60,000 visitors were counted.

The French national exhibition of 1849 precipitated the decision to organize the London exhibition of 1851 on an international basis. The Royal Society of Arts, of which Prince Albert (1818-61) was president, had been planning to hold a national industrial exhibition in 1851, and when Henry Cole, a member of the Society's exhibition committee, attended the 1849 Paris exhibition, he learned that a proposal to make the French exhibition an international one had been considered but turned down. When Cole made a similar proposal in his committee, Prince Albert accepted the idea with the comment that since "the productions of machinery, science, and taste . . . are the property of no country but belong as a whole to the civilized world, particular advantage to British industry might be derived from placing it in fair competition with that of other nations."

It is not surprising that French manufacturers, accustomed for a hundred years to taking second place to the British in technical prowess and foreign trade, had not been eager to invite the "nation of shopkeepers" to exhibit its wares in the middle of Paris. It is surprising, however, that the British manufacturer, secretive as he was and jealous of any privilege that maintained his advantage over others at home or abroad, should consent to display his wares openly not only to prospective buyers but to casual onlookers, and even to competitors. It can only be concluded that the British lion, conscious of its lead in technology over continental Europe and ignorant of the rest of the world except as a source of raw materials and an outlet for manufactured goods, was complacent in 1851. The foreign-born Prince Albert, who knew a great deal more about the Continent than did his British subjects, thought that a comparison of foreign with British goods might have a salutary effect upon industry and trade. Thus the proposed exhibition was given a royal commission and became the first international industrial exhibition.

THE CRYSTAL PALACE EXHIBITION: 1851

"The Great Exhibition of the Works of Industry of all Nations, 1851," was the official title of the undertaking. The "Crystal Palace" was the fanciful and wholly fitting name given by a writer in the magazine *Punch* to the building in which the Great Exhibition was held. A committee had been formed and a public competition held for a design for the exhibition building. After rejecting all of the 233 plans submitted, the building committee produced its own design. A vast and nondescript brick pavilion, covering an area of 18 acres, was to be capped by a monstrous iron dome, an idea of the audacious but outlandish engineer, Isambard Kingdom Brunel (1806-59). The exhibition building was to be located in Hyde Park, a wooded public place in London close by a residential district.

The Times, the leading London newspaper, waged a determined campaign to keep the Great Exhibition out of Hyde Park. The park and surrounding neighborhood would be overrun, it claimed, by visitors of the lowest classes. And a brick building, said *The Times*, certainly was not intended to be merely temporary. It would last for a hundred years, and worse than that, it would be as substantial as Buckingham Palace. Colonel Sibthorp, a member of Parliament, carried the fight to the House of Commons. He wanted to save the trees, and to keep out "all the bad characters at present scattered over the country who will be attracted to Hyde Park." Furthermore, he deplored the importation of "foreign stuff of every description—live and dead stock—without regard to quantity or quality." This could only lead to price reduction, he said; eventual ruin was inevitable.

Fortunately for the Great Exhibition, a last-minute solution of the building problem was presented when Joseph Paxton (1801-65), builder of greenhouses and manager of the Duke of Devonshire's estate, *Chatsworth*, suggested to Henry Cole a highly original design for the exhibition building in Hyde Park. Paxton's contribution was the "Crystal Palace," a fantasy of iron and glass that captured the public imagination even before the first structural member was set in place. Basing his design upon a small greenhouse he had built recently at *Chatsworth*, he extrapolated with unusual courage, skill, and taste to produce the great building for which he is particularly remembered.

The walls and roof of the Crystal Palace consisted of 22 acres of glass—18 acres in the roof and 4 in the walls. Supporting the glass were 3000 cast-iron columns, connected and stiffened by cast- and wrought-iron lattice girders. The framing employed modular units; aisles, for example, were either 24, 48, or 72 feet wide. A repetitive pattern throughout the building made possible the use of prefabricated components. The building could be dismantled readily, thus answering one of the objections to the Hyde Park site. A vaulted transept, or crossing, in the center of the building was added so that some of the trees might remain standing, without damage, inside the building. Another group of trees was cut down by stealth one night to make way for the building, prompting Colonel Sibthorp to cry out in anguish in the House of Commons.

Once the decision to build was made, the Crystal Palace grew with amazing speed. The contracting firm of Fox and Henderson received Paxton's plans around July 15, 1850, but the contract was not finally signed until October 30.

Fig. 44-1. The great Crystal Palace, London, 1851. The walls and roof consisted of 22 acres of glass. Supporting the glass were 3000 cast-iron columns, connected by cast- and wrought-iron lattice girders. (*The Bettmann Archive*)

Nevertheless, the site in Hyde Park was occupied on July 30; the first of the columns was erected on September 26. The whole building, 1850 feet long, 408 feet broad, and 63 feet high, rising to a height of 108 feet over the transept, was completed before the end of January 1851. The glass was all hand-blown; all of the nearly 300,000 panes, each 10 by 49 inches, were supplied by the Chance Brothers of Birmingham.

As the building rose in Hyde Park, new objections were voiced. Many self-styled experts agreed that the wind would blow the building down, so insubstantial did it appear to those familiar with brick, stone, and mortar. A glance at the balconies convinced others that they would never support the expected crowds. Although Paxton was no engineer, his defense was in keeping with the practice of his day: when the strength of galleries was questioned, an assembly of workmen jumped up and down on a sample section of the structure in question; a detail of soldiers marched, countermarched, and marked time; stacks of cannon balls in dollies were trundled over the floor, and the chief construction engineer offered to run a locomotive across the floor if the critics were not yet convinced. The soundness of the design was vindicated by time. After the Great Exhibition, the building was moved to another part of London and survived until fire destroyed it in 1936.

The popular view of the Crystal Palace and of the coming Great Exhibition was one of pride and expectancy. With appropriate ceremonies, the exhibition was opened on May 1, 1851, by Queen Victoria. The day reminded her of her coronation day. She wrote in her diary: "The tremendous cheering, the joy expressed in every face, the vastness of the building, with all its decorations and exhibits, the sound of the organ (with 200 instruments and 600 voices, which seemed nothing) and my beloved husband, the creator of this peace festival uniting the industry and art of all the nations of the earth, all this was indeed moving, and a day to live forever."

Throughout the summer and until October 15, the closing day, Hyde Park was jammed with throngs of visitors. Those who had objected to the exhibition were confounded by the uniformly good behavior of the visitors of all classes. Several insurance offices gave their clerks four days' holiday and £3 to visit the exhibition, and excursion trains brought thousands from all parts of England. The commandant of the Portsmouth dockyard gave his workmen six days' holiday and told them that if the excursion fares were not brought as low as he thought they should be, he would furnish a naval vessel to take the men to London. Organized parties of French workmen visited the Crystal Palace and wrote reports of their impressions.

By every standard that might be applied, the Great Exhibition was successful. The profusion of exhibits, the interest expressed by visitors; the holiday atmosphere of good-natured jostling; the approval of highly placed men of affairs who had earlier been skeptical (except Colonel Sibthorp, who would not enter the Crystal Palace, although he admitted it to be "a wonderful building externally")—all these were proof that Prince Albert's judgment had been sound. More than 100,000 exhibits were shown by 14,000 exhibitors, almost half of whom were foreigners. Six million visitors attended; as many as 90,000 were in the building at one time. Unlike most later exhibitions, the Great Exhibition was financially successful, taking in more money than was paid out.

Half of the Crystal Palace was taken up with British exhibits. In the foreign half, the United States had a respectable share of the space—too much, in fact, for the number of exhibits that could be assembled, and for a time the space was known as the "prairie ground." For while the United States government had furnished a vessel to take the exhibits to London, no financial provision had been made for unloading. George Peabody, an American banker who lived in London, finally advanced the money necessary to convey the exhibits to their place in Hyde Park.

Submerged in a building full of larger and more striking displays, the American exhibits were not particularly noticed until after the field-trials of agricultural machinery, at Tiptree Hall, had established the superiority of the McCormick reaper over the numerous British machines that were tried. After that, a number of novel American ideas were noted by those who were prepared to accept innovation not of British origin. For example, Samuel Colt's revolver and Robbins and Lawrence's military rifles were products of the "American system" of manufacture; the bank locks of Day and Newell resisted attempts at picking by English locksmiths, while Alfred Hobbs, the Day and Newell representative, went about opening English locks that had been assumed impregnable. Another event that attracted notice was the yacht race around the Isle of Wight, in which the yacht *America*, patterned after a New York pilot boat, swept a field of fifteen British vessels. "What's first?" asked a London writer of the day. "The *America*." "What's second?" "Nothing." The well-known America Cup, incidentally, was awarded to the winner of this race and named after it.

The effects of this Great Exhibition were unmistakable and far-reaching. Gratifying to the American representatives was the notice given to the United States by the British, who were more than a little dismayed to find that this new nation could, in the course of a generation or two, compete successfully in an industrial exhibition with the older, more experienced nations of Europe.

Some of the clear and direct results of the Exhibition can be enumerated. Within a few months of the closing date, one of the English centrifugal pumps shown at the Exhibition was at work draining English fen lands. On the strength of the newly revealed "American system" of manufacture, a delegation was sent in 1853 to the United States by the British Board of Ordnance to buy American machine tools to equip the new small-arms plant being built at Enfield Arsenal. A later comment by an English engineer on British developments in wood-working machinery referred to the Great Exhibition. "In 1851," he wrote, "we were taken by surprise with a few simple machines for working wood, which came from America, and from them, in a great measure, was derived the impetus that has pushed this branch up to its present condition." The German government commission published in 1852 a three-volume report of the Great Exhibition, which was carefully read by Germans who, though not yet politically united, were curious about their economic and industrial standing among the peoples of the world. The Royal Society of Arts in London held a series of lectures by prominent academic and practical men on the "Results of the Great Exhibition," and published the lectures in a volume of the same title. One of the commissioners of the exhibition published in 1855 a catalogue of some 500 works that he had personally collected "on, or having reference to, the Exhibition of 1851." Within a year of the Great Exhibition, the British patent system was completely reorganized as politicians came to recognize that the old system, with its absurdly complex and costly procedures, actually discouraged rather than fostered innovation.

The most significant revelation of the Exhibition, in the view of many British merchants and manufacturers, was the backward state of British "industrial design" compared to the artistic exhibits of the French and other continental nations. A much more significant conclusion might have been drawn regarding the deplorable state of primary and technical education in the United Kingdom, compared to that on the Continent and in the United States. Government schools of design, whose purpose was to apply art to industry, were strengthened; a Royal School of Mines was opened; and Hofmann's college of chemistry was taken over by the government. Nevertheless, the lesson had not been brought home with sufficient force to cause immediate changes in the general system of education that overlooked completely the future of the machine-tending and laboring classes. In the long run, however, the slight impetus given in 1851 to education has shown important results. With the profits of the season, the royal commission for the exhibition purchased the site in South Kensington on which are now located the Victoria and Albert Museum, the Science Museum and Library, and the Imperial College of Science and Technology.

After this first Great Exhibition, a nearly continuous succession of international exhibitions was held throughout the rest of the 19th century. Below is a list of the major international exhibitions, to which have been added the minor exhibitions of 1853, 1879, and 1880. It should be noted that "exhibition" was the original English designation, but these world's fairs are commonly spoken of today as "expositions," a word which entered the English language from the French.

Major International Exhibitions, 1851-1900

			NUMBER OF VISITORS (IN MILLIONS)
1851	London	The Great Exhibition	6.0
1853	New York	Exhibition of the Industry of all Nations	1.2
1855	Paris	Exposition Universelle	4.2
1862	London	International Exhibition	6.2
1867	Paris	Exposition Universelle	11.0
1873	Vienna	International Exhibition	7.3
1876	Philadelphia	Centennial International Exhibition	9.9
1878	Paris	Exposition Universelle	16.0
1879	Sydney	International Exhibition	1.0
1880	Melbourne	International Exhibition	1.0
1889	Paris	Exposition Universelle	32.4
1893	Chicago	World's Columbian Exposition	27.5
1900	Paris	Exposition Universelle	48.1

NEW YORK: 1853

While the success of the London Great Exhibition was still the talk of the Western world, a group of civic leaders and businessmen organized an international exhibition to be held in New York in 1853. Theodore Sedgwick (1811-59), a prominent lawyer, was made president of the proposed exhibition. After obtaining a charter from the state, the promoters asked permission of the city to erect a building alongside the reservoir, at 42nd Street and 6th Avenue. The City Council granted permission, but required that the building be of iron and glass and that no single admission charge be more than 50 cents.

The federal government made the exhibition building a bonded warehouse, which meant that foreign exhibitors would not have to pay tariff on their exhibits. The Secretary of State sent a circular letter to American consuls and ambassadors abroad, directing that they encourage foreign participation in the exhibition. Aside from these dispensations, however, and the attendance of President Franklin Pierce at the opening ceremonies, the federal government offered only moral encouragement. Even the effect of eliminating tariffs turned sour when, after the exhibition company went bankrupt, its creditors seized the foreign exhibitors' goods in order to recover their losses.

The building, which was called the Crystal Palace, was a frank imitation in spirit if not in detail of the London exhibition building. A design offered by Joseph Paxton was refused because it called for a long, narrow building, while the New York site was nearly square. The design adopted was from a New York firm of architects. A well-proportioned building, whose ground plan was octagonal, it had four valuted naves radiating from a central rotunda. An iron and glass dome, 100 feet in diameter, covered the rotunda. A long, low arcade containing machine exhibits was located on the east side of the square; boilers to supply steam to the machinery were located, for safety, across the street to the north of the machine arcade. In all, the buildings covered not quite four acres of ground; the London Crystal Palace had covered nearly nineteen.

The glass for the New York Crystal Palace was enamelled and the enamel was vitrified, giving the appearance of ground glass, so diffusing the sunlight that the great curtains of canvas, which had been employed in the London exhibition building, were unnecessary. Cast iron, employed in the New York Crystal Palace in much the same manner as in its London predecessor, had been used as a building material in the United States, and the New York iron-founder, James Bogardus, had erected a remarkable cast-iron office building and a shot-tower, both of which employed modular units. Despite these precedents, the new Crystal Palace taxed the resurces of its builders, and delays and difficulties marked its construction.

When the expected opening date, around the first of May 1853, arrived, the building was not sufficiently weather-tight to receive the exhibits. When a group of official British commissioners, including the machine-maker Joseph Whitworth, arrived in New York, on June 10, the interior of the building was so obstructed that the visiting dignitaries decided to tour various parts of the United States while waiting for the exhibition to open. Finally, on July 14, the exhibi-

Fig. 44-2. New York Crystal Palace, 1853. In frank imitation of the London Crystal Palace of 1851, this building housed the first international exhibition in the United States. (The J. Clarence Davies Collection, *Museum of the City of New York*)

tion was formally opened, although fewer than half of the exhibits were in place. It was not until September 5 that the managers were able to announce that the exhibition was complete.

The numerous delays in New York had been reported in the local press, and editorial comment was often sharply critical. When the reporters saw the exhibition in September, as it was meant to be, their tone changed to one of high praise. Unlike the London building, the New York Crystal Palace was artificially lighted, with gas, so that it might remain open after dark. "Seen at night," wrote the editor of *Harper's New Monthly Magazine*, "when it is lighted by only thirty less than the number of burners that light the streets of New York, it is a scene more gorgeous and graceful than the imagination of Eastern storytellers saw."

An impressive array of industrial exhibits was at length assembled. Well over half of the exhibits were from abroad. The United Kingdom occupied the largest space, nearly one-fourth of the building; Germany, France, and Belgium together occupied almost as much space; the remaining foreign ground was divided among exhibitors from a dozen other countries. The machine arcade was full of operating exhibits, powered by a Corliss and Nightingale steam engine.

The Belgian commissioner to the exhibition remarked upon the novelty, to a European, of many of the exhibits. In the American machinery section, for example, a considerable number of the 400 exhibitors displayed characteristically American special-purpose machines that were making obsolete many traditional skills. A gold-leaf beating machine; a new planing machine for fence pickets; a machine for cutting, dressing, and jointing barrel staves; a machine for pegging boots and shoes; a machine for making jack-chain links; a stone-dressing machine; and a machine for filing circular saws: these represented a logical extension of the ideas of division of labor and mechanization that guided British factory proprietors, but only in America had the ideas been so extensively developed. So impressed was the Belgian commissioner that he toured the United States before and after the exhibition, visiting iron works, mines, and centers of industry. When he returned home, he published an amply illustrated report of over 300 pages.

The editor of *Harper's* came close to the mark with his comment that "the true success of the Exhibition lies in its justification of the American pride. We have grown tired of hearing that we were such a great nation; but the Crystal Palace inclines us to tolerate the boast." There was much to be done, but much had already been accomplished. While sizable crowds had filled the New York Crystal Palace throughout the fall season, the exhibition could not regain the lost summer, and the enterprise was a failure financially. Nevertheless, the success or failure of the exhibition cannot be judged in terms of admissions revenues. If this criterion had been accepted at the time, most of the international exhibitions of the 19th century would never have been held, for none of the major exhibitions made money. Within a few years after this 1853 exhibition it would have been evident that such an enterprise had to be justified on some basis other than speculation. Private exhibitors found it profitable to pay heavily for the publicity that the international exhibitions afforded them, and governments, recognizing returns not immediately countable, became increasingly involved in the later exhibitions.

STILL MORE EXHIBITIONS

The next major international exhibition was held in Paris in 1855. An impressive industrial palace was built on the Champs Élysées, an art gallery arose nearby, and a machinery annex, three-quarters of a mile long and 89 feet wide, was erected along the right bank of the Seine. Although this exhibition was handicapped by its following too closely upon the heels of the Great Exhibition of 1851 and by the distractions of the then raging Crimean War, which diverted political attention in Paris and London to events on the Black Sea, this Universal Exposition attracted half again as many exhibitors as had the London exhibition; furthermore, the reputation of France for the princely hospitality with which she greeted foreign visitors was established during this exhibition. The four subsequent Paris expositions, held after 1867 at eleven-year intervals, were in many ways enlarged replicas of this first affair.

The London International Exhibition of 1862 was organized by the commissioners who had been responsible for the 1851 exhibition. An unfortunate choice of architect resulted in a very ordinary exhibition building. The death in 1861 of the consort, Prince Albert, who had been an enthusiastic promoter of the exhibition, cast a pall over the preparations. The Civil War in the United States reduced drastically the number of American exhibitors and caused the cancellation of an official American commission that had been expected to attend. In spite of difficulties, however, the 1862 exhibition drew as many visitors and more than twice as many exhibitors as there were in 1851. One tangible result was the establishment, in Berlin, of the Gewerbe-Museum, a commercial museum inspired directly by this exhibition. In view of the rule of the managers that British exhibitors would be permitted to show only objects that had been produced since 1850, there were undoubtedly many fresh and unproven innovations shown in 1862. Henry Bessemer, for example, had a very large display of objects illustrating his newly developed steel-making process. One visitor was struck by the "astonishing rapidity with which industry, in this country . . . particularly, assimilates the discoveries of science, gathering, so to speak, the ripe fruit of practical result from the seedling of theory only just sprung into existence."

An American novelty at the 1862 exhibition was the Porter-Allen high-speed steam engine, which Charles T. Porter, a gifted empirical designer, had taken to London in hopes of expanding his business. Whereas respectable English engines were running at 50 to 60 revolutions per minute, Porter's engine ran at 150, and did so more quietly and smoother than any other. Throughout the exhibition, Porter was mystified by the way his engine would attract visitors but no prospective customers. A visitor would ask him how he drove the air pump; he would answer that there was no air pump, because his engine was operating without a condenser. Thereupon the visitor would drift away, never to return. After the exhibition Porter learned that an air pump and a condenser were, in the mind of an Englishman, indispensable parts of a steam engine. A steam engine without a condenser was unthinkable—like Albert without Victoria.

Starting with the 1867 Paris Universal Exposition, the United States government, following the precedent of European nations, appointed an official commission to each of the major exhibitions; following each exhibition, an elaborate official report, usually five or six volumes in length, was published by the U.S. Government Printing Office. In the report of the 1873 Vienna International Exhibition, a section was devoted to the comments of foreign commissioners regarding the American exhibits. An English technologist, reporting on American machine tools, noted that the novelties exhibited in Vienna "were all improvements tending in a particular direction—precision and labor-saving and . . . advancing the gradual triumph of the human mind over matter." The Englishman did not belittle the solidity and rugged permanence of British machine tools, but he was impressed by the American innovations. A French observer commented upon the contrast between the decorated and embellished exhibits of other countries and the spare, utilitarian aspect of American exhibits. Everything in an American exhibit, he said, was expected to yield a return. Nothing was brought to Vienna but what would sell: furthermore, even the exhibits of American education were expected to interest prospective immigrants. "Except what will promote colonization or commerce," he wrote, "they have nothing." Commenting on the 1878 Paris Exposition, *The Times* of London made the point again in nearly the same language it had used in 1851, that "American inventive genius develops more that is new and practical in mechanism than all Europe combined."

THE CENTENNIAL: PHILADELPHIA, 1876

The 1876 Centennial International Exhibition, to celebrate the one hundredth anniversary of the Declaration of Independence, took place in Philadelphia less than a generation after the New York Crystal Palace Exhibition. Not only was it vastly larger than the earlier one—71 acres in Philadelphia compared with 4 in New York—but it was looked upon in a quite different light by the leaders of industry and politics in the United States. The federal government built and furnished an appropriately impressive building, some 480 feet long. The Smithsonian Institution and the Department of Interior each furnished a half-acre of exhibits; the Navy, war, agriculture, treasury, and post office departments all had separate exhibits. Nearly every state government had its own pavilion among the dozens of buildings erected in the extensive park surrounding the enormous central exhibition buildings.

Although there were more than enough non-mechanical exhibits in this exhibition to occupy the full time of the least technically inclined visitor, there was no question as to where the center of interest and purpose lay. Machinery Hall, which was itself nearly as large as the original Hyde Park Crystal Palace, was full of industrial exhibits; agricultural machinery was in Agriculture Hall, another vast and separate edifice; and numerous separate buildings for industrial exhibits were spread throughout the park—for example, the Campbell Printing Press Company, of New York, had erected a complete newspaper printing shop in a three-story building; a glass works was operated by a local firm; and there was a Singer Sewing Machine Building just opposite the art gallery.

Dominating Machinery Hall and, in many ways dominating the exhibition, was the great Corliss steam engine, which supplied power for the many working machinery exhibits. The engine was a double walking-beam engine, having two

Fig. 44-3. The great Corliss steam engine, 1876. Supplying power to the exhibits in Machinery Hall, the engine was the focal point of an exhibition that celebrated material progress. (*The Metropolitan Museum of Art*)

vertical cylinders, each of 40 inches bore and 10 feet stroke. The flywheel was a huge gear, 30 feet in diameter, which mated with a pinion on the main power shaft. As part of the opening ceremonies this impressive mechanical giant was started by the President of the United States and the Emperor of Brazil; its great overhead walking-beams began their deliberate bobbing—each making 34 bows per minute to the surrounding machinery—for the rest of the Centennial summer.

A correspondent of the *Atlantic Monthly*, who had made his way across the Centennial grounds, past the "horrid little restaurants" that diffused "a stale, sickening steamboat odor in their vicinity," came at length into the presence of the giant of Machinery Hall. It is instructive to read what he had to say about the engine, for his comment epitomizes the uncritical enthusiasm, bordering upon reverence, with which most Americans—if not most people of the world—regarded the wonders and novelties of technology, so dramatically exhibited here in a holiday atmosphere:

> The Corliss Engine does not lend itself to description; its personal acquaintance must be sought by those who would understand its vast and almost silent grandeur. It rises loftily in the centre of the huge structure, an athlete of steel and iron with not a superfluous ounce of metal on it; the mighty walking-beams plunge their pistons downward, the enormous flywheel revolves with a hoarded power that makes all tremble, the hundred life-like details do their office with unerring intelligence. In the midst of this ineffably strong mechanism is a chair where the engineer sits reading his newspaper, as in a peaceful bower. Now and then he lays down his paper and clambers up one of the stairways that cover the framework, and touches some irritated spot on the giant's body with a drop of oil, and goes down again and takes up his newspaper . . .

The German commissioner to the Centennial, Franz Reuleaux (1829-1905), saw the industrial exhibits in a much more practical light. Reuleaux, a prominent writer on machine theory, was a professor at the Royal Industrial Academy in Berlin and had been a commissioner to the exhibitions of 1862, 1867, and 1873. Upon comparing the German exhibits in Philadelphia with those of other nations, he wrote for publication a series of sharply critical "Letters from Philadelphia," in which he characterized the industrial production of Germany as *billig und schlecht* (cheap and ugly). These letters, which were eagerly read and passionately discussed throughout Germany, probably helped steer technical effort toward the manufacturing excellence that has long been associated with German industry.

Reuleaux was in 1879 and 1880 the manager of the German contributions to the Sydney and Melbourne international exhibitions. These audaciously conceived exhibitions, distant, as they were from world centers of industry, nevertheless put Australia on the commercial map, as they were intended to do. The German exhibitors, for example, opened German trade with Australia.

PARIS: 1889

The 1889 Paris Universal Exposition, celebrating the centennial of the French Revolution, was larger than any that had gone before and was the occasion for raising the Eiffel Tower above the skyline of Paris.

The idea of a tower, whose only purposes were to be conspicuous and to provide a bird's-eye view of the surroundings, was not new. Richard Trevithick, the noted English engineer, had proposed the erection in 1832 of a cast-iron tower, 1000 feet high, to celebrate the passage of the Reform Bill in Parliament, but it was not built. A wooden tower called the Latting Observatory, nearly 300 feet

Fig. 44-4. Eiffel Tower, Paris Universal Exposition, 1889. *La tour de trois cents mètres* (the 300-meter tower) was built as a spectacular visual focus of the 1889 exposition, whose buildings surround the base of the tower. The elevators to the second platform were supplied by Otis Brothers & Company, an American firm. (From *La Nature*, 1889)

high, had been built across the street from the New York Crystal Palace in 1853; a 1000-foot wrought-iron tower had been proposed for the Centennial Exhibition, but had not been built. The Eiffel Tower, 300 meters—nearly 1000 feet—high, was bitterly opposed by people who thought that the barbarous mass of iron and steel, like some great kitchen chimney, would dominate the city and overshadow the great monuments of the city's past. By the time the opposition was heard, however, work on the tower was under way.

Certainly it must be said that Gustave Eiffel built a tower that from an aesthetic point of view has worn rather well. The problem of the elevators, which had to operate within the sweeping curves of the tower legs, proved to be the occasion for a minor triumph of the Otis Elevator Company, an American firm. In all of France no reputable builder could be found who would undertake to design or build these elevators. In spite of the exhibition managers' prohibition of foreign material to build the tower, Eiffel was forced to accept Otis's proposal to supply and install elevators to carry visitors from the ground to the second balcony, some 380 feet high.

CHICAGO: 1893

The 1893 World's Columbian Exposition, to commemorate Columbus's discovery of America, opened a year later than originally planned. It was held in Chicago and was notable for the carefully planned site and for the architectural unity of the buildings.

The site-planning at this exhibition led directly to the City Beautiful Movement, which in turn produced Burnham's Chicago Plan of 1909, the first comprehensive metropolitan plan to be made in America. While its Neo-Classical buildings have been criticized as a retarding influence in American architecture, they did show the possibility of making a building of large roof-span look like something other than a train shed. The World's Columbian Exposition also brought Chicago to the attention of the world. An Englishman observed that before 1893 "Chicago was little more than a name on the map to nine-tenths of our manufacturers."

The first Ferris wheel, 250 feet in diameter and 28 feet wide, made its appearance at the Chicago exhibition. Designed and built by George Washington Gale Ferris (1859-96), a young American civil engineer, the great wheel was Ferris's reply to the challenge thrown out to American civil engineers to produce an object for the Chicago exhibition that would rival in interest the Eiffel Tower.

EXHIBITIONS AND EDUCATION

The Chicago exhibition, like the 1878 and 1889 Paris exhibitions, entertained a large number of international congresses, whose subject matter ranged over and beyond the physical sciences and technology, including such topics as agriculture, health, and religion. One of the most successful was an engineering congress, which encompassed the traditional branches and also resulted in a new Society for the Promotion of Engineering Education (now the American Society for Engineering Education).

Every one of the 19th-century international exhibitions, in fact, provided opportunities for fruitful discussions among men who had similar technical interests but who lived in different countries. The 1851 Great Exhibition was perhaps

Fig. 44-5. World's Columbian Exposition, Chicago, 1893. This nighttime view indicates the success of the alternating-current system of lighting. (*The Metropolitan Museum of Art*)

the first formal occasion in world history for any considerable number of technologists of different countries to come together with peaceful intent. In many of the reports of commissioners to subsequent exhibitions, particular mention was made of the valuable discussions that took place among the commissioners of different countries.

The effect of the 1851 exhibition on education has already been mentioned. A sequel to the flurry of concern that followed this first exhibition is to be found after the 1867 Paris exhibition, when John Scott Russell, English engineer and commissioner of the 1851 exhibition, observed that "by that [1867] Exhibition we were rudely awakened and thoroughly alarmed. We then learnt, not that we were equalled, but that we were beaten—not on some points, but by some nation or other on nearly all those points on which we had prided ourselves." It was evident to Russell, at least, that the reason for Britain's backward state was the absence of an educational system for workmen. Between 1851 and 1867 there had been time to train a whole new generation of workman, said Russell. On the Continent the opportunity had been seized; in Britain little had been done. While actual gains in elementary and technical education in the United Kingdom were discouragingly few, the succeeding international exhibitions provided those who were pushing for reform with clear evidence of need.

A direct, if faintly ludicrous, effect of the 1876 Centennial Exhibition on American engineering education resulted from the visit of John Runkle, president of the Massachusetts Institute of Technology, to the Russian section. In the exhibit of the St. Petersburg engineering college Runkle thought he had discovered the system of shop training that he had been searching for. Instead of using the college machine shops to build machinery for sale, the Russians had eliminated the commercial motive, and had started with elementary instruction in each of the skills comprising technical expertise. Thus, elementary courses in filing, forging, turning, and welding were required of a student before he could enter the shop in which general machine work was being done.

The Russian plan was quickly adopted at M.I.T. Before the year was out, an observer of the M.I.T. experiment had had "ocular demonstration of the results thus far achieved. The pupils, since the commencement of the lessons, had completed the course in 'filing,' and we saw in the results of only *eighty hours* of practice and instruction, such exquisite workmanship as could not be surpassed by an apprentice of two years' experience in an ordinary shop."

A number of the world's leading technical museums owe their existence to the international exhibitions that have been mentioned above. The 1851 Great Exhibition was directly responsible for the Victoria and Albert Museum, out of which evolved London's Science Museum. The original impetus for Vienna's Technisches Museum was supplied by the 1873 Vienna exhibition. The United States National Museum, part of the Smithsonian Institution in Washington, was put on a permanent footing as a result of its receiving a residue of objects from the 1876 Centennial Exhibition. The Museum of Science and Industry, in Chicago, is located in one of the 1893 exhibition buildings, which has since been rebuilt using permanent materials.

International exhibitions frequently played a crucial role in making innovations generally known. In a surprisingly large number of instances a new device or process was discovered by the technical world in an exhibition hall, even though it may have been in existence for many years. For example, the Bessemer steel exhibit at the 1862 London exhibition gave Germans their first knowledge of the Bessemer process, although it had been patented in 1856 and described by Bessemer in a paper before the British Association for the Advancement of Science meeting of 1856. A particular version of the Corliss engine valve gear was known in Germany as the "1867 valve gear," having been discovered by German visitors to the 1867 Paris exhibition; they later learned that this device had been patented in France in 1859. The conventional type of micrometer caliper, a workshop measuring instrument, was introduced by Brown and Sharpe into the United States in 1868 after Joseph R. Brown found at the 1867 Paris exhibition a micrometer caliper that had been patented in 1848 by Palmer, a Frenchman.

When in the early 1890's the question was raised as to how the power derived from Niagara Falls might be conveyed to Buffalo, about twenty miles away, an alternating-current system appeared to a small minority of engineers to be desirable, but nothing was known with any certainty about such a system. In 1891 there was exhibited at an electrical exhibition in Frankfurt, Germany, a long-distance alternating-current transmission system, conveying power from 100 miles away. In 1893, an alternating-current system was selected for the World's Columbian Exposition which demonstrated the suitability of alternating current to carry a large and variable light and power load. These two public demonstrations of alternating-current electricity helped give the Niagara Falls Power Company confidence to install an alternating-current system in the face of predicitions of dire consequences by such noted experts as Thomas Edison and Lord Kelvin.

THE BALANCE SHEET

During the latter half of the 19th century, when technology became a dominant factor in Western culture, international exhibitions were notably successful in diffusing and promoting technical ideas and objects. An English journalist who reported at great length on the mechanical exhibits in the 1893 Chicago exhibition thought that a much more systematic record of exhibitions should have been kept, in order to avoid past mistakes and to capitalize on forgotten successes. This is a familiar reaction to the insoluble problem of how to systematize knowledge so that it will come to the attention of those who need it or might use it, and how to retrieve information that has been already recorded. In spite of the haphazard and often purely fortuitous method of gaining knowledge by these exhibitions, it is doubtful that a more efficient promoter of technology could have been devised at the time.

The exhibitions were considerably less successful in accomplishing the fervent aim of Prince Albert to be "a happy means of promoting unity among nations, and peace and good will among the various races of mankind." Neither would the assertion of William Seward, American Secretary of State, that the exhibitions fostered "a better understanding between labor and capital," bear close examination.

There were the Colonel Sibthorps, who opposed this or that aspect of the various exhibitions, but their views were seldom popular. More frequently, the opponents appeared to be merely eccentric characters who would attempt to make more difficult the work of the world's benefactors. The strident, mechanical voice of progress echoed through the Crystal Palace and all subsequent exhibition halls, but it was an exceptional ear that could detect its presence. When steam-powered presses brought the news into the streets an hour earlier, there were few who asked why the haste; when Bell exhibited his telephone at the 1876 Centennial Exhibition, it was a renegade who wondered what people had to say that would not wait.

The task of international exhibitions was in the 20th century taken over by a rapidly expanding host of technical magazines, by catalogues and advertising pieces of all kinds, by the conventions of technical societies, by specialized exhibitions such as those showing only electrical goods or machine tools, and by the sober trade fairs that have become annual events in Milan, Hanover, and other cities throughout Europe. World's fairs linger on, and while they sometimes dazzle the public with their display of the "wonders" of science and technology, today's fairs are less substantial and less significant from the technical point of view than were their 19th-century predecessors.

45 / Technology and Culture at the End of the Nineteenth Century: The Will to Power

CHARLES L. SANFORD

The machine, by the end of the 19th century, dominated the imagination of all ranks of society in France, England, Germany, the United States, Japan, and to a lesser extent Russia, Italy, and southeastern Europe. Indeed, the machine was the master image of the day, whether expressed in metaphor and symbol or in the outpouring of new engineering marvels and the daily, familiar sight of factories, streetcars, steamships, locomotives, hydraulic pumps, dynamos, reapers, binders, elevators, printing presses, electrical appliances, sewing machines, telephones, and the machine tools and dies which stamped them out with uniform precision in prodigious number. Quite as important to the imagination as their physical presence, of course, was their incorporation into the vital functions of society. The major, underlying cultural issue of the period was not whether Western man belonged to the machine, but on what terms.

Perhaps the man who best represents the cultural tendencies of the end of the century was the German philosopher, Friedrich Nietzsche (1844-1900). He was one of those brilliant, intuitive, though unstable, fated geniuses in whom all the contradictory forces of a period gather for expression. In 1880, John Ruskin had complained that the "ferruginous temper" of the age was changing "merrie olde Englande" into the Man with the Iron Mask. Nietzsche personified the Man in the Iron Mask for all Western civilization, including its most distant outpost in Japan. He once wrote, "I belong to the machine which may fly to pieces." Indeed he went insane before his death; yet he was far more in harmony with the assertive energies of his time than, let us say, the self-proclaimed student of force, the American historian Henry Adams (1838-1918), who from a sense of failure withdrew into irony and self-contempt, or the English reformer William Morris (1834-96), who, looking backward, sought to revivify a medieval tradition of handicraft. Nietzsche's writings said "No" in thunder to all the things in which good people everywhere professed to have faith—liberal democracy, liberal Christianity, rationalism, and the cult of progress—and they were afraid.

Yet the heart of Nietzsche's message was affirmative, as was the main thrust of Western culture. Nietzsche was a nihilist (one who rejects customary religious beliefs and traditional morality) only insofar as he saw the necessity of discarding old lumber in order to build anew. In his radical innocence, his lyrical ecstasy, and even more than did Darwin and Freud, Niezsche shocked a generation which continued to justify its existence by sentimental illusions inherited from the past. In lectures on psychoanalysis after the turn of the century, Sigmund Freud (1856-1939) would point out that man's vanity had suffered three major insults from science: the first inflicted by the Copernican view that our earth was not the center of the universe; the second by the Darwinian hypothesis that men were descended from animals, possessing ineradicable animal instincts; the third—and one most resisted, from Freud's point of view —by Freud's own discovery that human will, that once proud monarch of the mind, took its marching orders from mysterious subconscious desires. Long before Freud, Nietzsche had startled others by uninhibitedly glorying in the subconscious forces which impelled men to power, and attempting to convert these into a source of joyous health.

THE POWER OF TECHNOLOGY

The dominant popular response to the question of man's relation to the machine can be called Nietzchean, though it did not always take the specific shape and form of which Nietzsche himself would have approved. It was Nietzschean because, in the main, consciously or unconsciously, it looked enthusiastically to machine technology as a field and instrument for the exercise of power—power transcending or nullifying the categories of good and evil, class power, national power, imperial power. It was the culmination of the Enlightenment's hopes of achieving mastery over Nature, if not of the Enlightenment's passion for rational understanding. Indeed, few men of the 1890's had moved much beyond the Enlightenment's conception of the universe as a vast machine. Throughout the 19th century this image was reinforced from all sides, and extended even to biology and the new sciences of man and society.

An Englishman, Edward Carpenter, who read a very different meaning into new scientific discoveries by Maxwell, Faraday, Röntgen, and the Curies complained that "mechanical ideas have come to colour all our conceptions of Science and the Universe. Modern Science holds it as a kind of ideal (even though finding it at times difficult to realise) to reduce everything to mechanical action, and to show each process of Nature intelligible in the same sense as a Machine is intelligible." He continued,

> It is curious that during this mechanical age of the last 100 years or so, we have not only come to regard Society in a mechanical light, as a concourse of separate individuals bound together by a mere cash-nexus, but have extended the same idea to the universe at large, which we look upon as a concourse of separate atoms, associated together by gravitation, or possibly by mere mutual impact.

Only a few members of the intelligentsia of the 1890's understood what the greatest artists and poets have always instinctively known, what Nietzsche knew as a poet if not as a thinker; namely, that any intelligible account of the universe must reckon with the irrational. But for the mass of people science remained mechanistic, and by science they usually meant a wonder-working *applied* science which would bear fruit in their own time.

Machine technology is, like any tool, morally neutral. When incorporated into the fabric of human society, however, it willy-nilly inculcates certain values associated with its leading characteristics. These are: causal sequence, order, functional efficiency, impersonality, uniformity, number, utility, power, and motion. The machine also whets the normal human appetite for creature comforts and pleasure formerly reserved to a small ruling elite, but now widely diffused. The enormous success of machine technology in all these respects begot by the end of the 19th century what the Spanish philosopher José Ortega y Gasset called, with aristocratic disdain, "the revolt of the masses." The image of the new mass is imprinted everywhere in the arts and history of the period, uprooted from the land, on the move, largely disrespectful of the old pieties, sensation-seeking, hungering for a cause, enormously attracted by the outward display of power in conspicuous consumption, glittering parades, the great international expositions, monumental public buildings, and the acquisition of colonial empires.

SOCIAL DARWINISM AND NATIONALISM

The idols of the masses, with whom they identified in the United States, were men like Andrew Carnegie, John D. Rockefeller, and Teddy Roosevelt—all of whom had risen to the leadership of vast aggregates of industrial power and endorsed the view of society as a Darwinian machine regulated by the principles of natural selection, adaptation, and the struggle for survival. This philosophy was extended to the international sphere, justifying America's venture into imperialism at the end of the century, competing with European powers.

Alfred Thayer Mahan's (1840-1914) primer for the United States, *The Influence of Sea Power* (1890), became an important text for the leading industrial nations of the earth by justifying imperialist expansion on the basis of profit. In it he wrote, "The instinct for commerce, bold enterprise in the pursuit of gain, and a keen scent for the trails that lead to it, all exist. . . ." For sentimental and tender-minded souls who winced at such bald statements of purpose, however, Rudyard Kipling (1865-1936) became the patron saint of colonial empire, equally popular in the United States and England.

The new mass man compensated for his loss of identity as an individual by rallying to the banners and march music of nationalism, which was an expression of himself written large with a central nervous system of telegraphs, railways, steamship lines, and printing presses. This expansion of consciousness helped also to fill the voids left by the decline of religious faith. Heralded by the American poet Walt Whitman (1819-92) in his *Song of Myself*, the transmutation of individualism into national egotism during the later years of the century is exemplified in the work of a French patriot, Maurice Barrès, who in the 1880's published a trilogy, *The Cult of Myself*, followed in the 1890's by another trilogy called *The Romance of National Energy*. Other writers, especially in Germany, brought to nationalism a kind of religious mystique of a super-state or race spirit, but in the main nationalism was materialistic, forged by machine technology and dedicated to the interest of the masses. As J. Holland Rose, the historian, put it:

> Functions of the state—now normally the national state—were being rapidly expanded to foster technological progress, to multiply public works and creature comforts, to cope with a wide latitude of economic and social problems issuing from the latest stage of industrialization, to elaborate and maintain vast systems of public education, to carry the advance of medical science into the big realm of public health. More and more, therefore, the national state was becoming, in the eyes of its citizens, a fairy godmother.

The wars of the second half of the 19th century resulted from an aggressive sensitive, competitive nationalism in which (despite the words of Alfred Mahan) the spirit of economic gain played a relatively subordinate role. Thus, a Cambridge don who wrote a very able, perceptive history of his own time blamed the Franco-Prussian War (1870-71) on national jealousy "fanned for four years by newspaper editors and popular speakers" who could not have prevailed without "the massing of mankind in great cities, where thought is superficial and feelings can quickly be stirred by a sensation-mongering press." Later historians have testified with a wealth of supporting documentation that the monied interests followed almost reluctantly the lead of adventurers, jingoistic publicists, patriotic professors, and public opinion into imperialism and war at the end of the century. Assuredly, the flag would not have followed trade without the support of the masses, who had no economic interests, but who devoured avidly the imperialistic harvests of writers like Kipling. The real stakes of the game were not capitalistic acquisition, but national power. So much for Nietzsche's characterization of the supposedly timid, leaderless, machine-like democracies which were supposed to have succeeded to a supposedly herd-like Christianity!

EXPOSITION OF TASTE

Yet Nietzsche had a point even here. Rose, the Cambridge don quoted above, although critical of the effect of imperialism upon subjugated peoples, let himself be beguiled by the sudden ascendancy of Western culture over the face of the globe. Musing upon this, however, he concluded that such ascendancy "has not been accomplished without losses, which are none the less serious because they cannot be assessed by financial standards. Chief among them is the tendency to excessive organization, drill, and strain observable in the Continental Governments of today. As a consequence, the freshness of individual life . . . has given place to a general sameness and respectable mediocrity, highly satisfactory to the drill-sergeant and tax-collector, but not fruitful in original achievements." Similar observations appeared elsewhere. A juror at the Paris International Exposition of 1889, for instance, drew upon a Darwinian analogy to describe the situation at the industrial exhibit:

> Each nation, gradually losing its prejudice, has adopted the improvements which have enabled another to take the lead in any particular direction, whether in matters of practical importance or of good taste. A general uniformity of excellence, amounting indeed to a certain lack of individuality in the design of the machinery of any one class, was a marked feature of the Exposition. . . . The doctrine of the survival of the fittest was here illustrated in its application to inanimate objects. (Reports of the United States Commissioners to the Universal Exposition of 1889 at Paris, III, p. 8.)

Other observers noted a greater uniformity in all departments, including the fine arts, than had prevailed at the earlier international expositions, which had begun at the London Crystal Palace in 1851, and not all agreed upon the uniformity of excellence.

The mass man was the creation of machine technology which transferred the will to power from the individual to the nation. The various international expositions, each one grander and more pretentious than the preceding one, were vast inventories of national achievement and power through which millions of admission-paying sightseers strolled in mingled awe and confusion, lingering longest before the wonders of science and technology, unaware that the most creative work of the age was not represented here because it was not yet known or valued. Let us pause and examine these international extravaganzas in the context of the general culture, for they present the age's most vivid and dramatic confrontation of the problem of men and machines—or, as one writer put it, they raised for the first time the question whether people knew where they and their civilization were driving with such headlong haste and apparent enthusiasm.

The original driving force behind these great spectacles had come from a young Englishman named Henry Cole, who became alarmed for the spiritual health of the individual as English society became mechanized ahead of other countries. Instead of joining the circle of reformers who preached a return to handicrafts, he organized a Society for the Encouragement of Arts, Manufactures and Commerce in an effort "to bridge the gaps between the artist, the manufacturer, and the designer." The purpose of the first international exposition in 1851, then, was to demonstrate a harmony of interest between artist and industrial engineer. Its creative center was to be the Crystal Palace, proof of a vision which dissolved "all traces of materiality." Prince Albert, president of the Society, was even more sanguine. Preaching a gospel of universal progress, he opened the Exhibition with a public address in which he said, "Nobody who has paid any attention to the peculiar features of our present era will doubt for a moment that we are living at a period of most wonderful transition, which tends rapidly to accomplish that great end to which indeed all history points —*the realisation of the unity of mankind.*"

THE NEW MASS MAN

Even at the first international exposition there were evidences of *disunity*, readily apparent, for example, in the sharp contrast between the blatancy of the machine-made carpets of Europe and the serenity of the primitive patterns from the Far East. The original undertaking was conceived and projected in the name of a humane culture sensitive to the needs of the whole man; but what began in the spirit of individualism was quickly taken over by the mass man, whose image even then was taking form in the pages of literature.

One of the earliest of the type came from the pen of the French novelist Gustave Flaubert (1821-80) in *Madame Bovary*. I have in mind primarily the fatuous would-be scientist, Homais, whose experiments at the expense of a poor beggar and a blind man only endeared him the more to a gullible, progress-loving public; but Flaubert's description of Charles Bovary's cap would do just as well as a parody of the mass man's ruling tastes at the expositions, being of "the composite order,"

> in which we can find traces of the bearskin, shako, billycock hat, sealskin cap, and cotton night-cap; one of those poor things, in fine, whose

> dumb ugliness has depths of expression, like an imbecile's face. Oval, stiffened with whalebone, it began with three round knobs; then came in succession lozenges of velvet and rabbit-skin separated by a red band; after that a sort of bag that ended in a cardboard polygon covered with complicated braiding, from which hung, at the end of a long, thin cord, small twisted gold threads in the manner of a tassel. The cap was new; its peak shone.

Offering much the same message, the American Henry Adams would describe the exhibits and architecture of the 1893 Chicago World's Fair as a "Babel of loose, ill-joined, vague, unrelated thoughts and half-thoughts and experimental outcries."

On the whole, however, American and Russian writers tended to have more sympathy for the "new man," delineated as William Dean Howells's Silas Lapham, who would "spiritualize mineral paint"; or as Mark Twain's champion of "hard unsentimental common sense," the character Hank Morgan from the novel *A Connecticut Yankee in King Arthur's Court*, whose failure ultimately to conquer ancient superstition Twain later attributed to the fact that although he was the boss of a machine shop, could build a locomotive or revolver, and run a telegraph line, this Yankee was "an ignoramus, nevertheless." The new type would reappear as the more prosaic hero of Edward Bellamy's utopian novel, *Looking Backward,* bringing the long heralded "brave new world" into being without a single jolt to human nature. What a showcase these writings would have made at the Paris Exposition of 1889, competing, for the attention of crowds, with such stellar attractions as night's myriad fountains, electrically illuminated, and the sky-soaring Eiffel Tower, eighth wonder of the civilized world! As a coherent, rational structure of human possibilities, the novel made far more sense than these marvels.

Then we come to the Russian novelists, who with a profound insight into human nature, conversant with the demonic as with mere unreason, were not unsympathetic to the new mass man. Ivan Turgenev (1818-83) gave us Bazarov; Feodor Dostoyevsky (1821-81) gave us Raskolnikov—both would-be supermen, believing like Bellamy's hero that environment was all, but differing in the bitter knowledge that violence was necessary to bring about the desired changes—to destroy in order to create, to do evil in behalf of a greater good; that was the irony, the paradox, the crucifixion.

In *Crime and Punishment* Dostoyevsky had Raskolnikov argue with an almost Marxist feeling for the dialectic that "if the discoveries of Kepler and Newton could not have been made known except by sacrificing the lives of one, a dozen, a hundred, or more men, Newton would have had the right, would indeed have been duty bound . . . for the sake of . . . humanity." But the character of Bazarov in Turgenev's *Fathers and Sons,* is more pertinent to our purpose. Bazarov comes nearer to being the archetypal figure to represent the forces and ideas in Western civilization for the late 19th-century international expositions, which in themselves symbolized the stirrings and counter-movements of the age. Bazarov was an avowed materialist and Russian nationalist who claimed to have no use for the so-called "finer things of life"—poetry, art, literature, religion, metaphysics, history, and the "Mother Russia" of the past. To him, these things were not merely frivolous; they meant aristocracy, elitism, injustice—until he encountered a superbly feminine, if high-strung, woman, Madame Odintsov. For he was, after all, intelligent, and humanly sensitive. Socially he was a nobody; she belonged to the lower aristocracy.

Turgenev records beautifully the anguish of Bazarov's awakening to the values of love and poetry and music, which he discovered to be inseparable and without which he was little more than a useful machine serving a shallow utilitarian morality. Unfortunately, his proud bad manners and Madame Odintsov's deep reserve maintained the barrier of class between them which had fallen for a few sweet, tempting moments. The proud superman who would have imposed his will on destiny later cut his finger by accident while doctoring his father's peasants, became infected, and died submissively, the ironic victim of fate; while his once submissive father rebelled against the injustice of the universe.

THE FATE OF THE ARTS

As if to rediscover or recover the inner human being from machine culture, the best art of the late 19th century went underground; the mediocre performances showed up at the international expositions. At the Paris Universal Exposition of 1878, the ruling taste of the mass man was already dominant. In America, meanwhile, the voice was feminine. The popular author William Dean Howells would call her "the Iron Madonna"; while Thomas Beer, the acidulous historian of popular taste, would rename her "the Titaness." We shall hear her voice at the fine arts section. At the 1878 Exposition, the United States spent $160,000, the bulk of it on building construction; in 1890, she would spend $250,000; in 1900, the enormous sum of $1,341,379.40—a rate of increase far exceeding, during the same period, the population growth or increase in national wealth. All the other nations raised the ante accordingly, a sure measure of the importance attached to national prestige.

What novelties captivated the 16 million people who attended the 1878 Paris Exposition? First came the electric light, then the new Bessemer products, followed by the industrial and education exhibits. Americans took ten grand prizes in the machine industry category—including Thomas Edison for his telephone and phonograph and Cyrus McCormick for his binding reaper; but France walked off with the majority of prizes in every other category. In a news article and editorial, the London *Times* for August 22, 1878, commended the United States for a mechanical genius "exceeding all Europe," but touched a sensitive nerve by declaring, "The mechanical development of the States belongs to a lower sphere than the splendours of Elizabethan literature or the glories of art in the Venetian and Dutch republics. . . ." French critics were equally patronizing. With James A. Whistler in mind, they thought it nice for an artist to be born in America, but to study and live abroad. The split personality of the modern world was now laid bare.

In his summary report, United States Art Commissioner William Wetmore Story, himself a distinguished sculptor and man of letters, raised a manly voice. Noting that "our needs and necessities have been amply supplied, but the heart and soul have been fed upon husks," he called for the founding of a national academy and government support of the arts, as in Europe. But, then, in his critical review of all the fine arts assembled at Paris, he lapsed into the feminine voice of the American Madonna. He was on good ground, of course, when he saw little differentiation among the national schools of art and architecture. Even with the French, who had raised an oriental palace in the Trocadero, imitativeness and eclecticism were the principles of the day. As Story said, "The modern facilities of travel have tended to draw all together into one great nationality of art." He liked French art best for its mastery of technique, but his American prudishness was offended by an "over-abundance of nudes" and "relish for brutality":

> The influence of France is greater than that of any other nation. It inoculates all the world with its disease, but nowhere is its contagion so deeply felt as in the United States. . . . The over-stress of the realistic school on the common, brutal, and even ugly, is of course the revolt against the artificially tawdry sentiment and effeminate feebleness of the vague idealistic school.

What he was complaining about here represented the very essence of machine culture, not yet well understood or appreciated by the masses, whose hunger for reality was divided between *things* and the kind of pseudo-reality fabricated in science fiction by Jules Verne or H. G. Wells, in detective stories by Arthur Conan Doyle, in the pallid adventures of Robert Louis Stevenson, in the exotic travel tales of Rudyard Kipling, Pierre Loti, and Alphonse Daudet.

The practitioners of the new realism, led by Emile Zola in France, self-consciously sought to interpret human life according to the principles of mechanism and science. They can be accused of being over-literal, even sensual (though sensual in an often healthy pagan sense, not in keeping with the decadent sensuality encouraged in the new mass man by the outpouring of machine-made goodies); on the other hand, very few of them exploited ugliness for the sake of mere sensation and ugliness. By "delineating the vulgarity of the present age," Hamlin Garland, an American naturalist, wrote, he wished to hasten into being an "age of beauty." Without being polemical reformers, that is, the realistic authors were sensitive to the human values which they felt were being eroded by urban industrialism.

EDUCATION AND REFORM

The voice of the mass man at the 1878 Exposition was heard most powerfully in connection with the education section. A Frenchman named Charles de Souches had recently published a tract entitled *Elementary Studies, Political, Social and Philosophical* dedicated to the French working class from which he sprang. He was quoted approvingly by the American commissioner of education, who had noted that sightseers had been much attracted to a legend inscribed over the American exhibit, "L'INSTRUCTION PUBLIQUE EST GRATUITE DANS TOUS LES ETATS DE L'UNION" (Public instruction is free in the United States). "We will have a notable revenge," Souches wrote:

> We will show to the world what a people can avail which rises above the low level of ignoble passions. We are resolved, cost what it may: we are ready. What will we do, is it asked? The most simple thing. For this famous revenge is already begun. It is born, full of life; and it will grow and flourish if we but give it faith and patriotism and resolution. . . . Our revenge is the REPUBLIC.

Another Frenchman, Jules Simon, emphasized that the objective of mass education was national aggrandizement, for "the people which has the best schools is the first people."

The theme of nationalism in education was to be pursued in a frightening manner by Americans at the 1900 Exposition. In a circular to all school officials in the United States, the director of the United States Commission appealed for evidence to prove to the world that the American system of education, and not mere fortuitous chance, had produced a new type of self-reliant citizen—"the man behind the gun." But what most interested Europeans about American education was its *practicality*, and they wanted to know more about the American technical and engineering schools and the application of science to the useful arts. They had had enough of German thoroughness, drill, and theory.

At the Paris Exposition of 1889, the United States won the largest number of awards, including 55 grand prizes, one this time in the fine arts, an oil painting by John Singer Sargent. A number of entirely new categories showed the drift of public taste. These reflected the vanity of the parvenu, pertaining to the adornment of the body for the most part: wearing apparel and accessories, jewelry, furniture and accessories, food products. United States Art Commissioner Rush Hawkins protested even more strongly the mass man's contempt for the fine arts, doubting even whether his report was worth writing, since it could not exercise any perceptible influence on the education of public taste; "The crushing power of unprecedented enormous wealth . . . and the conceit it engenders in the minds of its possessors, constitutes a sort of intangible Chinese wall. . . ." A crass public, he felt, wanted the "largest amount of show at the smallest possible price," while architects in America, lacking the backbone to rebel against enticements of the dollar, disdained simplicity and plain surfaces. He presented an exciting program of reform, based largely on the example of France—whose nudes and brutal realism, incidentally, he found refreshing. His proposals were more radical than Story's, including not only government support for the arts, but the building of model homes for slum people and a new concept for rural house-building. As will be seen, his frankness did not endear him to his compatriots.

The influence of the mass man was also beginning to be felt at this Exposition in the new concern for social welfare and reform. In an important address before one of the intellectual congresses convened to review the 100 years of enlightened progress since the French Revolution, the French minister of finance, economist, and author, Leon Say, denied that man is also a machine, to be driven ever faster and more efficiently, as the Americans were recommending for the operation of the French railroad system. He went on to endorse state regulation of business in the interest of law and justice, to "improve the relation between man and man." But the revolt of the masses was rising to full cry, with Populist unrest in the United States and labor troubles both there and in Europe, and at the Paris Exposition of 1900 this revolt was reflected in the exhibits devoted to social economy, labor, public health and welfare, and race relations in the United States. One of the United States jurors was Jane Addams, noted for her work in the settlement-house movement in Chicago.

But new categories at the 1900 Exposition also gave recognition to the military thrust of nationalism, with sections on colonial and military affairs. Also the full import of Japan's strides in education and artistic and material achievement were recognized for the first time. The accent on creature comforts and pleasures continued, with special attention given this time to decorations, sports, and recreation. One is almost reminded of the ancient Roman formula of "bread and circuses" to distract the masses from trouble-making. The Americans, however, had something far more important in mind, which was to spare their nation from the imputation of cultural barbarism which had developed in earlier expositions. Writing for home consumption, United States Commissioner General Ferdinand Peck claimed that the Exposition was especially significant for the "definite demonstration of the increasing prominence of the artistic in the United States." With a callow optimism characteristic of the time, he then concluded that the United States had succeeded in convincing the visiting public that "it promises fair to solve . . . that problem which has so long annoyed Europe and the world and all ages, viz., the practical and satisfying combination of the artistic with the utilitarian, the aesthetic with the enduring, the beautiful and graceful with the progressive and strong."

His subordinate in charge of art took pains to counteract the Hawkins indictment of 1889, discovering a prominent French museum director who was willing to testify to the presence of a strong new American school bearing a more distinctively national stamp. The American juror, F. D. Millet, having half Hawkins's brilliance and art knowledge, also repudiated Hawkins, sneering at the charge "which has become stale from frequent repetition, that the American pictures are echoes of the French school." The truth was, however, that there was very little change for the better, except for a few more medals, placing the United States second to France in total awards for fine arts. Once more, Whistler and Sargent carried the day with two grand prizes, the former an expatriate, the other no better than a superb illustrator. A British critic for the *Art Journal* in London wrote, "The American Section of the Exhibition is notable only because three painters who have made their homes in Europe have had their works included amongst the exhibits of their country of origin." Except for a belated recognition of French Impressionism and a magnificent tribute to Rodin, the best creative work of the period was overlooked at Paris.

Albert Pinkham Ryder, Thomas Eakins, Van Gogh, or Cezanne—also those architectural geniuses, Louis Sullivan and Frank Lloyd Wright, who were discovering aesthetic and cultural value in the aesthetic principle that "form follows function"—had to bow to the miracles of applied science. To Henry Adams, the dominant symbol of the 1900 Exposition was the dynamo, for him an expression of the acceleration of blind force in a strange, new, supersensual world in which he felt lost. For other Americans, however, the dynamo represented a triumph of heady national power. Thus, the decorative theme for the interior of the American pavilion was "America revealing her power and resources." At the center of a vast mural was the Spirit of America lifting her veil, flanked by a figure who represented "steam, the force of the past, generating electricity, the force of the future." This theme was wedded to a great ceremonial occasion in which all Paris participated to celebrate the centenary of an Age of Progress; and on July 4th, 1900, led by a parade of international dignitaries and the marching band of John Philip Sousa, America unveiled her gift to the French Republic, a gigantic statue of Lafayette on horseback.

FIN-DE-SIÈCLE

The French term *fin-de-siècle* has a connotation far beyond its literal meaning of "end of the century," for it refers to the decadence, languor, and boredom of those who rebelled against the dynamic materialism of an industrial society. They were, in modern slang, the "sick" generation of the close of the 19th century. Henry Adams might indeed speak of a sickliness which overspread "the dead-water of the fin-de-siècle"—and with some justice if one considered only the few academicians, aesthetes, and artists who saw life darkly through the glass of individual consciousness and temperament as a last refuge from the tyranny of things, the farce of force. The Italian philosopher, Benedetto Croce, spoke for them all when he announced at the end of the century:

> We no longer believe . . . like the Greeks, in happiness of life on earth; we no longer believe, like the Christians in happiness in an otherworldly life; we no longer believe, like the optimistic philosophers of the last century, in a happy future for the human race. . . . We no longer believe in anything of that, and *what we have alone retained is* the consciousness of ourselves. . . .

This was Nietzschean doctrine, shorn of the will of power. It helps to explain the shift at this time of Sigmund Freud and William James from physiology into psychology. It accounts in large part for the closing darkness of Mark Twain's life, for the melancholia of the Symbolists, the pessimism of Thomas Hardy and A. E. Housman in England or Gerhart Hauptmann in Germany, the skepticism of Georges Sorel in France. So ended for them the Enlightenment dreams of restoring man to an earthly paradise through science and technology.

But popular culture at the end of the century remained incurably and optimistically dynamic. Its faith in endless material progress under the auspices of national power had, if anything, established a quickened tempo to life, glorifying movement, motion, *change*. Mechanization, men were told, begins with motion which gradually supplants the human hand, until man arrives at the assembly line, already in full operation by the turn of the century. The idea of progress itself, says Lewis Mumford, associates value with movement in time.

It is not surprising, then, that the chief novelties for sightseers at the 1900 Exposition incorporated motion: the automobile, the Otis escalator (which won a grand prize), and, oddity of oddities, a French-designed "moving pavement" which silently wound through the entire exhibition on rollers powered by electricity. The latter was called "the most marked characteristic" of the Exposition as far as the general public was concerned, promising to revolutionize the future appearance of cities. An architectural over-view of the Exposition also suggested motion in the organization of line and mass by all things multiform, rococo, bizarre, writhing, dynamic.

The same principle was carried over into the "new art." A German credo of the 1890's claimed to be concerned with aspects of life "which are perpetually in motion" rather than static and fixed by formula, with "that which is still 'becoming'. . . ." Impressionist painting sought to capture the essence of the passing moment in a nervous, telegraphic style which dissolved harsh matter into soft fields of color charged with energy. It belonged to the age of steam, electricity, and radium. The movement called Expressionism was even more extreme. If Degas's impressionistic painting, "Rehearsal for the Ballet," made heads swim, the violently swirling skies and cypresses of Vincent Van Gogh left them numb.

The best engineering designs of the period also accentuated motion, as in John Roebling's design for the Brooklyn Bridge, completed in 1883. Its organization of arches, towers, and the diagonal trusses which carried the slender suspension suggested, according to a recent study, "not matter at rest, but well-balanced motion." The same principle was applied to the over-padded, over-stuffed, upholstered world of the body, as the new patent-furniture became increasingly convertible, collapsible, foldable, movable. But was all this "motion in a void"? The great public didn't think so, even if it didn't fully understand the significance of what was taking place.

The 19th century ended in joyous, triumphant, public celebration—dedicated ostensibly by a delirium of sight and sounds on parade to Teddy Roosevelt, the hero of San Juan Hill, to Queen Victoria on her Golden Jubilee, and to Lafayette and the French Republic; but secretly, the mass man was celebrating his own rise to power culminating a century of progress. The music was John Philip Sousa's. If one listened carefully, however, he could hear in the background the apocalyptic crescendos and ominous *leitmotivs* of Richard Wagner's "Twilight of the Gods."

Epilogue

CARROLL W. PURSELL, JR. AND MELVIN KRANZBERG

One of the hallmarks of the 19th century was a belief in progress, and more specifically in the progress of the material well-being of mankind through the benevolent agency of technology. And little wonder! As we have seen in the course of this volume, technology had, with but few setbacks, worked to increase man's control over his environment from his very beginnings. By the year 1900, man commanded hitherto unprecedented heights of physical power.

Even before our present human species, *homo sapiens,* had come into existence, our pre-human ancestors had begun to employ tools. The ever-increasing use of these tools became involved in the evolutionary process which helped to develop the species: *homo sapiens,* Man the Thinker, was also *homo faber,* Man the Maker. Man made tools, and tools helped make man.

By the time of the invention of writing, which separates prehistoric from historic times, men had developed an arsenal of tools and implements; had learned to use and control fire; had begun agriculture, which required settled communities and enabled the establishment of urban life; had domesticated animals; and had learned to use sails and wheels to assist him in his transportation. These innovations were basic to the development of civilized societies, which grew up first in the Western world in Mesopotamia and Egypt. These fundamental elements of technology, with but slight additions, provided the technological and economic foundations of the Classical civilizations of Greece and Rome.

Despite the great achievements of Classical antiquity in philosophy, art, literature, drama, religion, and science, there were few startling or major advances in the technical means which man had at his disposal. There were, of course, some innovations, but the great engineering feats of the Romans, for example, were largely the result of superior organization and mobilization of technological facilities rather than innovations in mechanical devices or changes in the energy sources, materials, or other technological requisites.

Surprisingly enough, it was in the Middle Ages, long denigrated by scholars as a thousand years of backwardness, stagnation, and decay, that major technical innovations took place: an agricultural revolution (use of the heavy-wheeled plow, the three-field system, and other agricultural innovations which remarkably increased the production of food and made possible once again the beginning of city life); a power revolution (water wheels and windmills, to take the burdens off the backs of men); and a transportation revolution (more effective use of horses through the development of the stirrup, the rigid horse-collar, horseshoes, and the tandem harness, as well as greatly improved means of transportation over water, through the sternpost rudder, the fore-and-aft sailing rig, the mariner's compass, and so forth). Thus, while scientific, philosophical, artistic, and literary efforts might have languished somewhat during the medieval centuries, and political and economic life might have been at a low level of activity, technological growth remains one of the bright spots of the Middle Ages.

On the other hand, the Renaissance, exalted by scholars as a time of "rebirth" for the human spirit and marked by developments in philosophy, literature, art, and religious thought, brought few fundamental innovations in terms of technology. Nevertheless, the Renaissance was able to exploit the new mechanical devices, such as gearing mechanisms and crank applications, which had grown out of the Middle Ages, to utilize the new navigational and ship improvements of medieval times to discover new continents, enlarge man's horizon, and to revolutionize commerce. This early modern period served to develop the secular outlook, the economic demand for new products and new ways of doing things, and the accumulation of vast riches which together provided the basis for the remarkable transformation of man's technological activity—and his entire society and culture—which began in the mid-18th century and which we term the Industrial Revolution.

The Industrial Revolution thoroughly transformed men's ways of working and living, and indeed, revolutionized every aspect of human life in slightly less than a century. Furthermore, it changed the tempo of existence, and pointed the way for even further and faster changes in the 20th century.

Transportation offers an obvious case in point. Until the late 18th century there had been little improvement in transport for hundreds of years, and that had been without much drama. Then, within a few decades, a transportation revolution had brought incredible changes. For most people these were summed up by the railroad—perhaps *the* symbol of the 19th century. Along with the railroad had come steamships and already, in the 1890's, automobiles. Groping for some expression which would sum up this change, it was most often said that time itself had been *annihilated*.

The same thing, of course, applied to communications. One of the great romantic episodes of the American West was the Pony Express which began in 1860. Yet this dramatic service was only the ultimate exploitation of a form of communication centuries old—a message carried by man on a horse. The trip from St. Joseph, Missouri, to San Francisco took ten and a half days—ten less than the best stagecoach time. One year later, however, the transcontinental telegraph was completed and the time for a message to reach the Pacific was again cut, this time to a fraction of a second. Within another 15 years, the telephone was a crude though a working reality, and the Atlantic Cable shortly after mid-century brought messages across the ocean in an instant. A rapidly growing and increasingly concerned public created the need for a better network of communication, which was served by the publication of books, magazines, and newspapers on steam presses.

Steam played an immense role in both transportation and communication, as well as in nearly every other phase of material life in the 19th century. It both epitomized and embodied the fundamental technological change of the period—an incredible jump in available energy. Since time immemorial man had tried to harness alien powers to his own ends. However, before the age of steam, the sum total of energy which he could effectively convert to his purposes through wind and water, horse and human power, was quite limited. Then, in the 19th century, he began to tap the power of fossil coal and liberate it through the medium of the steam engine on a scale never before possible. The new power, multiplied or divided almost at will, was applied to uncounted tasks. Before the century was out, electricity, produced from coal or water power, was coming into industrial use on a growing scale, and petroleum too showed promise of supplying great quantities of energy.

The large increase in the variety of energy sources was accompanied by a similar multiplication of available materials and processes. Both typical and most important was the change in the availability and usefulness of iron. Again, iron had been used for centuries but the ability to produce it on a large scale, and to work it once it was processed, was severely limited. In the late 18th and early 19th centuries the working of iron was revolutionized by the puddling furnace and the rolling mill, the hot blast and the steam hammer. Production was increased and fabrication facilitated by a whole new family of machine tools. The introduction of cheap steel at mid-century only intensified this change. By the end of the century aluminum was becoming available at a price which made it useful in manufacture. The importance of such developments, and countless others involving the chemical and other industries, was this: the best invention or manufacturing process was worthless if proper or adequate materials were not available to build the machines. In the 19th century the age-old technology of wood gave way to one of iron and steel.

The industrialization and urbanization which accompanied these transformations made new and unprecedented demands upon agriculture. For large numbers of farmers during the 19th century, agriculture ceased to be a trade and became a business: transportation linked his supply to world demands; communications placed him in a market larger and vastly more complex than any he had ever operated in before. The power of his own brawn was augmented by that of animals and, to a small degree, steam for a myriad of tasks (from breaking the soil to reaping the harvest) which had previously been beyond the reach of machines. Wooden plows were replaced by steel ones, the sickle was replaced by the reaper, and soil exhaustion was reversed by scientific analysis and fertilization. Both the cost and the rewards of agriculture rose rapidly.

Military technology was touched by the same spirit of technological progress. Despite holocausts of unprecedented carnage, like the American Civil War, the century was one of comparative peace—a Pax Britannica extended from the end of the Napoleonic Wars in 1815 to the beginning of the First World War in 1914. Such a climate was not conducive to the improvement of military weapons, but nonetheless great changes took place. On the oceans, the wooden sailing ships of the line were replaced within the century by steamships of steel. The cannon which brought down the feudal nobility gave way to the steel guns of Krupp. The musket fell to the rifle using cartridges and, in some cases, semiautomatic action. Railroads rushed troops to and from the scenes of battle. Balloons discovered the enemy's position, and the telegraph sent that intelligence to headquarters.

These great changes in the human condition were not universally applauded: the cult of progress had its detractors. Chief among these, perhaps, were those artists and writers who had never made their peace with the Industrial Revolution. The changes wrought by that upheaval had been psychological as well as material. Someone has written that technology is the knack of so ordering the world that we don't have to experience it. In this sense man's experience had, for a century, become more and more artificial, and less and less rooted in reality. Machine culture proved to have its own logic and imperatives: in America people were moved from the country to mill towns; in Italy religious holidays were curtailed so that factories could run efficiently; in Africa natives were forced to pay their taxes in cash rather than in traditional produce so that they had to live by the sweat of their brows rather than off the bounty of the land. These disruptions were felt by many, but the poets and painters and some philosophers felt them most acutely and communicated their anguish most passionately.

Thus, even when at the end of the 19th century most men were congratulating themselves over their increasing rational control over themselves and their environment, some poets and aesthetes viewed with alarm the "triumph" of technology, and such men as Freud and Nietzsche were rediscovering the basic irrationality of man and his fate.

But few men bothered with second thoughts. For most, the last decades of the 19th century had seen the ultimate triumph of the Enlightenment, which had for nearly two centuries given inspiration to change by its emphasis on the possibility of human material progress here on earth, to be achieved by the rational means of science and technology. Indeed, the machine and its improvement had transformed the Enlightenment from a dream and movement of the intellectuals to a redeemable promise for the masses. Furthermore, this progress was spatial as well as temporal. For many centuries Europe had trembled in fear of "barbarians" from the East—in 1683 (only four years before the publication of Newton's *Principia mathematica*) the Turks laid siege to the city of Vienna. Some 200 years later the vastly superior technology of Western Europe had enabled that area to place most of the rest of the world under its sway.

In short, technological advance was a tangible and measurable proof of man's progress. By the close of the 19th century, technology had manifested its power in transforming the human condition and—in most cases—helping to ameliorate it. There were some reservations, to be sure, but overcoming these were the high hopes held forth for continued progress.

As we shall see, the 20th century was to fulfill many of these expectations in more dramatic and accelerated form than had been imagined by even the most optimistic of prophets at the close of the 19th century. The technological changes which were to occur in the 20th century matched in magnitude—some might even say dwarfed—those of the preceding millennia, in terms of their impact on human life and society. The predictions of the science-fiction writers of the 19th century were more than fulfilled; at the same time, the reservations and qualifications which some critics at the close of the century had regarding the aim and purpose of technological advancement were also being heard. In our second volume we will see how technology in the 20th century built upon the past to produce our contemporary society with all its progress, promises, challenges—and threats.

Readings and References

Although systematic scholarly study of the history of technology is relatively recent in the United States, there already exists a large body of literature in this field. The titles listed below, compiled by the editors, do not comprise a comprehensive bibliography; instead, they represent a selected list of books and articles the authors found helpful in the preparation of their chapters, and others which they suggest to the serious reader and teacher for further reading on the many topics covered in this volume. In order to keep the bibliography within reasonable bounds, nearly all of the non-English works have been excluded. Attention is focused on works in English, many of which are available in college and public libraries.

General Works, Bibliographies, Periodicals

The basic and most comprehensive work in the history of technology is the encyclopedic *A History of Technology* (5 vols., Oxford, 1954-58), edited by Charles Singer, E. J. Holmyard, A. R. Hall, and Trevor I. Williams, and consisting of a series of articles by the world's leading authorities in the field. A one-volume condensation of this monumental work has been published as an introductory textbook: T. K. Derry and Trevor I. Williams, *A Short History of Technology from the Earliest Times to* A.D. *1900* (Oxford, 1961). There are several other textbooks which provide a useful introduction to the history of technology: R. J. Forbes, *Man the Maker* (London, 1958); Richard Shelton Kirby, Sidney Withington, Arthur Burr Darling, and Frederick Gridley Kilgour, *Engineering in History* (New York, 1956); Friedrich Klemm, *A History of Western Technology* (New York, 1959); James Kip Finch, *The Story of Engineering* (Garden City, N. Y., 1960); and W. H. G. Armytage, *A Social History of Engineering* (New York, 1961). A survey of American developments is given in John W. Oliver, *History of American Technology* (New York, 1956).

For the United States, the indispensable starting place is Brooke Hindle, *Technology in Early America: Needs and Opportunities for Study* (Chapel Hill, 1966). Two lavishly illustrated popularizations are Umberto Eco and G. B. Zorzoli, *The Picture History of Inventions* (New York, 1963) and Mitchell Wilson, *American Science and Invention* (New York, 1954). A major and pioneer effort to investigate the relation of technology to broad aspects of human culture is Lewis Mumford, *Technics and Civilization* (New York, 1934).

Eugene S. Ferguson's "Contributions to Bibliography in the History of Technology," which began appearing serially in Volume III (1962) of *Technology and Culture,* the international quarterly of the Society for the History of Technology, provides the best source for scholarly investigation of technological developments. These articles, which emphasize bibliographical works, indexes, and finding aids, are being revised and combined to form a volume in the Monograph Series of the Society for the History of Technology, scheduled for publication in late 1967 or early 1968. In addition, *Technology and Culture* has published annually since 1964 a "Current Bibliography in the History of Technology," compiled since 1965 by Jack Goodwin. Writings on the history of technology are also included in the "Critical Bibliography of the History of Science and Its Cultural Influences," published annually in *Isis,* the quarterly journal of the History of Science Society.

Articles in the history of technology regularly appear in *Technology and Culture* and two English publications, *Transactions of the Newcomen Society* and *Journal of Industrial Archaeology.* The Newcomen Society published cumulative indexes of its *Transactions.* Occasional articles on specialized aspects of the history of technology appear in *Isis, Business History Review, Journal of Economic History, Economic History Review, Chymia,* and other scholarly journals, as well as in trade and professional-society publications.

Prehistoric and Ancient Technology

The general background of prehistoric technology (Chapter 2) is in J. Hawkes and Sir Leonard Wooley, *Prehistory and the Beginnings of Civilization* (Vol. I in the UNESCO History of Mankind; New York, 1963). V. Gordon Childe is the author of two stimulating works on this early period: *What Happened in History* (New York, 1946) and *Man Makes Himself* (New York, 1951). Other important works include: Kenneth Oakley, *Man the Tool-Maker* (London, 1952); J. G. D. Clark, *Prehistoric Europe: The Economic Basis* (London, 1952); I. W. Cornwall, *The World of Ancient Man* (London, 1964). Julius Lips, *The Origin of Things* (London, 1949), tells of primitive and ancient technology from the standpoint of the ethnologist, while S. A. Semenov, *Prehistoric Technology* (London, 1964) is a fascinating account of the way in which early tools were made and used. R. J. Forbes has published nine volumes of his *Studies in*

Ancient Technology (Leiden, 1955-64) dealing with arts and crafts of the pre-Classical periods. Two more volumes are currently being prepared.

L. Sprague de Camp, *The Ancient Engineers* (New York, 1963) is a popular and entertaining account of engineering developments from prehistoric times to the Middle Ages, with emphasis on the earlier periods. Samuel Noah Kramer is one of the great authorities on Sumerian culture; his works include *The Sumerians* (Chicago, 1963) and *From the Tablets at Sumer,* 1956, published in a paperbound edition as *History Begins at Sumer* (Garden City, N.Y., 1959). An important specialized work on Mesopotamian technology (Chapter 3) is Martin Levey, *Chemistry and Chemical Technology in Ancient Mesopotamia* (Amsterdam, 1959).

Ancient Egyptian technology (Chapter 3) is dealt with at length in A. Lucas, *Ancient Egyptian Materials and Industries* (4th edn., London, 1962). Two excellent earlier works on the pyramids, Leonard Cottrell, *The Mountains of Pharaoh* (New York, 1956), and I. E. S. Edwards, *The Pyramids of Egypt* (Harmondsworth, England, n.d.), have been supplemented, if not superseded, by Ahmed Fakhry, *The Pyramids* (Chicago, 1961). For transportation, see Robert F. Heizer, "Ancient Heavy Transport: Methods and Achievements," *Science,* CLIII (Aug. 19, 1966), 821-30; and C. St. C. Davison, "Transporting Sixty-Ton Statues in Early Assyria and Egypt," *Technology and Culture,* II (1961), 11-16.

Some of the sources for the history of technology during Greek and Roman times (Chapter 4)—Vitruvius, Pliny, Frontinus, Aristotle—have been translated into English in the Loeb Classical Library, but the works of Heron and Philon and the writers on war engines are not to be found in that series. Heron's *Pneumatics* has been edited with an English translation by Bennet Woodcroft (London, 1851), but the rest of Heron's works are to be found only in French and German translations. A. G. Drachmann has written some very detailed works on aspects of Classical technology: *Ancient Oil Mills and Presses* (Copenhagen, 1932); *Ktesibios, Philon and Heron* (Copenhagen, 1948); and *The Mechanical Technology of Greek and Roman Antiquity* (Madison, Wisc., 1963).

The views of Benjamin Farrington are set forth in *Greek Science* (2 vols., Harmondsworth, England, 1944). There are specialized secondary works on various aspects of Roman technology, including the Roman aqueducts: T. Ashby, *The Aqueducts of Ancient Rome* (Oxford, 1935) and E. B. van Deman, *The Building of Roman Aqueducts* (Washington, 1943).

Medieval and Early Modern Technology

The appropriate volumes of the *Cambridge Economic History of Europe* provide an excellent introduction to these and other periods of technological development, but by far the best work on the technology of the Middle Ages (Chapter 5)—and, indeed, one of the finest works on any period of technological development—is Lynn White, jr., *Medieval Technology and Social Change* (Oxford, 1962) based on a close study of the sources, and notable also as a significant attempt to indicate the technological foundations of elements in medieval political, social, economic, and cultural life. A. C. Crombie has a lengthy chapter on "Technics and Science in the Middle Ages" in his *Augustine to Galileo: The History of Science,* A.D. *400-1650* (Cambridge, Mass., 1953), reprinted as *Medieval and Early Modern Science* (Garden City, N.Y., 1959). A monumental study of non-Western technology which deals in detail with this same chronological period is Joseph B. Needham's multi-volume *Science and Civilisation in China*; particularly valuable is Vol. 4, Part 2, *Mechanical Engineering* (Cambridge, 1966).

Two major primary sources are available for the study of medieval technology. The *Sketchbook* of Villard de Honnecourt has been edited and translated by T. Bowie (Bloomington, Ind., 1959), and the treatise of Theophilus, *On Divers Arts,* has been most recently translated and edited by John W. Hawthorne and Cyril Stanley Smith (Chicago, 1963).

Other important works for the history of medieval technology include: B. H. Slicher von Bath, *The Agrarian History of Western Europe,* A.D. *500-1850* (New York, 1963); J. Harvey, *Early Medieval Architects* (London, 1954); L. F. Salzman, *Building in England down to 1540* (Oxford, 1952); Rex Wailes, *The English Windmill* (London, 1954); and A. J. Holmyard, *Alchemy* (Hardmondsworth, 1957).

The invention of printing in the 15th century accounts for the larger number of sources available for the study of Renaissance technology (Chapter 6). Three major works on the history of metallurgy date from this period: Georgius Agricola (Georg Bauer), *De re metallica,* translated by Herbert C. Hoover and L. C. Hoover (1912; reprinted, New York, 1950); Vannoccio Biringuccio, *The Pirotechnia,* translated and edited by Cyril Stanley Smith and M. T. Gnudi (New York, 1959; paperbound edn., Cambridge, Mass. 1966); Lazarus Ercker, *Treatise on Ores and Assaying,* translated and edited by A. G. Sisco and Cyril Stanley Smith (Chicago, 1951). The mechanical treatises of Jacques Besson, Agostino Ramelli, and Vittorio Zonca, with their beautiful illustrations, have not yet appeared in English editions, but a selection of the illustrations appears in A. G. Keller, *A Theatre of Machines* (New York, 1965). Edward MacCurdy has edited *The Notebooks of Leonardo da Vinci* (London, 1948).

Other source materials for the early modern period include Francis Bacon, *The Advancement of Learning* and *The New Atlantis* (available in many editions); William Harrison, *An historicall description of the Islande of Britaigne* (London, 1577, and often republished); Thomas Tusser, *Five Hundreth Good pointes of Husbandry* (London, 1573).

Important secondary sources for the history of Renaissance technology deal with various aspects of development. Ship-building is dealt with in F. L. Lane, *Venetian Ships and Shipbuilding of the Renaissance* (Baltimore, 1934) and Sir Westcott S. Abell, *The Shipwright's Trade* (Cambridge, 1948). The development of printing is treated in T. F. Carter, *The Invention of Printing in China and Its Spread Westward* (revised edition by L. Carrington Goodrich, New York, 1955); P. Butler, *The Origin of Printing in Europe* (Chicago, 1940); and the sociocultural aspects are emphasized in Marshall McLuhan, *The Gutenberg Galaxy* (Toronto, 1962).

Other secondary works of importance are M. B. Donald, *Elizabethan Copper* (London, 1955); Charles Ffoulkes, *The Gunfounders of England* (Cambridge, England, 1937); A. Rupert Hall, *Ballistics in the Seventeenth Century* (Cambridge, England, 1952); John U. Nef, *The Rise of the British Coal Industry* (London, 1932); W. B. Parsons, *Engineers and Engineering in the Renaissance* (Baltimore, 1939); A. P. Usher, *A History of Mechanical Inventions* (Cambridge, Mass., 1954). Two works in French must also be mentioned: Maurice Daumas, *Les Instruments scientifiques aux XVIIe et XVIIIe siècles* (Paris, 1953) and Bertrand Gille, *Les Ingenieurs de la Renaissance* (Paris, 1964), translated as *The Renaissance Engineers* (London, 1966). Also of importance is A. Wolf, *A History of Science, Technology, and Philosophy in the 16th and 17th Centuries* (2nd edn., New York, 1950; paperbound edn., New York, 1959).

Background of the Industrial Revolution

The sociocultural foundations of the Industrial Revolution (Chapters 7 and 13) are treated in John U. Nef, *Cultural Foundations of Industrial Civilization* (Cambridge, 1958). One special aspect of this—the role of Protestantism in the development of capitalism—is dealt with in Max Weber, *The Protestant Ethic and the Spirit of Capitalism* (New York, 1952), and R. H. Tawney, *Religion and the Rise of Capitalism* (New York, 1954).

The economic background (Chapters 8 and 13) is dealt with in several works by Shepard B. Clough: *The Economic Development of Western Civilization* (New York, 1959); "Change and History," *The Social Sciences in Historical Study* (New York, 1954); and, with Carol Moodie, *European Economic History: Documents and Readings* (Princeton, 1965). See also T. S. Ashton, *Economic Fluctuations in England, 1700-1800* (Oxford, 1959).

Other elements of the economic background are covered in Carlo M. Cipolla, *Money, Prices, and Civilization in the Mediterranean: Fifth to Seventeenth Century* (Princeton, 1956); Earl J. Hamilton, *American Treasure and the Price Revolution in Spain, 1501-1650* (Cambridge, Mass., 1934); Eli Heckscher, *Mercantilism* (2 vols., London, 1934); John U. Nef, *The Conquest of the Mate-*

rial World (Chicago, 1964); J. H. Parry, *Europe and a Wider World* (London, 1949); Raymond de Roover, *The Rise and Fall of the Medici Bank, 1397-1494* (Cambridge, Mass., 1963).

For the technical background (Chapter 13), in addition to the works mentioned in other sections, there are A. Wolf, *A History of Science, Technology and Philosophy in the 18th Century* (2 vols., paperbound edn., New York, 1961); Shelby T. McCloy, *French Inventions of the 18th Century* (Lexington, Ky., 1952); and C. C. Gillispie (ed.), *A Diderot Pictorial Encyclopedia of Trades and Industry* (2 vols., New York, 1959).

The Agricultural Revolution

Two indispensable works for the study of the great agricultural transformations which took place up to and through the 18th century (Chapter 9) are B. H. Slicher von Bath, *The Agrarian History of Western Europe* (London, 1963) and N. S. B. Gras. *A History of Agriculture in Europe and America* (New York, 1954).

The standard work on the open-field system is C. S. and C. S. Orwin, *The Open Fields* (2nd edn., Oxford, 1954). A classic French work on developments in medieval agriculture is Marc Bloch, *Les Caractères originaux de l'histoire rurale française* (Paris, 1952; English trans., Berkeley, Calif., 1966). The strong influence of the Low Countries on British agriculture is treated in G. E. Fussell, "Low Countries' Influence on English Farming," *English Historical Review*, LXXIV (Oct. 1959), 611-22.

There are many works on the history of English farming, including R. E. Prothero (Lord Ernle), *English Farming, Past and Present* (New York, 1936) and C. S. Orwin, *A History of English Farming* (London, 1949). In addition to these general works, there are many studies of special topics: G. E. Mingay, *English Landed Society in the 18th Century* (London, 1963); Reginald Lennard, "English Agriculture under Charles II," *Economic History Review*, IV (1932), 23 ff.; Donald Grove Barnes, *A History of the English Corn Laws from 1660 to 1846* (London, 1930); A. H. Johnson, *The Disappearance of the Small Landowner* (2nd edn., London, 1963); Naomi Riches, *The Agricultural Revolution in Norfolk* (Chapel Hill, 1937). R. A. C. Parker, "Coke of Norfolk and the Agrarian Revolution," *Economic History Review*, 2nd series, VIII (1955), 160-65, offers a modern criticism of the popular over-estimation of Coke's proceedings; see also G. E. Fussell, "Norfolk Improvers, Their Farms and Methods—A Reassessment," *Norfolk Archaeology* (1964). For English agricultural influence on France, see Andre J. Bourde, *The Influence of England on the French Agronomes, 1750-1789* (Cambridge, 1953).

The outstanding historical work on farming implements is G. E. Fussell, *The Farmer's Tools, 1500-1900* (London, 1952). An interesting work, tying social to agricultural history, is Redcliffe N. Salaman, *The History and Social Influence of the Potato* (Cambridge, 1949).

In addition to the treatment of animal husbandry in general agricultural histories, there are some specialized works on this topic: H. Cecil Pawson, *Robert Bakewell, Pioneer Livestock Breeder* (London, 1957); H. B. Carter, *His Majesty's Spanish Flock* (Sydney and London, 1964), tells the story of the Merino sheep in detail; J. F. Smithcors, *Evolution of the Veterinary Art: a Narrative Account to 1850* (Kansas City, 1957); and a series of articles by G. E. Fussell, some co-authored with Constance Goodman, in *Agricultural History* (Oct. 1929, Oct. 1930, April and July 1937, Oct. 1941, Jan. 1942).

Metallurgical Developments

In addition to the sources mentioned in the chronological sections—Theophilus, Biringuccio, Agricola—there are several other primary sources for metallurgical developments (Chapters 10, 21, 36) during the early period. M. Jousse, *La Fidelle Ouverture de l'art de Serrurier* (La Flèche, 1627) has been partially translated by C. S. Smith and A. G. Sisco and published in *Technology and Culture,* II (1961), 131-45; Jousse, a blacksmith and locksmith, discusses the qualities of various kinds of iron and steel and their hardening procedure.

R. A. F. de Réaumur, *L'Art de convertir le fer forgé en acier . . .* (*The Art of Converting Wrought Iron into Steel . . .*) published in Paris, 1722, has been translated into English by A. G. Sisco as *Réaumur's Memoirs on Steel and Iron* (Chicago, 1956). Written by the inventor of malleable iron, this verbose but fascinating account of steel-making and of cast iron is the first significant scientific study of the hardening of steel. Benvenuto Cellini described his techniques in goldsmithing in his *Treatises on Goldsmithing and Sculpture* (Florence, 1568; English translation by C. R. Ashbee, London, 1898). An anthology of early papers on scientific metallurgy during the period 1532-1823, edited by Cyril Stanley Smith and entitled *Sources for the Science of Steel,* is to be published by the Society for the History of Technology in its Monograph Series in late 1967 or early 1968.

Among secondary works, a first-rate review of the whole history of metallurgy is to be found in Leslie Aitchison, *A History of Metals* (2 vols., London, 1960). For the very early period, excellent secondary materials include H. H. Coghlan, *Notes on Prehistoric Metallurgy of Copper and Bronze* (Oxford, 1951); R. J. Forbes, *Metallurgy in Antiquity* (2 vols., Leiden, 1950); C. S. Smith, "Materials and the Development of Civilization and Science," *Science,* 148 (1965), 908-17; T. A. Wertime, "Man's First Encounters with Metallurgy," *Science,* 146 (1964), 1257-67; and Joseph Needham, *The Development of Iron and Steel Technology in China* (Cambridge, England, 1964).

Two other excellent general works are H. R. Schubert, *A History of the Brit-*

ish Iron and Steel Industry, 450 B.C.-*1775* A.D. (London, 1957) and T. A. Wertime, *The Coming of the Age of Steel* (Chicago, 1962). For the middle period of the 19th century, J. S. Jeans, *Steel: Its History, Manufacture, and Uses* (London, 1880) retains its usefulness. For other metals there is Donald McDonald, *A History of Platinum* (London, 1960) and J. W. Richards, *Aluminum* (2nd edn., Philadelphia, 1890). E. N. Hartley, *Ironworks on the Saugus* (Norman, Okla., 1957) tells the story of America's first iron works.

Source materials for mid-19th century developments (Chapter 21) include a series of books by John Percy: *Metallurgy: Fuel . . . , Copper and Brass* (1861); *Iron and Steel* (1864); *Silver and Gold* (1870); *Lead* (1880). Others are W. C. Roberts-Austen, *Introduction to the Study of Metallurgy* (London, 1891) and Henry Bessemer, *Autobiography* (London, 1905).

For the iron and steel industry in the United States, one must still rely on the venerable James M. Swank, *History of the Manufacture of Iron in All Ages, and Particularly in the United States . . .* (Philadelphia, 1892). Arthur C. Bining, *Pennsylvania Iron Manufacture in the Eighteenth Century* (Harrisburg, Pa., 1938) is excellent on that period, and Peter Temin, *Iron and Steel in Nineteenth-Century America: An Economic Inquiry* (Cambridge Mass., 1964) is an attempt to apply econometrics to the problem. *The Autobiography of John Fritz* (New York, 1912) is the record of a major innovator.

For the scientific development of metallurgy (Chapter 36), source materials include W. Fairbairn, *Useful Information for Engineers* (3 vols., London, 1856); W. Kirkaldy, *Experiments on Wrought Iron and Steel* (Glasgow, 1862); *Report of the U.S. Board Appointed to Test Iron, Steel, and Other Metals* (2 vols., Washington, 1881); R. H. Thurston, "A Treatise on Iron and Steel," in his *Materials of Construction,* Part II (New York, 1883) and "Alloys, Brasses and Bronzes," *ibid.*, Part III (New York, 1884); I. L. Bell, *Principles of the Manufacture of Iron and Steel* (London, 1884); and H. M. Howe, *The Metallurgy of Steel* (New York, 1888). Secondary works of prime importance are Cyril Stanley Smith, *A History of Metallography* (Chicago, 1960); Cyril Stanley Smith (ed.), *Sorby Centennial Symposium on the History of Metallurgy* (New York, 1965); I. Todhunter and K. Pearson, *A History of Elasticity and the Strength of Materials* (2 vols. in 3, Cambridge, 1886-93).

The best book on mining in the United States is Rodman Wilson Paul, *Mining Frontiers of the Far West, 1848-1880* (New York, 1963).

Scientific Instruments

Some of the books mentioned below under the history of machine tools are essential for understanding the development of instrumentation (Chapter 11). Particularly important are the series of books on machine tools by Robert S. Woodbury. See also Edwin A. Battison, "Screw-Thread Cutting by the Master-Screw Method since 1480," Paper 37 in *Contributions from the Museum of History and Technology* (U.S. National Museum Bulletin 240, Washington, D.C., 1964).

For the history of scientific instruments themselves there are general works such as those of Maurice Daumas, *Les Instruments scientifiques aux XVIIe et XVIIIe siècles* (Paris, 1953); Francis Maddison, "Early Astronomical and Mathematical Instruments: A Brief Survey of Sources and Modern Studies," *History of Science,* II (1963), 17-50, the most comprehensive bibliography on the subject of scientific instrumentation; and E. G. R. Taylor, *The Mathematical Practitioners of Tudor and Stuart England* (Cambridge, England, 1954), a comprehensive biographical study of individual instrument-makers and practitioners and their published works. See also Silvio A. Bedini, *Early American Scientific Instruments and Their Makers* (U. S. National Museum Bulletin 231, Washington, D.C., 1964).

There are many works dealing with specialized fields of scientific instruments: Henry C. King, *History of the Telescope* (London, 1955); R. S. Clay and T. H. Court, *History of the Microscope* (London, 1932); W. Knowles Middleton, *The History of the Barometer* (Baltimore, 1964); Bern Dibner, *Early Electrical Machines* (Norwalk, Conn., 1957); and A. W. Richeson, *English Land Measuring to 1800: Instruments and Practice* (Cambridge, Mass., 1966). Silvio A. Bedini has written many articles on a variety of scientific instruments: "The Optical Workshop of Guiseppe Campani," *Journal of the History of Medicine and Allied Sciences,* XVI, 1 (Jan. 1961), 18-38; "On Making Telescope Tubes in the 17th Century," *Physis,* IV, 2 (1962), 110-16; "Seventeenth Century Italian Compound Microscopes," *Physis,* V, 4 (1963), 383-422; "Johann Philipp Treffler, Clockmaker of Augsburg," *Bulletin of the National Association of Watch and Clock Collectors* (1956); and many others.

The Industrial Revolution

Only a few of the many general works dealing with the early phases of the Industrial Revolution (Chapters 13-18) can be mentioned here. Many books dealing with specialized phases and topics of the Industrial Revolution are listed in the appropriate sections below.

The latest, and possibly the best, general study of the Industrial Revolution is the chapters by David S. Landes in Volume VI of the *Cambridge Economic History of Europe,* but there are many other important works in this field. Thomas S. Ashton's *Economic History of England in the Eighteenth Century* (London, 1955) presents a survey of the entire economy with little reference to technological change, but the same author's *The Industrial Revolution, 1760-1830* (New York, 1948) provides an excellent analysis of industrial development for the general reader. Phyllis Deane and W. A. Cole, *British Economic Growth, 1688-1959* (New York, 1962), provide a comprehensive survey of the statistical record with careful analysis and a new series on income and capital formation; their analysis of the 18th century shows the quantitative importance of trade and population growth as the background for technological change. J. L. and Barbara Hammond, *The Rise of Modern Industry* (London, 1925) is an important study of industrial development from the labor point of view. Not to be neglected is Phyllis Deane, *The First Industrial Revolution* (Cambridge, 1965).

Although later research has qualified many of its judgments, Paul Mantoux, *The Industrial Revolution in the Eighteenth Century* (London, 1907; paperbound edn., New York, 1962) remains an important part of the literature, as does Arnold Toynbee, *The Industrial Revolution* (paperbound edn., Boston, 1956).

Walther G. Hoffmann, *British Industry, 1700-1950* (Oxford, 1955) gives important new economic indexes for the 18th century, but the emphasis is mainly on the later period. An interesting case study is provided in Louis W. Moffit, *England on the Eve of the Industrial Revolution: A Study of Economic and Social Conditions, 1740-60, with Special Reference to Lancashire* (London, 1925). Essays covering recent research on the Industrial Revolution are to be found in Leslie S. Presnell (ed.), *Studies in the Industrial Revolution presented to T. S. Ashton* (London, 1960). Arthur L. Dunham, *The Industrial Revolution in France, 1815-1848* (New York, 1955) is a standard work in English on the topic. Sidney Pollard, in *The Genesis of Modern Management: A Study of the Industrial Revolution in Great Britain* (Cambridge, Mass., 1965), studies an important yet hitherto neglected aspect of industrialization during this period.

Biographical sketches of some of the leading figures of the Industrial Revolution are to be found in Samuel Smiles, *Lives of the Engineers,* first published in London in 1857 and collected in five volumes in 1904. Selections from Smiles have recently been edited by Thomas P. Hughes and published in a paperbound edition (Cambridge, Mass., 1966). Fascinating insights into some of the most important industrial leaders are to be found in Robert E. Schofield, *The Lunar Society of Birmingham* (Oxford, 1963).

The Textile Industry

Abbott Payson Usher, *A History of Mechanical Inventions* (paperbound edn., Boston, 1959) is a classic work in the history of technology and contains much important material on the Industrial Revolution in the textile industry (Chapter 14). A. P. Wadsworth and Julia de L. Mann, *The Cotton Trade and Industrial Lancashire, 1600-1780* (Manchester, 1931) is a comprehensive study of trade and industry showing how the techniques of calico printing and inventions in spinning and weaving built a new industry on the foundation created by the trade in raw cotton and East Indian fabrics.

Specific inventors and inventions are dealt with in many books. George

Unwin, *Samuel Oldknow and the Arkwrights* (Manchester, 1924) is an important study based on mill records and letters. *The Life of Robert Owen, written by Himself* (London, 1857) is important both for its account of the technical problems of the mills at New Lanark and for the idealistic effort to promote the welfare of the employees. E. Lipson, *The History of the Woolen and Worsted Industries* (London, 1928) affords a description of techniques of production and management which survived until late in the 18th century. Herbert Heaton, *The Yorkshire Woollen and Worsted Industries, from the earliest times up to the Industrial Revolution* (Oxford, 1920) is a fully documented study of a region where independent weavers maintained themselves to a late date. George W. Daniels, *The Early History of the Cotton Industry, with some unpublished letters of Samuel Crompton* (Manchester, 1920) is based on the business papers of M. Connel and Co., Ltd., for the period 1795-1835, and the source materials are carefully placed in their general setting.

B. P. Dobson tells *The Story of the Evolution of the Spinning Machine* (Manchester, 1910), while R. S. Fitton and A. P. Wadsworth provide an important study based on mill records and letters in *The Strutts and the Arkwrights, 1758-1830* (Manchester, 1958). The story of *James Hargreaves and the Spinning Jenny* is told by C. Aspin and S. D. Chapman (Helmshore, Lancashire, 1964). A description of looms and weave patterns well suited to the non-technical reader is provided in Luther Hooper *Handloom Weaving, Plain and Ornamental* (London, 1920).

The Steam Engine

The standard work on the development of the steam engine (Chapters 15, 17, 20) is Henry W. Dickinson, *A Short History of the Steam Engine* (Cambridge, 1938; 2nd edn., London, 1963). The relationship of the early steam engine to contemporary science is perceptively developed in D. S. L. Cardwell, *Steam Power in the Eighteenth Century* (London, 1963). In *Thomas Newcomen: The Prehistory of the Steam Engine* (London, 1963), L. T. C. Rolt brings together all that is known about the first modern steam engine (1712) and its builder.

An exhaustive study of the Watt engine, based on Boulton and Watt papers, is given in Henry W. Dickinson and Rhys Jenkins, *James Watt and the Steam Engine* (Oxford, 1927). Although his understanding of the early steam engine is not as sure as Dickinson's, Ivor B. Hart in *James Watt and the History of Steam Power* (New York, 1949) tells a story whose broad outlines are reasonably accurate. A re-evaluation of traditional beliefs concerning the Boulton and Watt "monopoly" (1775-1800) is provided in A. E. Musson and E. Robinson, "Early Growth of Steam Power," *Economic History Review*, 2nd series, II (1959), 418-39.

Other pioneers of the steam engine are dealt with in L. T. C. Rolt, *The*

Cornish Giant: The Story of Richard Trevithick, Father of the Steam Engine (London, 1960) and Greville and Dorothy Bathe, *Oliver Evans: A Chronicle of Early American Engineering* (Philadelphia, 1935).

Machine Tools, Machinery, and Manufacturing

A comprehensive account of machine-tool development (Chapters 16, 23, 38) from about 1775 is contained in L. T. C. Rolt, *A Short History of Machine Tools* (Cambridge, Mass., 1965). An older, but still useful account is Joseph M. Roe, *English and American Tool Builders* (New Haven, 1916).

Robert S. Woodbury has written monographic treatments of particular machine tools: *History of the Gear-Cutting Machine* (Cambridge, Mass., 1958); *History of the Grinding Machine* (Cambridge, Mass., 1959); *History of the Milling Machine* (Cambridge, Mass., 1960); *History of the Lathe to 1850* (Cleveland, 1961). A "must" for anyone interested in early American machine-building is Eugene S. Ferguson (ed.), *Early Engineering Reminiscences (1815-40) of George Escol Sellers* (U.S. National Museum Bulletin 238, Washington, D.C., 1964).

A standard and voluminous work on American manufacturing is Victor S. Clark, *History of the Manufactures of the United States* (3 vols., Washington, D.C., 1929). An interesting comparison of American and British technological differences, and the reasons therefore, is given in H. J. Habakkuk, *American and British Technology in the Nineteenth Century: The Search for Labour-Saving Inventions* (Cambridge, England, 1962). John A. Kouwenhoven, *Made in America: The Arts in Modern Civilization* (Garden City, N.Y., 1962) provides a fascinating and well-written study of the unique qualities of American technological development.

American leadership in the use of interchangeable parts—the basis for "the American system"—is seen through the eyes of important British contemporaries in Joseph Whitworth and George Wallis, *The Industry of the United States in Machinery, Manufactures, and Useful and Ornamental Arts* (London, 1854). An account of the process as found in the United States in 1880 is C. H. Fitch, "Interchangeable Mechanism," *United States Census: Report on the Manufactures of the United States,* Tenth Census (1880), II (Washington, D.C., 1883). Robert S. Woodbury has written a prize-winning article, "The Legend of Eli Whitney and Interchangeable Parts," *Technology and Culture,* I (Summer 1960), 235-53.

A survey of the field is provided in Aubrey F. Burstall, *A History of Mechanical Engineering* (London, 1963). Specialized fields of manufacturing are covered under the appropriate sections below.

For the development of the internal-combustion engine (Chapter 40), few works in English are available. Eugen Diesel, Gustav Goldbeck, and Friedrich

Schilderberger, *From Engines to Autos*, trans. Peter White (Chicago, 1960), is a survey of the internal-combustion engine in Germany, where most of the significant work was done: it contains chapters on Otto, Daimler, Benz, Diesel, and Bosch. Old textbooks are good sources of information about old engines: Bryan Donkin, Jr., *A Text Book on Gas, Oil, and Air Engines* (London, 1894) has an historical section. Lynwood Bryant has written an excellent study of the development of the first engine to operate on the four-stroke cycle: "The Silent Otto," *Technology and Culture*, VII, 2 (Spring 1966), 184-200.

Steamships and Railrods

The story of steam transportation on land and sea (Chapters 17 and 25) has fascinated enthusiasts as well as historians. As a result, there have probably been more books written on this topic than on any other aspect of technological development. Only a few can be singled out here for mention.

Edgar C. Smith, *A Short History of Naval and Marine Engineering* (Cambridge, England, 1938) provides a general survey of steam vessels, while James T. Flexner, *Steamboats Come True: American Inventors in Action* (New York, 1944) gives a full account of American developments from 1780 to Fulton's success in 1807. Louis C. Hunter, *Steamboats on the Western Rivers* (Cambridge, Mass., 1949) is undoubtedly the best book on steamboats ever written. Interesting for its focus on the many factors and problems besetting steamship "inventors" is S. Colum Gilfillan's *Inventing the Steamship* (Chicago, 1934). An older but still useful account is George Henry Preble, *A Chronological History of the Origin and Development of Steam Navigation* (Philadelphia, 1895). Thomas Boyd tells the story of *Poor John Fitch* (New York, 1935); Henry W. Dickinson, *Robert Fulton, Engineer and Artist* (London, 1913), primarily concerned with the technical aspects of Fulton's career, could not take into account the large amount of information that has come to light during the last 50 years.

The technical details of the early railroads are to be found in a carefully written and well-illustrated book by C. F. Dendy Marshall, *A History of British Railways down to the Year 1830* (London, 1938). C. H. Ellis, *British Railway History* (2 vols., London, 1954-60) is a basic and comprehensive account, noteworthy for its balance and insights. L. T. C. Rolt, *The Railway Revolution: George and Robert Stephenson* (New York, 1962) is a balanced retelling of the careers of the Stephensons, father and son, who were associated with the most important developments in English railway history during four decades, about 1815 to 1855. An older account of British railway development is W. M. Acworth, *The Railways of England* (5th edn., London, 1900).

Roger Burlingame's trilogy—*March of the Iron Men* (New York, 1938), *Engines of Democracy* (New York, 1940), and *Backgrounds of Power* (New

York, 1949)—deals with American transportation and other technological developments. Although written in a popular style for the non-technical reader, these books indicate the author's concern for historical scholarship. Other works on American railroads include E. P. Alexander, *Iron Horses* (New York, 1941) and John F. Stover, *American Railroads* (Chicago, 1961) which summarizes the history down to 1865. Forest G. Hill, *Roads, Rails, and Waterways: The Army Engineers and Early Transportation* (Norman, Okla., 1957) identifies the source of much of the engineering talent used in laying out American railroads and canals. The biography of one such army engineer (and the husband of "Whistler's Mother") may be found in Albert Parry, *Whistler's Father* (Indianapolis, 1939). The larger picture is set forth in Daniel H. Calhoun, *The American Civil Engineer: Origins and Conflict* (Cambridge, Mass., 1960). George W. Hilton and John F. Due have written *The Electric Interurban Railways in America* (Stanford, 1960) which provides much reference material on specific inter-urban companies, but which is not recommended for survey reading.

Seymour Dunbar, *A History of Travel in America* (New York, 1937) is a popular exposition of its subject. George R. Taylor, *The Transportation Revolution, 1815-1860* (New York, 1951) is a model work of scholarship tying together technological and economic developments.

Building and Construction

There are very few histories of structural techniques and materials (Chapters 12, 22, 37), and a comprehensive history of the whole domain has yet to be written. Nevertheless, there are some books which survey various fields within this area. These include Norman Davey, *A History of Building Materials* (London, 1961); H. Shirley-Smith, *The World's Great Bridges* (rev. edn., New York, 1965); David B. Steinman and Sara Ruth Watson, *Bridges and Their Builders* (rev. edn., New York, 1957); Hans Straub, *A History of Civil Engineering: An Outline from Ancient to Modern Times* (Cambridge, Mass., 1964); and Stephen P. Timoshenko, *History of Strength of Materials* (New York, 1953).

For the earlier period (Chapter 12), secondary works have been mentioned under the appropriate chronological periods above, as, for example, Parsons, *Engineers and Engineering of the Renaissance.* A few early texts, such as Alberti's *De re ædificatoria* (1485), the first modern printed book on architecture, are available in English translations. John Fitchen, *The Construction of Gothic Cathedrals* (Oxford, 1961) is an important contribution to our understanding of the most impressive structures of the Middle Ages. In 1925 the American Institute of Architects published *Old Bridges of France,* containing dimensioned drawings of many early French bridges, and Daniel Halevy's biography of Vauban has been translated into English as *Builder of Fortresses* (New York, 1925). James Kip Finch has published a series of articles on French engineers in *Consulting Engineer* as follows: Gautier (Oct. 1960); Bion (Nov. 1960); Belidor (Feb., March, May, and June 1961); Perronet (April 1962); Gauthey (Sept. 1962); and Girard (Aug. 1963).

Smiles, *Lives of the Engineers* contains biographies of some of the most important British civil engineers, such as Brindley, Rennie, Smeaton, and Telford. L. T. C. Rolt has published a readable and scholarly biography of *Thomas Telford* (London, 1958). Charles Hadfield has published a history of *British Canals* (London, 1950), and Eric de Maré has done the same for *The Bridges of Britain* (London, 1954).

For American developments (Chapters 22 and 37), the best general work is Carl W. Condit, *American Building Art: The Nineteenth Century* (New York, 1960), a well-written, finely illustrated, and comprehensive work which provides excellent discussions of the technical developments. Another important survey is James M. Fitch, *American Building: The Forces That Shape It* (Boston, 1948).

The history of bridges in America is covered in Llewellyn N. Edwards, *A Record of History and Evolution of Early American Bridges* (Orono, Maine, 1959); Robert Fletcher and J. P. Snow, "A History of the Development of Wooden Bridges," *Transactions, American Society of Civil Engineers,* XCIX (1934), 314-408; Theodore Cooper, "American Railroad Bridges," *Trans. ASCE,* XXI (1889), 1-60; and Richard S. Allen's finely illustrated volumes on covered bridges; *Covered Bridges of the Northeast* (Brattleboro, Vt., 1957) and *Covered Bridges of the Middle Atlantic States* (Brattleboro, Vt., 1959).

There are articles dealing specifically with the development of steel-frame structures: Turpin C. Bannister, "Bogardus Revisited," *Journal of the Society of Architectural Historians,* XV, 12-22; XVI, 11-19; and several by A. W. Skempton on some of the European background of this development: "Evolution of the Steel Frame Building," *The Guilds Engineer,* X, 37-51; "The Boat Store, Sheerness (1858-60), and Its Place in Structural History," *Transactions of the Newcomen Society,* XXXII, 57-78; and "The Origin of Iron Beams," *Actes du VIIIe Congrès International d'Histoire des Sciences* (Florence, 1956), 1029-39. Source material for later 19th-century developments can be found in two books by William H. Birkmire: *Architectural Iron and Steel, and Its Application in the Construction of Buildings* (New York, 1892) and *Skeleton Construction in Buildings* (New York, 1897). Another primary source is Joseph K. Freitag, *Architectural Engineering, with Special Reference to High Building Construction* (2nd edn., New York, 1901).

The growing use of concrete in buildings is seen in Peter Collins, *Concrete:*

The Vision of a New Architecture (London, 1959); Ernest L. Ransome and Alexis Saurbrey, *Reinforced Concrete Buildings* (New York, 1912); and Aly Ahmed Raafat, *Reinforced Concrete in Architecture* (New York, 1958).

Special attention should be given to the city of Chicago, where so much of modern building design and techniques originated. Frank A. Randall has written the *History of the Development of Building Construction in Chicago* (Urbana, Ill., 1949), and Carl W. Condit is the author of *The Chicago School of Architecture* (Chicago, 1964).

In addition, some specialized writings should be mentioned: Carroll L. V. Meeks has written an architectural history of *The Railroad Station* (New Haven, 1956); J. Carson Webster gives a brief summary of "The Skyscraper: Logical and Historical Considerations," *Journal of the Society of Architectural Historians*, XVIII, 126-39; Sigfried Giedion deals with the æsthetic and cultural values of architecture in *Space, Time and Architecture* (3rd edn., Cambridge, Mass., 1954); and Eduardo Torroja has written of the *Philosophy of Structures* (Berkeley, 1958).

Chemical Technology

Much of the literature on the history of industrial chemistry (Chapters 28, 41, and portions of Chapter 20) is uncritically antiquarian in nature, and despite a fairly extensive bibliography, this subject has been rather neglected in comparison with mechanical or electrical technology. The primary literature is, of course, vast, and occurs rather haphazardly in the older literatures of chemistry, economics, and such venerable technologies as metallurgy and textiles.

There are some general studies of varying quality and importance. F. S. Taylor, *A History of Industrial Chemistry* (New York, 1957) is too brief and diffuse to be of much use. Archibald and Nan Clow, *The Chemical Revolution* (London, 1952) is a definitive study of the British chemical industry into the first half of the 19th century; L. F. Haber, *The Chemical Industry during the Nineteenth Century* (Oxford, 1958) while not definitive, is informative, authoritative, and altogether the most readable book on this subject. William Haynes (ed.), *History of the American Chemical Industry* (6 vols., New York, 1945-54) is largely a history of companies. The student will find very helpful Aaron Ihde, *The Development of Modern Chemistry* (New York, 1964), and a compilation of historical articles, edited by Ihde and William F. Kieffer, from the *Journal of Chemical Education*, entitled *Selected Readings in the History of Chemistry* (Easton, Pa., 1965).

There has been a general literature of industrial chemistry since the mid-18th century, when chemistry scarcely concerned itself with any but "useful" materials, and when a textbook of chemistry was (from our modern point of view) more concerned with technology than science. From the mid-18th century, however, the main trend of chemical literature was in the direction which we today call scientific, while at the same time a few writers took a completely different tack and undertook to produce purely practical accounts of the "chemical arts." Among the most important British productions of this latter kind were Peter Shaw, *Chemical Lectures . . . for the improvement of arts, trades, and natural philosophy* (London, 1734); Samuel Parkes, *Chemical Essays . . .* (5 vols., London, 1815); and S. F. Gray, *The Operative Chemist* (London, 1828).

The character of mid-19th-century industrial chemistry is well revealed in such works as F. L. Knapp, *Lehrbuch der chemische Technologie* (2 vols., Braunschweig, 1847, and later editions, including an English translation); Andrew Ure, *Dictionary of Arts, Manufactures, and Mines* (London, 1839, and later editions); and Sheridan Muspratt, *Chemistry* (2 vols., Glasgow, 1860). A. W. Hofmann (ed.), *Bericht über die Entwicklung der chemischen Industrie während des letzten Jahrzehnts* (2 vols., Braunschweig, 1875) is interesting in that it shows the scientific chemist's views on the development of industrial chemistry.

The role of chemical science in industry at the turn of the century is seen in a textbook by F. M. Thorp, *Outlines of Industrial Chemistry* (New York, 1901).

Whereas the chemical industry of the 18th century was conducted in great secrecy—making information on the chamber sulfuric acid and Leblanc soda industries difficult to find—those industries which began in the 19th century immediately spawned a large independent literature. For example, A. Brandeln, *Die Operationen, Manipulationen, und Geratschaften der Electro-Chemie* (Weimar, 1849, but published earlier in French) was a book describing an industry which barely existed. The mere bibliography of such literature is not easy to assemble, and there is no more comprehensive history of industrial electrochemistry than the few pages in Wilhelm Ostwald, *L'Evolution de l'electrochimie* (Paris, 1912, translated from the German). One must resort to articles of reminiscence, such as August Eimer, "Early Days of the Carbide Industry," *Transactions of the American Electrochemical Society*, 51 (1927), 73-8; to the historical sections of such monographs as J. W. Richards, *Aluminum* (2nd edn., Philadelphia, 1890) and V. B. Lewes, *Acetylene* (Westminster and New York, 1900); and to such inclusive works as A. P. Bolley (ed.), *Handbuch der chemischen Technologie* (7 vols., Braunschweig, 1862-70) and J. W. Mellor, *A Comprehensive Treatise on Inorganic and Theoretical Chemistry* (16 vols., London, 1922-37). Some more recent studies of specialized aspects of chemical technology include Theodore J. Kreps, *The Economics of the Sulfuric Acid Industry* (Stanford, 1938) and R. Norris Shreve, *The Chemical Process Industries* (New York, 1945).

Early monographic literature has particular value not only because it is early, but because of a greater devotion to history than is fashionable today. Ample historical information (usually with citation of sources) appears in such books as those of Richards and Lewes, mentioned above. Works of similar value for the plastics industries are T. Seeligmann, G. Lamy Torrilhan, and H. Falconnet, *India Rubber and Gutta Percha* (London, 1903) and E. C. Worden, *Nitrocellulose Industry* (2 vols., New York, 1911). John Beer, *The Emergence of the German Dye Industry* (Urbana, Ill., 1959) is almost unique in being a treatment of industrial chemistry from the standpoint of the professional historian.

For the development of the petroleum industry (Chapter 41) there are three excellent historical works: Ralph W. and Muriel B. Hidy, *Pioneering in Big Business, 1882-1911: History of Standard Oil Company (New Jersey)* (New York, 1955); Arthur M. Johnson, *The Development of American Petroleum Pipelines: A Study in Private Enterprise and Public Policy* (Ithaca, N.Y., 1956); and Harold F. Williamson and Arnold R. Daum, *The American Petroleum Industry: The Age of Illumination, 1859-1899* (Evanston, Ill., 1959).

Electrical and Communications Technology

The early developments in the field of electricity belong more to the history of science than to the history of technology. However, the rapid rise of the electrical industry during the 19th century (Chapters 26, 27, 34, 35) has been well covered by a number of books which survey the field while emphasizing special aspects.

Harold I. Sharlin, *The Making of the Electrical Age* (New York, 1963) deals with the development of electrical technology with special stress on the factors required to make an invention a commercial success. Percy Dunsheath, *A History of Electrical Engineering* (London, 1962) provides descriptions of how electrical devices operate, for the technically competent student. Malcolm MacLaren, *The Rise of the Electrical Industry during the Nineteenth Century* (Princeton, 1943) provides a review of developments in lighting and communications, and has an excellent bibliography with lists of source materials. Notable for the wealth of its illustrative materials is W. James King, *The Development of Electrical Technology in the 19th Century* (Papers 28, 29, 30; pp. 231-407, of *Contributions from the Museum of History and Technology*, United States National Museum Bulletin 228, Washington, D.C., 1962).

Among the handsome books published by the Burndy Library (Norwalk, Conn.) are two by Bern Dibner which deal especially with the material covered in this volume: *Early Electrical Machines* (1957) and *The Atlantic Cable* (1959; paperbound edn., New York, 1964). E. A. Marland, *Early Electrical Communication* (London, 1964) covers the development of electricity from the earliest concepts through the development of the telephone by Alexander Graham Bell, but the greatest part is devoted to the development of the telegraph.

Carleton Mabee, *The American Leonardo: A Life of Samuel Morse* (New

York, 1943) is the standard biography of the inventor of the telegraph. Business and economic developments are stressed in Robert Luther Thompson, *Wiring a Continent: The History of the Telegraph Industry in the United States, 1832-1866* (Princeton, 1947).

The manufacturing end of the electrical industry has been the subject of several excellent investigations. Harold C. Passer, *The Electrical Manufacturers, 1875-1900* (Cambridge, Mass., 1953) is particularly good on the role of entrepreneurship in the development of technology and has an excellent discussion of the "Edison system." Edward Dean Adams, *Niagara Power: History of the Niagara Falls Power Company, 1886-1918* (Niagara Falls, N.Y., 1927) tells of the evolution of the central power station at the Falls and the development of the alternating current system there. Arthur A. Bright, Jr., *The Electric-Lamp Industry: Technological Change and Economic Development from 1800 to 1947* (New York, 1949) is very good on the early history of technological innovation in electrical illumination. John Winthrop Hammond, *Men and Volts: The Story of General Electric* (Philadelphia, 1941) covers the period from 1880 to 1939, beginning with the Thomson-Houston Co. to the formation of General Electric; it provides an excellent account of the economic growth of the General Electric Co. J. Scott, *Siemens Brothers, 1858-1958* (London, 1959) is a centennial account of the great Anglo-German electrical developers.

John Ambrose Fleming, an electrical engineer, wrote two works of historical interest: *The Alternate Current Transformer in Theory and Practice* (2 vols., New York, no date, but about 1900), which has good historical introductions in both volumes, plus descriptions of contemporary practice; and *Fifty Years of Electricity: The Memoirs of an Electrical Engineer* (London, no date, but about 1920), which covers telegraph, telephone, and motors, and provides a good description of electrical equipment in various stages of development. The best biography of Thomas A. Edison is Matthew Josephson's *Edison* (New York, 1959).

In addition to the works mentioned above, material on communications (Chapter 39) is to be found in J. Mellanby, *The History of Electric Wiring* (London, 1957); F. L. Rhodes, *The Beginning of Telephony* (New York, 1929); and F. W. Wile, *Emile Berliner, Maker of the Microphone* (Indianapolis, 1926).

M. W. Haynes, *The Student's History of Printing* (New York, 1930) covers the story from the beginning and includes important development in printing technology during the 19th century. The development of photography is recounted in J. M. Eder, *History of Photography* (New York, 1945) and in Helmut Gernsheim, *The History of Photography: From the Earliest Use of the Camera Obscura in the 11th Century up to 1914* (London, 1955). The book which most thoroughly covers the science, technology, and art of photography, with copious quotations from source materials and excellent bibliographies, is

Wolfgang Baier, *A Source Book of Photographic History* (New York, 1964) which, despite its English title, is almost wholly written in German. A stimulating and controversial book on the whole problem of communications technology is Marshall McLuhan, *Understanding Media: The Extensions of Man* (New York, 1964).

Agriculture and Food Technology in the Nineteenth Century

Unlike the materials listed above under the heading *The Agricultural Revolution,* which dealt mainly with European, especially English, developments, the history of agriculture and food technology in the 19th century (Chapter 24) focuses largely on American developments. There are two excellent and comprehensive surveys of this period in American agriculture: Paul W. Gates, *The Farmer's Age: Agriculture, 1815-1860* (New York, 1960) and Fred A. Shannon, *The Farmer's Last Frontier: Agriculture, 1860-97* (New York, 1945). A good selection of *Readings in the History of American Agriculture* (Urbana, Ill., 1960) has been edited by Wayne D. Rasmussen.

Walter P. Webb's classic work, *The Great Plains* (Boston, 1931), provides a provocative thesis for the understanding of American history and is notable for its emphasis on the role of the development of barbed wire in changing the agricultural pattern of the American plains area. Excellent studies of specialized topics are Clark C. Spence, *God Speed the Plow* (Urbana, Ill., 1960), which tells of the coming of steam cultivation to England, while Reynold M. Wik does the same for America in his *Steam Power on the American Farm* (Philadelphia, 1953).

Stewart Holbrook, *Machines of Plenty: Pioneering in American Agriculture* (New York, 1955) is a popular account of mechanization of farming during the 19th century. W. T. Hutchinson has written a two-volume biography of *Cyrus Hall McCormick* (New York, 1930), while Cyrus McCormick III has written *The Century of the Reaper* (Boston, 1931). Henry D. and Frances T. McCallum give a popular account of barbed wire in *The Wire That Fenced the West* (Norman, Okla., 1965). An old but still important general work is Leo Rogin, *The Introduction of Farm Machinery in Its Relation to the Productivity of Labor in the Agriculture of the United States in the Nineteenth Century* (Berkeley, 1931).

Among the very few books written on the development of food technology, including the preservation of food, are: C. Francis, *A History of Food and Its Preservation* (Princeton, 1937) and Oscar Anderson, *Refrigeration in America* (Princeton, 1957). Richard O. Cummings, *The American Ice Harvests: A Historical Study in Technology, 1800-1918* (Berkeley, 1949) tells an earlier story. Joe B. Frantz, *Gail Borden: Dairyman to a Nation* (Norman, Okla., 1951) relates the story of that man, famous for his inventions. Passing attention is given to Pasteur's work in food technology in René Dubos, *Louis Pasteur* (Boston, 1950).

Invention, Engineering Education, and Professionalization

On the subject of invention and inventors in the 19th and 20th centuries (Chapter 19) there is nothing to equal John Jewkes, David Sawers, and Richard Stillerman, *The Sources of Invention* (New York, 1958). Usher's *History of Mechanical Invention* (paper ed., 1959) is limited in its treatment of the modern period but has an excellent analysis of the nature of invention and technological change in Chapters 1-4. S. C. Gilfillan, *The Sociology of Invention* (Chicago, 1935) is an interesting and provocative study. H. Stafford Hatfield, *The Inventor and His World* (New York, 1933) is a survey of the field of invention by a man who is himself an eminent technologist.

The role of the entrepreneur in technical innovation is emphasized by Joseph Schumpeter, *The Theory of Economic Development* (Cambridge, Mass., 1934), pp. 74-94. A collection of articles on entrepreneurial activity is in H. G. J. Aitken (ed.), *Explorations in Enterprise* (Cambridge, Mass., 1965). W. Paul Strassman, *Risk and Technological Innovation: American Manufacturing Methods during the Nineteenth Century* (Ithaca, N.Y., 1959) shows the interaction of business enterprise and technological change in connection with innovation.

Two up-to-date studies of the engineer in relation to society and to his profession are W. H. G. Armytage, *A Social History of Engineering* (New York, 1961) and Daniel C. Calhoun, *The American Civil Engineer, Origins and Conflicts* (Cambridge, Mass., 1960). These two books devote much attention to the development of professionalization among engineers, and various engineering societies have published centennial or anniversary accounts of their own history. Engineering and technical education suffer from a lack of general and comprehensive studies, although some institutions have published their own histories for appropriate occasions. There is some material on technical education in France: Antoine Léon, *Histoire de l'education technique* (Paris, 1961) and Frederick B. Artz, *The Development of Technical Education in France, 1500-1850* (Cambridge, Mass., 1966). Eric Ashby, *Technology and the Academics: An Essay on Universities and the Scientific Revolution* (New York, 1958) concentrates on England. A recent attempt to describe the rise of the technical class is W. H. G. Armytage, *The Rise of the Technocrats: A Social History* (London, 1965).

Military and Political Impact of Industrialization

There are many general studies of the relationships between warfare and technology (Chapter 29), and the close ties between military and technological developments mean that no writers on military history can dare to ignore the technological factors in their studies.

There are a number of good general surveys of military developments, including Alfred Vagts, *A History of Militarism* (New York, 1937); Gordon Turner, *History of Military Affairs* (New York, 1953); Walter Millis, *Arms and Men* (New York, 1956); Theodore Ropp, *War in the Modern World* (Durham, N.C., 1959); and Hoffman Nickerson, *The Armed Horde, 1793-1939* (New York, 1940). Cyril Falls, *A Hundred Years of War, 1855-1954* (New York, 1953) presents a balanced account. One of the most prolific and influential of modern writers on military history was the British Brigadier J. F. C. Fuller; his *Armament and History* (New York, 1933) and *The Conduct of War, 1789-1961* (New Brunswick, N.J., 1961) deal most closely with technological developments.

Quincy Wright, *A Study of War* (2 vols., Chicago, 1942) considers many different aspects of warfare; there is much on the technological factor in Volume 1, pp. 291-328. Edward M. Earle, *Makers of Modern Strategy* (Princeton, 1943) tells of the impact of technological changes on strategical thought; the impact on military tactics is told in Tom Wintringham, *The Story of Weapons and Tactics* (Boston, 1943). A distinguished engineer who did much to mobilize American science and technology during World War II was Vannevar Bush, whose *Modern Arms and Free Men* (New York, 1948) touches on the same aspects of 19th-century developments.

There are some good studies on firearms. W. Y. Carman, *A History of Firearms* (London, 1955) tells the story from earliest times to the outbreak of World War I. Joseph W. Shields, *From Flintlock to M1* (New York, 1954) carries the story to World War II. Bernard and Fawn Brodie have written *From Crossbow to H-Bomb* (New York, 1962).

The technology of naval vessels and armaments is another major area within this general field. James P. Baxter III, *The Introduction of the Ironclad Warship* (Cambridge, Mass., 1933) is an excellent scholarly study of its subject. Other books on this topic are Frederick L. Robertson, *The Evolution of Naval Armament* (London, 1921) and Bernard Brodie, *Sea Power in the Machine Age* (London, 1943). David B. Tyler, *The American Clyde: A History of Iron and Steel Shipbuilding on the Delaware from 1840 to World War 1* (Newark, Del., 1958) should not be overlooked.

A study of a very specialized field is Thomas Weber, *The Northern Railroads in the Civil War, 1861-65* (New York, 1952). Robert V. Bruce, *Lincoln and the Tools of War* (Indianapolis, 1956) reports the failure of the Union to take advantage of potential new weapons.

The spread of the Industrial Revolution to countries outside Britain and the impact of technological change upon these nations (Chapter 30) are dealt with mainly in economic histories. W. Bowden, M. Karpovich, and A. P. Usher, *An*

Economic History of Europe since 1750 (New York, 1937) is a solid scholarly survey of the interactions among physical resources, technology, and social and economic institutions. Chapters 21-31 of Herbert Heaton, *An Economic History of Europe* (New York, 1948) deal with the story since 1750. Although an older account, J. H. Clapham, *The Economic Development of France and Germany, 1815-1914* (4th edn., Cambridge, 1936) remains one of the best books on these countries. An excellent scholarly study is R. E. Cameron, *France and the Economic Development of Europe* (Princeton, 1961).

Difficult because of their highly technical economics vocabulary, but important, are two excellent comparative studies: Charles P. Kindleberger, *Economic Growth in France and Britain, 1851-1950* (Cambridge, Mass., 1964) and H. J. Habbakuk, *American and British Technology in the 19th Century* (Cambridge, England, 1962).

Stuart Bruchey, *The Roots of American Economic Growth, 1607-1861* (New York, 1965) is a masterly critical presentation of recent research and interpretations of this formative period, as is Douglass C. North, *The Economic Growth of the United States, 1790-1860* (Englewood Cliffs, N.J., 1961; paperbound, New York, 1966). W. O. Henderson, *Britain and Industrial Europe, 1750-1870* (Liverpool, 1954) provides a good account of the transit of British technology to other portions of Europe. Thorstein Veblen, *Imperial Germany and the Industrial Revolution* (New York, 1915) remains unsurpassed in its perceptive insights into the coming of industrialism to Germany in the late 19th century.

The expansion of Europe—and Western technology—to non-Western areas of the world (Chapter 43) is a story which has often been told, but without very much attention paid to the technological aspects of Western imperialism. S. C. Easton, *The Rise and Fall of Western Colonialism* (New York, 1964) is a lucid and absorbing interpretation of colonial conquest and liberation from the perspective of the 1960's. J. T. Gallagher and R. I. Robinson, *Africa and the Victorians* (Oxford, 1962) is a novel and highly illuminating study of British expansion in Africa. Mary E. Townsend, *European Colonial Expansion since 1871* (Philadelphia, 1941) is a standard scholarly account of its subject up to World War II. Grover Clark, *The Balance Sheet of Imperialism* (New York, 1936) is a critical factual examination of the statistics of colonial trade and finance. For some scarcely remembered American efforts one may consult Merle Curti and Kendall Birr, *Prelude to Point Four: American Technical Missions Overseas, 1838-1938* (Madison, Wis., 1954).

The earlier period of colonial expansion has been studied recently by Carlo M. Cipolla in *Guns, Sails and Empires: Technological Innovation and the Early Phases of European Expansion, 1400-1700* (New York, 1966). His thesis is that naval armament was the major advantage held by the Europeans over the Oriental areas, and that this enabled them to gain a toehold in Asia during this early period.

The standard Marxist-Leninist interpretation of imperialism is V. I. Lenin, *Imperialism, the Highest Stage of Capitalism,* which was first published in 1915 but has had many editions since then. For discussion purposes it should be read in conjunction with J. A. Schumpeter, *Imperialism and Social Classes* (New York, 1949), a sociological explanation first published in 1919.

Economic Impact of Industrialization

Many of the works listed in the sections above—for example, Cameron, Clapham, Habakkuk, Kindleberger, Strassman, G. R. Taylor, and Usher—should be consulted in connection with the economic consequences of industrialization (Chapter 31). Other books which are important for understanding the interrelationships of economics and technology during the 19th century are: Phyllis Deane and W. A. Cole, *British Economic Growth, 1688-1959* (Cambridge, England, 1962) which contains excellent statistical series; L. Davis, J. Hughes, and D. McDougall, *American Economic History* (Homewood, Ill., 1961) for statistics on railway mileage; and National Bureau of Economic Research, *Trends in the American Economy in the Nineteenth Century* (Vol. 23 in Studies in Income and Wealth, Princeton, 1960).

N. Rosenberg has written an article of prime importance, "Technological Change in the Machine Tool Industry, 1840-1910," *Journal of Economic History,* XXIII (Dec. 1963), 414-43; its title belies its significance, for Rosenberg uses the machine-tool industry as the vehicle for tracing, with great perception and scholarly erudition, the complicated interrelationships between economic and technological developments on a very wide scale.

The effects of technology on distribution and marketing (Chapter 42) have been dealt with only incidentally, rather than directly, in histories of manufacturing and in economic histories. Hence many of the titles listed under other sections contain material which is applicable to this discussion.

Distribution is dealt with in such books as Harold Barger, *Distribution's Place in the American Economy since 1869* (Princeton, 1955); Chauncey Depew, *One Hundred Years of American Commerce* (New York, 1895); and Alfred Marshall, *Industry and Trade* (London, 1919). The changing character of marketing can be seen through the history of individual companies: Boris Emmet and John E. Jeuck, *Catalogues and Counters: A History of Sears, Roebuck and Company* (Chicago, 1950); Ralph M. Hower, *History of Macy's of New York* (Cambridge, Mass., 1943); and by the same author, *The History of an Advertising Agency* (Cambridge, Mass., 1949). A general account is given in George B. Hotchkiss, *Milestones of Marketing* (New York, 1938). Thomas D. Clark, *Pills, Petticoats and Plows* (Indianapolis, 1944) is a popular treatment.

Histories of various industries indicate the interrelationships among technological developments, enlarged-production factors, problems of mass distribution, and mass-consumption patterns. Among the most notable of such histories are Paul T. Cherington, *The Wool Industry* (New York, 1916); Rudolph A. Clemen, *The American Livestock and Meat Industry* (New York, 1923); Arthur H. Cole, *American Wool Manufacture* (2 vols., Cambridge, Mass., 1941); and, with Harold F. Williamson, *The American Carpet Manufacture* (Cambridge, Mass., 1941); M. T. Copeland, *The Cotton Manufacturing Industry of the United States* (Cambridge, Mass., 1923); Jesse Pope, *The Clothing Industry in New York* (Columbia, Mo., 1905); and B. M. Schleckman, *The Clothing and Textile Industries in New York* (New York, 1925).

A few scholarly studies of particular companies have been made by business historians; among these are Thomas R. Navin, *The Whitin Machine Works since 1831* (Cambridge, Mass., 1950) and George S. Gibb, *The Saco-Lowell Shops* (Cambridge, Mass., 1950).

Part III of Volume IX on "Manufactures" in the *Twelfth Census of the United States, 1900* contains important statistical material on distribution and marketing. An excellent selection of readings on various aspects of American economic development is to be found in Harold F. Williamson (ed.), *The Growth of the American Economy* (New York, 1951).

Social Impact of Industrialization

The condition of the factory workers during the early period of the Industrial Revolution in Britain (Chapter 18) has been the source of much contention among historians. Motivated by a zeal for social reform, many historians during the latter part of the 19th and early 20th centuries depicted the conditions of the works in the factories and mines in the blackest possible terms.

Representatives of this dismal picture of the social effects of industrialization are a series of works by J. L. and Barbara Hammond: *The Bleak Age* (London, 1934); *The Town Labourer* (London, 1919); *The Village Labourer* (London, 1920); *The Skilled Labourer* (London, 1919); and *The Age of the Chartists* (London, 1930). References to certain statistics and to official documents, such as the report of a parliamentary investigating committee (1832-33) headed by Michael Sadler, lent credence to such views. This "orthodox" interpretation has been extended by some major contemporary writers, most notably Lewis Mumford, in *Technics and Civilization* (New York, 1934), wherein he claims that industrialization befouled the environment and substituted false pecuniary values for human values, and Sigfried Giedion, *Mechanization Takes Command* (Oxford, 1948), with his thesis that the machine has become the master rather than the servant of man.

In the mid-20th century a group of economic historians have sought to revise the catastrophic interpretation of the Industrial Revolution, usually by reference to statistics, admittedly incomplete during the 18th and early 19th centuries, of real wages. Foremost among these "revisionists" have been Colin Clark, *The Conditions of Economic Progress* (London, 1940) and T. S. Ashton, *The Industrial Revolution* (Oxford, 1948). A good sampling of the arguments of the "revisionist" historians is to be found in F. A. Hayek (ed.), *Capitalism and the Historians* (Chicago, 1954). This controversy is by no means resolved.

A different type of approach to the problem of social changes wrought by the Industrial Revolution is provided by Neil J. Smelser, *Social Change in the Industrial Revolution: An Application of Theory to the British Cotton Industry, 1770-1840* (Chicago, 1959). Although difficult to read because of the highly abstract sociological analysis and vocabulary employed by the author, the book is a valuable attempt to arrive at a theoretical model of the relation of technological changes to social changes by reference to the cotton industry and the family system in Britain during this period.

A brief but perceptive historiographical survey of the problems, social and otherwise, posed by the Industrial Revolution is to be found in Eric E. Lampard, *Industrial Revolution: Interpretations and Perspectives* (Washington, D.C., 1957). The transformation of social conditions in England is treated in Sydney G. Checkland, *The Rise of Industrial Society in England, 1815-1885* (New York, 1964) and is handled exceptionally well by E. P. Thompson in a recent book which has already become a classic: *The Making of the English Working Class* (London, 1963; paperbound edn., New York, 1966). The European context is dealt with in Eric J. Hobsbawm, *The Age of Revolution: Europe, 1789-1848* (Cleveland, 1962). Some important aspects of the social impact of industrialization on America are treated in Stanley Lebergott, *Manpower in Economic Growth: The American Record since 1800* (New York, 1964), and Douglass C. North, *Growth and Welfare in the American Past: A New Economic History* (Englewood Cliffs, N.J., 1966).

Historians have only recently begun the serious scholarly investigation of population movements and urbanization; the study of the relation of these to technology (Chapter 32) is still in its infancy. Two books serve as the starting point for a study of American migrations: Bernard DeVoto, *Year of Decision: 1846* (Boston, 1942) and Oscar Handlin, *The Uprooted* (New York, 1951). DeVoto sees the westward movement in terms of the whole nation's experience, while Handlin gives a generalized picture of urban immigrant experience.

Following these generalized views, one should proceed to specialized studies. Robert Ernst, *Immigrant Life in New York City, 1825-63* (New York, 1949) gives a good account of the German experience; Oscar Handlin, *Boston's Immigrants* (rev. edn., Cambridge, Mass., 1959) concentrates on the Irish; Charlotte Erickson, *American Industry and the European Immigrant* (Cambridge, Mass., 1957) on labor importation. Particular ethnic groups have been studied by Robert F. Foerster, *The Italian Emigration of Our Times* (Cambridge, Mass.,

1924); Dorothy Swain Thomas, *Social and Economic Aspects of Swedish Population Movements* (New York, 1941); Emily G. Balch, *Our Slavic Fellow Citizens* (New York, 1910); Moses Rischiv, *The Promised City* (Cambridge, Mass., 1961) on New York's Jews; also, William E. B. DuBois, *The Philadelphia Negro* (Philadelphia, 1899) is a pioneer study containing material on 19th-century Negro migrations and city life.

Maurice R. Davie, *World Immigration with Special Reference to the United States* (New York, 1936) gives a brief account of all ethnic and national groups and includes bibliographic notes. The student should also be aware of the massive modern study of internal migration which is summarized in the concluding volume by Hope T. Eldridge and Dorothy Swain Thomas, *Population Redistribution and Economic Growth, U.S. 1870-1950* (Philadelphia, 1964). The best recent general study of historical demography is D. V. Glass and D. E. C. Eversley, *Population in History* (Chicago, 1965).

Writings referring specifically to the discussion in Chapter 32 are, for example, Ping-ti-Ho, *Studies on the Population of China, 1385-1953* (Harvard East Asian Studies 4, Cambridge, Mass., 1959), pp. 265-78. The factors in American population growth are discussed in Conrad and Irene B. Taeuber, *The Changing Population of the United States* (New York, 1958), pp. 292-313; and in Ansley J. Coale and Melvin Zelnik, *New Estimates of Fertility and Population in the United States* (Princeton, 1963), pp. 34-5. European migration to America is the subject of Marcus L. Hansen, *The Atlantic Migration, 1607-1860* (Cambridge, Mass., 1941), pp. 172-98. The factors involved in the migration of labor to the United States are discussed in Arthur Redford, *Labor Migration in England, 1800-1850* (Manchester, 1926), pp. 114-64; Hansen, *Atlantic Migration*, pp. 280-306; Mack Walker, *Germany and the Emigration of 1816-1885* (Cambridge, Mass., 1964), pp. 42-69, 153-94; and Rowland T. Berthoff, *British Immigrants in Industrial America, 1790-1950* (Cambridge, Mass., 1953), pp. 30-87. A graphic description of the role of the Irish famine in stimulating migration to America is given in Cecil B. Woodham-Smith, *The Great Hunger* (New York, 1962), pp. 9-96, 200-241.

The relationship between economic conditions, both European and American, and population migrations is dealt with in Brinley Thomas, *Migration and Economic Growth* (Cambridge, Mass., 1954) and Eric E. Lampard, "The Course of Industrial Urbanization, 1700-1900: An Unfinished Epic in Human Ecology," in Philip M. Hauser and Leo F. Schnore (eds.), *The Study of Urbanization* (New York, 1965).

For a general overview of the modern history of American and European cities, Lewis Mumford's classic *The Culture of Cities* (New York, 1938) is still the best, along with his newer and massive *The City in History: Its Origins, Its Transformations, and Its Prospects* (New York, 1961). A good visual supplement to Mumford would be Steen Eiler Rasmussen, *Towns and Buildings* (Cambridge, Mass., 1951 and John A. Kouwenhoven, *The Columbia Historical Portrait of New York* (Garden City, N.Y., 1953).

A few good American urban histories are: Bayard Still, *Mirror for Gotham* (rev. edn., New York, 1956); Robert G. Albion, *The Rise of New York Port, 1815-1860* (New York, 1959); an early industrial town is the subject of John Coolidge, *Mill and Mansion, A Study of Architecture and Society in Lowell, Massachusetts, 1820-1865* (New York, 1942); Gerald M. Capers, Jr., *The Biography of a River Town, Memphis: Its Heroic Age* (Chapel Hill, 1939); Bessie L. Pierce, *History of Chicago* (3 vols. in progress, New York, 1937-); and an antiquarian reconstruction of social history, Albert E. Fossier, *New Orleans, the Glamour Period, 1800-1840* (New Orleans, 1957).

An annotated list of books on urban history throughout the world is contained in Philip Dawson and Sam B. Warner, Jr., "A Selection of Works Relating to the History of Cities," in Oscar Handlin and John Burchard (eds.), *The Historian and the City* (Cambridge, Mass., 1963).

The nature of urban manufacturing in the United States during the first three-quarters of the 19th century is treated in Sam B. Warner, Jr., "Innovation and the Industrialization of Philadelphia, 1800-1850," in Handlin and Burchard (eds.), *The Historian and the City,* pp. 63-9. The reorganization of American urban labor along the lines of increasing specialization is told in John R. Commons (ed.), *History of Labor* (New York, 1935) Vol. 1 and in Norman Ware, *The Industrial Worker, 1840-1860* (Gloucester, Mass., 1959). The same development in England is treated by Eric E. Lampard, "History of Cities in Economically Advanced Areas," *Economic Development and Cultural Change,* III (Jan. 1955), 86-102; Smelser, *Social Change in the Industrial Revolution;* and Edward P. Thompson, *The Making of the English Working Class* (London, 1963).

The political basis of the racial and ethnic riots and turmoil in American cities in the decades immediately preceding the Civil War is discussed in Ray A. Billington, *The Protestant Crusade, 1800-1860* (New York, 1938); Robert Ernst, "Economic Nativism in New York City during the 1840's," *New York History,* XXIX (April 1948), 170-86; John Lardner, "The Martyrdom of Bill the Butcher," *New Yorker,* XXX (March 20, 27, 1954), 41-53, 38-59.

The development of the settlement pattern of the American industrial metropolis during the latter part of the 19th century receives thorough scholarly treatment in Sam B. Warner, Jr., *Streetcar Suburbs* (Cambridge, Mass., 1962). The biographies of famous municipal government "bosses" are related in Harold Zink, *City Bosses* (Durham, N.C., 1930).

The direct relation of technological change to domestic life (Chapter 33) remains unclear. Especially suggestive are two works by Lewis Mumford: *The Culture of Cities* and *Technics and Civilization.* Sigfried Giedion's *Mechaniza-*

tion Takes Command deals in part with the impact of mass production on home life. Margaret Mead, *Cultural Patterns and Technical Change* (New York, 1955) suggests the value of studying the adaptation by contemporary primitives of new techniques as models for the same process in Western civilization, and she includes a rich bibliography of such studies. The edition of G. M. Trevelyan's *English Social History* illustrated with perceptive drawings selected and analyzed by Ruth C. Wright (London, 1951) touches suggestively upon some of the social problems involved in industrialization. Phillippe Aries, *Centuries of Childhood* (New York, 1962) pioneers in the use of visual materials and erects very useful hypotheses which, though based on French data, have far wider implications for the analysis of the social impact of industrialism.

Specific connections between domestic life and technological developments are made, in narrow contexts, by J. C. Drummond and Ann Wilbraham, *The Englishman's Food* (London, 1939) and in W. M. Frazier, *A History of English Public Health, 1834-1939* (London, 1950).

Especially challenging in their implications are two very different works. David M. Potter in *People of Plenty: Abundance and the American Character* (Chicago, 1954) developed a hypothesis with historical methodology, that the American character and many aspects of family life were uniquely affected by American abundance. John M. W. Whiting and Irvin L. Child explore the possibility of comparative cultural study centering on the family in *Child Training and Personality* (New Haven, 1953); this technique would seem fundamental to a definitive study of the impact of technological change upon domestic life.

Cultural Impact of Industrialization

International exhibitions—world fairs—provided some element of public knowledge and understanding of technological developments (Chapter 44). Kenneth W. Luckhurst, *The Story of Exhibitions* (London, 1951) is a good account of exhibitions in general. A sprightly, heavily illustrated monograph on the Crystal Palace Exhibition, the first great exposition, is C. H. Gibbs-Smith, *The Great Exhibition of 1851* (London, 1950). Merle Curti, "America at the World Fairs, 1851-1893," *American Historical Review*, LV (July 1950), 833-56, provides a summary that may suggest additional reading. Eugene S. Ferguson has written on the influence of international expositions in the development of museums of science and technology: "Technical Museums and International Exhibitions," *Technology and Culture*, VI, 1 (Winter 1965), 30-46.

The cultural trends at the close of the 19th century, as these were related to technology (Chapter 45), reflect the widespread belief in "the idea of progress." A classic study of the origin and growth of the modern faith in progress is J. B. Bury, *The Idea of Progress* (paperbound edn., New York, 1955), with Chapter

18 the most relevant for the decade of the 1890's. For a delightful, but jaundiced view of American social and cultural life in this decade, one should consult Thomas Beer, *The Mauve Decade* (Garden City, N.Y., 1926). In *The Brown Decades* (New York, 1955), Lewis Mumford attempts to rescue what is most creative and enduring in the culture, while Arthur M. Schlesinger relates *The Rise of the City, 1878-1898* (New York, 1933) to the social and labor problems of the time. And, of course, no student of American cultural currents during this period should be without Henry Adams's *Education* (available in many editions).

The meaning of technology for early Americans is explored in two articles by Hugo A. Meier, "Technology and Democracy, 1800-1860," *Mississippi Valley Historical Review,* XLIII (March 1957), 618-40, and "American Technology and the Nineneeth-Century World," *American Quarterly,* X (Summer 1958), 116-30. The impact of technology on American and British writers is described in detail in Leo Marx, *The Machine in the Garden: Technology and the Pastoral Ideal in America* (New York, 1964). John A. Kouwenhoven, *Made in America: The Arts in Modern Civilization* (Garden City, N.Y. 1962) is perhaps the most provocative book on the subject.

The best attempt to tie together social, cultural, and political history during this era in Europe is Carlton J. H. Hayes, *A Generation of Materialism, 1871-1900* (New York, 1941). Two stimulating interpretations written by Europeans are Egon Friedell, *A Cultural History of the Modern Age* (New York, 1932), especially Vol. III, Book IV; and José Ortega y Gasset, *The Revolt of the Masses* (New York, 1932). In his patrician outlook, Ortega y Gasset reminds one of Henry Adams. A vast compendium of information bearing on all aspects of Western civilization are the various *Reports* of the U.S. Commissioners to the international expositions.

The great central issue of Western culture during the period—that of men and machines—is explored critically and pessimistically by the great Swiss architect-philosopher, Sigfried Giedion, in *Mechanization Takes Command,* and somewhat more optimistically by Lewis Mumford in *Technics and Civilization.* Advocating an organic philosophy, both works are full of insights. If one should need more help, valuable guides are Alan L. C. Bullock (ed.), *A Select List of Books on European History, 1815-1914* (Oxford, 1957); and Roy P. Basler *et al., A Guide to the Study of the United States of America: Books Reflecting the Development of American Life and Thought* (Washington, D.C., 1960).

Contributors

Silvio A. Bedini is Assistant Director of the Museum of History and Technology of the Smithsonian Institution. He is the author of many books and articles on the history of clocks and scientific instruments, including *The Scent of Time* and *Early American Scientific Instruments and Their Makers.*

Georg Borgstrom, Professor of Food Science and Geography at Michigan State University, is an authority on world food resources and food preservation. Among the best known of his books are *The Hungry Planet* and the four-volume *Fish as Food,* which he edited.

Lynwood Bryant is Associate Professor of History at the Massachusetts Institute of Technology. He has written on the history of the automobile and the automobile engine, particularly the early development of the internal-combustion engine.

Roger Burlingame is the author of a series of books on American technological history and industrial development. The best known of his books are *March of the Iron Men; Engines of Democracy; Background of Power;* and *Machines that Built America.*

Rondo Cameron is Professor of Economics and History at the University of Wisconsin. His publications include *France and the Economic Development of Europe, 1800-1914; The European World: A History;* and *Banking in the Early Stages of Industrialization.*

Shepard B. Clough, Professor of History at Columbia University, is vice president of the Economic History Association. Among the many books which he has written are the *Economic History of Europe* (with Charles W. Cole); *France, 1789-1939, A Study in National Economics;* and *The Economic History of Western Civilization.*

Carl W. Condit, Professor of Art at Northwestern University, is an editor of *Technology and Culture.* In addition to a number of articles on the development of structural techniques, he is the author of *The Rise of the Skyscraper;* the two-volume *American Building Art;* and *The Chicago School of Architecture.*

Bern Dibner is chairman of the Board of the Burndy Corporation and Director of the Burndy Library, which specializes in works in the history of science and technology. He has written many books and articles on the history of electricity and magnetism, including *Oersted; Alessandro Volta and the Electric Battery;* and *The Atlantic Cable.* Dr. Dibner is a vice president of the Society for the History of Technology.

Aage Gerhardt Drachmann retired from the University Library in Copenhagen in 1956 to pursue his studies in ancient technology. Among his books are *Ancient Oil Mills and Presses; Ktesibios, Philon and Heron;* and *The Mechanical Technology of Greek and Roman Antiquity.*

Eugene S. Ferguson, Professor of Mechanical Engineering at Iowa State University, was formerly curator of Mechanical and Civil Engineering in the Museum of History and Technology, Smithsonian Institution. An editor of *Technology and Culture,* he is the author of articles on the history of kinematics and thermodynamics and on American mechanical technology of the 19th century. Professor Ferguson's "Contributions to Bibliography in the History of Technology" will soon be published as a comprehensive bibliography in this field.

James Kip Finch, Dean Emeritus of the School of Engineering of Columbia University, is the author of a number of technical publications on hydraulics and structure. His historical studies include two books: *Engineering and Western Civilization* and *The Story of Engineering.*

Robert James Forbes served as a chemical engineer with the Royal Dutch/Shell Company while at the same time pursuing his interest in the history of ancient technology. Since 1947 he has been a profesor of the history of ancient science and technology at the University of Amsterdam. His publications include several volumes on the history of petroleum as well as the nine-volume *Studies in Ancient Technology.*

G. E. Fussell served for forty years in the English Ministry of Agriculture and Fishery. He founded the British Agricultural History Society and is a Fellow of the Royal Historical Society. His books on the history of agriculture include *Old English Farming: Fitzherbert to Tull, 1523-1730; The Farmer's Tools, 1500-1900;* and *The English Rural Labourer.* He is also the author of the forthcoming *The English Dairy Farmer, 1500-1900* and *Farming Techniques from Prehistoric to Modern Times.*

Anthony N. B. Garvan, Professor and Chairman of the Department of American Civilization at the University of Pennsylvania, served for a time as head curator of Civil History at the Smithsonian Institution and as an editor of *American Quarterly.* He has written many articles on the social aspects of American technology and is the author of *Architecture and Town Planning in Colonial Connecticut.*

Alfred Rupert Hall, Professor of the History of Science and Technology at the Imperial College of Science and Technology, University of London, was one of the editors of the five-volume *A History of Technology* along with Charles Singer, E. J. Holmyard, and T. I. Williams. In addition to many articles on the history of technology, he is the author of *The Scientific Revolution, 1500-1800* and *From Galileo to Newton*.

Herbert Heaton, Emeritus Professor of Economic History at the University of Minnesota and a former president of the Economic History Association, has written the *History of the Yorkshire Woolen and Worsted Industries from the Earliest Times to the Industrial Revolution; The Economic History of Europe;* and many other books and articles on economic and technological development.

Arthur M. Johnson, Professor of Business History at the Harvard Graduate School of Business Administration, has been editor of the *Business History Review* and associate editor of the Harvard Series in Business History. He is the author of *The Development of American Petroleum Pipelines* and *Government-Business Relations*. In 1954 he was awarded the Beveridge Prize of the American Historical Association.

Melvin Kranzberg is Professor of History at Case Institute of Technology. One of the founders of the Society for the History of Technology, he has served as editor-in-chief of its journal, *Technology and Culture*, since its inception. He is a vice president of the American Association for the Advancement of Science, chairman of the Historical Advisory Committee of the National Aeronautics and Space Administration, and has served as vice president of the Society for French Historical Studies. He is the author and editor of books and articles in general history and in the history of science and technology.

Eric E. Lampard is Professor of History and Director of the Graduate Program in Economic History, University of Wisconsin. He is the author of articles dealing with industrial urbanization. Among his publications is *Industrial Revolution: Interpretations and Perspectives*.

Robert Multhauf is Director of the Museum of History and Technology of the Smithsonian Institution and editor of *Isis*, the journal of the History of Science Society. He is the author of numerous articles on the history of technology and of a forthcoming book, *The Origins of Chemistry*. He has been an editor of *Technology and Culture* and is first vice president of the Society for the History of Technology.

Thomas A. Palmer, a lieutenant colonel in the United States Marine Corps, has seen duty throughout the world. He has served as a staff assistant to the Deputy Assistant Secretary of Defense for Education and has taught history and political science at the University of South Carolina. He has written many articles on the history of military technology.

Derek J. De Solla Price is Avalon Professor of the History of Science at Yale

University. His research interests range from the study of astrolabes, sundials, and other scientific instruments, to the scientific analysis of the modern growth and organization of scientific manpower and literature. Among his many publications are *Heavenly Clockwork* (with Joseph Needham and Ling Wang); *Science since Babylon;* and *Little Science, Big Science.*

Carroll W. Pursell, Jr., is Assistant Professor of History at the University of California, Santa Barbara. He is one of the editors of *The Politics of American Science: 1939 to the Present,* and has published articles on the history of American science and technology in many journals.

John B. Rae is Professor of History at Harvey Mudd College. He was formerly chairman of the Liberal Studies Division of the American Society for Engineering Education and is an advisory editor of *Technology and Culture.* His books include the prize-winning *American Automobile Manufacturers* and *The American Automobile.*

Nathan Rosenberg, Professor of Economics at Purdue University, is the author of *Economic Planning in the British Building Industry, 1945-1949,* and of many articles on the economic aspects of technological change.

Charles L. Sanford is Professor of Language and Literature at Rensselaer Polytechnic Institute. His publications include *Benjamin Franklin and the American Character; The Quest for Paradise: Europe and the American Moral Imagination;* and *Quest for America, 1810-1824.*

Harold I. Sharlin, Professor of History at Iowa State University, is the author of *The Making of the Electrical Age* and *The Convergent Century: The Unification of the Sciences in the 19th Century.*

Cyril Stanley Smith is Institute Professor, Professor of Metallurgy, and Professor of the History of Technology and Science at Massachusetts Institute of Technology. He was formerly Director of the Institute for the Study of Metals at the University of Chicago. He has published numerous papers in metallurgical and physical journals and has won many honors and distinctions as both a metallurgist and historian. His historical works include *A History of Metallography* and a series of annotated translations of classics in metallurgical history. Dr. Smith has served as an editor of *Technology and Culture* and as president of the Society for the History of Technology.

Thomas M. Smith is Associate Professor of the History of Science at the University of Oklahoma. His researches include studies of various communication and transportation developments. He is the co-editor, with Farrington Daniels, of *The Challenge of Our Times.*

Abbott Payson Usher, late Profesor Emeritus of Economics at Harvard University, was a pioneer investigator in the history of technology and in economic history. Among his many important books are *The Industrial History of England* and the classic *History of Mechanical Inventions.*

Sam Bass Warner, Jr., is an Associate Professor of History at Washington

University, St. Louis, and was formerly a research associate at the M.I.T.-Harvard Joint Center for Urban Studies. He is the author of *Streetcar Suburbs* and editor of *Planning for a Nation of Cities.*

Lynn White, Jr., Professor of Medieval History at the University of California, Los Angeles, served as president of Mills College for fifteen years. He is a Fellow of the Medieval Academy of America and past president of the Society for the History of Technology. He has received the Leonardo da Vinci Medal of that Society, and the Pfizer Award of the History of Science Society for his *Medieval Technology and Social Change.*

Harold F. Williamson, Professor of Economics at Northwestern University, is president of the Economic History Association, Director of the National Bureau of Economic Research, and secretary-treasurer of the American Economic Association. He has been author or co-author of many works in American economic history, including *The American Carpet Manufacture; The Growth of the American Economy;* and *The American Petroleum Industry.*

Robert S. Woodbury is Professor of the History of Technology at Massachusetts Institute of Technology. He has received the Usher Prize of the Society for the History of Technology and has been a Guggenheim Fellow. His publications include *History of the Milling Machine; History of the Gear-Cutting Machine; History of the Grinding Machine;* and *History of the Lathe to 1850.*

Subject Index

Name Index